MEMBRANE TECHNOLOGY AND APPLICATIONS

MEMBRANE TECHNOLOGY AND APPLICATIONS

SECOND EDITION

Richard W. Baker

Membrane Technology and Research, Inc.
Menlo Park, California

John Wiley & Sons, Ltd

First Edition published by McGraw-Hill, 2000. ISBN: 0 07 135440 9

Telephone (+44) 1243 779777

Email (for orders and customer service enquiries): cs-books@wiley.co.uk
Visit our Home Page on www.wileyeurope.com or www.wiley.com

Reprinted December 2004, February 2006, June 2007

Other Wiley Editorial Offices

John Wiley & Sons Inc., 111 River Street, Hoboken, NJ 07030, USA

Jossey-Bass, 989 Market Street, San Francisco, CA 94103-1741, USA

Wiley-VCH Verlag GmbH, Boschstr. 12, D-69469 Weinheim, Germany

John Wiley & Sons Australia Ltd, 33 Park Road, Milton, Queensland 4064, Australia

John Wiley & Sons (Asia) Pte Ltd, 2 Clementi Loop #02-01, Jin Xing Distripark, Singapore 129809

John Wiley & Sons Canada Ltd, 22 Worcester Road, Etobicoke, Ontario, Canada M9W 1L1

Wiley also publishes its books in a variety of electronic formats. Some content that appears in print may not be available in electronic books.

Library of Congress Cataloging-in-Publication Data

Baker, Richard W.
Membrane technology and applications / Richard W. Baker.—2nd ed.
p. cm.
Includes bibliographical references and index.
ISBN 0-470-85445-6 (Cloth : alk. paper)
1. Membranes (Technology) I. Title.
TP159.M4 B35 2004
660′.28424—dc22

2003021354

British Library Cataloguing in Publication Data

A catalogue record for this book is available from the British Library

ISBN 13: 978-0-470-85445-7 (HB)

Typeset in 10/12pt Times by Laserwords Private Limited, Chennai, India
Printed and bound in Great Britain by TJ International, Padstow, Cornwall
This book is printed on acid-free paper responsibly manufactured from sustainable forestry in which at least two trees are planted for each one used for paper production.

CONTENTS

PREFACE

My introduction to membranes was as a graduate student in 1963. At that time membrane permeation was a sub-study of materials science. What is now called membrane technology did not exist, nor did any large industrial applications of membranes. Since then, sales of membranes and membrane equipment have increased more than 100-fold and several tens of millions of square meters of membrane are produced each year—a membrane industry has been created.

This membrane industry is very fragmented. Industrial applications are divided into six main sub-groups: reverse osmosis; ultrafiltration; microfiltration; gas separation; pervaporation and electrodialysis. Medical applications are divided into three more: artificial kidneys; blood oxygenators; and controlled release pharmaceuticals. Few companies are involved in more than one sub-group of the industry. Because of these divisions it is difficult to obtain an overview of membrane science and technology; this book is an attempt to give such an overview.

The book starts with a series of general chapters on membrane preparation, transport theory, and concentration polarization. Thereafter, each major membrane application is treated in a single 20-to-40-page chapter. In a book of this size it is impossible to describe every membrane process in detail, but the major processes are covered. However, medical applications have been short-changed somewhat and some applications—fuel cell and battery separators and membrane sensors, for example—are not covered at all.

Each application chapter starts with a short historical background to acknowledge the developers of the technology. I am conscious that my views of what was important in the past differ from those of many of my academic colleagues. In this book I have given more credit than is usual to the engineers who actually made the processes work.

Readers of the theoretical section (Chapter 2) and elsewhere in the book will see that membrane permeation is described using simple phenomenological equations, most commonly, Fick's law. There is no mention of irreversible thermodynamics. The irreversible thermodynamic approach to permeation was very fashionable when I began to work with membranes in the 1960s. This approach has the appearance of rigor but hides the physical reality of even simple processes behind a fog of tough equations. As a student and young researcher, I struggled with irreversible thermodynamics for more than 15 years before finally giving up in the 1970s. I have lived happily ever after.

Finally, a few words on units. Because a great deal of modern membrane technology originated in the United States, the US engineering units—gallons, cubic feet, and pounds per square inch—are widely used in the membrane industry. Unlike the creators of the Pascal, I am not a worshipper of mindless uniformity. Metric units are used when appropriate, but US engineering units are used when they are the industry standard.

ACKNOWLEDGMENTS FOR THE FIRST EDITION

As a school boy I once received a mark of $\frac{1}{2}$ out of a possible 20 in an end-of-term spelling test. My spelling is still weak, and the only punctuation I ever really mastered was the period. This made the preparation of a polished final draft from my yellow notepads a major undertaking. This effort was headed by Tessa Ennals and Cindi Wieselman. Cindi typed and retyped the manuscript with amazing speed, through its numerous revisions, without complaint. Tessa corrected my English, clarified my language, unsplit my infinitives and added every semicolon found in this book. She also chased down a source for all of the illustrations used and worked with David Lehmann, our graphics artist, to prepare the figures. It is a pleasure to acknowledge my debt to these people. This book would have been far weaker without the many hours they spent working on it. I also received help from other friends and colleagues at MTR. Hans Wijmans read, corrected and made numerous suggestions on the theoretical section of the book (Chapter 2). Ingo Pinnau also provided data, references and many valuable suggestions in the area of membrane preparation and membrane material sciences. I am also grateful to Kenji Matsumoto, who read the section on Reverse Osmosis and made corrections, and to Heiner Strathmann, who did the same for Electrodialysis. The assistance of Marcia Patten, who proofed the manuscript, and Vivian Tran, who checked many of the references, is also appreciated.

ACKNOWLEDGMENTS FOR THE SECOND EDITION

Eighteen months after the first edition of this book appeared, it was out of print. Fortunately, John Wiley & Sons, Ltd agreed to publish a second edition, and I have taken the opportunity to update and revise a number of sections. Tessa Ennals, long-time editor at Membrane Technology and Research, postponed her retirement to help me finish the new edition. Tessa has the standards of an earlier time, and here, as in the past, she gave the task nothing but her best effort. I am indebted to her, and wish her a long and happy retirement. Marcia Patten, Eric Peterson, David Lehmann, Cindy Dunnegan and Janet Farrant assisted Tessa by typing new sections, revising and adding figures, and checking references, as well as helping with proofing the manuscript. I am grateful to all of these colleagues for their help.

1 OVERVIEW OF MEMBRANE SCIENCE AND TECHNOLOGY

Introduction

Membranes have gained an important place in chemical technology and are used in a broad range of applications. The key property that is exploited is the ability of a membrane to control the permeation rate of a chemical species through the membrane. In controlled drug delivery, the goal is to moderate the permeation rate of a drug from a reservoir to the body. In separation applications, the goal is to allow one component of a mixture to permeate the membrane freely, while hindering permeation of other components.

This book provides a general introduction to membrane science and technology. Chapters 2 to 4 cover membrane science, that is, topics that are basic to all membrane processes, such as transport mechanisms, membrane preparation, and boundary layer effects. The next six chapters cover the industrial membrane separation processes, which represent the heart of current membrane technology. Carrier facilitated transport is covered next, followed by a chapter reviewing the medical applications of membranes. The book closes with a chapter that describes various minor or yet-to-be-developed membrane processes, including membrane reactors, membrane contactors and piezodialysis.

Historical Development of Membranes

Systematic studies of membrane phenomena can be traced to the eighteenth century philosopher scientists. For example, Abbé Nolet coined the word 'osmosis' to describe permeation of water through a diaphragm in 1748. Through the nineteenth and early twentieth centuries, membranes had no industrial or commercial uses, but were used as laboratory tools to develop physical/chemical theories. For example, the measurements of solution osmotic pressure made with membranes by Traube and Pfeffer were used by van't Hoff in 1887 to develop his limit law, which explains the behavior of ideal dilute solutions; this work led directly to the

Membrane Technology and Applications R. W. Baker
 ISBN: 0-470-85445-6

van't Hoff equation. At about the same time, the concept of a perfectly selective semipermeable membrane was used by Maxwell and others in developing the kinetic theory of gases.

Early membrane investigators experimented with every type of diaphragm available to them, such as bladders of pigs, cattle or fish and sausage casings made of animal gut. Later, collodion (nitrocellulose) membranes were preferred, because they could be made reproducibly. In 1907, Bechhold devised a technique to prepare nitrocellulose membranes of graded pore size, which he determined by a bubble test [1]. Other early workers, particularly Elford [2], Zsigmondy and Bachmann [3] and Ferry [4] improved on Bechhold's technique, and by the early 1930s microporous collodion membranes were commercially available. During the next 20 years, this early microfiltration membrane technology was expanded to other polymers, notably cellulose acetate. Membranes found their first significant application in the testing of drinking water at the end of World War II. Drinking water supplies serving large communities in Germany and elsewhere in Europe had broken down, and filters to test for water safety were needed urgently. The research effort to develop these filters, sponsored by the US Army, was later exploited by the Millipore Corporation, the first and still the largest US microfiltration membrane producer.

By 1960, the elements of modern membrane science had been developed, but membranes were used in only a few laboratory and small, specialized industrial applications. No significant membrane industry existed, and total annual sales of membranes for all industrial applications probably did not exceed US$20 million in 2003 dollars. Membranes suffered from four problems that prohibited their widespread use as a separation process: They were too unreliable, too slow, too unselective, and too expensive. Solutions to each of these problems have been developed during the last 30 years, and membrane-based separation processes are now commonplace.

The seminal discovery that transformed membrane separation from a laboratory to an industrial process was the development, in the early 1960s, of the Loeb–Sourirajan process for making defect-free, high-flux, anisotropic reverse osmosis membranes [5]. These membranes consist of an ultrathin, selective surface film on a much thicker but much more permeable microporous support, which provides the mechanical strength. The flux of the first Loeb–Sourirajan reverse osmosis membrane was 10 times higher than that of any membrane then available and made reverse osmosis a potentially practical method of desalting water. The work of Loeb and Sourirajan, and the timely infusion of large sums of research and development dollars from the US Department of Interior, Office of Saline Water (OSW), resulted in the commercialization of reverse osmosis and was a major factor in the development of ultrafiltration and microfiltration. The development of electrodialysis was also aided by OSW funding.

Concurrently with the development of these industrial applications of membranes was the independent development of membranes for medical separation processes, in particular, the artificial kidney. W.J. Kolf [6] had demonstrated the first successful artificial kidney in The Netherlands in 1945. It took almost 20 years to refine the technology for use on a large scale, but these developments were complete by the early 1960s. Since then, the use of membranes in artificial organs has become a major life-saving procedure. More than 800 000 people are now sustained by artificial kidneys and a further million people undergo open-heart surgery each year, a procedure made possible by development of the membrane blood oxygenator. The sales of these devices comfortably exceed the total industrial membrane separation market. Another important medical application of membranes is for controlled drug delivery systems. A key figure in this area was Alex Zaffaroni, who founded Alza, a company dedicated to developing these products in 1966. The membrane techniques developed by Alza and its competitors are widely used in the pharmaceutical industry to improve the efficiency and safety of drug delivery.

The period from 1960 to 1980 produced a significant change in the status of membrane technology. Building on the original Loeb–Sourirajan technique, other membrane formation processes, including interfacial polymerization and multilayer composite casting and coating, were developed for making high-performance membranes. Using these processes, membranes with selective layers as thin as 0.1 μm or less are now being produced by a number of companies. Methods of packaging membranes into large-membrane-area spiral-wound, hollow-fine-fiber, capillary, and plate-and-frame modules were also developed, and advances were made in improving membrane stability. By 1980, microfiltration, ultrafiltration, reverse osmosis and electrodialysis were all established processes with large plants installed worldwide.

The principal development in the 1980s was the emergence of industrial membrane gas separation processes. The first major development was the Monsanto Prism® membrane for hydrogen separation, introduced in 1980 [7]. Within a few years, Dow was producing systems to separate nitrogen from air, and Cynara and Separex were producing systems to separate carbon dioxide from natural gas. Gas separation technology is evolving and expanding rapidly; further substantial growth will be seen in the coming years. The final development of the 1980s was the introduction by GFT, a small German engineering company, of the first commercial pervaporation systems for dehydration of alcohol. More than 100 ethanol and isopropanol pervaporation dehydration plants have now been installed. Other pervaporation applications are at the early commercial stage.

Types of Membranes

This book is limited to synthetic membranes, excluding all biological structures, but the topic is still large enough to include a wide variety of membranes that differ in chemical and physical composition and in the way they operate. In essence, a membrane is nothing more than a discrete, thin interface that moderates the

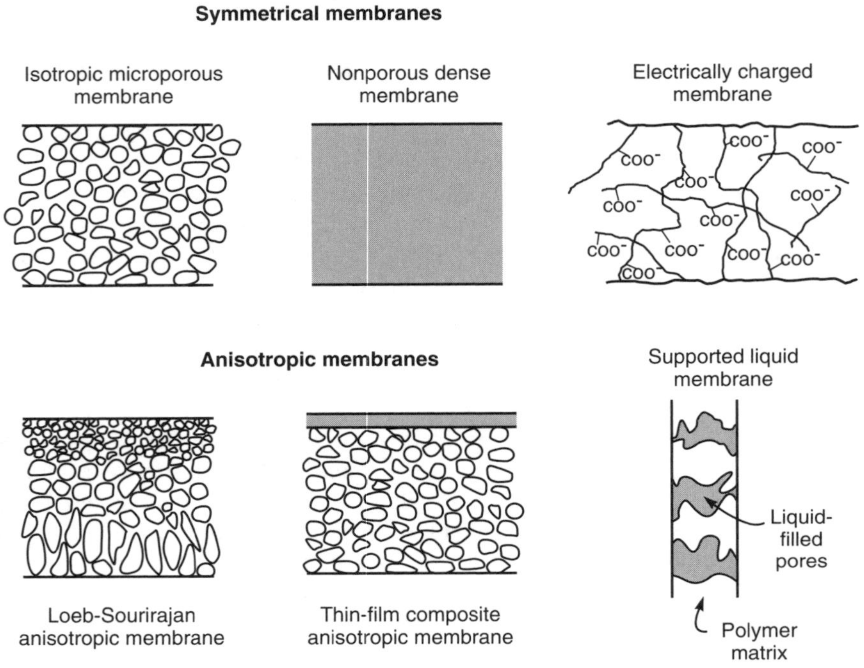

Figure 1.1 Schematic diagrams of the principal types of membranes

permeation of chemical species in contact with it. This interface may be molecularly homogeneous, that is, completely uniform in composition and structure, or it may be chemically or physically heterogeneous, for example, containing holes or pores of finite dimensions or consisting of some form of layered structure. A normal filter meets this definition of a membrane, but, by convention, the term filter is usually limited to structures that separate particulate suspensions larger than 1 to 10 μm. The principal types of membrane are shown schematically in Figure 1.1 and are described briefly below.

Isotropic Membranes

Microporous Membranes

A microporous membrane is very similar in structure and function to a conventional filter. It has a rigid, highly voided structure with randomly distributed, interconnected pores. However, these pores differ from those in a conventional filter by being extremely small, on the order of 0.01 to 10 μm in diameter. All particles larger than the largest pores are completely rejected by the membrane.

Particles smaller than the largest pores, but larger than the smallest pores are partially rejected, according to the pore size distribution of the membrane. Particles much smaller than the smallest pores will pass through the membrane. Thus, separation of solutes by microporous membranes is mainly a function of molecular size and pore size distribution. In general, only molecules that differ considerably in size can be separated effectively by microporous membranes, for example, in ultrafiltration and microfiltration.

Nonporous, Dense Membranes

Nonporous, dense membranes consist of a dense film through which permeants are transported by diffusion under the driving force of a pressure, concentration, or electrical potential gradient. The separation of various components of a mixture is related directly to their relative transport rate within the membrane, which is determined by their diffusivity and solubility in the membrane material. Thus, nonporous, dense membranes can separate permeants of similar size if their concentration in the membrane material (that is, their solubility) differs significantly. Most gas separation, pervaporation, and reverse osmosis membranes use dense membranes to perform the separation. Usually these membranes have an anisotropic structure to improve the flux.

Electrically Charged Membranes

Electrically charged membranes can be dense or microporous, but are most commonly very finely microporous, with the pore walls carrying fixed positively or negatively charged ions. A membrane with fixed positively charged ions is referred to as an anion-exchange membrane because it binds anions in the surrounding fluid. Similarly, a membrane containing fixed negatively charged ions is called a cation-exchange membrane. Separation with charged membranes is achieved mainly by exclusion of ions of the same charge as the fixed ions of the membrane structure, and to a much lesser extent by the pore size. The separation is affected by the charge and concentration of the ions in solution. For example, monovalent ions are excluded less effectively than divalent ions and, in solutions of high ionic strength, selectivity decreases. Electrically charged membranes are used for processing electrolyte solutions in electrodialysis.

Anisotropic Membranes

The transport rate of a species through a membrane is inversely proportional to the membrane thickness. High transport rates are desirable in membrane separation processes for economic reasons; therefore, the membrane should be as thin as possible. Conventional film fabrication technology limits manufacture of mechanically strong, defect-free films to about 20 μm thickness. The development of novel membrane fabrication techniques to produce anisotropic membrane structures was one of the major breakthroughs of membrane technology during the past 30 years. Anisotropic membranes consist of an extremely thin surface layer supported on a much thicker, porous substructure. The surface layer and its substructure may be formed in a single operation or separately. In composite membranes, the layers are usually made from different polymers. The separation properties and permeation rates of the membrane are determined exclusively by the surface layer; the substructure functions as a mechanical support. The advantages of the higher fluxes provided by anisotropic membranes are so great that almost all commercial processes use such membranes.

Ceramic, Metal and Liquid Membranes

The discussion so far implies that membrane materials are organic polymers and, in fact, the vast majority of membranes used commercially are polymer-based. However, in recent years, interest in membranes formed from less conventional materials has increased. Ceramic membranes, a special class of microporous membranes, are being used in ultrafiltration and microfiltration applications for which solvent resistance and thermal stability are required. Dense metal membranes, particularly palladium membranes, are being considered for the separation of hydrogen from gas mixtures, and supported liquid films are being developed for carrier-facilitated transport processes.

Membrane Processes

Six developed and a number of developing and yet-to-be-developed industrial membrane technologies are discussed in this book. In addition, sections are included describing the use of membranes in medical applications such as the artificial kidney, blood oxygenation, and controlled drug delivery devices. The status of all of these processes is summarized in Table 1.1.

The four developed industrial membrane separation processes are microfiltration, ultrafiltration, reverse osmosis, and electrodialysis. These processes are all well established, and the market is served by a number of experienced companies.

The range of application of the three pressure-driven membrane water separation processes—reverse osmosis, ultrafiltration and microfiltration—is illustrated in Figure 1.2. Ultrafiltration (Chapter 6) and microfiltration (Chapter 7) are basically similar in that the mode of separation is molecular sieving through increasingly fine pores. Microfiltration membranes filter colloidal particles and bacteria from 0.1 to 10 μm in diameter. Ultrafiltration membranes can be used to filter dissolved macromolecules, such as proteins, from solutions. The mechanism of separation by reverse osmosis membranes is quite different. In reverse osmosis membranes (Chapter 5), the membrane pores are so small, from 3 to 5 Å in diameter, that they are within the range of thermal motion of the polymer

Table 1.1 Membrane technologies addressed in this book

Category	Process	Status
Developed industrial membrane separation technologies	Microfiltration Ultrafiltration Reverse osmosis Electrodialysis	Well-established unit operations. No major breakthroughs seem imminent
Developing industrial membrane separation technologies	Gas separation Pervaporation	A number of plants have been installed. Market size and number of applications served are expanding
To-be-developed industrial membrane separation technologies	Carrier facilitated transport Membrane contactors Piezodialysis, etc.	Major problems remain to be solved before industrial systems will be installed on a large scale
Medical applications of membranes	Artificial kidneys Artificial lungs Controlled drug delivery	Well-established processes. Still the focus of research to improve performance, for example, improving biocompatibility

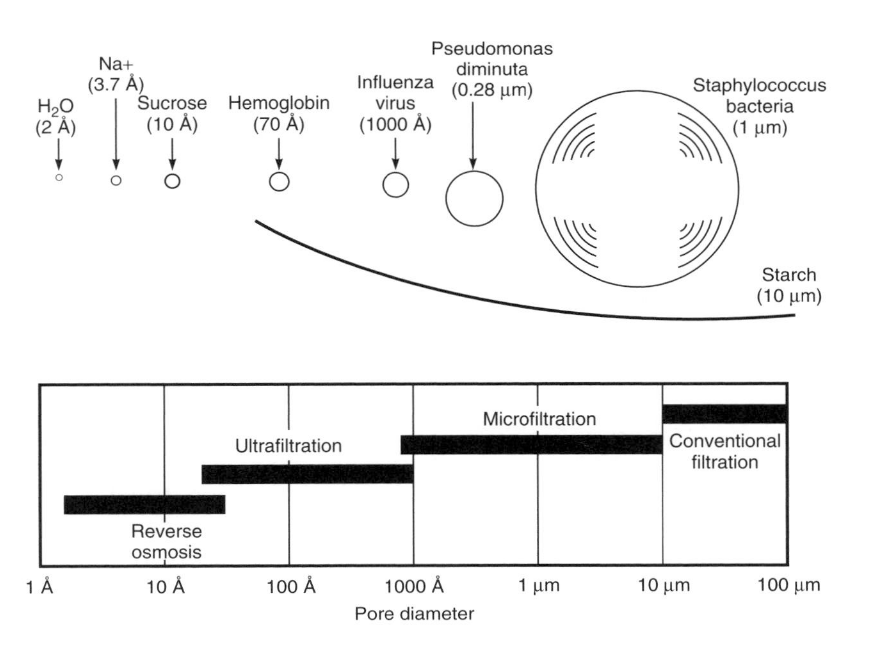

Figure 1.2 Reverse osmosis, ultrafiltration, microfiltration, and conventional filtration are related processes differing principally in the average pore diameter of the membrane filter. Reverse osmosis membranes are so dense that discrete pores do not exist; transport occurs via statistically distributed free volume areas. The relative size of different solutes removed by each class of membrane is illustrated in this schematic

chains that form the membrane. The accepted mechanism of transport through these membranes is called the solution-diffusion model. According to this model, solutes permeate the membrane by dissolving in the membrane material and diffusing down a concentration gradient. Separation occurs because of the difference in solubilities and mobilities of different solutes in the membrane. The principal application of reverse osmosis is desalination of brackish groundwater or seawater.

Although reverse osmosis, ultrafiltration and microfiltration are conceptually similar processes, the difference in pore diameter (or apparent pore diameter) produces dramatic differences in the way the membranes are used. A simple model of liquid flow through these membranes is to describe the membranes as a series of cylindrical capillary pores of diameter d. The liquid flow through a pore (q) is given by Poiseuille's law as:

$$q = \frac{\pi d^4}{128\ \mu\ell} \cdot \Delta p \tag{1.1}$$

where Δp is the pressure difference across the pore, μ is the liquid viscosity and ℓ is the pore length. The flux, or flow per unit membrane area, is the sum of all the flows through the individual pores and so is given by:

$$J = N \cdot \frac{\pi d^4}{128\ \mu\ell} \cdot \Delta p \tag{1.2}$$

where N is the number of pores per square centimeter of membrane.

For membranes of equal pore area and porosity (ε), the number of pores per square centimeter is proportional to the inverse square of the pore diameter. That is,

$$N = \varepsilon \cdot \frac{4}{\pi d^2} \tag{1.3}$$

It follows that the flux, given by combining Equations (1.2) and (1.3), is

$$J = \frac{\Delta p \varepsilon}{32\ \mu\ell} \cdot d^2 \tag{1.4}$$

From Figure 1.2, the typical pore diameter of a microfiltration membrane is 10 000 Å. This is 100-fold larger than the average ultrafiltration pore and 1000-fold larger than the (nominal) diameter of pores in reverse osmosis membranes. Because fluxes are proportional to the square of these pore diameters, the permeance, that is, flux per unit pressure difference ($J/\Delta p$) of microfiltration membranes is enormously higher than that of ultrafiltration membranes, which in turn is much higher than that of reverse osmosis membranes. These differences significantly impact the operating pressure and the way that these membranes are used industrially.

The fourth fully developed membrane process is electrodialysis (Chapter 10), in which charged membranes are used to separate ions from aqueous solutions under the driving force of an electrical potential difference. The process utilizes an electrodialysis stack, built on the filter-press principle and containing several hundred individual cells, each formed by a pair of anion and cation exchange membranes. The principal application of electrodialysis is the desalting of brackish groundwater. However, industrial use of the process in the food industry, for example, to deionize cheese whey, is growing, as is its use in pollution-control applications. A schematic of the process is shown in Figure 1.3.

Table 1.1 shows two developing industrial membrane separation processes: gas separation with polymer membranes (Chapter 8) and pervaporation (Chapter 9). Gas separation with membranes is the more advanced of the two techniques; at least 20 companies worldwide offer industrial, membrane-based gas separation systems for a variety of applications. Only a handful of companies currently offer industrial pervaporation systems. In gas separation, a gas mixture at an elevated pressure is passed across the surface of a membrane that is selectively permeable to one component of the feed mixture; the membrane permeate is enriched in this species. The basic process is illustrated in Figure 1.4. Major current applications

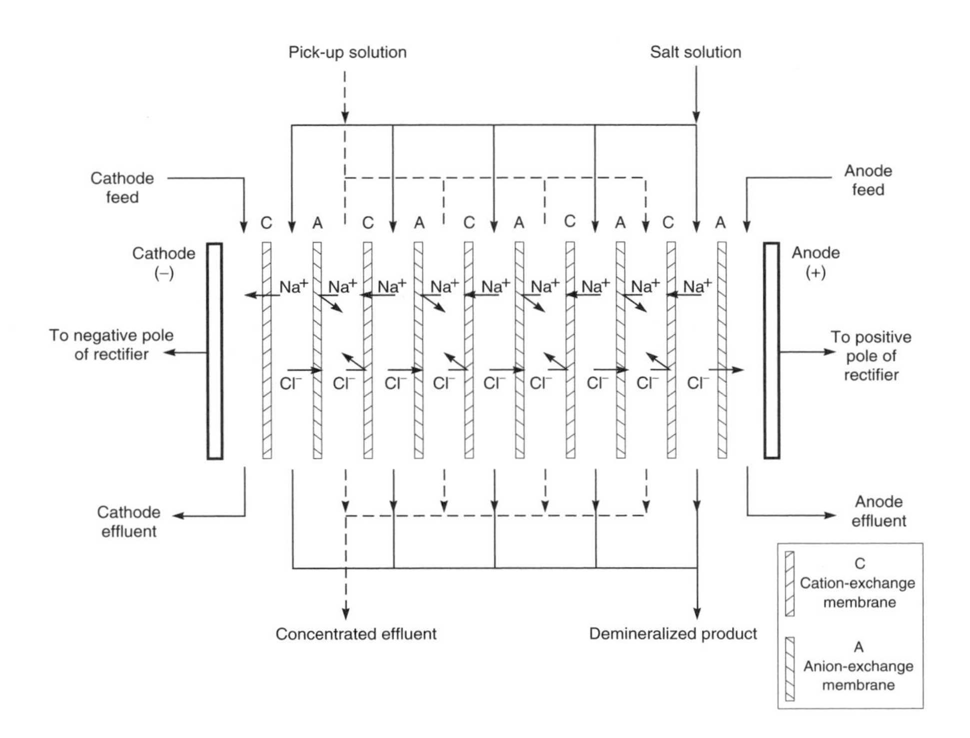

Figure 1.3 Schematic diagram of an electrodialysis process

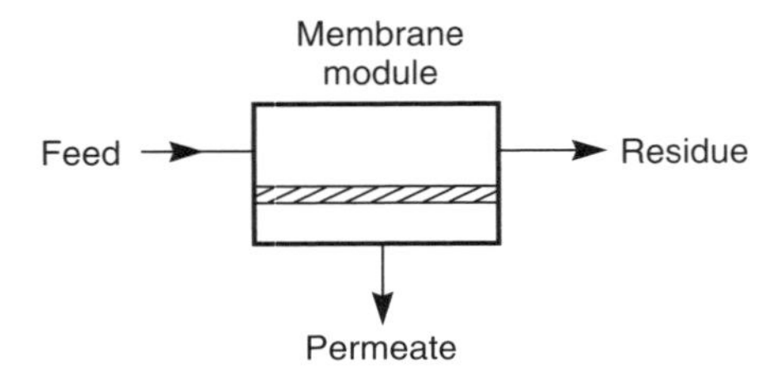

Figure 1.4 Schematic diagram of the basic membrane gas separation process

of gas separation membranes are the separation of hydrogen from nitrogen, argon and methane in ammonia plants; the production of nitrogen from air; and the separation of carbon dioxide from methane in natural gas operations. Membrane gas separation is an area of considerable current research interest, and the number of applications is expanding rapidly.

Pervaporation is a relatively new process that has elements in common with reverse osmosis and gas separation. In pervaporation, a liquid mixture contacts one side of a membrane, and the permeate is removed as a vapor from the other. The driving force for the process is the low vapor pressure on the permeate side of the membrane generated by cooling and condensing the permeate vapor. The attraction of pervaporation is that the separation obtained is proportional to the rate of permeation of the components of the liquid mixture through the selective membrane. Therefore, pervaporation offers the possibility of separating closely boiling mixtures or azeotropes that are difficult to separate by distillation or other means. A schematic of a simple pervaporation process using a condenser to generate the permeate vacuum is shown in Figure 1.5. Currently, the main industrial application of pervaporation is the dehydration of organic solvents, in particular, the dehydration of 90–95 % ethanol solutions, a difficult separation problem because of the ethanol–water azeotrope at 95 % ethanol. Pervaporation membranes that selectively permeate water can produce more than 99.9 % ethanol from these solutions. Pervaporation processes are also being developed for the removal of dissolved organics from water and for the separation of organic mixtures.

A number of other industrial membrane processes are placed in the category of to-be-developed technologies in Table 1.1. Perhaps the most important of these is carrier facilitated transport (Chapter 11), which often employs liquid membranes containing a complexing or carrier agent. The carrier agent reacts with one component of a mixture on the feed side of the membrane and then diffuses across the membrane to release the permeant on the product side of the membrane. The reformed carrier agent then diffuses back to the feed side of the membrane. Thus, the carrier agent acts as a shuttle to selectively transport one component from the feed to the product side of the membrane.

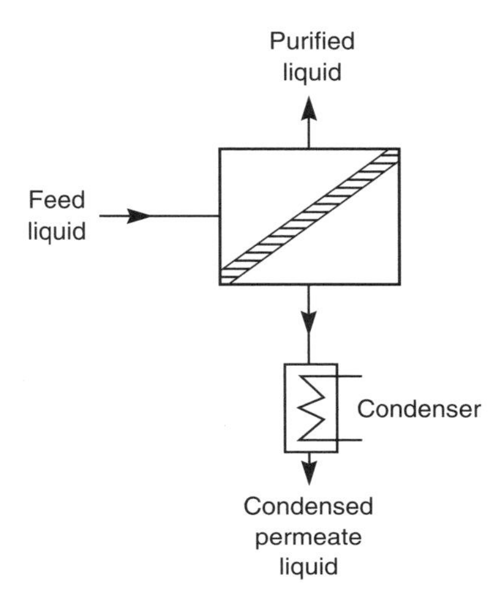

Figure 1.5 Schematic diagram of the basic pervaporation process

Facilitated transport membranes can be used to separate gases; membrane transport is then driven by a difference in the gas partial pressure across the membrane. Metal ions can also be selectively transported across a membrane, driven by a flow of hydrogen or hydroxyl ions in the other direction. This process is sometimes called coupled transport. Examples of carrier facilitated transport processes for gas and ion transport are shown in Figure 1.6.

Because the carrier facilitated transport process employs a reactive carrier species, very high membrane selectivities can be achieved. These selectivities are often far larger than the selectivities achieved by other membrane processes. This one fact has maintained interest in facilitated transport for the past 30 years, but no commercial applications have developed. The principal problem is the physical instability of the liquid membrane and the chemical instability of the carrier agent. In recent years a number of potential solutions to this problem have been developed, which may yet make carrier facilitated transport a viable process.

The membrane separation processes described above represent the bulk of the industrial membrane separation industry. Another process, dialysis, is not used industrially but is used on a large scale in medicine to remove toxic metabolites from blood in patients suffering from kidney failure. The first successful artificial kidney was based on cellophane (regenerated cellulose) dialysis membranes and was developed in 1945. Over the past 50 years, many changes have been made. Currently, most artificial kidneys are based on hollow-fiber membranes formed into modules having a membrane area of about 1 m^2; the process is illustrated in Figure 1.7. Blood is circulated through the center of the fiber, while isotonic

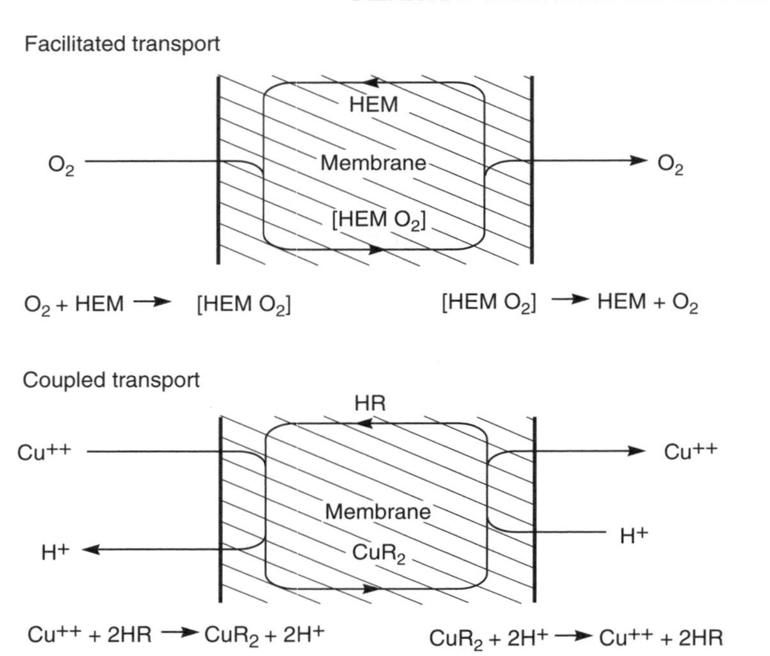

Figure 1.6 Schematic examples of carrier facilitated transport of gas and ions. The gas transport example shows the transport of oxygen across a membrane using hemoglobin as the carrier agent. The ion transport example shows the transport of copper ions across a membrane using a liquid ion-exchange reagent as the carrier agent

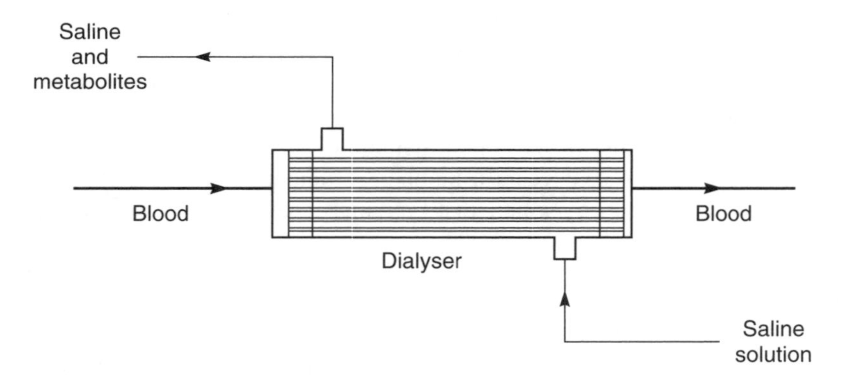

Figure 1.7 Schematic of a hollow fiber artificial kidney dialyser used to remove urea and other toxic metabolites from blood. About 100 million of these devices are used every year

saline, the dialysate, is pumped countercurrently around the outside of the fibers. Urea, creatinine, and other low-molecular-weight metabolites in the blood diffuse across the fiber wall and are removed with the saline solution. The process is quite slow, usually requiring several hours to remove the required amount of the metabolite from the patient, and must be repeated one or two times per week. In terms of membrane area used and dollar value of the membrane produced, artificial kidneys are the single largest application of membranes.

Following the success of the artificial kidney, similar devices were developed to remove carbon dioxide and deliver oxygen to the blood. These so-called artificial lungs are used in surgical procedures during which the patient's lungs cannot function. The dialysate fluid shown in Figure 1.7 is replaced with a carefully controlled sweep gas containing oxygen, which is delivered to the blood, and carbon dioxide, which is removed. These two medical applications of membranes are described in Chapter 12.

Another major medical use of membranes is in controlled drug delivery (Chapter 12). Controlled drug delivery can be achieved by a wide range of techniques, most of which involve membranes; a simple example is illustrated in Figure 1.8. In this device, designed to deliver drugs through the skin, drug is contained in a reservoir surrounded by a membrane. With such a system,

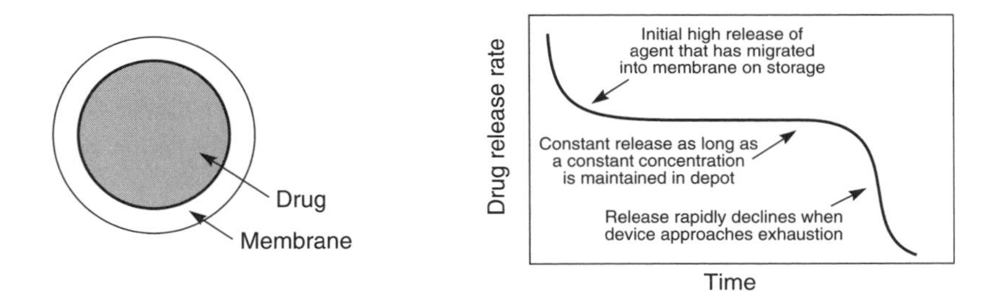

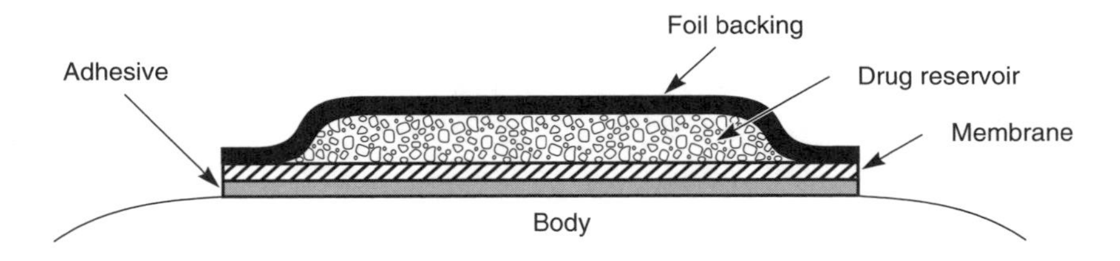

Figure 1.8 Schematic of transdermal patch in which the rate of delivery of drug to the body is controlled by a polymer membrane. Such patches are used to deliver many drugs including nitroglycerine, estradiol, nicotine and scopalamine

the release of drug is constant as long as a constant concentration of drug is maintained within the device. A constant concentration is maintained if the reservoir contains a saturated solution and sufficient excess of solid drug. Systems that operate using this principle are used to moderate delivery of drugs such as nitroglycerine (for angina), nicotine (for smoking cessation), and estradiol (for hormone replacement therapy) through the skin. Other devices using osmosis or biodegradation as the rate-controlling mechanism are also produced as implants and tablets. The total market of controlled release pharmaceuticals is comfortably above US$3 billion per year.

References

1. H. Bechhold, Kolloidstudien mit der Filtrationsmethode, *Z. Physik Chem.* **60**, 257 (1907).
2. W.J. Elford, Principles Governing the Preparation of Membranes Having Graded Porosities. The Properties of 'Gradocol' Membranes as Ultrafilters, *Trans. Faraday Soc.* **33**, 1094 (1937).
3. R. Zsigmondy and W. Bachmann, Über Neue Filter, *Z. Anorg. Chem.* **103**, 119 (1918).
4. J.D. Ferry, Ultrafilter Membranes and Ultrafiltration, *Chem. Rev.* **18**, 373 (1936).
5. S. Loeb and S. Sourirajan, Sea Water Demineralization by Means of an Osmotic Membrane, in *Saline Water Conversion–II, Advances in Chemistry Series Number 28*, American Chemical Society, Washington, DC, pp. 117–132 (1963).
6. W.J. Kolf and H.T. Berk, The Artificial Kidney: A Dialyzer with Great Area, *Acta Med Scand.* **117**, 121 (1944).
7. J.M.S. Henis and M.K. Tripodi, A Novel Approach to Gas Separation Using Composite Hollow Fiber Membranes, *Sep. Sci. Technol.* **15**, 1059 (1980).

2 MEMBRANE TRANSPORT THEORY

Introduction

The most important property of membranes is their ability to control the rate of permeation of different species. The two models used to describe the mechanism of permeation are illustrated in Figure 2.1. One is the solution-diffusion model, in which permeants dissolve in the membrane material and then diffuse through the membrane down a concentration gradient. The permeants are separated because of the differences in the solubilities of the materials in the membrane and the differences in the rates at which the materials diffuse through the membrane. The other model is the pore-flow model, in which permeants are transported by pressure-driven convective flow through tiny pores. Separation occurs because one of the permeants is excluded (filtered) from some of the pores in the membrane through which other permeants move. Both models were proposed in the nineteenth century, but the pore-flow model, because it was closer to normal physical experience, was more popular until the mid-1940s. However, during the 1940s, the solution-diffusion model was used to explain transport of gases through polymeric films. This use of the solution-diffusion model was relatively uncontroversial, but the transport mechanism in reverse osmosis membranes was a hotly debated issue in the 1960s and early 1970s [1–6]. By 1980, however, the proponents of solution-diffusion had carried the day; currently only a few die-hard pore-flow modelers use this approach to rationalize reverse osmosis.

Diffusion, the basis of the solution-diffusion model, is the process by which matter is transported from one part of a system to another by a concentration gradient. The individual molecules in the membrane medium are in constant random molecular motion, but in an isotropic medium, individual molecules have no preferred direction of motion. Although the average displacement of an individual molecule from its starting point can be calculated, after a period of time nothing can be said about the direction in which any individual molecule will move. However, if a concentration gradient of permeate molecules is formed in the medium, simple statistics show that a net transport of matter will occur

Membrane Technology and Applications R. W. Baker
© 2004 John Wiley & Sons, Ltd ISBN: 0-470-85445-6

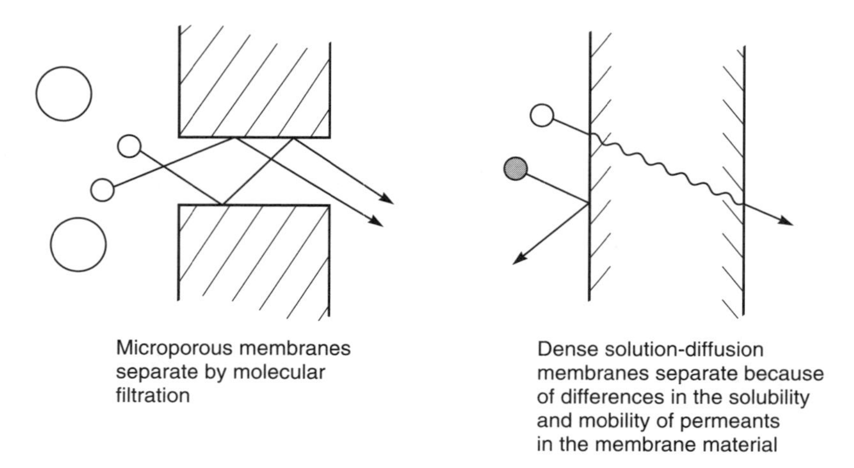

Figure 2.1 Molecular transport through membranes can be described by a flow through permanent pores or by the solution-diffusion mechanism

from the high concentration to the low concentration region. For example, when two adjacent volume elements with slightly different permeant concentrations are separated by an interface, then simply because of the difference in the number of molecules in each volume element, more molecules will move from the concentrated side to the less concentrated side of the interface than will move in the other direction. This concept was first recognized by Fick theoretically and experimentally in 1855 [7]. Fick formulated his results as the equation now called Fick's law of diffusion, which states

$$J_i = -D_i \frac{dc_i}{dx} \tag{2.1}$$

where J_i is the rate of transfer of component i or flux (g/cm^2 · s) and dc_i/dx *is* the concentration gradient of component i. The term D_i is called the diffusion coefficient (cm^2/s) and is a measure of the mobility of the individual molecules. The minus sign shows that the direction of diffusion is down the concentration gradient. Diffusion is an inherently slow process. In practical diffusion-controlled separation processes, useful fluxes across the membrane are achieved by making the membranes very thin and creating large concentration gradients in the membrane.

Pressure-driven convective flow, the basis of the pore flow model, is most commonly used to describe flow in a capillary or porous medium. The basic equation covering this type of transport is Darcy's law, which can be written as

$$J_i = K' c_i \frac{dp}{dx} \tag{2.2}$$

where dp/dx is the pressure gradient existing in the porous medium, c_i is the concentration of component i in the medium and K' is a coefficient reflecting the nature of the medium. In general, convective-pressure-driven membrane fluxes are high compared with those obtained by simple diffusion.

The difference between the solution-diffusion and pore-flow mechanisms lies in the relative size and permanence of the pores. For membranes in which transport is best described by the solution-diffusion model and Fick's law, the free-volume elements (pores) in the membrane are tiny spaces between polymer chains caused by thermal motion of the polymer molecules. These volume elements appear and disappear on about the same timescale as the motions of the permeants traversing the membrane. On the other hand, for a membrane in which transport is best described by a pore-flow model and Darcy's law, the free-volume elements (pores) are relatively large and fixed, do not fluctuate in position or volume on the timescale of permeant motion, and are connected to one another. The larger the individual free volume elements (pores), the more likely they are to be present long enough to produce pore-flow characteristics in the membrane. As a rough rule of thumb, the transition between transient (solution-diffusion) and permanent (pore-flow) pores is in the range 5–10 Å diameter.

The average pore diameter in a membrane is difficult to measure directly and must often be inferred from the size of the molecules that permeate the membrane or by some other indirect technique. With this caveat in mind membranes can be organized into the three general groups shown in Figure 2.2:

- Ultrafiltration, microfiltration and microporous Knudsen-flow gas separation membranes are all clearly microporous, and transport occurs by pore flow.
- Reverse osmosis, pervaporation and polymeric gas separation membranes have a dense polymer layer with no visible pores, in which the separation occurs. These membranes show different transport rates for molecules as small as 2–5 Å in diameter. The fluxes of permeants through these membranes are also much lower than through the microporous membranes. Transport is best described by the solution-diffusion model. The spaces between the polymer chains in these membranes are less than 5 Å in diameter and so are within the normal range of thermal motion of the polymer chains that make up the membrane matrix. Molecules permeate the membrane through free volume elements between the polymer chains that are transient on the timescale of the diffusion processes occurring.
- Membranes in the third group contain pores with diameters between 5 Å and 10 Å and are intermediate between truly microporous and truly solution-diffusion membranes. For example, nanofiltration membranes are intermediate between ultrafiltration membranes and reverse osmosis membranes. These membranes have high rejections for the di- and trisaccharides sucrose and raffinose with molecular diameters of 10–13 Å, but freely pass the monosaccharide fructose with a molecular diameter of about 5–6 Å.

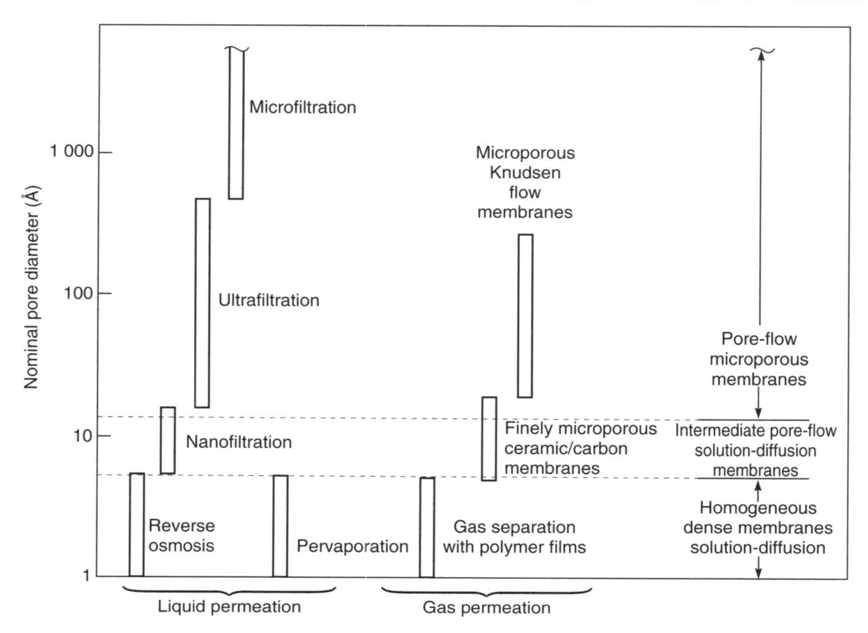

Figure 2.2 Schematic representation of the nominal pore size and best theoretical model for the principal membrane separation processes

In this chapter, permeation through dense nonporous membranes is covered first; this includes permeation in reverse osmosis, pervaporation, and gas separation membranes. Transport occurs by molecular diffusion and is described by the solution-diffusion model. The predictions of this model are in good agreement with experimental data, and a number of simple equations that usefully rationalize the properties of these membranes result. In the second part of the chapter, transport in microporous ultrafiltration and microfiltration membranes is covered more briefly. Transport through these membranes occurs by convective flow with some form of sieving mechanism producing the separation. However, the ability of theory to rationalize transport in these membranes is poor. A number of factors concurrently affect permeation, so a simple quantitative description of the process is not possible. Finally, a brief discussion of membranes that fall into the 'intermediate' category is given.

Solution-diffusion Model

Molecular Dynamics Simulations

The solution-diffusion model applies to reverse osmosis, pervaporation and gas permeation in polymer films. At first glance these processes appear to be very different. Reverse osmosis uses a large pressure difference across the membrane to separate water from salt solutions. In pervaporation, the pressure difference across the membrane is small, and the process is driven by the vapor pressure difference between the feed liquid and the low partial pressure of the permeate vapor. Gas permeation involves transport of gases down a pressure or concentration gradient. However, all three processes involve diffusion of molecules in a dense polymer. The pressure, temperature, and composition of the fluids on either side of the membrane determine the concentration of the diffusing species at the membrane surface in equilibrium with the fluid. Once dissolved in the membrane, individual permeating molecules move by the same random process of molecular diffusion no matter whether the membrane is being used in reverse osmosis, pervaporation, or gas permeation. Often, similar membranes are used in very different processes. For example, cellulose acetate membranes were developed for desalination of water by reverse osmosis, but essentially identical membranes have been used in pervaporation to dehydrate alcohol and are widely used in gas permeation to separate carbon dioxide from natural gas. Similarly, silicone rubber membranes are too hydrophobic to be useful in reverse osmosis but are used to separate volatile organics from water by pervaporation and organic vapors from air in gas permeation.

The advent of powerful computers has allowed the statistical fluctuations in the volumes between polymer chains due to thermal motion to be calculated. Figure 2.3 shows the results of a computer molecular dynamics simulation calculation for a small-volume element of a polymer. The change in position of individual polymer molecules in a small-volume element can be calculated at short enough time intervals to represent the normal thermal motion occurring in a polymeric matrix. If a penetrant molecule is placed in one of the small-free-volume microcavities between polymer chains, its motion can also be calculated. The simulated motion of a carbon dioxide molecule in a 6FDA-4PDA polyimide matrix is shown in Figure 2.3 [8]. During the first 100 ps of the simulation, the carbon dioxide molecule bounces around in the cavity where it has been placed, never moving more than about 5 Å, the diameter of the microcavity. After 100 ps, however, a chance thermal motion moves a segment of the polymer chains sufficiently for the carbon dioxide molecule to jump approximately 10 Å to an adjacent cavity where it remains until another movement of the polymer chains allows it to jump to another cavity. By repeating these calculations many times and averaging the distance moved by the gas molecule, its diffusion coefficient can be calculated.

An alternative method of representing the movement of an individual molecule by computational techniques is shown in Figure 2.4 [9]. This figure shows the movement of three different permeate molecules over a period of 200 ps in a silicone rubber polymer matrix. The smaller helium molecule moves more frequently and makes larger jumps than the larger methane molecule. Helium, with a molecular diameter of 2.55 Å, has many more opportunities to move from one

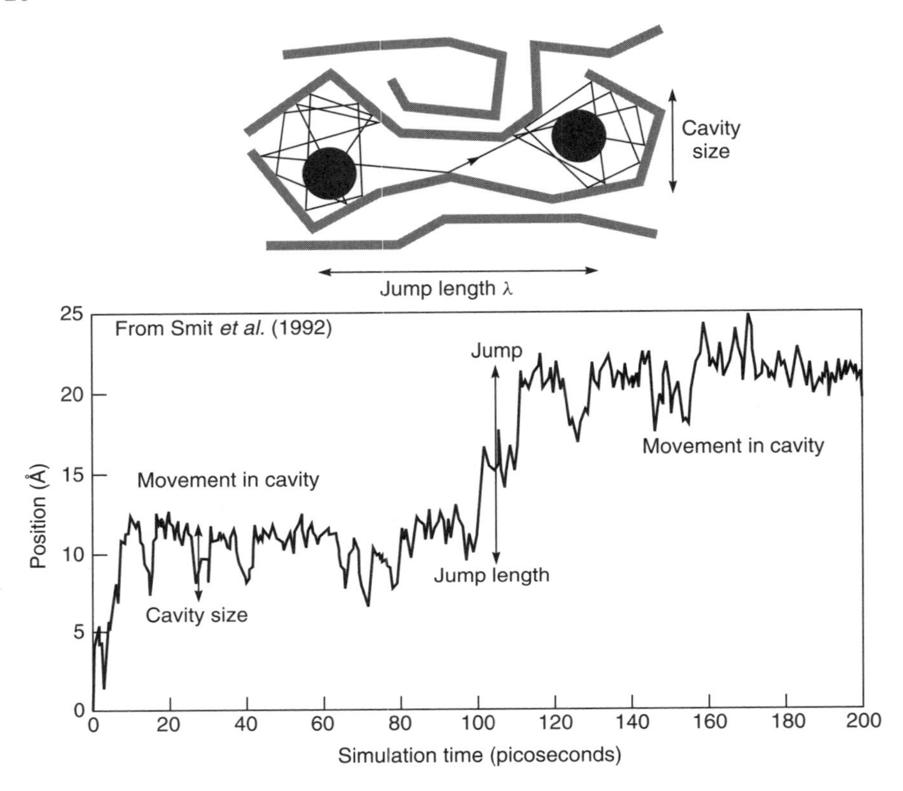

Figure 2.3 Motion of a carbon dioxide molecule in a 6FDA-4PDA polymer matrix [8]. Reprinted from *J. Membr. Sci.* **73**, E. Smit, M.H.V. Mulder, C.A. Smolders, H. Karrenbeld, J. van Eerden and D. Feil, Modeling of the Diffusion of Carbon Dioxide in Polyimide Matrices by Computer Simulation, p. 247, Copyright 1992, with permission from Elsevier

position to another than methane, with a molecular diameter of 3.76 Å. Oxygen, with a molecular diameter of 3.47 Å, has intermediate mobility. The effect of polymer structure on diffusion can be seen by comparing the distance moved by the gas molecules in the same 200-ps period in Figures 2.3 and 2.4. Figure 2.3 simulates diffusion in a glassy rigid-backbone polyimide. In 200 ps, the permeate molecule has made only one large jump. Figure 2.4 simulates diffusion in silicone rubber, a material with a very flexible polymer backbone. In 200 ps, all the permeants in silicone rubber have made a number of large jumps from one microcavity to another.

Molecular dynamics simulations also allow the transition from the solution-diffusion to the pore-flow transport mechanism to be seen. As the microcavities become larger, the transport mechanism changes from the diffusion process

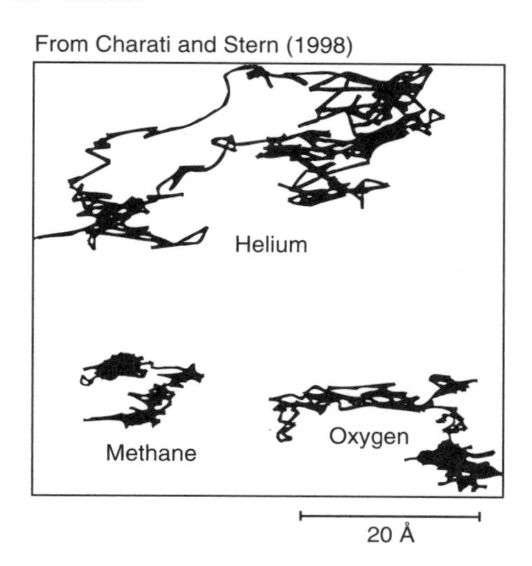

Figure 2.4 Simulated trajectories of helium, oxygen and methane molecules during a 200-ps time period in a poly(dimethylsiloxane) matrix [9]. Reprinted with permission from S.G. Charati and S.A. Stern, Diffusion of Gases in Silicone Polymers: Molecular Dynamic Simulations, *Macromolecules* **31**, 5529. Copyright 1998, American Chemical Society

simulated in Figures 2.3 and 2.4 to a pore-flow mechanism. Permanent pores form when the microcavities are larger than about 10 Å in diameter.

However, molecular dynamics calculations are at an early stage of development. Current estimates of diffusion coefficients from these simulations are generally far from matching the experimental values, and enormous computing power and a better understanding of the interactions between the molecules of polymer chains will be required to produce accurate predictions. Nonetheless, the technique demonstrates the qualitative basis of the solution-diffusion model in a very graphic way. Currently, the best quantitative description of permeation uses phenomenological equations, particularly Fick's law. This description is given in the section that follows, which outlines the mathematical basis of the solution-diffusion model. Much of this section is adapted from a 1995 *Journal of Membrane Science* article written with my colleague, Hans Wijmans [10].

Concentration and Pressure Gradients in Membranes

The starting point for the mathematical description of diffusion in membranes is the proposition, solidly based in thermodynamics, that the driving forces of pressure, temperature, concentration, and electrical potential are interrelated and that the overall driving force producing movement of a permeant is the gradient in its chemical potential. Thus, the flux, J_i (g/cm^2 · s), of a component, i, is described by the simple equation

$$J_i = -L_i \frac{d\mu_i}{dx} \tag{2.3}$$

where $d\mu_i/dx$ is the chemical potential gradient of component i and L_i is a coefficient of proportionality (not necessarily constant) linking this chemical potential driving force to flux. Driving forces, such as gradients in concentration, pressure, temperature, and electrical potential can be expressed as chemical potential gradients, and their effect on flux expressed by this equation. This approach is extremely useful, because many processes involve more than one driving force, for example, both pressure and concentration in reverse osmosis. Restricting the approach to driving forces generated by concentration and pressure gradients, the chemical potential is written as

$$d\mu_i = RT\,d\ln(\gamma_i n_i) + \upsilon_i dp \tag{2.4}$$

where n_i is the mole fraction (mol/mol) of component i, γ_i is the activity coefficient (mol/mol) linking mole fraction with activity, p is the pressure, and υ_i is the molar volume of component i.

In incompressible phases, such as a liquid or a solid membrane, volume does not change with pressure. In this case, integrating Equation (2.4) with respect to concentration and pressure gives

$$\mu_i = \mu_i^o + RT\ln(\gamma_i n_i) + \upsilon_i(p - p_i^o) \tag{2.5}$$

where μ_i^o is the chemical potential of pure i at a reference pressure, p_i^o.

In compressible gases, the molar volume changes with pressure. Using the ideal gas laws in integrating Equation (2.4) gives

$$\mu_i = \mu_i^o + RT\ln(\gamma_i n_i) + RT\ln\frac{p}{p_i^o} \tag{2.6}$$

To ensure that the reference chemical potential μ_i^o is identical in Equations (2.5) and (2.6), the reference pressure p_i^o is defined as the saturation vapor pressure of i, $p_{i_{sat}}$. Equations (2.5) and (2.6) can then be rewritten as

$$\mu_i = \mu_i^o + RT\ln(\gamma_i n_i) + \upsilon_i(p - p_{i_{sat}}) \tag{2.7}$$

for incompressible liquids and the membrane phase, and as

$$\mu_i = \mu_i^o + RT\ln(\gamma_i n_i) + RT\ln\frac{p}{p_{i_{sat}}} \tag{2.8}$$

for compressible gases.

Several assumptions must be made to define any permeation model. Usually, the first assumption governing transport through membranes is that the fluids on either side of the membrane are in equilibrium with the membrane material at the interface. This assumption means that the gradient in chemical potential from one side of the membrane to the other is continuous. Implicit in this assumption is that the rates of absorption and desorption at the membrane interface are much higher than the rate of diffusion through the membrane. This appears to be the case in almost all membrane processes, but may fail in transport processes involving chemical reactions, such as facilitated transport, or in diffusion of gases through metals, where interfacial absorption can be slow.

The second assumption concerns the pressure and concentration gradients in the membrane. The solution-diffusion model assumes that when pressure is applied across a dense membrane, the pressure throughout the membrane is constant at the highest value. This assumes, in effect, that solution-diffusion membranes transmit pressure in the same way as liquids. Consequently, the solution-diffusion model assumes that the pressure within a membrane is uniform and that the chemical potential gradient across the membrane is expressed only as a concentration gradient [5,10]. The consequences of these two assumptions are illustrated in Figure 2.5, which shows pressure-driven permeation of a one-component solution through a membrane by the solution-diffusion mechanism.

In the solution-diffusion model, the pressure within the membrane is constant at the high-pressure value (p_o), and the gradient in chemical potential across the membrane is expressed as a smooth gradient in solvent activity ($\gamma_i n_i$). The flow that occurs down this gradient is expressed by Equation (2.3), but because no pressure gradient exists within the membrane, Equation (2.3) can be rewritten by combining Equations (2.3) and (2.4). Assuming γ_i is constant, this gives

$$J_i = -\frac{RTL_i}{n_i} \cdot \frac{\mathrm{d}n_i}{\mathrm{d}x} \tag{2.9}$$

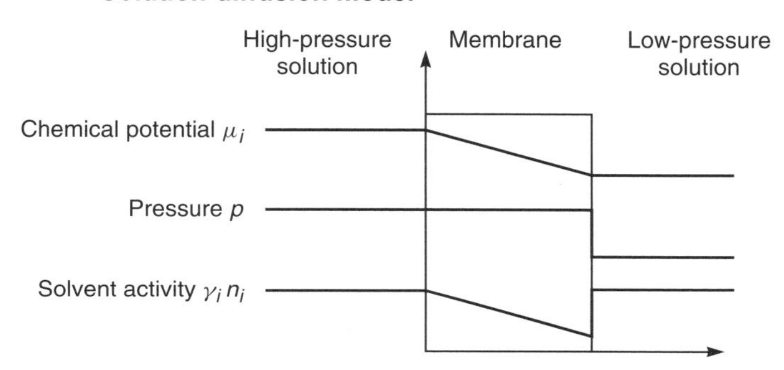

Figure 2.5 Pressure driven permeation of a one-component solution through a membrane according to the solution-diffusion transport model

In Equation (2.9), the gradient of component i across the membrane is expressed as a gradient in mole fraction of component i. Using the more practical term concentration c_i (g/cm^3) defined as

$$c_i = m_i \rho n_i \tag{2.10}$$

where m_i is the molecular weight of i (g/mol) and ρ is the molar density (mol/cm^3), Equation (2.9) can be written as

$$J_i = -\frac{RTL_i}{c_i} \cdot \frac{\mathrm{d}c_i}{\mathrm{d}x} \tag{2.11}$$

Equation (2.11) has the same form as Fick's law in which the term RTL_i/c_i can be replaced by the diffusion coefficient D_i. Thus,

$$J_i = -D_i \frac{\mathrm{d}c_i}{\mathrm{d}x} \tag{2.12}$$

Integrating over the thickness of the membrane then gives[1]

$$J_i = \frac{D_i (c_{i_{o(m)}} - c_{i_{\ell(m)}})}{\ell} \tag{2.13}$$

By using osmosis as an example, concentration and pressure gradients according to the solution-diffusion model can be discussed in a somewhat more complex situation. The activity, pressure, and chemical potential gradients within this type of membrane are illustrated in Figure 2.6.

Figure 2.6(a) shows a semipermeable membrane separating a salt solution from the pure solvent. The pressure is the same on both sides of the membrane. For simplicity, the gradient of salt (component j) is not shown in this figure, but the membrane is assumed to be very selective, so the concentration of salt within the membrane is small. The difference in concentration across the membrane results in a continuous, smooth gradient in the chemical potential of the water (component i) across the membrane, from μ_{i_ℓ} on the water side to μ_{i_o} on the salt side. The pressure within and across the membrane is constant (that is, $p_o = p_m = p_\ell$) and the solvent activity gradient ($\gamma_{i_{(m)}} n_{i_{(m)}}$) falls continuously from the pure water (solvent) side to the saline (solution) side of the membrane. Consequently, water passes across the membrane from right to left.

Figure 2.6(b) shows the situation at the point of osmotic equilibrium, when sufficient pressure has been applied to the saline side of the membrane to bring the flow across the membrane to zero. As shown in Figure 2.6(b), the pressure

[1] In the equations that follow, the terms i and j represent components of a solution, and the terms o and ℓ represent the positions of the feed and permeate interfaces, respectively, of the membrane. Thus the term c_{i_o} represents the concentration of component i in the fluid (gas or liquid) in contact with the membrane at the feed interface. The subscript m is used to represent the membrane phase. Thus, $c_{i_{o(m)}}$ is the concentration of component i in the membrane at the feed interface (point o).

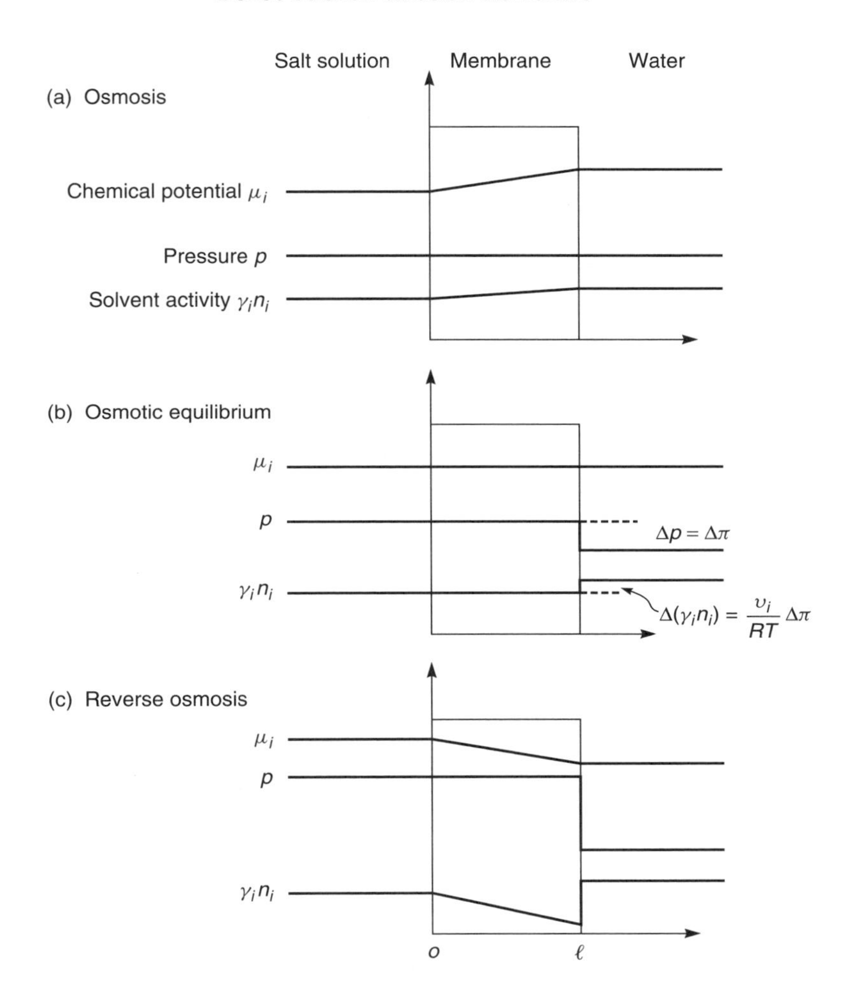

Figure 2.6 Chemical potential, pressure, and solvent activity profiles through an osmotic membrane following the solution-diffusion model. The pressure in the membrane is uniform and equal to the high-pressure value, so the chemical potential gradient within the membrane is expressed as a concentration gradient

within the membrane is assumed to be constant at the high-pressure value (p_o). There is a discontinuity in pressure at the permeate side of the membrane, where the pressure falls abruptly from p_o to p_ℓ, the pressure on the solvent side of the membrane. This pressure difference ($p_o - p_\ell$) can be expressed in terms of the chemical potential difference between the feed and permeate solutions.

The membrane in contact with the permeate-side solution is in equilibrium with this solution. Thus, Equation (2.7) can be used to link the two phases in terms of their chemical potentials, that is

$$\mu_{i_{\ell(m)}} = \mu_{i_\ell} \tag{2.14}$$

and so

$$RT\ln(\gamma_{i_{\ell(m)}} n_{i_{\ell(m)}}) + \upsilon_i p_o = RT\ln(\gamma_{i_\ell} n_{i_\ell}) + \upsilon_i p_\ell \tag{2.15}$$

On rearranging, this gives

$$RT\ln(\gamma_{i_{\ell(m)}} n_{i_{\ell(m)}}) - RT\ln(\gamma_{i_\ell} n_{i_\ell}) = -\upsilon_i(p_o - p_\ell) \tag{2.16}$$

At osmotic equilibrium $\Delta(\gamma_i n_i)$ can also be defined by

$$\Delta(\gamma_i n_i) = \gamma_{i_\ell} n_{i_\ell} - \gamma_{i_{\ell(m)}} n_{i_{\ell(m)}} \tag{2.17}$$

and, since $\gamma_{i_\ell} n_{i_\ell} \approx 1$, it follows, on substituting Equation (2.17) into (2.16), that

$$RT\ln[1 - \Delta(\gamma_i n_i)] = -\upsilon_i(p_o - p_\ell) \tag{2.18}$$

Since $\Delta(\gamma_i n_i)$ is small, $\ln[1 - \Delta(\gamma_i n_i)] \approx \Delta(\gamma_i n_i)$, and Equation (2.18) reduces to

$$\Delta(\gamma_i n_i) = \frac{-\upsilon_i(p_o - p_\ell)}{RT} = \frac{-\upsilon_i \Delta\pi}{RT} \tag{2.19}$$

Thus, the pressure difference, $(p_o - p_\ell) = \Delta\pi$, across the membrane balances the solvent activity difference $\Delta(\gamma_i n_i)$ across the membrane, and the flow is zero.

If a pressure higher than the osmotic pressure is applied to the feed side of the membrane, as shown in Figure 2.6(c), then the solvent activity difference across the membrane increases further, resulting in a flow from left to right. This is the process of reverse osmosis.

The important conclusion illustrated by Figures 2.5 and 2.6 is that, although the fluids on either side of a membrane may be at different pressures and concentrations, within a perfect solution-diffusion membrane, there is no pressure gradient—only a concentration gradient. Flow through this type of membrane is expressed by Fick's law, Equation (2.13).

Application of the Solution-diffusion Model to Specific Processes

In this section the solution-diffusion model is used to describe transport in dialysis, reverse osmosis, gas permeation and pervaporation membranes. The resulting equations, linking the driving forces of pressure and concentration with flow, are then shown to be consistent with experimental observations.

The general approach is to use the first assumption of the solution-diffusion model, namely, that the chemical potential of the feed and permeate fluids are in equilibrium with the adjacent membrane surfaces. From this assumption, the chemical potential in the fluid and membrane phases can be equated using the appropriate expressions for chemical potential given in Equations (2.7) and (2.8). By rearranging these equations, the concentrations of the different species in the membrane at the fluids interface ($c_{i_{o(m)}}$ and $c_{i_{\ell(m)}}$) can be obtained in terms of the pressure and composition of the feed and permeate fluids. These values for $c_{i_{o(m)}}$ and $c_{i_{\ell(m)}}$ can then be substituted into the Fick's law expression, Equation (2.13), to give the transport equation for the particular process.

Dialysis

Dialysis is the simplest application of the solution-diffusion model because only concentration gradients are involved. In dialysis, a membrane separates two solutions of different compositions. The concentration gradient across the membrane causes a flow of solute and solvent from one side of the membrane to the other.

Following the general procedure described above, equating the chemical potentials in the solution and membrane phase at the feed-side interface of the membrane gives

$$\mu_{i_o} = \mu_{i_{o(m)}} \tag{2.20}$$

Substituting the expression for the chemical potential of incompressible fluids from Equation (2.7) gives

$$\mu_i^o + RT\ln(\gamma_{i_o}^{\mathrm{L}} n_{i_o}) + \upsilon_i(p_o - p_{i_{\mathrm{sat}}}) = \mu_i^o + RT\ln(\gamma_{i_{o(m)}} n_{i_{o(m)}}) + \upsilon_{i_{o(m)}}(p_o - p_{i_{\mathrm{sat}}}) \tag{2.21}$$

which leads to[2]

$$\ln(\gamma_{i_o}^{\mathrm{L}} n_{i_o}) = \ln(\gamma_{i_{o(m)}} n_{i_{o(m)}}) \tag{2.22}$$

and thus

$$n_{i_{o(m)}} = \frac{\gamma_{i_o}^{\mathrm{L}}}{\gamma_{i_{o(m)}}} \cdot n_{i_o} \tag{2.23}$$

or from Equation (2.10)

$$c_{i_{o(m)}} = \frac{\gamma_{i_o} \rho_m}{\gamma_{i_{o(m)}} \rho_o} \cdot c_{i_o} \tag{2.24}$$

Hence, defining a sorption coefficient K_i^{L} as

$$K_i^{\mathrm{L}} = \frac{\gamma_{i_o} \rho_m}{\gamma_{i_{o(m)}} \rho_o} \tag{2.25}$$

Equation (2.24) becomes

$$c_{i_{o(m)}} = K_i^{\mathrm{L}} \cdot c_{i_o} \tag{2.26}$$

[2]The superscripts G and L are used here and later to distinguish between gas phase and liquid phase activity coefficients, sorption coefficients and permeability coefficients.

From Theeuwes et al. (1978)

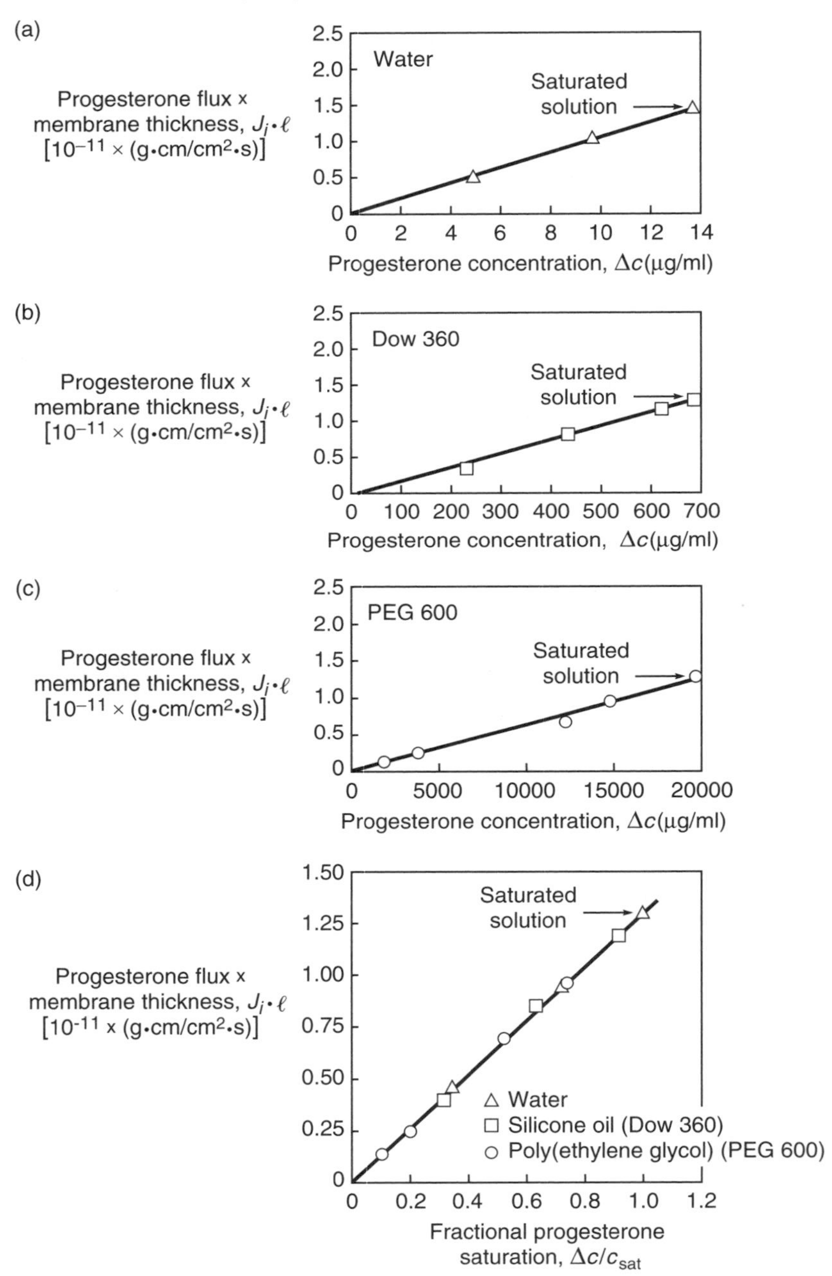
(a)
Progesterone flux x membrane thickness, J_i • ℓ [10^-11 × (g•cm/cm^2•s)]
Water
Saturated solution
2.5
2.0
1.5
1.0
0.5
0
0 2 4 6 8 10 12 14
Progesterone concentration, Δc(μg/ml)
(b)
Progesterone flux x membrane thickness, J_i • ℓ [10^-11 × (g•cm/cm^2•s)]
Dow 360
Saturated solution
2.5
2.0
1.5
1.0
0.5
0
0 100 200 300 400 500 600 700
Progesterone concentration, Δc(μg/ml)
(c)
Progesterone flux x membrane thickness, J_i • ℓ [10^-11 × (g•cm/cm^2•s)]
PEG 600
Saturated solution
2.5
2.0
1.5
1.0
0.5
0
0 5000 10000 15000 20000
Progesterone concentration, Δc(μg/ml)
(d)
Progesterone flux x membrane thickness, J_i • ℓ [10^-11 x (g•cm/cm^2•s)]
Saturated solution
1.50
1.25
1.00
0.75
0.50
0.25
0
△ Water
□ Silicone oil (Dow 360)
○ Poly(ethylene glycol) (PEG 600)
0 0.2 0.4 0.6 0.8 1.0 1.2
Fractional progesterone saturation, Δc/c_sat

On the permeate side of the membrane, the same procedure can be followed, leading to an equivalent expression

$$c_{i_{\ell(m)}} = K_i^{\mathrm{L}} \cdot c_{i_\ell} \tag{2.27}$$

The concentrations of permeant within the membrane phase at the two interfaces can then be substituted from Equations (2.26) and (2.27) into the Fick's law expression, Equation (2.13), to give the familiar expression describing permeation through dialysis membranes:

$$J_i = \frac{D_i K_i^{\mathrm{L}}}{\ell}(c_{i_o} - c_{i_\ell}) = \frac{P_i^{\mathrm{L}}}{\ell}(c_{i_o} - c_{i_\ell}) \tag{2.28}$$

The product $D_i K_i^{\mathrm{L}}$ is normally referred to as the permeability coefficient, P_i^{L}. For many systems, D_i, K_i^{L}, and thus P_i^{L} are concentration dependent. Thus, Equation (2.28) implies the use of values for D_i, K_i^{L}, and P_i^{L} that are averaged over the membrane thickness.

The permeability coefficient P_i^{L} is often treated as a pure materials constant, depending only on the permeant and the membrane material, but in fact the nature of the solvent used in the liquid phase is also important. From Equations (2.28) and (2.25), P_i^{L} can be written as

$$P_i^{\mathrm{L}} = D_i \cdot \gamma_i^{\mathrm{L}} / \gamma_{i(m)} \cdot \frac{\rho_m}{\rho_o} \tag{2.29}$$

The presence of the term γ_i^{L} makes the permeability coefficient a function of the solvent used as the liquid phase. Some experimental data illustrating this effect are shown in Figure 2.7 [11], which is a plot of the product of the progesterone flux and the membrane thickness, $J_i \ell$ against the concentration difference across the membrane, $(c_{i_o} - c_{i_\ell})$. From Equation (2.28), the slope of this line is the permeability, P_i^L. Three sets of dialysis permeation experiments are reported, in which the solvent used to dissolve the progesterone is water, silicone oil and poly(ethylene glycol) MW 600 (PEG 600), respectively. The permeability calculated from these plots varies from 9.5×10^{-7} cm^2/s for water to 6.5×10^{-10} cm^2/s for PEG 600. This difference reflects the activity term γ_i^{L} in Equation (2.28). However, when the driving force across the membrane is

---

Figure 2.7 Permeation of progesterone through polyethylene vinyl acetate films. The thickness-normalized progesterone flux $(J_i \cdot \ell)$ is plotted against the progesterone concentration across the membrane, Δc [11]. The solvents for the progesterone are (a) water, (b) silicone oil and (c) poly(ethylene glycol) (PEG 600). Because of the different solubilities of progesterone in these solvents, the permeabilities calculated from these data through Equation (2.28) vary 1000-fold. All the data can be rationalized onto a single curve by plotting the thickness-normalized flux against fractional progesterone saturation as described in the text and shown in (d). The slope of this line, $P_i^{\mathrm{L}} c_{\mathrm{sat}}$ or $D_i m_i \rho_m / \gamma_{i_{(m)}}$ is a materials property dependent only on the membrane material and the permeant and independent of the solvent

represented not as a difference in concentration but as a difference in fractional saturation between the feed and permeate solution, all the data fall on a single line as shown in Figure 2.7(d). The slope of this line is the term $P_i^L c_{sat}$. This result is also in agreement with Equation (2.29), which when combined with the approximation that, for dilute solutions, the activity of component i can be written as

$$\gamma_i^L = \frac{1}{n_{i_{sat}}} = \frac{m_i \rho_o}{c_{i_{sat}}} \tag{2.30}$$

yields

$$P_i^L c_{i_{sat}} = \frac{D_i m_i \rho_m}{\gamma_{i_{(m)}}} \tag{2.31}$$

The terms $D_i m_i \rho_m / \gamma_{i_{(m)}}$ and, therefore, $P_i^L c_{sat}$ are determined solely by the permeant and the membrane material and are thus independent of the liquid phase surrounding the membrane.

Reverse Osmosis

Reverse osmosis and normal osmosis (dialysis) are directly related processes. In simple terms, if a selective membrane (i.e., a membrane freely permeable to water, but much less permeable to salt) separates a salt solution from pure water, water will pass through the membrane from the pure water side of the membrane into the side less concentrated in water (salt side) as shown in Figure 2.8. This process is called normal osmosis. If a hydrostatic pressure is applied to the salt side of the membrane, the flow of water can be retarded and, when the applied pressure is sufficient, the flow ceases. The hydrostatic pressure required to stop

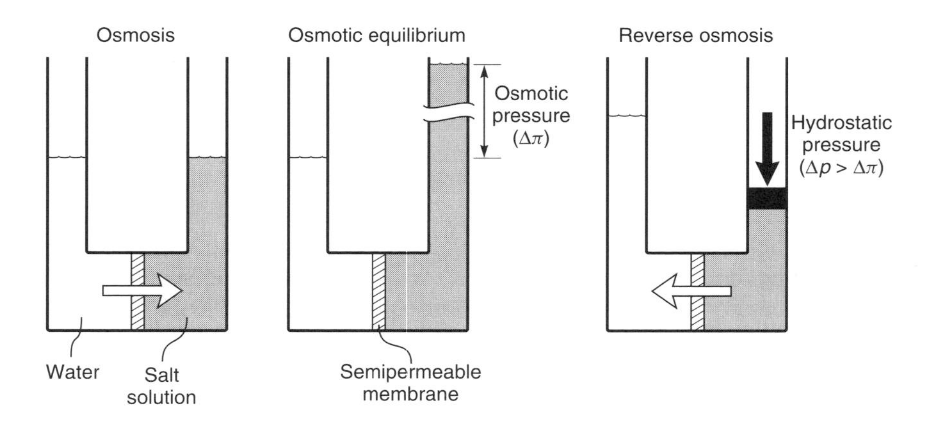

Figure 2.8 A schematic illustration of the relationship between osmosis (dialysis), osmotic equilibrium and reverse osmosis

the water flow is called the osmotic pressure ($\Delta\pi$). If pressures greater than the osmotic pressure are applied to the salt side of the membrane, then the flow of water is reversed, and water begins to flow from the salt solution to the pure water side of the membrane. This process is called reverse osmosis, which is an important method of producing pure water from salt solutions.

Reverse osmosis usually involves two components, water (i) and salt (j). Following the general procedure, the chemical potentials at both sides of the membrane are first equated. At the feed interface, the pressure in the feed solution and within the membrane are identical (as shown in Figure 2.6c). Equating the chemical potentials at this interface gives the same expression as in dialysis [cf. Equation (2.26)]

$$c_{i_{o(m)}} = K_i^{\mathrm{L}} \cdot c_{i_o} \tag{2.32}$$

A pressure difference exists at the permeate interface (as shown in Figure 2.6c) from p_o within the membrane to p_ℓ in the permeate solution. Equating the chemical potentials across this interface gives

$$\mu_{i_\ell} = \mu_{i_{\ell(m)}} \tag{2.33}$$

Substituting the appropriate expression for the chemical potential of an incompressible fluid to the liquid and membrane phases [Equation (2.7)] yields

$$\mu_i^o + RT\ln(\gamma_{i_\ell}^{\mathrm{L}} n_{i_\ell}) + \upsilon_i(p_\ell - p_{i_{\mathrm{sat}}}) = \mu_i^o + RT\ln(\gamma_{i_{\ell(m)}} n_{i_{\ell(m)}}) + \upsilon_i(p_o - p_{i_{\mathrm{sat}}}) \tag{2.34}$$

which leads to

$$\ln(\gamma_{i_\ell}^{\mathrm{L}} n_{i_\ell}) = \ln(\gamma_{i_{\ell(m)}}^{\mathrm{L}} n_{i_{\ell(m)}}) + \frac{\upsilon_i(p_o - p_\ell)}{RT} \tag{2.35}$$

Rearranging and substituting for the sorption coefficient, K_i^{L} [Equations (2.10) and (2.25)], gives the expression

$$c_{i_{\ell(m)}} = K_i^{\mathrm{L}} \cdot c_{i_\ell} \cdot \exp\left[\frac{-\upsilon_i(p_o - p_\ell)}{RT}\right] \tag{2.36}$$

The expressions for the concentrations within the membrane at the interface in Equations (2.32) and (2.36) can now be substituted into the Fick's law expression, Equation (2.13), to yield

$$J_i = \frac{D_i K_i^{\mathrm{L}}}{\ell}\left\{c_{i_o} - c_{i_\ell}\exp\left[\frac{-\upsilon_i(p_o - p_\ell)}{RT}\right]\right\} \tag{2.37}$$

Equation (2.37) and the equivalent expression for component j give the water flux and the salt flux across the reverse osmosis membrane in terms of the pressure and concentration difference across the membrane. There is an analytical expression for Equation (2.37) for a two-component feed mixture that allows the performance of the membrane to be calculated for known permeabilities, $D_i K_i^{\mathrm{L}}/\ell$ and $D_j K_j^{\mathrm{L}}/\ell$, and feed concentrations, c_{i_o} and c_{j_o}. However, more commonly

Equation (2.37) is simplified by assuming that the membrane selectivity is high, that is, $D_i K_i^L/\ell \gg D_j K_j^L/\ell$. This is a good assumption for most of the reverse osmosis membranes used to separate salts from water. Consider the water flux first. At the point at which the applied hydrostatic pressure balances the water activity gradient, that is, the point of osmotic equilibrium in Figure 2.6(b), the flux of water across the membrane is zero. Equation (2.37) becomes

$$J_i = 0 = \frac{D_i K_i^L}{\ell}\left\{c_{i_o} - c_{i_\ell}\exp\left[\frac{-\upsilon_i(\Delta\pi)}{RT}\right]\right\} \tag{2.38}$$

and, on rearranging

$$c_{i_\ell} = c_{i_o}\exp\left[\frac{\upsilon_i(\Delta\pi)}{RT}\right] \tag{2.39}$$

At hydrostatic pressures higher than $\Delta\pi$, Equations (2.37) and (2.39) can be combined to yield

$$J_i = \frac{D_i K_i^L c_{i_o}}{\ell}\left(1 - \exp\left\{\frac{-\upsilon_i[(p_o - p_\ell) - \Delta\pi]}{RT}\right\}\right) \tag{2.40}$$

or

$$J_i = \frac{D_i K_i^L c_{i_o}}{\ell}\left\{1 - \exp\left[\frac{-\upsilon_i(\Delta p - \Delta\pi)}{RT}\right]\right\} \tag{2.41}$$

where Δp is the difference in hydrostatic pressure across the membrane $(p_o - p_\ell)$. A trial calculation shows that the term $-\upsilon_i(\Delta p - \Delta\pi)/RT$ is small under the normal conditions of reverse osmosis. For example, in water desalination, when $\Delta p = 100$ atm, $\Delta\pi = 10$ atm, and $\upsilon_i = 18$ cm^3/mol, the term $\upsilon_i(\Delta p - \Delta\pi)/RT$ is about 0.06.

Under these conditions, the simplification $1 - \exp(x) \rightarrow x$ as $x \rightarrow 0$ can be used, and Equation (2.41) can be written to a very good approximation as

$$J_i = \frac{D_i K_i^L c_{i_o}\upsilon_i(\Delta p - \Delta\pi)}{\ell RT} \tag{2.42}$$

This equation can be simplified to

$$J_i = A(\Delta p - \Delta\pi) \tag{2.43}$$

where A is a constant equal to the term $D_i K_i^L c_{i_o}\upsilon_i/\ell RT$. In the reverse osmosis literature, the constant A is usually called the *water permeability constant*.

Similarly, a simplified expression for the salt flux, J_j, through the membrane can be derived, starting with the equivalent to Equation (2.37)

$$J_j = \frac{D_j K_j^L}{\ell}\left\{c_{j_o} - c_{j_\ell}\exp\left[\frac{-\upsilon_j(p_o - p_\ell)}{RT}\right]\right\} \tag{2.44}$$

Because the term $-\upsilon_j(p_o - p_\ell)/RT$ is small, the exponential term in Equation (2.44) is close to one, and Equation (2.44) can then be written as

$$J_j = \frac{D_j K_j^{\mathrm{L}}}{\ell}(c_{j_o} - c_{j_\ell}) \tag{2.45}$$

or

$$J_j = B(c_{j_o} - c_{j_\ell}) \tag{2.46}$$

where B is usually called the *salt permeability constant* and has the value

$$B = \frac{D_j K_j^{\mathrm{L}}}{\ell} \tag{2.47}$$

Predictions of salt and water transport can be made from this application of the solution-diffusion model to reverse osmosis (first derived by Merten and co-workers) [12,13]. According to Equation (2.43), the water flux through a reverse osmosis membrane remains small up to the osmotic pressure of the salt solution and then increases with applied pressure, whereas according to Equation (2.46), the salt flux is essentially independent of pressure. Some typical results are shown in Figure 2.9. Also shown in this figure is a term called the rejection coefficient, $\mathbb{R}$, which is defined as

$$\mathbb{R} = \left(1 - \frac{c_{j_\ell}}{c_{j_o}}\right) \times 100\,\% \tag{2.48}$$

The rejection coefficient is a measure of the ability of the membrane to separate salt from the feed solution.

For a perfectly selective membrane the permeate salt concentration, $c_{j_\ell} = 0$ and $\mathbb{R} = 100\,\%$, and for a completely unselective membrane the permeate salt concentration is the same as the feed salt concentration, $c_{j_\ell} = c_{j_o}$ and $\mathbb{R} = 0\,\%$. The rejection coefficient increases with applied pressure as shown in Figure 2.9, because the water flux increases with pressure, but the salt flux does not.

Hyperfiltration

By convention, the term reverse osmosis is used to describe the separation of an aqueous salt solution by pressure-driven flow through a semipermeable membrane. Recently, the same type of process has been applied to the separation of organic mixtures. For example, Mobil Oil has installed a large plant to separate methyl ethyl ketone (MEK) from MEK–oil mixtures created in the production of lubricating oil [14] as described in Chapter 5. Separation of this type of mixture is probably best called hyperfiltration.

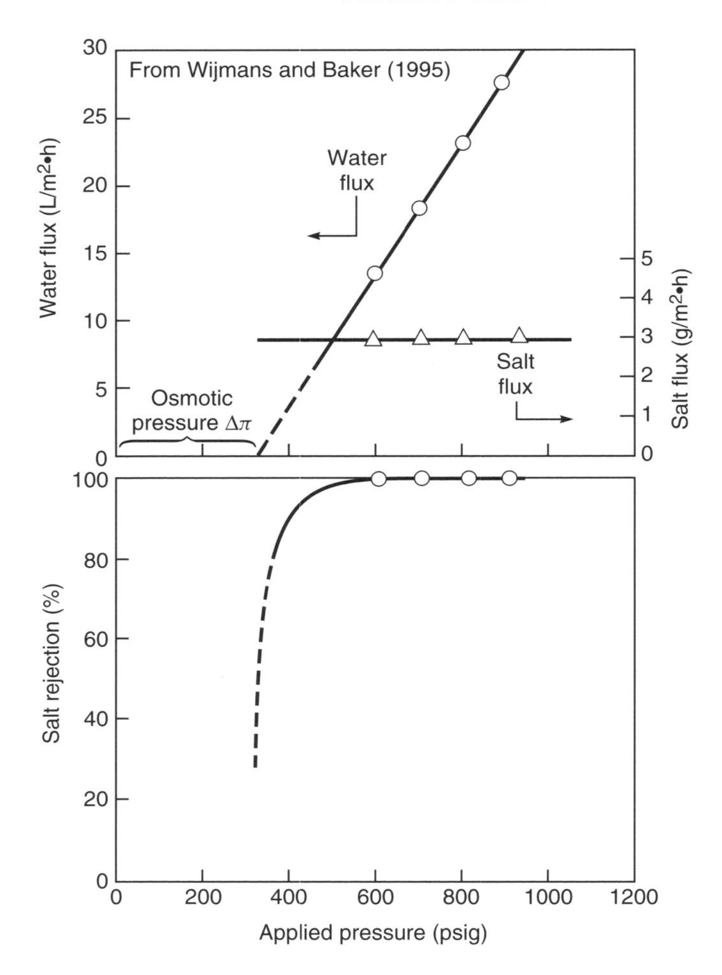

Figure 2.9 Flux and rejection data for a model seawater solution (3.5 % sodium chloride) in a good quality reverse osmosis membrane (FilmTec Corp. FT 30 membrane) as a function of pressure [10]. The salt flux, in accordance with Equation (2.44), is essentially constant and independent of pressure. The water flux, in accordance with Equation (2.43), increases with pressure, and, at zero flux, meets the pressure axis at the osmotic pressure of seawater ~350 psi

The mathematical description of this process is identical to that of reverse osmosis given in Equations (2.37) and (2.44) and leads to expressions for the solute and solvent fluxes

$$J_i = \frac{D_i K_i^{\mathrm{L}}}{\ell}\left\{c_{i_o} - c_{i_\ell}\exp\left[\frac{-\upsilon_i(p_o - p_\ell)}{RT}\right]\right\} \tag{2.49}$$

and

$$J_j = \frac{D_j K_j^{\mathrm{L}}}{\ell} \left\{ c_{j_o} - c_{j_\ell} \exp\left[\frac{-\upsilon_j (p_o - p_\ell)}{RT} \right] \right\} \qquad (2.50)$$

With the advent of the personal computer, the numerical solution to these equations is straightforward even for multicomponent mixtures. Figure 2.10 shows an example calculation for the separation of a 20 wt% solution of *n*-decane in MEK. In these calculations, the ratio of the permeabilities of MEK

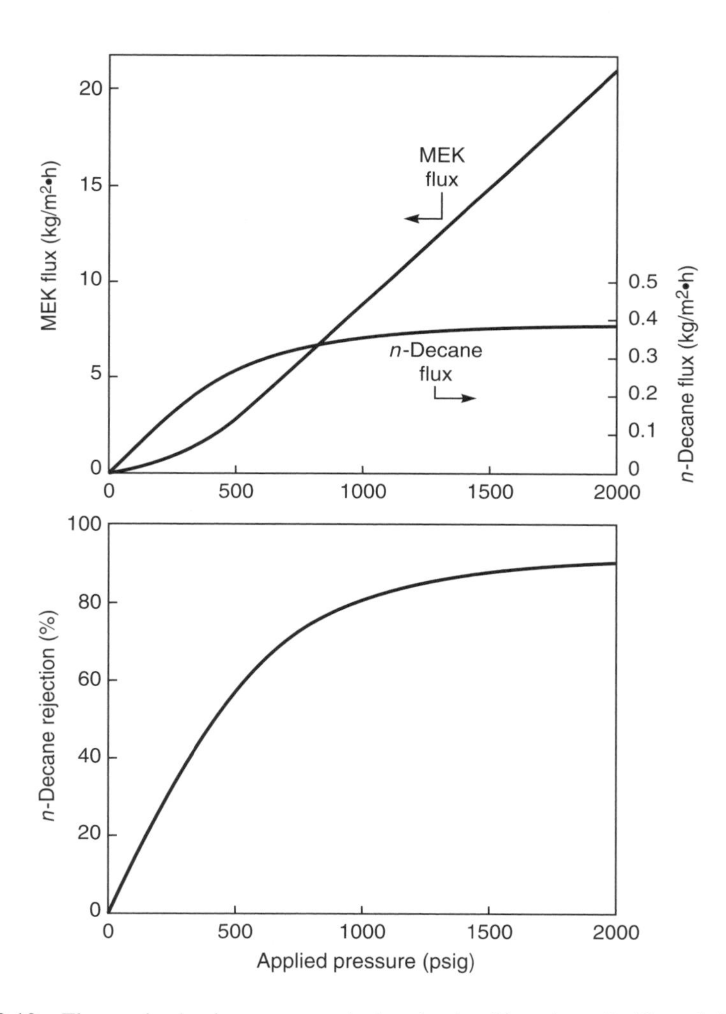

Figure 2.10 Flux and rejection curves calculated using Equations (2.49) and (2.50) for a 20 wt % *n*-decane solution in methyl ethyl ketone (MEK). MEK is assumed to be 10 times more permeable than *n*-decane

and n-decane, $D_i K_i / D_j K_j$, is set at 10. The curves have essentially the same form as the salt solution flux data in Figure 2.9. At high pressures, the rejection approaches a limiting value of 90 %, and the limiting Equations (2.43) for the solvent (MEK) flux and (2.47) for the solute flux apply.

Gas Separation

In gas separation, a gas mixture at a pressure p_o is applied to the feed side of the membrane, while the permeate gas at a lower pressure (p_ℓ) is removed from the downstream side of the membrane. As before, the starting point for the derivation of the gas separation transport equation is to equate the chemical potentials on either side of the gas/membrane interface. This time, however, the chemical potential for the gas phase is given by Equation (2.8) for a compressible fluid, whereas Equation (2.7) for an incompressible medium is applied to the membrane phase. Substitution of these equations into Equation (2.20) at the gas/membrane feed interface yields[3]

$$\mu_i^o + RT\ln(\gamma_{i_o}^{G} n_{i_o}) + RT\ln\frac{p_o}{p_{i_{\mathrm{sat}}}} = \mu_i^o + RT\ln(\gamma_{i_{o(m)}} n_{i_{o(m)}}) + \upsilon_i(p_o - p_{i_{\mathrm{sat}}}) \tag{2.51}$$

which rearranges to

$$n_{i_{o(m)}} = \frac{\gamma_{i_o}^{G}}{\gamma_{i_{o(m)}}} \cdot \frac{p_o}{p_{i_{\mathrm{sat}}}} \cdot n_{i_o} \exp\left[\frac{-\upsilon_i(p_o - p_{i_{\mathrm{sat}}})}{RT}\right] \tag{2.52}$$

Because the exponential term is again very close to one,[4] even for very high pressures (p_o), Equation (2.52) reduces to

$$n_{i_{o(m)}} = \frac{\gamma_{i_o}^{G} n_{i_o}}{\gamma_{i_{o(m)}}} \cdot \frac{p_o}{p_{i_{\mathrm{sat}}}} \tag{2.53}$$

The term $n_{i_o} p_o$ is the partial pressure of i in the feed gas, p_{i_o}. Equation (2.53) then simplifies to

$$n_{i_{o(m)}} = \frac{\gamma_{i(o)}^{G}}{\gamma_{i_{o(m)}}} \cdot \frac{p_{i_o}}{p_{i_{\mathrm{sat}}}} \tag{2.54}$$

or

$$c_{i_{o(m)}} = m_i \rho_m \frac{\gamma_{i_o}^{G} p_{i_o}}{\gamma_{i_{o(m)}} p_{i_{\mathrm{sat}}}} \tag{2.55}$$

[3] At this point the superscript G is introduced to denote the gas phase. For example γ_i^{G}, the activity of component i in the gas phase, and K_i^{G}, the sorption coefficient of component i between the gas and membrane phases [Equation (2.56)].

[4] In evaluating this exponential term (the Poynting correction), it is important to recognize that υ_i is not the molar volume of i in the gas phase, but the molar volume of i dissolved in the membrane material, which is approximately equal to the molar volume of liquid i.

By defining a gas phase sorption coefficient K_i^G as

$$K_i^G = \frac{m_i \rho_m \gamma_{i_o}^G}{\gamma_{i_{o(m)}} p_{i_{sat}}} \tag{2.56}$$

the concentration of component i at the feed interface of the membrane can be written as

$$c_{i_{o(m)}} = K_i^G \cdot p_{i_o} \tag{2.57}$$

In exactly the same way, the concentration of component i at the membrane/permeate interface can be shown to be

$$c_{i_{\ell(m)}} = K_i^G \cdot p_{i_\ell} \tag{2.58}$$

Combining Equations (2.57) and (2.58) with the Fick's law expression, Equation (2.13), gives

$$J_i = \frac{D_i K_i^G (p_{i_o} - p_{i_\ell})}{\ell} \tag{2.59}$$

The product $D_i K_i^G$ is often abbreviated to a permeability coefficient, P_i^G, to give the familiar expression

$$J_i = \frac{P_i^G (p_{i_o} - p_{i_\ell})}{\ell} \tag{2.60}$$

Equation (2.60) is widely used to accurately and predictably rationalize the properties of gas permeation membranes.

The derivation of Equation (2.60) might be seen as a long-winded way of arriving at a trivial result. However, this derivation explicitly clarifies the assumptions behind this equation. First, a gradient in concentration occurs within the membrane, but there is no gradient in pressure. Second, the absorption of a component into the membrane is proportional to its activity (partial pressure) in the adjacent gas, but is independent of the total gas pressure. This is related to the approximation made in Equation (2.52), in which the Poynting correction was assumed to be 1.

The permeability coefficient, P_i^G, equal to the product $D_i K_i^G$ can be expressed from the definition of K_i^G in Equation (2.56) as

$$P_i^G = \frac{\gamma_i^G D_i m_i \rho_m}{\gamma_{i_{(m)}} \cdot p_{i_{sat}}} \tag{2.61}$$

In Equation (2.60) the membrane flux, J_i, is a mass flux (g/cm$^2 \cdot$ s), whereas the gas separation literature predominantly uses a molar flux, typically expressed in the units cm^3(STP)/cm$^2 \cdot$ s. The molar flux, j_i, can be linked to the mass flux, J_i, by the expression

$$j_i = J_i \frac{\upsilon_i^G}{m_i} \tag{2.62}$$

where υ_i^G is the molar volume of gas i (cm^3(STP)/mol). Similarly the mass permeability unit P_i^G, defined in Equation (2.60), can be linked to the molar gas permeability $\mathcal{P}_i^G$, usually in the units cm^3(STP) · cm/cm^2 · s · cmHg, as

$$\mathcal{P}_i^G = \frac{P_i^G \upsilon_i^G}{m_i} \tag{2.63}$$

Equation (2.60) can then be written as

$$j_i = \frac{\mathcal{P}_i^G}{\ell}(p_{i_o} - p_{i_\ell}) \tag{2.64}$$

and combining Equations (2.61) and (2.63) gives

$$\mathcal{P}_i^G = \frac{\gamma_i^G D_i \upsilon_i^G \rho_{(m)}}{\gamma_{i_{(m)}} p_{i_{sat}}} \tag{2.65}$$

Equation (2.65) is not commonly used as an expression for gas-phase membrane permeability, but is of interest because it shows that large permeability coefficients are obtained for compounds with a large diffusion coefficient (D_i), a limited affinity for the gas phase (large γ_i^G), a high affinity for the membrane material (small $\gamma_{i(m)}$), and a low saturation vapor pressure ($p_{i_{sat}}$). The molar gas permeation permeability ($\mathcal{P}_i^G$) is close to being a materials constant, relatively independent of the composition and pressure of the feed and permeate gases. This is in sharp contrast to the permeability constant for liquids as described in the discussion centered on Figure 2.7 earlier, but, even for gases, the concept of permeability as a materials constant must be treated with caution. For example, the permeability of vapors at partial pressures close to saturation often increases substantially with increasing partial pressure. This effect is commonly ascribed to plasticization and other effects of the permeant on the membrane, changing D_i and $\gamma_{i_{(m)}}$ in Equation (2.65). However, significant deviations of the vapor's activity coefficient, γ_i^G, from ideality can also occur at high partial pressures.

Equation (2.65) is also a useful way to rationalize the effect of molecular weight on permeability. The permeant's saturation vapor pressure ($p_{i_{sat}}$) and diffusion coefficient both decrease with increasing molecular weight, creating competing effects on the permeability coefficient. In glassy polymers, the decrease in diffusion coefficient far outweighs other effects, and permeabilities fall significantly as molecular weight increases [15]. In rubbery polymers, on the other hand, the two effects are more balanced. For molecular weights up to 100, permeability generally increases with increasing molecular weight because $p_{i_{sat}}$ is the dominant term. Above molecular weight 100, the molecular weight term gradually becomes dominant, and permeabilities fall with increasing molecular weight of the permeant. Some data illustrating this behavior for permeation of simple alkanes in silicone rubber membranes are shown in Figure 2.11. As the molecular weight increases from CH_4 to C_5H_{12}, the effect of the decrease in $p_{i_{sat}}$

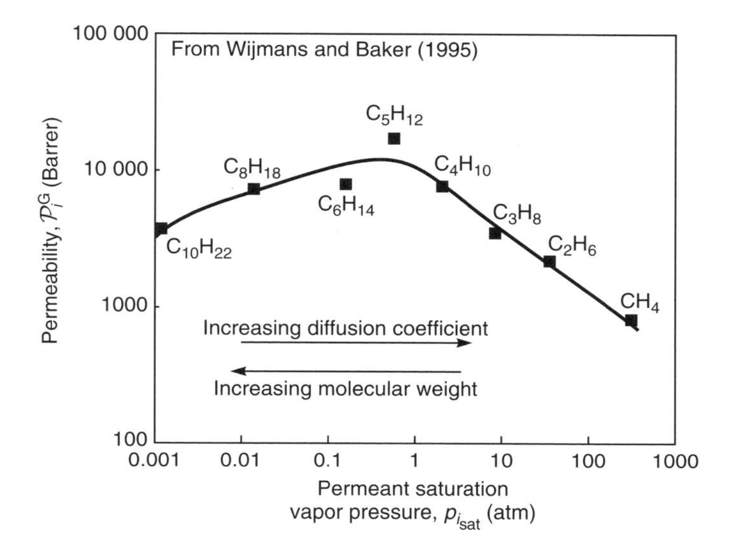

Figure 2.11 Permeability coefficient, $\mathcal{P}_i^G$, of n-alkanes in poly(dimethylsiloxane) as a function of saturation pressure ($p_{i_{sat}}$)

is larger than the effect of increasing size or D_i. Above pentane, however, the trend is reversed.

Pervaporation

Pervaporation is a separation process in which a multicomponent liquid is passed across a membrane that preferentially permeates one or more of the components. A partial vacuum is maintained on the permeate side of the membrane, so that the permeating components are removed as a vapor mixture. Transport through the membrane is induced by maintaining the vapor pressure of the gas on the permeate side of the membrane at a lower vapor pressure than the feed liquid. The gradients in chemical potential, pressure, and activity across the membrane are illustrated in Figure 2.12.

At the liquid solution/membrane feed interface, the chemical potential of the feed liquid is equilibrated with the chemical potential in the membrane at the same pressure. Equation (2.7) then gives

$$\mu_i^o + RT\ln(\gamma_{i_o}^L n_{i_o}) + \upsilon_i(p_o - p_{i_{sat}}) = \mu_i^o + RT\ln(\gamma_{i_{o(m)}} n_{i_{o(m)}}) + \upsilon_i(p_o - p_{i_{sat}}) \tag{2.66}$$

which leads to an expression for the concentration at the feed side interface

$$c_{i_{o(m)}} = \frac{\gamma_{i_o}^L \rho_m}{\gamma_{i_{o(m)}} \rho_o} \cdot c_{i_o} = K_i^L \cdot c_{i_o} \tag{2.67}$$

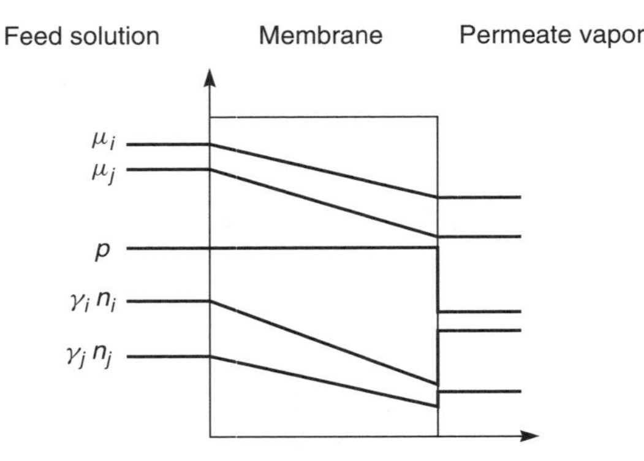

Figure 2.12 Chemical potential, pressure, and activity profiles through a pervaporation membrane following the solution-diffusion model

where K_i^{L} is the liquid-phase sorption coefficient defined in Equation (2.26) in the dialysis section.

At the permeate gas/membrane interface, the pressure drops from p_o in the membrane to p_ℓ in the permeate vapor. The equivalent expression for the chemical potentials in each phase is then

$$\mu_i^o + RT\ln(\gamma_{i_\ell}^{\mathrm{G}} n_{i_\ell}) + RT\ln\left(\frac{p_\ell}{p_{i_{\mathrm{sat}}}}\right) = \mu_i^o + RT\ln(\gamma_{i_{\ell(m)}} n_i) + \upsilon_i(p_o - p_{i_{\mathrm{sat}}}) \tag{2.68}$$

Rearranging Equation (2.68) gives

$$n_{i_{\ell(m)}} = \frac{\gamma_{i_\ell}^{\mathrm{G}}}{\gamma_{i_{\ell(m)}}} \cdot \frac{p_\ell}{p_{i_{\mathrm{sat}}}} \cdot n_{i_\ell} \cdot \exp\left[\frac{-\upsilon_i(p_o - p_{i_{\mathrm{sat}}})}{RT}\right] \tag{2.69}$$

As before, the exponential term is close to unity; thus, the concentration at the permeate side interface is

$$n_{i_{\ell(m)}} = \frac{\gamma_{i_\ell}^{\mathrm{G}}}{\gamma_{i_{\ell(m)}}} \cdot n_{i_\ell} \cdot \frac{p_\ell}{p_{i_{\mathrm{sat}}}} \tag{2.70}$$

The product $n_{i_\ell} p_\ell$ can be replaced by the partial pressure term p_{i_ℓ}, thus

$$n_{i_{\ell(m)}} = \frac{\gamma_{i_\ell}^{\mathrm{G}}}{\gamma_{i_{\ell(m)}}} \cdot \frac{p_{i_\ell}}{p_{i_{\mathrm{sat}}}} \tag{2.71}$$

or, substituting concentration for mole fraction from Equation (2.10),

$$c_{i_{\ell(m)}} = m_i \rho_m \cdot \frac{\gamma_{i_\ell}^{\mathrm{G}} p_{i_\ell}}{\gamma_{i_{\ell(m)}} p_{i_{\mathrm{sat}}}} = K_i^{\mathrm{G}} p_{i_\ell} \tag{2.72}$$

where K_i^G is the gas-phase sorption coefficient defined in Equation (2.56) in the gas separation section.

The concentration terms in Equations (2.67) and (2.72) can be substituted into Equation (2.13) (Fick's law) to obtain an expression for the membrane flux.

$$J_i = \frac{D_i(K_i^L c_{i_o} - K_i^G p_{i_\ell})}{\ell} \tag{2.73}$$

However, the sorption coefficient in Equation (2.67) is a liquid-phase coefficient, whereas the sorption coefficient in Equation (2.72) is a gas-phase coefficient. The interconversion of these two coefficients can be handled by considering a hypothetical vapor in equilibrium with a feed solution. This vapor–liquid equilibrium can then be written

$$\mu_i^o + RT\ln(\gamma_i^L n_i^L) + \upsilon_i(p - p_{i_{sat}}) = \mu_i^o + RT\ln(\gamma_i^G n_i^G) + RT\ln\left(\frac{p}{p_{i_{sat}}}\right) \tag{2.74}$$

where the superscripts L and G represent the liquid and the gas phases. By following the same steps as were taken from Equation (2.68) to (2.72), Equation (2.74) becomes

$$n_i^L = \frac{\gamma_i^G p_i}{\gamma_i^L p_{i_{sat}}} \tag{2.75}$$

Substituting for concentration with Equation (2.10) gives

$$c_i^L = m_i \rho \frac{\gamma_i^G p_i}{\gamma_i^L p_{i_{sat}}} \tag{2.76}$$

and so

$$c_i^L = \frac{K_i^G}{K_i^L} \cdot p_i \tag{2.77}$$

This expression links the concentration of component i in the liquid phase, c_i^L with p_i, the partial vapor pressure of i in equilibrium with the liquid. Substitution of Equation (2.77) into Equation (2.73) yields

$$J_i = \frac{D_i K_i^G (p_{i_o} - p_{i_\ell})}{\ell} \tag{2.78}$$

where p_{i_o} and p_{i_ℓ} are the partial vapor pressures of component i on either side of the membrane. Equation (2.67) can also be written as

$$J_i = \frac{P_i^G}{\ell}(p_{i_o} - p_{i_\ell}) \tag{2.79}$$

This equation explicitly expresses the driving force in pervaporation as the vapor pressure difference across the membrane, a form of the pervaporation process derived first by Kataoka *et al.* [16]. Equation (2.77) links the concentration of a sorbed vapor in the liquid phase ($c_{i_o}^{\mathrm{L}}$) with the equilibrium partial pressure of the vapor. This is known as Henry's law and is usually written as[5]

$$H_i \cdot c_i^{\mathrm{L}} = p_i \tag{2.80}$$

From Equations (2.77) and (2.80) it follows that the Henry's law coefficient, H_i, can be written as

$$H_i = \frac{K_i^{\mathrm{L}}}{K_i^{\mathrm{G}}} = \frac{\gamma_i^{\mathrm{L}} p_{i_{\mathrm{sat}}}}{m_i \rho \gamma_i^{\mathrm{G}}} \tag{2.81}$$

These expressions can be used to write Equation (2.73) as

$$J_i = \frac{P_i^{\mathrm{G}}}{\ell}(c_{i_o} H_i - p_{i_\ell}) \tag{2.82}$$

or

$$J_i = \frac{P_i^{\mathrm{L}}}{\ell}(c_{i_o} - p_{i_\ell}/H_i) \tag{2.83}$$

Equation (2.79) expresses the driving force in pervaporation in terms of the vapor pressure. The driving force could equally well have been expressed in terms of concentration differences, as in Equation (2.83). However, in practice, the vapor pressure expression provides much more useful results and clearly shows the connection between pervaporation and gas separation, Equation (2.60). Also, P_i^{G}, the gas phase coefficient, is much less dependent on temperature than P_i^{L}. The reliability of Equation (2.79) has been amply demonstrated experimentally [17,18]. Figure 2.13, for example, shows data for the pervaporation of water as a function of permeate pressure. As the permeate pressure (p_{i_ℓ}) increases, the water flux falls, reaching zero flux when the permeate pressure is equal to the feed-liquid vapor pressure ($p_{i_{\mathrm{sat}}}$) at the temperature of the experiment. The straight lines in Figure 2.13 indicate that the permeability coefficient (P_i^{G}) of water in silicone rubber is constant, as expected in this and similar systems in which the membrane material is a rubbery polymer and the permeant swells the polymer only moderately.

Greenlaw *et al.* [18] have studied the effect of feed and permeate pressure on pervaporation flux in some detail; some illustrative results are shown in

[5] In Equation (2.80), the Henry's law coefficient H_i has the units atm $\cdot$ cm^3/g. More commonly, Henry's law is written in terms of mole fraction:

$$H_i' \cdot n_i^{\mathrm{L}} = p_i$$

where H_i' has the units atm/mol fraction. Using Equation (2.10), the two coefficients are linked by the expression

$$H_i = \frac{H_i'}{m_i \cdot p_i}$$

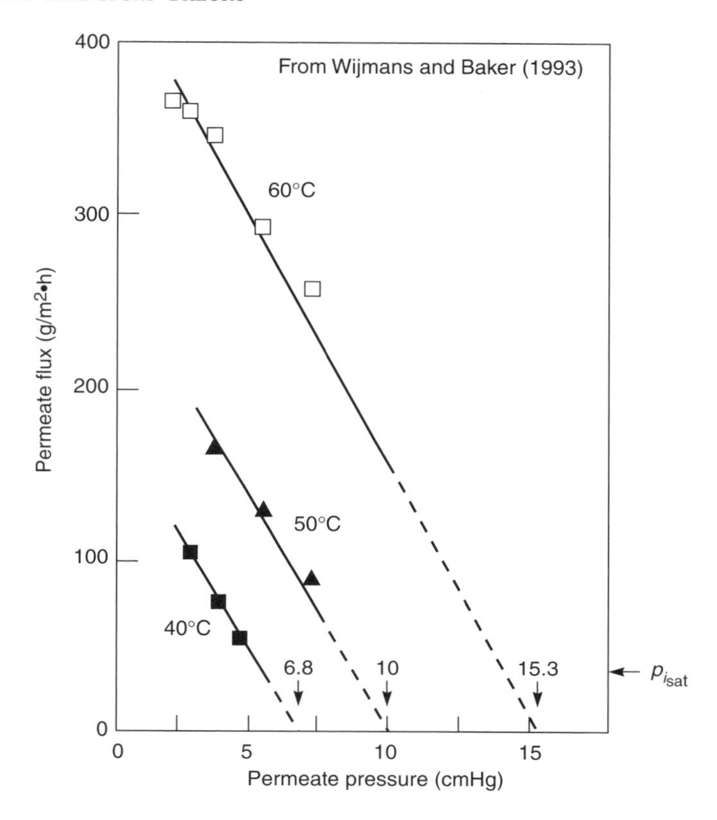

Figure 2.13 The effect of permeate pressure on the water flux through a silicone rubber pervaporation membrane. The arrows on the lower axis represent the saturation vapor pressures of the feed solution at the temperature of these experiments as predicted by Equation (2.79) [15]

Figure 2.14. As Figure 2.14(a) shows, the dependence of flux on permeate pressure in pervaporation is in accordance with Equation (2.79). The flux decreases with increasing permeate pressure, reaching a minimum value when the permeate pressure equals the saturation vapor pressure of the feed. The curvature of the line in Figure 2.14(a) shows that the permeability coefficient decreases with decreasing permeate pressure, that is, P^{G}_{hexane} decreases as hexane concentration in the membrane decreases. This behavior is typical of membranes that are swollen significantly by the permeant. If, on the other hand, as shown in Figure 2.14(b), the permeate pressure is fixed at a low value, the hydrostatic pressure of the feed liquid can be increased to as much as 20 atm without any significant change in the flux. This is because increased hydrostatic pressure produces a minimal change in the partial pressure of the feed liquid partial pressure (p_{i_o}), the true

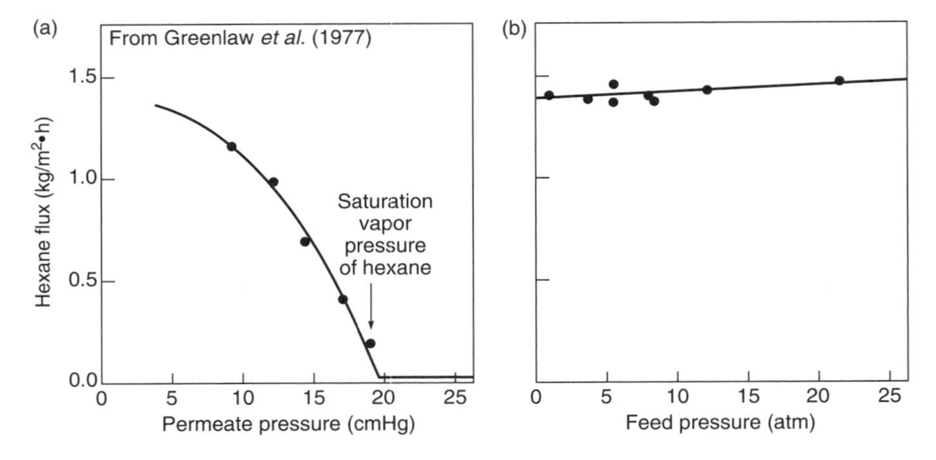

Figure 2.14 The effect of feed and permeate pressure on the flux of hexane through a rubbery pervaporation membrane. The flux is essentially independent of feed pressure up to 20 atm but is extremely sensitive to permeate pressure [18]. The explanation for this behavior is in the transport equation (2.79). Reprinted from *J. Membr. Sci.* **2**, F.W. Greenlaw, W.D. Prince, R.A. Shelden and E.V. Thompson, Dependence of Diffusive Permeation Rates by Upstream and Downstream Pressures, p. 141, Copyright 1977, with permission from Elsevier

driving force shown in Equation (2.79). Thus, the properties of pervaporation membranes illustrated in Figures 2.13 and 2.14 are easily rationalized by the solution-diffusion model but are much more difficult to explain by a pore-flow mechanism, although this has been tried.

Evidence for the Solution-diffusion Model

In the discussion above, the solution-diffusion model was used to derive equations that predict the experimentally observed performance of the membrane processes of dialysis, gas separation, reverse osmosis, and pervaporation. It was not necessary to resort to any additional process-specific model to obtain these results. This agreement between theory and experiment is good evidence for the validity of the solution-diffusion model. Moreover, the large body of permeability, diffusion, and partition coefficient data obtained over the past 20 years for these membrane processes are in good numerical agreement with one another. This universality and the simplicity of the solution-diffusion model are its most useful features and are a strong argument for the validity of the model. Finally, a number of direct experimental measurements can be made to distinguish between the solution-diffusion model and other models, such as the pore-flow model.

One prediction of the solution-diffusion model, controversial during the 1970s, is that the action of an applied pressure on the feed side of the membrane is to *decrease* the concentration of the permeant on the low pressure side of the membrane. This counterintuitive effect is illustrated by Figures 2.5 and 2.6. A number of workers have verified this prediction experimentally with a variety of polymer membranes, ranging from diffusion of water in glassy cellulose acetate membranes to diffusion of organics in swollen rubbers [19–21]. Convincing examples of this type of experiment are the results of Rosenbaum and Cotton shown in Figure 2.15 [20]. In these experiments, four thin cellulose acetate films were laminated together, placed in a high pressure reverse osmosis cell, and subjected

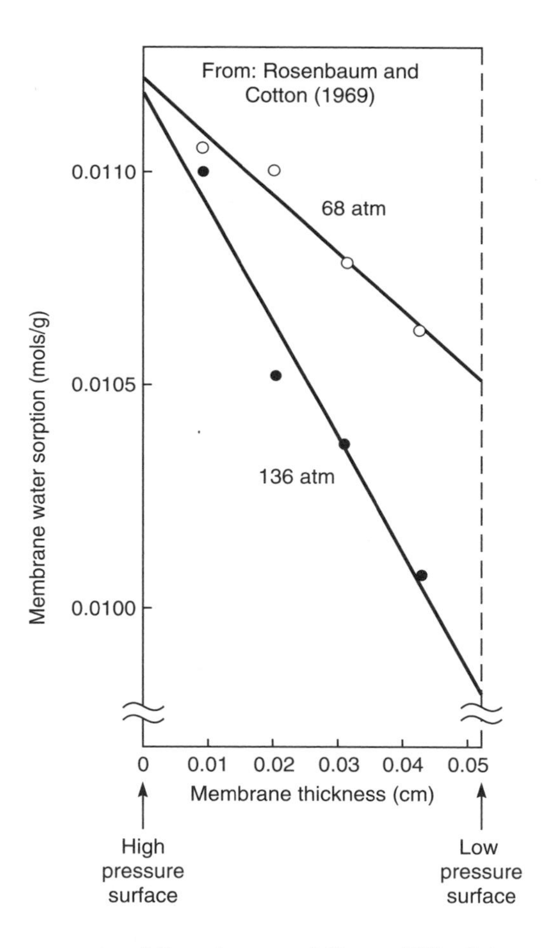

Figure 2.15 Measurements of Rosenbaum and Cotton [20] of the water concentration gradients in a laminated reverse osmosis cellulose acetate membrane under applied pressures of 68 and 136 atm. Reprinted from Steady-state Distribution of Water in Cellulose Acetate Membrane, S. Rosenbaum and O. Cotton, *J. Polym. Sci.* **7**, 101; Copyright © 1969. This material is used by permission of John Wiley & Sons, Inc.

to feed pressures of 68 or 136 atm. The permeate was maintained at atmospheric pressure. After permeation through the membrane laminate had reached a steady state, the membrane was quickly removed from the cell, and the water concentration in each laminate measured. As predicted by the solution-diffusion model and shown in Figure 2.15, the applied pressure decreases the concentration of water on the permeate side of the membrane. Also, the concentration difference across the membrane at 136 atm applied pressure is about twice that observed at 68 atm, and the measured concentration on the permeate side is within 20 % of the expected value calculated from Equation (2.36).

Another series of papers by Paul and co-workers [4–6,19,22] focuses on the same phenomenon using rubbery membranes and permeation of organic solvents such as hexane, benzene and carbon tetrachloride. Such membranes are highly swollen by the organic solvents and, when operated in reverse osmosis mode,

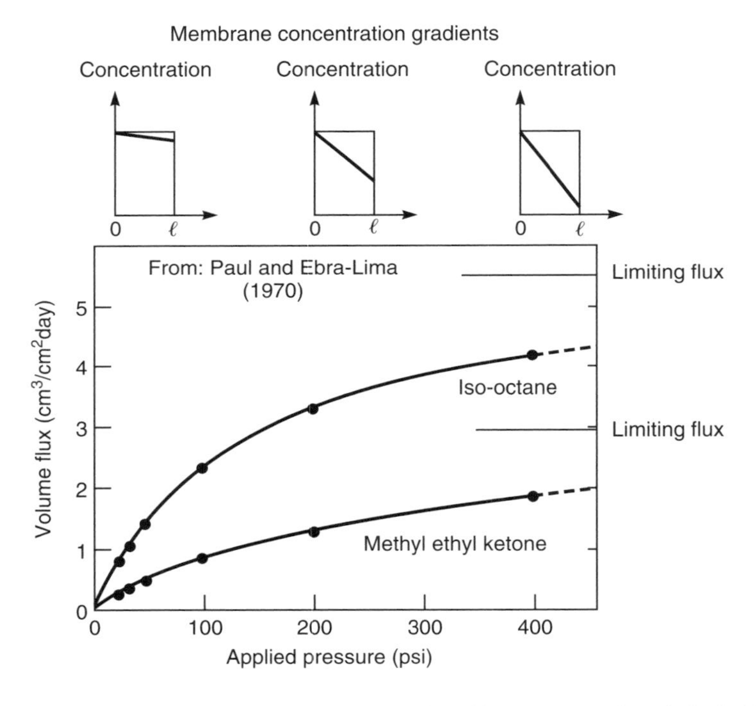

Figure 2.16 Pressure permeation (reverse osmosis) of iso-octane and methyl ethyl ketone through crosslinked 265-μm-thick natural rubber membranes. The change in the concentration gradient in the membrane as the applied pressure is increased is illustrated by the inserts. At high applied pressures, the concentration gradient and the permeation fluxes approach their limiting values [4]. Reprinted from Pressure-induced Diffusion of Organic Liquids Through Highly Swollen Polymer Membranes," D.R. Paul and O.M. Ebra-Lima, *J. Appl. Polym. Sci.* **14**, 2201; Copyright © 1970. This material is used by permission of John Wiley & Sons, Inc.

large concentration gradients develop through the membrane even at relatively modest applied pressures. This means that the concentration in the membrane on the permeate side approaches zero and the flux through the membrane reaches a limiting value as the feed pressure is increased. Representative data are shown in Figure 2.16.

Paul and Paciotti [19] took this work a step further by measuring the flux of a liquid (hexane) through a membrane both in pervaporation experiments with atmospheric pressure on the feed side of the membrane and a vacuum on the permeate side, and in reverse osmosis experiments with liquid at elevated pressures on the feed side and at atmospheric pressure on the permeate side. The hexane flux obtained in these two sets of experiments is plotted in Figure 2.17 against the hexane concentration difference in the membrane $(c_{i_{o(m)}} - c_{i_{\ell(m)}})$. The concentrations, $c_{i_{o(m)}}$ and $c_{i_{\ell(m)}}$, were calculated from Equations (2.26), (2.36) and (2.72).

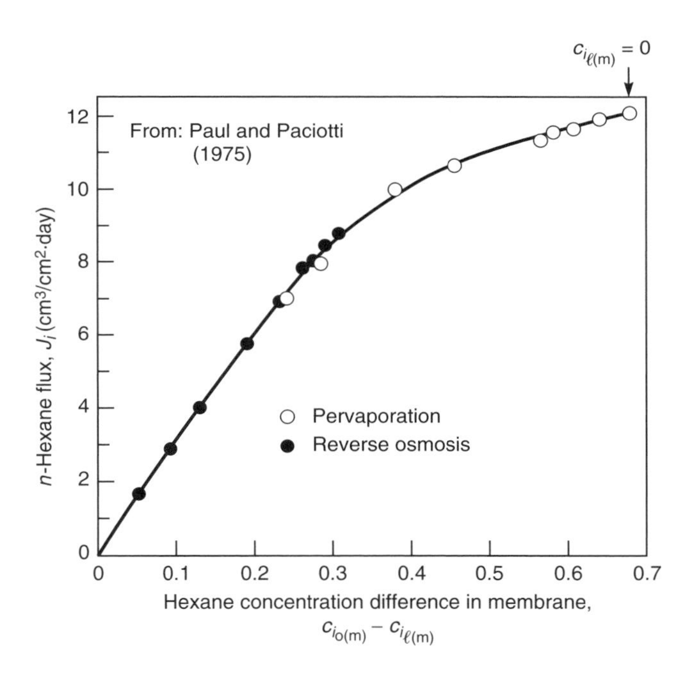

Figure 2.17 Flux of n-hexane through a rubbery membrane as a function of the hexane concentration difference in the membrane. Data taken from both reverse osmosis (●) and pervaporation (○) experiments. Feed-side and permeate-side membrane concentrations, $c_{i_{o(m)}}$ and $c_{i_{\ell(m)}}$, calculated from the operating conditions through Equations (2.26), (2.36) and (2.76). Maximum flux is obtained at the maximum concentration difference, when the permeate-side membrane concentration $(c_{i_{\ell(m)}})$, equals zero [19]. Reprinted from Driving Force for Hydraulic and Pervaporation Transport in Homogeneous Membranes, D.R. Paul and D.J. Paciotti, *J. Polym. Sci., Polym. Phys. Ed.* **13**, 1201; Copyright © 1975. This material is used by permission of John Wiley & Sons, Inc.

Sorption data were used to obtain values for K_i^L. As pointed out by Paul and Paciotti, the data in Figure 2.17 show that reverse osmosis and pervaporation obey one unique transport equation—Fick's law. In other words, transport follows the solution-diffusion model. The slope of the curve decreases at the higher concentration differences, that is, at smaller values for $c_{i_{\ell(m)}}$ because of decreases in the diffusion coefficient, as the swelling of the membrane decreases.

The results illustrated in Figure 2.16 show that the solvent flux tends towards a limiting value at very high pressures. This value is reached when the concentration of sorbed solvent at the permeate side of the membrane reaches zero, the limiting value.

Structure–Permeability Relationships in Solution-diffusion Membranes

In the preceding section the effect of concentration and pressure gradient driving forces on permeation through membranes was described in terms of the solution-diffusion model and Fick's law. The resulting equations all contain a permeability term, P, that must be experimentally determined. This section describes how the nature of the membrane material affects permeant diffusion and sorption coefficients, which in turn determine membrane permeability. This is a difficult subject. By analyzing the factors that determine membrane permeability, useful correlations and rules of thumb can be derived to guide the selection of membrane materials with the optimum flux and selectivity properties. Most of the experimental data in this area have been obtained with gas-permeable membranes. However, the same general principles apply to all polymeric solution-diffusion membranes.

The problem of predicting membrane permeability can be divided into two parts because permeability is the product of the diffusion coefficient and the sorption coefficient:

$$P = D \cdot K \tag{2.84}$$

The sorption coefficient (K) in Equation (2.84) is the term linking the concentration of a component in the fluid phase with its concentration in the membrane polymer phase. Because sorption is an equilibrium term, conventional thermodynamics can be used to calculate solubilities of gases in polymers to within a factor of two or three. However, diffusion coefficients (D) are kinetic terms that reflect the effect of the surrounding environment on the molecular motion of permeating components. Calculation of diffusion coefficients in liquids and gases is possible, but calculation of diffusion coefficients in polymers is much more difficult. In the long term, the best hope for accurate predictions of diffusion in polymers is the molecular dynamics calculations described in an earlier section. However, this technique is still under development and is currently limited to calculations of the diffusion of small gas molecules in amorphous polymers; the

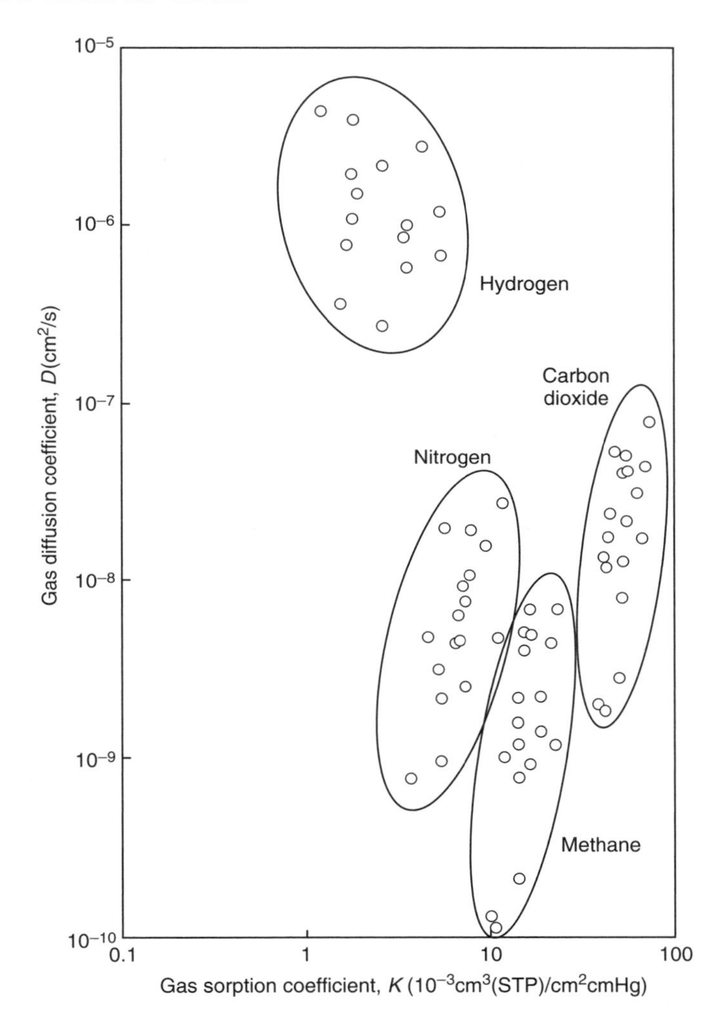

Figure 2.18 Diffusion and sorption coefficients plotted for gases in a family of 18 related polyimides. Data of Tanaka *et al.* [23]

agreement between theory and experiment is modest. In the meantime, simple correlations based on polymer free volume must be used.

As a general rule, membrane material changes affect the diffusion coefficient of a permeant much more than the sorption coefficient. For example, Figure 2.18 shows some typical gas permeation data taken from a paper of Tanaka *et al.* [23]. The diffusion and sorption coefficients of four gases in a family of 18 related polyimides are plotted against each other. Both sorption and diffusion coefficients are fairly well grouped for each gas. However, for any one gas the difference in diffusion coefficient from the highest to lowest value is approximately 100-fold, whereas the spread in sorption coefficient is only 2- to 4-fold. Changes in polymer chemistry affect both the sorption and diffusion coefficients, but the effect on the diffusion coefficient is much more profound.

More detailed examination of the data shown in Figure 2.18 shows that the relative position of each polymer within the group of 18 is approximately the same for all gases. That is, the polymer with the highest diffusion coefficient for methane also has the highest diffusion coefficient for nitrogen, carbon dioxide and hydrogen. The trend for the solubility coefficients is similar. As a general rule, changes in polymer chemistry and structure that change the diffusion coefficient or sorption coefficient of one gas change the properties of other gases in the same way. This is why membrane permeabilities can be easily varied by orders of magnitude by changing the membrane material, whereas changing membrane selectivities (proportional to the ratio of permeabilities) by more than a factor of two or three is difficult.

In the following sections the factors that determine the magnitude of diffusion and solubility coefficients in polymers are discussed.

Diffusion Coefficients

The Fick's law diffusion coefficient of a permeating molecule is a measure of the frequency with which the molecule moves and the size of each movement. Therefore, the magnitude of the diffusion coefficient is governed by the restraining forces of the medium on the diffusing species. Isotopically labeled carbon in a diamond lattice has a very small diffusion coefficient. The carbon atoms of diamond move infrequently, and each movement is very small—only 1 to 2 Å. On the other hand, isotopically labeled carbon dioxide in a gas has an extremely large diffusion coefficient. The gas molecules are in constant motion and each jump is of the order of 1000 Å or more. Table 2.1 lists some representative values of diffusion coefficients in different media.

Table 2.1 Typical diffusion coefficients in various media (25 °C)

Permeant/material	Diffusion coefficient, D (cm^2/s)
Oxygen in air (atmospheric pressure)	1×10^{-1}
Salt in water	1.5×10^{-5}
Albumin (MW 60 000) in water	6×10^{-7}
Oxygen in silicone rubber	1×10^{-5}
Oxygen in polysulfone	4×10^{-8}
Sodium atoms in sodium chloride crystals	1×10^{-20}
Aluminum atoms in metallic copper	1×10^{-30}

The main observation from Table 2.1 is the enormous range of values of diffusion coefficients—from 10^{-1} to 10^{-30} cm^2/s. Diffusion in gases is well understood and is treated in standard textbooks dealing with the kinetic theory of gases [24,25]. Diffusion in metals and crystals is a topic of considerable interest to the semiconductor industry but not to membrane permeation. This book focuses principally on diffusion in liquids and polymers in which the diffusion coefficient can vary from about 10^{-5} to about 10^{-10} cm^2/s.

Diffusion in Liquids

Liquids are simple, well defined systems and provide the starting point for modern theories of diffusion. An early and still fundamentally sound equation was derived by Einstein who applied simple macroscopic hydrodynamics to diffusion at the molecular level. He assumed the diffusing solute to be a sphere moving in a continuous fluid of solvent, in which case it can be shown that

$$D = \frac{kT}{6\pi a \eta} \tag{2.85}$$

where k is Boltzmann's constant, a is the radius of the solute and η is the solution viscosity. This is known as the Stokes–Einstein equation. The equation is a good approximation for large solutes with radii greater than 5–10 Å. But, as the solute becomes smaller, the approximation of the solvent as a continuous fluid becomes less valid. In this case there may be slip of solvent at the solute molecule's surface. A second limiting case assumes complete slip at the surface of the solute sphere; in this case

$$D = \frac{kT}{4\pi a \eta} \tag{2.86}$$

Thus, the Stokes–Einstein equation is perhaps best expressed as

$$D = \frac{kT}{n\pi a \eta} \qquad 4 \leq n \leq 6 \tag{2.87}$$

An important conclusion to be drawn from the Stokes–Einstein equation is that the diffusion coefficient of solutes in a liquid only changes slowly with molecular weight, because the diffusion coefficient is proportional to the reciprocal of the radius, which in turn is approximately proportional to the cube root of the molecular weight.

Application of the Stokes–Einstein equation requires a value for the solute radius. A simple approach is to assume the molecule to be spherical and to calculate the solute radius from the molar volume of the chemical groups making up the molecule. Using values for the solute radius calculated this way along with measured and known diffusion coefficients of solutes in water, Edward [26]

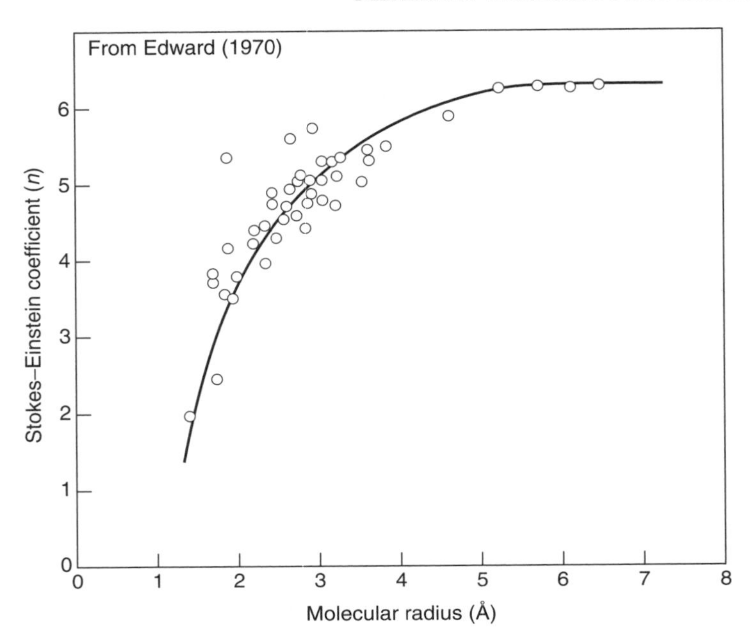

Figure 2.19 Value of the coefficient n in the Stokes–Einstein equation [Equation (2.87)] required to achieve agreement between calculation and experimental solute diffusion coefficients in water. [26]. Reprinted with permission from the *Journal of Chemical Education* **47**, No. 4, 1970, pp. 261–270, Figure 12, copyright © 1970, Division of Chemical Education, Inc.

constructed the graph of the coefficient n in the Stokes–Einstein equation, Equation (2.87), as a function of solute radius as shown in Figure 2.19. With large solutes, n approaches 6, that is, Einstein's application of normal macroscopic fluid dynamics at the molecular level is a valid approximation. However, when the solute radius falls below about 4 Å water can no longer be regarded as a continuous fluid, and n falls below 6. Nonetheless, that an equation based on macroscopic hydrodynamic theory applies to molecules to the 4 Å level is an interesting result.

The Stokes–Einstein equation works well for diffusion of solutes in simple liquids but fails in more complex fluids, such as a solution of a high-molecular-weight polymer. Dissolving a polymer in a liquid increases the solvent's viscosity, but the solute's diffusion coefficient is not significantly affected. For example, as the concentration of poly(vinyl pyrrolidone) dissolved in water is changed from 0 to 20 wt %, the viscosity of the solution increases by several orders of magnitude. However, the diffusion coefficient of sucrose in these solutions only changes by a factor of four [27]. The long polymer chains of dissolved poly(vinyl pyrrolidone) molecules link distant parts of the aqueous solution and change the viscosity of the fluid substantially, but, in the fluid immediately surrounding the diffusing sucrose molecule, the effect of polymer chain length is much less noticeable. This result illustrates the difference between the microscopic viscosity in the immediate environment of the diffusing solute and the macroscopic viscosity measured by conventional viscometers. In simple liquids the macroscopic and microscopic viscosities are the same, but in liquids containing dissolved macromolecules, or in gels and polymer films, the microscopic viscosity and the macroscopic viscosity differ significantly.

Diffusion in Polymers

The concept that the local environment around the permeating molecule determines the permeate's diffusion coefficient is key to understanding diffusion in polymer membranes. Polymers can be divided into two broad categories—rubbery and glassy. In a rubbery polymer, segments of the polymer backbone can rotate freely around their axis; this makes the polymer soft and elastic. Thermal motion of these segments also leads to high permeant diffusion coefficients. In a glassy polymer, steric hindrance along the polymer backbone prohibits rotation of polymer segments; the result is a rigid, tough polymer. Thermal motion in this type of material is limited, so permeant diffusion coefficients are low. If the temperature of a glassy polymer is raised, a point is reached at which the increase in thermal energy is sufficient to overcome the steric hindrance restricting rotation of polymer backbone segments. At this temperature, called the *glass transition temperature* (T_g), the polymer changes from a glass to a rubber.

Figure 2.20 shows a plot of diffusion coefficient as a function of molecular weight for permeants diffusing through a liquid (water), two soft rubbery polymers (natural rubber and silicone rubber), and a hard, stiff glassy polymer (polystyrene) [28]. For very small molecules, such as helium and hydrogen, the diffusion coefficients in all of the media are comparable, differing by no more than a factor of two or three. These very small molecules only interact with one or two atoms in their immediate proximity. The local environment for these small solutes in the three polymers is not radically different to that in a liquid such as water. On the other hand, larger diffusing solutes with molecular weights of 200 to 300 and above have molecular diameters of 6 to 10 Å. Such solutes are in quite different local environments in the different media. In water, the Stokes–Einstein equation applies, and the resistance to movement of the solute is not much larger than that of a very small solute. In polymer membranes, however, several segments of the polymer chain are involved in each movement of the diffusing species. This type of cooperative movement is statistically unlikely; consequently, diffusion coefficients are much smaller than in liquid water. Moreover, the differences between the motion of polymer segments in the flexible rubbery membranes and in the stiff polystyrene membrane are large. The polymer chains in rubbers are considerably more flexible and rotate more easily than

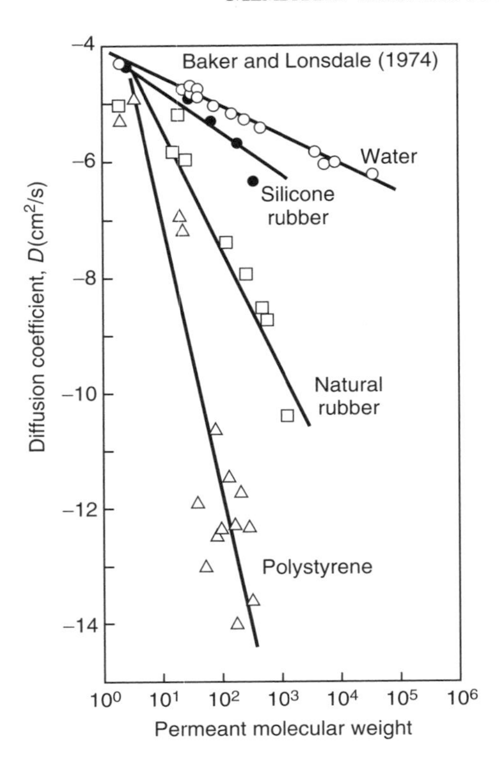

Figure 2.20 Permeant diffusion coefficient as a function of permeant molecular weight in water, natural rubber, silicone rubber and polystyrene. Diffusion coefficients of solutes in polymers usually lie between the value in natural rubber, an extremely permeable polymer, and the value in polystyrene, an extremely impermeable material [28]

those in polystyrene. One manifestation of this difference in chain flexibility is the difference in elastic properties; another is the difference in diffusion coefficient.

An example of the change in diffusion coefficient as the matrix material changes is illustrated by Figure 2.21. In this example, the polymer matrix material is changed by plasticization of the polymer, ethyl cellulose, by the permeant, dichloroethane [29]. The resulting change in the diffusion coefficient is shown in the figure. The concentration of dichloroethane in the polymer matrix increases from very low levels (<1 % dichloroethane) to very high levels (>90 % dichloroethane). As the concentration of dichloroethane increases, the polymer changes from a glassy polymer to a rubbery polymer, to a solvent-swollen gel, and finally to a dilute polymer solution. Ethyl cellulose is a glassy polymer with a glass transition of about 45–50 °C. At low concentrations of dichloroethane (below about 5 vol %) in the polymer, the ethyl cellulose matrix is glassy, and the dichloroethane diffusion coefficient is in the range 1 to 5×10^{-9} cm^2/s. As

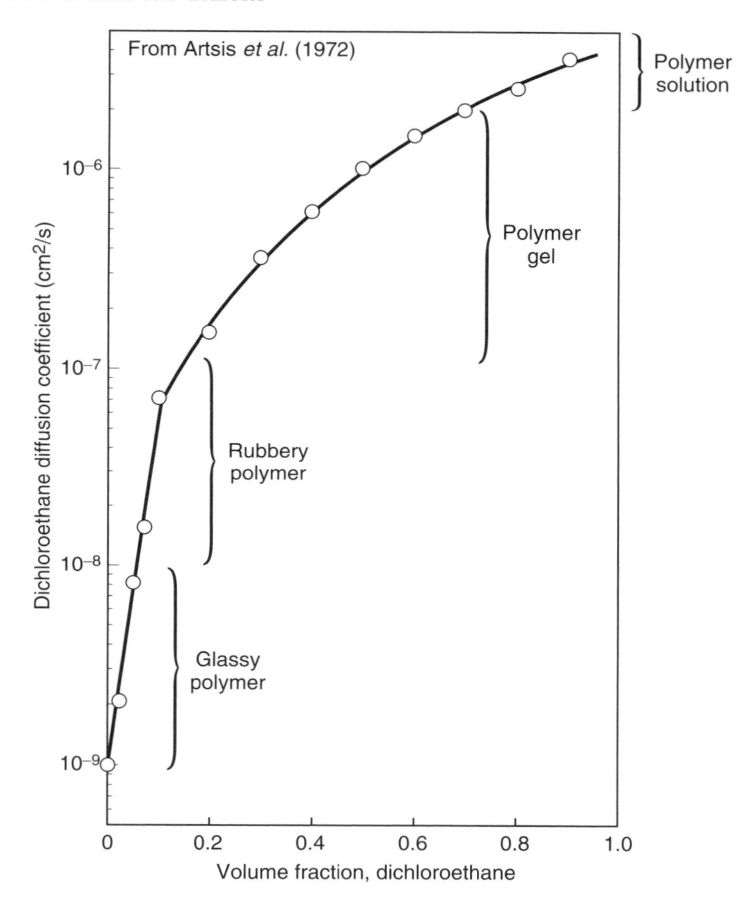

Figure 2.21 Changes in the diffusion coefficient of dichloroethane in ethyl cellulose as a function of the volume fraction of dichloroethane dissolved in the polymer matrix. Data of Artsis *et al.* [29]

the dichloroethane concentration increases to above 5 vol %, enough solvent has dissolved in the polymer to reduce the glass transition temperature to below the temperature of the experiment. The polymer chains then have sufficient freedom to rotate, and the polymer becomes rubbery. As the dichloroethane concentration increases further, the polymer chain mobility also increases as does the diffusion coefficient of dichloroethane. At 20 % dichloroethane, the diffusion coefficient is 1×10^{-7} cm^2/s, 100 times greater than the diffusion coefficient in the glassy polymer. Above 20 vol % dichloroethane, sufficient solvent is present to allow relatively large segments of the polymer chain to move. In this range, between 20 and 70 vol % dichloroethane, the matrix is best characterized as a solvent-swollen gel, and the diffusion coefficient of dichloroethane increases from 1×10^{-7} to 2×10^{-6} cm^2/s. Finally, at dichloroethane concentrations above 70 vol %, sufficient solvent is present for the matrix to be characterized as a polymer solution. In this final solvent concentration range, the increase in diffusion coefficient with further increases in dichloroethane concentration is relatively small.

Figures 2.20 and 2.21 show the significant difference between diffusion in liquids and in rubbery and glassy polymers. A great deal of work has been performed over the last two decades to achieve a quantitative link between the structure of polymers and their permeation properties. No such quantitative structure–property relationship is at hand or even in sight. What has been achieved is a set of semiempirical rules that allow the permeation properties of related families of polymers to be correlated based on small changes in their chemical structures. The correlating tool most generally used is the polymer's *fractional free volume* v_f (cm^3/cm^3), usually defined as

$$v_f = \frac{v - v_o}{v} \tag{2.88}$$

where v is the specific volume of the polymer (cm^3/g), that is, the reciprocal of the polymer density, and v_o is the volume occupied by the molecules themselves (cm^3/g). The free volume of a polymer is the sum of the many small spaces between the polymer chains in these amorphous, noncrystalline materials.

The free volume of a polymer can be determined by measuring the polymer's specific volume, then calculating the occupied volume (v_o) of the groups that form the polymer. Tables of the molar volume of different chemical groups have been prepared by Bondi [30] and Van Krevelen [31]. By summing the molar volume of all the groups in the polymer repeat unit, the occupied molar volume of the polymer can be calculated. The occupied volume obtained in this way is about 1.3 times larger than the Van der Waals volume of the groups. The factor of 1.3 occurs because some unoccupied space is inevitably present even in crystals at 0 K. The fractional free volumes of a number of important membrane materials are given in Table 2.2.

The concept of polymer free volume is illustrated in Figure 2.22, which shows polymer specific volume (cm^3/g) as a function of temperature. At high temperatures the polymer is in the rubbery state. Because the polymer chains do not pack perfectly, some unoccupied space—free volume—exists between the polymer chains. This free volume is over and above the space normally present between molecules in a crystal lattice; free volume in a rubbery polymer results from its amorphous structure. Although this free volume is only a few percent of the total volume, it is sufficient to allow some rotation of segments of the polymer backbone at high temperatures. In this sense a rubbery polymer, although solid at the macroscopic level, has some of the characteristics of a liquid. As the temperature of the polymer decreases, the free volume also decreases. At the glass transition temperature, the free volume is reduced to a point at which the

Table 2.2 Calculated fractional free volume for representative membrane materials at ambient temperatures (Bondi method)

Polymer	Polymer type	Glass transition temperature, T_g (°C)	Fractional free volume (cm^3/cm^3)
Silicone rubber	Rubber	−129	0.16
Natural rubber	Rubber	−73	0.16
Polycarbonate	Glass	150	0.16
Poly(phenylene oxide)	Glass	167	0.20
Polysulfone	Glass	186	0.16
6FDA-ODA polyimide	Glass	300	0.16
Poly(4-methyl-2-pentyne) (PMP)	Glass	>250	0.28
Poly(1-trimethylsilyl-1-propyne) (PTMSP)	Glass	>250	0.34

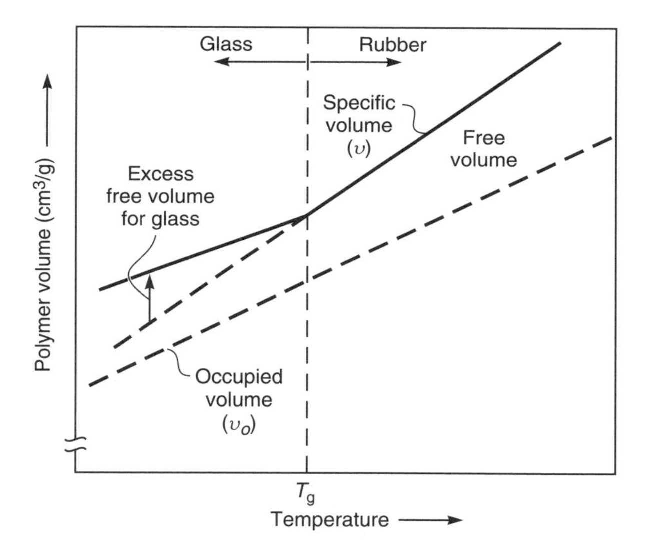

Figure 2.22 The change in specific volume as a function of temperature for a typical polymer

polymer chains can no longer rotate freely. Segmental motion then ceases, and the remaining free volume elements between the polymer chains are essentially frozen into the polymer matrix. As the polymer temperature is reduced further, its occupied volume will continue to decrease as the vibrational energy of the groups forming the polymer decreases, but the free volume elements remain essentially constant. Therefore, a glassy polymer contains both the normal free volume elements caused by the incomplete packing of the groups making up the polymer chains and the excess free volume elements frozen into the polymer matrix because the polymer chains cannot rotate.

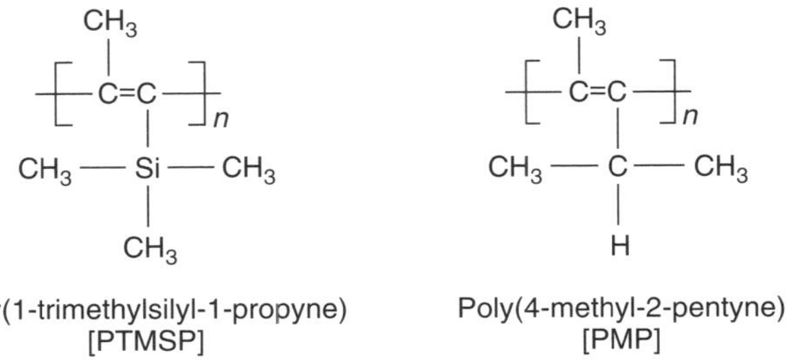

Poly(1-trimethylsilyl-1-propyne) [PTMSP]

Poly(4-methyl-2-pentyne) [PMP]

Figure 2.23 Structure of two high-free-volume substituted polyacetylenes, PTMSP and PMP. The carbon–carbon double bond is completely rigid, and depending on the size of the substituents, rotation around the carbon–carbon single bond can be very restricted also. The result is very stiff-backboned, rigid polymer chains which pack very poorly, leading to unusually high fractional free volumes

The fractional free volume of most materials is quite small and the value depends on the methods used for the calculation. For rubbers, the volume calculated by the Bondi method is generally about 10 to 15 % and for glassy polymers slightly higher, generally in the range 15 to 20 % because of the excess free volume contribution. Recently, a number of substituted polyacetylene polymers with extraordinarily rigid polymer backbones have been prepared. The structures of two such polymers are shown in Figure 2.23. Their glass transition temperatures are very high, and their free volumes are correspondingly unusually high—as much as 25 to 35 % of the polymers' volume is unoccupied space.

Correlation of the permeation properties of a wide variety of polymers with their free volume is not possible [32]. But, within a single class of materials, there is a correlation between the free volume of polymers and gas diffusion coefficients; an example is shown in Figure 2.24 [33]. The relationship between the free volume and the sorption and diffusion coefficients of gases in polymers, particularly glassy polymers, has been an area of a great deal of experimental and theoretical work. The subject has recently been reviewed in detail by Petropoulos [34] and by Paul and co-workers [35,36].

Sorption Coefficients in Polymers

The second key factor determining permeability in polymers is the sorption coefficient. The data in Figure 2.18 show that sorption coefficients for a particular gas are relatively constant within a single family of related materials. In fact, sorption coefficients of gases in polymers are relatively constant for a wide range of chemically different polymers. Figure 2.25 plots sorption and diffusion coefficients of methane in Tanaka's fluorinated polyimides [23], carboxylated polyvinyl trimethylsiloxane [37] and substituted polyacetylenes [38], all amorphous glassy polymers, and a variety of substituted siloxanes [39], all rubbers. The diffusion

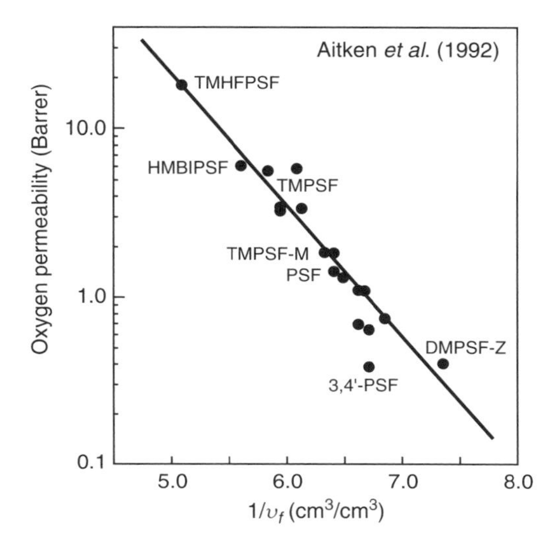

Figure 2.24 Correlation of the oxygen permeability coefficient for a family of related polysulfones with inverse fractional free volume (calculated using the Bondi method) [33]. Reprinted with permission from C.L. Aitken, W.J. Koros and D.R. Paul, Effect of Structural Symmetry on Gas Transport Properties of Polysulfones, *Macromolecules* **25**, 3424. Copyright 1992, American Chemical Society

coefficients of methane in the different polymers vary by more than 100 000, showing the extraordinary sensitivity of the permeant diffusion coefficients to changes in the packing of the polymer chains and to their flexibility. In contrast, sorption coefficients vary by only a factor of 10 around a mean value of about 15×10^{-3} cm^3(STP)/cm$^3 \cdot$ cmHg.

The sorption coefficients of gases in polymers remain relatively constant because sorption in polymers behaves as though the polymers were ideal fluids. Gas sorption in a polymer is expressed from Equation (2.57) as

$$c_{i_{(m)}} = K_i^{\mathrm{G}} p_i \tag{2.89}$$

By substituting for the sorption coefficient K_i^{G} from Equation (2.56), Equation (2.89) can be written as

$$c_{i_{(m)}} = m_i \rho_m \frac{\gamma_i^{\mathrm{G}} p_i}{\gamma_{i_{(m)}} p_{i_{\mathrm{sat}}}} \tag{2.90}$$

From the conversion of concentration to mole fraction [Equation (2.10)], it follows that

$$c_{i_{(m)}} = m_i \rho_m n_{i_{(m)}} \tag{2.91}$$

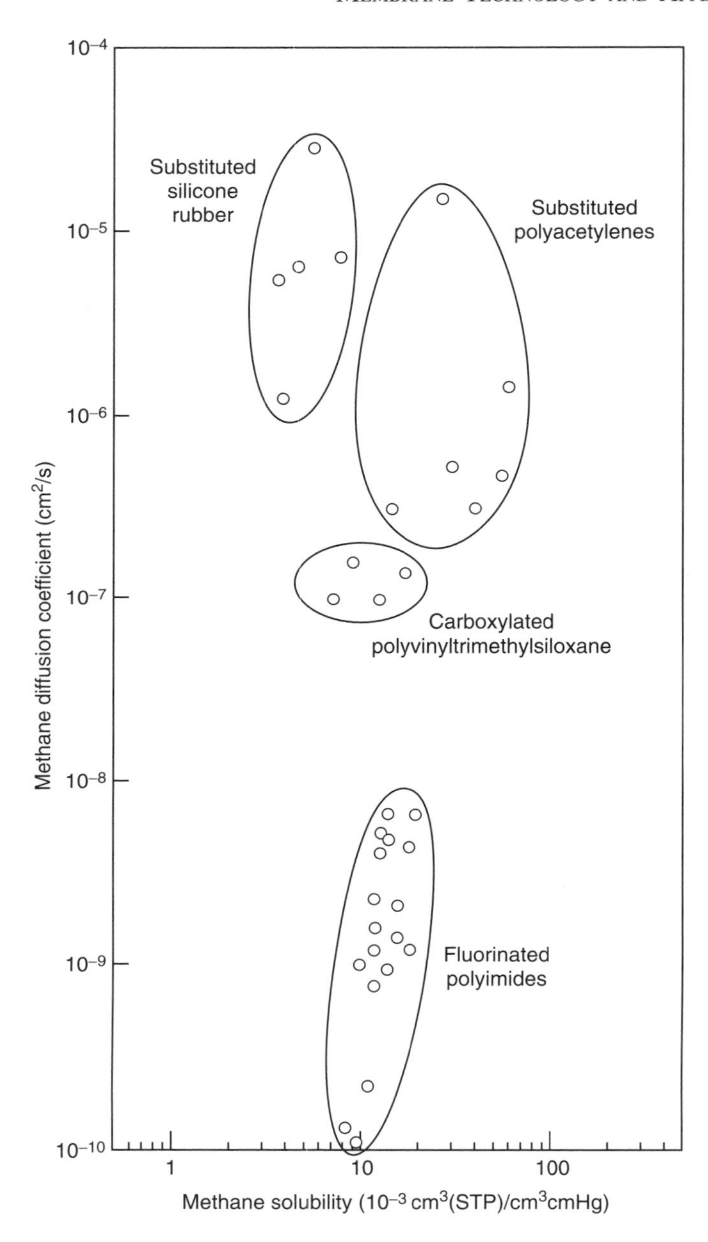

Figure 2.25 Diffusion and sorption coefficients of methane in different families of polymer materials. Diffusion coefficients change over a wide range but sorption coefficients are relatively constant. Data from references [23,35–37]

and so Equation (2.90) can be written as

$$\frac{c_{i_{(m)}}}{\rho_m m_i} = n_{i_{(m)}} = \frac{\gamma_i^{\mathrm{G}} p_i}{\gamma_{i_{(m)}} p_{i_{\mathrm{sat}}}} \tag{2.92}$$

For an ideal gas dissolving in an ideal liquid, γ_i^{G} and $\gamma_{i_{(m)}}$ are both unity, so Equation (2.92) can be written as

$$n_{i_{(m)}} = \frac{p_i}{p_{i_{\mathrm{sat}}}} \tag{2.93}$$

where $n_{i_{(m)}}$ is the mole fraction of the gas sorbed in the liquid, p_i is the partial pressure of the gas, and $p_{i_{\mathrm{sat}}}$ is the saturation vapor pressure at the pressure and temperature of the liquid. To apply Equation (2.93), the gas saturation vapor pressure must be determined. This can be done by extrapolating from available vapor pressure data to the ambient range using the Clausius–Clapeyron equation. For some gases the vapor pressure thereby obtained does not correspond to a stable gas–liquid equilibrium because the gas is supercritical at ambient temperatures. However, the calculated value is adequate to calculate the sorption coefficient using Equation (2.93) [40]. At 25 °C the saturation vapor pressure of methane extrapolated in this way is 289 atm. Thus, from Equation (2.93) the mole fraction of methane dissolved in an ideal liquid is 1/289 or 0.0035. The ideal solubility and measured solubilities of methane in a number of common liquids are given in Table 2.3. Although there is some spread in the data, particularly for small polar solvent molecules such as water or methanol, the overall agreement is remarkably good. A more detailed discussion of the solubility of gases in liquids is given in the book by Fogg and Gerrard [41].

To apply the procedure outlined above to a polymer, it is necessary to use the Flory–Huggins theory of polymer solution, which takes into account the entropy of mixing of solutes in polymers caused by the large difference in molecular size

Table 2.3 Mole fraction of methane in various solvents at 25 °C and 1 atm. The solubility of methane in an ideal liquid under these conditions is 0.0035 [40]

Liquid	Methane solubility (mole fraction)
Ethyl ether	0.0045
Cyclohexane	0.0028
Carbon tetrachloride	0.0029
Acetone	0.0022
Benzene	0.0021
Methanol	0.0007
Water	0.00002

between the two components. The Flory–Huggins expression for the free energy of mixing of a gas in polymer solution can be written [42]

$$\Delta G = RT \ln \frac{p_i}{p_{i_{sat}}} = RT\left[\ln V_i + \left(1 - \frac{\upsilon_i}{\upsilon_j}\right)(1 - V_i)\right] \tag{2.94}$$

where υ_i and υ_j are the molar volumes of the gas i and the polymer j respectively, and V_i the volume fraction of the polymer j occupied by the sorbed gas i. When $\upsilon_i \approx \upsilon_j$, that is, the gas and polymer molecules are approximately the same size, Equation (2.94) reduces to Equation (2.93), the ideal liquid case. When $\upsilon_i \ll \upsilon_j$, that is, when the molar volume of a gas (υ_i) is much smaller than the molar volume of the polymer (υ_j), then $\upsilon_i/\upsilon_j \rightarrow 0$ and Equation (2.94) becomes

$$\ln \frac{p_i}{p_{i_{sat}}} = \ln V_i + (1 - V_i) \tag{2.95}$$

Equation (2.95) can be rearranged to

$$V_i = \frac{p_i/p_{i_{sat}}}{\exp(1 - V_i)} \tag{2.96}$$

and since V_i is small, $\exp(1 - V_1)$ is approximately $\exp(1) \approx 2.72$, Equation (2.95) then becomes

$$V_i = \frac{p_i/p_{i_{sat}}}{2.72} \tag{2.97}$$

Comparing Equations (2.93) and (2.97), we see that the volume fraction of gas sorbed by an ideal polymer is 1/2.72 of the mole fraction of a gas sorbed in an ideal liquid.[6]

The results of such a calculation are shown in Table 2.4. In Figure 2.26, the calculated sorption coefficients in an ideal polymer from Table 2.4 are plotted against the average sorption coefficients of the same gases in Tanaka's polyimides [23]. The calculated values are within a factor of two of the experimental values, which is extremely good agreement considering the simplicity of Equation (2.97). A more detailed discussion of sorption of gases in polymers is given in a review by Petropoulos [34].

As shown above, thermodynamics can qualitatively predict the sorption of simple gases in polymers to within a factor of two. Moreover, Equation (2.97) predicts that all polymers should have about the same sorption for the same gas and that sorption is inversely proportional to saturation vapor pressure.

Another way of showing the same effect is to plot gas sorption against some convenient measure of saturation vapor pressure, such as the gas boiling point

---

[6] V_i is the volume fraction of the gas sorbed in the polymer. To calculate the amount of gas sorbed in cm^3 (STP)/cm^3, the molar density of the sorbed gas must be known. We assume this density is 1/MW (mol/cm^3).

Table 2.4 Solubility of gases in an ideal liquid and an ideal polymer (35 °C)

Gas	Calculated saturation vapor pressure, $p_{i_{sat}}$ (atm)	Ideal solubility in a liquid at 1 atm (mole fraction) [Equation (2.93)]	Ideal solubility in a polymer [10^{-3} cm^3(STP)/cm^3 · cmHg] [Equation (2.97)]
N_2	1400	0.0007	2.6
O_2	700	0.0014	4.8
CH_4	366	0.0027	18.4
CO_2	79.5	0.0126	29.5

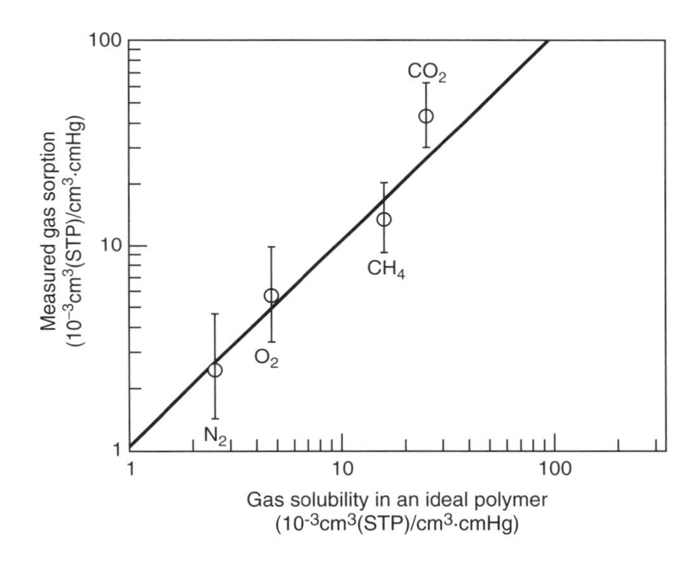

Figure 2.26 Average sorption coefficients of simple gases in a family of 18 related polyimides plotted against the expected sorption in an ideal polymer calculated using Equation (2.97). Data from Tanaka *et al.* [23]

or critical temperature. Figure 2.27 shows a plot of this type for a typical glassy polymer (polysulfone), a typical rubber (silicone rubber), and the values for the ideal solubility of a gas in a polymer calculated using Equation (2.97) [43]. The figure shows that the difference in gas sorptions of polymers is relatively small and the values are grouped around the calculated value.

Although all of these predictions are qualitatively correct, the differences between the behavior of an ideal polymer and an actual polymer are important in selecting the optimum material for a particular separation. The usual starting point for this fine-tuning is the dual-sorption model originally proposed by Barrer *et al.* [44]. This model has since been extended by Michaels *et al.* [45], Paul *et al.* [46], Koros *et al.* [47] and many others.

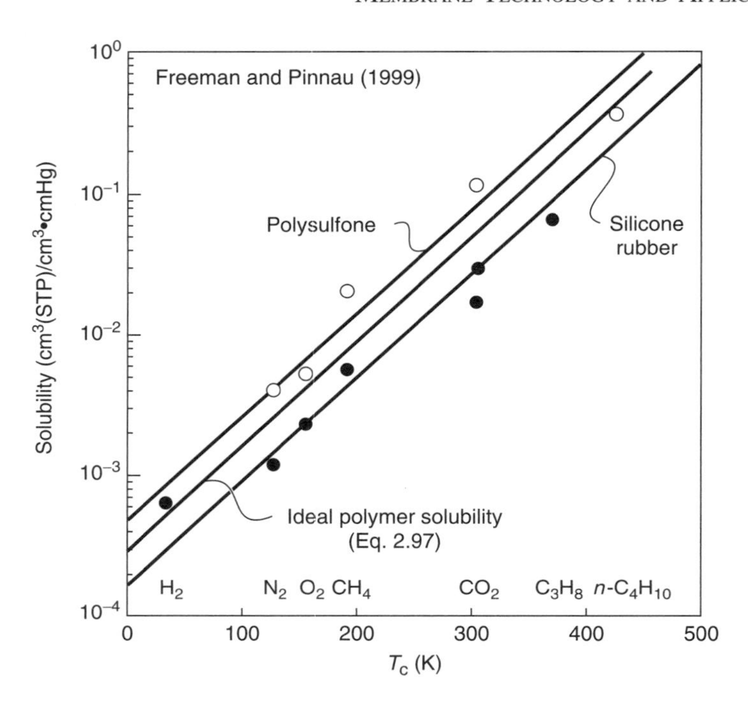

Figure 2.27 Solubilities as a function of critical temperature (T_c) for a typical glassy polymer (polysulfone) and a typical rubbery polymer (silicone rubber) compared with values for the ideal solubility calculated from Equation (2.97)[43]

According to the dual-sorption model, gas sorption in a polymer (c_m) occurs in two types of sites. The first type is filled by gas molecules dissolved in the equilibrium free volume portion of material (concentration c_H). In rubbery polymers this is the only population of dissolved gas molecules, but in glassy polymers a second type of site exists. This population of dissolved molecules (concentration c_D) is dissolved in the excess free volume of the glassy polymer. The total sorption in a glassy polymer is then

$$c_m = c_D + c_H \tag{2.98}$$

The number of molecules (c_D) dissolved in the equilibrium free volume portion of the polymer will behave as in normal sorption in a liquid and can be related to the pressure in the surrounding gas by a linear expression equivalent to Equation (2.89)

$$c_D = K_D p \tag{2.99}$$

This fraction of the total sorption is equivalent to the value calculated in Equation (2.97). The other fraction (c_H) is assumed to be sorbed into the excess free volume elements, which are limited, so sorption will cease when all the sites are filled. Sorption in these sites is best approximated by a Langmuir-type absorption isotherm

$$c_H = \frac{c'_H bp}{1 + bp} \tag{2.100}$$

At high pressures $c_H \rightarrow c'_H$, where c'_H is the saturation sorption concentration at which all excess free volume sites are filled.

From Equations (2.99) and (2.100) it follows that the total sorption can be written as

$$c_m = K_D p + \frac{c'_H bp}{1 + bp} \tag{2.101}$$

The form of the sorption isotherm predicted from the dual sorption model is shown in Figure 2.28. Because the expressions for sorption contain three adjustable parameters, good agreement between theory and experiment is obtained.

Often, much is made of the particular values of the constants c'_H, b, and K. However, these constants should be treated with caution because they depend totally on the starting point of the curve-fitting exercise. That is, starting with an arbitrary value of c'_H, the other constants b and K can usually be adjusted to obtain good agreement of Equation (2.101) with experiment. If the starting value for c'_H is changed, then equally good agreement between theory and experiment can still be obtained but with different values of b and K [48].

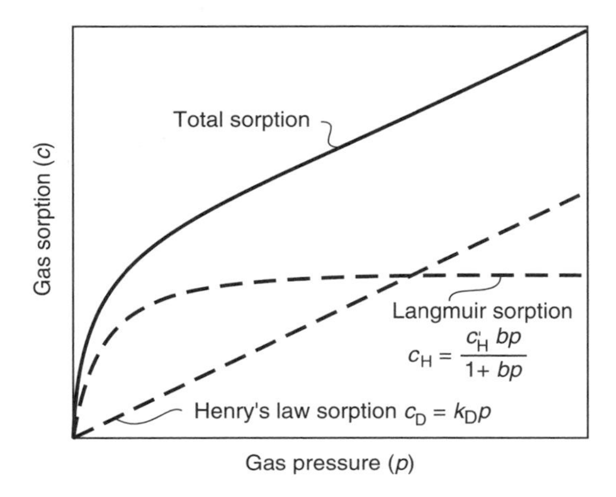

Figure 2.28 An illustration of the two components that contribute to gas sorption in a glassy polymer according to the dual sorption model. Henry's law sorption occurs in the equilibrium free volume portion of the polymer. Langmuir sorption occurs in the excess free volume between polymer chains that exists in glassy polymers

Permeation of gases in glassy polymers can also be described in terms of the dual sorption model. One diffusion coefficient (D_D) is used for the portion of the gas dissolved in the polymer according to the Henry's law expression and a second, somewhat larger, diffusion coefficient (D_H) for the portion of the gas contained in the excess free volume. The Fick's law expression for flux through the membrane has the form

$$J = -D_D \frac{\mathrm{d}c_D}{\mathrm{d}x} - D_H \frac{\mathrm{d}c_H}{\mathrm{d}x} \tag{2.102}$$

Pore-flow Membranes

The creation of a unified theory able to rationalize transport in the dense membranes used in reverse osmosis, pervaporation and gas separation occurred over a 20-year period from about 1960 to 1980. Development of this theory was one of the successes of membrane science. The theory did not form overnight as the result of one single breakthrough but rather as the result of a series of incremental steps. The paper of Lonsdale *et al.* [12] applying the solution-diffusion model to reverse osmosis for the first time was very important.[7] Also important was the series of papers by Paul and co-workers showing the connection between hydraulic permeation (reverse osmosis) and pervaporation [4–6, 19] and providing the experimental support for the solution-diffusion model as applied to these processes. Unfortunately no equivalent unified theory to describe transport in microporous membranes has been developed. Figure 2.29 illustrates part of the problem, namely the extremely heterogeneous nature of microporous membranes. All of the microporous membranes shown in this figure perform approximately the same separation, but their porous structure and the mechanism of the separation differ significantly. The nucleation track membrane (Figure 2.29a) and the asymmetric Loeb–Sourirajan membrane (Figure 2.29d) both separate particles by molecular sieving. The cellulose acetate/cellulose nitrate membrane (Figure 2.29c) is a depth filter which captures particles within the interior of the membrane by adsorption. The expanded film membrane (Figure 2.29b) captures particles by both methods. The materials from which these membranes are made also differ, ranging from polyethylene and polysulfone, both hydrophobic, low-surface-energy materials, to cellulose acetate, a hydrophilic material that often carries charged surface groups.

The parameters available to characterize the complexity of microporous membranes are also imperfect. Some widely used parameters are illustrated in Figure 2.30. The membrane porosity (ε) is the fraction of the total membrane

[7]This paper was initially submitted by its three industrial authors for publication to the *Journal of Physical Chemistry* and was rejected as insufficiently fundamental. More than 30 years after it was finally published in the *Journal of Applied Polymer Science*, it remains one of the most cited papers on membrane transport theory.

(a) Track etch

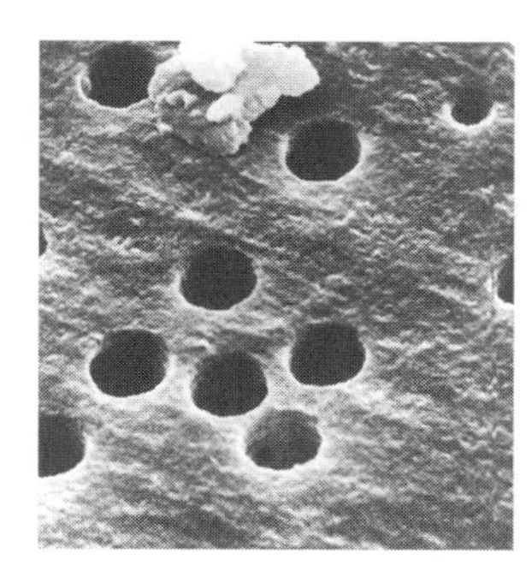

(b) Expanded film

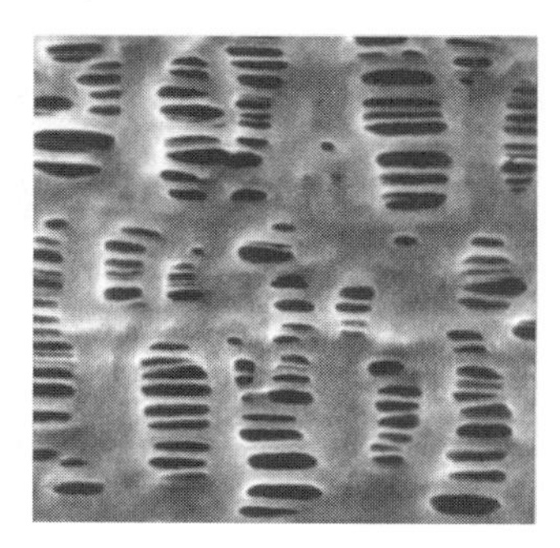

(c) Phase separation

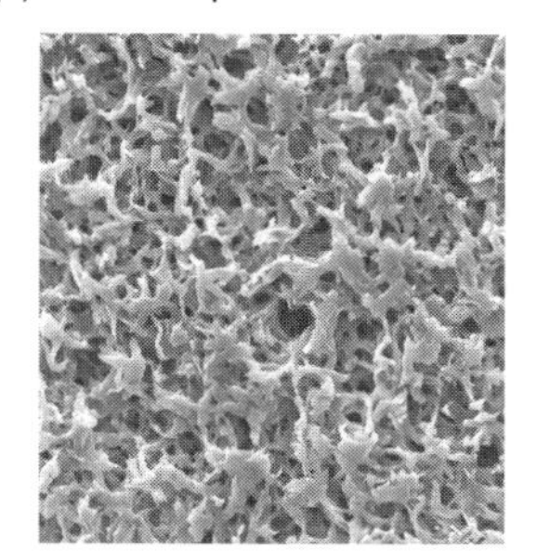

(d) Loeb–Sourirajan

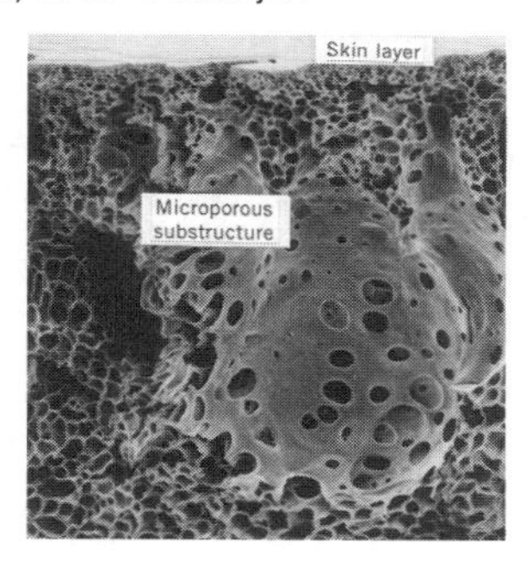

Figure 2.29 Scanning electron micrographs at approximately the same magnification of four microporous membranes having approximately the same particle retention. (a) Nuclepore (polycarbonate) nucleation track membrane; (b) Celgard® (polyethylene) expanded film membrane; (c) Millipore cellulose acetate/cellulose nitrate phase separation membrane made by water vapor imbibition (Courtesy of Millipore Corporation, Billerica, MA); (d) anisotropic polysulfone membrane made by the Loeb–Sourirajan phase separation process

volume that is porous. Typical microporous membranes have average porosities in the range 0.3–0.7. This number can be obtained easily by weighing the membrane before and after filling the pores with an inert liquid. The average porosity obtained this way must be treated with caution, however, because the porosity of a membrane can vary from place to place. For example, anisotropic membranes, such as the Loeb–Sourirajan phase separation membrane shown in Figure 2.29(d), often have an average porosity of 0.7–0.8, but the porosity of the skin layer that performs the actual separation may be as low as 0.05.

The membrane tortuosity (τ) reflects the length of the average pore compared to the membrane thickness. Simple cylindrical pores at right angles to the membrane surface have a tortuosity of one, that is, the average length of the pore is the

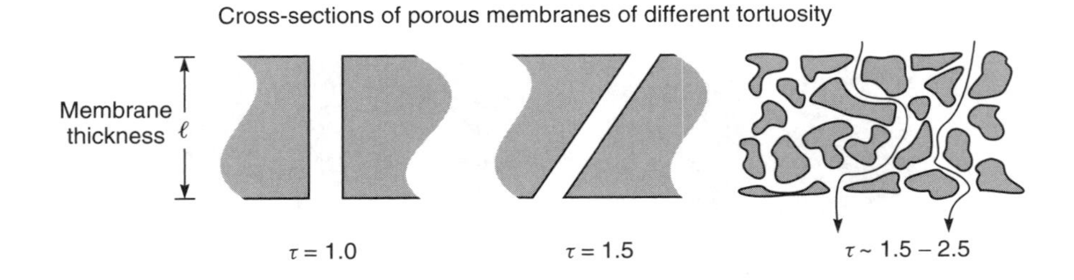

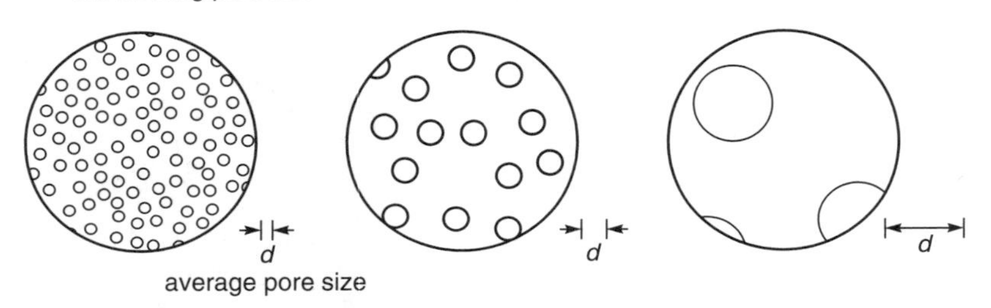

Figure 2.30 Microporous membranes are characterized by their tortuosity (τ), their porosity (ε), and their average pore diameter (d)

membrane thickness. Usually pores take a more meandering path through the membrane, so typical tortuosities are in the range 1.5–2.5.

The most important property characterizing a microporous membrane is the pore diameter (d). Some of the methods of measuring pore diameters are described in Chapter 7. Although microporous membranes are usually characterized by a single pore diameter value, most membranes actually contain a range of pore sizes. In ultrafiltration, the pore diameter quoted is usually an average value, but to confuse the issue, the pore diameter in microfiltration is usually defined in terms of the largest particle able to penetrate the membrane. This nominal pore diameter can be 5 to 10 times smaller than the apparent pore diameter based on direct microscopic examination of the membrane.

Permeation in Ultrafiltration and Microfiltration Membranes

Microporous ultrafiltration and microfiltration membranes used to filter particulates from liquids fall into the two general categories illustrated in Figure 2.31. The first category (a) is the surface or screen filter; such membranes contain surface pores smaller than the particles to be removed. Particles in the permeating

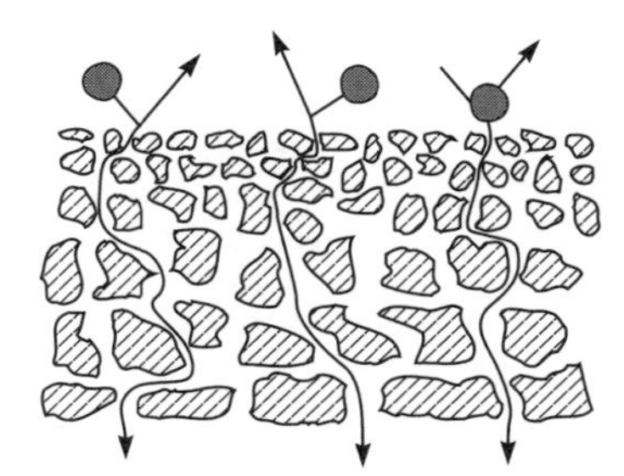

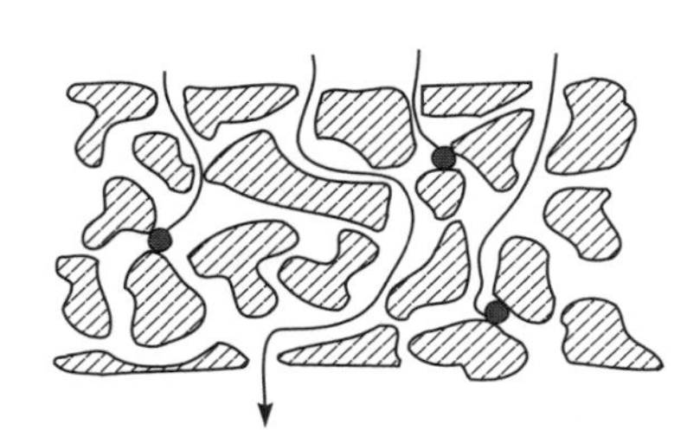

Figure 2.31 Separation of particulates can take place at the membrane surface according to a screen filtration mechanism (a) or in the interior of the membrane by a capture mechanism as in depth filtration (b)

fluid are captured and accumulate on the surface of the membrane. These membranes are usually anisotropic, with a relatively finely microporous surface layer on a more open microporous support. Particles small enough to pass through the surface pores are not normally captured in the interior of the membrane. Most ultrafiltration membranes are screen filters.

The second category of microporous membranes is the depth filter (b), which captures the particles to be removed in the interior of the membrane. The average pore diameter of a depth filter is often 10 times the diameter of the smallest particle able to permeate the membrane. Some particles are captured at small constrictions within the membrane, others by adsorption as they permeate the membrane by a tortuous path. Depth filters are usually isotropic, with a similar pore structure throughout the membrane. Most microfiltration membranes are depth filters.

Screen Filters

The mechanism of particle filtration by screen filters has been the subject of many studies because it is relatively easily described mathematically; Bungay has published a review of this work [49]. Ferry [50] was the first to model membrane retention by a screen filter; in his model pores were assumed to be equal circular capillaries with a large radius, r, compared to the solvent molecule radius. Therefore, the total area of the pore is available for transport of solvent. A solute molecule whose radius, a, is an appreciable fraction of the pore radius cannot approach nearer than one molecular radius of the pore overall. The model is illustrated in Figure 2.32.

The area, A, of the pore available for solute transport is given by the equation

$$\frac{A}{A_o} = \frac{(r-a)^2}{r^2} \tag{2.103}$$

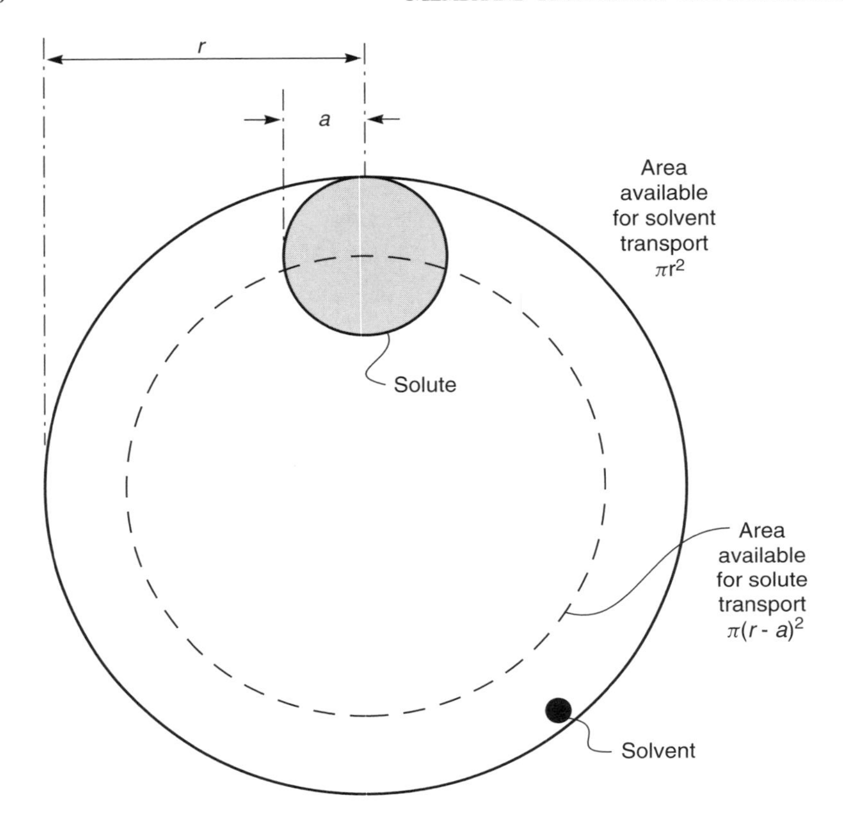

Figure 2.32 Illustration of the Ferry mechanical exclusion model of solute transport in small pores

where A_o is the area of the pore available for solvent molecules. Later, Renkin [51] showed that Equation (2.103) has to be modified to account for the parabolic velocity profile of the fluid as it passes through the pore. The effective fractional pore area available for solutes in this case is

$$\left(\frac{A}{A_o}\right)' = 2\left(1 - \frac{a}{r}\right)^2 - \left(1 - \frac{a}{r}\right)^4 \tag{2.104}$$

where $(A/A_o)'$ is equal to the ratio of the solute concentration in the filtrate (c_ℓ) to the concentration in the feed (c_o), that is,

$$\left(\frac{A}{A_o}\right)' = \left(\frac{c_\ell}{c_o}\right) \tag{2.105}$$

It follows that from Equation (2.105) and the definition of solution rejection [Equation (2.48)] that the rejection of the membrane is

$$\mathbb{R} = \left[1 - 2\left(1 - \frac{a}{r}\right)^2 + \left(1 - \frac{a}{r}\right)^4\right] \times 100\,\% \tag{2.106}$$

The Ferry–Renkin equation can be used to estimate the pore size of ultrafiltration membranes from the membrane's rejection of a solute of known radius. The rejections of globular proteins by four typical ultrafiltration membranes plotted against the cube root of the protein molecular weight (an approximate measure of the molecular radius) are shown in Figure 2.33(a). The theoretical curves

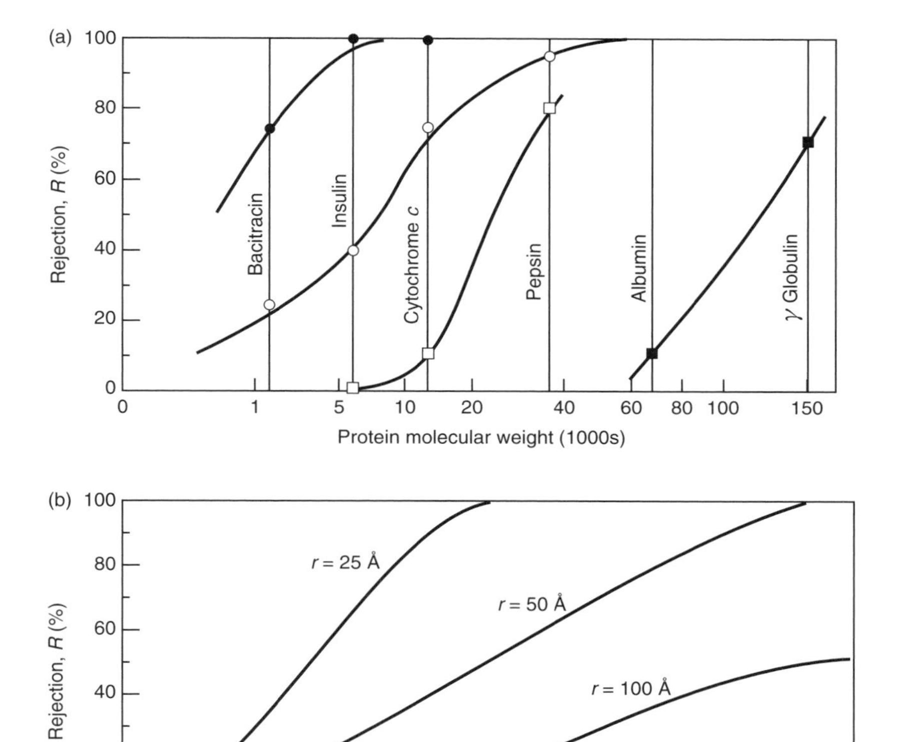

Figure 2.33 (a) Rejection of globular proteins by ultrafiltration membranes of increasing pore size; (b) calculated rejection curves from the Ferry–Renkin equation (2.106) plotted on the same scale [52]

Table 2.5 Marker molecules used to characterize ultrafiltration membranes

Species	Molecular weight (×1000)	Estimated molecular diameter (Å)
Sucrose	0.34	11
Raffinose	0.59	13
Vitamin B_{12}	1.36	17
Bacitracin	1.41	17
Insulin	5.7	27
Cytochrome *c*	13.4	38
Myoglobin	17	40
α-Chymotrysinogene	25	46
Pepsin	35	50
Ovalbumin	43	56
Bovine albumin	67	64
Aldolase	142	82
γ-Globulin	150	84

calculated from Equation (2.106) are shown directly in Figure 2.33(b) [52]. The abscissae of both figures have been made comparable because the radius of gyration of albumin is approximately 30 Å. A pore size that appears to be reasonable can then be obtained by comparing the two graphs. This procedure for obtaining an approximate pore size from membrane retention measurements shown in Figure 2.33 has been widely used. Globular proteins are usually the basis for this work because their molecular weights and molecular diameter can be calculated precisely. A list of some commonly used molecular markers is given in Table 2.5.

Depth Filters

The mechanism of particle capture by depth filtration is more complex than for screen filtration. Simple capture of particles by sieving at pore constructions in the interior of the membrane occurs, but adsorption of particles on the interior surface of the membrane is usually at least as important. Figure 2.34 shows four mechanisms that contribute to particle capture in depth membrane filters. The most obvious mechanism, simple sieving and capture of particles at constrictions in the membrane, is often a minor contributor to the total separation. The three other mechanisms, which capture particles by adsorption, are inertial capture, Brownian diffusion and electrostatic adsorption [53,54]. In all cases, particles smaller than the diameter of the pore are captured by adsorption onto the internal surface of the membrane.

In inertial capture, relatively large particles in the flowing liquid cannot follow the fluid flow lines through the membrane's tortuous pores. As a result, such particles are captured as they impact the pore wall. This capture mechanism is

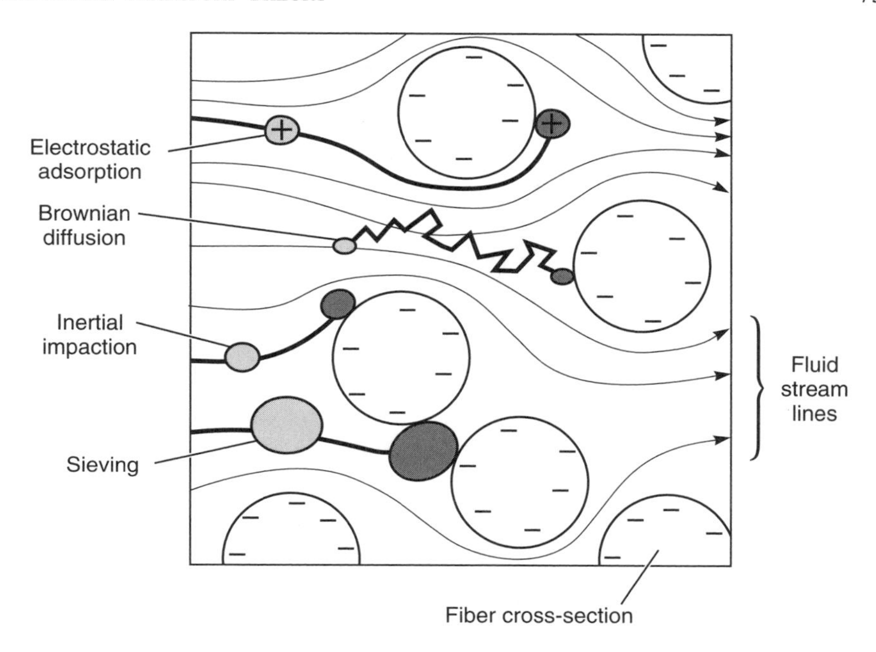

Figure 2.34 Particle capture mechanism in filtration of liquid solutions by depth microfilters. Four capture mechanisms are shown: simple sieving; electrostatic adsorption; inertial impaction; and Brownian diffusion

more frequent for larger diameter particles. In experiments with colloidal gold particles and depth filtration membranes with tortuous pores approximately 5 μm in diameter, Davis showed that 60 % of 0.05-μm-diameter particles were captured [55]. Nucleation track membranes with 5-μm, almost straight-through pores and no tortuosity retained less than 1 % of the particles. The retention of the small particles by the depth filter was caused by the greater tortuosity which led to inertial capture.

The second mechanism is capture by Brownian diffusion, which is more of a factor for smaller particles. Small particles are easily carried along by the moving fluid. However, because the particles are small, they are subject to random Brownian motion that periodically brings them into contact with the pore walls. When this happens, capture by surface adsorption occurs.

The third mechanism is capture of charged particles by membranes having surface-charged groups. Many common colloidal materials carry a slight negative charge, so membranes containing an excess of positive groups can provide enhanced removals. Several microfiltration membrane manufacturers produce this type of charged membrane. One problem is that the adsorption capacity of the charged group is exhausted as filtration proceeds, and the retention falls.

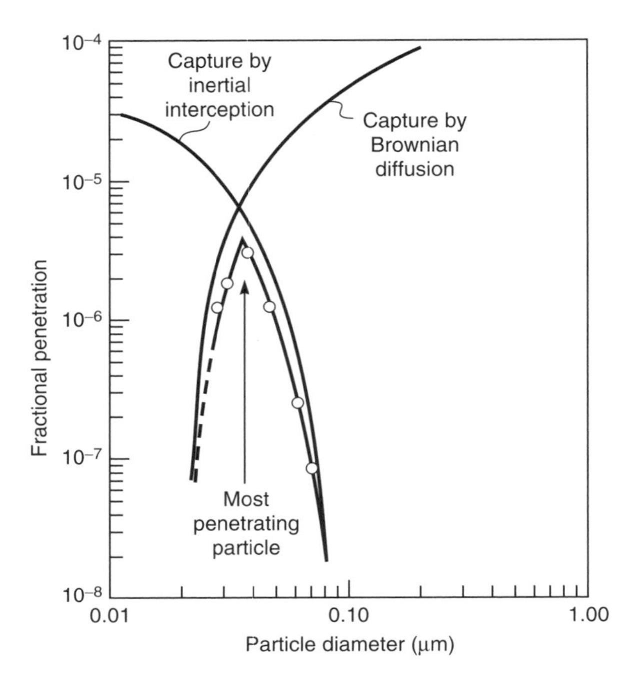

Figure 2.35 Gas-borne particle penetration through an ultrathin PVDF membrane [55,56]

In filtration of gas-borne aerosol particles by microfiltration membranes, capture by adsorption is usually far more important than capture by sieving. This leads to the paradoxical result that the most penetrating particle may not be the smallest one. This is because capture by inertial interception is most efficient for larger particles, whereas capture by Brownian motion is most efficient for smaller particles. As a result the most penetrating particle has an intermediate diameter, as shown in Figure 2.35 [55,56].

Knudsen Diffusion and Surface Diffusion in Microporous Membranes

Essentially all industrial gas separation membranes involve permeation through dense polymeric membranes. But the study of gas permeation through finely microporous membranes has a long history dating back to Graham's work in the 1850s. To date, the only application of these membranes has been the separation of $U^{235}F_6$ and $U^{238}F_6$ in the Manhattan project. More recently finely microporous membranes made by carbonizing poly(vinylidene chloride) and other polymers have been developed and taken to the pilot plant scale.

If the pores of a microporous membrane are 0.1 μm or larger, gas permeation will take place by normal convective flow described by Poiseuille's law. As the pore radius (r) decreases it can become smaller than the mean free path (λ) of the gas. (At atmospheric pressure the mean free path of common gases is in the range 500–2000 Å.) When this occurs the ratio of the pore radius to the gas mean free path (r/λ) is less than one. Diffusing gas molecules then have more collisions with the pore walls than with other gas molecules. Gas permeation in this region is called Knudsen diffusion. At every collision with the pore walls, the gas molecules are momentarily adsorbed and then reflected in a random direction. Molecule–molecule collisions are rare, so each gas molecule moves independently of all others. Hence with gas mixtures in which the different species move at different average velocities, a separation is possible. The gas flow in a membrane made of cylindrical right capillaries for Knudsen diffusion is given by Equation (2.103)

$$j = \frac{4r\varepsilon}{3} \cdot \left(\frac{2RT}{\pi m}\right)^{1/2} \cdot \frac{p_o - p_\ell}{\ell \cdot RT} \tag{2.107}$$

where m is the molecular weight of the gas, j is the flux in gmol/cm$^2 \cdot$ s, ε is the porosity of the membrane, r is the pore radius, ℓ is the pore length and p_o and p_ℓ are the absolute pressures of the gas species at the beginning of the pore ($x = 0$) and at the end ($x = \ell$).

The equivalent equation for permeation by Poiseuille flow is

$$j = \frac{r^2\varepsilon}{8\eta} \cdot \frac{[p_o - p_\ell][p_o + p_\ell]}{\ell \cdot RT} \tag{2.108}$$

where η is the viscosity of the gas. Equation (2.108) differs from the more familiar Poiseuille equation for liquids by the additional term $[p_o + p_\ell]$ which arises from the expansion of a gas as it moves down the pressure gradient.

Figure 2.36 shows the effect of the ratio r/λ on the relative proportions of Knudsen to Poiseuille flow in a cylindrical capillary [57]. When r/λ is greater than one, Poiseuille flow predominates. Because the mean free path of gases at atmospheric pressure is in the range of 500–2000 Å, for Knudsen flow to predominate and a separation to be obtained, the membrane pore radius must be less than 500 Å.

It follows from Equation (2.107) that the permeability of a gas (i) through a Knudsen diffusion membrane is proportional to $1/\sqrt{m_i}$. The selectivity of this membrane ($\alpha_{i/j}$), proportional to the ratio of gas permeabilities, is given by the expression

$$\alpha_{i/j} = \sqrt{\frac{m_j}{m_i}} \tag{2.109}$$

This result was first observed experimentally by Graham and is called Graham's law of diffusion. Knudsen diffusion membranes have been used to separate gas isotopes that are difficult to separate by other methods, for example tritium from

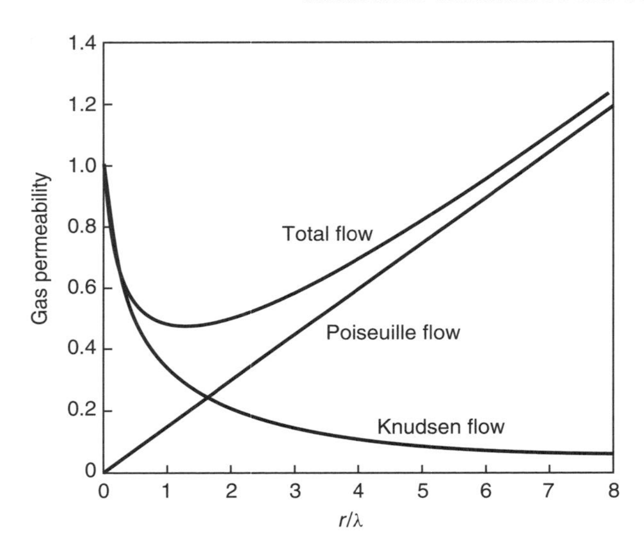

Figure 2.36 Illustration of the proportion of Knudsen to Poiseuille flow as a function of r/λ (after Barrer) [57]

hydrogen, $C^{12}H_4$ from $C^{14}H_4$ and most importantly $U^{235}F_6$ from $U^{238}F_6$. The membrane selectivity for $U^{235}F_6/U^{238}F_6$ mixtures is only 1.0043, so hundreds of separation stages are required to produce a complete separation. Nevertheless, at the height of the Cold War, the US Atomic Energy Commission operated three plants fitted with microporous metal membranes that processed almost 20 000 tons/year of uranium.

When the pore diameter of a microporous membrane decreases to the 5–10 Å range, the pores begin to separate gases by a molecular sieving effect. The difficulty of making these membranes defect-free has so far prevented their application to industrial separations. However, in the laboratory, spectacular separations have been reported for gases that differ in size by only 0.1 Å. Figure 2.37 shows some data for permeation through microporous silica membranes [58]. No polymeric membranes can match this separation.

Surface adsorption and diffusion add a second contribution to gas permeation that can occur in small-pore-diameter membranes. This phenomenon is shown schematically in Figure 2.38. Adsorption onto the walls of the small pores becomes noticeable when the pore diameter drops below about 100 Å. At this pore diameter the surface area of the pore walls is in the range 100 m^2/cm^3 of material. Significant amounts of gas then adsorb onto the pore walls, particularly if the gas is condensable. Often the amount of gas sorbed on the pore walls is much greater than the amount of nonsorbed gas. Sorbed gas molecules are mobile and can move by a process of surface diffusion through the membrane according

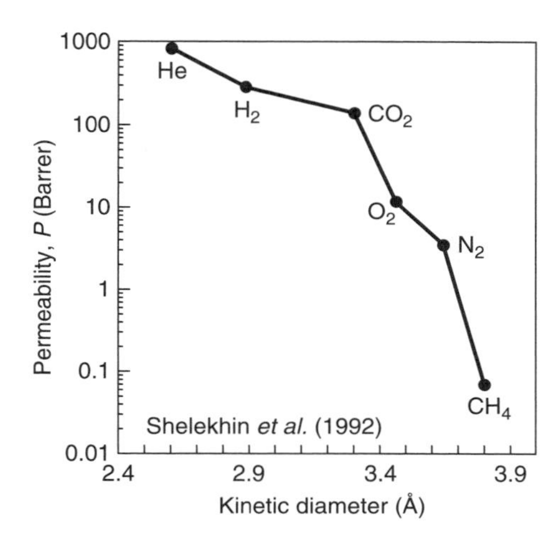

Figure 2.37 Permeability coefficients as a function of the gas kinetic diameter in microporous silica hollow fine fibers [58]. Reprinted from *J. Membr. Sci.* **75**, A.B. Shelekhin, A.G. Dixon and Y.H. Ma, Adsorption, Permeation, and Diffusion of Gases in Microporous Membranes, 233, Copyright 1992, with permission from Elsevier

to a Fick's law type of expression

$$J_s = -D_s \frac{\mathrm{d}c_s}{\mathrm{d}x} \tag{2.110}$$

where J_s is the contribution to permeation by surface diffusion of the sorbed gas c_s and D_s is a surface diffusion coefficient. At room temperature, typical surface diffusion coefficients are in the range 1×10^{-3}–1×10^{-4} cm^2/s, intermediate between the diffusion coefficients of molecules in gases and liquids [59]. Although these coefficients are less than the diffusion coefficients for nonsorbed gas, surface diffusion still makes a significant contribution to total permeation.

Some typical results illustrating the effect of surface diffusion are shown in Figure 2.39 for permeation of gases through microporous glass [60]. The expected permeability normalized for gas molecular weight, $\mathcal{P}\sqrt{m}$, is constant, but only the very low boiling gases, helium, hydrogen and neon, approach this value. As the condensability of the gas increases (as measured by boiling point or critical temperature) the amount of surface adsorption increases and the contribution of surface diffusion to gas permeation increases. For butane, for example, 80 % of the total gas permeation is due to surface diffusion.

In experiments with mixtures of condensable and noncondensable gases, adsorption of the condensable gas component can restrict or even completely block permeation of the noncondensable gas [61,62]. This effect was first noticed by Barrer and Pope in experiments with sulfur dioxide/hydrogen mixtures [63]; some of the data are shown in Figure 2.40. Sorption of sulfur dioxide on the pore walls

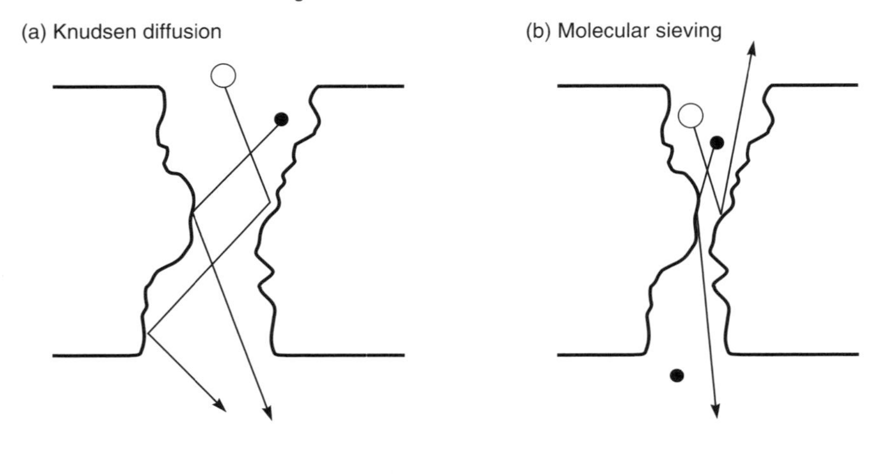

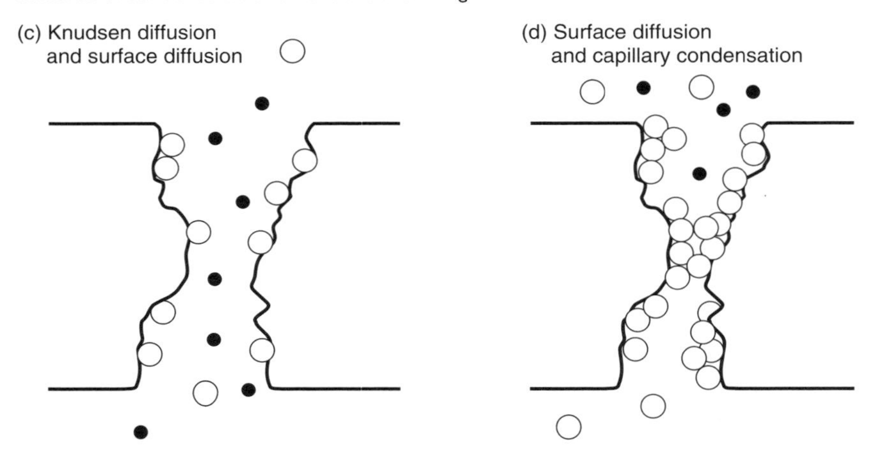

Figure 2.38 Permeation of noncondensable and condensable gas mixtures through finely microporous membranes. With noncondensable gases molecular sieving occurs when the pore wall reaches the 5- to 10-Å diameter range. With gas mixtures containing condensable gases surface diffusion increases as the pore diameter decreases and the temperature decreases (increasing adsorption)

of the microporous carbon membrane inhibits the flow of hydrogen. If adsorption is increased by increasing the sulfur dioxide partial pressure or by lowering the temperature, sufficient sulfur dioxide is adsorbed to cause capillary condensation of sulfur dioxide in the membrane pores, completely blocking permeation of hydrogen. At this point the membrane only permeates sulfur dioxide.

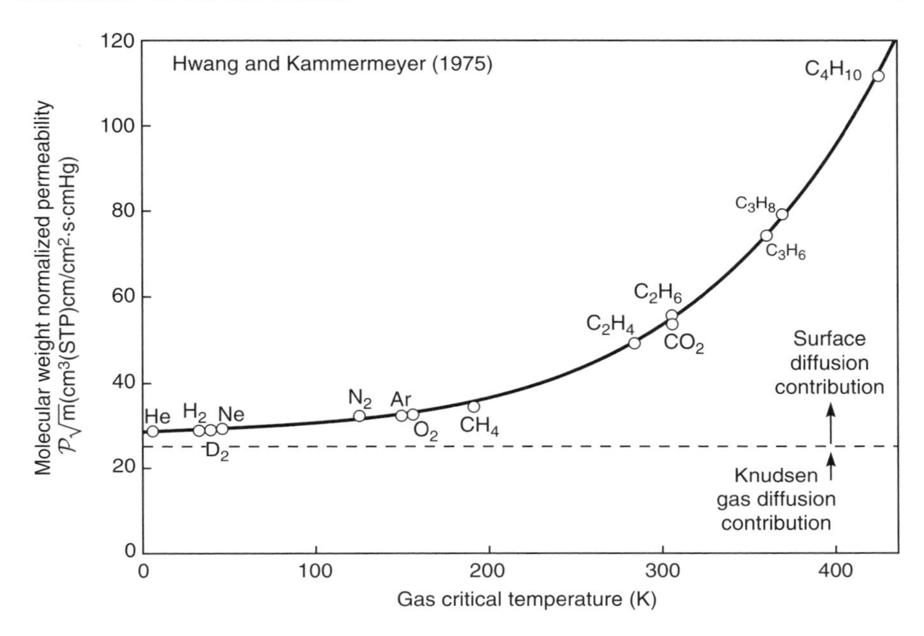

Figure 2.39 Molecular-weight-normalized permeability of gases through Vycor microporous glass membranes [60]. Reprinted from *Techniques of Chemistry, Vol. VII, Membranes in Separations*, S.T. Hwang and K. Kammermeyer; A. Weissberger (ed.); Copyright © 1975. This material is used by permission of John Wiley & Sons, Inc.

Microporous carbonized hollow fibers were developed over a period of 20 years by Soffer, Koresh, and others (64) at Carbon Membranes Ltd. and were brought to the small module scale. Spectacular separations were reported, but the membranes were difficult to make defect-free and were relatively sensitive to fouling and breaking. More recently, Rao, Sirkar, and others at Air Products tried to use microporous membranes to separate hydrogen/light hydrocarbon gas mixtures found in refinery waste gas streams [65,66]. They also used microporous carbon membranes, this time formed by vacuum carbonization of polymer films cast onto microporous ceramic supports. The adsorbed hydrocarbons permeate the membranes by surface diffusion while permeation of hydrogen in the gas phase is blocked by capillary condensation in the membrane pores. The process was tried at the pilot-plant scale, but eventually abandoned in part because of blocking of the membranes by permanently adsorbed higher hydrocarbons in the feed gas.

Despite these failures, microporous carbon membranes continue to be a subject of research by a number of groups [67–70]. The selectivities obtained are often very good, even for simple gas mixtures such as oxygen/nitrogen or carbon dioxide/methane. However long-term, it is difficult to imagine carbon membranes

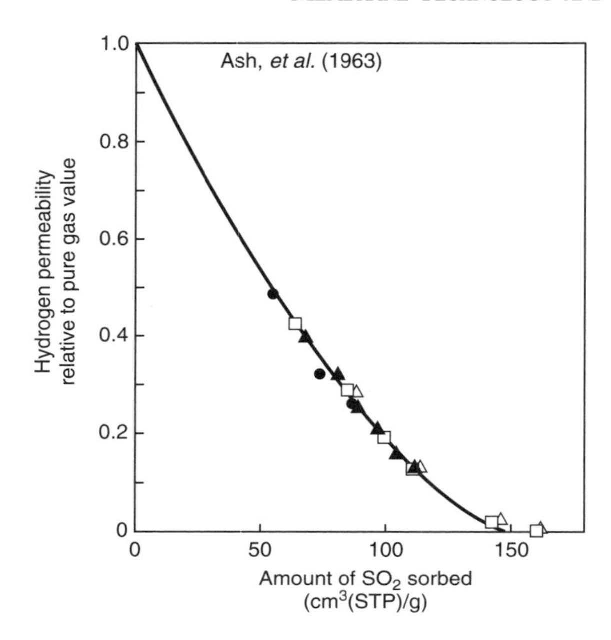

Figure 2.40 Blocking of hydrogen in hydrogen/sulfur dioxide gas mixture permeation experiments with finely microporous membranes [63] as a function of the amount of sulfur dioxide adsorbed by the membrane. As sulfur dioxide sorption increases the hydrogen permeability is reduced until at about 140 $cm^3(SO_2)(STP)/g$, the membrane is completely blocked and only sulfur dioxide permeates. Data obtained at several temperatures fall on the same master curve (●, 0 °C; ▲, −10 °C; □, −20.7 °C; △, −33.6 °C). Reprinted from R. Ash, R.M. Barrer and C.G. Pope, Flow of Adsorbable Gases and Vapours in Microporous Medium, *Proc. R. Soc. London, Ser. A*, **271**, 19 (1963) with permission from The Royal Society

competing with polymeric membranes for these separations. Carbon membranes are likely to be 10 to 100 times more expensive than equivalent polymeric membranes. This cost differential can only be tolerated in applications in which polymeric membranes completely fail to make the separation. Such applications might be the high-temperature separation of hydrocarbon vapor/vapor mixtures; the chemical and physical stability of ceramic and carbon membranes is a real advantage in this type of separation.

Although the literature of gas separation with microporous membranes is dominated by inorganic materials, polymer membranes have also been tried with some success. The polymers used are substituted polyacetylenes, which can have an extraordinarily high free volume, on the order of 25 vol %. The free volume is so high that the free volume elements in these polymers are probably interconnected. Membranes made from these polymers appear to function as finely microporous materials with pores in the 5 to 15 Å diameter range [71,72]. The two most widely studied polyacetylenes are poly(1-trimethylsilyl-1-propyne) (PTMSP) and poly(4-methyl-2-pentyne) (PMP), with the structures shown in Figure 2.23. Gas permeabilities in these materials are orders of magnitude higher than those of conventional, low-free-volume glassy polymers, and are even substantially higher than those of poly(dimethylsiloxane), for many years the most permeable polymer known. The extremely high free volume provides a sorption capacity as much as 10 times that of a conventional glassy polymer. More dramatically, diffusion coefficients are 10^3 to 10^6 times greater than those observed in conventional glassy polymers. This combination of extraordinarily high permeabilities, together with the very high free volume, hints at a pore-flow contribution. Nonetheless, the ratio of the diffusion coefficients of oxygen and nitrogen (D_{O_2}/D_{N_2}) is 1.4, a small value for a glassy polymer membrane but still more than would be expected for a simple Knudsen diffusion membrane.

These high-free-volume polymers also have unusual permeability characteristics with mixtures of condensable and noncondensable gases. For example, in the presence of as little as 1200 ppm of a condensable vapor such as the perfluorocarbon FC-77 (a perfluoro octane-perfluoro decane mixture), the nitrogen permeability of PTMSP is 20 times lower than the pure nitrogen permeability [71], as shown in Figure 2.41. When the condensable vapor is removed from the feed gas the nitrogen permeability rapidly returns to its original value. The best

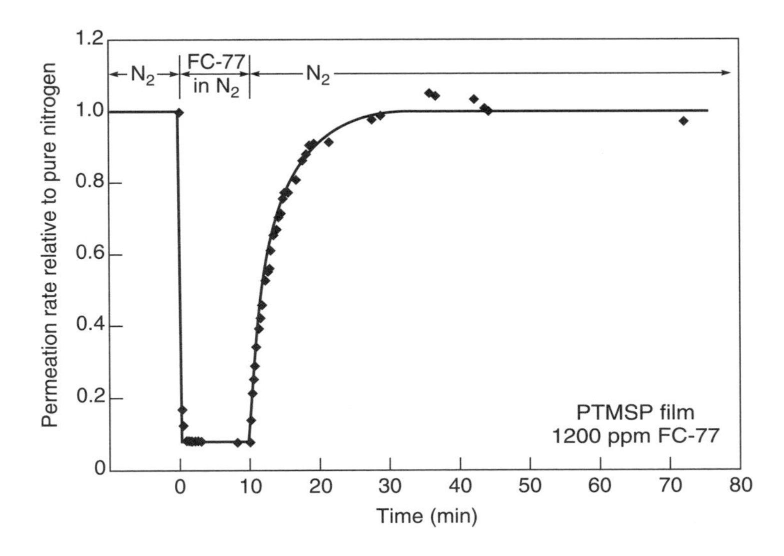

Figure 2.41 The change in nitrogen flux through a PTMSP membrane caused by the presence of a condensable vapor in the feed gas [71]. This behavior is characteristic of extremely finely porous microporous ceramic or ultrahigh-free-volume polymeric membranes such as PTMSP. The condensable vapor adsorbs in the 5- to 15-Å-diameter pores of the membrane, blocking the flow of the noncondensable nitrogen gas

explanation for these unusual vapor permeation properties is that PTMSP, because of its very high free volume, is an ultra-microporous membrane in which pore-flow transport occurs. The FC-77 vapor causes capillary condensation in which the pores are partially or completely blocked by the adsorbed vapor, preventing the flow of noncondensed gases (nitrogen) through the membrane.

The Transition Region

The transition between pore-flow and solution-diffusion transport seems to occur with membranes having very small pores. Ultrafiltration membranes that reject sucrose and raffinose but pass all micro-ions are clearly pore-flow membranes, whereas desalination-grade sodium-chloride-rejecting reverse osmosis membranes clearly follow the solution-diffusion model. Presumably, the transition is in the nanofiltration range, with membranes having good rejections to divalent ions and most organic solutes, but rejection of monovalent ions in the 20–70 % range. The performance of a family of nanofiltration membranes of this type is illustrated in Table 2.6 [73]. The FT30 membrane is clearly a good reverse osmosis membrane, whereas the XP-20 is a very small pore flow ultrafiltration membrane. The XP-45 membrane is intermediate in character.

The transition between reverse osmosis membranes with a salt rejection of more than 95 % and molecular weight cutoffs below 50 and ultrafiltration membranes with a salt rejection of less than 10 % and a molecular weight cutoff of more than 1000 is shown in Figure 2.42 [74]. The very large change in the pressure-normalized flux of water that occurs as the membranes become more retentive is noteworthy. Because these are anisotropic membranes, the thickness of the separating layer is difficult to measure, but clearly the permeability of

Table 2.6 Rejection of microsolutes by nanofiltration membranes (FilmTec data) [73]. Reprinted from *Desalination*, **70**, J. Cadotte, R. Forester, M. Kim, R. Petersen and T. Stocker, Nanofiltration Membranes Broaden the Use of Membrane Separation Technology, p. 77, Copyright 1988, with permission from Elsevier

Solute	Solute rejection (%)		
	FT-30	XP-45	XP-20
NaCl	99.5	50	20
$MgCl_2$	>99.5	83	—
$MgSO_4$	>99.5	97.5	85
$NaNO_3$	90	<20	0
Ethylene glycol	70	24	11
Glycerol	96	44	15
Glucose	99	95	60
Sucrose	100	100	89

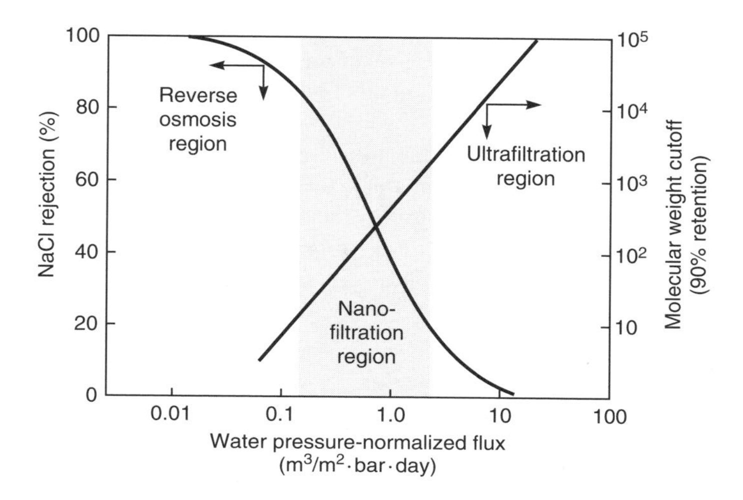

Figure 2.42 Diagram of the region of nanofiltration membrane performance relative to reverse osmosis and ultrafiltration membranes [74]

water through the pores of ultrafiltration membranes is orders of magnitude higher than permeability through dense solution-diffusion reverse osmosis membranes. Gas permeation also places high-free-volume substituted polyacetylene polymer membranes in the transition area between solution-diffusion and pore flow.

Conclusions and Future Directions

During the last 30 years the basis of permeation through membranes has become much clearer. This is particularly true for reverse osmosis, gas permeation and pervaporation for which the solution-diffusion model is now almost universally accepted and well-supported by a body of experimental evidence. This model provides simple equations that accurately link the driving forces of concentration and pressure with flux and selectivity. The solution-diffusion model has been less successful at providing a link between the nature of the membrane material and the membrane permeation properties. This link requires an ability to calculate membrane diffusion and sorption coefficients. These calculations require knowledge of the molecular level of interactions of permeant molecules and their motion in the polymer matrix that is not yet available. Only semiempirical correlations such as the dual sorption model or free volume correlations are available. The best hope for future progress towards *a priori* methods of calculating permeant sorption and diffusion coefficients lies in computer-aided molecular dynamic simulations, but accurate predictions using this technique are years—perhaps decades—away.

The theory of permeation through microporous membranes in ultrafiltration and microfiltration is much less developed and it is difficult to see a clear path forward. Permeation through these membranes is affected by a variety of hard-to-compute effects and is also very much a function of membrane structure and composition. Measurements of permeation through ideal uniform-pore-diameter membranes made by the nucleation track method are in good agreement with theory. Unfortunately, industrially useful membranes have nonuniform tortuous pores and are often anisotropic as well. Current theories cannot predict the permeation properties of these membranes.

References

1. S. Sourirajan, *Reverse Osmosis*, Academic Press, New York (1970).
2. H. Yasuda and A. Peterlin, Diffusive and Bulk Flow Transport in Polymers, *J. Appl. Polym. Sci.* **17**, 433 (1973).
3. P. Meares, On the Mechanism of Desalination by Reversed Osmotic Flow Through Cellulose Acetate Membrane, *Eur. Polym. J.* **2**, 241 (1966).
4. D.R. Paul and O.M. Ebra-Lima, Pressure-induced Diffusion of Organic Liquids Through Highly Swollen Polymer Membranes, *J. Appl. Polym. Sci.* **14**, 2201 (1970).
5. D.R. Paul, Diffusive Transport in Swollen Polymer Membranes, in *Permeability of Plastic Films and Coatings*, H.B. Hopfenberg (ed.), Plenum Press, New York, pp. 35–48 (1974).
6. D.R. Paul, The Solution-diffusion Model for Swollen Membranes, *Sep. Purif. Meth.* **5**, 33 (1976).
7. A. Fick, Über Diffusion, *Poggendorff's Annal. Physik Chem.* **94**, 59 (1855).
8. E. Smit, M.H.V. Mulder, C.A. Smolders, H. Karrenbeld, J. van Eerden and D. Feil, Modeling of the Diffusion of Carbon Dioxide in Polyimide Matrices by Computer Simulation, *J. Membr. Sci.* **73**, 247 (1992).
9. S.G. Charati and S.A. Stern, Diffusion of Gases in Silicone Polymers: Molecular Dynamic Simulations, *Macromolecules* **31**, 5529 (1998).
10. J.G. Wijmans and R.W. Baker, The Solution-diffusion Model: A Review, *J. Membr. Sci.* **107**, 1 (1995).
11. F. Theeuwes, R.M. Gale and R.W. Baker, Transference: a Comprehensive Parameter Governing Permeation of Solutes Through Membranes, *J. Membr. Sci.* **1**, 3 (1976).
12. H.K. Lonsdale, U. Merten and R.L. Riley, Transport Properties of Cellulose Acetate Osmotic Membranes, *J. Appl. Polym. Sci.* **9**, 1341 (1965).
13. U. Merten, Transport Properties of Osmotic Membranes, in *Desalination by Reverse Osmosis*, U. Merten (ed.), MIT Press, Cambridge, MA, pp. 15–54 (1966).
14. N. Bhore, R.M. Gould, S.M. Jacob, P.O. Staffeld, D. McNally, P.H. Smiley and C.R. Wildemuth, New Membrane Process Debottlenecks Solvent Dewaxing Unit, *Oil Gas J.* **97**, 67 (1999).
15. R.W. Baker and J.G. Wijmans, Membrane Separation of Organic Vapors from Gas Streams, in *Polymeric Gas Separation Membranes*, D.R. Paul and Y.P. Yampol'skii (eds), CRC Press, Boca Raton, FL, pp. 353–398 (1994).
16. T. Kataoka, T. Tsuro, S.-I. Nakao and S. Kimura, Membrane Transport Properties of Pervaporation and Vapor Permeation in an Ethanol–Water System Using Polyacrylonitrile and Cellulose Acetate Membranes, *J. Chem. Eng. Jpn* **24**, 326 (1991).
17. J.G. Wijmans and R.W. Baker, A Simple Predictive Treatment of the Permeation Process in Pervaporation, *J. Membr. Sci.* **79**, 101 (1993).

18. F.W. Greenlaw, W.D. Prince, R.A. Shelden and E.V. Thompson, Dependence of Diffusive Permeation Rates by Upstream and Downstream Pressures, *J. Membr. Sci.* **2**, 141 (1977).
19. D.R. Paul and D.J. Paciotti, Driving Force for Hydraulic and Pervaporation Transport in Homogeneous Membranes, *J. Polym. Sci., Polym. Phys. Ed.* **13**, 1201 (1975).
20. S. Rosenbaum and O. Cotton, Steady-state Distribution of Water in Cellulose Acetate Membrane, *J. Polym. Sci.* **7**, 101 (1969).
21. S.N. Kim and K. Kammermeyer, Actual Concentration Profiles in Membrane Permeation, *Sep. Sci.* **5**, 679 (1970).
22. D.R. Paul, J.D. Paciotti and O.M. Ebra-Lima, Hydraulic Permeation of Liquids Through Swollen Polymeric Networks, *J. Appl. Polym. Sci.* **19**, 1837 (1975).
23. K. Tanaka, H. Kita, M. Okano and K. Okamoto, Permeability and Permselectivity of Gases in Fluorinated and Non-fluorinated Polyimides, *Polymer* **33**, 585 (1992).
24. J.O. Hirschfelder, C.F. Curtis and R.B. Bird, *Molecular Theory of Gases and Liquids*, John Wiley, New York (1954).
25. R.C. Reid, J.M. Prausnitz and B.E. Poling, *The Properties of Gases and Liquids*, 4th Edn, McGraw Hill, New York (1987).
26. J.T. Edward, Molecular Volumes and the Stokes–Einstein Equation, *J. Chem. Ed.* **47**, 261 (1970).
27. Y. Nishijima and G. Oster, Diffusion in Concentrated Polymer Solutions, *J. Polym. Sci.* **19**, 337 (1956).
28. R.W. Baker and H.K. Lonsdale, Controlled Release: Mechanisms and Rates, in *Controlled Release of Biological Active Agents*, A.C. Tanquary and R.E. Lacey (eds), Plenum Press, New York, pp. 15–72 (1974).
29. M. Artsis, A.E. Chalykh, N.A. Khalturinskii, W. Moiseev and G.E. Zaikov, Diffusion of Organic Diluents into Ethyl Cellulose, *Eur. Polym. J.* **8**, 613 (1972).
30. A. Bondi, *Physical Properties of Molecular Crystals, Liquids, and Glasses*, Wiley, New York (1968).
31. D.W. Van Krevelen, *Properties of Polymers*, Elsevier, Amsterdam (1990).
32. Y. Kirayama, T. Yoshinaga, Y. Kusuki, K. Ninomiya, T. Sakakibara and T. Tameri, Relation of Gas Permeability with Structure of Aromatic Polyimides, *J. Membr. Sci.* **111**, 169 (1996).
33. C.L. Aitken, W.J. Koros and D.R. Paul, Effect of Structural Symmetry on Gas Transport Properties of Polysulfones, *Macromolecules* **25**, 3424 (1992).
34. J.H. Petropoulos, Mechanisms and Theories for Sorption and Diffusion of Gases in Polymers, in *Polymeric Gas Separation Membranes*, D.R. Paul and Y.P. Yampol'skii (eds), CRC Press, Boca Raton, FL, pp. 17–82 (1994).
35. M.R. Paxton and D.R. Paul, Relationship Between Structure and Transport Properties for Polymers with Aromatic Backbones, in *Polymeric Gas Separation Membranes*, D.R. Paul and Y.P. Yampol'skii (eds), CRC Press, Boca Raton, FL, pp. 83–154 (1994).
36. J.Y. Park and D.R. Paul, Correlation and Prediction of Gas Permeability in Glassy Polymer Membrane Materials via a Modified Free Volume Based Group Contribution Method, *J. Membr. Sci.* **125**, 29 (1997).
37. N.A. Platé and Y.P. Yampol'skii, Relationship between Structure and Transport Properties for High Free Volume Polymeric Materials, in *Polymeric Gas Separation Membranes*, D.R. Paul and Y.P. Yampol'skii (eds), CRC Press, Boca Raton, FL, pp. 155–208 (1994).
38. T. Masuda, Y. Iguchi, B.-Z. Tang and T. Higashimura, Diffusion and Solution of Gases in Substituted Polyacetylene Membranes, *Polymer* **29**, 2041 (1988).
39. S.A. Stern, V.M. Shah and B.J. Hardy, Structure–Permeability Relationships in Silicone Polymers, *J. Polym. Sci., Polym. Phys. Ed.* **25**, 1263 (1987).

40. K. Denbigh, *The Principles of Chemical Equilibrium*, Cambridge University Press, Cambridge (1961).
41. P.G.T. Fogg and W. Gerrard, *Solubility of Gases in Liquids*, Wiley, Chichester (1991).
42. P.J. Flory, *Principles of Polymer Chemistry*, Cornell University Press, Ithaca, NY, p. 511 (1953).
43. B.D. Freeman and I. Pinnau, Polymer Membranes for Gas Separation, *ACS Symp. Ser.* **733**, 6 (1999).
44. R.M. Barrer, J.A. Barrie and J. Slater, Sorption and Diffusion in Ethyl Cellulose, *J. Polym. Sci.* **27**, 177 (1958).
45. A.S. Michaels, W.R. Vieth and J.A. Barrie, Solution of Gases in Polyethylene Terephthalate, *J. Appl. Phys.* **34**, 1 (1963).
46. K. Toi, G. Morel and D.R. Paul, Gas Sorption in Poly(phenylene oxide) and Comparisons with Other Glassy Polymers, *J. Appl. Sci.* **27**, 2997 (1982).
47. W.J. Koros, A.H. Chan and D.R. Paul, Sorption and Transport of Various Gases in Polycarbonate, *J. Membr. Sci.* **2**, 165 (1977).
48. A. Morisato, B.D. Freeman, I. Pinnau and C.G. Casillas, Pure Hydrocarbon Sorption Properties of Poly(1-trimethylsilyl-1-propyne) [PTMSP] and Poly(1-phenyl-1-propyne) [PPP] and PTMSP/PPP Blends, *J. Polym. Sci., Polym. Phys. Ed.* **34**, 1925 (1996).
49. P.M. Bungay, Transport Principles–Porous Membranes, in *Synthetic Membranes: Science Engineering and Applications*, P.M. Bungay, H.K. Lonsdale and M.N. dePintio (eds), D. Reidel, Dordrecht, pp. 109–154 (1986).
50. J.D. Ferry, Ultrafilter Membranes and Ultrafiltration, *Chem. Rev.* **18**, 373 (1936).
51. E.M. Renkin, Filtration, Diffusion and Molecular Sieving Through Porous Cellulose Membranes, *J. Gen. Physiol.* **38**, 225 (1955).
52. R.W. Baker and H. Strathmann, Ultrafiltration of Macromolecular Solutions with High-flux Membranes, *J. Appl. Polym. Sci.* **14**, 1197 (1970).
53. R.C. Lukaszewicz, G.B. Tanny and T.H. Meltzer, Membrane-filter Characterizations and their Implications for Particle Retention, *Pharm. Tech.* **2**, 77 (1978).
54. T.H. Meltzer, *Filtration in the Pharmaceutical Industry*, Marcel Dekker, New York (1987).
55. R.H. Davis and D.C. Grant, Theory for Dead End Microfiltration, in *Membrane Handbook*, W.S.W. Ho and K.K. Sirkar (eds), Van Nostrand Reinhold, New York, pp. 461–479 (1992).
56. D.C. Grant, B.Y.H. Liu, W.G. Fischer and R.A. Bowling, Particle Capture Mechanisms in Gases and Liquids: an Analysis of Operative Mechanisms, *J. Environ. Sci.* **42**, 43 (1989).
57. R.M. Barrer, Diffusion in Porous Media, *Appl. Mater. Res.* **2**, 129 (1963).
58. A.B. Shelekhin, A.G. Dixon and Y.H. Ma, Adsorption, Permeation, and Diffusion of Gases in Microporous Membranes, *J. Membr. Sci.* **75**, 233 (1992).
59. R. Ash, R.M. Barrer and P. Sharma, Sorption and Flow of Carbon Dioxide and Some Hydrocarbons in a Microporous Carbon Membrane, *J. Membr. Sci.* **1**, 17 (1976).
60. S.T. Hwang and K. Kammermeyer, *Techniques of Chemistry, Vol. VII, Membranes in Separations*, Wiley, New York (1975).
61. K. Keizer, A.J. Burggraaf, Z.A.E.P. Vroon and H. Verweij, Two Component Permeation Through Thin Zeolite MFI Membranes, *J. Membr. Sci.* **147**, 159 (1998).
62. M.H. Hassan, J.D. Way, P.M. Thoen and A.C. Dillon, Single Component and Mixed Gas Transport in a Silica Fiber Membrane, *J. Membr. Sci.* **104**, 27 (1995).
63. R. Ash, R.M. Barrer and C.G. Pope, Flow of Adsorbable Gases and Vapours in Microporous Medium, *Proc. R. Soc. London, Ser. A* **271**, 19 (1963).
64. A. Soffer, J.E. Koresh and S. Saggy, Separation Device, US Patent 4,685,940, August 1987.

65. M.B. Rao and S. Sirkar, Nanoporous Carbon Membranes for Separation of Gas Mixtures by Selective Surface Flow, *J. Membr. Sci.* **85**, 253 (1994).
66. M.B. Rao and S. Sirkar, Performance and Pore Characterization of Nanoporous Carbon Membranes for Gas Separation, *J. Membr. Sci.* **110**, 109 (1996).
67. D.Q. Vu, W.J. Koros and S.J. Miller, High Pressure CO_2/CH_4 Separations Using Carbon Molecular Sieve Hollow Fiber Membranes, *Ind. Eng. Chem. Res.* **41**, 367 (2002).
68. N. Tanihara, H. Shimazaki, Y. Hirayama, N. Nakanishi, T. Yoshinaga and Y. Kusuki, Gas Permeation Properties of Asymmetric Carbon Hollow Fiber Membranes Prepared from Asymmetric Polymer Hollow Fibers, *J. Membr. Sci.* **160**, 179 (1999).
69. A.B. Fuertes, Adsorption-selective Carbon Membranes for Gas Separation, *J. Membr. Sci.* **177**, 9 (2000).
70. H. Kita, H. Maeda, K. Tanaka and K. Okamoto, Carbon Molecular Sieve Membranes Prepared from Phenolic Resin, *Chem. Lett.* **179** (1997).
71. I. Pinnau and L.G. Toy, Transport of Organic Vapors through Poly[1-(trimethylsilyl)-1-propyne], *J. Membr. Sci.* **116**, 199 (1996).
72. R. Srinivasan, S.R. Auvil and P.M. Burban, Elucidating the Mechanism(s) of Gas Transport in Poly[1-(trimethylsilyl)-1-propyne] (PTMSP) Membranes, *J. Membr. Sci.* **86**, 67 (1994).
73. J. Cadotte, R. Forester, M. Kim, R. Petersen and T. Stocker, Nanofiltration Membranes Broaden the Use of Membrane Separation Technology, *Desalination* **70**, 77 (1988).
74. S. Egli, A. Ruf and F. Widmer, Entwicklung und Charakterisierung von Kompositmembranen für die Nano- und Ultrafiltration, *Swiss Chem.* **11**(9), 53 (1989).

3 MEMBRANES AND MODULES

Introduction

The surge of interest in membrane separation processes that began in the late 1960s was prompted by two developments: first, the ability to produce high flux, essentially defect-free membranes on a large scale and second, the ability to form these membranes into compact, high-surface-area, economical membrane modules. These breakthroughs in membrane technology took place in the 1960s to early 1970s, as part of the development of reverse osmosis and ultrafiltration. Adaptation of the technology to other membrane processes took place in the 1980s.

Several factors contribute to the successful fabrication of a high-performance membrane module. First, membrane materials with the appropriate chemical, mechanical and permeation properties must be selected; this choice is very process-specific. However, once the membrane material has been selected, the technology required to fabricate this material into a robust, thin, defect-free membrane and then to package the membrane into an efficient, economical, high-surface-area module is similar for all membrane processes. Therefore, this chapter focuses on methods of forming membranes and membrane modules. The criteria used to select membrane materials for specific processes are described in the chapters covering each application.

In this chapter membrane preparation techniques are organized by membrane structure: isotropic membranes, anisotropic membranes, ceramic and metal membranes, and liquid membranes. Isotropic membranes have a uniform composition and structure throughout; such membranes can be porous or dense. Anisotropic (or asymmetric) membranes, on the other hand, consist of a number of layers each with different structures and permeabilities. A typical anisotropic membrane has a relatively dense, thin surface layer supported on an open, much thicker microporous substrate. The surface layer performs the separation and is the principal barrier to flow through the membrane. The open support layer provides mechanical strength. Ceramic and metal membranes can be either isotropic or anisotropic.

Membrane Technology and Applications R. W. Baker
 ISBN: 0-470-85445-6

However, these membranes are grouped separately from polymeric membranes because their preparation methods are so different.

Liquid membranes are the final membrane category. The selective barrier in these membranes is a liquid phase, usually containing a dissolved carrier that selectively reacts with a specific permeant to enhance its transport rate through the membrane. Liquid membranes are used almost exclusively in carrier facilitated transport processes, so preparation of these membranes is covered in that chapter (Chapter 11).

The membrane classification scheme described above works fairly well. However, a major membrane preparation technique, phase separation, also known as phase inversion, is used to make both isotropic and anisotropic membranes. This technique is covered under anisotropic membranes.

Isotropic Membranes

Isotropic Nonporous Membranes

Dense nonporous isotropic membranes are rarely used in membrane separation processes because the transmembrane flux through these relatively thick membranes is too low for practical separation processes. However, they are widely used in laboratory work to characterize membrane properties. In the laboratory, isotropic (dense) membranes are prepared by solution casting or thermal melt-pressing. The same techniques can be used on a larger scale to produce, for example, packaging material.

Solution Casting

Solution casting is commonly used to prepare small samples of membrane for laboratory characterization experiments. An even film of an appropriate polymer solution is spread across a flat plate with a casting knife. The casting knife consists of a steel blade, resting on two runners, arranged to form a precise gap between the blade and the plate onto which the film is cast. A typical hand-held knife is shown in Figure 3.1. After casting, the solution is left to stand, and the solvent evaporates to leave a thin, uniform polymer film. A detailed description of many types of hand casting knives and simple casting machines is given in the book by Gardner and Sward [1].

The polymer solution used for solution casting should be sufficiently viscous to prevent it from running over the casting plate, so typical polymer concentrations are in the range 15–20 wt%. Preferred solvents are moderately volatile liquids such as acetone, ethyl acetate and cyclohexane. Films cast from these solutions are dry within a few hours. Solvents with high boiling points such as dimethyl formamide or *N*-methyl pyrrolidone are unsuitable for solution casting, because their low volatility requires long evaporation times. During an extended solvent evaporation time, the cast film can absorb sufficient atmospheric water to

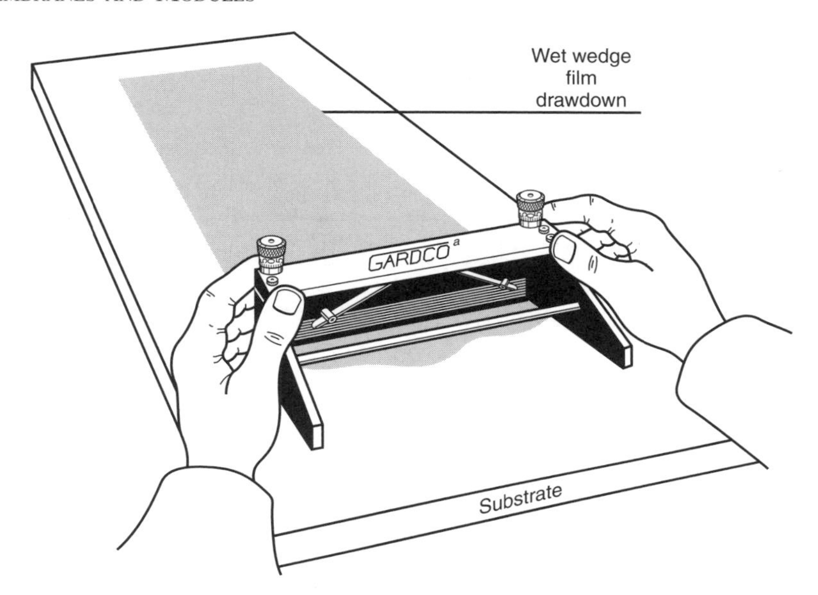

Figure 3.1 A typical hand-casting knife. (Courtesy of Paul N. Gardner Company, Inc., Pompano Beach, FL)

precipitate the polymer, producing a mottled, hazy surface. Very volatile solvents such as methylene chloride can also cause problems. Rapid evaporation of the solvent cools the casting solution, causing gelation of the polymer. The result is a film with a mottled, orange-peel-like surface. Smooth films can be obtained with rapidly evaporating solvents by covering the cast film with a glass plate raised 1 to 2 cm above the film to slow evaporation. When the solvent has completely evaporated the dry film can be lifted from the glass plate. If the cast film adheres to the plate, soaking in a swelling non-solvent such as water or alcohol will usually loosen the film.

Solution-cast film is produced on a larger scale for medical applications, battery separators, or other specialty uses with machinery of the type shown in Figure 3.2 [2]. Viscous film is made by this technique. The solution is cast onto the surface of a rotating drum or a continuous polished stainless steel belt. These machines are generally enclosed to control water vapor pickup by the film as it dries and to minimize solvent vapor losses to the atmosphere.

Melt Extruded Film

Many polymers, including polyethylene, polypropylene, and nylons, do not dissolve in appropriate solvents at room temperature, so membranes cannot be made by solution casting. To prepare small pieces of film, a laboratory press

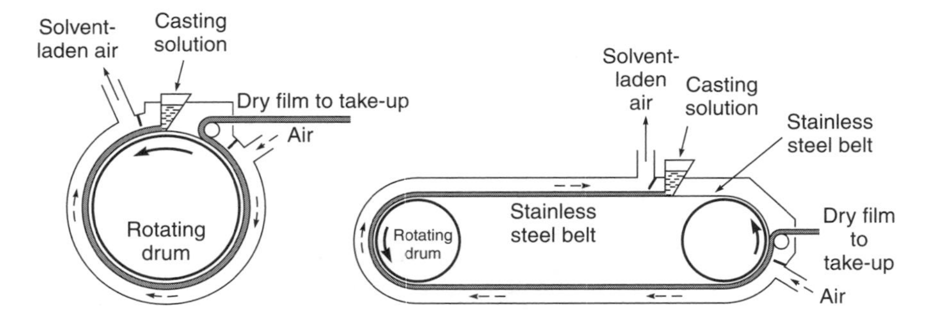

Figure 3.2 Machinery used to make solution-cast film on a commercial scale

as shown in Figure 3.3 can be used. The polymer is compressed between two heated plates. Typically, a pressure of 2000–5000 psi is applied for 1–5 min, at a plate temperature just below the melting point of the polymer. Melt extrusion is also used on a very large scale to make dense films for packaging applications, either by extrusion as a sheet from a die or as blown film. Detailed descriptions of this equipment can be found in specialized monographs. A good overview is given in the article by Mackenzie in the *Encyclopedia of Chemical Technology* [2].

Isotropic Microporous Membranes

Isotropic microporous membranes have much higher fluxes than isotropic dense membranes and are widely used as microfiltration membranes. Further significant uses are as inert spacers in battery and fuel cell applications and as the rate-controlling element in controlled drug delivery devices.

The most important type of microporous membrane is formed by one of the phase separation techniques discussed in the next section; about half of the isotropic microporous membrane used is made in this way. The remaining types are made by various proprietary techniques, the more important of which are described below.

Track-etch Membranes

Track-etch membranes were developed by the General Electric Corporation Schenectady Laboratory [3]. The two-step preparation process is illustrated in Figure 3.4. First, a thin polymer film is irradiated with fission particles from a nuclear reactor or other radiation source. The massive particles pass through the film, breaking polymer chains and leaving behind a sensitized track of damaged polymer molecules. These tracks are much more susceptible to chemical attack than the base polymer material. So when the film is passed through a solution

Figure 3.3 A typical laboratory press used to form melt-pressed membranes. (Courtesy of Carver, Inc., Wabash, IN)

that etches the polymer, the film is preferentially etched along the sensitized nucleation tracks, thereby forming pores. The exposure time of the film to radiation determines the number of membrane pores; the etch time determines the pore diameter [4]. A feature of the track-etch preparation technique is that the pores are uniform cylinders traversing the membrane at right angles. The

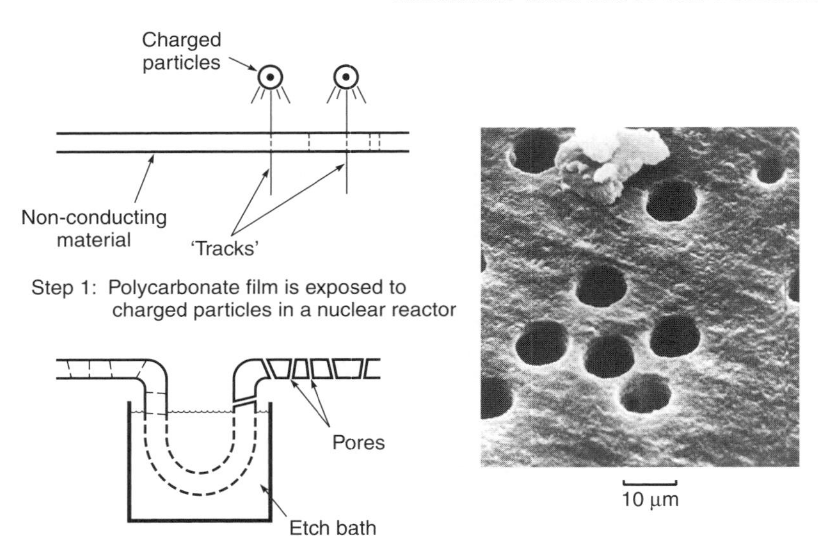

Figure 3.4 Diagram of the two-step process to manufacture nucleation track membranes [4] and photograph of resulting structure. (Photograph courtesy of Whatman plc, Maidstone, Kent, UK)

membrane tortuosity is, therefore, close to one, and all pores have the same diameter. These membranes are almost a perfect screen filter; therefore, they are widely used to measure the number and type of suspended particles in air or water. A known volume of fluid is filtered through the membrane, and all particles larger than the pore diameter are captured on the surface of the membrane so they can be easily identified and counted. To minimize the formation of doublet holes produced when two nucleation tracks are close together, the membrane porosity is usually kept relatively low, about 5 % or less. This low porosity results in low fluxes. General Electric, the original developers of these membranes, assigned the technology to a spin-off company, the Nuclepore Corporation, in 1972 [5]. Nuclepore® membranes remain the principal commercially available track-etch membranes. Polycarbonate or polyester films are usually used as the base membrane material and sodium hydroxide as the etching solution. Other materials can also be used; for example, etched mica has been used in research studies.

Expanded-film Membranes

Expanded-film membranes are made from crystalline polymers by an orientation and annealing process. A number of manufacturers produce porous membranes by this technique. The original development was due to a group at Celanese, which made microporous polypropylene membranes by this process under the trade name Celgard® [6]. In the first step of the process, a highly oriented film is produced by extruding polypropylene at close to its melting point coupled with a very rapid drawdown. The crystallites in the semi-crystalline polymer are then aligned in the direction of orientation. After cooling and annealing, the film is stretched a second time, up to 300 %. During this second elongation the amorphous regions between the crystallites are deformed, forming slit-like voids, 200 to 2500 Å wide, between the polymer crystallites. The pore size of the membrane is controlled by the rate and extent of the second elongation step. The formation process is illustrated in Figure 3.5. This type of membrane is also made from poly(tetrafluoroethylene) film by W.L. Gore and sold under the

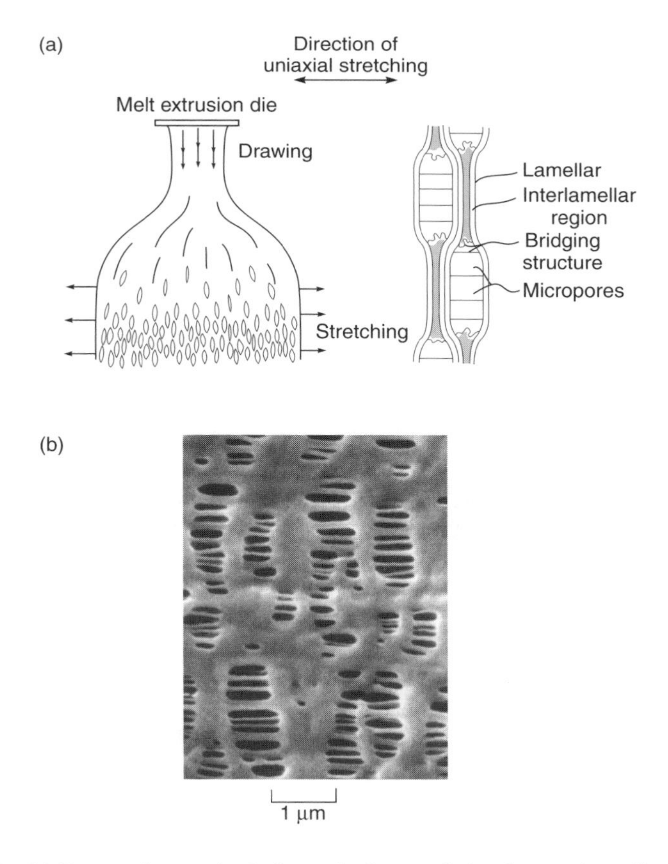

Figure 3.5 (a) Preparation method of a typical expanded polypropylene film membrane, in this case Celgard®. (b) Scanning electron micrograph of the microdefects formed on uniaxial stretching of films [6]

trade name Gore-Tex® [7]. Expanded film membrane was originally produced as rolled flat sheets. More recently the process has also been adapted to the production of hollow fibers [8]; Membrana produces this type of fiber on a large scale for use in blood oxygenator equipment (Chapter 12) and membrane contactors (Chapter 13). Gore-Tex poly(tetrafluoroethylene) film is widely used as a water-vapor-permeable (that is, breathable) but liquid-water-impermeable fabric. The commercial success of these membranes has motivated a number of other companies to produce similar materials [9,10].

Template Leaching

Template leaching is another method of producing isotropic microporous membranes from insoluble polymers such as polyethylene, polypropylene and poly(tetrafluoroethylene). In this process a homogeneous melt is prepared from a mixture of the polymeric membrane matrix material and a leachable component. To finely disperse the leachable component in the polymer matrix, the mixture is often homogenized, extruded, and pelletized several times before final extrusion as a thin film. After formation of the film, the leachable component is removed with a suitable solvent, and a microporous membrane is formed [11–13]. The leachable component can be a soluble, low-molecular-weight solid, a liquid such as liquid paraffin, or even a polymeric material such as polystyrene. A drawing of a template leaching membrane production machine is shown in Figure 3.6.

Anisotropic Membranes

Anisotropic membranes are layered structures in which the porosity, pore size, or even membrane composition change from the top to the bottom surface of the membrane. Usually anisotropic membranes have a thin, selective layer supported on a much thicker, highly permeable microporous substrate. Because the selective layer is very thin, membrane fluxes are high. The microporous substrate

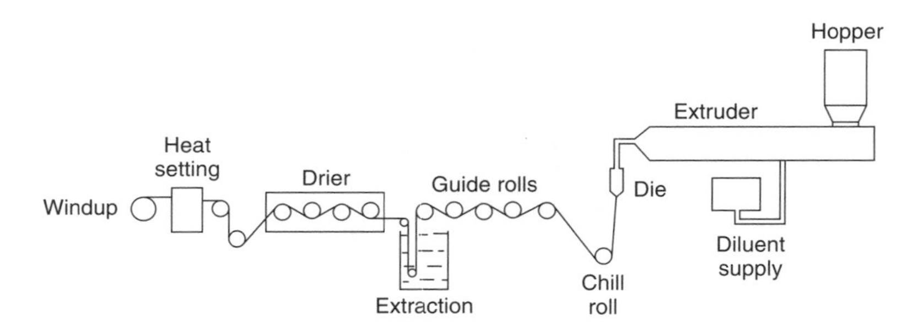

Figure 3.6 Flow schematic of a melt extruder system used to make polypropylene membranes by template leaching [13]

provides the strength required for handling the membrane. The importance of anisotropic membranes was not recognized until Loeb and Sourirajan prepared the first high-flux, anisotropic reverse osmosis membranes by what is now known as the Loeb–Sourirajan technique [14]. Hindsight makes it clear that some of the membranes produced in the 1930s and 1940s were also anisotropic. Loeb and Sourirajan's discovery was a critical breakthrough in membrane technology. Their anisotropic reverse osmosis membranes were an order of magnitude more permeable than the isotropic membranes produced previously from the same materials. For a number of years the Loeb–Sourirajan technique was the only method of making anisotropic membranes, but the demonstrated benefits of the anisotropic structure encouraged the development of other methods. Improvements in anisotropic membrane preparation methods and properties were accelerated by the availability in the late 1960s of the scanning electron microscope (SEM), which allowed the effects of changes in the membrane formation process on structure to be assessed easily.

Membranes made by the Loeb–Sourirajan process consist of a single membrane material, but the porosity and pore size change in different layers of the membrane. Anisotropic membranes made by other techniques and used on a large scale often consist of layers of different materials which serve different functions. Important examples are membranes made by the interfacial polymerization process discovered by Cadotte [15] and the solution-coating processes developed by Ward [16], Francis [17] and Riley [18]. The following sections cover four types of anisotropic membranes:

- *Phase separation membranes.* This category includes membranes made by the Loeb–Sourirajan technique involving precipitation of a casting solution by immersion in a nonsolvent (water) bath. Also covered are a variety of related techniques such as precipitation by solvent evaporation, precipitation by absorption of water from the vapor phase, and precipitation by cooling.
- *Interfacial polymerization membranes.* This type of anisotropic membrane is made by polymerizing an extremely thin layer of polymer at the surface of a microporous support polymer.
- *Solution-coated composite membranes.* To prepare these membranes, one or more thin, dense polymer layers are solution coated onto the surface of a microporous support.
- *Other anisotropic membranes.* This category covers membranes made by a variety of specialized processes, such as plasma deposition, in the laboratory or on a small industrial scale to prepare anisotropic membranes for specific applications.

Phase Separation Membranes

The Loeb–Sourirajan technique is now recognized as a special case of a more general class of membrane preparation process, best called the phase separation

Table 3.1 Phase separation membrane preparation procedures

Procedure	Process
Water precipitation (the Loeb–Sourirajan process)	The cast polymer solution is immersed in a nonsolvent bath (typically water). Absorption of water and loss of solvent cause the film to rapidly precipitate from the top surface down
Water vapor absorption	The cast polymer solution is placed in a humid atmosphere. Water vapor absorption causes the film to precipitate
Thermal gelation	The polymeric solution is cast hot. Cooling causes precipitation
Solvent evaporation	A mixture of solvents is used to form the polymer casting solution. Evaporation of one of the solvents after casting changes the solution composition and causes precipitation

process, but sometimes called the phase inversion process or the polymer precipitation process. The term phase separation describes the process most clearly, namely, changing a one-phase casting solution into two separate phases. In all phase separation processes, a liquid polymer solution is precipitated into two phases: a solid, polymer-rich phase that forms the matrix of the membrane and a liquid, polymer-poor phase that forms the membrane pores.

Precipitation of the cast liquid polymer solution to form the anisotropic membrane can be achieved in several ways, as summarized in Table 3.1. Precipitation by immersion in a bath of water was the technique discovered by Loeb and Sourirajan, but precipitation can also be caused by absorption of water from a humid atmosphere. A third method is to cast the film as a hot solution. As the cast film cools, a point is reached at which precipitation occurs to form a microporous structure; this method is called thermal gelation. Finally, evaporation of one of the solvents in the casting solution can be used to cause precipitation. In this technique the casting solution consists of a polymer dissolved in a mixture of a volatile good solvent and a less volatile nonsolvent (typically water or alcohol). When a film of the solution is cast and allowed to evaporate, the volatile good solvent evaporates first, the film then becomes enriched in the nonvolatile nonsolvent, and finally precipitates. Many combinations of these processes have also been developed. For example, a cast film placed in a humid atmosphere can precipitate partly because of water vapor absorption but also because of evaporation of one of the more volatile components.

Polymer Precipitation by Water (the Loeb–Sourirajan Process)

The first phase separation membrane was developed at UCLA from 1958 to 1960 by Sidney Loeb, then working on his Master's degree, and Srinivasa Sourirajan, then a post-doctoral researcher. In their process, now called the Loeb–Sourirajan technique, precipitation is induced by immersing the cast film of polymer solution in a water bath. In the original Loeb–Sourirajan process, a solution containing 20 to 25 wt% cellulose acetate dissolved in a water-miscible solvent was cast as a thin film on a glass plate. The film was left to stand for 10–100 s to allow some of the solvent to evaporate, after which the film was immersed in a water bath to precipitate the film and form the membrane. The membrane was usually post-treated by annealing in a bath of hot water. The steps of the process are illustrated in Figure 3.7.

The Loeb–Sourirajan process remains by far the most important membrane-preparation technique. The process is part of the overall membrane preparation procedure for almost all reverse osmosis and ultrafiltration and for many gas separation membranes. Reverse osmosis and gas separation membranes made by this technique consist of a completely dense top surface layer (the skin) on top of a microporous support structure. Ultrafiltration membranes, support membranes for solution coating, and interfacial polymerization membranes have the same general anisotropic structure, but the skin layer is very finely microporous, typically with pores in the 10- to 200-Å diameter range. Also, the porous substrate of ultrafiltration membranes is usually more open, often consisting of large finger-like cavities extending from just under the selective skin layer to the bottom surface of the membrane. Scanning electron micrographs of typical sponge-structure reverse-osmosis type and finger-structure ultrafiltration-type membranes are shown in Figure 3.8 [19]. These photographs show how small

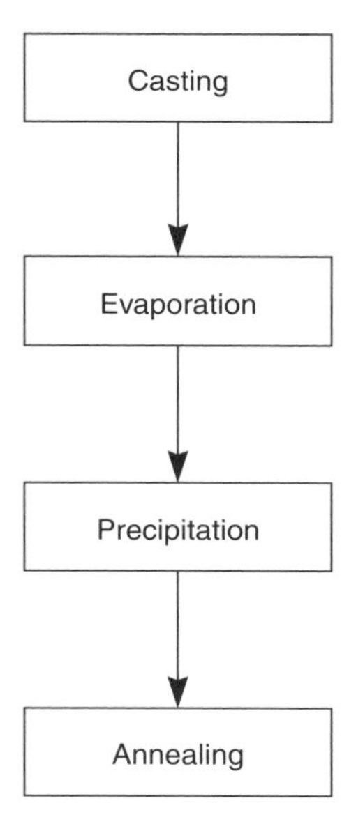

Figure 3.7 Process scheme used to form Loeb–Sourirajan water precipitation phase separation membranes [14]

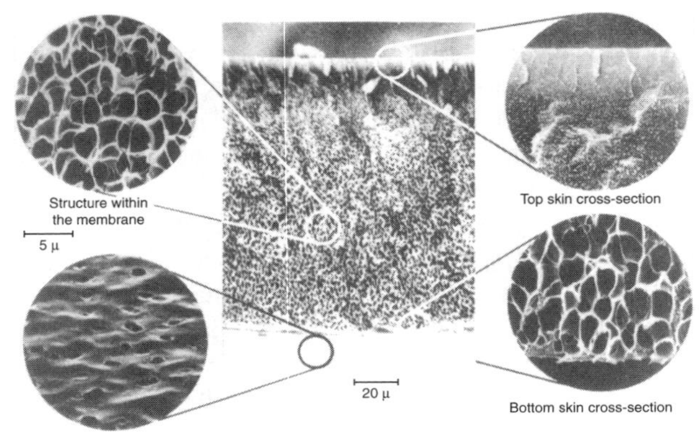

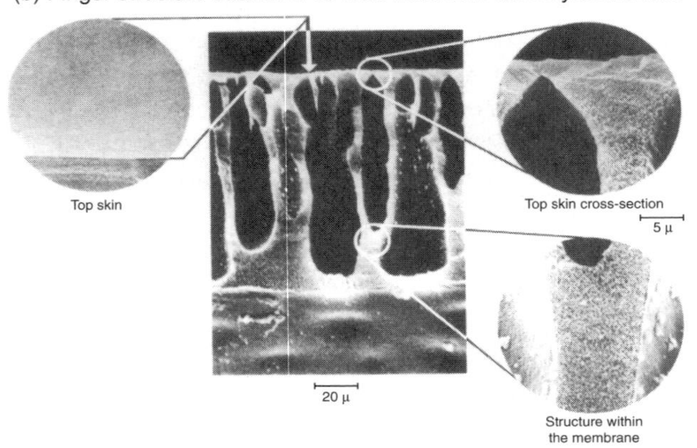

Figure 3.8 Scanning electron micrographs of aromatic polyamide (Nomex, Du Pont) Loeb–Sourirajan membranes cast from 22 and 18 wt% Nomex in dimethylacetamide [19]

changes in the casting solution can produce major differences in membrane properties. Both membranes are prepared from a Nomex® (DuPont, Wilmington, DE) polyamide-dimethylacetamide casting solution, but the polymer concentration in the solutions is different.

The Loeb–Sourirajan water precipitation membranes shown in Figure 3.8 were made by casting the membranes onto glass plates. This procedure is still used in the laboratory, but for commercial production large casting machines produce rolls of membrane up to 5000 m long and 1 to 2 m wide. A diagram of a small casting machine is shown in Figure 3.9. The polymer solution is cast onto a moving nonwoven paper web. The cast film is then precipitated by immersion in a water bath. The water precipitates the top surface of the cast film rapidly,

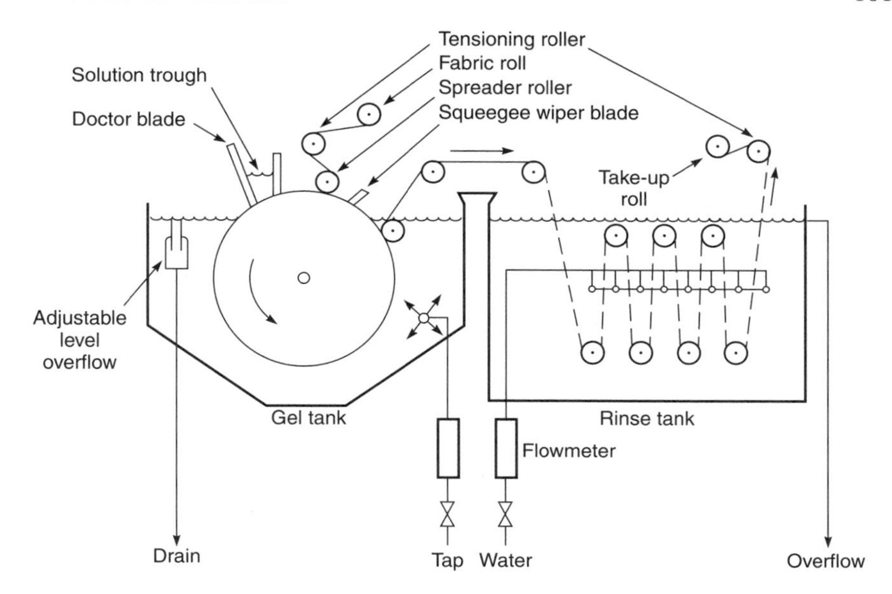

Figure 3.9 Schematic of Loeb–Sourirajan membrane casting machine used to prepare reverse osmosis or ultrafiltration membranes. A knife and trough are used to coat the casting solution onto a nonwoven paper web. The coated web then enters the water-filled gel tank, where the casting solution precipitates. After the membrane has formed, it is washed thoroughly to remove residual solvent before being wound up on the take-up roll

forming the dense, selective skin. This skin slows entry of water into the underlying polymer solution, which precipitates much more slowly and forms a more porous substructure. Depending on the polymer, the casting solution, and other parameters, the thickness of the dense skin varies from 0.1 to 1.0 μm. Casting machine speeds vary from as low as 1 to 2 m/min for slowly precipitating casting solutions, such as cellulose acetate, to 10 m/min for rapidly precipitating casting solutions, such as polysulfone. A listing of some typical casting solutions and precipitation conditions for membranes made by the Loeb–Sourirajan technique is given in Table 3.2 [14,20–23].

Since the discovery of the Loeb–Sourirajan technique in the 1960s, development of the technology has proceeded on two fronts. Industrial users of the technology have generally taken an empirical approach, making improvements in the technique based on trial and error experience. Concurrently, theories of membrane formation based on fundamental studies of the precipitation process have been developed. These theories originated with the early industrial developers of membranes at Amicon [19,22,24] and were then taken up at a number of academic centers. Unfortunately, much of the recent academic work is so complex that many industrial producers of phase separation membranes no longer follow this literature.

Table 3.2 Historically important examples of conditions for preparation of solution-precipitation (Loeb–Sourirajan) membranes

Casting solution composition	Precipitation conditions	Application and comments
22.2 wt% cellulose acetate (39.8 wt% acetyl polymer) 66.7 wt% acetone 10.0 wt% water 1.1 wt% magnesium perchlorate	3 min evaporation, precipitate into 0 °C water, anneal for 5 min at 65–85 °C	The first Loeb–Sourirajan reverse osmosis membrane [14]
25 wt% cellulose acetate (39.8 wt% acetyl polymer) 45 wt% acetone 30 wt% formamide	0.5–2 min evaporation, cast into 0 °C water, anneal for 5 min at 65–85 °C	The Manjikian formulation widely used in early 1970s for reverse osmosis membranes [20]
8.2 wt% cellulose acetate (39.8 wt% acetyl polymer) 8.2 wt% cellulose triacetate (43.2 wt% acetyl polymer) 45.1 wt% dioxane 28.7 wt% acetone 7.4 wt% methanol 2.5 wt% maleic acid	Up to 3 min evaporation at −10 °C, precipitation into an ice bath, anneal at 85–90 °C for 3 min	A high-performance reverse osmosis cellulose acetate blend membrane [21]
15 wt% polysulfone (Udell P 1700) 85 wt% *N*-methyl-2-pyrrolidone	Cast into 25 °C water bath. No evaporation or annealing step necessary	An early ultrafiltration membrane formulation [22]. Similar polysulfone-based casting solutions are still widely used
20.9 wt% polysulfone 33.2 wt% dimethyl formamide 33.2 wt% tetrahydrofuran 12.6 wt% ethanol	Forced evaporation with humid air 10–15 s. Precipitate into 20 °C water	A high-performance gas separation membrane with a completely dense nonporous skin ∼1000Å thick [23]

Empirical Approach to Membrane Formation by Water Precipitation

Over the years several rules of thumb have developed to guide producers of solution precipitation membranes. These rules can be summarized as follows:

Choice of Polymer. The ideal polymer is a tough, amophorous, but not brittle thermoplastic with a glass transition temperature more than 50 °C above the expected use temperature. A high molecular weight is important. Commercial polymers made for injection molding have molecular weights in the 30 000 to 40 000 Dalton range, but, for solution precipitation, polymers with higher molecular weights are usually preferable. If the polymer is crystalline or a rigid glass, the resulting membrane may be too brittle and will break if bent during later handling. The polymer must also be soluble in a suitable water-miscible solvent. Polymers that meet these specifications include cellulose acetate, polysulfone, poly(vinylidine fluoride), polyetherimide and aromatic polyamides.

Choice of Casting Solution Solvent. Generally the best casting solution solvents are aprotic solvents such as dimethyl formamide, *N*-methyl pyrrolidone and dimethyl acetamide. These solvents dissolve a wide variety of polymers, and casting solutions based on these solvents precipitate rapidly when immersed in water to give porous, very anisotropic membranes. Casting solutions using low-solubility-parameter solvents, such as tetrahydrofuran, acetone, dioxane and ethyl formate, are generally not appropriate. Such casting solutions precipitate slowly and give relatively nonporous membranes. However, small amounts of these solvents may be added as casting solution modifiers (see below). Figure 3.10 illustrates the apparent correlation between solvent solubility parameter and membrane porosity as demonstrated by So *et al.* [25].

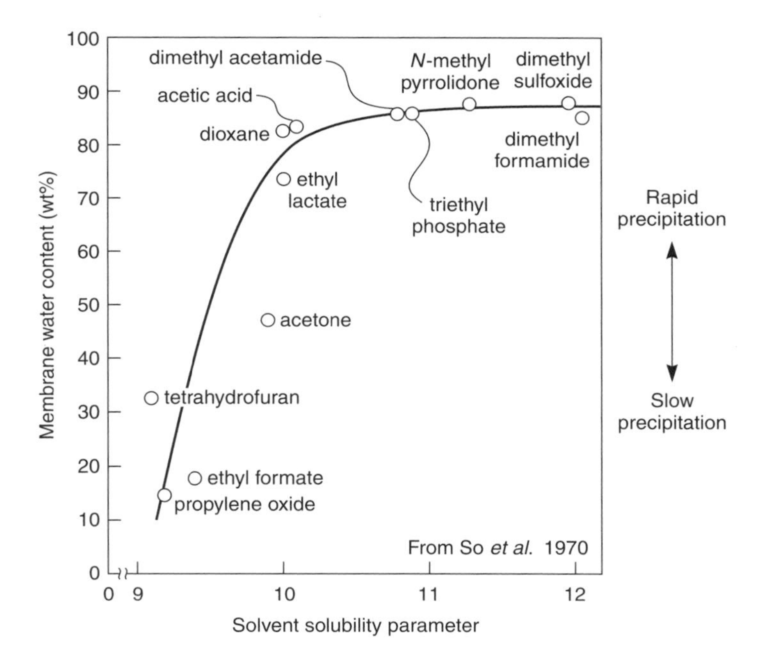

Figure 3.10 The porosity of cellulose acetate membranes cast from 15-wt% solutions with various solvents. The same trend of high porosity and rapid precipitation with high solubility-parameter solvents was seen with a number of other membrane materials [25]

Increasing the polymer casting solution concentration always reduces the porosity and flux of the membrane. Typical concentrations for porous ultrafiltration membranes are in the range 15–20 wt%. Polymer casting solution concentrations for reverse osmosis or gas separation membranes are higher, generally about 25 wt%, and casting solutions used to make hollow fiber membranes by spinning a hot solution at 60 to 80 °C may contain as much as 35 % polymer.

Precipitation Medium. Water is almost always the casting solution precipitation medium. Some work has been done with organic solvents, particularly to form hollow fiber membranes for which the mechanical and safety problems of handling an organic solvent precipitation bath and limiting atmospheric emissions are more easily controlled than in flat sheet casting. In general, the results obtained with nonaqueous precipitation baths have not justified the increased complexity of the process. Organic-based solvent precipitation media such as methanol or isopropanol almost always precipitate the casting solution more slowly than water, and the resulting membranes are usually denser, less anisotropic, and lower flux than membranes precipitated with water.

The temperature of the water used to precipitate the casting solution is important; this temperature is controlled in commercial membrane plants. Generally low-temperature precipitation produces lower flux, more retentive membranes. For this reason chilled water is frequently used to prepare cellulose acetate reverse osmosis membranes.

Casting Solution Modifiers. Membrane properties are often tailored by adding small amounts of modifiers to the casting solution. The casting solutions shown in Table 3.2 contain two to four components, but modern commercial casting solutions may be more complex. Even though the solution may contain only 5 to 20 wt% modifiers, these modifiers can change the membrane performance significantly. This aspect of membrane preparation is a black art, and most practitioners have their preferred ingredients. Addition of low solubility solvents such as acetone, tetrahydrofuran or dioxane will normally produce denser, more retentive membranes. Increasing the polymer concentration of the casting solution will also make the membrane more dense. Addition of salts such as zinc chloride and lithium chloride usually gives more open membranes. Polymeric additives may also be used—commonly poly(vinyl pyrrolidone) and poly(ethylene glycol); generally these polymers make the membrane more porous. Also, although most of these water-soluble polymers and salts are removed during precipitation and washing of the membrane, a portion remains trapped, making the final membrane more hydrophilic.

When developing membranes from a new polymer, practitioners of the empirical approach usually prepare a series of trial casting solutions based on past experience with similar polymers. Membrane films are made by casting onto glass plates and precipitation in a water bath. The casting solutions most likely to yield good membranes are often immediately apparent. The rate of precipitation is important. Slow precipitation produces dense, more isotropic membranes; rapid precipitation produces porous, anisotropic membranes. The appearance and mechanical properties of the membrane surface—shine, brittleness and thickness—compared to casting solution thickness also provide clues to the membrane structure. Based on these trials one or more casting solutions will be selected for systematic parametric development.

Theoretical Approach to Membrane Formation

Over the years several approaches have been used to rationalize the formation of Loeb–Sourirajan (solution precipitation) and other phase inversion membranes. Most have involved the polymer–solvent–precipitation medium phase diagrams popularized by Michaels [22], Strathmann [19,24,26] and Smolders [27–29]. In this approach the change in composition of the casting solution as membrane formation takes place is tracked as a path through the phase diagram. The path starts at a point representing the original casting solution and finishes at a point representing the composition of the final membrane. The casting solution composition moves to the final membrane composition by losing solvent and gaining water.

A typical three-component phase diagram for the components used to prepare Loeb–Sourirajan membranes is shown in Figure 3.11. The corners of the

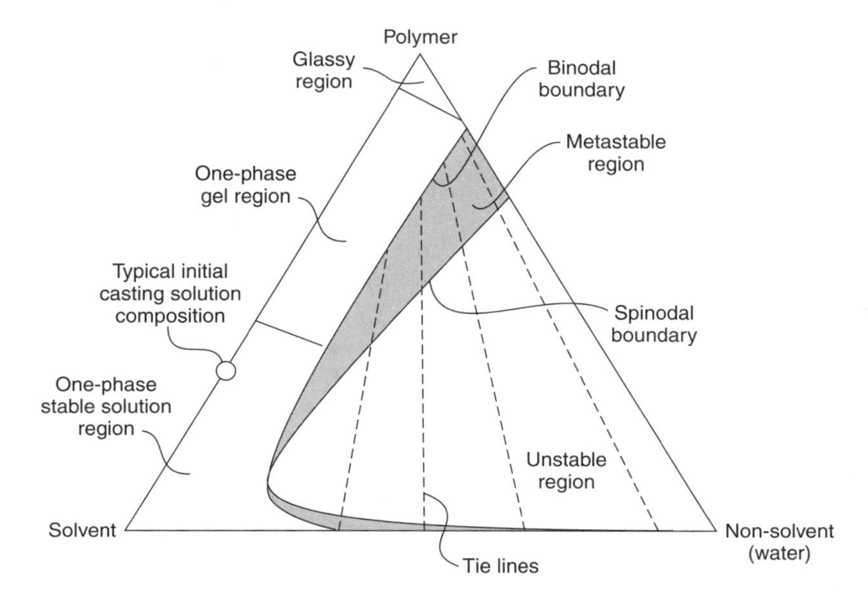

Figure 3.11 Schematic of the three-component phase diagram often used to rationalize the formation of water-precipitation phase separation membranes

triangle represent the three pure components—polymer, solvent, and nonsolvent (water); points within the triangle represent mixtures of the three components. The diagram has two principal regions: a one-phase region, in which all components are miscible; and a two-phase region, in which the system separates into a solid (polymer-rich) phase and a liquid (polymer-poor) phase. During precipitation of the membrane casting solution, the solution loses solvent and gains water. The casting solution moves from a composition in the one-phase region to a composition in the two-phase region.

Although the one-phase region in the phase diagram is thermodynamically continuous, for practical purposes it can be conveniently subdivided into a liquid polymer solution region, a polymer gel region, and a glassy solid polymer region. Thus, in the low-polymer-concentration region, typical of the original casting solution, the compositions are viscous liquids. But, if the concentration of polymer is increased, the viscosity of compositions in the one-phase region increases rapidly, reaching such high values that the system can be regarded as a solid gel. The transition between the liquid and gel regions is arbitrary but can be placed at a polymer concentration of 30 to 40 wt%. If the one-phase solution contains more than 90 wt% polymer, the swollen polymer gel may become so rigid that the polymer chains can no longer rotate. The polymer gel then becomes a solid polymer glass.

During the precipitation process, the casting solution enters the two-phase region of the phase diagram by crossing the so-called binodal boundary. This brings the casting solution into a metastable two-phase region. Polymer solution compositions in this region are thermodynamically unstable but will not normally precipitate unless well nucleated. The metastable region in the phase diagrams of low-molecular-weight materials is very small, but can be large for high-molecular-weight materials. As more solvent leaves the casting solution and water enters the solution, the composition crosses into another region of the phase diagram in which a one-phase solution is always thermodynamically unstable. In this region, polymer solutions spontaneously separate into two phases with compositions linked by tie lines. The boundary between the metastable and unstable regions is called the spinodal boundary.

Thus, the membrane precipitation process is a series of steps. First, solvent exchange with the precipitation medium occurs. Then, as the composition enters the two-phase region of the phase diagram, phase separation or precipitation begins. The time taken for solvent–water exchange before precipitation occurs can be measured because the membrane turns opaque as soon as precipitation begins. Depending on the casting solution composition, the time to first precipitation may be almost instantaneous to as long as 30–60 s. Initially, the polymer phase that separates on precipitation may be a liquid or semi-liquid gel, and the precipitation domains may be able to flow and agglomerate at this point. In the final step of the precipitation process, desolvation of the polymer phase converts the polymer to a relatively solid gel phase, and the membrane structure is fixed. The solid polymer phase forms the matrix of the final membrane, and the liquid solvent–nonsolvent phase forms the pores. The precipitation behavior of polymer–solvent mixtures is further complicated by slow kinetics caused by the viscosity of polymer solutions and by thermodynamic effects that allow metastable solutions to exist for a prolonged time without precipitating. Much has been made of these effects in a number of theoretical papers, but application to concretely predicting membrane permeation properties has proved difficult.

The original approach of Strathmann *et al.* [24] was to present the process of membrane formation as a line through the phase diagram. This approach is shown in Figure 3.12. During membrane formation, the composition changes from a composition A, which represents the initial casting solution composition, to a composition D, which represents the final membrane composition. At composition D, the two phases are in equilibrium: a solid (polymer-rich) phase, which forms the matrix of the final membrane, represented by point S, and a liquid (polymer-poor) phase, which constitutes the membrane pores filled with precipitant, represented by point L. The position of composition D on the line S-L determines the overall porosity of the membrane. The entire precipitation process is represented by the path A-D, along which the solvent is exchanged by the precipitant. The point B along the path is the concentration at which the

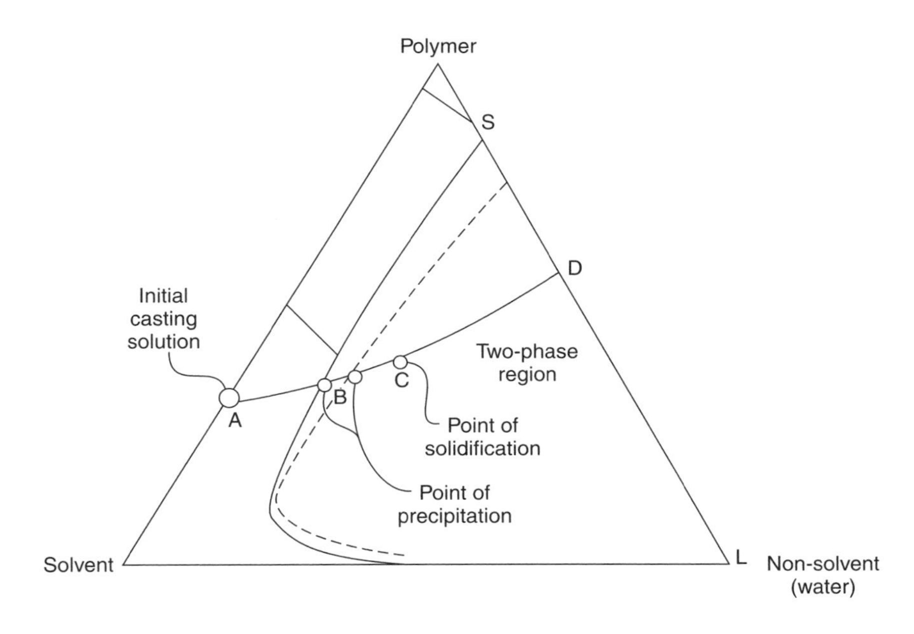

Figure 3.12 Membrane formation in water-precipitation membranes was first rationalized as a path through the three-component phase diagram from the initial polymer casting solution (A) to the final membrane (D) [24]

polymer initially precipitates. As precipitation proceeds, more solvent is lost, and precipitant is imbibed by the polymer-rich phase, raising the viscosity. At some point, the viscosity is high enough for the precipitated polymer to be regarded as a solid. This composition is at C in Figure 3.12. Once the precipitated polymer solidifies, further bulk movement of the polymer is hindered.

The precipitation path in Figure 3.12 is shown as a single line representing the average composition of the whole membrane. In fact, the rate of precipitation and the precipitation path through the phase diagram differ at different points in the membrane. When the cast film of polymer solution is exposed to the precipitation medium, the top surface begins to precipitate first. This surface layer precipitates rapidly, so the two phases formed on precipitation do not have time to agglomerate. The resulting structure is finely microporous. However, the precipitated surface layer then becomes a barrier that slows further loss of solvent and imbibition of nonsolvent by the cast film. The result is increasingly slow precipitation from the top surface to the bottom surface of the film. As precipitation slows, the average pore size increases because the two phases formed on precipitation have more time to separate. The differences between the precipitation rates and the pathway taken by different places in the casting solution mean that the precipitation process is best represented by the movement of a line through the phase diagram rather than a single point. This concept was developed in a series of papers on phase-separation membranes by Smolders and co-workers at Twente University [27–29]. The movement of this line is illustrated in Figure 3.13 [27]. At time t_2, for example, a few seconds after the precipitation process has begun, the top surface of the polymer film has almost completely precipitated, and the composition of this surface layer is close to the polymer nonsolvent axis. On the other hand, at the bottom surface of the film where precipitation has only just begun, the composition is close to that of the original casting solution.

In Figure 3.13 the precipitation pathway enters the two-phase region of the phase diagram above the critical point at which the binodal and spinodal lines intersect. This is important because it means that precipitation will occur as a liquid droplet in a continuous polymer-rich phase. If dilute casting solutions are used, in which the precipitation pathway enters the two-phase region of the phase diagram below the critical point, precipitation produces polymer gel particles in a continuous liquid phase. The membrane that forms is then weak and powdery.

The simplified treatment of membrane formation using the three-component phase diagram given above is about as far as this approach can be usefully taken. Experimental measurement of the path taken by the membrane during the formation process is difficult. Recently, much effort has been made to calculate these pathways through the phase diagrams and to use them to predict the effect of membrane formation variables on the fine membrane structure. As quantitative predictors of membrane performance this approach has failed. However, as a tool to qualitatively rationalize the complex interplay of factors determining membrane performance, the phase diagram approach has proved useful. Many of the

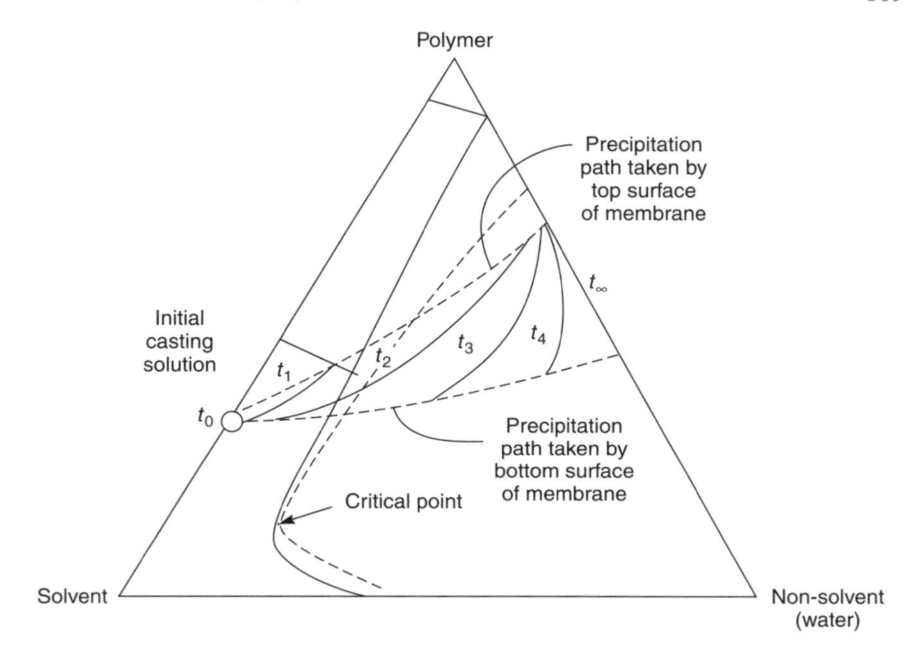

Figure 3.13 The surface layer of water-precipitation membranes precipitates faster than the underlying substrate. The precipitation pathway is best represented by the movement of a line through the three-component phase diagram [27]

recent papers describing the application of the phase diagram approach to membrane formation are a heavy read for industrial membrane producers faced with real-world problems. This literature is reviewed in detail elsewhere [27,30–32].

Polymer Precipitation by Cooling

Perhaps the simplest solution-precipitation membrane preparation technique is thermal gelation, in which a film is cast from a hot, one-phase polymer/solvent solution. As the cast film cools, the polymer precipitates, and the solution separates into a polymer matrix phase containing dispersed pores filled with solvent. Because cooling is usually uniform throughout the cast film, the resulting membranes are relatively isotropic microporous structures with pores that can be controlled within 0.1–10 μm.

The precipitation process that forms thermal gelation membranes can be represented by the phase diagram shown in Figure 3.14 and described in an early Akzo patent of Castro [33]. This is a simplified drawing of the actual phase diagram, which was described later in papers by Lloyd *et al.* [34], Vadalia *et al.* [35] and Caneba and Soong [36]. The phase diagram shows the metastable region

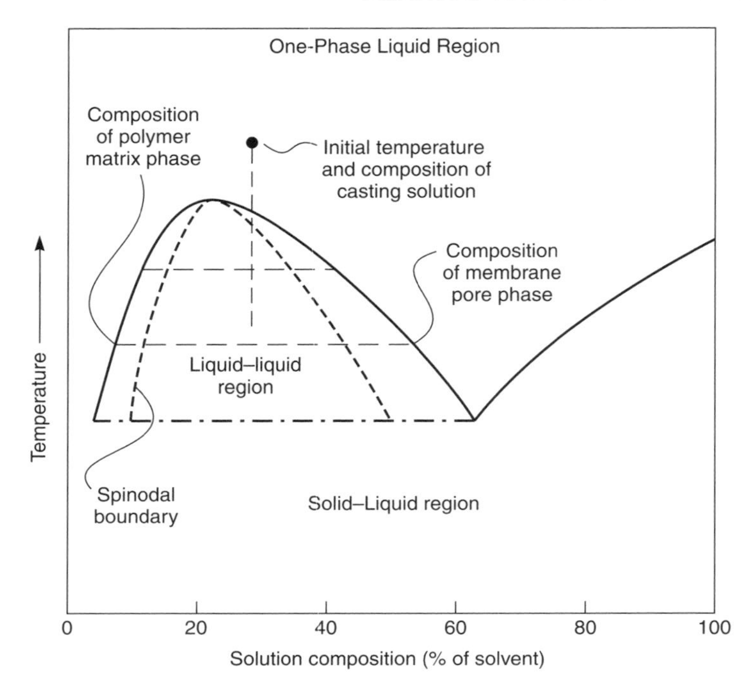

Figure 3.14 Phase diagram showing the composition pathway traveled by the casting solution during precipitation by cooling

between the binodal and spinodal phase boundaries discussed in reference to Figure 3.11, with additional complications caused by the crystalline nature of many of the polymers used to form thermal phase-separation membranes. The pore volume in the final membrane is determined mainly by the initial composition of the solution, because this determines the ratio of the polymer to liquid phase in the cooled film. However, the spatial distribution and size of the pores are determined largely by the rate of cooling and hence, precipitation of the film. In general, more rapid cooling produces smaller membrane pores and greater membrane anisotropy [37,38]. Membrane preparation by thermal gelation is possible with many polymers, but the technique is used mainly to make membranes from polyethylene and polypropylene, which cannot be formed into microporous membranes by standard solution-casting methods.

Polymer precipitation by cooling to produce microporous membranes was first developed and commercialized by Akzo [33,37], which continues to market microfiltration polypropylene and poly(vinylidene fluoride) membranes produced by this technique under the trade name Accurel®. Flat sheet and hollow fiber membranes are made. Polypropylene membranes are prepared from a solution of polypropylene in *N,N*-bis(2-hydroxyethyl)tallowamine. The amine and polypropylene form a clear solution at temperatures above 100–150 °C. Upon cooling, the solvent and polymer phases separate to form a microporous structure. If the solution is cooled slowly, an open cell structure of the type shown in Figure 3.15(a) results. The interconnecting passageways between cells are generally in the micrometer range. If the solution is cooled and precipitated

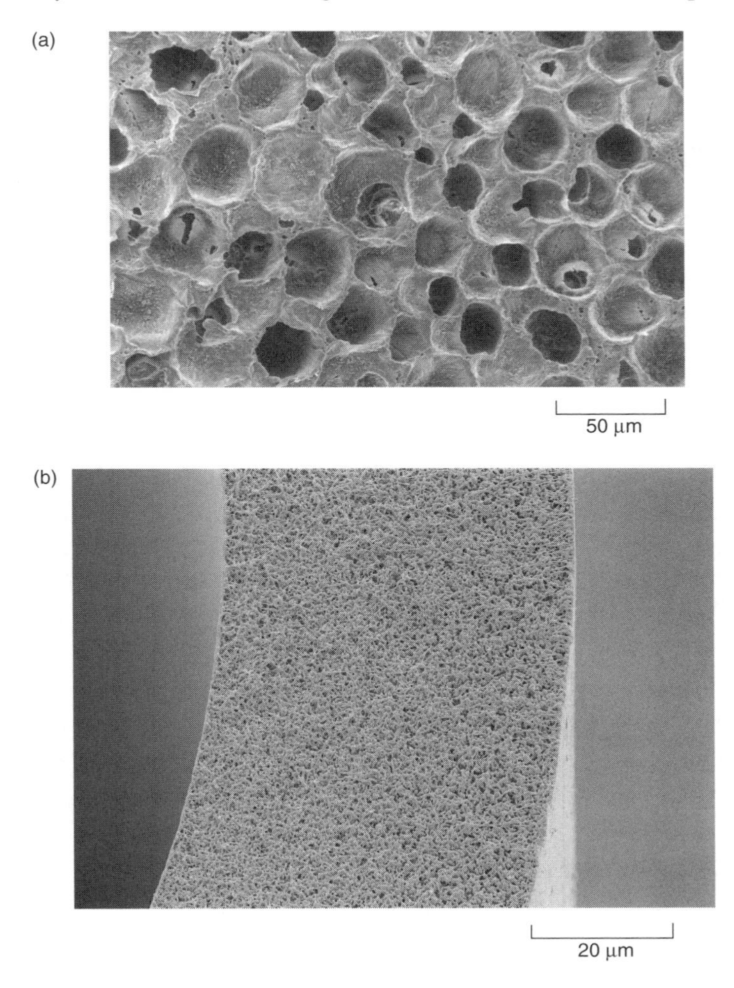

Figure 3.15 Polypropylene structures. (a) Type I: open cell structure formed at low cooling rates. (b) Type II: fine structure formed at high cooling rates [37]. Reprinted with permission from W.C. Hiatt, G.H. Vitzthum, K.B. Wagener, K. Gerlach and C. Josefiak, Microporous Membranes via Upper Critical Temperature Phase Separation, in *Materials Science of Synthetic Membranes*, D.R. Lloyd (ed.), ACS Symposium Series Number 269, Washington, DC. Copyright 1985, American Chemical Society and American Pharmaceutical Association

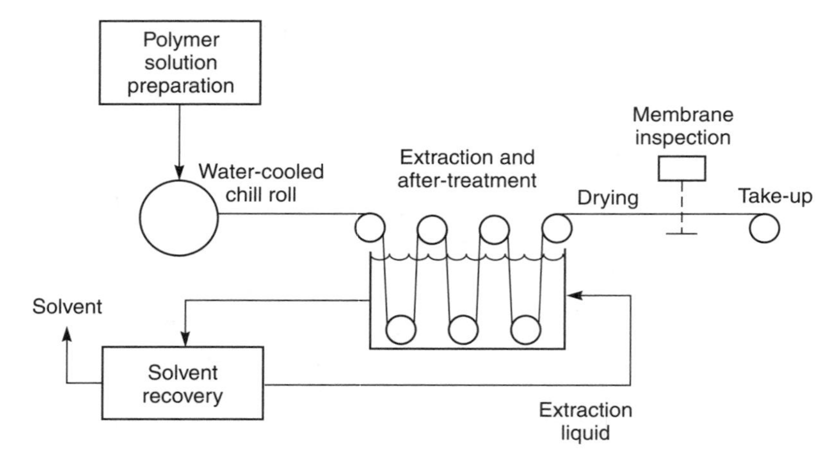

Figure 3.16 Equipment to prepare microporous membranes by the polymer precipitation by cooling technique [37]. Reprinted with permission from W.C. Hiatt, G.H. Vitzthum, K.B. Wagener, K. Gerlach and C. Josefiak, Microporous Membranes via Upper Critical Temperature Phase Separation, in *Materials Science of Synthetic Membranes*, D.R. Lloyd (ed.), ACS Symposium Series Number 269, Washington, DC. Copyright 1985, American Chemical Society and American Pharmaceutical Association

rapidly, a much finer structure is formed, as shown in Figure 3.15(b). The rate of cooling is, therefore, a key parameter determining the final structure of the membrane. The anisotropy of the membranes can be increased by cooling the top and bottom surface of the cast film at different rates.

A schematic diagram of a commercial-scale thermal gelation polymer precipitation process is shown in Figure 3.16. The hot polymer solution is cast onto a water-cooled chill roll, which cools the solution, causing the polymer to precipitate. The precipitated film is passed through an extraction tank containing methanol, ethanol or isopropanol to remove the solvent. Finally, the membrane is dried, sent to a laser inspection station, trimmed and rolled up.

Polymer Precipitation by Solvent Evaporation

This technique, one of the earliest methods of making microporous membranes, was used by Bechhold, Elford, Pierce, Ferry and others in the 1920s and 1930s [39–43]. In the simplest form of the method, a polymer is dissolved in a two-component solvent mixture consisting of a volatile solvent, such as methylene chloride or acetone, in which the polymer is readily soluble, and a less volatile nonsolvent, typically water or an alcohol. The polymer solution is cast onto a glass plate. As the volatile solvent evaporates, the casting solution is enriched in the nonvolatile solvent, so the polymer precipitates, forming the membrane structure. The process can be continued until the membrane has completely formed, or it can be stopped, and the membrane structure fixed, by immersing the cast film in a precipitation bath of water or other nonsolvent. The precipitation process used to form these membranes is much slower than precipitation by immersion into liquid water (the Loeb–Sourirajan process). As a result membranes formed by solvent evaporation are only modestly anisotropic and have large pores. Scanning electron micrographs of some membranes made by this process are shown in Figure 3.17 [44].

Many factors determine the porosity and pore size of membranes formed by the solvent evaporation method. As Figure 3.17 shows, if the membrane is immersed in a nonsolvent after a short evaporation time, the resulting structure will be finely microporous. If the evaporation step is prolonged before fixing the structure by immersion in water, the average pore size will be larger. In general, increasing the nonsolvent content of the casting solution, or decreasing the polymer concentration, increases porosity. It is important for the nonsolvent to be completely incompatible with the polymer. If partially compatible nonsolvents

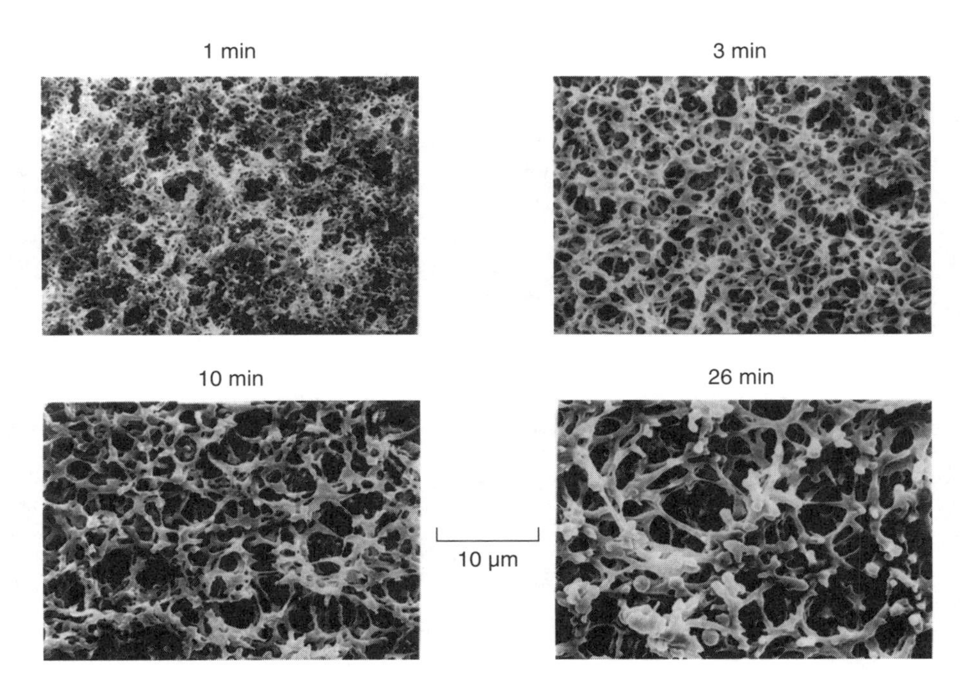

Figure 3.17 SEM photomicrographs of the bottom surface of cellulose acetate membranes cast from a solution of acetone (volatile solvent) and 2-methyl-2,4-pentanediol (nonvolatile nonsolvent). The evaporation time before the structure is fixed by immersion in water is shown [44]. Reprinted from *J. Membr. Sci.*, **87**, L. Zeman and T. Fraser, Formation of Air-cast Cellulose Acetate Membranes, p. 267, Copyright 1994, with permission from Elsevier

are used, the precipitating polymer phase contains sufficient residual solvent to allow it to flow, and the pores will collapse as the solvent evaporates. The result is a dense rather than a microporous film.

Polymer Precipitation by Absorption of Water Vapor

Preparation of microporous membranes by solvent evaporation alone is not widely practiced. However, a combination of solvent evaporation and absorption of water vapor from a humid atmosphere is an important method of making microfiltration membranes. The processes involve proprietary casting formulations not normally disclosed by membrane developers. However, during the development of composite membranes at Gulf General Atomic, Riley *et al.* prepared this type of membrane and described the technology in some detail in a series of Office of Saline Water Reports [45]. These reports remain the best published description of the technique. Casting solutions used to prepare these membranes are complex and often contain 5 to 10 components. For example, a typical casting solution composition taken from Riley's report [45] comprises 8.1 wt% cellulose nitrate, 1.3 wt% cellulose acetate, 49.5 wt% acetone (a volatile good solvent), 22.3 wt% ethanol and 14.7 wt% *n*-butanol (nonvolatile poor solvents), 2.6 wt% water (a nonsolvent), 0.5 wt% Triton X-100 (a surfactant solution modifier), and 1.2 wt% glycerin (a polymer plasticizer).

The type of equipment used by Riley *et al.* is shown in Figure 3.18. The casting solution is cast onto a moving stainless steel belt. The cast film then passes through a series of environmental chambers. Warm, humid air is usually circulated through the first chamber, where the film loses the volatile solvent by evaporation and simultaneously absorbs water. A key issue is to avoid formation of a dense surface skin on the air side of the membrane. Dense skin formation is

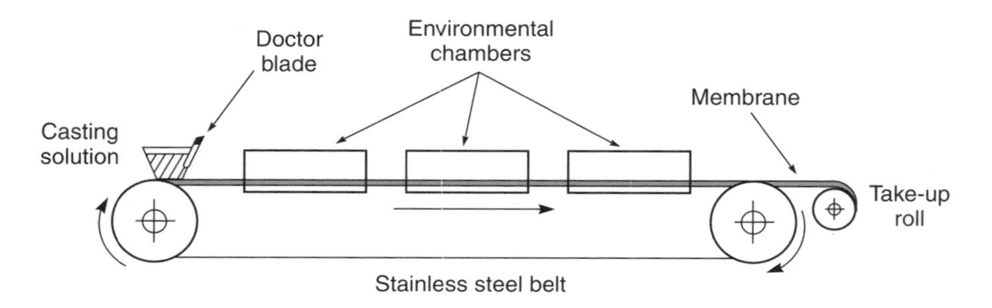

Figure 3.18 Schematic of casting machine used to make microporous membranes by water vapor absorption. A casting solution is deposited as a thin film on a moving stainless steel belt. The film passes through a series of humid and dry chambers, where the solvent evaporates from the solution, and water vapor is absorbed. This precipitates the polymer, forming a microporous membrane that is taken up on a collection roll [45]

generally prevented by incorporating sufficient polymer nonsolvent in the casting solution. Polymer precipitation and formation of two phases then occur when even a small portion of the volatile solvent component in the mixture evaporates. The total precipitation process is slow, taking about 10–30 min to complete. Typical casting speeds are of the order of 1 to 5 ft/min. To allow higher casting speeds the casting machine must be very long—commercial machines can be up to 100 feet. The resulting membrane structure is more isotropic and more microporous than membranes precipitated by immersion in water. After precipitation in the environmental chambers, the membrane passes to a second oven, through which hot, dry air is circulated to evaporate the remaining solvent and dry the film. The formed membrane is then wound onto a take-up roll. This type of membrane is widely used in microfiltration. Membranes made by the water vapor absorption-solvent evaporation precipitation process often have the characteristic nodular form shown in Figure 3.19. A discussion of some of the

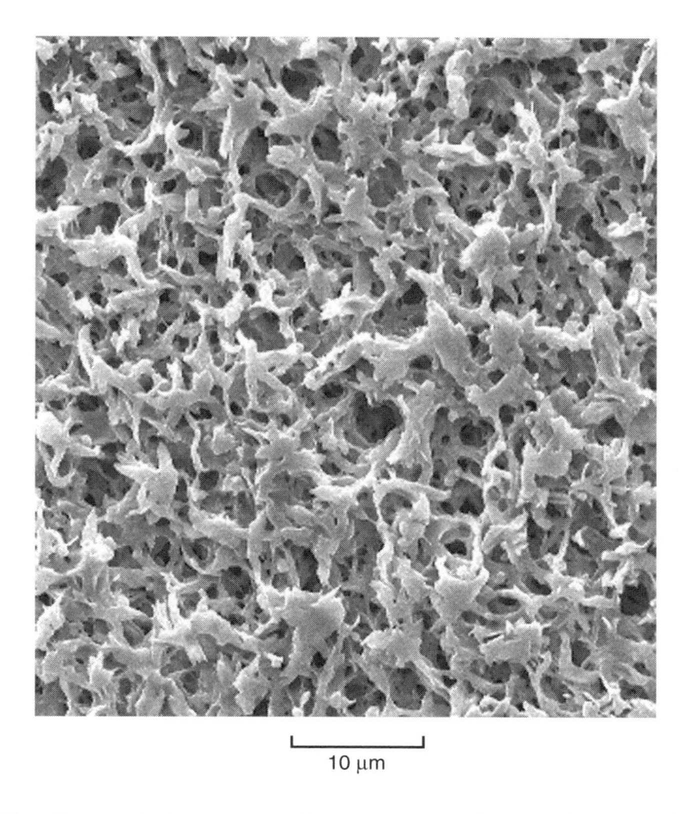

Figure 3.19 Characteristic structure of a phase-separation membrane made by water vapor absorption and solvent evaporation. (Courtesy of Millipore Corporation, Billerica, MA)

practical considerations involved in making this type of membrane is given in a recent book by Zeman and Zydney [46].

Interfacial Polymerization Membranes

The production by Loeb and Sourirajan of the first successful anisotropic membranes spawned numerous other techniques in which a microporous membrane is used as a support for a thin, dense separating layer. One of the most important of these was interfacial polymerization, an entirely new method of making anisotropic membranes developed by John Cadotte, then at North Star Research. Reverse osmosis membranes produced by this technique had dramatically improved salt rejections and water fluxes compared to those prepared by the Loeb–Sourirajan process. Almost all reverse osmosis membranes are now made by the interfacial polymerization process, illustrated in Figure 3.20. In this method, an aqueous solution of a reactive prepolymer, such as a polyamine, is first deposited in the pores of a microporous support membrane, typically a polysulfone ultrafiltration membrane. The amine-loaded support is then immersed in a water-immiscible solvent solution containing a reactant, such as a diacid chloride in hexane. The amine and acid chloride react at the interface of the two immiscible

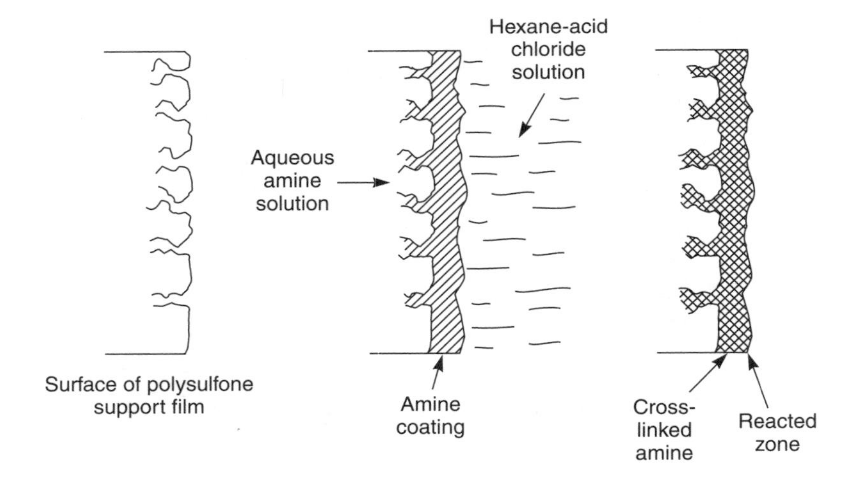

Figure 3.20 Schematic of the interfacial polymerization process. The microporous film is first impregnated with an aqueous amine solution. The film is then treated with a multivalent crosslinking agent dissolved in a water-immiscible organic fluid, such as hexane or Freon-113. An extremely thin polymer film forms at the interface of the two solutions [47]. Reprinted from L.T. Rozelle, J.E. Cadotte, K.E. Cobian, and C.V. Knopp, Jr, Nonpolysaccharide Membranes for Reverse Osmosis: NS-100 Membranes, in *Reverse Osmosis and Synthetic Membranes*, S. Sourirajan (ed.), National Research Council Canada, Ottawa, Canada (1977) by permission from NRC Research Press

solutions to form a densely crosslinked, extremely thin membrane layer. The first membrane made by Cadotte was based on polyethyleneimine crosslinked with toluene-2,4-diisocyanate, to form the structure shown in Figure 3.21 [47]. The process was later refined by Cadotte *et al.* at FilmTec Corp. [15,48], Riley *et al.* at UOP [49], and Kamiyama *et al.* [50] at Nitto in Japan.

Membranes made by interfacial polymerization have a dense, highly crosslinked polymer layer formed on the surface of the support membrane at the interface of the two solutions. A less crosslinked, more permeable hydrogel layer forms under this surface layer and fills the pores of the support membrane. The dense, crosslinked polymer layer, which can only form at the interface, is extremely thin, on the order of 0.1 μm or less, so the membrane permeability is high. Because the polymer is highly crosslinked, its selectivity is also high. Although the crosslinked interfacial polymer layer determines membrane selectivity, the nature of the microporous support film affects membrane flux

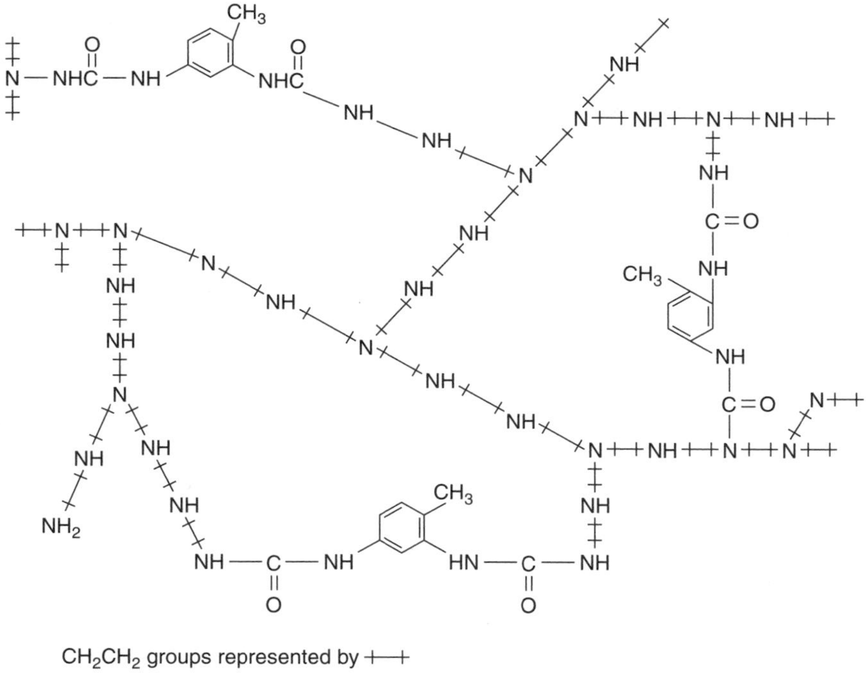

CH_2CH_2 groups represented by ┼─┼

Figure 3.21 Idealized structure of polyethyleneimine crosslinked with toluene 2,4-diisocyanate. This was called the NS-100 membrane. The chemistry was first developed by Cadotte to make interfacial reverse osmosis membranes with almost twice the water flux and one-fifth the salt leakage of the best reverse osmosis membranes then available. Even better membranes have since been developed by Cadotte and others [47]

significantly. The film has to be very finely porous to withstand the high pressures applied but must also have a high surface porosity so it is not a barrier to flow. The first reverse osmosis membranes made by the interfacial polymerization method were five times less salt-permeable than the best cellulose acetate Loeb–Sourirajan membranes but had better water fluxes. Since then interfacial polymerization chemistry has been refined. The first membrane produced by this method (and shown in Figure 3.21) was based on the reaction of a polyethyleneimine (in water) and toluene-2,4-diisocyanate or isophthaloyl chloride (in hexane). These NS-100 membranes had very good permeation properties but were very sensitive to even trace amounts (ppb levels) of chlorine commonly used as an antibacterial agent in water. The chlorine caused chain cleavage of the polymer at the amide bonds resulting in loss of salt rejection. A number of other chemistries have been developed over the years; the FT-30 membrane produced by reaction of phenylenediamine with trimesoyl chloride, also developed by Cadotte when at FilmTec (Dow Chemical), is particularly important. This membrane, which has a high water flux and consistent salt rejections of greater than 99.5 % with seawater [51], made single-pass seawater desalination with anisotropic membranes possible. A more detailed description of the chemistry of interfacial composite membranes is given in the discussion of reverse osmosis membranes in Chapter 5 and in a review by Petersen [48].

Production of interfacial polymerization membranes in the laboratory is relatively easy, but development of equipment to produce these membranes on a large scale required some ingenuity. The problem is the fragility of the interfacial surface film, which cannot be handled once formed. One solution to this problem is illustrated in Figure 3.22. The polysulfone or other material used as the support film is first immersed in an aqueous amine bath. On leaving this bath the membrane passes to a second organic acid chloride bath and then through a drying/curing oven. The transfer rollers are arranged so that the surface layer of the polymer on which the membrane forms never contacts a roller. On leaving the oven, the interfacial membrane is completely formed. This membrane is then coated with a protective solution of a water-soluble polymer such as poly(vinyl alcohol). When this solution is dried, the membrane is wound onto a take-up roll. The poly(vinyl alcohol) layer protects the membrane from damage during subsequent handling as it is formed into spiral-wound modules. When the module is used for the first time, the feed water washes off the water-soluble poly(vinyl alcohol) layer to expose the interfacial polymerized membrane, and the module is ready for use.

Interfacial polymerization membranes are widely used in reverse osmosis and nanofiltration but not for gas separation because of the water-swollen hydrogel that fills the pores of the support membrane. In reverse osmosis, this layer is hydrated and offers little resistance to water flow, but when the membrane is dried for use in gas separation the gel becomes a rigid glass with very low gas permeability. This glassy polymer fills the membrane pores and, as a result, defect-free

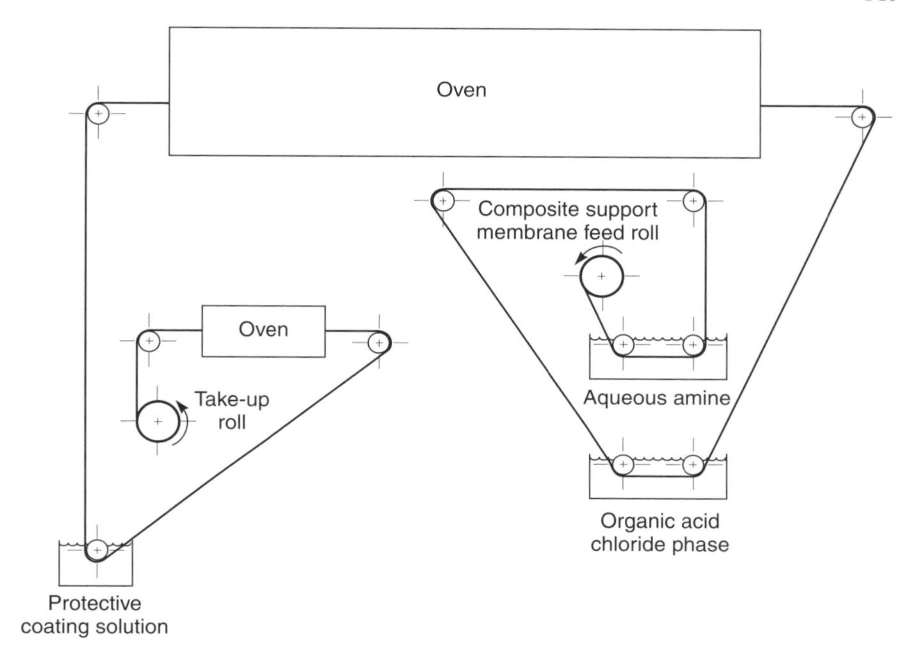

Figure 3.22 Schematic of the type of machinery used to make interfacial composite membranes

interfacial composite membranes usually have low gas fluxes, although their selectivities can be good.

Solution-coated Composite Membranes

Another important group of anisotropic composite membranes is formed by solution-coating a thin (0.5–2.0 μm) selective layer on a suitable microporous support. Membranes of this type were first prepared by Ward, Browall, and others at General Electric [52] and by Forester and Francis at North Star Research [17,53] using a type of Langmuir trough system. In this system, a dilute polymer solution in a volatile water-insoluble solvent is spread over the surface of a water-filled trough.

The apparatus used to make small sections of water-cast composite membranes is shown in Figure 3.23. The dilute polymer solution is cast on the surface between two Teflon rods. The rods are then moved apart to spread the film. The thin polymer film formed on the water surface is picked up on a microporous support. The main problem with this method is the transfer of the fragile, ultrathin film onto the microporous support. This is usually done by sliding the support

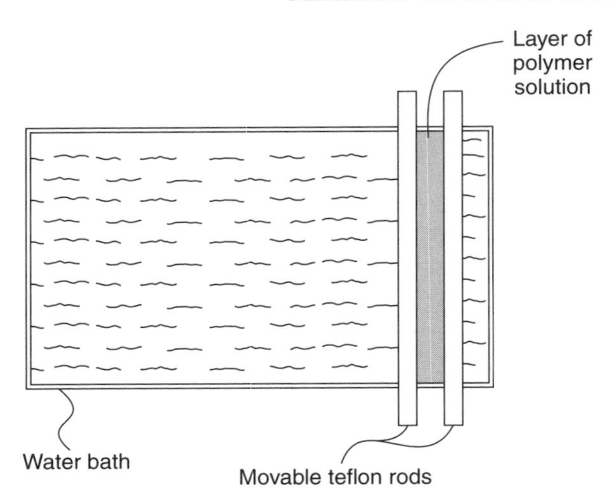

Figure 3.23 Schematic of the apparatus developed by Ward *et al.* [52] to prepare water-cast composite membranes. Reprinted from *J. Membr. Sci.*, **1**, W.J. Ward, III, W.R. Browall and R.M. Salemme, Ultrathin Silicone Rubber Membranes for Gas Separations, p. 99, Copyright 1976, with permission from Elsevier

membrane under the spread film. With care, small pieces of membrane as thin as 200 Å can be made.

The water-casting procedure was scaled up to produce continuous composite membrane films at General Electric. Figure 3.24 shows a schematic of the system used to produce composite polycarbonate-silicone copolymer membranes for small air separation units to produce oxygen-enriched air for medical use. The polymer casting solution added to the surface of the water bath spreads as a thin film and is picked up on the moving microporous support membrane. Membranes as thin as 0.1–0.2 μm can be made. This water-casting technique was used at General Electric and its spinoff, the Oxygen Enrichment Company, to make gas separation membranes for several years in the 1970s. The technique has also been adapted to coat hollow fiber membranes for gas separation applications [54,55].

Currently, most solution-coated composite membranes are prepared by the method first developed by Riley and others [45,56,57]. In this technique, a polymer solution is cast directly onto the microporous support. The support must be clean, defect-free and very finely microporous, to prevent penetration of the coating solution into the pores. If these conditions are met, the support can be coated with a liquid layer 50–100 μm thick, which after evaporation leaves a thin selective film 0.5–2 μm thick. A schematic drawing of the meniscus-coating technique is shown in Figure 3.25 [58]. Obtaining defect-free films by this technique requires considerable attention to the preparation procedure and the coating solution.

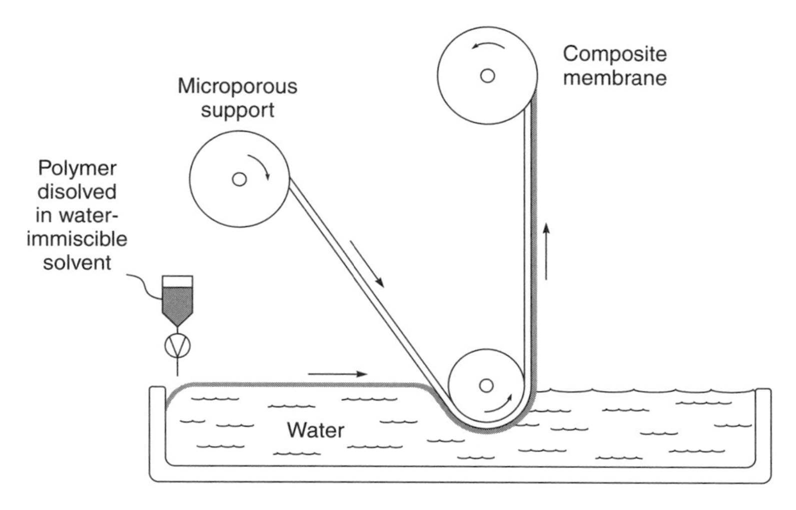

Figure 3.24 Method developed by Ward, Browall and others at General Electric to make multilayer composite membranes by the water casting technique [55]

The characteristics of the microporous support are very important. Because the selective layer is extremely thin, the support layer can contribute significantly to the total resistance to transport through the membrane. Not only does the resistance of the support decrease the flux through the membrane, but it can affect the separation [32,59]. To achieve the intrinsic selectivity of the selective membrane layer, the flux of the uncoated support material must be at least 10 times that of the coated support. This ensures that more than 90 % of the resistance to flow lies within the selective coating layer. As well as having a high flux, the surface layer of the microporous support material must also be very finely microporous. The pores must be small enough to support the thin selective layer under high pressure, and must also be close together so the permeating components do not take a long tortuous path to reach the pore. When the selective layer is only a few tenths of a micrometer thick this requirement may be difficult to meet. One solution to the problem is an intermediate gutter layer of a highly permeable polymer between the microporous support and the selective layer. The gutter layer material is much more permeable than the thin selective layer and acts as a conduit to transport material to the support membrane pores. Finally, because the selective layer of the composite membrane is often very thin and correspondingly delicate, such membranes are often protected by a sealing layer, also formed from a highly permeable material, to protect the membrane from damage during handling. A schematic of a multilayer composite membrane of this type is shown in Figure 3.26 together with a scanning electron micrograph. A discussion of the issues involved in preparing this type of membrane is found in the review by Koros and Pinnau [32].

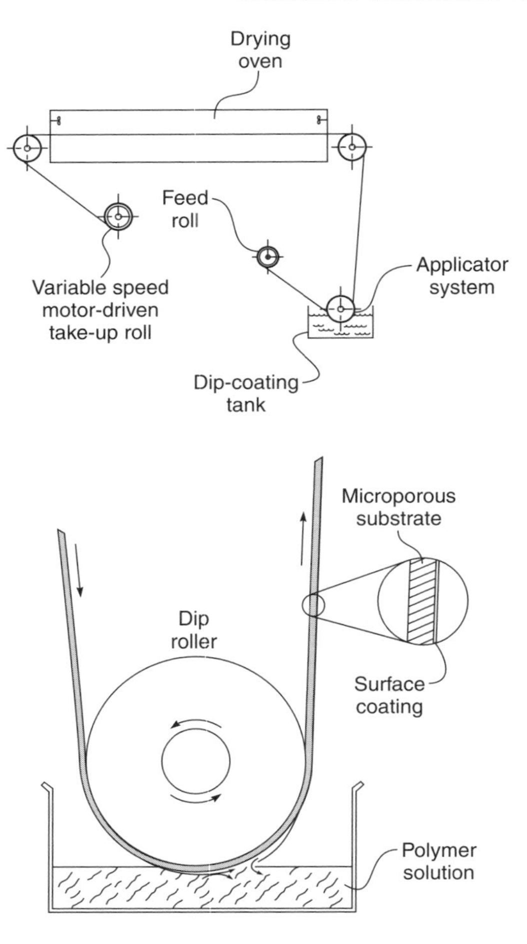

Figure 3.25 Schematic diagram of a film coating apparatus [58]

Other Anisotropic Membranes

Most anisotropic membranes are produced by solution precipitation, interfacial polymerization or solution coating. A number of other techniques developed in the laboratory are reviewed briefly below; none are used on a large scale.

Plasma Polymerization Membranes

Plasma polymerization of films was first used to form electrical insulation and protective coatings, but a number of workers have also prepared selective membranes by this method [60–63]. A simple plasma polymerization apparatus is

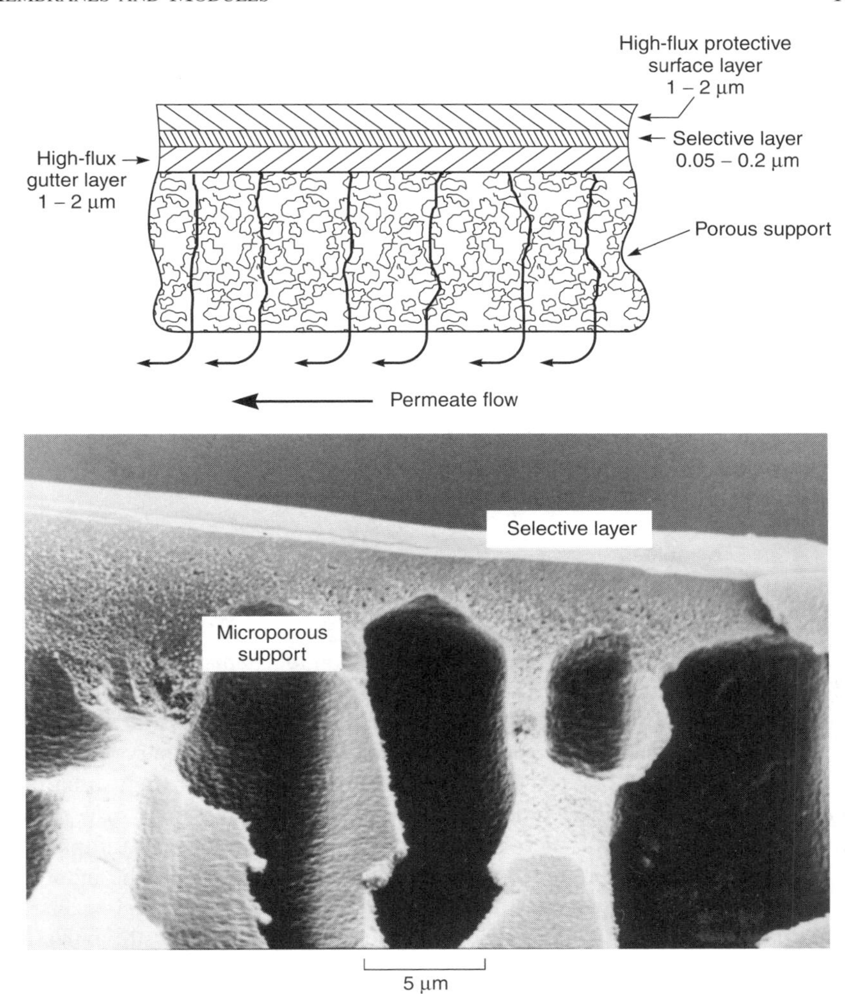

Figure 3.26 Schematic and scanning electron micrograph of a multilayer composite membrane on a microporous support. (Courtesy of Membrane Technology and Research, Inc.)

shown in Figure 3.27. Most workers used radio frequency fields at frequencies of 2–50 MHz to generate the plasma. In a typical plasma experiment helium, argon, or another inert gas is introduced at a pressure of 50–100 mTorr and a plasma is initiated. Monomer vapor is then introduced to bring the total pressure to 200–300 mTorr. These conditions are maintained for a period of 1–10 min, during which a thin polymer film is deposited on the membrane sample held in the plasma field.

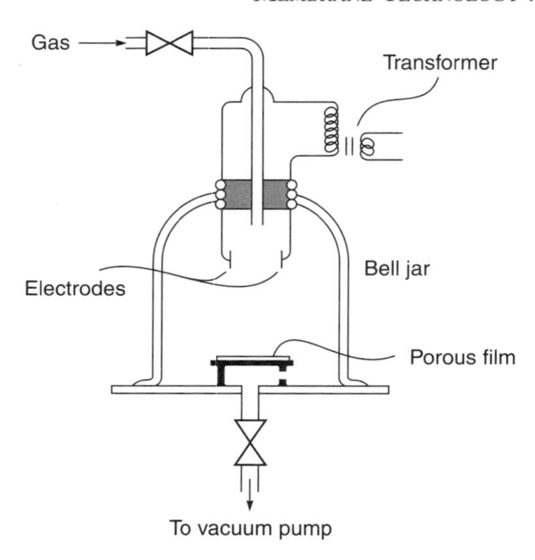

Figure 3.27 Simple bell jar plasma coating apparatus

Monomer polymerization proceeds by a complex mechanism involving ionized molecules and radicals and is completely different from conventional polymerization reactions. In general, the polymer films are highly crosslinked and may contain radicals that slowly react on standing. The stoichiometry of the film may also be quite different from the original monomer due to fragmentation of monomer molecules during the plasma polymerization process. The susceptibility of monomers to plasma polymerization or the characteristics of the resulting polymer film are difficult to predict. For example, many vinyl and acrylic monomers polymerize very slowly, whereas unconventional monomers such as benzene and hexane polymerize readily. The vapor pressure of the monomers, the power and voltage used in the discharge reaction, and the type and temperature of the substrate all affect the polymerization reaction. The inert gas used in the plasma may also enter into the reaction. Nitrogen and carbon monoxide, for example, are particularly reactive. In summary, the products of plasma polymerization are ill-defined and vary according to the experimental procedures. However, the resulting films can be very thin and have been shown to be quite selective.

The most extensive studies of plasma-polymerized membranes were performed in the 1970s and early 1980s by Yasuda, who tried to develop high-performance reverse osmosis membranes by depositing plasma films onto microporous polysulfone films [60,61]. More recently other workers have studied the gas permeability of plasma-polymerized films. For example, Stancell and Spencer [62] were able to obtain a gas separation plasma membrane with a hydrogen/methane selectivity of almost 300, and Kawakami *et al.* [63] have reported plasma membranes with an oxygen/nitrogen selectivity of 5.8. Both selectivities are good compared to those of other membranes, and the plasma films were also quite thin. However, in both cases the plasma film was formed on a substrate made from a thick (25–100 μm), dense polymer film, so the flux through the composite membrane was still low. Scale-up of plasma polymerization is likely to prove difficult, so the process will remain a laboratory technique until membranes with unique properties are produced.

Dynamically Formed Membranes

In the late 1960s and early 1970s, much attention was devoted to preparing dynamically formed anisotropic membranes, principally by Johnson, Kraus and others at Oak Ridge National Laboratory [64,65]. The general procedure is to form a layer of inorganic or polymeric colloids on the surface of a microporous support membrane by filtering a solution containing suspended colloid through the support membrane. A thin colloidal layer is laid down on the membrane surface and acts as a semipermeable membrane. Over time the colloidal surface layer is lost, and membrane performance falls. The support membrane is then cleaned, and a new layer of colloid is deposited. In the early development of this technique a wide variety of support membranes were used. Recently, microporous ceramic or porous carbon tubes have become the most commonly used materials. Typical colloidal materials used to make the selective membrane layer are polyvinyl methyl ether, acrylic acid copolymers or hydrated metal oxides such as zirconium hydroxide.

Dynamically formed membranes were pursued for many years for reverse osmosis because of their high water fluxes and relatively good salt rejection, especially with brackish water feeds. However, the membranes proved to be unstable and difficult to reproduce reliably and consistently. For these reasons, and because high-performance interfacial composite membranes were developed in the meantime, dynamically formed reverse osmosis membranes fell out of favor. A small application niche in high-temperature nanofiltration and ultrafiltration remains, and Rhône Poulenc continues their production. The principal application is poly(vinyl alcohol) recovery from hot wash water produced in textile dyeing operations.

Reactive Surface Treatment

Recently several groups have tried to improve the properties of anisotropic gas separation membranes by chemically modifying the surface selective layer. For example, Langsam at Air Products and Paul *et al.* at the University of Texas, Austin have treated films and membranes with dilute fluorine gas [66–71]. In this treatment fluorine chemically reacts with the polymer structure. By careful

Table 3.3 Effect of fluorination on the carbon dioxide/methane selectivity of various glassy membrane materials

Base polymer	Carbon dioxide/methane selectivity	
	Before fluorination	After fluorination
Poly(1-trimethylsilyl-1-propyne) (PTMSP) [69]	2.0	48
Poly(phenylene oxide) [71]	15	50–60
Poly(4-methyl-1-pentene) [70]	5.4	30–40

control of the process conditions, the reaction can be limited to a 100- to 200-Å surface layer. The dramatic improvements in selectivity produced by this surface treatment are illustrated by the data in Table 3.3. Scaling up this process for safe operation on a large scale will be difficult, but several groups are studying the approach. Ozone has also been suggested as a possible reactive surface treatment agent [72].

Repairing Membrane Defects

In preparing anisotropic membranes, the goal is to make the selective layer that performs the separation as thin as possible, but still defect free. Over the past 20 years, a great deal of work has been devoted to understanding the factors that determine the properties and thickness of the selective layer. The selective layer can be dense, as in reverse osmosis or gas separation membranes, or finely microporous with pores in the 100- to 500-Å diameter range, as in ultrafiltration membranes. In good quality membranes a thickness as low as 500–1000 Å can be achieved, but with layers as thin as this, formation of minute membrane defects is a problem. The defects, caused by gas bubbles, dust particles and support fabric imperfections, can be very difficult to eliminate. Such defects may not significantly affect the performance of anisotropic membranes used in liquid separation processes, such as ultrafiltration and reverse osmosis, but can be disastrous in gas separation applications. Browall [55] solved this problem by overcoating defective solution-cast composite membranes with a second thin coating layer of a highly permeable polymer to seal defects, as shown in Figure 3.28.

Later Henis and Tripodi [73] showed that membrane defects in anisotropic Loeb–Sourirajan membranes could be overcome in a similar way by coating the membrane with a thin layer of a relatively permeable material such as silicone rubber. A sufficiently thin coating does not change the properties of the underlying selective layer but does plug defects, through which simple convective gas flow can occur. Henis and Tripodi's membrane is illustrated in Figure 3.29. The silicone rubber layer is many times more permeable than the selective layer and

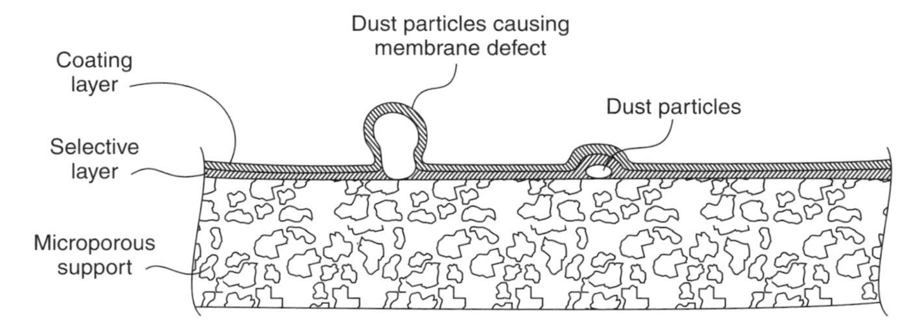

Figure 3.28 Method developed by Ward, Browall and others at General Electric to seal membrane defects in composite membranes made by the water coating technique [55]

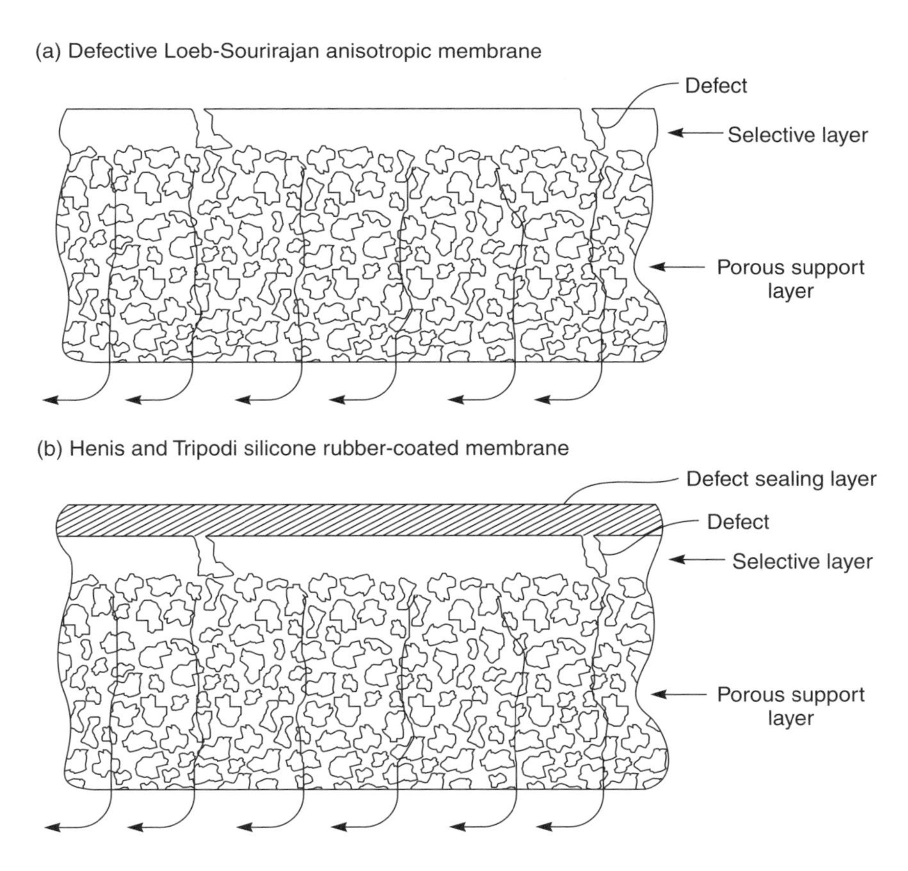

Figure 3.29 Schematic of (a) Loeb–Sourirajan and (b) Henis and Tripodi gas separation membranes [73]

does not function as a selective barrier but rather plugs defects, thereby reducing non-diffusive gas flow. The flow of gas through the portion of the silicone rubber layer over the pore is high compared to the flow through the defect-free portion of the membrane. However, because the area of membrane defects is very small, the total gas flow through these plugged defects is negligible. When this coating technique is used, the polysulfone skin layer of the Loeb–Sourirajan membrane no longer has to be completely defect free; therefore, the membrane can be made with a thinner skin than is possible with an uncoated membrane. The increase in flux obtained by decreasing the thickness of the selective skin layer more than compensates for the slight reduction in flux due to the silicone rubber sealing layer.

Metal Membranes and Ceramic Membranes

Metal Membranes

Metal membranes, particularly palladium-based, have been considered for hydrogen separation for a long time. In the 1950s and 1960s, Union Carbide installed and operated a palladium membrane plant to separate hydrogen from a refinery off-gas stream [74]. The plant produced 99.9 % pure hydrogen in a single pass through 25-μm-thick palladium membranes. However, even at a feed pressure of 450 psi, the membranes had to be operated at 370 °C to obtain a useful trans-membrane hydrogen flux. A further problem was the very high membrane cost; a 25-μm-thick palladium membrane requires approximately 250 g palladium/m^2 of membrane. At current palladium costs of US$20/g, the metal cost alone is US$5000/m^2 of membrane, which is 50 times the total cost of typical polymeric membranes used for gas separations. Small-scale palladium membrane systems, to produce ultrapure hydrogen for specialized applications, are marketed by Johnson Matthey and Company. These systems use palladium/silver alloy membranes based on those developed by Hunter [75,76].

If noble metal membranes are ever to be used on a large scale their cost must be reduced. One approach [77,78] is to sputter-coat a 500- to 1000-Å film of the metal on a polymer support. Because the film is extremely thin these membranes have extremely high hydrogen fluxes even at room temperature. Another approach, used by Buxbaum [79,80], is to coat a thin layer of palladium on a tantalum or vanadium support film. Tantalum and vanadium are also quite permeable to hydrogen and much less expensive than palladium. These metals cannot be used alone because they easily form an impenetrable oxide surface film. However, protected by a thin palladium layer, these membranes are quite permeable at high temperatures. Edlund [81,82] is pursuing a similar approach. A detailed discussion of hydrogen permeation in metals is given in the book by Alefeld and Völkl [83].

Ceramic Membranes

Metal Oxide Membranes

Several companies have developed inorganic ceramic membranes for ultrafiltration and microfiltration. These microporous membranes are made from aluminum, titanium or silica oxides. Ceramic membranes have the advantages of being chemically inert and stable at high temperatures, conditions under which polymer membranes fail. This stability makes ceramic microfiltration/ultrafiltration membranes particularly suitable for food, biotechnology and pharmaceutical applications in which membranes require repeated steam sterilization and cleaning with aggressive solutions. Pore diameters in ceramic membranes for microfiltration and ultrafiltration range from 0.01 to 10 μm; these membranes are generally made by a slip coating-sintering procedure. Other techniques, particularly sol-gel methods, are used to produce membranes with pores from 10 to 100 Å. Sol-gel membranes are the subject of considerable research interest particularly for gas separation applications, but so far have found only limited commercial use. A number of reviews covering the general area of ceramic membrane preparation and use have appeared recently [84,85].

In the slip coating-sintering process a porous ceramic support tube is made by pouring a dispersion of a fine-grain ceramic material and a binder into a mold and sintering at high temperature. The pores between the particles that make up this support tube are large. One surface of the tube is then coated with a suspension of finer particles in a solution of a cellulosic polymer or poly(vinyl alcohol) which acts as a binder and viscosity enhancer to hold the particles in suspension. This mixture is called a slip suspension; when dried and sintered at high temperatures, a finely microporous surface layer remains. Usually several slip-coated layers are applied in series, each layer being formed from a suspension of progressively finer particles and resulting in an anisotropic structure. Most commercial ceramic ultrafiltration membranes are made this way, generally in the form of tubes or perforated blocks. A scanning electron micrograph of the surface of this type of multilayer membrane is shown in Figure 3.30.

The slip coating-sintering procedure can be used to make membranes with pore diameters down to about 100–200 Å. More finely porous membranes are made by sol-gel techniques. In the sol-gel process slip coating is taken to the colloidal level. Generally the substrate to be coated with the sol-gel is a microporous ceramic tube formed by the slip coating-sintering technique. The solution coated onto this support is a colloidal or polymeric gel of an inorganic hydroxide. These solutions are prepared by controlled hydrolysis of metal salts or metal alkoxides to hydroxides.

Sol-gel methods fall into two categories, depending on how the colloidal coating solution is formed. The processes are shown schematically in Figure 3.31

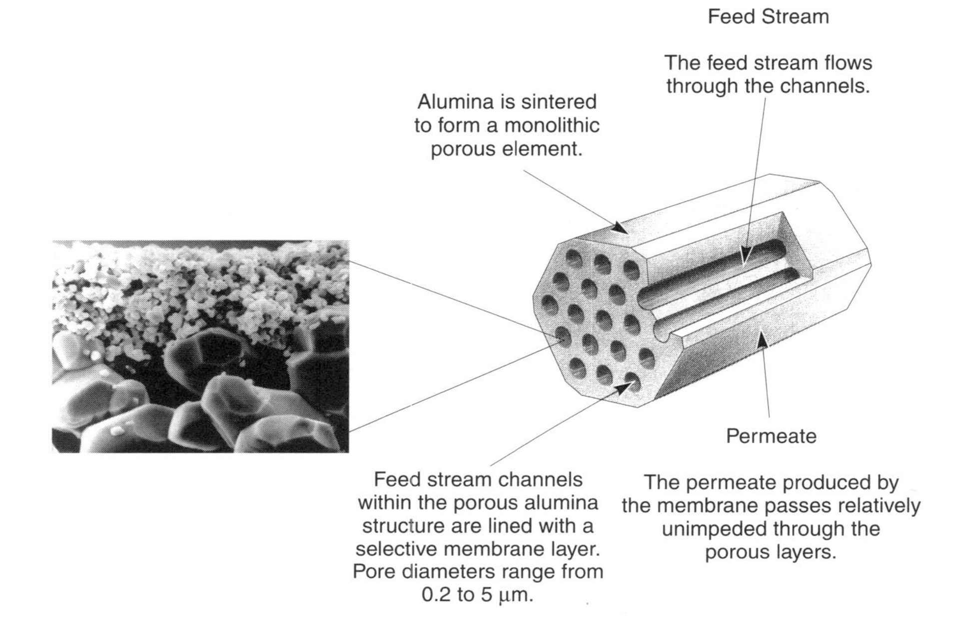

Figure 3.30 Cross-sectional scanning electron micrograph of a three-layered alumina membrane/support (pore sizes of 0.2, 0.8 and 12 μm, respectively). (Courtesy of Pall Corporation, Filterite Division, Timonium, MD)

[86–88]. In the particulate-sol method a metal alkoxide dissolved in alcohol is hydrolyzed by addition of excess water or acid. The precipitate that results is maintained as a hot solution for an extended period during which the precipitate forms a stable colloidal solution. This process is called peptization from the Greek pep—to cook (not a misnomer; many descriptions of the sol-gel process have a strong culinary flavor). The colloidal solution is then cooled and coated onto the microporous support membrane. The layer formed must be dried carefully to avoid cracking the coating. In the final step the film is sintered at 500–800 °C. The overall process can be represented as:

Precipitation: $Al(OR)_3 + H_2O \rightarrow Al(OH)_3$

Peptization: $Al(OH)_3 \rightarrow \gamma\text{-}Al_2O_3 \cdot H_2O$ (Böhmite) or $\delta\text{-}Al_2O_3 \cdot 3H_2O$ (Bayerite)

Sintering: $\gamma\text{-}Al_2O_3 \cdot H_2O \rightarrow \gamma\text{-}Al_2O_3 + H_2O$

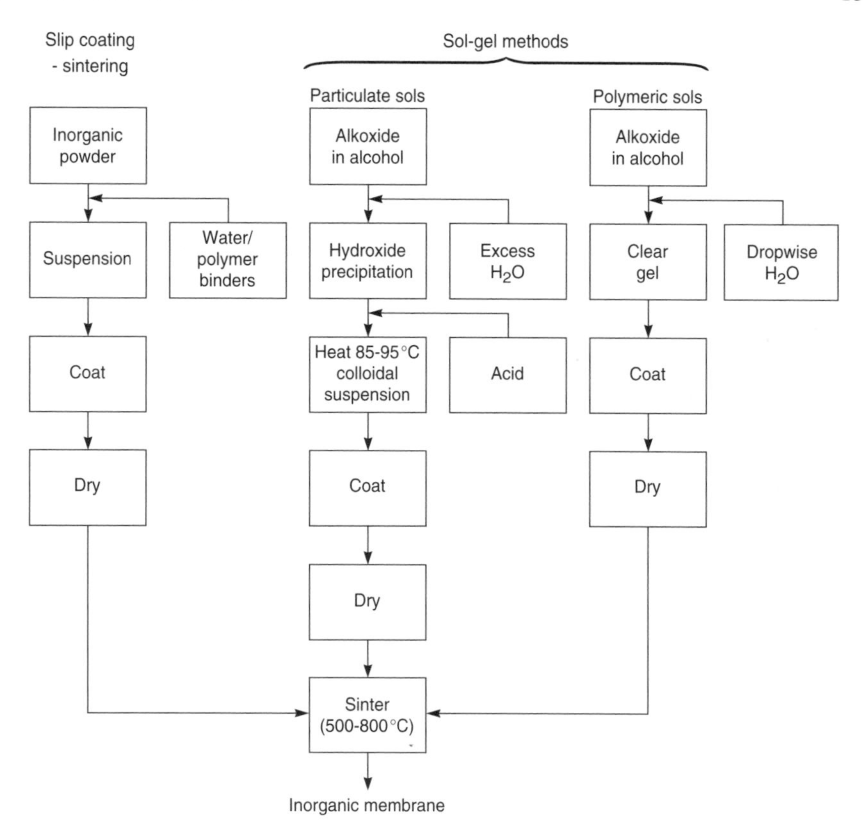

Figure 3.31 Slip coating-sintering and sol-gel processes used to make ceramic membranes

In the polymeric sol-gel process, partial hydrolysis of a metal alkoxide dissolved in alcohol is accomplished by adding the minimum of water to the solution. The active hydroxyl groups on the alkoxides then react to form an inorganic polymer molecule that can then be coated onto the ceramic support. On drying and sintering, the metal oxide film forms. Chemically the polymeric sol-gel process can be represented as:

$$\text{Hydrolysis:} \quad Ti(OR)_4 + H_2O \longrightarrow Ti(OR)_2(OH)_2 + ROH$$

$$\text{Polymerization:} \quad nTi(OR)_2(OH)_2 \longrightarrow [Ti(OR)_2 - O]_n + H_2O$$

$$\text{Crosslinking:} \quad [Ti(OR)_2 - O]_n \longrightarrow [Ti(OH)_2 - O][\underset{|}{\overset{\text{O}}{}}Ti(OH) - O]$$

Depending on the starting material and the coating procedure, a wide range of membranes can be made by the sol-gel process. The problem of cracking the films on drying and sintering can be alleviated by adding small amounts of a polymeric binder to the coating solution. The coating process may also be repeated several times to give a defect-free film. With care, membranes with pore sizes in the 10- to 100-Å range can be made by this method. In principle these membranes could be useful in a number of processes—membrane reactors, for example. Currently the technology is still at the laboratory stage.

Microporous Carbon Membranes

The first microporous carbon membranes were produced by Barrer in the 1950s and 1960s by compressing high-surface-area carbon powders at very high pressures [89,90]. The resulting porous plugs had pores of 5- to 30-Å diameter and were used to study diffusion of gases and vapors. More recently, practical ways of producing microporous carbon membranes have been developed by Koresh and Soffer [91], Hayashi *et al.* [92] and at Air Products by Rao and Sirkar [93]. All three groups are producing extremely finely microporous carbon membranes by pyrolizing preformed polyimide or polyacrylonitrile membranes in an inert atmosphere or vacuum at 500 to 800 °C. Under these conditions the polymer is converted to carbon. Permeation properties show that the carbon membranes have pores from 10 to 20 Å diameter. The Air Products membranes are made as thin films coated onto a ceramic support. The membranes of Koresh and Soffer are made as hollow fine fibers. These membranes are brittle and difficult to produce on a large scale, but have exceptional separation properties for some gas mixtures.

Microporous Glass Membranes

Microporous glass membranes in the form of tubes and fibers have been made by Corning, PPG, and Schott. Currently only the Corning membranes are still available, under the trade name Vycor®. The leaching process used to make this type of membrane has been described by Beaver [94]. The starting material is a glass containing 30–70 % silica, as well as oxides of zirconium, hafnium or titanium and extractable materials. The extractable materials comprise one or more boron-containing compounds and alkali metal oxides and/or alkaline earth metal oxides. Glass hollow fibers produced by melt extrusion are treated with dilute hydrochloric acid at 90 °C for 2–4 h to leach out the extractable materials, washed to remove residual acid, and then dried.

Liquid Membranes

Liquid membranes containing carriers to facilitate selective transport of gases or ions were the subject of a considerable research effort in the 1970s and 1980s.

A number of published reviews summarize this work [95,96]. Although these membranes are still being studied in a number of laboratories, improvements in selective conventional polymer membranes have diminished interest in processes using liquid membranes. The preparation and use of these membranes are described in Chapter 11.

Hollow Fiber Membranes

The membrane preparation techniques described so far were developed to produce flat-sheet membranes. However, these techniques can be adapted to produce membranes in the form of thin tubes or fibers. An important advantage of hollow fiber membranes is that compact modules with very high membrane surface areas can be formed. However, this advantage is offset by the generally lower fluxes of hollow fiber membranes compared to flat-sheet membranes made from the same materials. Nonetheless, the development of hollow fiber membranes by Mahon and the group at Dow Chemical in 1966 [97] and their later commercialization by Dow, Monsanto, Du Pont, and others represents one of the major events in membrane technology. A good review of the early development of hollow fiber membranes is given by Baum *et al.* [98]. Reviews of more recent developments are given by Moch [99] and McKelvey *et al.* [100].

The diameter of hollow fibers varies over a wide range, from 50 to 3000 μm. Fibers can be made with a uniformly dense structure, but preferably are formed as a microporous structure having a dense selective layer on either the outside or the inside surface. The dense surface layer can be either integral with the fiber or a separate layer coated onto the porous support fiber. Many fibers must be packed into bundles and potted into tubes to form a membrane module; modules with a surface area of even a few square meters require many kilometers of fibers. Because a module must contain no broken or defective fibers, hollow fiber production requires high reproducibility and stringent quality control.

The types of hollow fiber membranes in production are illustrated in Figure 3.32. Fibers of 50- to 200-μm diameter are usually called hollow fine fibers. Such fibers can withstand very high hydrostatic pressures applied from the outside, so they are used in reverse osmosis or high-pressure gas separation applications in which the applied pressure can be 1000 psig or more. The feed fluid is applied to the outside (shell side) of the fibers, and the permeate is removed down the fiber bore. When the fiber diameter is greater than 200–500 μm, the feed fluid is commonly applied to the inside bore of the fiber, and the permeate is removed from the outer shell. This technique is used for low-pressure gas separations and for applications such as hemodialysis or ultrafiltration. Fibers with a diameter greater than 500 μm are called capillary fibers.

Two methods are used to prepare hollow fibers: solution spinning and melt spinning [98,99]. The most common process is solution spinning or wet spinning, in which a 20–30 wt% polymer solution is extruded and precipitated

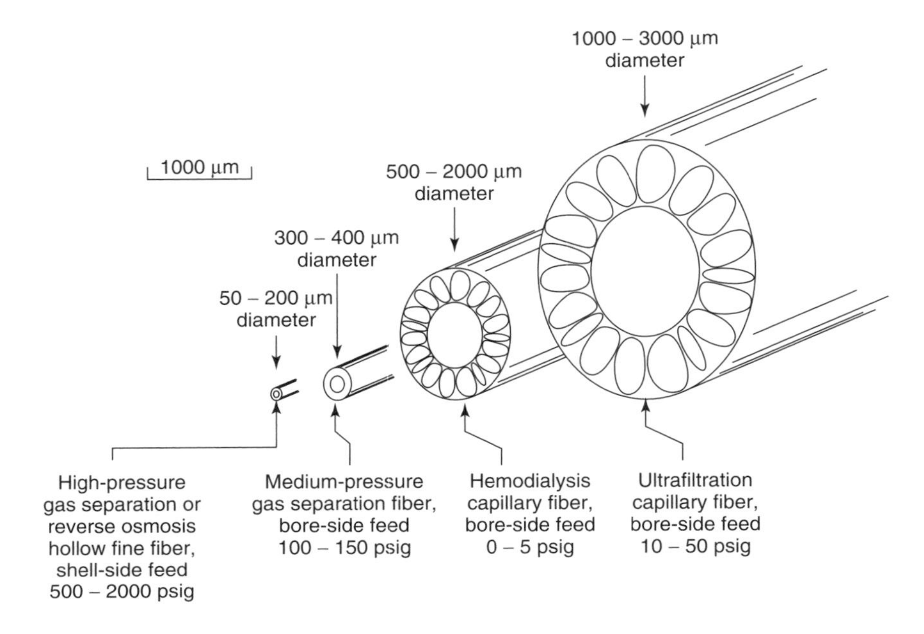

Figure 3.32 Schematic of the principal types of hollow fiber membranes

into a nonsolvent, generally water. Fibers made by solution spinning have the anisotropic structure of Loeb–Sourirajan membranes. This technique is generally used to make relatively large, porous hemodialysis and ultrafiltration fibers. In the alternative technique of melt spinning, a hot polymer melt is extruded from an appropriate die and is then cooled and solidified in air prior to immersion in a quench tank. Melt-spun fibers are usually denser and have lower fluxes than solution-spun fibers, but, because the fiber can be stretched after it leaves the die, very fine fibers can be made. Melt-spun fibers can also be produced at high speeds. The technique is usually used to make hollow fine fibers for high-pressure reverse osmosis and gas separation applications and is also used with polymers such as poly(trimethylpentene), which are not soluble in convenient solvents and are difficult to form by wet spinning. The distinction between solution spinning and melt spinning has gradually faded over the years. To improve fluxes, solvents and other additives are generally added to melt spinning dopes so spinning temperatures have fallen considerably. Many melt-spun fibers are now produced from spinning dopes containing as much as 30 to 60 wt% solvent, which requires the spinner to be heated to only 70–100 °C to make the dope flow. These fibers are also often cooled and precipitated by spinning into a water bath, which also helps to form an anisotropic structure.

The first hollow fiber spinneret system was devised by Mahon at Dow [97]. Mahon's spinneret consists of two concentric capillaries, the outer capillary having a diameter of approximately 400 μm, and the central capillary having an outer diameter of approximately 200 μm and an inner diameter of 100 μm. Polymer solution is forced through the outer capillary, while air or liquid is forced through the inner one. The rate at which the core fluid is injected into the fibers relative to the flow of polymer solution governs the ultimate wall thickness of the fiber. Figure 3.33 shows a cross-section of this type of spinneret, which is widely used to produce the large-diameter fibers used in ultrafiltration. Experimental details of this type of spinneret can be found elsewhere [101–103]. A complete hollow fiber spinning system is shown in Figure 3.34.

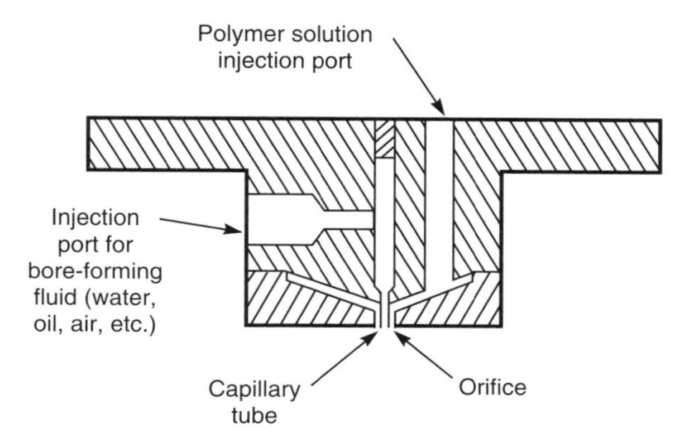

Figure 3.33 Twin-orifice spinneret design used in solution-spinning of hollow fiber membranes. Polymer solution is forced through the outer orifice, while bore-forming fluid is forced through the inner capillary

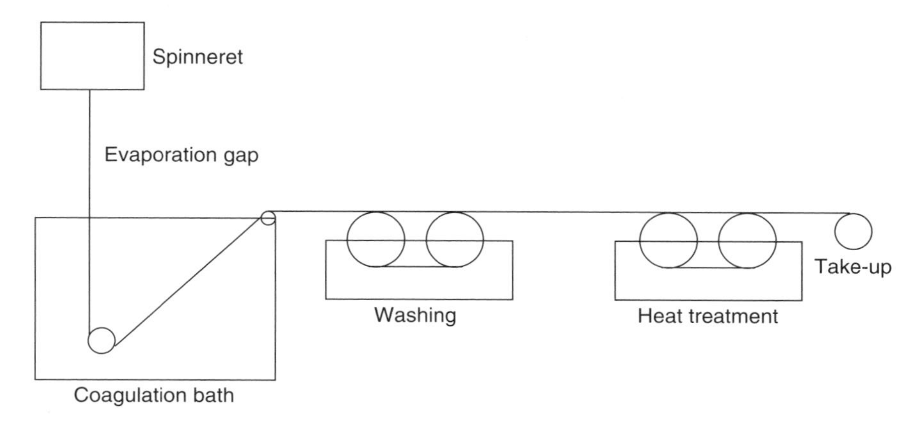

Figure 3.34 A complete hollow fiber solution-spinning system. The fiber is spun into a coagulation bath, where the polymer spinning solution precipitates forming the fiber. The fiber is then washed, dried, and taken up on a roll

The evaporation time between the solution exiting the spinneret and entering the coagulation bath is a critical variable, as are the compositions of the bore fluid and the coagulation bath. The position of the dense anisotropic skin can be adjusted by varying the bath and bore solutions. For example, if water is used as the bore fluid and the coagulation bath contains some solvent, precipitation will occur first and most rapidly on the inside surface of the fiber. If the solutions are reversed so that the bore solution contains some solvent and the coagulation bath is water, the skin will tend to be formed on the outside surface of the fiber, as shown in Figure 3.35. In many cases precipitation will begin on both surfaces of the fiber, and a dense layer will form on both inside and outside surfaces. This ability to manipulate the position of the dense skin is important because the skin should normally face the feed fluid.

Generally, the spinning dope used in solution spinning has a higher polymer concentration and is more viscous than the casting solutions used to form equivalent flat sheet membranes. This is because hollow fiber membranes must be able not only to perform the separation required but also to withstand the applied pressure of the process without collapsing. The mechanical demands placed on the microporous substructure of hollow fiber membranes are more demanding than for their flat-sheet equivalents. Consequently, a finer, stronger, and higher density microporous support structure is required. Because more concentrated casting solutions are used, the thickness of the skin layer of hollow fiber membranes is also greater than their flat-sheet equivalents. Usually lower membrane fluxes result. However, the low cost of producing a large membrane area in hollow fiber form compensates for the poorer performance.

Hollow fiber spinning dopes and preparation procedures vary over a wider range than their flat-sheet equivalents, but some representative dopes and spinning conditions taken from the patent literature [98,103,104] are given in Table 3.4.

Recently some interest in forming more complex hollow fibers has developed; for example, composite hollow fibers in which the microporous shell of the fiber provides the mechanical strength, but the selective layer is a coating of a

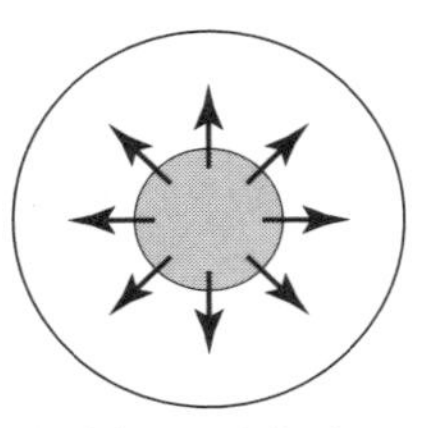

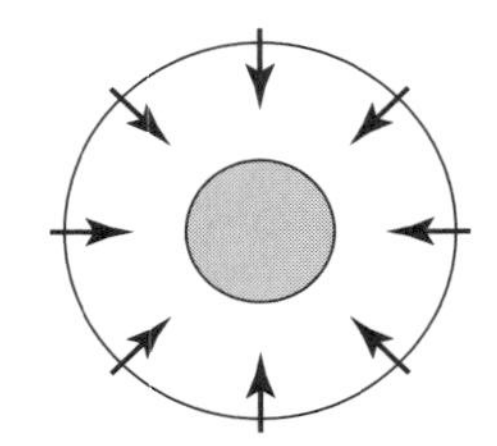

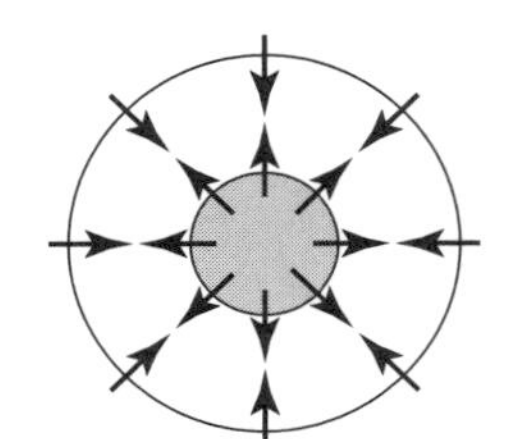

Figure 3.35 Depending on the bore fluid and the composition of the coagulation bath, the selective skin layer can be formed on the inside, the outside or both sides of the hollow fiber membrane

Table 3.4 Preparation parameters for various hollow fiber membranes

Casting dope	Bore fluid	Precipitation bath	Membrane type
37 wt% polysulfone (Udel P3500) 36 wt% *N*-methyl pyrrolidone 27 wt% propionic acid (spun at 15–100 °C)	Water	Water 25–50 °C	Gas separation fiber $\alpha O_2/N_2$ 5.2, ≈50 μm diameter, anisotropic outside-skinned fibers, finely microporous substrate [103]
25 wt% polyacrylonitrile-vinyl acetate copolymer 68 wt% dimethyl formamide 7 wt% formamide (spun at 65 °C)	10 wt% dimethyl formamide in water	40 wt% dimethyl formamide in water 4 °C	Ultrafiltration capillary membrane, inside skin, 98 % rejection to 110 000 MW dextran [104]
69 wt% cellulose triacetate (spun at 200 °C) 17.2 wt% sulfolane 13.8 wt% poly(ethylene glycol) (MW 400)	Air	No precipitation bath used; fiber forms on cooling. Solvents removed in later extraction step	Early (Dow) 80-μm-diameter fine fiber reverse osmosis membrane [98]

different material. Ube, Praxair, Air Products and Medal all produce this type of fiber for gas separation applications. Various techniques are described in the patent literature [105–107]. A device proposed by Air Products is shown in Figure 3.36. The preformed hollow fiber support membrane is drawn through a volatile solution of the coating polymer. The thickness of the film formed on the outer surface of the fiber is controlled by the concentration of polymer in the casting solution and the diameter of the orifice in the coating die. The solvent is evaporated and the fiber wound up.

Another method of producing composite hollow fibers, described by Kusuki *et al.* at Ube [108] and Kopp *et al.* at Memtec [109], is to spin double-layered fibers with a double spinneret of the type shown in Figure 3.37. This system allows different spinning solutions to be used for the outer and inner surface of the fibers and gives more precise control of the final structure. Often, two different polymers are incorporated into the same fiber. The result is a hollow fiber composite membrane equivalent to the flat sheet membrane shown in Figure 3.26. A reason for the popularity of composite hollow fiber membranes is that different polymers can be used to form the mechanically strong support and the selective layer. This can reduce the amount of selective polymer required. The tailor-made polymers developed for gas separation applications can cost as much as

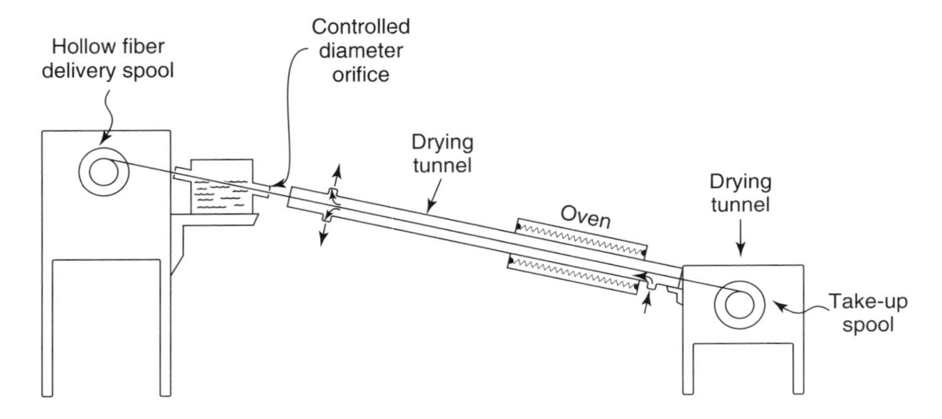

Figure 3.36 Apparatus to make composite hollow fiber membranes by coating a hollow fiber support membrane with a thin selective coating [105]

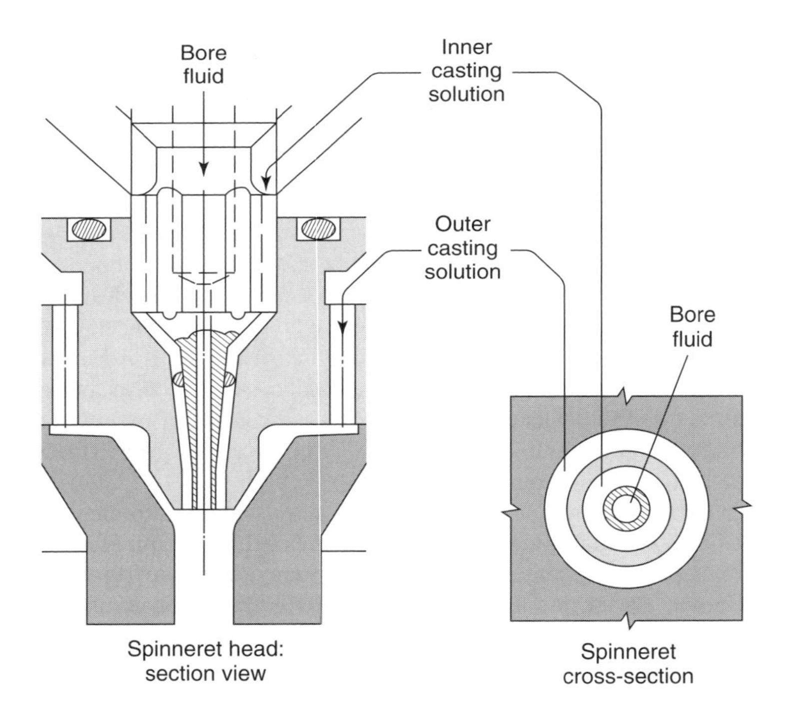

Figure 3.37 A double capillary spinneret sometimes used to produce two-layer hollow fibers. After Kopp *et al.* [109]

US$5–10/g. Single-layer hollow fiber membranes contain 10–20 g of polymer per square meter of membrane, for a material cost alone of US$50–200/$m^2$. Thus, using a composite structure consisting of a relatively inexpensive core polymer material coated with a thin layer of the expensive selective polymer reduces the overall membrane material cost significantly.

Membrane Modules

Industrial membrane plants often require hundreds to thousands of square meters of membrane to perform the separation required on a useful scale. Before a membrane separation can be performed industrially, therefore, methods of economically and efficiently packaging large areas of membrane are required. These packages are called membrane modules. The development of the technology to produce low-cost membrane modules was one of the breakthroughs that led to commercial membrane processes in the 1960s and 1970s. The earliest designs were based on simple filtration technology and consisted of flat sheets of membrane held in a type of filter press: these are called plate-and-frame modules. Membranes in the form of 1- to 3-cm-diameter tubes were developed at about the same time. Both designs are still used, but because of their relatively high cost they have been largely displaced in most applications by two other designs—the spiral-wound module and the hollow fiber module.

Despite the importance of membrane module technology, many researchers are astonishingly uninformed about module design issues. In part this is because module technology has been developed within companies, and developments are only found in patents, which are ignored by many academics. The following sections give an overview of the principal module types, followed by a section summarizing the factors governing selection of particular types for different membrane processes. Cost is always important, but perhaps the most important issues are membrane fouling and concentration polarization. This is particularly true for reverse osmosis and ultrafiltration systems, but concentration polarization issues also affect the design of gas separation and pervaporation modules.

Plate-and-frame Modules

Plate-and-frame modules were one of the earliest types of membrane system. A plate-and-frame design proposed by Stern [110] for early Union Carbide plants to recovery helium from natural gas is shown in Figure 3.38. Membrane, feed spacers, and product spacers are layered together between two end plates. The feed mixture is forced across the surface of the membrane. A portion passes through the membrane, enters the permeate channel, and makes its way to a central permeate collection manifold.

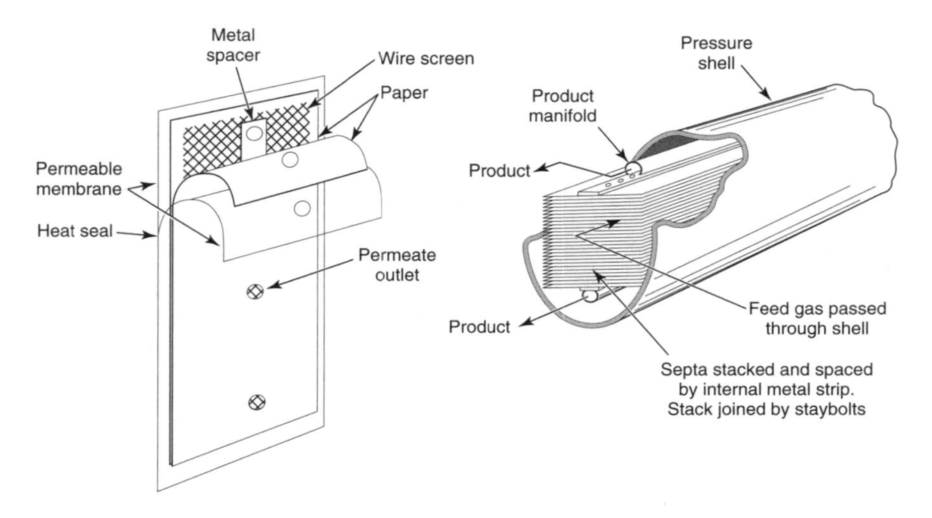

Figure 3.38 Early plate-and-frame design developed by Stern *et al.* [110] for the separation of helium from natural gas. Reprinted with permission from S.A. Stern, T.F. Sinclaire, P.J. Gareis, N.P. Vahldieck and P.H. Mohr, Helium Recovery by Permeation, *Ind. Eng. Chem.* **57**, 49. Copyright 1965, American Chemical Society and American Pharmaceutical Association

Plate-and-frame units have been developed for some small-scale applications, but these units are expensive compared to the alternatives, and leaks through the gaskets required for each plate are a serious problem. Plate-and-frame modules are now only used in electrodialysis and pervaporation systems and in a limited number of reverse osmosis and ultrafiltration applications with highly fouling feeds. An example of one of these reverse osmosis units is shown in Figure 3.39 [111].

Tubular Modules

Tubular modules are now generally limited to ultrafiltration applications, for which the benefit of resistance to membrane fouling due to good fluid hydrodynamics outweighs their high cost. Typically, the tubes consist of a porous paper or fiberglass support with the membrane formed on the inside of the tubes, as shown in Figure 3.40.

The first tubular membranes were between 2 and 3 cm in diameter, but more recently, as many as five to seven smaller tubes, each 0.5–1.0 cm in diameter, are nested inside a single, larger tube. In a typical tubular membrane system a large number of tubes are manifolded in series. The permeate is removed from each tube and sent to a permeate collection header. A drawing of a 30-tube system is shown in Figure 3.41. The feed solution is pumped through all 30 tubes connected in series.

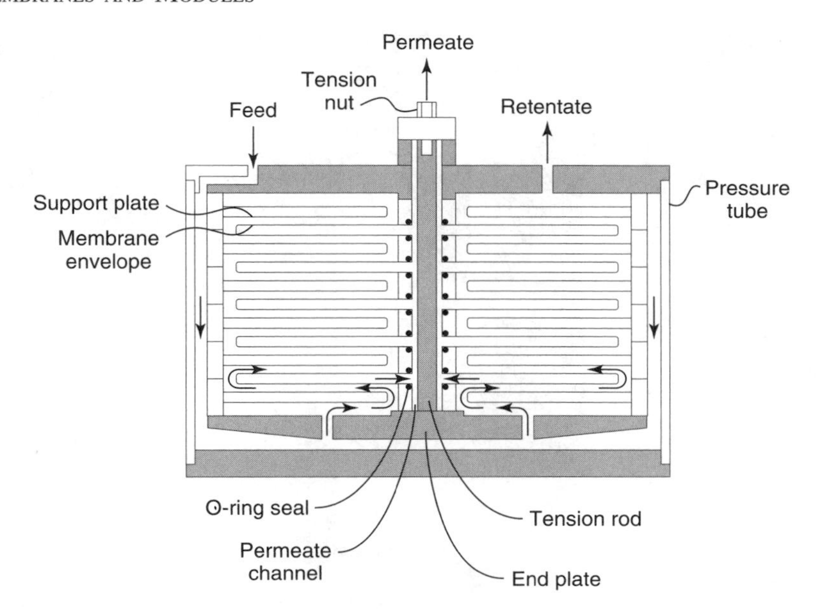

Figure 3.39 Schematic of a plate-and-frame module. Plate-and-frame modules provide good flow control on both the permeate and feed side of the membrane, but the large number of spacer plates and seals lead to high module costs. The feed solution is directed across each plate in series. Permeate enters the membrane envelope and is collected through the central permeate collection channel [111]

Spiral-wound Modules

Spiral-wound modules were used in a number of early artificial kidney designs, but were fully developed for industrial membrane separations by Gulf General Atomic (a predecessor of Fluid Systems, Inc.). This work, directed at reverse osmosis membrane modules, was carried out under the sponsorship of the Office of Saline Water [112–114]. The design shown in Figure 3.42 is the simplest, consisting of a membrane envelope of spacers and membrane wound around a perforated central collection tube; the module is placed inside a tubular pressure vessel. Feed passes axially down the module across the membrane envelope. A portion of the feed permeates into the membrane envelope, where it spirals towards the center and exits through the collection tube.

Small laboratory spiral-wound modules consist of a single membrane envelope wrapped around the collection tube, as shown in Figure 3.42. The membrane area of these modules is typically 0.2 to 1.0 m^2. Industrial-scale modules contain several membrane envelopes, each with an area of 1–2 m^2, wrapped around the central collection pipe. The multi-envelope design developed at Gulf General Atomic by Bray [113] and others is illustrated in Figure 3.43. Multi-envelope

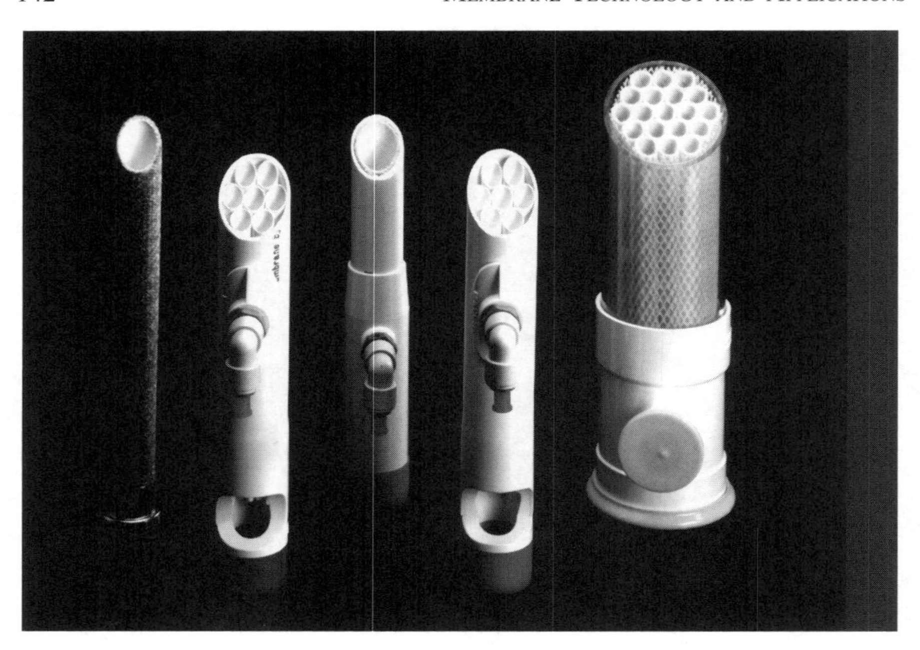

Figure 3.40 Typical tubular ultrafiltration module design. The membrane is usually cast on a porous fiberglass or paper support, which is then nested inside a plastic or steel support tube. In the past, each plastic housing contained a single 2- to 3-cm-diameter tube. More recently, several 0.5- to 1.0-cm-diameter tubes, nested inside single housings, have been introduced. (Courtesy of Koch Membrane Systems)

designs minimize the pressure drop encountered by the permeate fluid traveling towards the central pipe. If a single membrane envelope were used in a large-membrane-area module, the path taken by the permeate to reach the central collection pipe would be several meters long, depending on the module diameter. Such a long permeate path would result in a large pressure drop in the permeate collection channel. By using multiple short envelopes the pressure drop in any one envelope is kept at a manageable level. The standard industrial spiral-wound module has an 8-in. diameter and is 40 in. long. Twelve-inch-diameter modules up to 60 in. long have been made and offer some economy of scale. There is, therefore, a trend towards increasing the module diameter for larger plants. The approximate membrane area and number of membrane envelopes used in industrial 40-in.-long spiral-wound modules are given in Table 3.5.

Four to six spiral-wound membrane modules are normally connected in series inside a single pressure vessel (tube). A typical 8-in.-diameter tube containing six modules has 100–200 m^2 of membrane area. An exploded view of a membrane tube containing two modules is shown in Figure 3.44 [115]. The end of each module is fitted with an anti-telescoping device (ATD) which is designed to

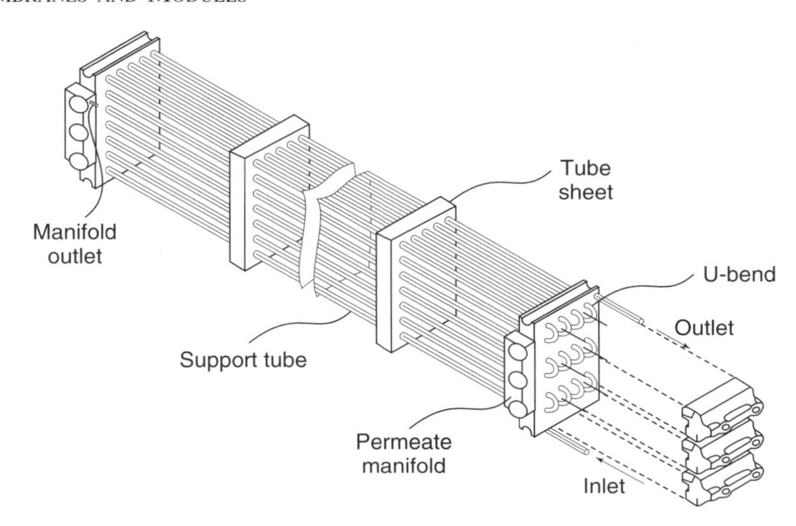

Figure 3.41 Exploded view of a tubular ultrafiltration system in which 30 tubes are connected in series. Permeate from each tube is collected in the permeate manifold

prevent the module leaves shifting under the feed-to-residue pressure difference required to force feed fluid through the module. The ATD is also fitted with a rubber seal to form a tight connection between the module and the pressure vessel. This seal prevents fluid bypassing the module in the gap between the module and the vessel wall.

In some applications of reverse osmosis and ultrafiltration spiral-wound modules in the food industry, it may be desirable to allow a small portion of the feed solution to bypass the module to prevent bacteria growing in the otherwise stagnant fluid. One way of achieve this bypass is by perforating the ATD as illustrated in Figure 3.45 [115].

Hollow Fiber Modules

Hollow fiber membrane modules are formed in two basic geometries. The first is the shell-side feed design illustrated in Figure 3.46(a) and used, for example, by Monsanto in their hydrogen separation systems and by Du Pont in their reverse osmosis systems. In such a module, a loop or a closed bundle of fibers is contained in a pressure vessel. The system is pressurized from the shell side; permeate passes through the fiber wall and exits through the open fiber ends. This design is easy to make and allows very large membrane areas to be contained in an economical system. Because the fiber wall must support considerable hydrostatic pressure, the fibers usually have small diameters and thick walls, typically 50-μm internal diameter and 100- to 200-μm outer diameter.

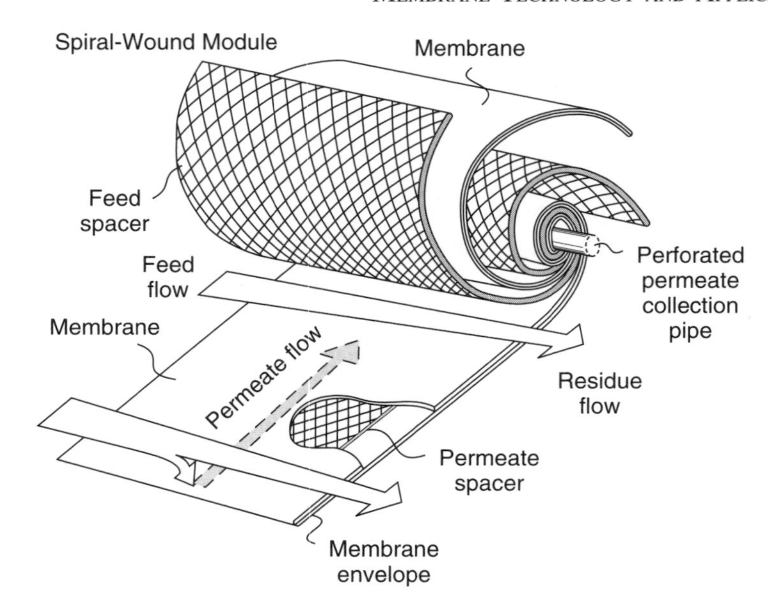

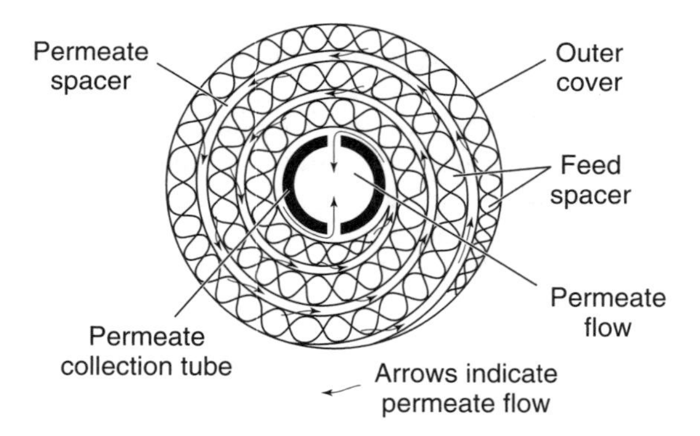

Figure 3.42 Exploded view and cross-section drawings of a spiral-wound module. Feed solution passes across the membrane surface. A portion passes through the membrane and enters the membrane envelope where it spirals inward to the central perforated collection pipe. One solution enters the module (the feed) and two solutions leave (the residue and the permeate). Spiral-wound modules are the most common module design for reverse osmosis and ultrafiltration as well as for high-pressure gas separation applications in the natural gas industry

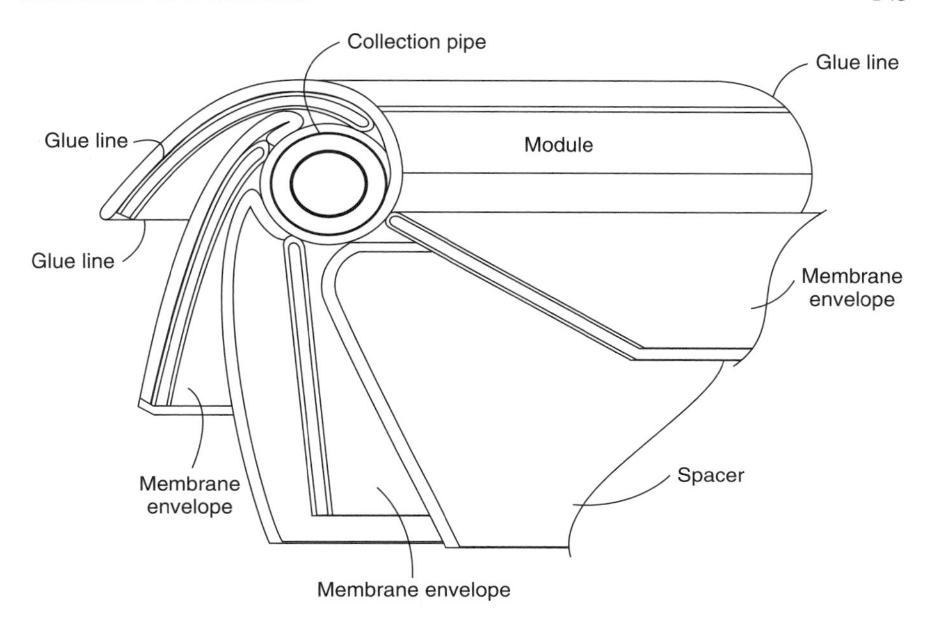

Figure 3.43 Multi-envelope spiral-wound module [113], used to avoid excessive pressure drops on the permeate side of the membrane. Large-diameter modules may have as many as 30 membrane envelopes, each with a membrane area of 1–2 m^2

Table 3.5 Typical membrane area and number of membrane envelopes for 40-in.-long industrial spiral-wound modules. The thickness of the membrane spacers used for different applications causes the variation in membrane area

Module diameter (in.)	4	6	8	12
Number of membrane envelopes	4–6	6–10	15–30	30–40
Membrane area (m^2)	3–6	6–12	20–40	30–60

The second type of hollow fiber module is the bore-side feed type illustrated in Figure 3.46(b). The fibers in this type of unit are open at both ends, and the feed fluid is circulated through the bore of the fibers. To minimize pressure drop inside the fibers, the diameters are usually larger than those of the fine fibers used in the shell-side feed system and are generally made by solution spinning. These so-called capillary fibers are used in ultrafiltration, pervaporation, and some low- to medium-pressure gas applications. Feed pressures are usually limited to below 150 psig in this type of module.

In bore-side feed modules, it is important to ensure that all of the fibers have identical fiber diameters and permeances. Even fiber variation as small as ±10 % from the average fiber can lead to large variations in module performance [116,117]. The flow of fluid through the fiber bore is proportional to the fiber

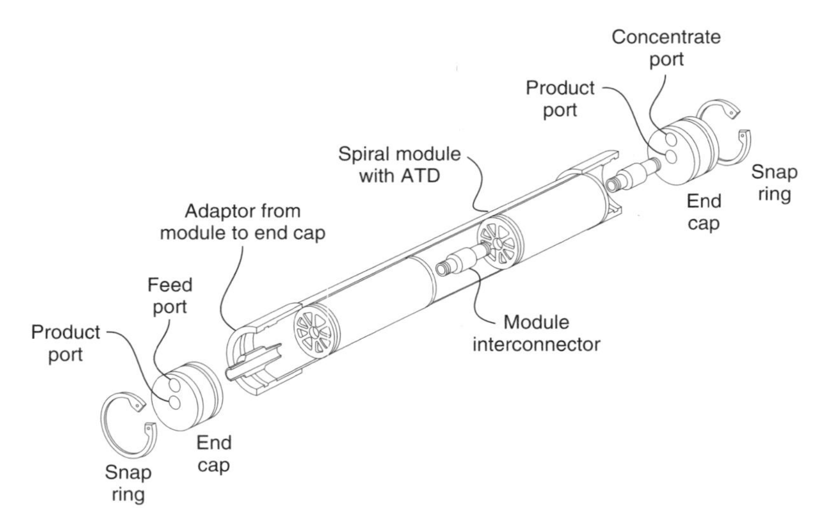

Figure 3.44 Schematic of a spiral-wound module [115] installed in a multimodule pressure vessel. Typically four to six modules are installed in a single pressure vessel. Reprinted from *Reverse Osmosis Technology*, B.S. Parekh (ed.), Marcel Dekker, New York (1988), p. 81, by courtesy of Marcel Dekker, Inc.

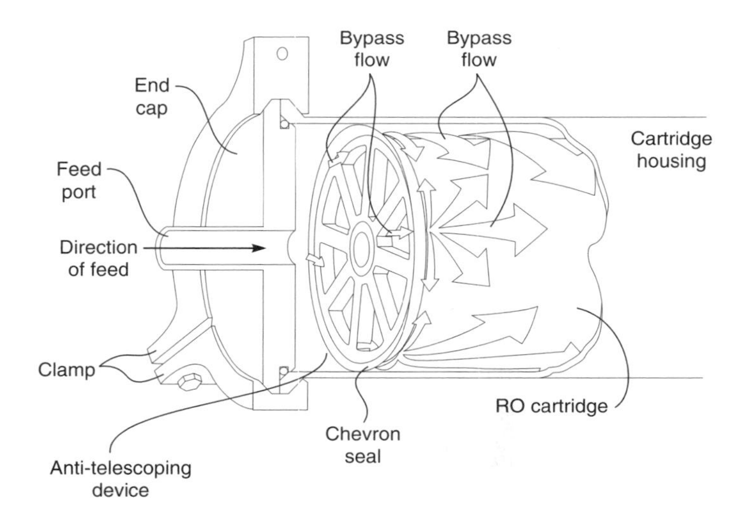

Figure 3.45 By perforating the antitelescoping device, a small controlled bypass of fluid past the module seal is achieved to eliminate the stagnant area between the reverse osmosis module and the pressure vessel walls. This device is used in food and other sanitary applications of spiral-wound modules [115]. Reprinted from *Reverse Osmosis Technology*, B.S. Parekh (ed.), Marcel Dekker, New York (1988), p. 359, by courtesy of Marcel Dekker, Inc.

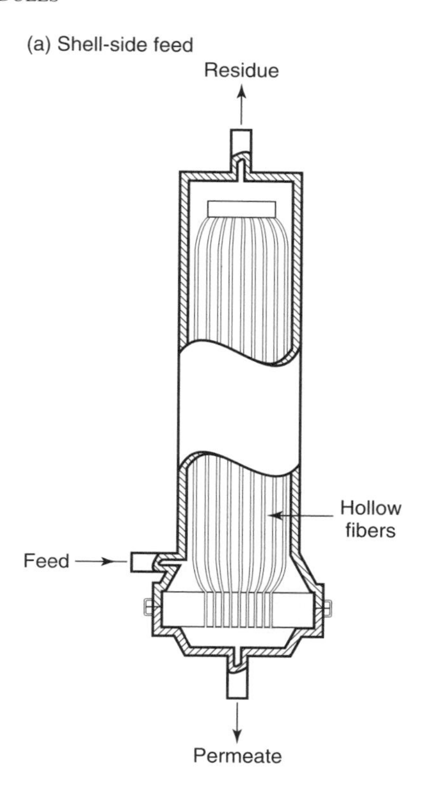

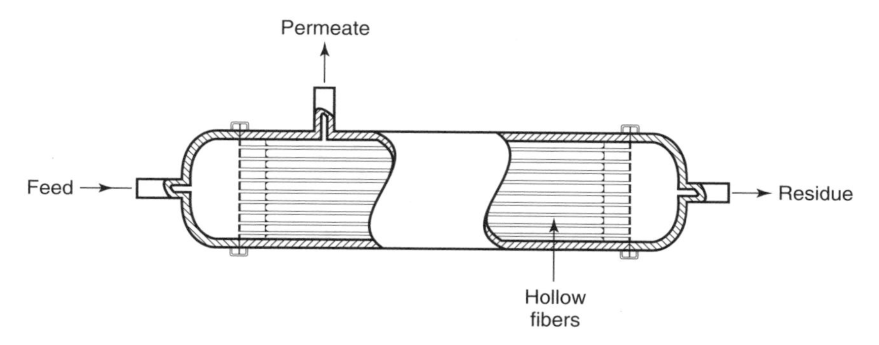

Figure 3.46 Two types of hollow-fiber modules used for gas separation, reverse osmosis, and ultrafiltration applications. Shell-side feed modules are generally used for high-pressure applications up to 1000 psig. Fouling on the feed side of the membrane can be a problem with this design, and pretreatment of the feed stream to remove particulates is required. Bore-side feed modules are generally used for medium-pressure feed streams up to 150 psig, for which good flow control to minimize fouling and concentration polarization on the feed side of the membrane is desired

diameter to the fourth power, whereas the membrane area only changes by the second power. The effect is particularly important in the production of nitrogen from air and in hollow-fiber kidney modules, in which high levels of removal of the permeable component in a single pass are desired. If the fibers have different diameters, a few overly large or overly small fibers can significantly affect the removal achieved by the module.

Concentration polarization is well controlled in bore-side feed modules. The feed solution passes directly across the active surface of the membrane, and no stagnant dead spaces are produced. This is far from the case in shell-side feed modules in which flow channeling and stagnant areas between fibers, which cause significant concentration polarization problems, are difficult to avoid [118]. Any suspended particulate matter in the feed solution is easily trapped in these stagnant areas, leading to irreversible fouling of the membrane. Baffles to direct the feed flow have been tried [119,120], but are not widely used. A more common method of minimizing concentration polarization is to direct the feed flow normal to the direction of the hollow fibers as shown in Figure 3.47. This produces a cross-flow module with relatively good flow distribution across the fiber surface. Several membrane modules may be connected in series, so high feed solution velocities can be used. A number of variants on this basic design have been patented [121,122] and are reviewed by Koros and Fleming [123].

A second problem in shell-side feed hollow fine fibers is permeate side parasitic pressure drops. The permeate channel in these fibers is so narrow, and presents such a resistance to fluid passage, that a significant pressure drop develops along the length of the permeate channel, reducing the pressure difference across the membrane that provides the driving force for permeation. In applications involving separation of mixtures of relatively impermeable components, such as oxygen and nitrogen in air, the pressure drop that develops is small

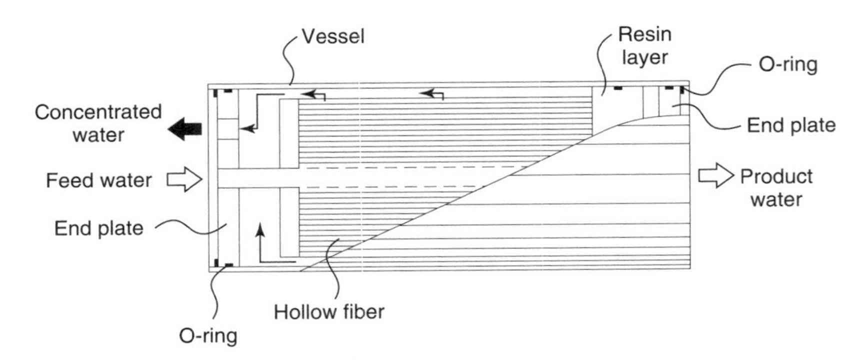

Figure 3.47 A cross-flow hollow fiber module used to obtain better flow distribution and reduce concentration polarization (the Tyobo Hollosep reverse osmosis module). Feed enters through the perforated central pipe and flows towards the module shell

and unimportant. But in separations of more permeable gas mixtures, such as hydrogen or carbon dioxide from methane, the pressure drop can be a significant fraction of the total applied pressure. Permeate-side pressure drops also tend to develop in spiral-wound modules. However, because the permeate channels are wider in this type of module, pressure drops are usually smaller and less significant.

The greatest single advantage of hollow fiber modules is the ability to pack a very large membrane area into a single module. The magnitude of this advantage can be gauged by the membrane area per module data shown in Table 3.6. This table shows the calculated membrane area contained in an 8-in.-diameter, 40-in.-long module; a spiral-wound module of this size would contain about 20–40 m^2 of membrane area. The equivalent hollow fiber module, filled with fibers with a diameter of 100 μm, will contain approximately 300 m^2 of membrane area, 10 times the area in a spiral-wound module. As the diameter of the fibers in the module increases, the membrane area decreases. Capillary ultrafiltration membrane modules have almost the same area as equivalent-sized spiral-wound modules.

Table 3.6 also shows the huge numbers of hollow fibers required for high-surface-area modules. A hollow fine fiber module with an area of 300 m^2 will contain 1000 km of fiber. Expensive, sophisticated, high-speed automated spinning and fiber handling and module fabrication equipment is required to produce these modules. A typical hollow fiber spinning operation will have 50–100 spinnerets. In general the capital investment for a hollow fine fiber production plant is so large that the technology can only be considered when large numbers of modules are being produced on a round-the-clock basis. The technology is maintained as a trade secret within the handful of companies that produce this type of module. A clue to the type of machinery involved can be obtained from the patent literature. Figure 3.48, for example, shows a module winding machine from an old Du Pont patent [124]. Fibers from several bobbins are wound around a porous paper sheet, laying down the bundle that ultimately becomes the module insert.

Table 3.6 Effect of fiber diameter on membrane area and the number of fibers in a module 20 cm (8 in.) in diameter and 1 m (40 in.) long. Twenty-five percent of the module volume is filled with fiber. A spiral-wound module of this size contains approximately 20–40 m^2 of membrane area and has a packing density of 6–13 cm^2/cm^3

Module use	High-pressure reverse osmosis and gas separation	Low-pressure gas separation		Ultrafiltration	
Fiber diameter (μm)	100	250	500	1000	2000
Number of fibers/module (thousands)	1000	250	40	10	2.5
Membrane area (m^2)	315	155	65	32	16
Packing density (cm^2/cm^3)	100	50	20	10	5

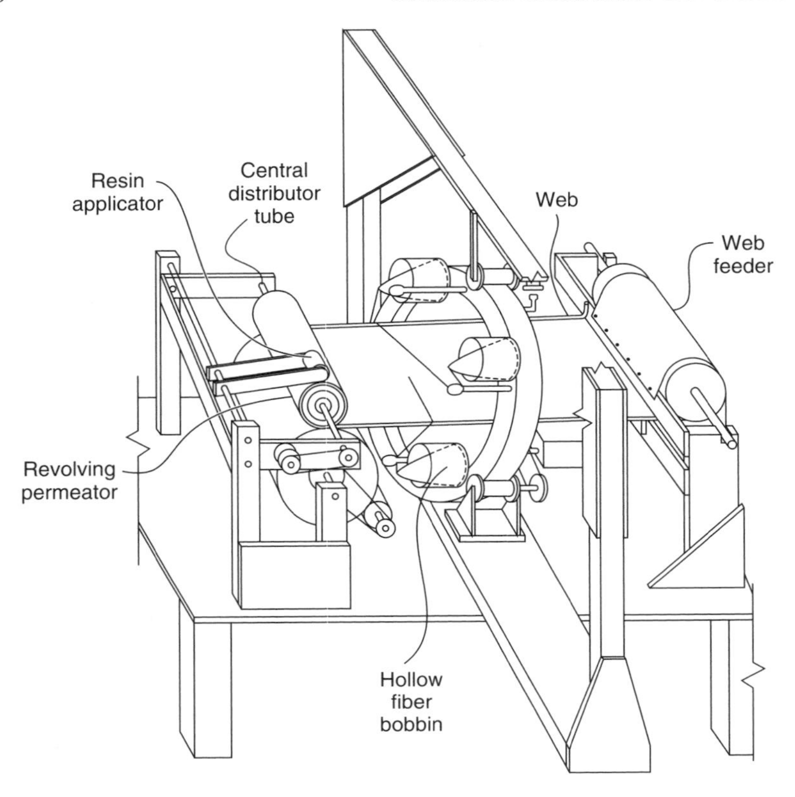

Figure 3.48 Hollow fiber module winding apparatus from a 1972 Du Pont patent [124]. Machines of this general type are still used to produce hollow fiber modules

Vibrating and Rotating Modules

In all of the module designs described thus far, the fluid to be separated (gas or liquid) is pumped across the surface of the membrane at high velocity to control concentration polarization. A few vibrating or rotating modules, in which the membrane moves, and moves much faster than the fluid flowing across its surface, have been developed. One such design, a vibrating module, from New Logic International, is shown in Figure 3.49 [125,126]. Vibration of the membrane at high speed creates interior agitation directly at the membrane surface. These modules have proved to be able to ultrafilter extremely concentrated, viscous solutions that could not be treated by conventional module designs. Currently the modules are extremely expensive—in the range US\$2000–5000/m^2 membrane—compared to alternative designs. This limits their application to high-value separations that cannot be performed by other processes.

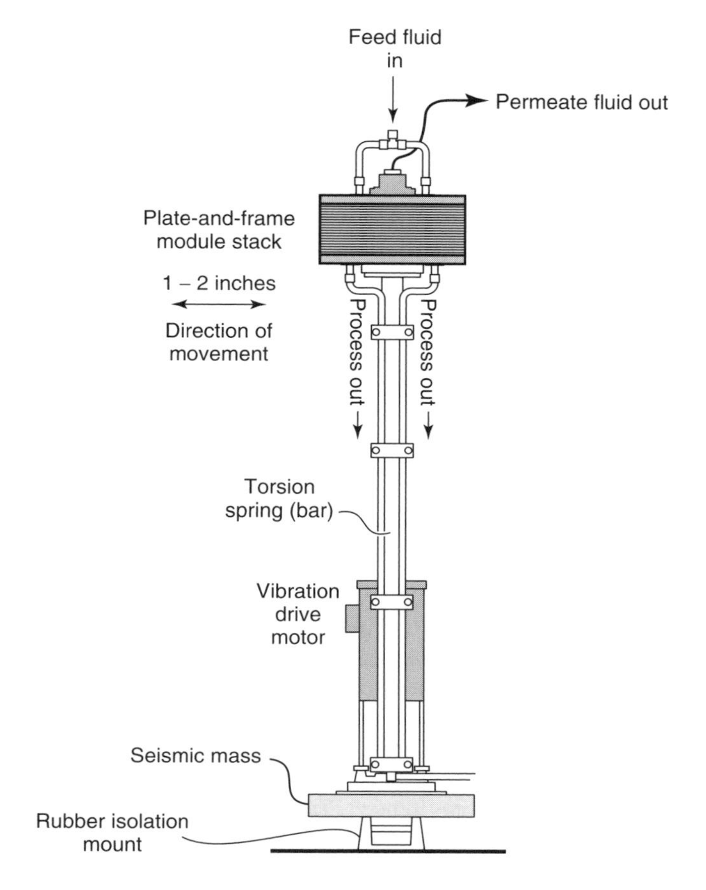

Figure 3.49 New Logic International vibrating plate-and-frame module design [125]. A motor taps a metal plate (the seismic mass) supported by a rubber mount at 60 times/s. A bar that acts as a torsion spring connects the vibrating mass to a plate-and-frame membrane module, which then vibrates by 1–2 in. at the same frequency. By shaking the membrane module, high turbulence is induced in the pressurized feed fluid flowing through the module. The turbulence occurs directly at the membrane surface, providing good control of membrane fouling

Module Selection

The choice of the most suitable membrane module type for a particular membrane separation must balance a number of factors. The principal module design parameters that enter into the decision are summarized in Table 3.7.

Cost, always important, is difficult to quantify because the actual selling price of the same module design varies widely, depending on the application. Generally, high-pressure modules are more expensive than low-pressure or vacuum

Table 3.7 Parameters for membrane module design

Parameter	Hollow fine fibers	Capillary fibers	Spiral-wound	Plate-and-frame	Tubular
Manufacturing cost (US$/m^2)	5–20	10–50	5–100	50–200	50–200
Concentration polarization fouling control	Poor	Good	Moderate	Good	Very good
Permeate-side pressure drop	High	Moderate	Moderate	Low	Low
Suitability for high-pressure operation	Yes	No	Yes	Yes	Marginal
Limited to specific types of membrane material	Yes	Yes	No	No	No

modules. The total volume of product likely to be produced to satisfy a particular application is a key issue. For example, spiral-wound modules for reverse osmosis are produced by three or four manufacturers in large volumes, resulting in severe competition and low prices. Similar modules used in ultrafiltration are produced in much lower numbers and so are much more expensive. Hollow fiber modules are significantly cheaper, per square meter of membrane, than spiral-wound or plate-and-frame modules but can only be economically produced for very high volume applications that justify the expense of developing and building the spinning and module fabrication equipment. This cost advantage is often offset by the lower fluxes of the membranes compared with their flat-sheet equivalents. An estimate of module manufacturing cost is given in Table 3.7; the selling price is typically two to five times higher.

Two other major factors determining module selection are concentration polarization control and resistance to fouling. Concentration polarization control is a particularly important issue in liquid separations such as reverse osmosis and ultrafiltration. In gas separation applications, concentration polarization is more easily controlled but is still a problem with high-flux, highly selective membranes. Hollow fine fiber modules are notoriously prone to fouling and concentration polarization and can be used in reverse osmosis applications only when extensive, costly feed solution pretreatment removes all particulates. These fibers cannot be used in ultrafiltration applications at all.

Another factor is the ease with which various membrane materials can be fabricated into a particular module design. Almost all membranes can be formed into plate-and-frame, spiral-wound and tubular modules, but many membrane materials cannot be fabricated into hollow fine fibers or capillary fibers. Finally, the suitability of the module design for high-pressure operation and the relative magnitude of pressure drops on the feed and permeate sides of the membrane can be important factors.

The types of modules generally used in some of the major membrane processes are listed in Table 3.8.

In reverse osmosis, the commonly used modules are spiral-wound. Plate-and-frame and tubular modules are limited to a few applications in which membrane fouling is particularly severe, for example, in food applications or processing heavily contaminated industrial wastewater. The hollow fiber reverse osmosis modules used in the past have now been almost completely displaced by spiral-wound modules, which are inherently more fouling resistant, and require less feed pretreatment.

For ultrafiltration applications, hollow fine fibers have never been seriously considered because of their susceptibility to fouling. If the feed solution is extremely fouling, tubular systems are still used. Recently, however, spiral-wound modules with improved resistance to fouling have been developed; these modules are increasingly displacing the more expensive tubular systems. This is particularly the case with clean feed solutions, for example, in the ultrafiltration of boiler feed water or municipal water to make ultrapure water for the electronics industry. Capillary systems are also used in some ultrafiltration applications.

Table 3.8 Module designs most commonly used in the major membrane separation processes

Application	Module type
Reverse osmosis: seawater	Spiral-wound modules. Only one hollow fiber producer remains
Reverse osmosis: industrial and brackish water	Spiral-wound modules used almost exclusively; fine fibers too susceptible to scaling and fouling
Ultrafiltration	Tubular, capillary and spiral-wound modules all used. Tubular generally limited to highly fouling feeds (automotive paint), spiral-wound to clean feeds (ultrapure water)
Gas separation	Hollow fibers for high volume applications with low flux, low selectivity membranes in which concentration polarization is easily controlled (nitrogen from air)
	Spiral-wound when fluxes are higher, feed gases more contaminated and concentration polarization a problem (natural gas separations, vapor permeation)
Pervaporation	Most pervaporation systems are small so plate-and-frame systems were used in the first systems
	Spiral-wound and capillary modules being introduced

For high-pressure gas separation applications, hollow fine fibers have a major segment of the market. Hollow fiber modules are clearly the lowest cost design per unit membrane area, and their poor resistance to fouling is not a problem in many gas separation applications because gaseous feed streams can easily be filtered. Also, gas separation membrane materials are often rigid glassy polymers such as polysulfones, polycarbonates and polyimides, which are easily formed into hollow fine fibers. Spiral-wound modules are used to process natural gas streams, which are relatively dirty, often containing oil mist and condensable components that would foul hollow fine fiber modules rapidly.

Spiral-wound modules are much more commonly used in low-pressure or vacuum gas separation applications, such as the production of oxygen-enriched air or the separation of organic vapors from air. In these applications, the feed gas is at close to ambient pressure, and a vacuum is drawn on the permeate side of the membrane. Parasitic pressure drops on the permeate side of the membrane and the difficulty in making high-performance hollow fine fiber membranes from the rubbery polymers used to make them both work against hollow fine fiber modules for such applications.

Pervaporation operates under constraints similar to those for low-pressure gas separation. Pressure drops on the permeate side of the membrane must be small, and many pervaporation membrane materials are rubbery, so both spiral-wound modules and plate-and-frame systems are in use. Plate-and-frame systems are competitive in this application despite their high cost, primarily because they can be operated at high temperatures with relatively aggressive feed solutions, conditions under which spiral-wound modules might fail.

Conclusions and Future Directions

The technology to fabricate ultrathin high-performance membranes into high-surface-area membrane modules has steadily improved during the modern membrane era. As a result the inflation-adjusted cost of membrane separation processes has decreased dramatically over the years. The first anisotropic membranes made by Loeb–Sourirajan processes had an effective thickness of 0.2–0.4 μm. Currently, various techniques are used to produce commercial membranes with a thickness of 0.1 μm or less. The permeability and selectivity of membrane materials have also increased two to three fold during the same period. As a result, today's membranes have 5 to 10 times the flux and better selectivity than membranes available 30 years ago. These trends are continuing. Membranes with an effective thickness of less than 0.05 μm have been made in the laboratory using advanced composite membrane preparation techniques or surface treatment methods.

As a result of these improvements in membrane performance, the major factors determining system performance have become concentration polarization and membrane fouling. All membrane processes are affected by these problems, so membrane modules with improved fluid flow to minimize concentration polarization and modules formed from membranes that can be easily cleaned if fouled are likely to become increasingly important development areas.

References

1. H.A. Gardner and G.G. Sward, *Physical and Chemical Examination of Paints, Varnishes, Lacquers, and Colors*, 11th Edn, H.A. Gardner Laboratory, Maryland (1950).
2. K.J. Mackenzie, Film and Sheeting Material, in *Encyclopedia of Chemical Technology*, 4th Edn, Vol. 10, p. 761 (1992).
3. R.L. Fleischer, H.W. Alter, S.C. Furman, P.B. Price and R.M. Walker, Particle Track Etching, *Science* **172**, 225 (1972).
4. M.C. Porter, A Novel Membrane Filter for the Laboratory, *Am. Lab.* November (1974).
5. J.E. Gingrich, The Nuclepore Story, *The 1988 Sixth Annual Membrane Technology/Planning Conference*, Cambridge, MA (1988).
6. H.S. Bierenbaum, R.B. Isaacson, M.L. Druin and S.G. Plovan, Microporous Polymeric Films, *Ind. Eng. Chem. Proc. Res. Dev.* **13**, 2 (1974).
7. R.W. Gore, Porous Products and Process Therefor, US Patent 4,187,390 (February, 1980).
8. J.J. Kim, T.S. Jang, Y.D. Kwon, U.Y. Kim and S.S. Kim, Structural Study of Microporous Polypropylene Hollow Fiber Membranes Made by the Melt-Spinning and Cold-Stretching Method, *J. Membr. Sci.* **93**, 209 (1994).
9. K. Okuyama and H. Mizutani, Process for Preparing Air-permeable Film, US Patent 4,585,604 (April, 1986).
10. Y. Mizutani, S. Nakamura, S. Kaneko and K. Okamura, Microporous Polypropylene Sheets, *Ind. Eng. Chem. Res.* **32**, 221 (1993).
11. C.C. Chau and J.-H. Im, Process of Making a Porous Membrane, US Patent 4,874, 568 (October, 1989).
12. T. Ichikawa, K. Takahara, K. Shimoda, Y. Seita and M. Emi, Hollow Fiber Membrane and Method for Manufacture Thereof, US Patent 4,708,800 (November, 1987).
13. G. Lopatin, L.Y. Yen and R.R. Rogers, Microporous Membranes from Polypropylene, US Patent 4,874,567 (October, 1989).
14. S. Loeb and S. Sourirajan, Sea Water Demineralization by Means of an Osmotic Membrane, in *Saline Water Conversion-II*, Advances in Chemistry Series Number 28, American Chemical Society, Washington, DC, pp. 117–132 (1963).
15. J.E. Cadotte, Evolution of Composite Reverse Osmosis Membranes, in *Materials Science of Synthetic Membranes*, D.R. Lloyd (ed.), ACS Symposium Series Number 269, American Chemical Society, Washington, DC, pp. 273–294 (1985).
16. W.J. Ward, III, W.R. Browall and R.M. Salemme, Ultrathin Silicone Rubber Membranes for Gas Separations, *J. Membr. Sci.* **1**, 99 (1976).
17. P.S. Francis, Fabrication and Evaluation of New Ultrathin Reverse Osmosis Membranes, *Offices of Saline Water Report*, NTIS# PB-177083 (February, 1966).
18. R.L. Riley, H.K. Lonsdale, C.R. Lyons and U. Merten, Preparation of Ultrathin Reverse Osmosis Membranes and the Attainment of Theoretical Salt Rejection, *J. Appl. Polym. Sci.* **11**, 2143 (1967).
19. H. Strathmann, K. Kock, P. Amar and R.W. Baker, The Formation Mechanism of Anisotropic Membranes, *Desalination* **16**, 179 (1975).
20. S. Manjikian, Desalination Membranes from Organic Casting Solutions, *Ind. Eng. Chem. Prod. Res. Dev.* **6**, 23 (1967).

21. C.W. Saltonstall, Jr, Development and Testing of High-retention Reverse-osmosis Membranes, *International Conference PURAQUA*, Rome, Italy (February, 1969).
22. A.S. Michaels, High Flow Membrane, US Patent 3,615,024 (October, 1971).
23. I. Pinnau and W.J. Koros, Defect Free Ultra High Flux Asymmetric Membranes, US Patent 4,902,422 (February, 1990).
24. H. Strathmann, P. Scheible and R.W. Baker, A Rationale for the Preparation of Loeb–Sourirajan-type Cellulose Acetate Membranes, *J. Appl. Polym. Sci.* **15**, 811 (1971).
25. M.T. So, F.R. Eirich, H. Strathmann and R.W. Baker, Preparation of Anisotropic Loeb–Sourirajan Membranes, *Polym. Lett.* **11**, 201 (1973).
26. H. Strathmann and K. Kock, The Formation Mechanism of Phase Inversion Membranes, *Desalination* **21**, 241 (1977).
27. J.G. Wijmans and C.A. Smolders, Preparation of Anisotropic Membranes by the Phase Inversion Process, in *Synthetic Membranes: Science, Engineering, and Applications*, P.M. Bungay, H.K. Lonsdale and M.N. de Pinho (eds), D. Reidel, Dordrecht, pp. 39–56 (1986).
28. F.W. Altena and C.A. Smolders, Calculation of Liquid–Liquid Phase Separation in a Ternary System of a Polymer in a Mixture of a Solvent and Nonsolvent, *Macromolecules* **15**, 1491 (1982).
29. A.J. Reuvers, J.W.A. Van den Berg and C.A. Smolders, Formation of Membranes by Means of Immersion Precipitation, *J. Membr. Sci.* **34**, 45 (1987).
30. M. Mulder, *Basic Principles of Membrane Technology*, Kluwer Academic Publishers, Dordrecht (1991).
31. I. Pinnau and W.J. Koros, A Qualitative Skin Layer Formation Mechanism for Membranes Made by Dry/Wet Phase Inversion, *J. Polym. Sci., Part B: Polym. Phys.* **31**, 419 (1993).
32. W.J. Koros and I. Pinnau, Membrane Formation for Gas Separation Processes, in *Polymeric Gas Separation Membranes*, D.R. Paul and Y.P. Yampol'skii (eds), CRC Press, Boca Raton, FL, pp. 209–272 (1994).
33. A.J. Castro, Methods for Making Microporous Products, US Patent 4,247,498 (January, 1981).
34. D.R. Lloyd, J.W. Barlow and K.E. Kinzer, Microporous Membrane Formation via Thermally-induced Phase, in *New Membrane Materials and Processes for Separation*, K.K. Sirkar and D.R. Lloyd (eds), AIChE Symposium Series 261, AIChE, New York, NY, p. 84 (1988).
35. H.J.C. Vadalia, H.K. Lee, A.S. Meyerson and K. Levon, Thermally Induced Phase Separations in Ternary Crystallizable Polymer Solutions, *J. Membr. Sci.* **89**, 37 (1994).
36. G.T. Caneba and D.S. Soong, Polymer Membrane Formation Through the Thermal-inversion Process, *Macromolecules* **18**, 2538 (1985).
37. W.C. Hiatt, G.H. Vitzthum, K.B. Wagener, K. Gerlach and C. Josefiak, Microporous Membranes via Upper Critical Temperature Phase Separation, in *Materials Science of Synthetic Membranes*, D.R. Lloyd (ed.), ACS Symposium Series Number 269, American Chemical Society, Washington, DC, p. 229 (1985).
38. D.R. Lloyd, S.S. Kim and K.E. Kinzer, Microporous Membrane Formation in Thermally Induced Phase Separation, *J. Membr. Sci.* **64**, 1 (1991).
39. H. Bechhold, Kolloidstudien mit der Filtationmethode, *Z. Physik. Chem.* **60**, 257 (1907).
40. W.J. Elford, Principles Governing the Preparation of Membranes Having Graded Porosities. The Properties of 'Gradocol' Membranes as Ultrafilters, *Trans. Faraday Soc.* **33**, 1094 (1937).

41. H.F. Pierce, Nitrocellulose Membranes of Graded Permeability, *J. Biol. Chem.* **75**, 795 (1927).
42. J.D. Ferry, Ultrafilter Membranes and Ultrafiltration, *Chem. Rev.* **18**, 373 (1936).
43. R. Zsigmondy and W. Bachmann, Uber Neue Filter, *Z. Anorg. Chem.* **103**, 119 (1918).
44. L. Zeman and T. Fraser, Formation of Air-cast Cellulose Acetate Membranes, *J. Membr. Sci.* **87**, 267 (1994).
45. R.L. Riley, H.K. Lonsdale, L.D. LaGrange and C.R. Lyons, Development of Ultrathin Membranes, *Office of Saline Water Research and Development Progress Report No. 386*, PB# 207036 (January, 1969).
46. L.J. Zeman and A.L. Zydney, *Microfiltration and Ultrafiltration*, Marcel Dekker, New York (1996).
47. L.T. Rozelle, J.E. Cadotte, K.E. Cobian and C.V. Kopp, Jr, Nonpolysaccharide Membranes for Reverse Osmosis: NS-100 Membranes, in *Reverse Osmosis and Synthetic Membranes*, S. Sourirajan (ed.), National Research Council Canada, Ottawa, Canada, pp. 249–262 (1977).
48. R.J. Petersen, Composite Reverse Osmosis and Nanofiltration Membranes, *J. Membr. Sci.* **83**, 81 (1993).
49. R.L. Riley, R.L. Fox, C.R. Lyons, C.E. Milstead, M.W. Seroy and M. Tagami, Spiral-wound Poly(ether/amide) Thin-film Composite Membrane System, *Desalination* **19**, 113 (1976).
50. Y. Kamiyama, N. Yoshioka, K. Matsui and E. Nakagome, New Thin-film Composite Reverse Osmosis Membranes and Spiral Wound Modules, *Desalination* **51**, 79 (1984).
51. R.E. Larson, J.E. Cadotte and R.J. Petersen, The FT-30 Seawater Reverse Osmosis Membrane-element Test Results, *Desalination* **38**, 473 (1981).
52. W.J. Ward, III, W.R. Browall and R.M. Salemme, Ultrathin Silicone Rubber Membranes for Gas Separations, *J. Membr. Sci.* **1**, 99 (1976).
53. R.H. Forester and P.S. Francis, Method of Producing an Ultrathin Polymer Film Laminate, US Patent 3,551,244 (December, 1970).
54. M. Haubs and W. Hassinger, Coated Fibers, US Patent 5,344,702 (September, 1994).
55. W.R. Browall, Method for Sealing Breaches in Multi-layer Ultrathin Membrane Composites, US Patent 3,980,456 (September, 1976).
56. R.L. Riley, H.K. Lonsdale and C.R. Lyons, Composite Membranes for Seawater Desalination by Reverse Osmosis, *J. Appl. Polym. Sci.* **15**, 1267 (1971).
57. F.-J. Tsai, D. Kang and M. Anand, Thin Film Composite Gas Separation Membranes: On the Dynamics of Thin Film Formation Mechanism on Porous Substrates, *Sep. Sci. Technol.* **30**, 1639 (1995).
58. I. Pinnau, Ultrathin Ethyl Cellulose/Poly(4-methyl pentene-1) Permselective Membranes, US Patent 4,871,378 (October 1989).
59. I. Pinnau, J.G. Wijmans, I. Blume, T. Kuroda and K.-V. Peinemann, Gas Separation Through Composite Membranes, *J. Membr. Sci.* **37**, 81 (1988).
60. H. Yasuda, Plasma Polymerization for Protective Coatings and Composite Membranes, *J. Membr. Sci.* **18**, 273 (1984).
61. H. Yasuda, Composite Reverse Osmosis Membranes Prepared by Plasma Polymerization, in *Reverse Osmosis and Synthetic Membranes*, S. Sourirajan (ed.), National Research Council Canada, Ottawa, Canada, pp. 263–294 (1977).
62. A.R. Stancell and A.T. Spencer, Composite Permselective Membrane by Deposition of an Ultrathin Coating from a Plasma, *J. Appl. Polym. Sci.* **16**, 1505 (1972).
63. M. Kawakami, Y. Yamashita, M. Iwamoto and S. Kagawa, Modification of Gas Permeabilities of Polymer Membranes by Plasma Coating, *J. Membr. Sci.* **19**, 249 (1984).

64. K.A. Kraus, A.J. Shor and J.S. Johnson, Hyperfiltration Studies, *Desalination* **2**, 243 (1967).
65. J.S. Johnson, K.A. Kraus, S.M. Fleming, H.D. Cochran and J.J. Perona, Hyperfiltration Studies, *Desalination* **5**, 359 (1968).
66. M. Langsam, Fluorinated Polymeric Membranes for Gas Separation Processes, US Patent 4,657,564 (April, 1987); M. Langsam and C.L. Savoca, Polytrialkylgermylpropyne Polymers and Membranes, US Patent 4,759,776 (July, 1988).
67. M. Langsam, M. Anand and E.J. Karwacki, Substituted Propyne Polymers I. Chemical Surface Modification of Poly[1-(trimethylsilyl)propyne] for Gas Separation Membranes, *Gas Sep. Purif.* **2**, 162 (1988).
68. J.M. Mohr, D.R. Paul, I. Pinnau and W.J. Koros, Surface Fluorination of Polysulfone Anisotropic Membranes and Films, *J. Membr. Sci.* **56**, 77 (1991).
69. J.M. Mohr, D.R. Paul, Y. Taru, T. Mlsna and R.J. Lagow, Surface Fluorination of Composite Membranes, *J. Membr. Sci.* **55**, 149 (1991).
70. M. Langsam, Fluorinated Polymeric Membranes for Gas Separation Processes, US Patent 4,657,564 (April, 1987).
71. J.D. Le Roux, D.R. Paul, M.F. Arendt, Y. Yuan and I. Cabasso, Surface Fluorination of Poly(phenylene oxide) Composite Membranes, *J. Membr. Sci.* **90**, 37 (1994).
72. P.W. Kramer, M.K. Murphy, D.J. Stookey, J.M.S. Henis and E.R. Stedronsky, Membranes Having Enhanced Selectivity and Methods of Producing Such Membranes, US Patent 5,215,554 (June, 1993).
73. J.M.S. Henis and M.K. Tripodi, A Novel Approach to Gas Separations Using Composite Hollow Fiber Membranes, *Sep. Sci. Technol.* **15**, 1059 (1980).
74. R.B. McBride and D.L. McKinley, A New Hydrogen Recovery Route, *Chem. Eng. Prog.* **61**, 81 (1965).
75. J.B. Hunter, Silver–Palladium Film for Separation and Purification of Hydrogen, US Patent 2,773,561 (December, 1956).
76. J.B. Hunter, A New Hydrogen Purification Process, *Platinum Met. Rev.* **4**, 130 (1960).
77. R.W. Baker, J. Louie, P.H. Pfromm and J.G. Wijmans, Ultrathin Composite Metal Membranes, US Patent 4,857,080 (August, 1989).
78. A.L. Athayde, R.W. Baker and P. Nguyen, Metal Composite Membranes for Hydrogen Separation, *J. Membr. Sci.* **94**, 299 (1994).
79. R.E. Buxbaum, Composite Metal Membrane for Hydrogen Extraction, US Patent 4,215,729 (June, 1993).
80. R.E. Buxbaum and T.L. Marker, Hydrogen Transport Through Non-porous Membranes of Palladium-coated Niobium, Tantalum, and Vanadium, *J. Membr. Sci.* **85**, 29 (1993).
81. D.J. Edlund and D.T. Friesen, Hydrogen-permeable Composite Metal Membrane and Uses Thereof, US Patent 5,217,506 (June, 1993).
82. D.J. Edlund, D. Friesen, B. Johnson and W. Pledger, Hydrogen-permeable Metal Membranes for High-temperature Gas Separations, *Gas Sep. Purif.* **8**, 131 (1994).
83. G. Alefeld and J. Völkl (eds), *Hydrogen in Metals—Basic Properties*, Springer-Verlag, Berlin (1978).
84. R.R. Bhave (ed.), *Inorganic Membranes: Synthesis Characterization and Applications*, Chapman Hall, New York (1991).
85. K. Keizer, R.J.R. Uhlhorn and T.J. Burggraaf, Gas Separation Using Inorganic Membranes, in *Membrane Separation Technology, Principles and Applications*, R.D. Nobel and S.A. Stern (eds), Elsevier, Amsterdam, pp. 553–584 (1995).
86. T.J. Burggraaf and K. Keizer, Synthesis of Inorganic Membranes, in *Inorganic Membranes Synthesis, Characteristics, and Applications*, R.R. Bhave (ed.), Chapmann Hall, New York, pp. 10–63 (1991).

87. A. Larbot, J.P. Fabre, C. Guizard and L. Cot, Inorganic Membranes Obtained by Sol-Gel Techniques, *J. Membr. Sci.* **39**, 203 (1988).
88. M.A. Anderson, M.J. Gieselmann and Q. Xu, Titania and Alumina Ceramic Membranes, *J. Membr. Sci.* **39**, 243 (1988).
89. R. Ash, R.M. Barrer and C.G. Pope, Flow of Adsorbable Gases and Vapours in a Microporous Medium, *Proc. R. Soc. London, Ser. A* **271**, 19 (1963).
90. R.M. Barrer and T. Gabor, Sorption and Diffusion of Simple Gases in Silica–Aluminum Cracking Catalyst, *Proc. R. Soc. London, Ser. A* **265**, 267 (1960).
91. J.E. Koresh and A. Soffer, Molecular Sieve Carbon Selective Membrane, *Sep. Sci. Technol.* **18**, 723 (1983).
92. J. Hayashi, H. Mizuta, M. Yamamoto, K. Kusakabe and S. Morooka, Pore Size Control of Carbonized BPDA-pp′ODA Polyimide Membrane by Chemical Vapor Deposition of Carbon, *J. Membr. Sci.* **124**, 243 (1997).
93. M.B. Rao and S. Sircar, Nanoporous Carbon Membranes for Separation of Gas Mixtures by Selective Surface Flow, *J. Membr. Sci.* **85**, 253 (1993).
94. R.P. Beaver, Method of Production Porous Hollow Silica-rich Fibers, US Patent 4,778,499 (October, 1988).
95. R.W. Baker and I. Blume, Coupled Transport Membranes, in *Handbook of Industrial Membrane Technology*, M.C. Porter (ed.), Noyes Publications, Park Ridge, NJ, pp. 511–558 (1990).
96. E.L. Cussler, Facilitated and Active Transport, in *Polymeric Gas Separation Membranes*, D.R. Paul and Y.P. Yampol'skii (eds), CRC Press, Boca Raton, FL, pp. 273–300 (1994).
97. H.I. Mahon, Permeability Separatory Apparatus, Permeability Separatory Membrane Element, Method of Making the Same and Process Utilizing the Same, US Patent 3,228,876 (January, 1966).
98. B. Baum, W. Holley, Jr and R.A. White, Hollow Fibres in Reverse Osmosis, Dialysis, and Ultrafiltration, in *Membrane Separation Processes*, P. Meares (ed.), Elsevier, Amsterdam, pp. 187–228 (1976).
99. I. Moch, Jr, Hollow Fiber Membranes, in *Encyclopedia of Chemical Technology*, 4th Edn, John Wiley-InterScience Publishing, New York, Vol. 13, p. 312 (1995).
100. S.A. McKelvey, D.T. Clausi and W.J. Koros, A Guide to Establishing Fiber Macroscopic Properties for Membrane Applications, *J. Membr. Sci.* **124**, 223 (1997).
101. T. Liu, D. Zhang, S. Xu and S. Sourirajan, Solution-spun Hollow Fiber Polysulfone and Polyethersulfone Ultrafiltration Membranes, *Sep. Sci. Technol.* **27**, 161 (1992).
102. S.H. Lee, J.J. Kim, S.S. Kim and U.Y. Kim, Morphology and Performance of a Polysulfone Hollow Fiber Membrane, *J. Appl. Polym. Sci.* **49**, 539 (1993).
103. R.F. Malon and C.A. Cruse, Anisotropic Gas Separation Membranes Having Improved Strength, US Patent 5,013,767 (May, 1991).
104. S. Takao, Process for Producing Acrylonitrile Separation Membranes in Fibrous Form, US Patent 4,409,162 (October, 1983).
105. P.S. Puri, Continuous Process for Making Coated Composite Hollow Fiber Membranes, US Patent 4,863,761 (September, 1989).
106. H.-D. Sluma, R. Weizenhofer, A. Leeb and K. Bauer, Method of Making a Multilayer Capillary Membrane, US Patent 5,242,636 (September, 1993).
107. M. Haubs and W. Hassinger, Method and Apparatus for Applying Polymeric Coating, US Patent 5,156,888 (October, 1992).
108. Y. Kusuki, T. Yoshinaga and H. Shimazaki, Aromatic Polyimide Double Layered Hollow Filamentary Membrane and Process for Producing Same, US Patent 5,141, 642 (August, 1992).
109. C.V. Kopp, R.J.W. Streeton and P.S. Khoo, Extrusion Head for Forming Polymeric Hollow Fiber, US Patent 5,318,417 (June, 1994).

110. S.A. Stern, T.F. Sinclaire, P.J. Gareis, N.P. Vahldieck and P.H. Mohr, Helium Recovery by Permeation, *Ind. Eng. Chem.* **57**, 49 (1965).
111. R. Günther, B. Perschall, D. Reese and J. Hapke, Engineering for High Pressure Reverse Osmosis, *J. Membr. Sci.* **121**, 95 (1996).
112. J.C. Westmoreland, Spirally Wrapped Reverse Osmosis Membrane Cell, US Patent 3,367,504 (February, 1968).
113. D.T. Bray, Reverse Osmosis Purification Apparatus, US Patent 3,417,870 (December, 1968).
114. S.S. Kremen, Technology and Engineering of ROGA Spiral-wound Reverse Osmosis Membrane Modules, in *Reverse Osmosis and Synthetic Membranes*, S. Sourirajan (ed.), National Research Council Canada, Ottawa, Canada, pp. 371–386 (1977).
115. B.S. Parekh (ed.), *Reverse Osmosis Technology*, Marcel Dekker, New York (1988).
116. R.O. Crowder and E.L. Cussler, Mass Transfer in Hollow-fiber Modules with Nonuniform Fibers, *J. Membr. Sci.* **134**, 235 (1997).
117. J. Lemanski and G.G. Lipscomb, Effect of Fiber Variation on the Performance of Counter-current Hollow-fiber Gas Separation Modules, *J. Membr. Sci.* **167**, 241 (2000).
118. J. Lemanski and G.G. Lipscomb, Effect of Shell-side Flows on the Performance of Hollow-fiber Gas Separation Modules, *J. Membr. Sci.* **195**, 215 (2002).
119. R. Prasad, C.J. Runkle and H.F. Shuey, Spiral-wound Hollow Fiber Cartridge and Modules Having Flow Directing Baffles, US Patent 5,352,361 (October, 1994).
120. R.C. Schucker, C.P. Darnell and M.M. Hafez, Hollow Fiber Module Using Fluid Flow Control Baffles, US Patent 5,169,530 (December, 1992).
121. T.J. Eckman, Hollow Fiber Cartridge, US Patent 5,470,469 (November, 1995).
122. R.P. de Filippi and R.W. Pierce, Membrane Device and Method, US Patent 3,536, 611 (October, 1970).
123. W.J. Koros and G.K. Fleming, Membrane Based Gas Separation, *J. Membr. Sci.* **83**, 1 (1993).
124. P.R. McGinnis and G.J. O'Brien, Permeation Separation Element, US Patent 3,690, 465 (September, 1972).
125. B. Culkin, A. Plotkin and M. Monroe, Solve Membrane Fouling with High-shear Filtration, *Chem. Eng. Prog.* **94**, 29 (1998).
126. O. Al Akoum, M.Y. Jaffrin, L. Ding, P. Paullier and C. Vanhoutte, An Hydrodynamic Investigation of Microfiltration and Ultrafiltration in a Vibrating Membrane Module, *J. Membr. Sci.* **197**, 37 (2002).

4 CONCENTRATION POLARIZATION

Introduction

In membrane separation processes, a gas or liquid mixture contacts the feed side of the membrane, and a permeate enriched in one of the components of the mixture is withdrawn from the downstream side of the membrane. Because the feed mixture components permeate at different rates, concentration gradients form in the fluids on both sides of the membrane. The phenomenon is called concentration polarization. Figure 4.1 illustrates a dialysis experiment in which a membrane separates two solutions containing different concentrations of dissolved solute. Solute (i) diffuses from right to left; solvent (j) diffuses from left to right. Unless the solutions are extremely well stirred, concentration gradients form in the solutions on either side of the membrane. The same phenomenon occurs in other processes that involve transport of heat or mass across an interface. Mathematical descriptions of these processes can be found in monographs on heat and mass transfer, for example, the books by Carslaw and Jaeger [1], Bird *et al.* [2] and Crank [3].

The layer of solution immediately adjacent to the membrane surface becomes depleted in the permeating solute on the feed side of the membrane and enriched in this component on the permeate side. Equivalent gradients also form for the other component. This concentration polarization reduces the permeating component's concentration difference across the membrane, thereby lowering its flux and the membrane selectivity. The importance of concentration polarization depends on the membrane separation process. Concentration polarization can significantly affect membrane performance in reverse osmosis, but it is usually well controlled in industrial systems. On the other hand, membrane performance in ultrafiltration, electrodialysis, and some pervaporation processes is seriously affected by concentration polarization.

Figure 4.1 also shows the formation of concentration polarization gradients on both sides of the membrane. However, in most membrane processes there is a bulk flow of liquid or gas through the membrane, and the permeate-side composition

Membrane Technology and Applications R. W. Baker
 ISBN: 0-470-85445-6

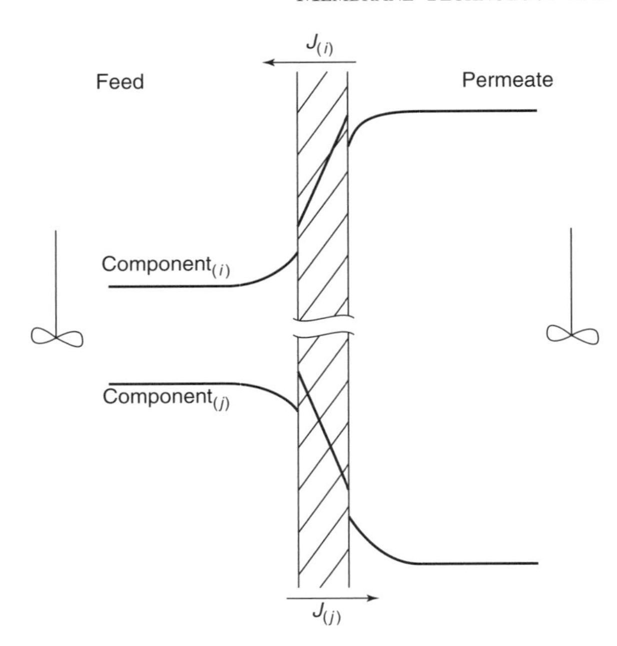

Figure 4.1 Concentration gradients formed when a dialysis membrane separates two solutions of different concentrations

depends only on the ratio of the components permeating the membrane. When this is the case, concentration gradients only form on the feed side of the membrane.

Two approaches have been used to describe the effect of concentration polarization. One has its origins in the dimensional analysis used to solve heat transfer problems. In this approach the resistance to permeation across the membrane and the resistance in the fluid layers adjacent to the membrane are treated as resistances in series. Nothing is assumed about the thickness of the various layers or the transport mechanisms taking place.

Using this model and the assumption that concentration polarization occurs only on the feed side of the membrane, the flux J_i across the combined resistances of the feed side boundary layer and the membrane can be written as

$$J_i = k_{ov}(c_{i_b} - c_{i_p}) \tag{4.1}$$

where k_{ov} is the overall mass transfer coefficient, c_{i_b} is the concentration of component i in the bulk feed solution, and c_{i_p} is the concentration of component i in the bulk permeate solution. Likewise, the flux across the boundary layer is also J_i and can be written as

$$J_i = k_{b\ell}(c_{i_b} - c_{i_o}) \tag{4.2}$$

where $k_{b\ell}$ is the fluid boundary layer mass transfer coefficient, and c_{i_o} is the concentration of component i in the fluid at the feed/membrane interface, and the flux across the membrane can be written as

$$J_i = k_m(c_{i_o} - c_{i_p}) \tag{4.3}$$

where k_m is the mass transfer coefficient of the membrane.

Since the overall concentration drop $(c_{i_b} - c_{i_p})$ is the sum of the concentration drops across the boundary layer and the membrane, a simple restatement of the resistances-in-series model using the terms of Equations (4.1–4.3) is

$$\frac{1}{k_{ov}} = \frac{1}{k_m} + \frac{1}{k_{b\ell}} \tag{4.4}$$

When the fluid layer mass-transfer coefficient ($k_{b\ell}$) is large, the resistance $1/k_{b\ell}$ of this layer is small, and the overall resistance is determined only by the membrane. When the fluid layer mass-transfer coefficient is small, the resistance term $1/k_{b\ell}$ is large, and becomes a significant fraction of the total resistance to permeation. The overall mass transfer coefficient (k_{ov}) then becomes smaller, and the flux decreases. The boundary layer mass transfer coefficient is thus an arithmetical fix used to correct the membrane permeation rate for the effect of concentration polarization. Nothing is revealed about the causes of concentration polarization.

The boundary layer mass-transfer coefficient is known from experiment to depend on many system properties; this dependence can be expressed as an empirical relationship of the type

$$k_{b\ell} = \text{constant } Q^{\alpha} h^{\beta} D^{\gamma} T^{\delta} \ldots.. \tag{4.5}$$

where, for example, Q is the fluid velocity through the membrane module, h is the feed channel height, D is the solute diffusion coefficient, T is the feed solution temperature, and so on. Empirical mass-transfer correlations obtained this way can be used to estimate the performance of a new membrane unit by extrapolation from an existing body of experimental data [4–7]. However, these correlations have a limited range of applicability and cannot be used to obtain a priori estimates of the magnitude of concentration polarization. This approach also does not provide insight into the dependence of concentration polarization on membrane properties. A more detailed and more sympathetic description of the mass-transfer approach is given in Cussler's monograph [8].

The second approach to concentration polarization, and the one used in this chapter, is to model the phenomenon by assuming that a thin layer of unmixed fluid, thickness δ, exists between the membrane surface and the well-mixed bulk solution. The concentration gradients that control concentration polarization form in this layer. This boundary layer film model oversimplifies the fluid hydrodynamics occurring in membrane modules and still contains one adjustable parameter, the boundary layer thickness. Nonetheless this simple model can explain most of the experimental data.

Boundary Layer Film Model

The usual starting point for the boundary layer film model is illustrated in Figure 4.2, which shows the velocity profile in a fluid flowing through the channel of a membrane module. The average velocity of the fluid flowing down the channel is normally of the order 1–5 m/s. This velocity is far higher than the average velocity of the fluid flowing at right angles through the membrane, which is typically 10–20 μm/s. However, the velocity in the channel is not uniform. Friction at the fluid–membrane surface reduces the fluid velocity next to the membrane to essentially zero; the velocity increases as the distance from the membrane surface increases. Thus, the fluid flow velocity in the middle of the channel is high, the flow there is often turbulent, and the fluid is well mixed. The velocity in the boundary layer next to the membrane is much lower, flow is laminar, and mixing occurs by diffusion. Concentration gradients due to concentration polarization are assumed to be confined to the boundary layer.

Figure 4.1 shows the concentration gradients that form on either side of a dialysis membrane. However, dialysis differs from most membrane processes in that the volume flow across the membrane is usually small. In processes such as reverse osmosis, ultrafiltration, and gas separation, the volume flow through the membrane from the feed to the permeate side is significant. As a result the permeate concentration is typically determined by the ratio of the fluxes of the components that permeate the membrane. In these processes concentration polarization gradients form only on the feed side of the membrane, as shown in Figure 4.3. This simplifies the description of the phenomenon. The few membrane processes in which a fluid is used to sweep the permeate side of the membrane,

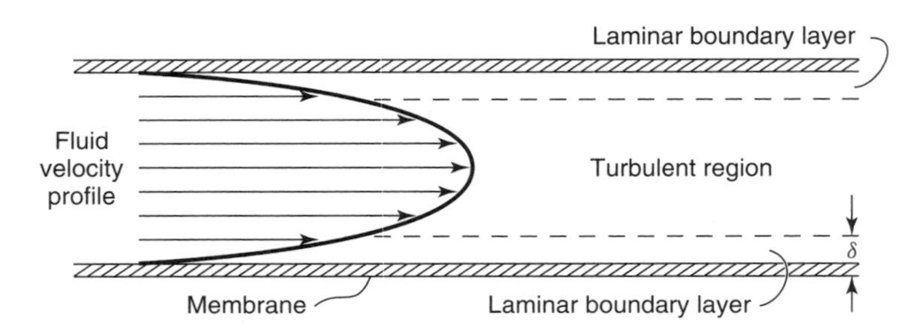

Figure 4.2 Fluid flow velocity through the channel of a membrane module is nonuniform, being fastest in the middle and essentially zero adjacent to the membrane. In the film model of concentration polarization, concentration gradients formed due to transport through the membrane are assumed to be confined to the laminar boundary layer

to change the permeate-side concentration from the value set by the ratio of permeating components, are discussed in the final section of this chapter.

In any process, if one component is enriched at the membrane surface, then mass balance dictates that a second component is depleted at the surface. By convention, concentration polarization effects are described by considering the concentration gradient of the minor component. In Figure 4.3(a), concentration polarization in reverse osmosis is represented by the concentration gradient of salt, the minor component rejected by the membrane. In Figure 4.3(b), which illustrates dehydration of aqueous ethanol solutions by pervaporation, concentration polarization is represented by the concentration gradient of water, the minor component that preferentially permeates the membrane.

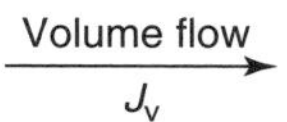

(a) Component enriched at membrane surface
(for example, salt in desalination of water by reverse osmosis)

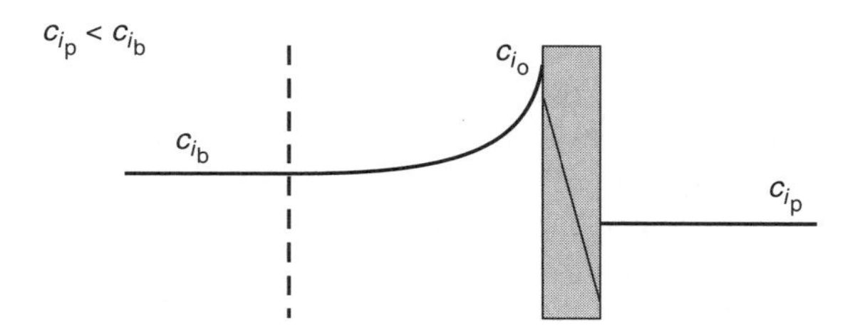

(b) Component depleted at membrane surface
(for example, water in dehydration of ethanol by pervaporation)

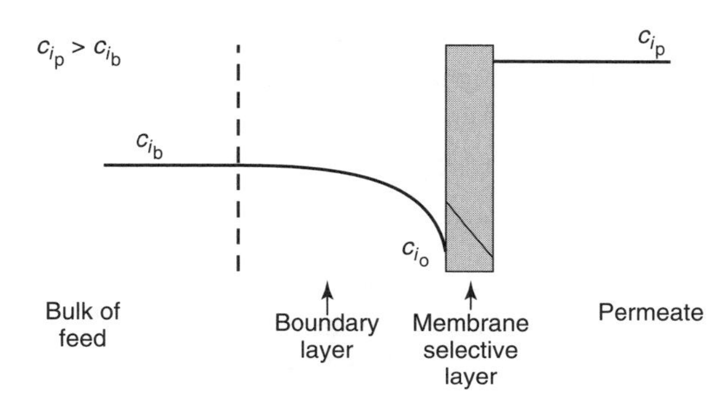

Figure 4.3 Concentration gradients formed as a result of permeation through a selective membrane. By convention, concentration polarization is usually represented by the gradient of the minor component—salt in the reverse osmosis example and water in the pervaporation example (dehydration of an ethanol solution)

In the case of desalination of water by reverse osmosis illustrated in Figure 4.3(a), the salt concentration c_{i_o} adjacent to the membrane surface is higher than the bulk solution concentration c_{i_b} because reverse osmosis membranes preferentially permeate water and retain salt. Water and salt are brought toward the membrane surface by the flow of solution through the membrane J_v.[1] Water and a little salt permeate the membrane, but most of the salt is rejected by the membrane and retained at the membrane surface. Salt accumulates at the membrane surface until a sufficient gradient has formed to allow the salt to diffuse to the bulk solution. Steady state is then reached.

In the case of dehydration of ethanol by pervaporation illustrated in Figure 4.3(b), the water concentration c_{i_o} adjacent to the membrane surface is lower than the bulk solution concentration c_{i_b} because the pervaporation membrane preferentially permeates water and retains ethanol. Water and ethanol are brought towards the membrane surface by the flow of solution through the membrane. Water and a little ethanol permeate the membrane, but most of the ethanol is retained at the membrane surface. Ethanol accumulates at the membrane surface until a sufficient gradient has formed to allow it to diffuse back to the bulk solution. An equal and opposite water gradient must form; thus, water becomes depleted at the membrane surface.

The formation of these concentration gradients can be expressed in mathematical form. Figure 4.4 shows the steady-state salt gradient that forms across a reverse osmosis membrane.

The salt flux through the membrane is given by the product of the permeate volume flux J_v and the permeate salt concentration c_{i_p}. For dilute liquids the permeate volume flux is within 1 or 2 % of the volume flux on the feed side of the membrane because the densities of the two solutions are almost equal. This means that, at steady state, the net salt flux at any point within the boundary layer must also be equal to the permeate salt flux $J_v c_{i_p}$. In the boundary layer this net salt flux is also equal to the convective salt flux towards the membrane $J_v c_i$ minus the diffusive salt flux away from the membrane expressed by Fick's law $(D_i \mathrm{d}c_i/\mathrm{d}x)$. So, from simple mass balance, transport of salt at any point within the boundary layer can be described by the equation

$$J_v c_i - D_i \mathrm{d}c_i/\mathrm{d}x = J_v c_{i_p} \tag{4.6}$$

where D_i is the diffusion coefficient of the salt, x is the coordinate perpendicular to the membrane surface, and J_v is the volume flux in the boundary layer generated by permeate flow through the membrane. The mass balance equation (4.6) can be integrated over the thickness of the boundary layer to give the well-known polarization equation first derived by Brian [9] for reverse osmosis:

$$\frac{c_{i_o} - c_{i_p}}{c_{i_b} - c_{i_p}} = \exp(J_v \delta / D_i) \tag{4.7}$$

[1] In this chapter, the term J_v is the volume flux ($\mathrm{cm^3/cm^2 \cdot s}$) through the membrane measured at the feed-side conditions of the process.

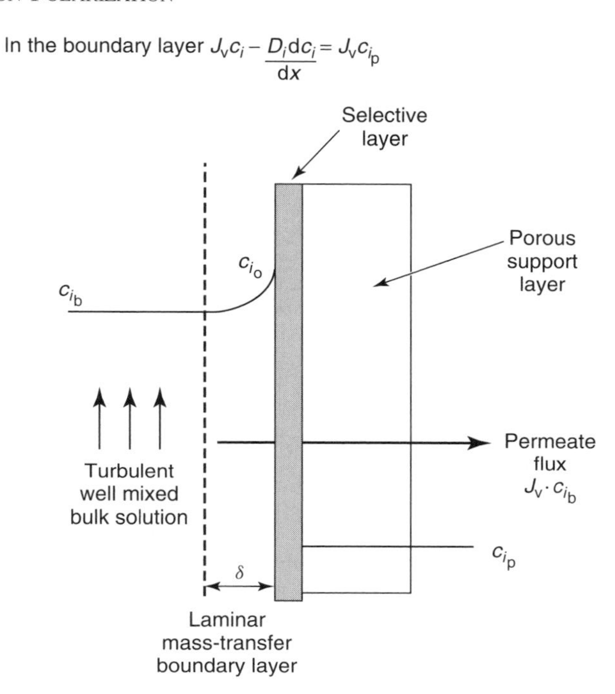

Figure 4.4 Salt concentration gradients adjacent to a reverse osmosis desalination membrane. The mass balance equation for solute flux across the boundary layer is the basis of the film model description of concentration polarization

In this equation, c_{i_o} is the concentration of solute in the feed solution at the membrane surface, and δ is the thickness of the boundary layer. An alternative form of Equation (4.7) replaces the concentration terms by an enrichment factor E, defined as c_{i_p}/c_{i_b}. The enrichment obtained in the absence of a boundary layer E_o is then defined as c_{i_p}/c_{i_o}, and Equation (4.7) can be written as

$$\frac{1/E_o - 1}{1/E - 1} = \exp(J_v \delta / D_i) \tag{4.8}$$

In the case of reverse osmosis, the enrichment factors (E and E_o) are less than 1.0, typically about 0.01, because the membrane rejects salt and permeates water. For other processes, such as dehydration of aqueous ethanol by pervaporation, the enrichment factor for water will be greater than 1.0 because the membrane selectively permeates the water.

The increase or decrease of the permeate concentration at the membrane surface c_{i_o}, compared to the bulk solution concentration c_{i_b}, determines the extent of concentration polarization. The ratio of the two concentrations, c_{i_o}/c_{i_b} is called the concentration polarization modulus and is a useful measure of the extent of concentration polarization. When the modulus is 1.0, no concentration polarization occurs, but as the modulus deviates farther from 1.0, the effect of concentration polarization on membrane selectivity and flux becomes increasingly important. From the definitions of E and E_o, the concentration polarization modulus is equal to E/E_o and, from Equations (4.7) and (4.8), the modulus can be written as

$$\frac{c_{i_o}}{c_{i_b}} = \frac{\exp(J_v\delta/D_i)}{1 + E_o[\exp(J_v\delta/D_i) - 1]} \tag{4.9}$$

Depending on the enrichment term (E_o) of the membrane, the modulus can be larger or smaller than 1.0. For reverse osmosis E_o is less than 1.0, and the concentration polarization modulus is normally between 1.1 and 1.5; that is, the concentration of salt at the membrane surface is 1.1 to 1.5 times larger than it would be in the absence of concentration polarization. The salt leakage through the membrane and the osmotic pressure that must be overcome to produce a flow of water are increased proportionately. Fortunately, modern reverse osmosis membranes are extremely selective and permeable, and can still produce useful desalted water under these conditions. In other membrane processes, such as pervaporation or ultrafiltration, the concentration polarization modulus may be as large as 5 to 10 or as small as 0.2 to 0.1, and may seriously affect the performance of the membrane.

Equation (4.9) shows the factors that determine the magnitude of concentration polarization, namely the boundary layer thickness δ, the membrane enrichment E_o, the volume flux through the membrane J_v, and the diffusion coefficient of the solute in the boundary layer fluid D_i. The effect of changes in each of these parameters on the concentration gradients formed in the membrane boundary layer are illustrated graphically in Figure 4.5 and discussed briefly below.

Of the four factors that affect concentration polarization, the one most easily changed is the boundary layer thickness δ. As δ decreases, Equation (4.9) shows that the concentration polarization modulus becomes exponentially smaller. Thus, the most straightforward way of minimizing concentration polarization is to reduce the boundary layer thickness by increasing turbulent mixing at the membrane surface. Factors affecting turbulence in membrane modules are described in detail in the review of Belfort *et al.* [10]. The most direct technique to promote mixing is to increase the fluid flow velocity past the membrane surface. Therefore, most membrane modules operate at relatively high feed fluid velocities. Membrane spacers are also widely used to promote turbulence by disrupting fluid flow in the module channels, as shown in Figure 4.6 [11]. Pulsing the feed fluid flow through the membrane module is another technique [12]. However, the energy consumption of the pumps required and the pressure drops produced place a practical limit to the turbulence that can be obtained in a membrane module.

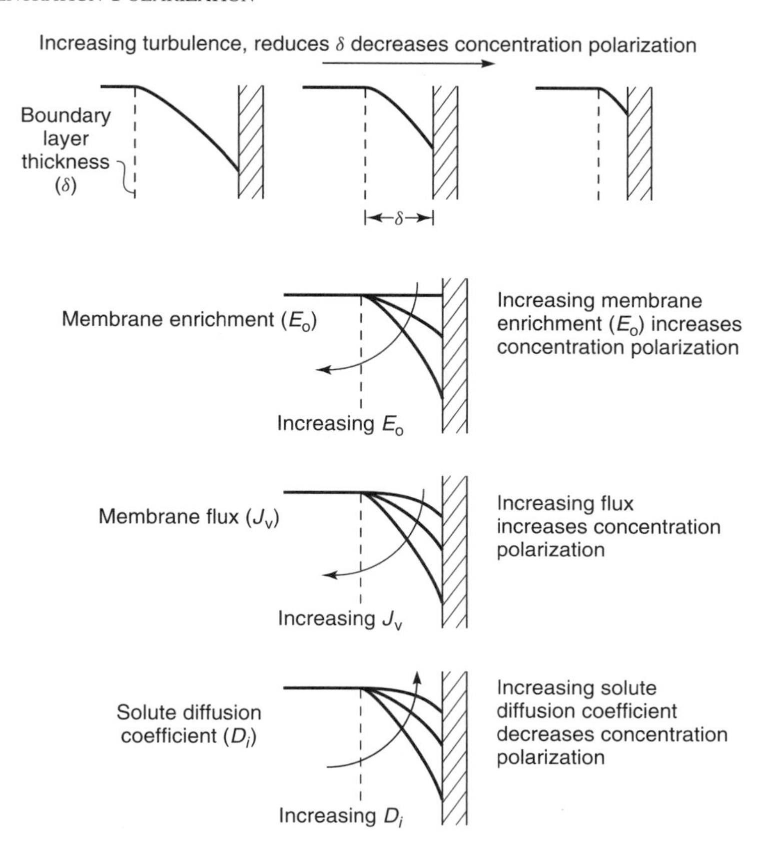

Figure 4.5 The effect of changes in boundary layer thickness δ, membrane enrichment E_o, membrane flux J_v, and solute diffusion D_i on concentration gradients in the stagnant boundary layer

The membrane's intrinsic enrichment E_o also affects concentration polarization. If the membrane is completely unselective, $E_o = 1$. The relative concentrations of the components passing through the membrane do not change, so concentration gradients are not formed in the boundary layer. As the difference in permeability between the more permeable and less permeable components increases, the intrinsic enrichment E_o achieved by the membrane increases, and the concentration gradients that form become larger. As a practical example, in pervaporation of organics from water, concentration polarization is much more important when the solute is toluene (with an enrichment E_o of 5000 over water) than when the solute is methanol (with an enrichment E_o less than 5).

Another important characteristic of Equation (4.9) is that the enrichment E_o produced by the membrane, not the intrinsic selectivity α, determines the

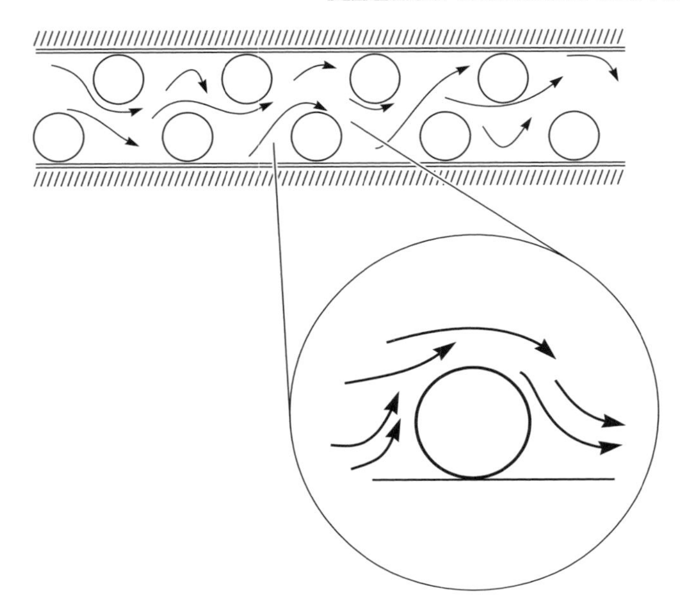

Figure 4.6 Flow dynamics around the spacer netting often used to promote turbulence in a membrane module and reduce concentration polarization

membrane separation performance and the concentration polarization modulus. Enrichment and intrinsic selectivity are linked but are not identical. This distinction is illustrated by the separation of hydrogen from inert gases in ammonia plant purge gas streams, which typically contain 30 % hydrogen. Hydrogen is 100 to 200 times more permeable than the inert gases nitrogen, methane, and argon, so the intrinsic selectivity of the membrane is very high. The high selectivity means that the membrane permeate is 97 % hydrogen; even so, the enrichment E_o is only 97/30, or 3.3, so the concentration polarization modulus is small. On the other hand, as hydrogen is removed, its concentration in the feed gas falls. When the feed gas contains 5 % hydrogen, the permeate will be 90 % hydrogen and the enrichment 90/5 or 18. Under these conditions, concentration polarization can affect the membrane performance.

Equation (4.9) shows that concentration polarization increases exponentially as the total volume flow J_v through the membrane increases. This is one of the reasons why modern spiral-wound reverse osmosis membrane modules are operated at low pressures. Modern membranes have two to five times the water permeability, at equivalent salt selectivities, of the first-generation cellulose acetate reverse osmosis membranes. If membrane modules containing these new membranes were operated at the same pressures as early cellulose acetate modules, two to five times the desalted water throughput could be achieved with the same number of modules. However, at such high fluxes, spiral-wound modules suffer from excessive concentration polarization, which leads to increased salt leakage and scale formation. For this reason, modern, high permeability modules are operated at about the same volume flux as the early modules, but at lower applied pressures. This reduces energy costs.

The final parameter in Equation (4.9) that determines the value of the concentration polarization modulus is the diffusion coefficient D_i of the solute away from the membrane surface. The size of the solute diffusion coefficient explains why concentration polarization is a greater factor in ultrafiltration than in reverse osmosis. Ultrafiltration membrane fluxes are usually higher than reverse osmosis fluxes, but the difference between the values of the diffusion coefficients of the retained solutes is more important. In reverse osmosis the solutes are dissolved salts, whereas in ultrafiltration the solutes are colloids and macromolecules. The diffusion coefficients of these high-molecular-weight components are about 100 times smaller than those of salts.

In Equation (4.9) the balance between convective transport and diffusive transport in the membrane boundary layer is characterized by the term $J_v\delta/D_i$. This dimensionless number represents the ratio of the convective transport J_v and diffusive transport D_i/δ and is commonly called the Peclet number. When the Peclet number is large ($J_v \gg D_i/\delta$), the convective flux through the membrane cannot easily be balanced by diffusion in the boundary layer, and the concentration polarization modulus is large. When the Peclet number is small ($J_v \ll D_i/\delta$), convection is easily balanced by diffusion in the boundary layer, and the concentration polarization modulus is close to unity.

Wijmans *et al.* [13] calculated the concentration polarization modulus using Equation (4.9) as a function of the Peclet number $J_v\delta/D_i$ that is, the varying ratio of convection to diffusion. The resulting, very informative plot is shown in Figure 4.7. This figure is divided into two regions depending on whether the concentration polarization modulus, c_{i_o}/c_{i_b}, is smaller or larger than 1. The polarization modulus is smaller than 1 when the permeating minor component is enriched in the permeate. In this case, the component becomes depleted in the boundary layer, for example, in the dehydration of ethanol by pervaporation shown in Figure 4.3(b). The polarization modulus is larger than 1 when the permeating minor component is depleted in the permeate. In this case, the component is enriched in the boundary layer, for example, in the reverse osmosis of salt solutions shown in Figure 4.3(a). As might be expected, the concentration polarization modulus deviates increasingly from unity as the Peclet number increases. At high values of the ratio $J_v\delta/D_i$, the exponential term in Equation (4.9) increases toward infinity, and the concentration polarization modulus c_{i_o}/c_{i_b} approaches a limiting value of $1/E_o$.

A striking feature of Figure 4.7 is its asymmetry with respect to enrichment and rejection of the minor component by the membrane. This means that, under comparable conditions, concentration polarization is much larger when the minor

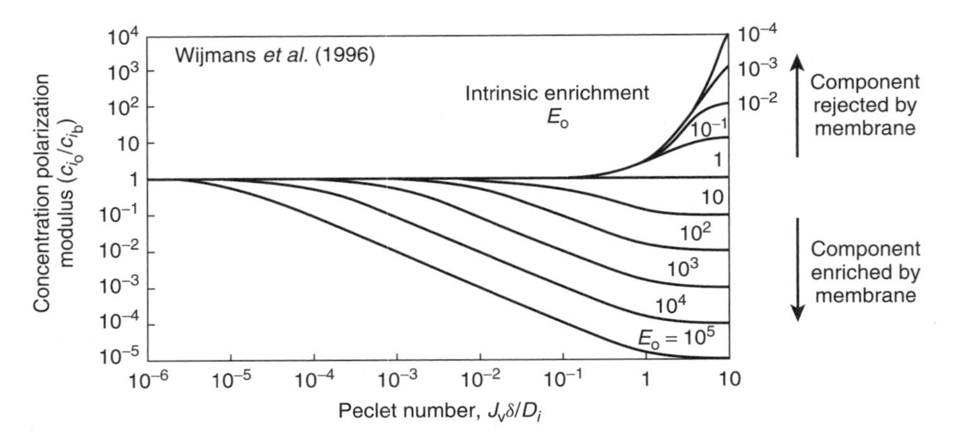

Figure 4.7 Concentration polarization modulus c_{i_o}/c_{i_b} as a function of the Peclet number $J_v\delta/D_i$ for a range of values of the intrinsic enrichment factor E_o. Lines calculated through Equation (4.9). This figure shows that components that are enriched by the membrane ($E_o > 1$) are affected more by concentration polarization than components that are rejected by the membrane ($E_o < 1$) [13]

component of the feed is preferentially permeated by the membrane than when it is rejected. This follows from the form of Equation (4.9). Consider the case when the Peclet number $J_v\delta/D_i$ is 1. The concentration polarization modulus expressed by Equation (4.9) then becomes

$$\frac{c_{i_o}}{c_{i_b}} = \frac{\exp(1)}{1 + E_o[\exp(1) - 1]} = \frac{2.72}{1 + E_o(1.72)} \tag{4.10}$$

For components rejected by the membrane ($E_o \leq 1$) the enrichment E_o produced by the membrane lies between 1 and 0. The concentration polarization modulus c_{i_o}/c_{i_b} then lies between 1 (no concentration polarization) and a maximum value of 2.72. That is, the flux of the less permeable component cannot be more than 2.72 times higher than that in the absence of concentration polarization. In contrast, for a component enriched by the membrane in the permeate ($E_o \geq 1$), no such limitation on the magnitude of concentration polarization exists. For dilute solutions (c_{i_b} small) and selective membranes, the intrinsic enrichment can be 100 to 1000 or more. The concentration polarization modulus can then change from 1 (no concentration polarization) to close to zero (complete concentration polarization). These two cases are illustrated in Figure 4.8.

Determination of the Peclet Number

Equation (4.9) and Figure 4.7 are powerful tools to analyze the importance of concentration polarization in membrane separation processes. However, before

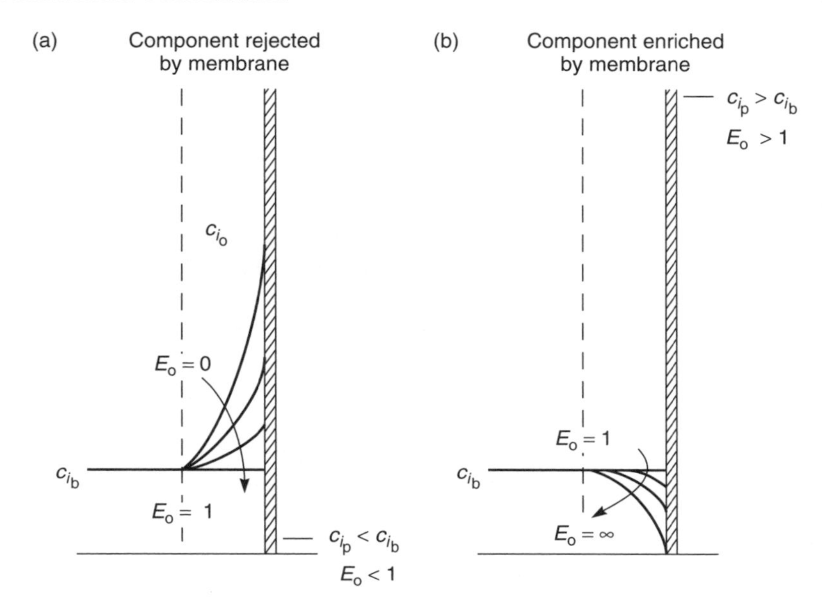

Figure 4.8 Concentration gradients that form adjacent to the membrane surface for components (a) rejected or (b) enriched by the membrane. The Peclet number, characterizing the balance between convection and diffusion in the boundary layer, is the same $J_v\delta/D_i = 1$. When the component is rejected, the concentration at the membrane surface c_{i_o} cannot be greater than $2.72\,c_{i_b}$, irrespective of the membrane selectivity. When the minor component permeates the membrane, the concentration at the membrane surface can decrease to close to zero, so the concentration polarization modulus becomes very small

these tools can be used, the appropriate value to be assigned to the Peclet number $J_v\delta/D_i$ must be determined. The volume flux J_v through the membrane is easily measured, so determining the Peclet number then becomes a problem of measuring the coefficient D_i/δ.

One approach to the boundary layer problem is to determine the ratio D_i/δ experimentally. This can be done using a procedure first proposed by Wilson [14]. The starting point for Wilson's approach is Equation (4.8), which can be written as

$$\ln\left(1 - \frac{1}{E}\right) = \ln\left(1 - \frac{1}{E_o}\right) - J_v\delta/D_i \tag{4.11}$$

The boundary layer thickness δ in Equation (4.11) is a function of the feed solution velocity u in the module feed flow channel; thus, the term δ/D_i can be expressed as

$$\frac{D_i}{\delta} = k_o u^n \tag{4.12}$$

where u is the superficial velocity in the feed flow channel and k_o and n are constants. Equation (4.11) can then be rewritten as

$$\ln\left(1 - \frac{1}{E}\right) = \ln\left(1 - \frac{1}{E_o}\right) - \frac{J_v}{k_o u^n} \tag{4.13}$$

Equation (4.13) can be used to calculate the dependence of pervaporation system performance on concentration polarization. One method is to use data obtained with a single module operated at various feed solution velocities. A linear regression analysis is used to fit data obtained at different feed velocities to obtain an estimate for k_o and E_o; the exponent n is adjusted to minimize the residual error. Figure 4.9 shows some data obtained in pervaporation experiments with dilute aqueous toluene solutions and silicone rubber membranes [15]. Toluene is considerably more permeable than water through these membranes. In Figure 4.9, when the data were regressed, the best value for n was 0.96. The values of E_o, the intrinsic enrichment of the membrane, and k_o obtained by regression analysis are 3600 and 7.1×10^{-4}, respectively. The boundary layer coefficient, D_i/δ is given by

$$\frac{D_i}{\delta} = 7.1 \times 10^{-4} u^{0.96} \tag{4.14}$$

where u is the superficial velocity in the module.

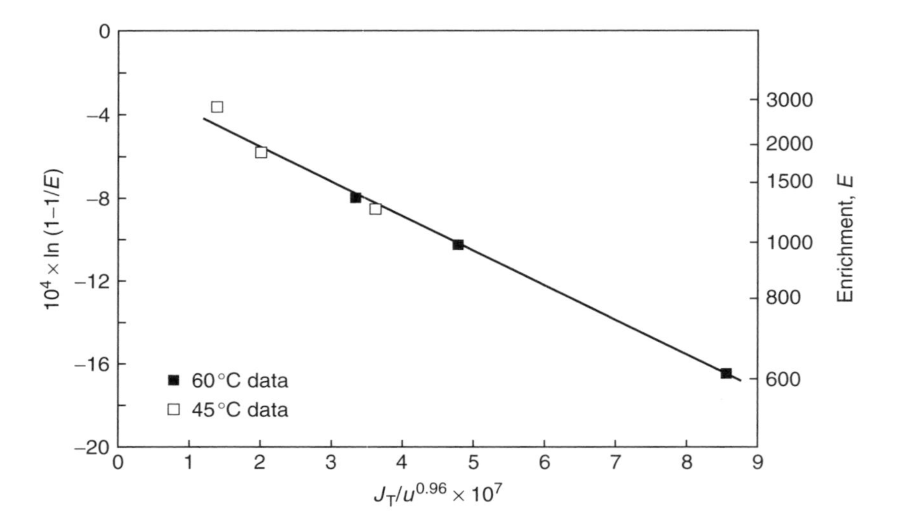

Figure 4.9 Derivation of the mass transfer coefficient by Wilson's method. Toluene/water enrichments are plotted as a function of feed solution superficial velocity in pervaporation experiments. Enrichments were measured at different feed solution superficial velocities with spiral-wound membrane modules [15]

A second method of determining the coefficient (D_i/δ) and the intrinsic enrichment of the membrane E_o is to use Equation (4.11). The term $\ln(1 - 1/E)$ is plotted against the permeate flux measured at constant feed solution flow rates but different permeate pressures or feed solution temperatures. This type of plot is shown in Figure 4.10 for data obtained with aqueous trichloroethane solutions in pervaporation experiments with silicone rubber membranes.

The coefficients D_i/δ obtained at each velocity in Figure 4.10 can then be plotted as a function of the feed superficial velocity. The data show that the ratio D_i/δ varies with the superficial velocity according to the equation

$$D_i/\delta = 9 \times 10^{-4} u^{0.8} \tag{4.15}$$

From Equations (4.14) and (4.15), the value of the term D_i/δ at a fluid velocity of 30 cm/s is $1.6–1.8 \times 10^{-2}$ cm/s. Based on a trichloroethane diffusion coefficient in the boundary layer of 2×10^{-5} cm^2/s, this yields a boundary layer thickness of 10–15 μm. This boundary layer thickness is in the same range as values calculated for reverse osmosis with similar modules.

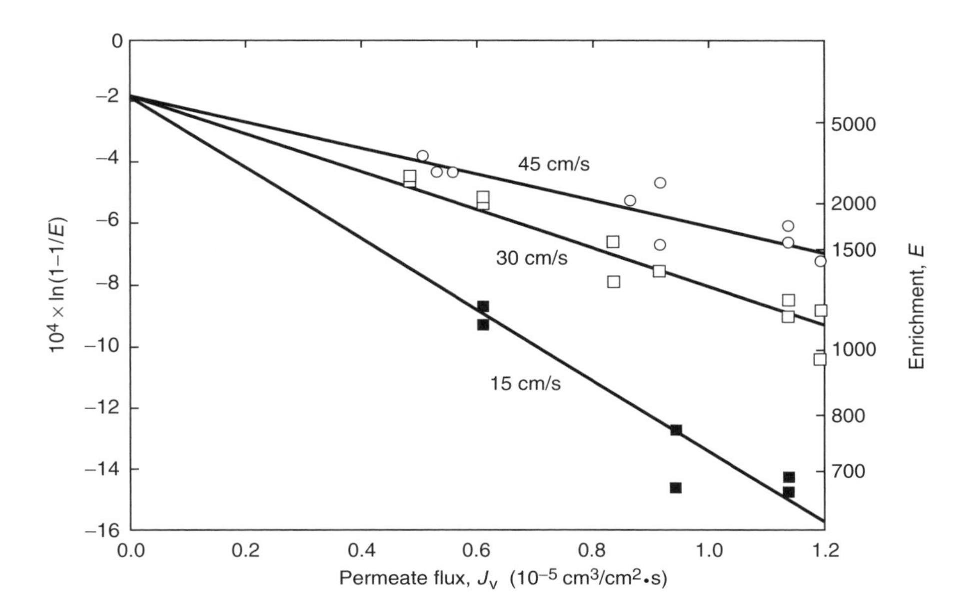

Figure 4.10 Trichloroethane enrichment $[\ln(1 - 1/E)]$ as a function of permeate flux J_v in pervaporation experiments with silicone rubber membranes in spiral-wound modules using solutions of 100 ppm trichloromethane in water [15]. Feed solution flow rates are shown

Concentration Polarization in Liquid Separation Processes

The effect of concentration polarization on specific membrane processes is discussed in the individual application chapters. However, a brief comparison of the magnitude of concentration polarization is given in Table 4.1 for processes involving liquid feed solutions. The key simplifying assumption is that the boundary layer thickness is 20 μm for all processes. This boundary layer thickness is typical of values calculated for separation of solutions with spiral-wound modules in reverse osmosis, pervaporation, and ultrafiltration. Tubular, plate-and-frame, and bore-side feed hollow fiber modules, because of their better flow velocities, generally have lower calculated boundary layer thicknesses. Hollow fiber modules with shell-side feed generally have larger calculated boundary layer thicknesses because of their poor fluid flow patterns.

Table 4.1 shows typical enrichments and calculated Peclet numbers for membrane processes with liquid feeds. In this table it is important to recognize the difference between enrichment and separation factor. The enrichments shown are calculated for the minor component. For example, in the dehydration of ethanol, a typical feed solution of 96 % ethanol and 4 % water yields a permeate containing about 80 % water; the enrichment, that is, the ratio of the permeate to feed concentration, is about 20. In Figure 4.11, the calculated Peclet numbers and enrichments shown in Table 4.1 are plotted on the Wijmans graph to show the relative importance of concentration polarization for the processes listed.

Table 4.1 Representative values of the concentration polarization modulus calculated for a variety of liquid separation processes. For these calculations a boundary layer thickness of 20 μm, typical of that in most spiral-wound membrane modules, is assumed

Process	Typical enrichment, E_o	Typical flux [in engineering units and as J_v (10^{-3} cm/s)]	Diffusion coefficient (10^{-6} cm^2/s)	Peclet number, $J_v\delta/D_i$	Concentration polarization modulus [Equation (4.9)]
Reverse osmosis					
Seawater desalination	0.01	30 gal/ft^2 · day(1.4)	10	0.28	1.3
Brackish water desalination	0.01	50 gal/ft^2 · day(2.3)	10	0.46	1.5
Ultrafiltration					
Protein separation	0.01	30 gal/ft^2 · day(1.4)	0.5	5.6	70
Pervaporation					
Ethanol dehydration	20	0.1 kg/m^2 · h(0.003)	20	0.0003	1.0
VOC from water	2000	1.0 kg/m^2 · h(0.03)	20	0.003	0.14
Coupled transport					
Copper from water	1000	60 mg/cm^2 · min(0.001)	10	0.0002	0.8

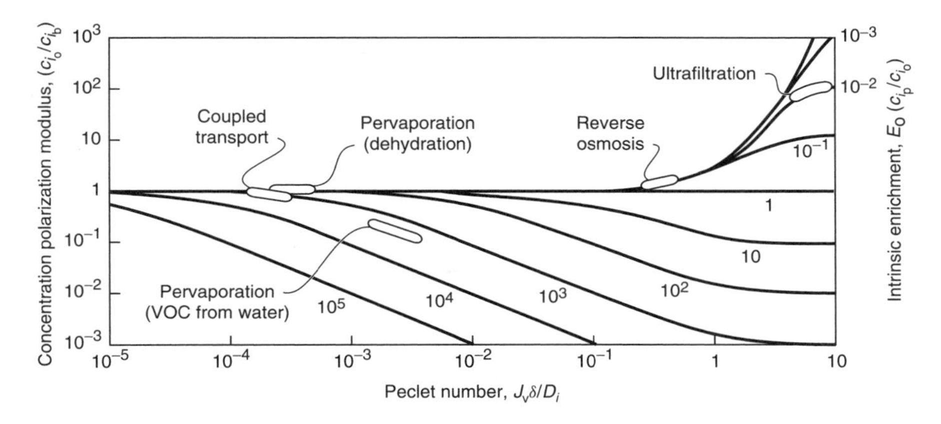

Figure 4.11 Peclet numbers and intrinsic enrichments for the membrane separation processes shown in Table 4.1 superimposed on the concentration polarization plot of Wijmans *et al.* [13]

In coupled transport and solvent dehydration by pervaporation, concentration polarization effects are generally modest and controllable, with a concentration polarization modulus of 1.5 or less. In reverse osmosis, the Peclet number of 0.3–0.5 was calculated on the basis of typical fluxes of current reverse osmosis membrane modules, which are 30- to 50-gal/ft^2 · day. Concentration polarization modulus values in this range are between 1.0 and 1.5.

Figure 4.11 shows that ultrafiltration and pervaporation for the removal of organic solutes from water are both seriously affected by concentration polarization. In ultrafiltration, the low diffusion coefficient of macromolecules produces a concentration of retained solutes 70 times the bulk solution volume at the membrane surface. At these high concentrations, macromolecules precipitate, forming a gel layer at the membrane surface and reducing flux. The effect of this gel layer on ultrafiltration membrane performance is discussed in Chapter 6.

In the case of pervaporation of dissolved volatile organic compounds (VOCs) from water, the magnitude of the concentration polarization effect is a function of the enrichment factor. The selectivity of pervaporation membranes to different VOCs varies widely, so the intrinsic enrichment and the magnitude of concentration polarization effects depend strongly on the solute. Table 4.2 shows experimentally measured enrichment values for a series of dilute VOC solutions treated with silicone rubber membranes in spiral-wound modules [15]. When these values are superimposed on the Wijmans plot as shown in Figure 4.12, the concentration polarization modulus varies from 1.0, that is, no concentration polarization, for isopropanol, to 0.1 for trichloroethane, which has an enrichment of 5700.

Table 4.2 Enrichment factors measured for the pervaporation of VOCs from dilute solutions with silicone rubber spiral-wound modules

Solute	Enrichment (E_o)
Trichloroethylene	5700
Toluene	3600
Ethyl acetate	270
Isopropanol	18

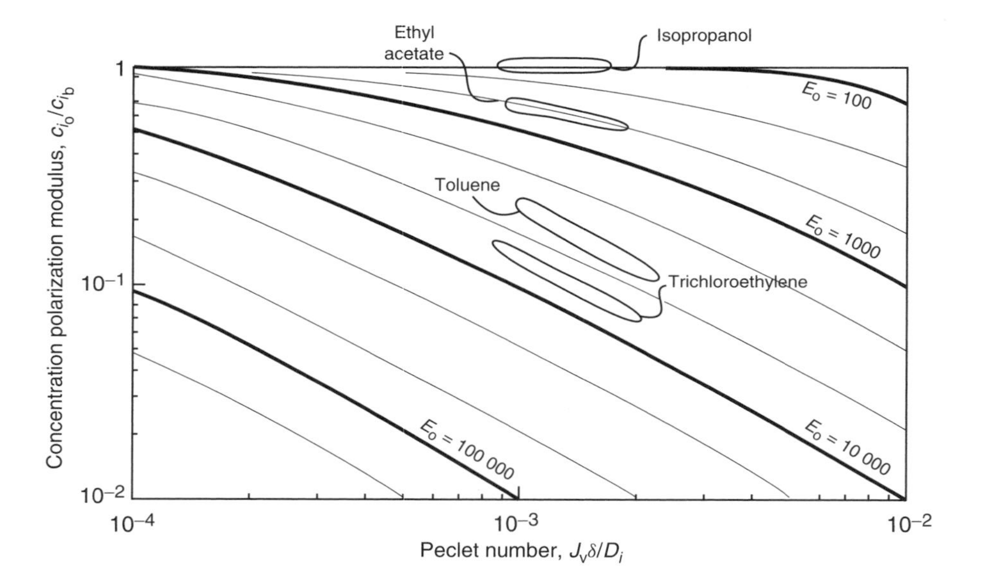

Figure 4.12 A portion of the Wijmans plot shown in Figure 4.7 expanded to illustrate concentration polarization in pervaporation of dilute aqueous organic solutions. With solutes such as toluene and trichloroethylene, high intrinsic enrichments produce severe concentration polarization. Concentration polarization is much less with solutes such as ethyl acetate (enrichment 270), and is essentially eliminated with isopropanol (enrichment 18) [15]

Concentration Polarization in Gas Separation Processes

Concentration polarization in gas separation processes has not been widely studied, and the effect is often assumed to be small because of the high diffusion coefficients of gases. However, the volume flux of gas through the membrane is also high, so concentration polarization effects are important for several processes.

In calculating the expression for the concentration polarization modulus of gases, the simplifying assumption that the volume fluxes on each side of the membrane are equal cannot be made. The starting point for the calculation is the mass-balance Equation (4.6), which for gas permeation is written

$$J_{v_f} c_i - \frac{D_i \mathrm{d}c_i}{\mathrm{d}x} = J_{v_p} c_{i_p} \tag{4.16}$$

where J_{v_f} is the volume flux of gas on the feed side of the membrane and J_{v_p} is the volume flux on the permeate side. These volume fluxes ($\mathrm{cm}^3/\mathrm{cm}^2 \cdot \mathrm{s}$) can be linked by correcting for the pressure on each side of the membrane using the expression

$$J_{v_f} p_o = J_{v_p} p_\ell \tag{4.17}$$

where p_o and p_ℓ are the gas pressures on the feed and permeate sides of the membrane. Hence,

$$J_{v_f} \frac{p_o}{p_\ell} = J_{v_f} \varphi = J_{v_p} \tag{4.18}$$

where φ is the pressure ratio p_o/p_ℓ across the membrane. Substituting Equation (4.18) into Equation (4.16) and rearranging gives

$$-D_i \frac{\mathrm{d}c_i}{\mathrm{d}x} = J_{v_f} (\varphi\, c_{i_p} - c_i) \tag{4.19}$$

Integrating across the boundary layer thickness, as before, gives

$$\frac{c_{i_o}/\varphi - c_{i_p}}{c_{i_b}/\varphi - c_{i_p}} = \exp\left(\frac{J_{v_f} \delta}{D}\right) \tag{4.20}$$

For gases, the enrichment terms, E and E_o, are most conveniently expressed in volume fractions, so that

$$E_o = \frac{c_{i_p}}{p_\ell} \frac{p_o}{c_{i_o}} = \frac{c_{i_p}}{c_{i_o}} \varphi \tag{4.21}$$

and

$$E = \frac{c_{i_p}}{p_\ell} \cdot \frac{p_o}{c_{i_b}} = \frac{c_{i_p}}{c_{i_b}} \cdot \varphi \tag{4.22}$$

Equation (4.20) can then be written as

$$\exp\left(\frac{J_{v_f} \delta}{D_i}\right) = \frac{1 - 1/E_o}{1 - 1/E} \tag{4.23}$$

which on rearranging gives

$$E/E_o = c_{i_o}/c_{i_b} = \frac{\exp(J_{v_f} \delta/D_i)}{1 + E_o[\exp(J_{v_f} \delta/D_i) - 1]} \tag{4.24}$$

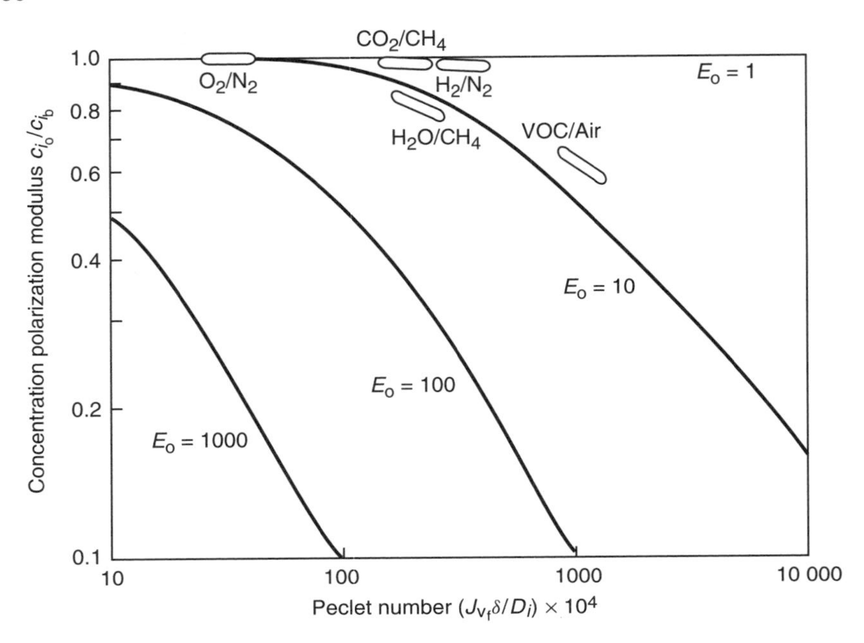

Figure 4.13 A portion of the Wijmans plot shown in Figure 4.7 expanded to illustrate concentration polarization in some important gas separation applications

Equation (4.24) has the same form as the expression for the concentration polarization modulus of liquids, Equation (4.9).

Equation (4.24) can be used to calculate the expected concentration polarization modulus for some of the better-known gas separation applications. The results of the calculations are tabulated in Table 4.3 and shown on a Wijmans plot in Figure 4.13. To obtain agreement between these calculations and industrial experience [16,17], it is necessary to assume the boundary layer thickness in gases is far greater than in liquids. In the calculations of Peclet numbers listed in Table 4.3 a boundary layer thickness of 2000 μm is used, 100 times larger than the value used in similar calculations for the Peclet number for liquid separation processes given in Table 4.1. A boundary layer thickness this large does not seem physically reasonable and in some cases is more than the membrane channel width. The reason for this huge difference between gases and liquids may be related to the difference in the densities of these fluids. Channeling, in which a portion of the feed gas completely bypasses contact with the membrane through some flow maldistribution in the module, can also reduce module efficiency in a way that is difficult to separate from concentration polarization. Channeling is much more noticeable in gas permeation modules than in liquid permeation modules.

Table 4.3 Calculated Peclet numbers for several important gas separations. The boundary layer thickness is assumed to be 2000 μm. Permeant diffusion coefficients in the gas boundary layer are taken from tables for ambient pressure diffusion coefficients in Cussler [8] and corrected for pressure. Membrane enrichments E_o are calculated using Equation (4.21)

Gas separation process	Pressure-normalized flux, P/ℓ [10^{-6} cm^3(STP)/cm$^2 \cdot$ s $\cdot$ cmHg]	Pressure feed/permeate (atm/atm)	Volume flux at feed pressure, J_{v_f} (10^{-3} cm^3/cm$^2 \cdot$ s)	Membrane selectivity, α	Enrichment, E_o	Feed gas diffusion coefficient at feed pressure (10^{-3} cm^2/s)	Peclet number, $J_{v_f}\delta/D_i (\times 10^4)$	Concentration polarization modulus [Equation (4.24)]
O_2/N_2 in air (5 % O_2)	10	10/1	0.68	7	3.9	20	68	0.98
VOC from air (1 % VOC)	100	10/1	6.8	50	8.3	10	1360	0.58
H_2/N_2 ammonia purge gas (20 % H_2)	50	40/10	2.9	100	3.6	20	286	0.93
CO_2/CH_4 natural gas (5 % CO_2)	5	35/5	0.32	20	4.5	5	126	0.96
H_2O/CH_4 natural gas (0.05 % H_2O)	10	30/1	0.74	500	18.2	10	148	0.80

Cross-flow, Co-flow and Counter-flow

In the discussion of concentration polarization to this point, the assumption is made that the volume flux through the membrane is large, so the concentration on the permeate side of the membrane is determined by the ratio of the component fluxes. This assumption is almost always true for liquid separation processes, such as ultrafiltration or reverse osmosis, but must be modified in a few gas separation and pervaporation processes. In these processes, a lateral flow of gas is sometimes used to change the composition of the gas on the permeate side of the membrane. Figure 4.14 illustrates a laboratory gas permeation experiment using this effect. As the pressurized feed gas mixture is passed over the membrane surface, certain components permeate the membrane. On the permeate side of the membrane, a lateral flow of helium or other inert gas sweeps the permeate from the membrane surface. In the absence of the sweep gas, the composition of the gas mixture on the permeate side of the membrane is determined by the flow of components from the feed. If a large flow of sweep gas is used, the partial

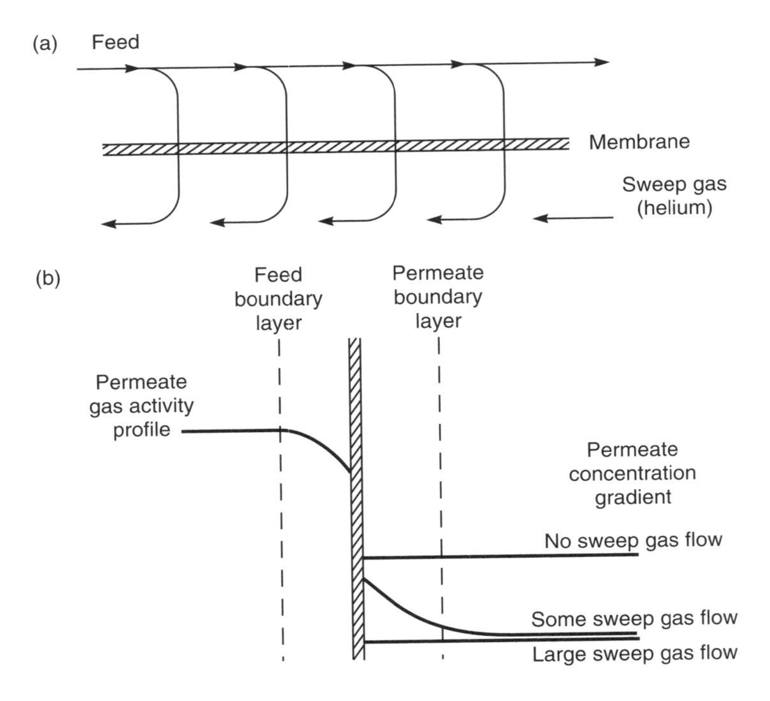

Figure 4.14 (a) Flow schematic of permeation using a permeate-side sweep gas sometimes used in laboratory gas separation and pervaporation experiments. (b) The concentration gradients that form on the permeate side of the membrane depend on the volume of sweep gas used. In laboratory experiments a large sweep-gas-to-permeate-gas flow ratio is used, so the concentration of permeate at the membrane surface is very low

pressure of the permeating components on the permeate side of the membrane is reduced to a low value. The difference in partial pressure of the permeating gases from the feed to the permeate side of the membrane is thereby increased, and the flow across the membrane increases proportionately. Sweep gases are sometimes used in gas permeation and pervaporation laboratory experiments. The sweep gas is generally helium and the helium/permeate gas mixture is fed to a gas chromatograph for analysis.

The drawback of using an external permeate-side sweep gas to lower the partial pressure on the permeate side of the membrane for an industrial process is that the sweep gas and permeating component must subsequently be separated. In some cases this may not be difficult; some processes that have been suggested but rarely used are shown in Figure 4.15. In these examples, the separation of the sweep gas and the permeating component is achieved by condensation. If the permeating gas is itself easily condensed, an inert gas such as nitrogen can be used as the sweep [18]. An alternative is a condensable vapor such as steam [19–21].

In the examples illustrated in Figures 4.14 and 4.15, the sweep gas is an inert gas that is unlike the permeating components. However, the sweep gas could be a mixture of the permeating components at a different composition. An example of this type of process is shown schematically in Figure 4.16, which illustrates the separation of nitrogen from air using a membrane that preferentially permeates oxygen. The feed air, containing approximately 20 vol% oxygen, is introduced under pressure at one end of the module. The permeate gas at this end of the module typically contains about 50 vol% oxygen (at a lower pressure). As the feed gas travels down the membrane module it becomes increasingly depleted in oxygen (enriched in nitrogen) and leaves the module as a residue gas containing 99 % nitrogen. The permeate gas at this end of the module contains about 5 vol% oxygen and 95 vol% nitrogen. If this gas is directed to flow counter to the incoming feed gas, as shown in Figure 4.16, the effect is to sweep the permeate side of the membrane with a flow of oxygen-depleted, nitrogen-enriched gas. This is beneficial because the oxygen gradient through the membrane is increased, which increases its flux through the membrane. Simultaneously the nitrogen gradient is decreased, which decreases its flux through the membrane. An opposite negative result would result if the permeate gas were moved in the same direction as the feed gas (that is co-flow). This would have the effect of sweeping the permeate side of the membrane with oxygen-enriched gas.

The cross-, co- and counter-flow schemes are illustrated in Figure 4.17, together with the concentration gradient across a median section of the membrane. It follows from Figure 4.17 that system performance can be improved by operating a module in an appropriate flow mode (generally counter-flow). However, such improvements require that the concentration at the membrane permeate surface equals the bulk concentration of the permeate at that point. This condition cannot be met with processes such as ultrafiltration or reverse osmosis in which the permeate is a liquid. In these processes, the selective side of the membrane faces the

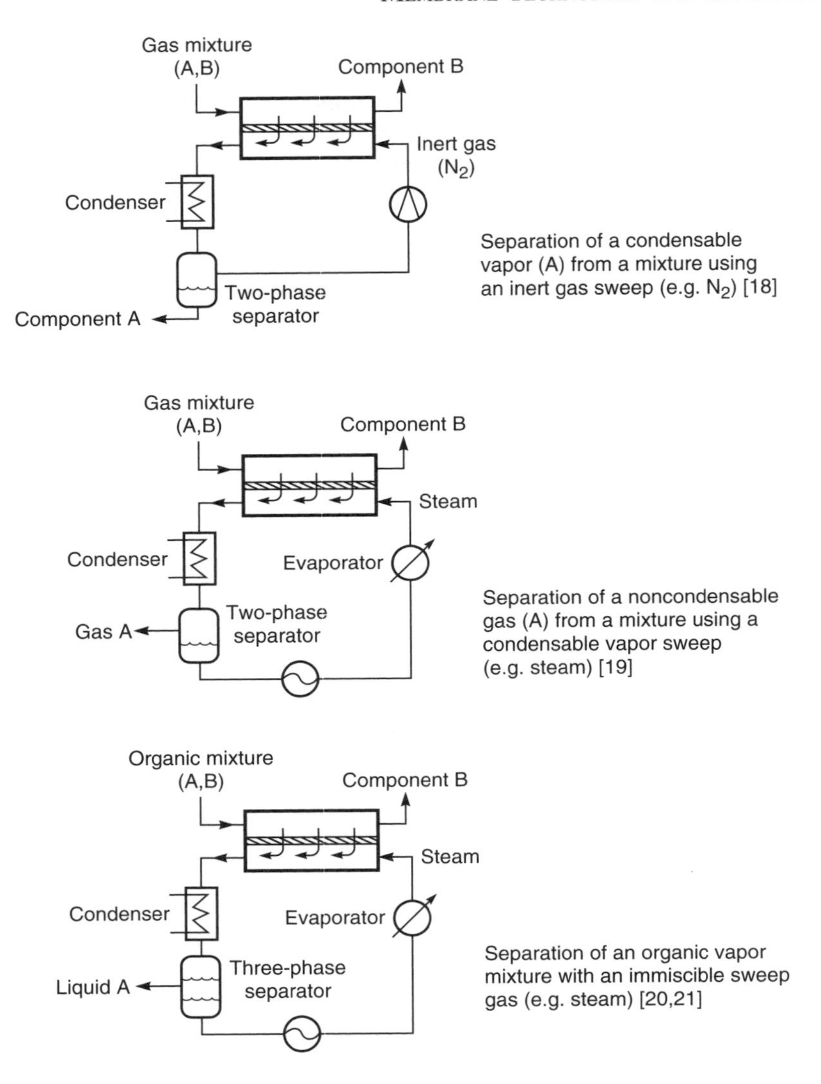

Figure 4.15 Sweep gas systems proposed for industrial processes

feed solution, and a microporous support layer faces the permeate. Concentration gradients easily build up in this boundary layer, completely outweighing the benefit of counter-flow. Thus, counter-flow (sweep) module designs are limited to gas separation and pervaporation processes. In these processes the permeate is a gas, and permeate-side concentration gradients are more easily controlled because diffusion coefficients in gases are high.

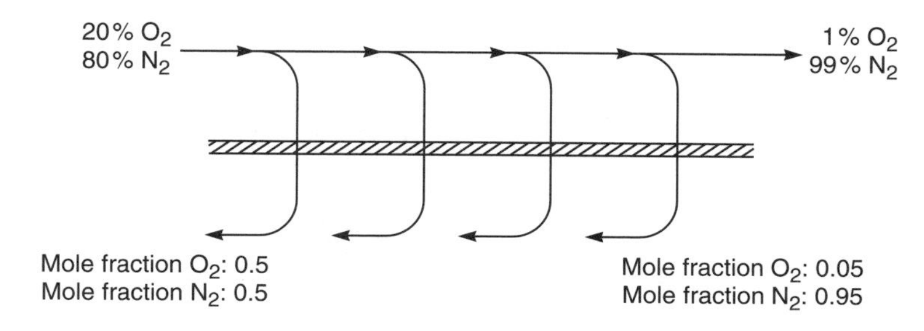

Figure 4.16 An illustration of a counter-flow module for the separation of nitrogen from air. Directing the permeate to flow counter to the feed sweeps the permeate side of the membrane with a flow of oxygen-depleted gas. This increases the oxygen flux and decreases the nitrogen flux through the membrane

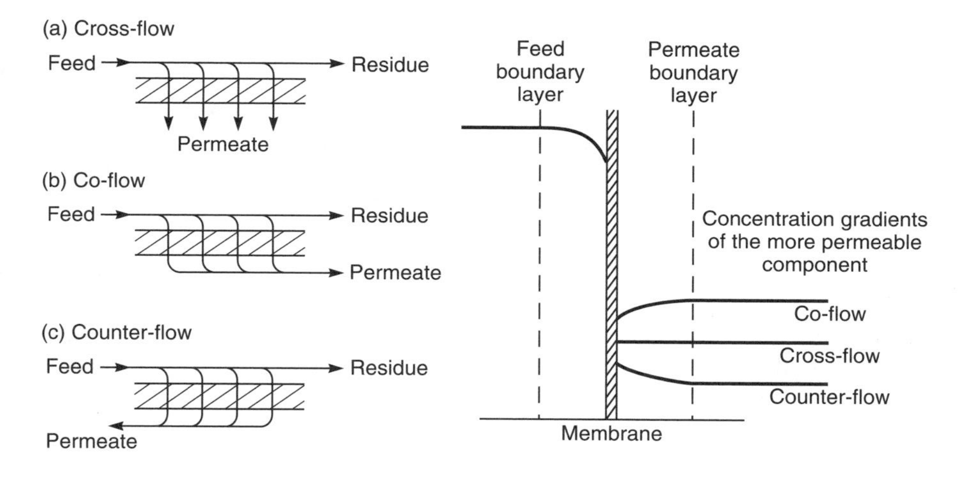

Figure 4.17 (a) Cross-, (b) co- and (c) counter-flow schemes in a membrane module and the changes in the concentration gradients that occur across a median section of the membrane

The benefit obtained from counter-flow depends on the particular separation, but it can often be substantial, particularly in gas separation and pervaporation processes. A comparison of cross-flow, counter-flow, and counter-flow/sweep for the same membrane module used to dehydrate natural gas is shown in Figure 4.18. Water is a smaller molecule and much more condensable than methane, the main component of natural gas, so membranes with a water/methane selectivity of 400–500 are readily available. In the calculations shown in Figure 4.18, the membrane is assumed to have a pressure-normalized

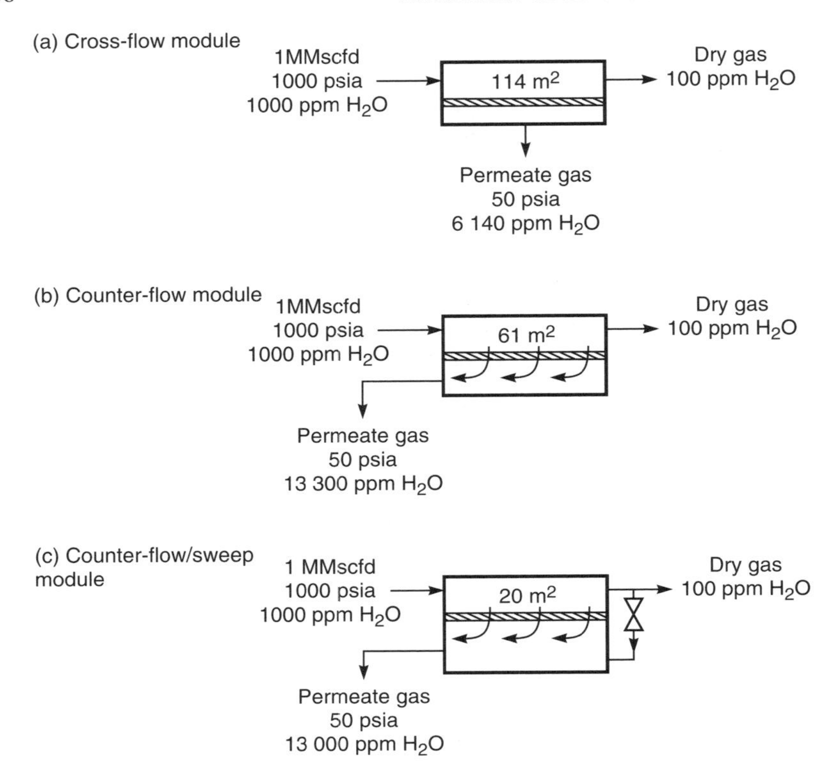

Figure 4.18 Comparison of (a) cross-flow, (b) counter-flow and (c) counter-flow sweep module performance for the separation of water vapor from natural gas. Pressure-normalized methane flux: 5×10^{-6}cm^3(STP)/cm$^2 \cdot$ s $\cdot$ cmHg; membrane selectivity, water/methane: 200

methane flux of 5×10^{-6} cm^3(STP)/cm$^2 \cdot$ s $\cdot$ cmHg and a water/methane selectivity of 200. Counter-flow/sweep modules have a substantial advantage in this separation because the separation is completely pressure-ratio-limited.[2]

[2]The importance of the pressure ratio in separating gas mixtures can be illustrated by considering the separation of a gas mixture with component concentrations (mol%) n_{i_o} and n_{j_o} at a feed pressure of p_o. A flow of component across the membrane can only occur if the partial pressure of component i on the feed side of the membrane, $n_{i_o} p_o$, is greater than the partial pressure of component i on the permeate side of the membrane, $n_{i_\ell} p_\ell$. That is

$$n_{i_o} p_o > n_{i_\ell} p_\ell$$

It follows that the maximum enrichment achieved by the membrane can be expressed as

$$\frac{n_{i_\ell}}{n_{i_o}} \leq \frac{p_o}{p_\ell}$$

In the cross-flow module illustrated in Figure 4.18(a) the average concentration of water on the feed side of the membrane as it decreases from 1000 to 100 ppm is 310 ppm (the log mean). The pooled permeate stream has a concentration of 6140 ppm. The counter-flow module illustrated in Figure 4.18(b) performs substantially better, providing a pooled permeate stream with a concentration of 13 300 ppm. Not only does the counter-flow module perform a two-fold better separation, it also requires only about half the membrane area.

In the case of the counter-flow/sweep membrane module illustrated in Figure 4.18(c) a portion of the dried residue gas stream is expanded across a valve and used as the permeate-side sweep gas. The separation obtained depends on how much gas is used as a sweep. In the calculation illustrated, 5 % of the residue gas is used as a sweep; even so the result is dramatic. The concentration of water vapor in the permeate gas is 13 000 ppm, almost the same as the perfect counter-flow module shown in Figure 4.18(b), but the membrane area required to perform the separation is one-third of the counter-flow case. *Mixing separated residue gas with the permeate gas improves the separation!* The cause of this paradoxical result is illustrated in Figure 4.19 and discussed in a number of papers by Cussler *et al.* [16].

Figure 4.19(a) shows the concentration of water vapor on the feed and permeate sides of the membrane module in the case of a simple counter-flow module. On the high-pressure side of the module, the water vapor concentration in the feed gas drops from 1000 ppm to about 310 ppm halfway through the module and to 100 ppm at the residue end. The graph directly below the module drawing shows the theoretical maximum concentration of water vapor on the permeate side of the membrane. This maximum is determined by the feed-to-permeate pressure ratio of 20 as described in the footnote to page 186. The actual calculated permeate-side concentration is also shown. The difference between these two lines is a measure of the driving force for water vapor transport across the membrane. At the feed end of the module, this difference is about 1000 ppm, but at the permeate end the difference is only about 100 ppm.

Figure 4.19(b) shows an equivalent figure for a counter-flow module in which 5 % of the residue gas containing 100 ppm water vapor is expanded to 50 psia and introduced as a sweep gas. The water vapor concentration in the permeate gas at the end of the membrane then falls from 1900 ppm to 100 ppm, producing a dramatic increase in water vapor permeation through the membrane at the residue end of the module. The result is a two-thirds reduction in the size of the module.

This means that the enrichment can never exceed the pressure ratio of p_o/p_ℓ, no matter how selective the membrane. In the example above, the maximum water vapor enrichment across the membrane is 20 (1000 psia/50 psia) even though the membrane is 200 times more permeable to water than methane.

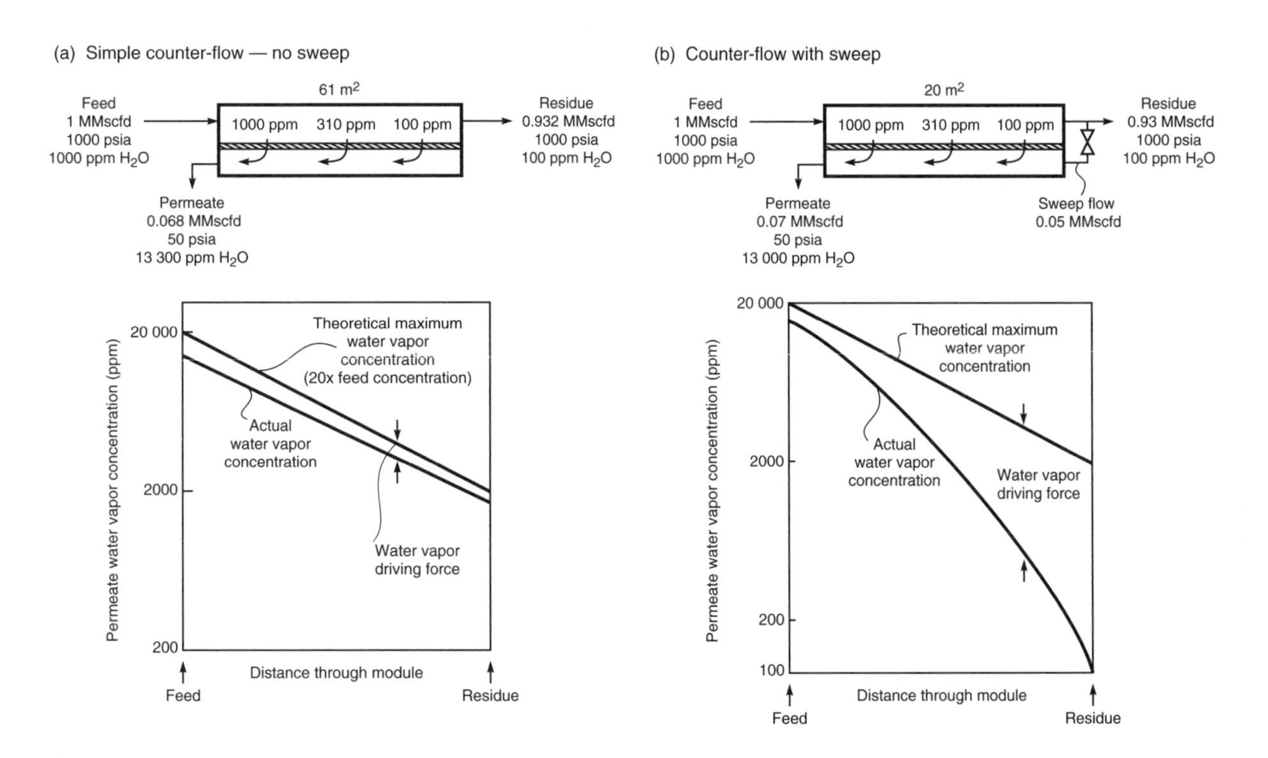

Figure 4.19 The effect of a small permeate-side, counter-flow sweep on the water vapor concentration on the permeate side of a membrane. In this example calculation, the sweep flow reduces the membrane area by two-thirds

Conclusions and Future Directions

Few membrane processes are unaffected by concentration polarization, and the effect is likely to become more important as membrane materials and membrane fabrication techniques improve. As membrane flux and selectivity increase concentration polarization effects become exponentially larger. In the laboratory, concentration polarization is controlled by increasing the turbulence of the feed fluid. However, in industrial systems this approach has practical limits. In ultrafiltration and electrodialysis, for example, liquid recirculation pumps are already a major portion of the plant's capital cost and consume 20 to 40 % of the power used for the separation. The best hope for minimizing concentration polarization effects lies in improving membrane module design, understanding the basis for the choice of channel spacer materials, and developing methods of controlling the feed fluid flow in the module. Unfortunately, this type of work is generally performed in membrane system manufacturing companies and is not well covered in the open literature.

References

1. H.S. Carslaw and J.C. Jaeger, *Conduction of Heat in Solids*, Oxford University Press, London (1947).
2. R.B. Bird, W.E. Stewart and E.N. Lightfoot, *Transport Phenomena*, Wiley, New York (1960).
3. J. Crank, *The Mathematics of Diffusion*, Oxford University Press, London (1956).
4. M.C. Porter, Concentration Polarization with Membrane Ultrafiltration, *Ind. Eng. Chem. Prod. Res. Dev.* **11**, 234 (1972).
5. J.V. Lepore and R.C. Ahlert, Fouling in Membrane Processes, in *Reverse Osmosis Technology*, B.S. Parekh (ed.), Marcel Dekker, New York, pp. 141–184 (1988).
6. S.R. Wickramasinghe, M.J. Semmens and E.L. Cussler, Mass Transfer in Various Hollow Fiber Geometries, *J. Membr. Sci.* **69**, 235 (1992).
7. L. Mi and S.T. Hwang, Correlation of Concentration Polarization and Hydrodynamic Parameters in Hollow Fiber Modules, *J. Membr. Sci.* **159**, 143 (1999).
8. E.L. Cussler, *Diffusion Mass Transfer in Fluid Systems*, 2nd Edn, Cambridge University Press, New York, NY and Cambridge, UK (1997).
9. P.L.T. Brian, Mass Transport in Reverse Osmosis, in *Desalination by Reverse Osmosis*, U. Merten (ed.), MIT Press, Cambridge, MA, pp. 161–292 (1966).
10. G. Belfort, R.H. Davis and A.L. Zydney, The Behavior of Suspensions and Macromolecular Solutions in Cross Flow Microfiltration, *J. Membr. Sci.* **1**, 96 (1994).
11. A.R. Da Costa, A.G. Fane and D.E. Wiley, Spacer Characterization and Pressure Drop Modeling in Spacer-filled Channels for Ultrafiltration, *J. Membr. Sci.* **87**, 79 (1994).
12. M.Y. Jaffrin, B.B. Gupta and P. Paullier, Energy Savings Pulsatile Mode Crossflow Filtration, *J. Membr. Sci.* **86**, 281 (1994).
13. J.G. Wijmans, A.L. Athayde, R. Daniels, J.H. Ly, H.D. Kamaruddin and I. Pinnau, The Role of Boundary Layers in the Removal of Volatile Organic Compounds from Water by Pervaporation, *J. Membr. Sci.* **109**, 135 (1996).
14. E.E. Wilson, A Basis for Rational Design of Heat Transfer Apparatus, *Trans. ASME* **37**, 47 (1915).

15. R.W. Baker, J.G. Wijmans, A.L. Athayde, R. Daniels, J.H. Ly and M. Le, The Effect of Concentration Polarization on the Separation of Volatile Organic Compounds from Water by Pervaporation, *J. Membr. Sci.* **137**, 159 (1997).
16. K.L. Wang, S.H. McCray, D.N. Newbold and E.L. Cussler, Hollow Fiber Air Drying, *J. Membr. Sci.* **72**, 231 (1992).
17. O. Lüdtke, R.D. Behling and K. Ohlrogge, Concentration Polarization in Gas Permeation, *J. Membr. Sci.* **146**, 145 (1998).
18. F.E. Frey, Process for Concentrating Hydrocarbons, US Patent 2,159,434 (May, 1939).
19. R.J. Walker, C.J. Drummond and J.M. Ekmann, Evaluation of Advanced Separation Techniques for Application to Flue Gas Cleanup Processes for the Simultaneous Removal of Sulfur Dioxide and Nitrogen Oxides, *Department of Energy Report, DE85102227* (May, 1985).
20. A.E. Robertson, Separation of Hydrocarbons, US Patent 2,475,990 (July, 1969).
21. R.W. Baker, E.L. Cussler, W. Eykamp, W.J. Koros, R.L. Riley and H. Strathmann, *Membrane Separation Systems*, Noyes Data Corp., Park Ridge, NJ, p. 155 (1991).

5 REVERSE OSMOSIS

Introduction and History

Reverse osmosis is a process for desalting water using membranes that are permeable to water but essentially impermeable to salt. Pressurized water containing dissolved salts contacts the feed side of the membrane; water depleted of salt is withdrawn as a low-pressure permeate. The ability of membranes to separate small solutes from water has been known for a very long time. Pfeffer, Traube and others studied osmotic phenomena with ceramic membranes as early as the 1850s. In 1931 the process was patented as a method of desalting water, and the term reverse osmosis was coined [1]. Modern interest dates from the work of Reid and Breton, who in 1959 showed that cellulose acetate films could perform this type of separation [2]. Their films were 5–20 μm thick so fluxes were very low but, by pressurizing the feed salt solution to 1000 psi, they obtained salt removals of better than 98 % in the permeate water. The breakthrough discovery that made reverse osmosis a practical process was the development of the Loeb–Sourirajan anisotropic cellulose acetate membrane [3]. This membrane had 10 times the flux of the best membrane of Reid and Breton and equivalent rejections. With these membranes, water desalination by reverse osmosis became a potentially practical process, and within a few years small demonstration plants were installed. The first membrane modules were tubular or plate-and-frame systems, but Westmoreland, Bray, and others at the San Diego Laboratories of Gulf General Atomics (the predecessor of Fluid Systems Inc.) soon developed practical spiral-wound modules [4,5]. Later, Du Pont [6], building on the earlier work of Dow, introduced polyaramide hollow fine fiber reverse osmosis modules under the name Permasep®.

Anisotropic cellulose acetate membranes were the industry standard through the 1960s to the mid-1970s, until Cadotte, then at North Star Research, developed the interfacial polymerization method of producing composite membranes [7]. Interfacial composite membranes had extremely high salt rejections, combined with good water fluxes. Fluid Systems introduced the first commercial interfacial composite membrane in 1975. The construction of a large seawater desalination plant at Jiddah, Saudi Arabia using these membranes was a milestone in reverse

Membrane Technology and Applications R. W. Baker
 ISBN: 0-470-85445-6

osmosis development [8]. Later, at FilmTec, Cadotte developed a fully aromatic interfacial composite membrane based on the reaction of phenylene diamine and trimesoyl chloride [9,10]. This membrane has become the new industry standard. The most recent development, beginning in the mid-1980s, was the introduction of low-pressure nanofiltration membranes by all of the major reverse osmosis companies [11,12]. These membranes are used to separate trace amounts of salts and other dissolved solutes from already good-quality water to produce ultrapure water for the electronics industry. An important recent advance by Grace Davison working with Mobil Oil, now ExxonMobil, is the development of a reverse osmosis (hyperfiltration) process to separate a solution of methyl ethyl ketone and lube oil. A plant installed at a Beaumont, Texas, refinery in 1998 was the first large-scale use of pressure-driven membranes to separate organic solvent mixtures.

Currently, approximately one billion gal/day of water are desalted by reverse osmosis. Half of this capacity is installed in the United States, Europe, and Japan, principally to produce ultrapure industrial water. The remainder is installed in the Middle East and other desert regions to produce municipal drinking water from brackish groundwater or seawater. In recent years, the interfacial composite membrane has displaced the anisotropic cellulose acetate membrane in most applications. Interfacial composite membranes are supplied in spiral-wound module form; the market share of hollow fiber membranes is now less than

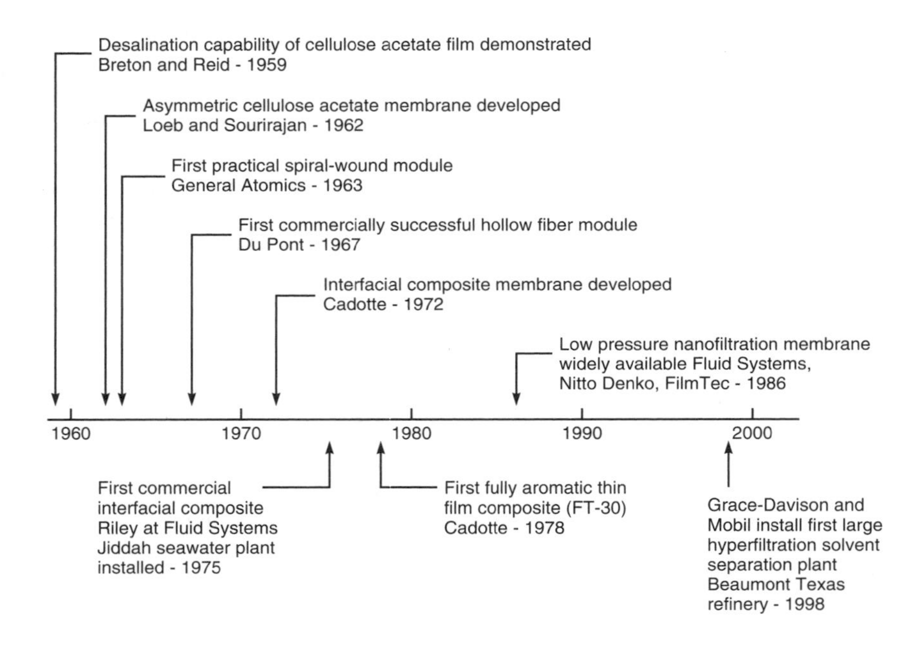

Figure 5.1 Milestones in the development of reverse osmosis

10 % of new installed capacity and shrinking [13]. Tubular and plate-and-frame systems, which are only competitive for small niche applications involving particularly highly fouling water, have less than 5 % of the market. Some of the milestones in the development of the reverse osmosis industry are summarized in Figure 5.1.

Theoretical Background

Salt and water permeate reverse osmosis membranes according to the solution-diffusion transport mechanism are described in Chapter 2. The water flux, J_i, is linked to the pressure and concentration gradients across the membrane by the equation

$$J_i = A(\Delta p - \Delta\pi) \tag{5.1}$$

where Δp is the pressure difference across the membrane, $\Delta\pi$ is the osmotic pressure differential across the membrane, and A is a constant. As this equation shows, at low applied pressure, when $\Delta p < \Delta\pi$, water flows from the dilute to the concentrated salt-solution side of the membrane by normal osmosis. When $\Delta p = \Delta\pi$, no flow occurs, and when the applied pressure is higher than the osmotic pressure, $\Delta p > \Delta\pi$, water flows from the concentrated to the dilute salt-solution side of the membrane.

The salt flux, J_j, across a reverse osmosis membrane is described by the equation

$$J_j = B(c_{j_o} - c_{j_\ell}) \tag{5.2}$$

where B is the salt permeability constant and c_{j_o} and c_{j_ℓ}, respectively, are the salt concentrations on the feed and permeate sides of the membrane. The concentration of salt in the permeate solution (c_{j_ℓ}) is usually much smaller than the concentration in the feed (c_{j_o}), so equation (5.2) can be simplified to

$$J_j = Bc_{j_o} \tag{5.3}$$

It follows from these two equations that the water flux is proportional to the applied pressure, but the salt flux is independent of pressure. This means that the membrane becomes more selective as the pressure increases. Selectivity can be measured in a number of ways, but conventionally, it is measured as the salt rejection coefficient $\mathbb{R}$, defined as

$$\mathbb{R} = \left[1 - \frac{c_{j_\ell}}{c_{j_o}}\right] \times 100\,\% \tag{5.4}$$

The salt concentration on the permeate side of the membrane can be related to the membrane fluxes by the expression

$$c_{j_\ell} = \frac{J_j}{J_i} \times \rho_i \tag{5.5}$$

where ρ_i is the density of water (g/cm^3). By combining equations (5.1) to (5.5), the membrane rejection can be expressed as

$$\mathbb{R} = \left[1 - \frac{\rho_i \cdot B}{A(\Delta p - \Delta\pi)}\right] \times 100\,\% \tag{5.6}$$

The effects of the most important operating parameters on membrane water flux and salt rejection are shown schematically in Figure 5.2 [14]. The effect of feed pressure on membrane performance is shown in Figure 5.2(a). As predicted by Equation (5.1), at a pressure equal to the osmotic pressure of the feed (350 psi), the water flux is zero; thereafter, it increases linearly as the pressure is increased. The salt rejection also extrapolates to zero at a feed pressure of 350 psi as predicted by Equation (5.6), but increases very rapidly with increased pressure to reach salt rejections of more than 99 % at an applied pressure of 700 psi (twice the feed solution osmotic pressure).

The effect of increasing the concentration of salt in the feed solution on membrane performance is illustrated in Figure 5.2(b). Increasing the salt concentration effectively increases the osmotic pressure term in Equation (5.1); consequently, at a constant feed pressure, the water flux falls with increasing salt concentration at a feed pressure of 1000 psi. The water flux approaches zero when the salt concentration is about 10 wt%, at which point the osmotic pressure equals the applied hydrostatic pressure. The salt rejection also extrapolates to zero rejection at this point but increases rapidly with decreasing salt concentration. Salt rejections of more than 99 % are reached at salt concentrations below 6 %, corresponding to a net applied pressure of about 400 psi.

The effect of temperature on salt rejection and water flux illustrated in Figure 5.2(c) is more complex. Transport of both salt and water represented by Equations (5.1) and (5.3) is an activated process, and both increase exponentially with increasing temperature. As Figure 5.2(c) shows, the effect of temperature on the water flux of membranes is quite dramatic: the water flux doubles as the temperature is increased by 30 °C. However, the effect of temperature on the salt flux is even more marked. This means that the salt rejection coefficient, proportional to the ratio B/A in Equation (5.6), actually declines slightly as the temperature increases.

Measurements of the type shown in Figure 5.2 are typically obtained with small laboratory test cells. A typical test system is illustrated in Figure 5.3. Such systems are often used in general membrane quality control tests with a number of cells arranged in series through which fluid is pumped. The system is usually operated with a test solution of 0.2 to 1.0 % sodium chloride at pressures ranging from 150 to 600 psi. The storage tank and flow recirculation rate are made large enough that changes in concentration of the test solution due to loss of permeate can be ignored.

Some confusion can occur over the rejection coefficients quoted by membrane module manufacturers. The intrinsic rejection of good quality membranes measured in a laboratory test system might be in the range 99.5 to 99.7 %, whereas

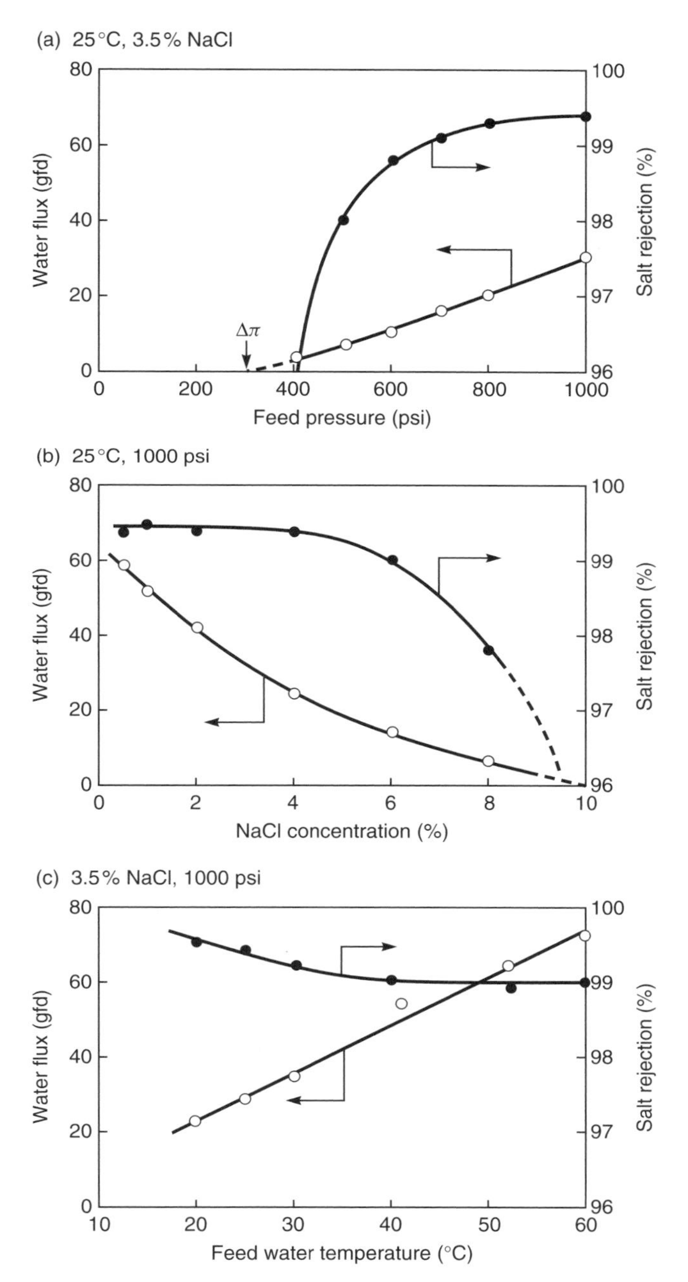

Figure 5.2 Effect of pressure, feed salt concentration and feed temperature on the properties of good quality seawater desalination membranes (SW-30) [14]

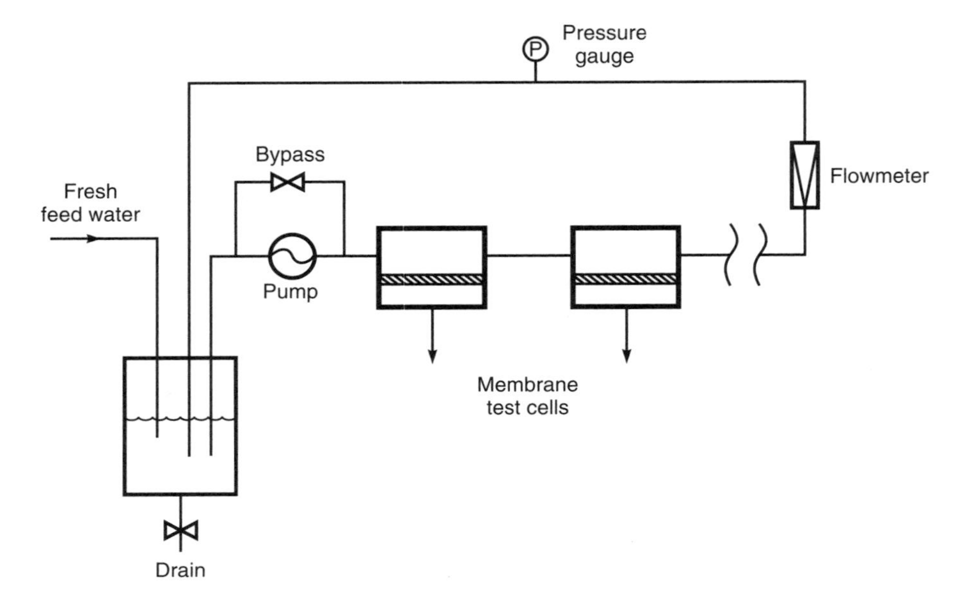

Figure 5.3 Flow schematic of a high-pressure laboratory reverse osmosis test system

the same membrane in module form may have a salt rejection of 99.4 to 99.5 %. This difference is due to small membrane defects introduced during module production and to concentration polarization, which has a small but measurable effect on module rejection. Manufacturers call the module value the nominal rejection. However, manufacturers will generally only guarantee a lower figure, for example, 99.3 % for the initial module salt rejection to take into account variations between modules. To complicate matters further, module performance generally deteriorates slowly during the 1- to 3-year guaranteed module lifetime due to membrane compaction, membrane fouling, and membrane degradation from hydrolysis, chlorine attack, or membrane cleaning. A decrease in the membrane flux by 20 % over the 3-year lifetime of typical modules is not unusual, and the rejection can easily fall by 0.2–0.3 %. Reverse osmosis system manufacturers allow for this decline in performance when designing systems.

Membranes and Materials

A number of membrane materials and membrane preparation techniques have been used to make reverse osmosis membranes. The target of much of the early work was seawater desalination (approximately 3.5 wt% salt), which requires membranes with salt rejections of greater than 99.3 % to produce an acceptable permeate containing less than 500 ppm salt. Early membranes could only meet this target performance when operated at very high pressures, up to 1500 psi. As membrane performance has improved, this pressure has dropped to 800–1000 psi. Recently, the need for desalination membranes has shifted more to brackish water feeds with salt concentrations of 0.2–0.5 wt%. For this application, membranes are typically operated at pressures in the 150–400 psi range with a target salt rejection of about 99 %. With the growth of the electronics industry the demand for ultrapure water to wash silicon wafers has increased. The feed to an ultrapure water reverse osmosis plant is often municipal drinking water, which may only contain 100 to 200 ppm dissolved salts, mostly divalent ions. The target membrane performance in this case may be 98–99 % sodium chloride rejection but more than 99.5 % divalent ion rejection. These membranes are operated at low pressures, typically in the 100–200 psi range. Many manufacturers tailor the properties of a single membrane material to meet the requirements of different applications. Invariably a significant trade-off between flux and rejection is involved.

A brief description of the commercially important membranes in current use follows. More detailed descriptions can be found in specialized reviews [13,15,16]. Petersen's review on interfacial composite membranes is particularly worth noting [17].

Cellulosic Membranes

Cellulose acetate was the first high-performance reverse osmosis membrane material discovered. The flux and rejection of cellulose acetate membranes have now been surpassed by interfacial composite membranes. However, cellulose acetate membranes still maintain a small fraction of the market because they are easy to make, mechanically tough, and resistant to degradation by chlorine and other oxidants, a problem with interfacial composite membranes. Cellulose acetate membranes can tolerate up to 1 ppm chlorine, so chlorination can be used to sterilize the feed water, a major advantage with feed streams having significant bacterial loading.

The water and salt permeability of cellulose acetate membranes is extremely sensitive to the degree of acetylation of the polymer used to make the membrane [2,18,19]. The effect of degree of acetylation on salt and water permeability is illustrated in Figure 5.4 [20]. Fully substituted cellulose triacetate (44.2 wt% acetate) has an extremely high water-to-salt permeability ratio, reflecting its very high selectivity. Unfortunately the water permeability is low so these membranes have low water fluxes. Nonetheless, cellulose triacetate hollow fine fiber membranes are still produced for some seawater desalination plants because salt rejections of about 99.5 % with a seawater feed are attainable. However, most commercial cellulose acetate membranes use a polymer containing about 40 wt% acetate with a degree of acetylation of 2.7. These membranes generally achieve 98–99 % sodium chloride rejection and have reasonable fluxes. The permeability

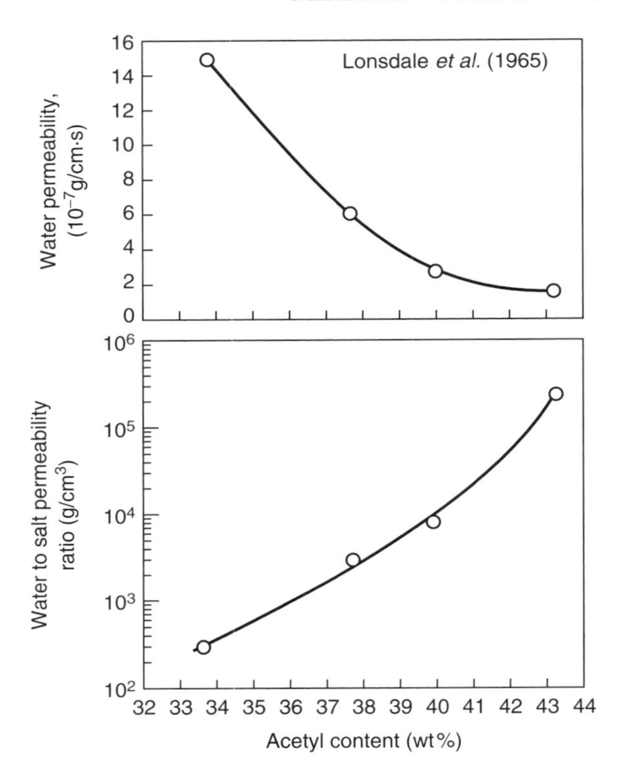

Figure 5.4 Permeabilities of cellulose acetate to water and sodium chloride as a function of acetyl content at 25 °C. Data from Lonsdale *et al.* [20]

data shown in Figure 5.4 can be replotted to show expected salt rejections, as shown in Figure 5.5.

The data in Figure 5.5 show that thick films of cellulose acetate made from 39.8 wt% acetate polymer should reject 99.5 % sodium chloride. In practice, this theoretical rejection is very difficult to obtain with practical thin membranes [21]. Figure 5.6 shows the salt rejection properties of 39.8 wt% acetate membranes made by the Loeb–Sourirajan process [22]. The freshly formed membranes have very high water fluxes of almost 200 gal/ft^2 · day (gfd) but almost no rejection of sodium chloride. The membranes appear to have a finely microporous structure and are permeable to quite large solutes such as sucrose. The rejection of these membranes can be greatly improved by heating in a bath of hot water for a few minutes. Apparently, this annealing procedure, used with all cellulose acetate membranes, modifies the salt rejection layer of the membrane by eliminating the micropores and producing a denser, more salt-rejecting skin. The water flux decreases, and the sodium chloride rejection increases. The temperature of this

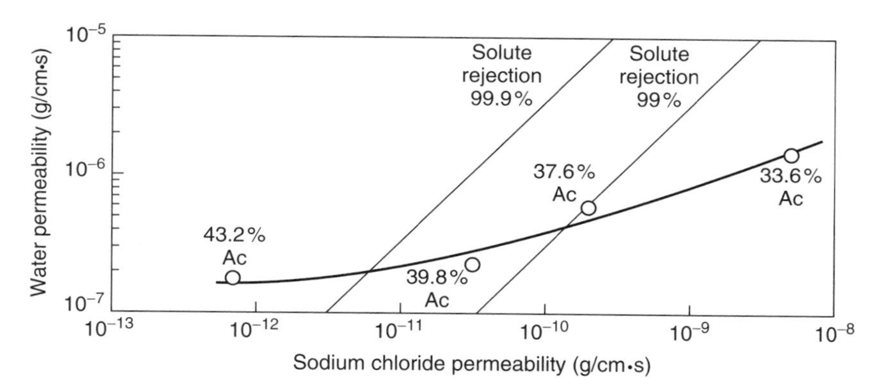

Figure 5.5 Water permeability as a function of sodium chloride permeability for membranes made from cellulose acetate of various degrees of acetylation. The expected rejection coefficients for these membranes, calculated for dilute salt solutions using Equation (5.6), are shown

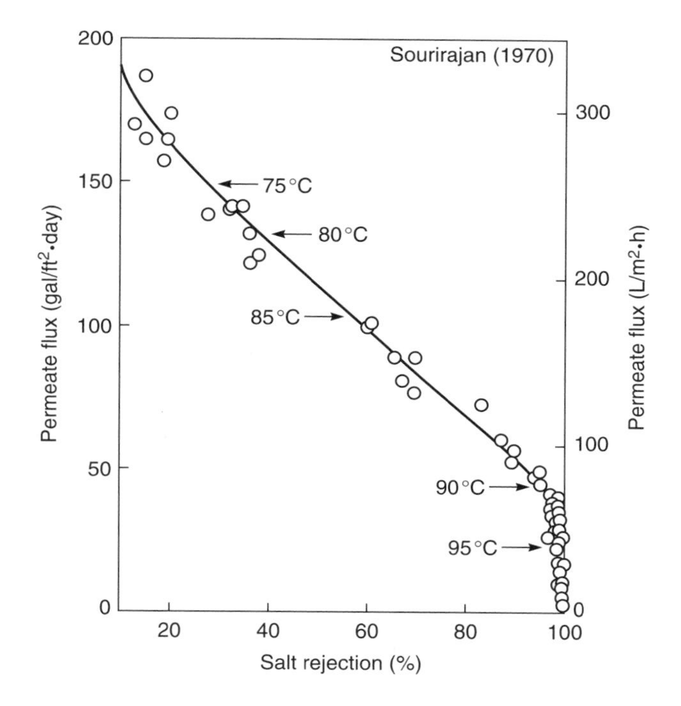

Figure 5.6 The effect of annealing temperatures on the flux and rejection of cellulose acetate membranes. The anealing temperature is shown on the figure. (Cellulose diacetate membranes tested at 1500 psig with 0.5 M NaCl) [22]. Reprinted with permission from Elsevier

annealing step determines the final properties of the membrane. A typical rejection/flux curve for various annealed membranes is shown in Figure 5.6. Because their properties change on heating, cellulose acetate membranes are generally not used above about 35 °C. The membranes also slowly hydrolyze over time, so the feed water is usually adjusted to pH 4–6, the range in which the membranes are most stable [23].

Throughout the 1960s considerable effort was expended on understanding the Loeb–Sourirajan membrane production process to improve the quality of the membranes produced. The casting solution composition is critically important. Other important process steps are the time of evaporation before precipitation, the temperature of the precipitation bath, and the temperature of the annealing step. Most of the early membranes were made of 39.8 wt% acetate polymer because this material was readily available and had the most convenient solubility properties. By the 1970s, however, a number of workers, particularly Saltonstall and others at Envirogenics, had developed better membranes by blending the 39.8 wt% acetate polymer with small amounts of triacetate polymer (44.2 wt% acetate) or other cellulose esters such as cellulose acetate butyrate [24]. These blends are generally used to form current cellulose acetate membranes. Good-quality blend membranes with seawater salt rejections of 99.0–99.5 %, close to the theoretical salt rejection determined by thick film measurements, can be made, but the flux of these membranes is modest. However, most applications of cellulose acetate membranes do not require such high salt rejections, so the typical commercial cellulose acetate membrane has good fluxes and a sodium chloride rejection of about 96 %.

Noncellulosic Polymer Membranes

During the 1960s and 1970s the Office of Saline Water sponsored development of noncellulosic reverse osmosis membranes. Many polymers were evaluated as Loeb–Sourirajan membranes but few matched the properties of cellulose acetate. Following the development of interfacial composite membranes by Cadotte, this line of research was abandoned by most commercial membrane producers.

Nonetheless a few commercially successful noncellulosic membrane materials were developed. Polyamide membranes in particular were developed by several groups. Aliphatic polyamides have low rejections and modest fluxes, but aromatic polyamide membranes were successfully developed by Toray [25], Chemstrad (Monsanto) [26] and Permasep (Du Pont) [27], all in hollow fiber form. These membranes have good seawater salt rejections of up to 99.5 %, but the fluxes are low, in the 1 to 3 gal/ft^2 · day range. The Permasep® membrane, in hollow fine fiber form to overcome the low water permeability problems, was produced under the names B-10 and B-15 for seawater desalination plants until the year 2000. The structure of the Permasep B-15 polymer is shown in Figure 5.7. Polyamide membranes, like interfacial composite membranes, are susceptible to degradation by chlorine because of their amide bonds.

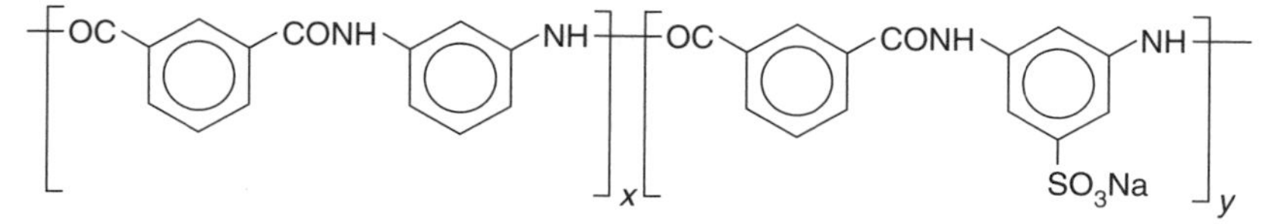

Figure 5.7 Aromatic polyamide used by Du Pont in its Permasep B-15 hollow fine fibers [27]

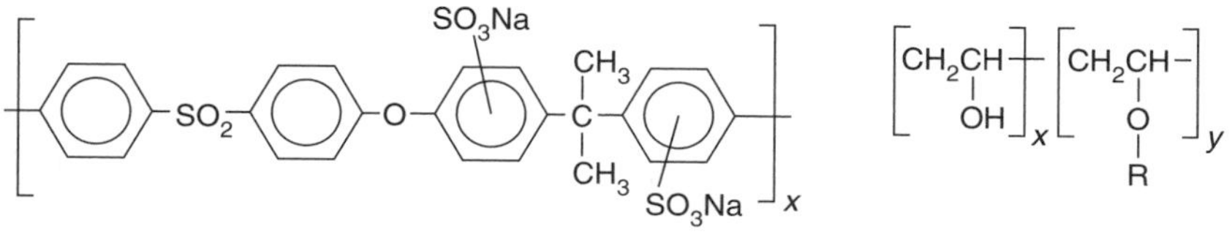

Figure 5.8 Membranes based on sulfonated polysulfone and substituted poly(vinyl alcohol) are produced by Hydranautics (Nitto) for nanofiltration applications

Loeb–Sourirajan membranes based on sulfonated polysulfone and substituted poly(vinyl alcohol) produced by Hydranautics (Nitto) have also found a commercial market as high-flux, low-rejection membranes in water softening applications because their divalent ion rejection is high. These membranes are also chlorine-resistant and have been able to withstand up to 40 000 ppm · h of chlorine exposure without degradation.[1] The structures of the polymers used by Hydranautics are shown in Figure 5.8.

Interfacial Composite Membranes

Since the discovery by Cadotte and his co-workers that high-flux, high-rejection reverse osmosis membranes can be made by interfacial polymerization [7,9,10], this method has become the new industry standard. Interfacial composite membranes have significantly higher salt rejections and fluxes than cellulose acetate membranes. The first membranes made by Cadotte had salt rejections in tests with 3.5 % sodium chloride solutions (synthetic seawater) of greater than 99 % and fluxes of 18 gal/ft^2 · day at a pressure of 1500 psi. The membranes could also be operated at temperatures above 35 °C, the temperature ceiling for Loeb–Sourirajan cellulose acetate membranes. Today's interfacial composite membranes are significantly better. Typical membranes, tested with 3.5 % sodium chloride solutions,

[1]The ability of a reverse osmosis membrane to withstand chlorine attack without showing significant loss in rejection is measured in ppm · h. This is the product of chlorine exposure expressed in ppm and the length of exposure expressed in hours. Thus, 1000 ppm · h is 1 ppm chlorine for 1000 h or 10 ppm chlorine for 100 h or 1000 ppm chlorine for 1 h, and so on.

have a salt rejection of 99.5 % and a water flux of 30 gal/ft^2 · day at 800 psi; this is less than half the salt passage of the cellulose acetate membranes and twice the water flux. The rejection of low-molecular-weight dissolved organic solutes by interfacial membranes is also far better than cellulose acetate. The only drawback of interfacial composite membranes, and a significant one, is the rapid, permanent loss in selectivity that results from exposure to even ppb levels of chlorine or hypochlorite disinfectants [28]. Although the chlorine resistance of interfacial composite membranes has been improved, these membranes still cannot be used with feed water containing more than a few ppb of chlorine.

The chemistry and properties of some of the important interfacial composite membranes developed over the past 25 years are summarized in Table 5.1 [10,12,29,30]. The chemistry of the FT-30 membrane, which has an all-aromatic structure based on the reaction of phenylene diamine and trimesoyl chloride, is widely used. This chemistry, first developed by Cadotte [9] and shown in Figure 5.9, is now used in modified form by all the major reverse osmosis membrane producers.

For a few years after the development of the first interfacial composite membranes, it was believed that the amine portion of the reaction chemistry had to be polymeric to obtain good membranes. This is not the case, and the monomeric amines, piperazine and phenylenediamine, have been used to form membranes with very good properties. Interfacial composite membranes based on urea or amide bonds are subject to degradation by chlorine attack, but the rate of degradation of the membrane is slowed significantly if tertiary aromatic amines are used and the membranes are highly crosslinked. Chemistries based on all-aromatic or piperazine structures are moderately chlorine tolerant and can withstand very low level exposure to chlorine for prolonged periods or exposure to ppm levels

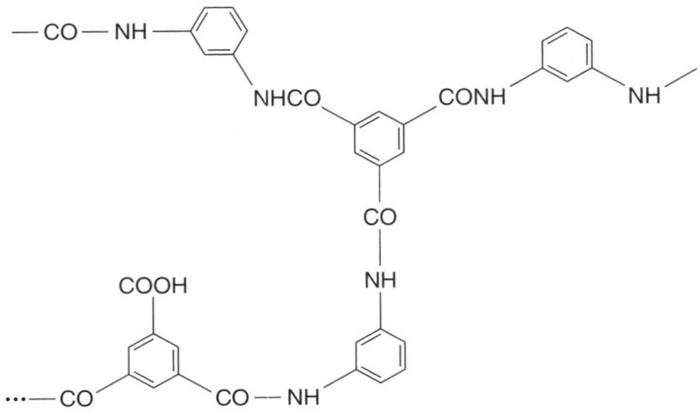

Figure 5.9 Chemical structure of the FT-30 membrane developed by Cadotte using the interfacial reaction of phenylene diamine with trimesoyl chloride

Table 5.1 Characteristics of major interfacial polymerization reverse osmosis membranes

Membrane	Developer	Properties
NS100 Polyethylenimine crosslinked with toluene 2,4-diisocyanate	Cadotte [29] North Star Research	The first interfacial composite membrane achieved seawater desalination characteristics of >99 % rejection, 18 gal/ft^2 · day at 1500 psi with seawater
PA 300/RC-100 Epamine (epichlorohydrin-ethylenediamine adduct) crosslinked with isophthalyl or toluene 2,4-diisocyanate	Riley *et al.* [30] Fluid Systems, San Diego	The PA 300, based on isophthalyl chloride, was introduced first but RC-100, based on toluene 2,4-diisocyanate, proved more stable. This membrane was used at the first large reverse osmosis seawater desalination plant (Jiddah, Saudia Arabia)
NF40 and NTR7250 Piperazine crosslinked with trimesoyl chloride	Cadotte FilmTec [10] and Kamiyama Nitto Denko [12]	The first all-monomeric interfacial membrane. Only modest seawater desalination properties but is a good brackish water membrane. More chlorine-tolerant than earlier membranes because of the absence of secondary amine bonds
FT-30/SW-30 *m*-Phenylenediamine crosslinked with trimesoyl chloride	Cadotte FilmTec [10]	An all-aromatic, highly crosslinked structure giving exceptional salt rejection and very high fluxes. By tailoring the preparation techniques, brackish water or seawater membranes can be made. Seawater version has a rejection 99.3–99.5 % at 800 psi. Brackish water version has >99 % salt rejection at 25 gal/ft^2 · day and 225 psi. All the major reverse osmosis companies produce variations of this membrane

for a few days. Early interfacial composite membranes such as the NS100 or PA300 membrane showed significant degradation at a few hundred ppm · h. Current membranes, such as the fully aromatic FilmTec FT-30 or the Hydranautics ESPA membrane, can withstand up to 1000 ppm · h chlorine exposure. A number of chlorine tolerance studies have been made over the years; a discussion of the literature has been given by Glater *et al.* [31]. Heavy metal ions such as iron appear to strongly catalyze chlorine degradation. For example, the FT-30 fully aromatic membrane is somewhat chlorine resistant in heavy-metal-free water, but in natural waters, which normally contain heavy metal ions, chlorine resistance is low. The rate of chlorine attack is also pH sensitive.

Other Membrane Materials

An interesting group of composite membranes with very good properties is produced by condensation of furfuryl alcohol with sulfuric acid. The first membrane of this type was made by Cadotte at North Star Research and was known as the NS200 membrane [32]. These membranes are not made by the interfacial composite process; rather a polysulfone microporous support membrane is contacted first with an aqueous solution of furfuryl alcohol and then with sulfuric acid. The coated support is then heated to 140 °C. The furfuryl alcohol forms a polymerized, crosslinked layer on the polysulfone support; the membrane is completely black. The chemistry of condensation and reaction is complex, but a possible polymerization scheme is shown in Figure 5.10.

These membranes have exceptional properties, including seawater salt rejections of up to 99.6 % and fluxes of 23 $gal/ft^2 \cdot day$ at 800 psi. Unfortunately, they are even more sensitive to oxidants such as chlorine or dissolved oxygen than the polyamide/polyurea interfacial composites. The membranes lose their excellent properties after a few hundred hours of operation unless the feed water is completely free of dissolved chlorine and oxygen. A great deal of work was devoted to stabilizing this membrane, with little success.

Later, Kurihara and co-workers [33] at Toray produced a related membrane, using 1,3,5-tris(hydroxy ethyl) isocyanuric acid as a comonomer. A possible reaction scheme is shown in Figure 5.11. This membrane, commercialized by Toray under the name PEC-1000, has the highest rejection of any membrane developed,

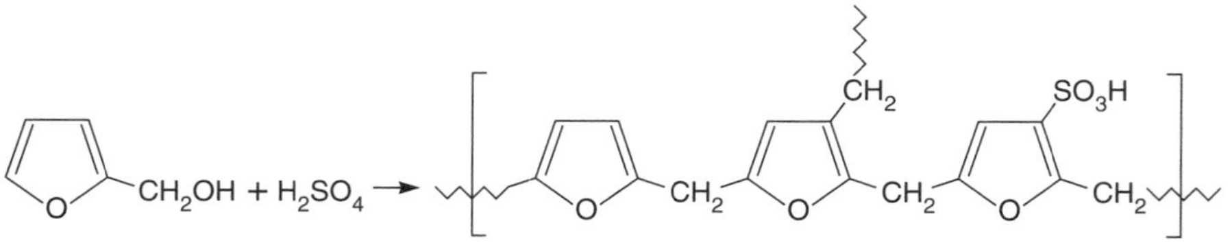

Figure 5.10 Formation of the NS200 condensation membrane

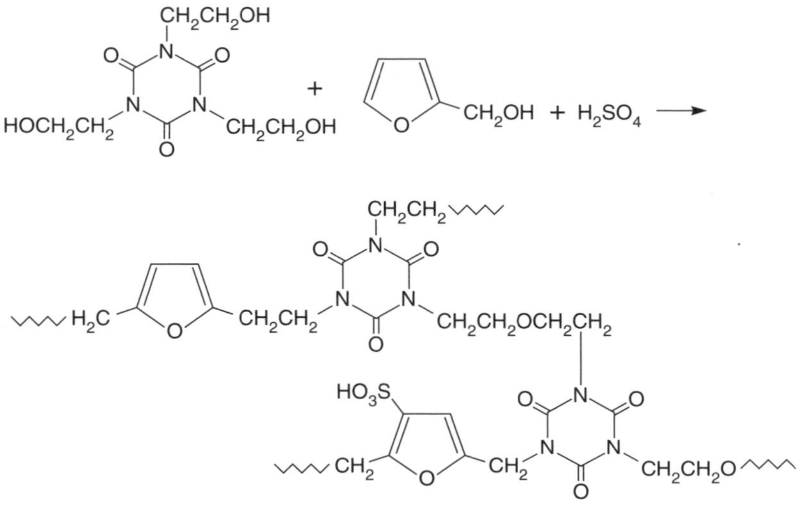

Figure 5.11 Reaction sequence for Toray's PEC-1000 membrane

with seawater rejections of 99.9 % and fluxes of 12 gal/ft^2 · day at 1000 psi. The membrane also shows the highest known rejections to low-molecular-weight organic solutes, typically more than 95 % from relatively concentrated feed solutions [34]. Unfortunately these exceptional selectivities are accompanied by the same sensitivity to dissolved oxidants as the NS200 membrane. This problem was never completely solved, so the PEC-1000 membrane, despite its unsurpassed properties, is no longer commercially available.

Reverse Osmosis Membrane Categories

Reverse osmosis membranes can be grouped into three main categories:

- Seawater and brackish water desalination membranes operated with 0.5 to 5 wt% salt solutions at pressures of 200–1000 psi.
- Low-pressure nanofiltration membranes operated with 200–5000 ppm salt solutions at pressures of 100–200 psi.
- Hyperfiltration membranes used to separate solutes from organic solvent solutions.

Seawater and Brackish Water Desalination Membranes

The relative performances of membranes produced for the desalination market are shown in Figure 5.12, a plot of sodium chloride rejection as a function of

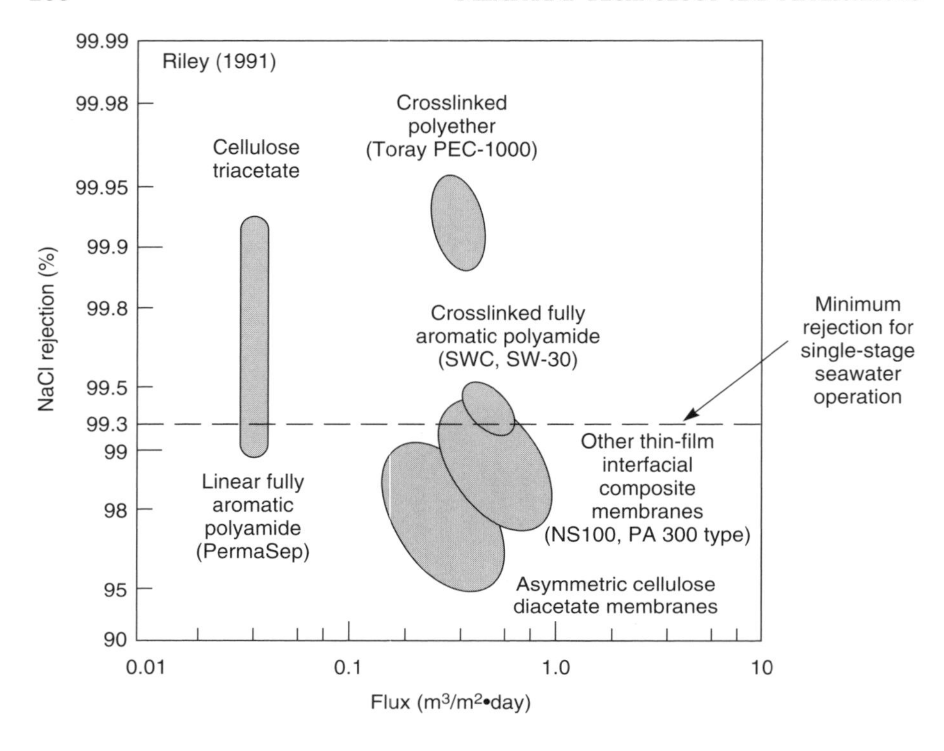

Figure 5.12 Performance characteristics of membranes operating on seawater at 56 kg/cm^2 (840 psi) and 25 °C [13]

membrane flux. The figure is divided into two sections by a dotted line at a rejection of 99.3 %. This salt rejection is generally considered to be the minimum sodium chloride rejection that can produce potable water from seawater in a practical single-stage reverse osmosis plant. Membranes with lower sodium chloride rejections can be used to desalinate seawater, but at least a portion of the product water must be treated in a second-stage operation to achieve the target average permeate salt concentration of less than 500 ppm. Two-stage operation is generally not competitive with alternative desalination technologies.

As Figure 5.12 shows, Toray's PEC-1000 crosslinked furfuryl alcohol membrane has by far the best sodium chloride rejection combined with good fluxes. This explains the sustained interest in this membrane despite its extreme sensitivity to dissolved chlorine and oxygen in the feed water. Hollow fine fiber membranes made from cellulose triacetate by Toyobo or aromatic polyamides by Permasep (Du Pont) are also comfortably in the one-stage seawater desalination performance range, but the water fluxes of these membranes are low. However, because large-surface-area, hollow fine fiber reverse osmosis modules can be produced very economically, these membranes remained competitive until 2000, when DuPont finally ceased production. Currently, all new seawater desalination plants are based on interfacial composite membranes of the fully aromatic type, such as the SW-30 membrane of FilmTec (Dow) or the SWC membrane of Hydranautics (Nitto). Even the best Loeb–Sourirajan cellulose acetate membranes are not suitable for one-stage seawater desalination because their maximum salt rejection is less than 99 %.

Brackish water generally has a salt concentration in the 2000–10 000 ppm range. Groundwater aquifers with these salt levels must be treated to make the water useful. The objective of the desalination plant is to convert 80–90 % of the feed water to a desalted permeate containing 200–500 ppm salt and a concentrated brine that is reinjected into the ground, sent to an evaporation pond, or discharged to the sea. In this application, membranes with 95–98 % sodium chloride rejection are usually adequate. For this reason some brackish water plants still use cellulose acetate membranes with salt rejections of 96–98 %, although interfacial composite membranes are more common. The fluxes and rejections of the composite membranes at the same operating pressures are usually greater than those of cellulose acetate membranes. Therefore, composite membranes are always preferred for large operations such as municipal drinking water plants, which can be built to handle the membrane's chlorine sensitivity. Some small system operators, on the other hand, still prefer cellulose acetate membranes because of their greater stability. The membranes are often operated at higher pressures to obtain the required flux and salt rejection.

The comparative performance of high-pressure, high-rejection reverse osmosis membranes, medium-pressure brackish water desalting membranes, and low-pressure nanofiltration membranes is shown in Table 5.2. Generally, the performance of a membrane with a particular salt can be estimated reliably once the

Table 5.2 Properties of current good-quality commercial membranes

Parameter	Seawater membrane (SW-30)	Brackish water membrane (CA)	Nanofiltration membrane (NTR-7250)
Pressure (psi)	800–1000	300–500	100–150
Solution concentration (%)	1–5	0.2–0.5	0.05
Rejection (%)			
NaCl	99.5	97	60
$MgCl_2$	99.9	99	89
$MgSO_4$	99.9	99.9	99
Na_2SO_4	99.8	99.1	99
$NaNO_3$	90	90	45
Ethylene glycol	70	—	—
Glycerol	96	—	—
Ethanol	—	20	20
Sucrose	100	99.9	99.0

performance of the membrane with one or two marker salts, such as sodium chloride and magnesium sulfate, is known. The rejection of dissolved neutral organic solutes is less predictable. For example, the PEC-1000 membrane had rejections of greater than 95 % for almost all dissolved organics, but the rejections of even the best cellulose acetate membrane are usually no greater than 50–60 %.

Nanofiltration Membranes

The goal of most of the early work on reverse osmosis was to produce desalination membranes with sodium chloride rejections greater than 98 %. More recently membranes with lower sodium chloride rejections but much higher water permeabilities have been produced. These membranes, which fall into a transition region between pure reverse osmosis membranes and pure ultrafiltration membranes, are called loose reverse osmosis, low-pressure reverse osmosis, or more commonly, nanofiltration membranes. Typically, nanofiltration membranes have sodium chloride rejections between 20 and 80 % and molecular weight cutoffs for dissolved organic solutes of 200–1000 dalton. These properties are intermediate between reverse osmosis membranes with a salt rejection of more than 90 % and molecular weight cut-off of less than 50 and ultrafiltration membranes with a salt rejection of less than 5 %.

Although some nanofiltration membranes are based on cellulose acetate, most are based on interfacial composite membranes. The preparation procedure used to form these membranes can result in acid groups attached to the polymeric backbone. Neutral solutes such as lactose, sucrose and raffinose are not affected by the presence of charged groups and the membrane rejection increases in proportion to solute size. Nanofiltration membranes with molecular weight cut-offs to neutral solutes between 150 and 1500 dalton are produced. Typical rejection curves for low molecular weight solutes by two representative membranes are shown in Figure 5.13 [35].

The rejection of salts by nanofiltration membranes is more complicated and depends on both molecular size and Donnan exclusion effects caused by the acid groups attached to the polymer backbone. The phenomenon of Donnan exclusion is described in more detail in Chapter 10. In brief, charged groups tend to exclude ions of the same charge, particularly multivalent ions while being freely permeable to ions of the opposite charge, particularly multivalent ions.

Some results obtained by Peters *et al.* that illustrate the type of results that can be produced are shown in Figure 5.14 [36], in which the permeation properties of neutral, positively charged and negatively charged membranes are compared.

The neutral nanofiltration membrane rejects the various salts in proportion to molecular size, so the order of rejection is simply

$$Na_2SO_4 > CaCl_2 > NaCl$$

The anionic nanofiltration membrane has positive groups attached to the polymer backbone. These positive charges repel positive cations, particularly divalent

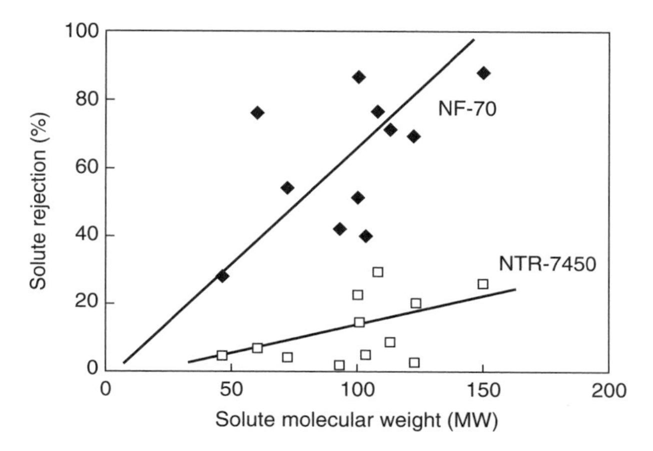

Figure 5.13 Rejection of neutral solutes by two membrane types spanning the range of commonly available nanofiltration membranes [35]

cations such as Ca^{2+}, while attracting negative anions, particularly divalent anions such as SO_4^{2-}. The result is an order of salt rejection

$$CaCl_2 > NaCl > Na_2SO_4$$

The cationic nanofiltration membrane has negative groups attached to the polymer backbone. These negative charges repel negative anions, such as SO_4^{2-}, while attracting positive cations, particularly divalent cations such as Ca^{2+}. The result is an order of salt rejection

$$Na_2SO_4 > NaCl > CaCl_2$$

Many nanofiltration membranes follow these rules, but oftentimes the behavior is more complex. Nanofiltration membranes frequently combine both size and Donnan exclusion effects to minimize the rejection of all salts and solutes. These so-called low-pressure reverse osmosis membranes have very high rejections and high permeances of salt at low salt concentrations, but lose their selectivity at salt concentrations above 1000 or 2000 ppm salt in the feed water. The membranes are therefore used to remove low levels of salt from already relatively clean water. The membranes are usually operated at very low pressures of 50–200 psig.

Hyperfiltration Organic Solvent Separating Membranes

A promising new application of reverse osmosis in the chemical industry is the separation of organic/organic mixtures. These separations are difficult because of the high osmotic pressures that must be overcome and because they require

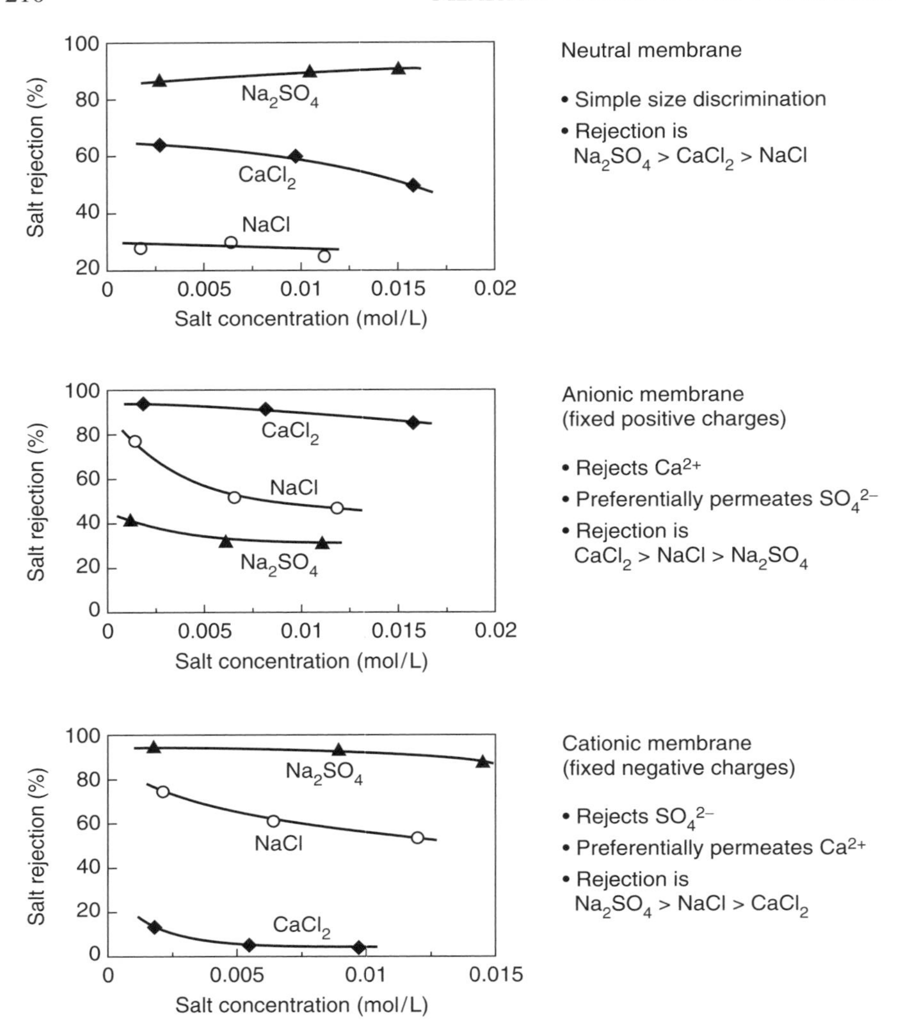

Figure 5.14 Salt rejection with neutral, anionic and cationic nanofiltration membranes showing the effect of Donnan exclusion and solute size on relative rejections. Data of Peters *et al.* [36]

membranes that are sufficiently solvent-resistant to be mechanically stable, but are also sufficiently permeable for good fluxes to be obtained. Nonetheless this is an area of keen industrial interest, and from 1988 to 2002 more than 70 US patents covering membranes and membrane systems for these applications were issued.

Developing membranes for processing organic solvent solutions is more difficult than conventional reverse osmosis because different membranes must be developed for each category of solvent. In the 1980s, Nitto Denko developed polyimide-based ultrafiltration membranes that found a small use in the recovery of acetone, toluene, ethyl acetate and hexane and other solvents from waste paint and polymer solutions [37]. These were microporous membranes with a molecular weight cut-off of 2000–6000. The first dense, solution-diffusion, hyperfiltration membranes did not appear until the late 1990s. Kiryat Weitzman, Ltd, now part of Koch (Abcor), produced crosslinked silicone composite membranes that have some uses in the hyperfiltration of nonpolar solvents [38,39]. The flux of different simple solvents through these membranes is shown in Figure 5.15. These membranes can be used as nanofiltration membranes to separate large dyes or catalyst solutes from solvents. However, because the membranes are made from rubbers that are easily swollen and plasticized by most solvents, they show poor selectivity when used to separate simple solvent mixtures.

The first, and currently only, successful solvent-permeable hyperfiltration membrane is the Starmem® series of solvent-resistant membranes developed by W.R. Grace [40]. These are asymmetric polyimide phase-inversion membranes prepared from Matrimid® (Ciba-Geigy) and related materials. The Matrimid polyimide structure is extremely rigid with a T_g of 305 °C and the polymer remains glassy and unswollen even in aggressive solvents. These membranes found their first large-scale commercial use in Mobil Oil's processes to separate lube oil from methyl ethyl ketone–toluene solvent mixtures [41–43]. Scarpello *et al.* [44] have also achieved rejections of >99 % when using these membranes to separate dissolved phase transfer catalysts (MW ~ 600) from tetrahydrofuran and ethyl acetate solutions.

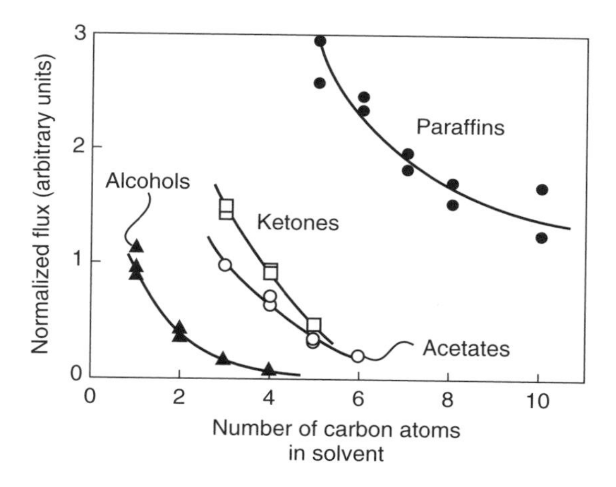

Figure 5.15 Normalized flux of homologous solvent series versus the number of carbon atoms in the solvent molecules (MFP-60 Kiryat Weitzman membranes) [39]

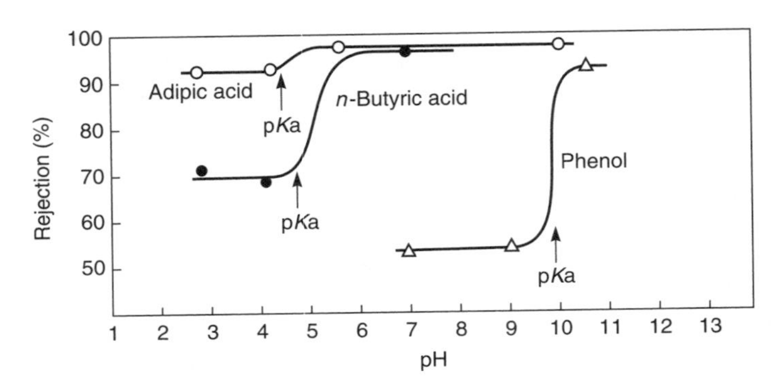

Figure 5.16 Effect of pH on rejection of organic acids. Solute rejection increases at the pKa as the acid converts to the ionized form. Data from Matsuura and Sourirajan [46,47]

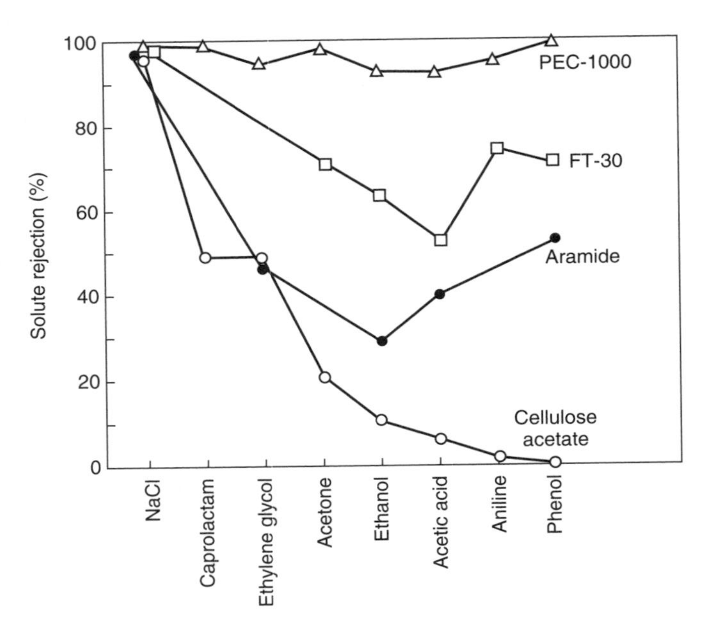

Figure 5.17 Organic rejection data for the PEC-1000 membrane compared to FT-30, anisotropic aramide and anisotropic cellulose membranes [28]

Membrane Selectivity

Rautenbach and Albrecht [45] have proposed some general guidelines for membrane selectivity that can be summarized as follows:

1. Multivalent ions are retained better than monovalent ions. Although the absolute values of the salt rejection vary over a wide range, the ranking for the

different salts is the same for all membranes. In general, the order of rejection of ions by reverse osmosis membranes is as shown below.
For cations:

$$Fe^{3+} > Ni^{2+} \approx Cu^{2+} > Mg^{2+} > Ca^{2+} > Na^{+} > K^{+} \quad (5.7)$$

For anions:

$$PO_4{}^{3-} > SO_4{}^{2-} > HCO_3{}^{-} > Br^{-} > Cl^{-} > NO_3{}^{-} \approx F^{-} \quad (5.8)$$

2. Dissolved gases such as ammonia, carbon dioxide, sulfur dioxide, oxygen, chlorine and hydrogen sulfide always permeate well.
3. Rejection of weak acids and bases is highly pH dependent. When the acid or base is in the ionized form the rejection will be high, but in the nonionized form rejection will be low [46,47]. Data for a few weak acids are shown in Figure 5.16. At pHs above the acid pKa, the solute rejection rises significantly, but at pHs below the pKa, when the acid is in the neutral form, the rejection falls.
4. Rejection of neutral organic solutes generally increases with the molecular weight (or diameter) of the solute. Components with molecular weights above 100 are well rejected by all reverse osmosis membranes. Although differences between the rejection of organic solutes by different membranes are substantial, as the data in Figure 5.17 show, the rank order is generally consistent between membranes. Caprolactam rejection, for example, is better than ethanol rejection for all reverse osmosis membranes. The dependence of solute rejection on molecular weight is shown for three different membranes in Figure 5.18.
5. Negative rejection coefficients, that is, a higher concentration of solute in the permeate than in the feed are occasionally observed, for example, for phenol and benzene with cellulose acetate membranes [48].

Membrane Modules

Currently, 8-in.-diameter, 40-in.-long spiral-wound modules are the type most commonly used for reverse osmosis. Five to seven modules are housed inside a filament-wound, fiber-glass-reinforced plastic tube. Larger modules, up to 12 in. diameter and 60 in. length, are produced by some manufacturers but have not been widely adopted. The module elements can be removed from the pressure vessels and exchanged as needed. A photograph of a typical skid-mounted system is shown in Figure 5.19. A typical spiral-wound 8-in.-diameter membrane module will produce 8000–10 000 gal/day of permeate, so the 75-module plant shown in Figure 5.19 has a capacity of about 700 000 gal/day.

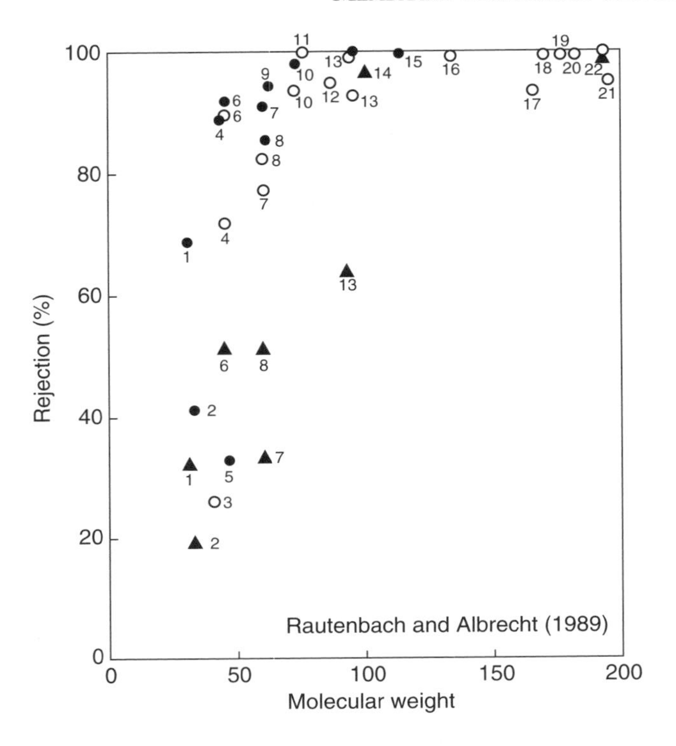

Number	Name	Number	Name
1	Formaldehyde	12	Ethyl acetate
2	Methanol	13	Phenol
3	Acetonitrile	14	*n*-Methyl-2-pyrrolidone
4	Acetaldehyde	15	ε-Caprolactam
5	Formic acid	16	D,L-aspartic acid
6	Ethanol	17	Tetrachloroethylene
7	Acetic acid	18	*o*-Phenyl phenol
8	Urea	19	Butyl benzonate
9	Ethylene glycol	20	Trichlorobenzene
10	Methyl ethyl ketone	21	Dimethyl phthalate
11	Glycine	22	Citric acid

Figure 5.18 Organic solute rejection as a function of solute molecular weight for three representative reverse osmosis membranes [45]: the interfacial composite membranes, (○) PA300 (UOP) and (▲) NTR 7197 (Nitto), and the crosslinked furfuryl alcohol membrane (●) PEC-1000 (Toray). Reprinted from R. Rautenbach and R. Albrecht, *Membrane Processes*, Copyright © 1989. This material is used by permission of John Wiley & Sons, Inc.

Figure 5.19 Skid-mounted reverse osmosis plant able to produce 700 000 gal/day of desalted water. Courtesy of Christ Water Technology Group

Hollow fine fiber modules made from cellulose triacetate or aromatic polyamides were produced in the past for seawater desalination. These modules incorporated the membrane around a central tube, and feed solution flowed rapidly outward to the shell. Because the fibers were extremely tightly packed inside the pressure vessel, flow of the feed solution was quite slow. As much as 40–50 % of the feed could be removed as permeate in a single pass through the module. However, the low flow and many constrictions meant that extremely good pretreatment of the feed solution was required to prevent membrane fouling from scale or particulates. A schematic illustration of such a hollow fiber module is shown in Figure 3.47.

Membrane Fouling Control

Membrane fouling is the main cause of permeant flux decline and loss of product quality in reverse osmosis systems, so fouling control dominates reverse osmosis system design and operation. The cause and prevention of fouling depend greatly on the feed water being treated, and appropriate control procedures must be devised for each plant. In general, sources of fouling can be divided into four principal categories: scale, silt, bacteria, and organic. More than one category may occur in the same plant.

Fouling control involves pretreatment of the feed water to minimize fouling as well as regular cleaning to handle any fouling that still occurs. Fouling by particulates (silt), bacteria and organics such as oil is generally controlled by a suitable pretreatment procedure; this type of fouling affects the first modules in the plant the most. Fouling by scaling is worse with more concentrated feed solutions; therefore, the last modules in the plant are most affected because they are exposed to the most concentrated feed water.

Scale

Scale is caused by precipitation of dissolved metal salts in the feed water on the membrane surface. As salt-free water is removed in the permeate, the concentration of ions in the feed increases until at some point the solubility limit is exceeded. The salt then precipitates on the membrane surface as scale. The proclivity of a particular feed water to produce scale can be determined by performing an analysis of the feed water and calculating the expected concentration factor in the brine. The ratio of the product water flow rate to feed water flow rate is called the recovery rate, which is equivalent to the term stage-cut used in gas separation.

$$\text{Recovery Rate} = \frac{\text{product flow rate}}{\text{feed flow rate}} \tag{5.9}$$

Assuming all the ions remain in the brine solution, the concentration factor is given by

$$\text{Concentration factor} = \frac{1}{1 - \text{recovery rate}} \tag{5.10}$$

The relationship between brine solution concentration factor and water recovery rate is shown in Figure 5.20. With plants that operate below a concentration factor of 2, that is, 50 % recovery rate, scaling is not normally a problem. However, many brackish water reverse osmosis plants operate at recovery rates of 80 or 90 %. Salt concentrations on the brine side of the membrane may then be far above the solubility limit. In order of importance, the salts that most commonly form scale are:

- calcium carbonate;
- calcium sulfate;
- silica complexes;
- barium sulfate;

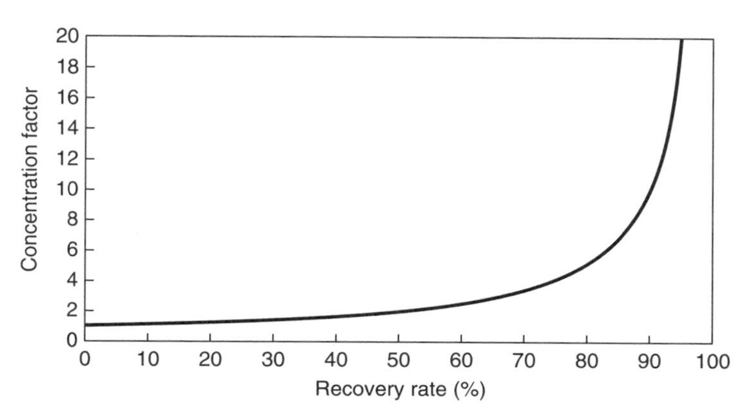

Figure 5.20 The effect of water recovery rate on the brine solution concentration factor

- strontium sulfate;
- calcium fluoride.

Scale control is complex; the particular procedure depends on the composition of the feed water. Fortunately, calcium carbonate scale, by far the most common problem, is easily controlled by acidifying the feed or by using an ion exchange water softener to exchange calcium for sodium. Alternatively, an antiscalant chemical such as sodium hexametaphosphate can be added. Antiscalants interfere with the precipitation of the insoluble salt and maintain the salt in solution even when the solubility limit is exceeded. Polymeric antiscalants may also be used, sometimes in combination with a dispersant to break up any flocs that occur.

Silica can be a particularly troublesome scalant because no effective antiscalant or dispersant is available. The solubility of silica is a strong function of pH and temperature, but in general the brine should not exceed 120 ppm silica. Once formed, silica scale is difficult to remove.

Silt

Silt is formed by suspended particulates of all types that accumulate on the membrane surface. Typical sources of silt are organic colloids, iron corrosion products, precipitated iron hydroxide, algae, and fine particulate matter. A good predictor of the likelihood of a particular feed water to produce fouling by silt is the silt density index (SDI) of the feed water. The SDI, an empirical measurement (ASTM Standard D-4189-82, 1987), is the time required to filter a fixed volume of

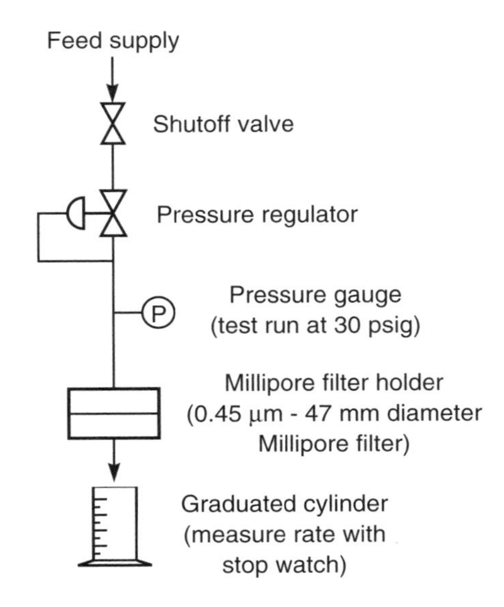

(1) Measure the amount of time required for 500 ml of feed water to flow through a 0.45 micrometer Millipore filter (47 mm in diameter) at a pressure of 30 psig.

(2) Allow the feed water to continue flowing at 30 psig applied pressure and measure the time required for 500 ml to flow through the filter after 5, 10 and 15 minutes.

(3) After completion of the test, calculate the SDI by using the equation below.

$$SDI = \frac{100\,(1 - T_i/T_f)}{T_t}$$

where SDI = Silt Density Index
T_t = Total elapsed test time (either 5, 10 or 15 minutes)
T_i = Initial time in seconds required to collect the 500 ml sample
T_f = Time in seconds required to collect the second 500 ml sample after test time T_t (normally after 15 minutes).

Figure 5.21 The silt density index (SDI) test [49]. Reprinted with permission from Noyes Publications

water through a standard 0.45-μm pore size microfiltration membrane. Suspended material in the feed water that plugs the microfilter increases the sample filtration time, giving a higher SDI. The test procedure is illustrated in Figure 5.21 [49].

An SDI of less than 1 means that the reverse osmosis system can run for several years without colloidal fouling. An SDI of less than 3 means that the system can run several months between cleaning. An SDI of 3–5 means that particulate fouling is likely to be a problem and frequent, regular cleaning will be needed. An SDI of more than 5 is unacceptable and indicates that additional pretreatment is required to bring the feed water into an acceptable range. The maximum tolerable SDI also varies with membrane module design. Spiral-wound modules generally require an SDI of less than 5, whereas hollow fine fiber modules are more susceptible to fouling and require an SDI of less than 3.

To avoid fouling by suspended solids, some form of feed water filtration is required. All reverse osmosis units are fitted with a 0.45-μm cartridge filter in front of the high-pressure pump, but a sand filter, sometimes supplemented by addition of a flocculating chemical such as alum or a cationic polymer, may be required. The target SDI after filtration is normally less than 3–5. Groundwaters usually have very low SDI values, and cartridge filtration is often sufficient. However, surface or seawater may have an SDI of up to 200, requiring flocculation, coagulation, and deep-bed multimedia filtration before reverse osmosis treatment.

Biofouling

Biological fouling is the growth of bacteria on the membrane surface. The susceptibility of membranes to biological fouling is a strong function of the membrane composition. Cellulose acetate membranes are an ideal nutrient for bacteria and can be completely destroyed by a few weeks of uncontrolled bacterial attack. Therefore, feed water to cellulose acetate membranes must always be sterilized. Polyamide hollow fibers are also somewhat susceptible to bacterial attack, but thin-film composite membranes are generally quite resistant. Periodic treatment of such membranes with a bactericide usually controls biological fouling. Thus, control of bacteria is essential for cellulose acetate membranes and desirable for polyamides and composite membranes. Because cellulose acetate can tolerate up to 1 ppm chlorine, sufficient chlorination is used to maintain 0.2 ppm free chlorine. Chlorination can also be used to sterilize the feed water to polyamide and interfacial composite membranes, but residual chlorine must then be removed because the membranes are chlorine-sensitive. Dechlorination is generally achieved by adding sodium metabisulfate. In ultrapure water systems, water sterility is often maintained by UV sterilizers.

Organic Fouling

Organic fouling is the attachment of materials such as oil or grease onto the membrane surface. Such fouling may occur accidentally in municipal drinking

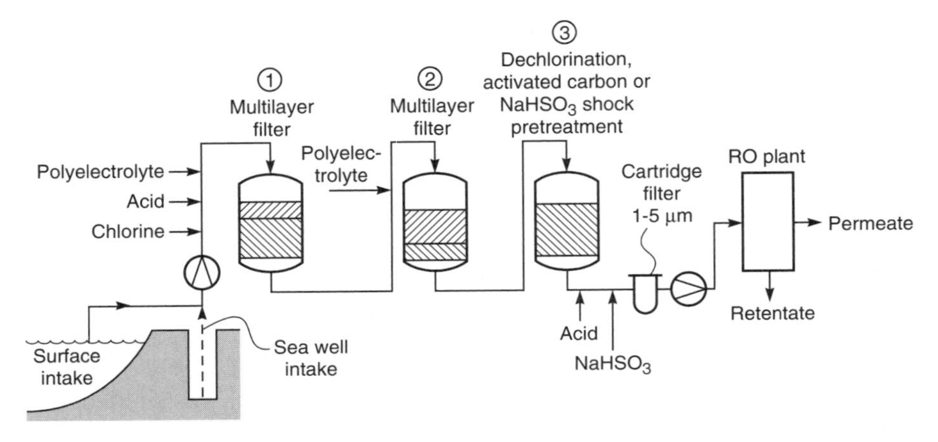

Figure 5.22 Flow scheme showing the pretreatment steps in a typical seawater reverse osmosis system [50]

water systems, but is more common in industrial applications in which reverse osmosis is used to treat a process or effluent stream. Removal of the organic material from the feed water by filtration or carbon adsorption is required.

An example of a complete pretreatment flow scheme for a seawater reverse osmosis plant is shown in Figure 5.22 [50]. The water is controlled for pH, scale, particulates and biological fouling. The feed water is first treated with chlorine to sterilize the water and to bring it to a pH of 5–6. A polyelectrolyte is added to flocculate suspended matter, and two multilayer depth filters then remove suspended materials. The water is dechlorinated by dosing with sodium bisulfite followed by passage through an activated carbon bed. As a final check the pH is adjusted a second time, and the water is filtered through a 1- to 5-μm cartridge filter before being fed to the reverse osmosis modules. Obviously, such pretreatment is expensive and may represent as much as one-third of the operating and capital cost of the plant; however, it is essential for reliable long-term operation.

Membrane Cleaning

A good pretreatment system is essential to achieve a long reverse osmosis membrane life, but pretreatment must be backed up by an appropriate cleaning schedule. Generally this is done once or twice a year, but more often if the feed is a problem water. As with pretreatment, the specific cleaning procedure is a function of the feed water chemistry, the type of membrane, and the type of fouling. A typical cleaning regimen consists of flushing the membrane modules by recirculating the cleaning solution at high speed through the module, followed by a soaking period, followed by a second flush, and so on. The chemical cleaning agents commonly used are acids, alkalis, chelatants, detergents, formulated products, and sterilizers.

Acid cleaning agents such as hydrochloric, phosphoric, or citric acids effectively remove common scaling compounds. With cellulose acetate membranes the pH of the solution should not go below 2.0 or else hydrolysis of the membrane will occur. Oxalic acid is particularly effective for removing iron deposits. Acids such as citric acid are not very effective with calcium, magnesium, or barium sulfate scale; in this case a chelatant such as ethylene diamine tetraacetic acid (EDTA) may be used.

To remove bacteria, silt or precipitates from the membrane, alkalis combined with surfactant cleaners are often used. Biz® and other laundry detergents containing enzyme additives are useful for removing biofoulants and some organic foulants. Most large membrane module producers now distribute formulated products, which are a mixture of cleaning compounds. These products are designed for various common feed waters and often provide a better solution to membrane cleaning than devising a cleaning solution for a specific feed.

Sterilization of a membrane system is also required to control bacterial growth. For cellulose acetate membranes, chlorination of the feed water is sufficient to control bacteria. Feed water to polyamide or interfacial composite membranes need not be sterile, because these membranes are usually fairly resistant to biological attack. Periodic shock disinfection using formaldehyde, peroxide or peracetic acid solutions as part of a regular cleaning schedule is usually enough to prevent biofouling.

Repeated cleaning gradually degrades reverse osmosis membranes. Most manufacturers now supply membrane modules with a 1- to 2-year limited warranty depending on the application. Well designed and maintained plants with good feed water pretreatment can usually expect membrane lifetimes of 3 years, and lifetimes of 5 years or more are not unusual. As membranes approach the end of their useful life, the water flux will normally have dropped by at least 20 %, and the salt rejection will have begun to fall. At this point operators may try to 'rejuvenate' the membrane by treatment with a dilute polymer solution. This surface treatment plugs microdefects and restores salt rejection [51]. Typical polymers are poly(vinyl alcohol)/vinyl acetate copolymers or poly(vinyl methyl ether). In this procedure the membrane modules are carefully cleaned and then flushed with dilute solutions of the rejuvenation polymer. The exact mechanism of rejuvenation is unclear.

Applications

Approximately one-half of the reverse osmosis systems currently installed are desalinating brackish or seawater. Another 40 % are producing ultrapure water for the electronics, pharmaceutical, and power generation industries. The remainder are used in small niche applications such as pollution control and food processing. A review of reverse osmosis applications has been done by Williams *et al.* [52].

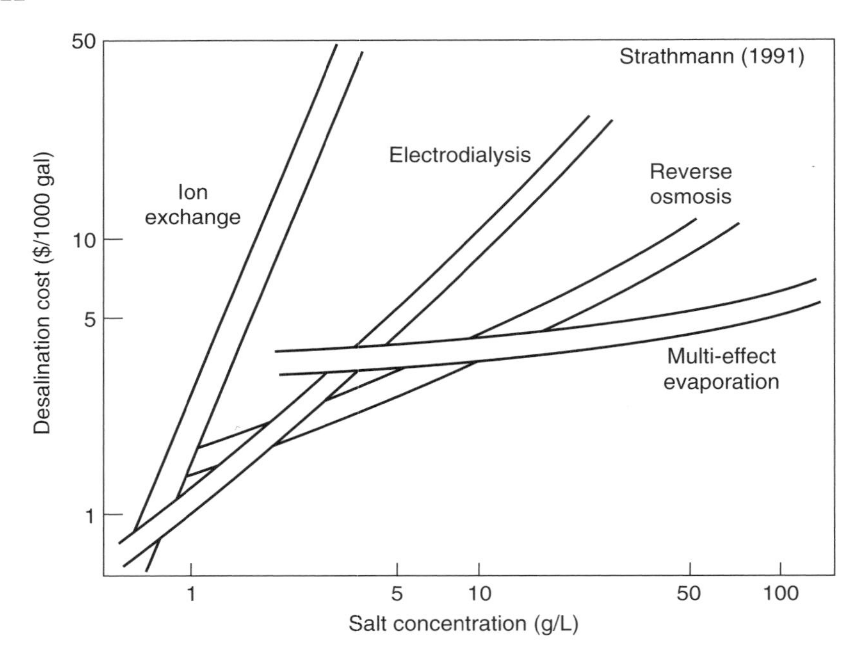

Figure 5.23 Comparative costs of the major desalination technologies as a function of salt concentration. These costs should be taken as a guide only; site-specific factors can affect costs significantly [53]

The relative cost of reverse osmosis compared with other desalting technologies (ion exchange, electrodialysis, and multi-effect evaporation) is shown in Figure 5.23. The operating costs of electrodialysis and ion exchange scale almost linearly in proportion to the salt concentration in the feed. Therefore, these technologies are best suited to low-salt-concentration feed streams. On the other hand, the cost of multi-effect evaporation is relatively independent of the salt concentration and is mainly proportional to the mass of water to be evaporated. Thus, desalination by evaporation is best performed with concentrated salt solution feeds. Reverse osmosis costs increase significantly with salt concentration but at a lower rate than electrodialysis does. The result is that reverse osmosis is the lowest-cost process for streams containing between 3000 and 10 000 ppm salt. However, site-specific factors or plant size often make the technology the best approach for more dilute feed water or for streams as concentrated as seawater (35 000 ppm salt).

The approximate operating costs for brackish and seawater reverse osmosis plants are given in Table 5.3. These numbers are old, but improvements in membrane technology have kept pace with inflation so the costs remain reasonably current.

Table 5.3 Operating costs for large brackish water and seawater reverse osmosis plants [49]. Capital costs are approximately US$1.25 per gal/day capacity for the brackish water plant and US$4–5 per gal/day capacity for the seawater plant

	Brackish water (US$/1000 gal product)	Seawater (US$/1000 gal product)
Energy (US$0.06/kWh)	0.36	1.80
Chemicals	0.09	0.14
Labor	0.12	0.19
Maintenance	0.05	0.22
Membrane replacement	0.10	0.90
Amortization (12 %/20 years)	0.48	1.75
Total	1.20	5.00

Brackish Water Desalination

The salinity of brackish water is usually between 2000 and 10 000 mg/L. The World Health Organization (WHO) recommendation for potable water is 500 mg/L, so up to 90 % of the salt must be removed from these feeds. Early cellulose acetate membranes could achieve this removal easily, so treatment of brackish water was one of the first successful applications of reverse osmosis. Several plants were installed as early as the 1960s.

The osmotic pressure of brackish water is approximately 11 psi per 1000 ppm salt, so osmotic pressure effects do not generally limit water recovery significantly. Limitations are generally due to scaling. Typical water recoveries are in the 70–90 % range, which means the brine stream leaving the system is up to 10 times more concentrated in calcium, sulfate and silica ions present in the feed. If scaling occurs, the last modules in the system must be replaced first.

A simplified flow scheme for a brackish water reverse osmosis plant is shown in Figure 5.24. In this example, it is assumed that the brackish water is heavily contaminated with suspended solids, so flocculation followed by a sand filter and a cartridge filter is used to remove particulates. The pH of the feed solution might be adjusted, followed by chlorination to sterilize the water to prevent bacterial growth on the membranes and addition of an anti-scalant to inhibit precipitation of multivalent salts on the membrane. Finally, if chlorine-sensitive interfacial composite membranes are used, sodium sulfite is added to remove excess chlorine before the water contacts the membrane. Generally, more pretreatment is required in plants using hollow fiber modules than in plants using spiral-wound modules. This is one reason why hollow fiber modules have been displaced by spiral-wound systems for most brackish water installations.

A feature of the system design shown in Figure 5.24 is the staggered arrangement of the module pressure vessels. As the volume of the feed water is reduced as water is removed in the permeate, the number of modules arranged in parallel is also reduced. In the example shown, the feed water passes initially through

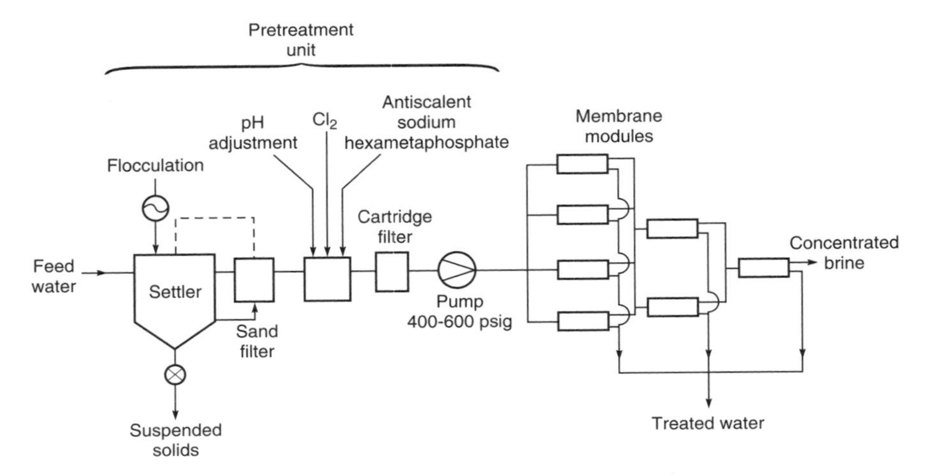

Figure 5.24 Flow schematic of a typical brackish water reverse osmosis plant. The plant contains seven pressure vessels each containing six membrane modules. The pressure vessels are in a 'Christmas tree' array to maintain a high feed velocity through the modules

four modules in parallel, then through two, and finally through a single module in series. This is called a 'Christmas tree' or 'tapered module' design and provides a high average feed solution velocity through the modules. As the volume of the feed water is reduced by removing water as permeate, the number of modules arranged in parallel is reduced also.

The operating pressure of brackish water reverse osmosis systems has gradually fallen over the past 20 years as the permeability and rejections of membranes have steadily improved. The first plants operated at pressures of 800 psi, but typical brackish water plants now operate at pressures in the 200- to 300-psi range. Capital costs of brackish water plants have stayed remarkably constant for almost 20 years; the rule of thumb of US$1.00 per gal/day capacity is still true. Accounting for inflation, this reflects a very large reduction in real costs resulting from the better performance of today's membranes.

Seawater Desalination

Seawater has a salt concentration of 3.2–4.0 %, depending on the region of the world. Because of this high salinity, only membranes with salt rejections of 99.3 % or more can produce potable water in a single pass. Application to seawater desalination of the first-generation cellulose acetate membranes, with rejections of 97–99 %, was limited. With the development of the polyamide hollow fine fibers and interfacial composites, suitable seawater membranes became available, and many plants have been installed. In general, membranes are not

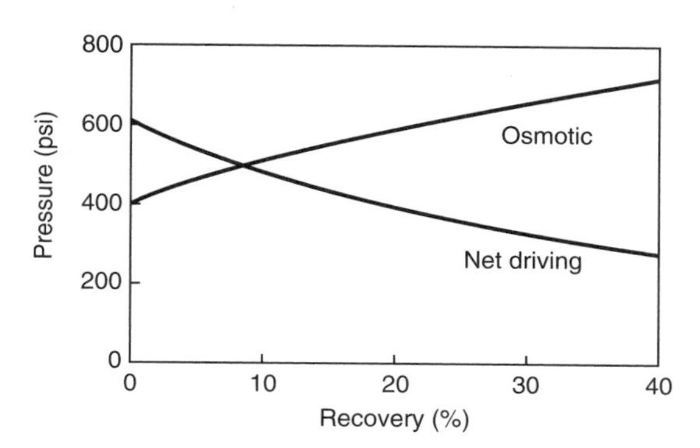

Figure 5.25 Effect of water recovery on the seawater feed osmotic pressure and net driving pressure of a plant operating at 1000 psi

competitive for large seawater desalination plants—multistage flash evaporation is usually used for plants larger than 10 million gal/day capacity. Often these plants are powered by waste steam from an adjacent electric power generation unit. A number of these very large plants have been installed in the Middle East. In the 1–10 million gal/day range membranes are more competitive, and the flexibility of membrane systems as well as their easy start-up/shut-down and turndown capability are advantages.

Early seawater reverse osmosis plants operated at very high pressures, up to 1500 psi, but as membranes improved, operating pressures dropped to 800–1000 psi. The osmotic pressure of seawater is about 350 psi, and the osmotic pressure of the rejected brine can be as much as 600 psi, so osmotic pressure affects the net operating pressure in a plant markedly. This effect is illustrated in Figure 5.25. Typical seawater plants do not operate at a recovery rate of more than 35–45 % because of the high brine osmotic pressure; at this modest recovery rate, more than half of the feed water leaves the plant as pressurized brine. Because of the high pressures involved in seawater desalination, recovery of compression energy from the high-pressure brine stream is almost always worthwhile. This can be achieved with a hydro-turbine linked to the high-pressure feed water pump, lowering total power costs by as much as 30 %.

Raw seawater requires considerable pretreatment before it can be desalinated (Figure 5.22), but these pretreatment costs can be reduced by using shallow sea-front wells as the water source. The SDI of this water is usually quite low, and little more than a sand filter may be required for particulate control. However, sterilization of the water and addition of antiscalants will still be necessary.

Ultrapure Water

Production of ultrapure water for the electronics industry is an established and growing application of reverse osmosis [54,55]. The usual feed is municipal

Table 5.4 Ultrapure water specifications for typical wafer manufacturing process and levels normally found in drinking water

	Ultrapure water	Typical drinking water
Resistivity at 25 °C (megohm-cm)	18.2	—
TOC (ppb)	<5	5000
Particles/L by laser >0.1 μm	<100	—
Bacteria/100 mL by culture	<0.1	<30
Silica, dissolved (ppb)	<3	3000
Boron (ppb)	<1	40
Ions (ppb)		
Na^+	<0.01	3000
K^+	<0.02	2000
Cl^-	<0.02	10 000
Br^-	<0.02	—
NO_3^-	<0.02	—
SO_4^{2-}	<0.02	15 000
Total ions	<0.1	<100 000

drinking water, which often contains less than 200 ppm dissolved solids. However, the electronics industry requires water of extraordinarily high purity for wafer production, so extensive treatment of municipal water is required. Table 5.4 shows the target water quality required by a modern electronics plant compared to that of typical municipal drinking water.

The first ultrapure water reverse osmosis system was installed at a Texas Instruments plant in 1970 as a pretreatment unit to an ion exchange process. These systems have increased in complexity as the needs of the industry for better quality water have increased. The flow scheme for a typical modern ultrapure water treatment system is shown in Figure 5.26. The plant comprises a complex array of operations, each requiring careful maintenance to achieve the necessary water quality. As the key part of the process, the reverse osmosis plant typically removes more than 98 % of all the salts and dissolved particulates in the feed water. Because the feed water is dilute, these systems often operate at very high recovery rates—90 % or more. Carbon adsorption then removes dissolved organics, followed by ion exchange to remove final trace amounts of ionic impurities. Bacterial growth is a major problem in ultrapure water systems; sterility is maintained by continuously recirculating the water through UV sterilizers and cartridge microfilters.

Wastewater Treatment

In principle, pollution control should be a major application for reverse osmosis. In practice, membrane fouling, causing low plant reliability, has inhibited its widespread use in this area. The most common applications are special situations

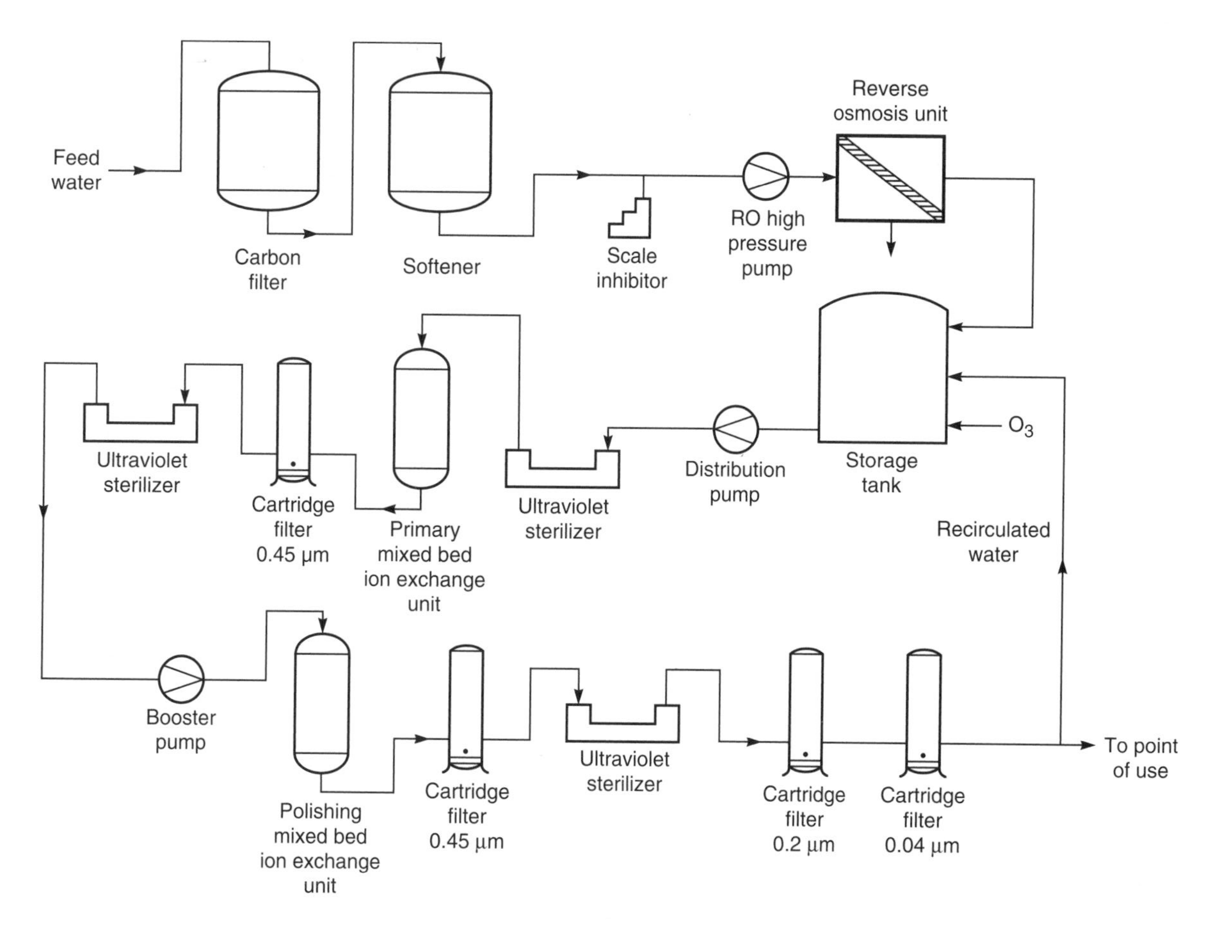

Figure 5.26 Flow schematic of an ultrapure water treatment system [54]

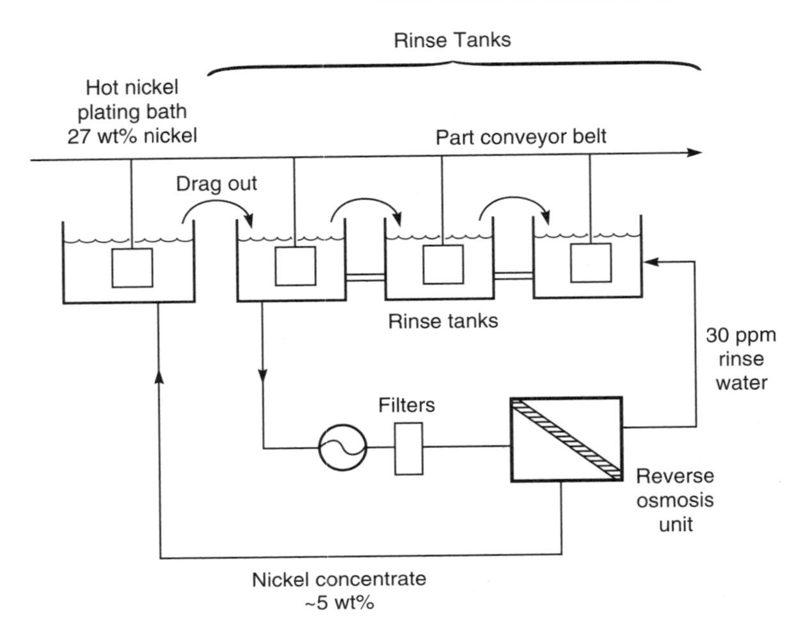

Figure 5.27 Flow scheme showing the use of a reverse osmosis system to control nickel loss from rinse water produced in a countercurrent electroplating rinse tank

in which the chemicals separated from the water are valuable. An example is the recovery of nickel from nickel-plating rinse tanks, shown schematically in Figure 5.27. Watts nickel-plating baths contain high concentrations of nickel and other plating chemicals. After plating, a conveyor belt moves the parts through a series of connected rinse tanks. Water circulates through these tanks to rinse the part free of nickel for the next plating operating. A typical countercurrent rinse tank produces a waste stream containing 2000–3000 ppm nickel; the water is a pollution problem and valuable material is lost. This is an ideal application for reverse osmosis because the rinse water is at nearly neutral pH, in contrast to many plating rinse waters which are very acidic [56,57]. The reverse osmosis unit produces permeate water containing only 20–50 ppm nickel that can be reused and a small nickel concentrate stream that can be sent to the plating tank. Although the concentrate is more dilute than the plating tank drag-out, evaporation from the hot plating bath tank compensates for the extra water.

In the early days of membrane development, membranes were expected to be widely used in the tertiary treatment of water to produce drinking water from sewage. At a cost of US$2–3 per 1000 gal, this idea makes good economic sense in many water-limited regions of the world. However, psychological barriers have inhibited its widespread adoption. A few small plants have been introduced in Japan and at least one large plant in the US. This plant, called Water Factory 21, is in Orange County, California, an arid region where the principal local surface water source, the Colorado River, has a total salinity of 750 ppm. Operation of this 5-million-gal/day system is described in detail by Nusbaum and Argo [58]. The system treats secondary sewage to produce good-quality water, which is reinjected into the aquifer below the county. The water is then mixed with natural groundwater before being removed and used as a drinking water supply elsewhere in the county. Apparently, confusing the source of the water supply in this way makes the process acceptable.

Nanofiltration

Nanofiltration membrane usually have high rejections to most dissolved organic solutes with molecular weights above 100–200 and good salt rejection at salt concentrations below 1000–2000 ppm salt. The membranes are also two- to five-fold more permeable than brackish and sea water reverse osmosis membranes, so they can be operated at pressures as low as 50–150 psig and still produce useful fluxes. For these reasons, their principal application has been in the removal of low levels of contaminants from already relatively clean water. For example, nanofiltration membranes are widely used as point-of-use drinking water treatment units in southern California and the southwestern United States. The water in this region contains on the order of 700 ppm dissolved salt and trace amounts of agricultural run-off contaminants. Many households use small 0.5-m^2 spiral-wound nanofiltration modules (under-the-sink modules) to filter this water using the 30- to 50-psig tap water pressure to provide the driving force. On a larger scale, similar membranes are used to soften municipal water by removing sulfate and divalent cations or as an initial pretreatment unit for an ultrapure water treatment plant.

Organic Solvent Separation

The use of membranes to separate organic solvent solutions is still at a very early stage. One application that has already become commercial is the separation of small solvent molecules from larger hydrocarbons in mixtures resulting from the extraction of vacuum resid oil in refineries [41–43]. Figure 5.28(a) shows a simplified flow diagram of a refining lube oil separation process–these operations are very large. In a 100 000–200 000-barrel/day refinery, about 15 000–30 000 barrels/day of the oil entering the refinery remain as residual oil. A large fraction of this oil is sent to the lube oil plant, where the heavy oil is mixed with 3–10 volumes of a solvent such as methyl ethyl ketone and toluene. On cooling the mixture, the heavy wax components precipitate out and are removed by a drum filter. The light solvent is then stripped from the lube oil by vacuum distillation and recycled through the process. The vacuum distillation step is very energy intensive because of the high solvent-to-oil ratios employed.

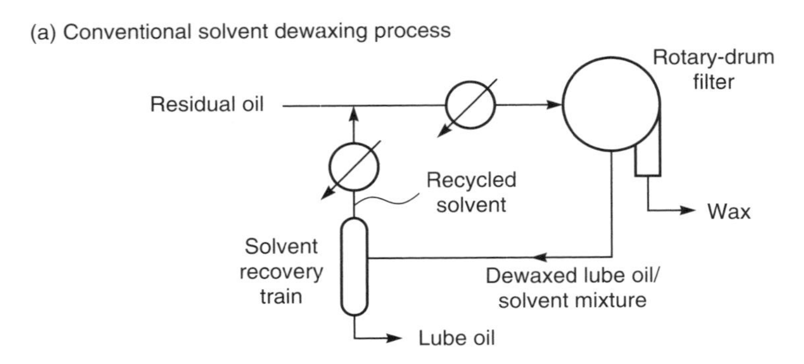

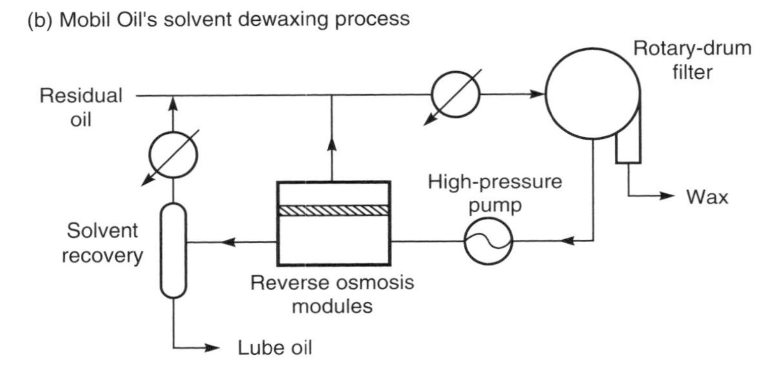

Figure 5.28 Simplified flow schemes of (a) a conventional and (b) Mobil Oil's membrane solvent dewaxing processes. Refrigeration economizers are not shown. The first 3 million gallon/day commercial unit was installed at Mobil's Beaumont refinery in 1998. Polyimide membranes in spiral-wound modules were used [41–43]

A hyperfiltration process developed by Mobil Oil, now ExxonMobil, for this separation is illustrated in Figure 5.28(b). Polyimide membranes formed into spiral-wound modules are used to separate up to 50 % of the solvent from the dewaxed oil. The membranes have a flux of 10–20 gal/ft^2 day at a pressure of 450–650 psi. The solvent filtrate bypasses the distillation step and is recycled directly to the incoming oil feed. The net result is a significant reduction in the refrigeration load required to cool the oil and in the size and energy consumption of the solvent recovery vacuum distillation section.

ExxonMobil is now licensing this technology to other refineries. Development of similar applications in other operations is likely. Initially, applications will probably involve relatively easy separations such as the separation of methyl ethyl ketone/toluene from lube oil described above or soybean oil from hexane in food oil production. Long-term, however, the technology may become sufficiently advanced to be used in more important refining operations, such as fractionation of linear from branched paraffins, or the separation of benzene and other aromatics from paraffins and olefins in the gasoline pool.

Conclusions and Future Directions

The reverse osmosis industry is now well established. The market is divided between three or four large manufacturers, who between them produce 70 % of the membrane modules, and a much larger number of system builders. The system builders buy modules almost as commodities from the various suppliers according to their particular needs. A handful of companies serving various niche markets produce both modules and systems. Total membrane module sales in 1998 were about US$200 million worldwide; system sales were another US$200 million. Short-term prospects for future growth are good. The demand for reverse osmosis systems to produce ultrapure water for the electronics and pharmaceutical industries is very strong. Municipalities in arid regions of the world are also continuing to buy brackish water and some seawater desalination units.

The industry is extremely competitive, with the manufacturers producing similar products and competing mostly on price. Many incremental improvements have been made to membrane and module performance over the past 20 years, resulting in steadily decreasing water desalination costs in inflation-adjusted dollars. Some performance values taken from a paper by Furukawa are shown in Table 5.5. Since 1980, just after the introduction of the first interfacial composite membranes, the cost of spiral-wound membrane modules on a per square meter basis has decreased seven-fold. At the same time the water flux has doubled, and the salt permeability has decreased seven-fold. Taking these improvements into account, today's membranes are almost 100 times better than those of the 1980s. This type of incremental improvement is likely to continue for some time.

The key short-term technical issue is the limited chlorine resistance of interfacial composite membranes. A number of incremental steps made over the past 10–15 years have improved resistance, but current chlorine-resistant interfacial composites do not have the rejection and flux of the best conventional membranes.

Table 5.5 Advances in spiral-wound module reverse osmosis performance

Year	Cost normalized (1980 US$)	Productivity normalized (to 1980)	Reciprocal salt passage normalized (to 1980)	Figure of merit[a]
1980	1.00	1.00	1.00	1.0
1985	0.65	1.10	1.56	2.6
1990	0.34	1.32	2.01	7.9
1995	0.19	1.66	3.52	30.8
2000	0.14	1.94	7.04	99.3

[a]Figure of merit = (productivity) × (reciprocal salt passage/cost).
Source: Dave Furukawa.

All the major membrane manufacturers are working on this problem, which is likely to be solved in the next few years. Three longer-term, related technical issues are fouling resistance, pretreatment, and membrane cleaning. Current membrane modules are subject to fouling by particulates and scale; this fouling can only be controlled by good (and expensive) feed water pretreatment and by membrane cleaning. In some large potential reverse osmosis markets, such as municipal wastewater reclamation and industrial process water treatment, the complexity, expense, and low reliability due to membrane fouling limit expansion significantly.

A further long-term area of research is likely to be the development of reverse osmosis membranes to recover organic solutes from water. This chapter has focused almost entirely on the separation of ionic solutes from water, but some membranes (such as the PEC-1000 membrane) have excellent organic solute rejections also. The PEC-1000 membrane was chemically unstable, but it demonstrated what is achievable with membranes. A stable membrane with similar properties could be used in many wastewater applications.

References

1. A.G. Horvath, Water Softening, US Patent 1,825,631 (September, 1931).
2. C.E. Reid and E.J. Breton, Water and Ion Flow Across Cellulosic Membranes, *J. Appl. Polym. Sci.* **1**, 133 (1959).
3. S. Loeb and S. Sourirajan, Sea Water Demineralization by Means of an Osmotic Membrane, in *Saline Water Conversion II*, R.F. Gould (ed.), Advances in Chemistry Series Number 38, American Chemical Society, Washington, DC, pp. 117–132 (1963).
4. J.C. Westmoreland, Spirally Wrapped Reverse Osmosis Membrane Cell, US Patent 3,367,504 (February, 1968).
5. D.T. Bray, Reverse Osmosis Purification Apparatus, US Patent 3,417,870 (December, 1968).
6. C.P. Shields, Five Years Experience with Reverse Osmosis Systems using DuPont Permasep Permeators, *Desalination* **28**, 157 (1979).
7. J.E. Cadotte, Evaluation of Composite Reverse Osmosis Membrane, in *Materials Science of Synthetic Membranes*, D.R. Lloyd (ed.), ACS Symposium Series Number 269, American Chemical Society, Washington, DC (1985).
8. A. Muirhead, S. Beardsley and J. Aboudiwan, Performance of the 12,000 m^3/day Sea Water Reverse Osmosis Desalination Plant at Jeddah, Saudi Arabia (Jan. 1979–Jan. 1981), *Desalination* **42**, 115 (1982).
9. R.E. Larson, J.E. Cadotte and R.J. Petersen, The FT-30 Seawater Reverse Osmosis Membrane-element Test Results, *Desalination* **38**, 473 (1981).
10. J.E. Cadotte, Interfacially Synthesized Reverse Osmosis Membrane, US Patent 4,277,344 (July, 1981).
11. P. Eriksson, Water and Salt Transport Through Two Types of Polyamide Composite Membranes, *J. Membr. Sci.* **36**, 297 (1988).
12. Y. Kamiyama, N. Yoshioki, K. Matsui and K. Nakagome, New Thin Film Composite Reverse Osmosis Membranes and Spiral Wound Molecules, *Desalination* **51**, 79 (1984).

13. R.L. Riley, Reverse Osmosis, in *Membrane Separation Systems*, R.W. Baker, E.L. Cussler, W. Eykamp, W.J. Koros, R.L. Riley and H. Strathmann (eds), Noyes Data Corp., Park Ridge, NJ, pp. 276–328 (1991).
14. J.E. Cadotte, R.J. Petersen, R.E. Larson and E.E. Erickson, A New Thin Film Sea Water Reverse Osmosis Membrane, Presented at the 5th Seminar on Membrane Separation Technology, Clemson University, Clemson, SC (1980).
15. Z. Amjad (ed.), *Reverse Osmosis,* Van Nostrand Reinhold, New York (1993).
16. B. Parekh (ed.), *Reverse Osmosis Technology*, Marcel Dekker, New York (1988).
17. R.J. Petersen, Composite Reverse Osmosis and Nanofiltration Membranes, *J. Membr. Sci.* **83**, 81 (1993).
18. S. Rosenbaum, H.I. Mahon and O. Cotton, Permeation of Water and Sodium Chloride Through Cellulose Acetate, *J. Appl. Polym. Sci.* **11**, 2041 (1967).
19. H.K. Lonsdale, Properties of Cellulose Acetate Membranes, in *Desalination by Reverse Osmosis*, M. Merten (ed.), MIT Press, Cambridge, MA, pp. 93–160 (1966).
20. H.K. Lonsdale, U. Merten and R.L. Riley, Transport Properties of Cellulose Acetate Osmotic Membranes, *J. Appl. Polym. Sci.* **9**, 1341 (1965).
21. R.L. Riley, H.K. Lonsdale, C.R. Lyons and U. Merten, Preparation of Ultrathin Reverse Osmosis Membranes and the Attainment of Theoretical Salt Rejection, *J. Appl. Polym. Sci.* **11**, 2143 (1967).
22. S. Sourirajan, *Reverse Osmosis*, Academic Press, New York (1970).
23. K.D. Vos, F.O. Burris, Jr and R.L. Riley, Kinetic Study of the Hydrolysis of Cellulose Acetate in the pH Range of 2–10, *J. Appl. Polym. Sci.* **10**, 825 (1966).
24. W.M. King, D.L. Hoernschemeyer and C.W. Saltonstall, Jr, Cellulose Acetate Blend Membranes, in *Reverse Osmosis Membrane Research*, H.K. Lonsdale and H.E. Podall (eds), Plenum Press, New York, pp. 131–162 (1972).
25. R. Endoh, T. Tanaka, M. Kurihara and K. Ikeda, New Polymeric Materials for Reverse Osmosis Membranes, *Desalination* **21**, 35 (1977).
26. R. McKinney and J.H. Rhodes, Aromatic Polyamide Membranes for Reverse Osmosis Separations, *Macromolecules* **4**, 633 (1971).
27. J.W. Richter and H.H. Hoehn, Selective Aromatic Nitrogen-containing Polymeric Membranes, US Patent 3,567,632 (March, 1971).
28. R.J. Petersen and J.E. Cadotte, Thin Film Composite Reverse Osmosis Membranes, in *Handbook of Industrial Membrane Technology*, M.C. Porter (ed.), Noyes Publications, Park Ridge, NJ, pp. 307–348 (1990).
29. J.E. Cadotte, Reverse Osmosis Membrane, US Patent 4,039,440 (August, 1977).
30. R.L. Riley, C.E. Milstead, A.L. Lloyd, M.W. Seroy and M. Takami, Spiral-wound Thin Film Composite Membrane Systems for Brackish and Seawater Desalination by Reverse Osmosis, *Desalination* **23**, 331 (1977).
31. J. Glater, S.K. Hong and M. Elimelech, The Search for a Chlorine-Resistant Reverse Osmosis Membrane, *Desalination* **95**, 325 (1994).
32. J.E. Cadotte, Reverse Osmosis Membrane, US Patent 3,926,798 (December, 1975).
33. M. Kurihara, N. Harumiya, N. Kannamaru, T. Tonomura and M. Nakasatomi, Development of the PEC-1000 Composite Membrane for Single Stage Sea Water Desalination and the Concentration of Dilute Aqueous Solutions Containing Valuable Materials, *Desalination* **38**, 449 (1981).
34. Y. Nakagawa, K. Edogawa, M. Kurihara and T. Tonomura, Solute Separation and Transport Characteristics Through Polyether Composite (PEC)-1000 Reverse-Osmosis Membranes, in *Reverse Osmosis and Ultrafiltration*, S. Sourirajan and T. Matsuura (eds), ACS Symposium Series Number 281, American Chemical Society, Washington, DC, pp. 187–200 (1985).

35. B. Van der Bruggen, J. Schaep, D. Wilms and C. Vandecasteele, Influence of Molecular Size, Polarity and Charge on the Retention of Organic Molecules by Nanofiltration, *J. Membr. Sci.* **156**, 29 (1999).
36. J.M.M. Peters, J.P. Boom, M.H.V. Mulder and H. Strathmann, Retention Measurements of Nanofiltration Membranes with Electrolyte Solutions, *J. Membr. Sci.* **145**, 199 (1998).
37. A. Iwama and Y. Kazuse, New Polyimide Ultrafiltration Membranes for Organic Use, *J. Membr. Sci.* **11**, 279 (1982).
38. C. Linder, M. Nemas, M. Perry and R. Katraro, Silicone-derived Solvent Stable Membranes, US Patent 5,265,734 (November 1993).
39. D.R. Machado, D. Hasson and R. Semiat, Effect of Solvent Properties on Permeate Flow Through Nanofiltration Membranes. Part I: Investigation of Parameters Affecting Solvent Flux, *J. Membr. Sci.* **163**, 93 (1999).
40. L.S. White, I.-F. Wang and B.S. Minhas, Polyimide Membranes for Separation of Solvents from Lube Oil, US Patent 5,264,166 (November 1993).
41. L.S. White and A.R. Nitsch, Solvent Recovery from Lube Oil Filtrates with Polyimide Membranes, *J. Membr. Sci.* **179**, 267 (2000).
42. R.M. Gould and A.R. Nitsch, Lubricating Oil Dewaxing with Membrane Separation of Cold Solvent, US Patent 5,494,566 (February 1996).
43. N. Bhore, R.M. Gould, S.M. Jacob, P.O. Staffield, D. McNally, P.H. Smiley and C.R. Wildemuth, New Membrane Process Debottlenecks Solvent Dewaxing Unit, *Oil Gas J.* **97**, 67 (1999).
44. J.T. Scarpello, D. Nair, L.M. Freitas dos Santos, L.S. White and A.G. Livingston, The Separation of Homogenous Organometallic Catalysts using Solvent Resistant Nanofiltration, *J. Membr. Sci.* **203**, 71 (2002).
45. R. Rautenbach and R. Albrecht, *Membrane Processes*, John Wiley & Sons, Inc. Chichester (1989).
46. T. Matsuura and S. Sourirajan, Reverse Osmosis Separation of Phenols in Aqueous Solutions Using Porous Cellulose Acetate Membranes, *J. Appl. Polym. Sci.* **15**, 2531 (1972).
47. T. Matsuura and S. Sourirajan, Physiochemical Criteria for Reverse Osmosis Separation of Alcohols, Phenols, and Monocarboxylic Acid in Aqueous Solutions Using Porous Cellulose Acetate Membranes, *J. Appl. Polym. Sci.* **15**, 2905 (1971).
48. H.K. Lonsdale, U. Merten and M. Tagami, Phenol Transport in Cellulose Acetate Membranes, *J. Appl. Polym. Sci.* **11**, 1877 (1967).
49. R.G. Sudak, Reverse Osmosis, in *Handbook of Industrial Membrane Technology*, M.C. Porter (ed.), Noyes Publications, Park Ridge, NJ, pp. 260–306 (1990).
50. K. Marquardt, Sea Water Desalination by Reverse Osmosis, *GVC/VDI Gesellschaft Verfahrenstechnik und Chemieingenieurwesen Seawater Desalination-Water Pretreatment and Conditioning*, VDI Verlag, Düsseldorf (1981).
51. A. Ko and D.B. Guy, Brackish and Seawater Desalting, in *Reverse Osmosis Technology*, B.S. Parekh (ed.), Marcel Dekker, New York, pp. 141–184 (1988).
52. M.E. Williams, D. Bhattacharyya, R.J. Ray and S.B. McCray, Selected Applications of Reverse Osmosis, in *Membrane Handbook*, W.S.W. Ho and K.K. Sirkar (eds), Van Nostrand Reinhold, New York, pp. 312–354 (1992).
53. H. Strathmann, Electrodialysis in Membrane Separation Systems, in *Membrane Separation Systems*, R.W. Baker, E.L. Cussler, W. Eykamp, W.J. Koros, R.L. Riley and H. Strathmann (eds), Noyes Data Corp., Park Ridge, NJ, pp. 396–420 (1991).
54. G.A. Pittner, High Purity Water Production Using Reverse Osmosis Technology, in *Reverse Osmosis*, Z. Amjad (ed.), Van Nostrand Reinhold, New York (1993).

55. C.F. Frith, Jr, Electronic-grade Water Production Using Reverse Osmosis Technology, in *Reverse Osmosis Technology*, B.S. Parekh (ed.), Marcel Dekker, New York, pp. 279–310 (1988).
56. A. Golomb, Application of Reverse Osmosis to Electroplating Waste Treatment, in *Reverse Osmosis and Synthetic Membranes*, S. Sourirajan (ed.), National Research Council Canada, Ottawa, Canada, pp. 481–494 (1977).
57. A. Golomb, Applications of Reverse Osmosis to Electroplating Waste Treatment, *AES Research Project* **31**, 376 (1970).
58. I. Nusbaum and D.G. Argo, Design and Operation of a 5-mgd Reverse Osmosis Plant for Water Reclamation, in *Synthetic Membrane Processes*, G. Belfort (ed.), Academic Press, Orlando, FL, pp. 377–436 (1984).

6 ULTRAFILTRATION

Introduction and History

Ultrafiltration uses a finely porous membrane to separate water and microsolutes from macromolecules and colloids. The average pore diameter of the membrane is in the 10–1000 Å range. The first synthetic ultrafiltration membranes were prepared by Bechhold from collodion (nitro cellulose) [1]. Bechhold was probably the first to measure membrane bubble points, and he also coined the term 'ultrafilter'. Other important early workers were Zsigmondy and Bachmann [2], Ferry [3] and Elford [4]. By the mid-1920s, collodion ultrafiltration and microfiltration membranes were commercially available for laboratory use. Although collodion membranes were widely used in laboratory studies, no industrial applications existed until the 1960s. The crucial breakthrough was the development of the anisotropic cellulose acetate membrane by Loeb and Sourirajan in 1963 [5]. Their goal was to produce high-flux reverse osmosis membranes, but others, particularly Michaels at Amicon, realized the general applicability of the technique. Michaels and his coworkers [6] produced ultrafiltration membranes from cellulose acetate and many other polymers including polyacrylonitrile copolymers, aromatic polyamides, polysulfone and poly(vinylidene fluoride). These materials are still widely used to fabricate ultrafiltration membranes.

In 1969, Abcor (now a division of Koch Industries) installed the first commercially successful industrial ultrafiltration system equipped with tubular membrane modules [7] to recover electrocoat paint from automobile paint shop rinse water. The economics were compelling, and within a few years many similar systems were installed. Shortly thereafter (1970), the first cheese whey ultrafiltration system was installed. Within a decade, 100 similar systems had been sold worldwide. These early systems used tubular or plate-and-frame modules, which were relatively expensive, but lower cost designs were gradually introduced. Hollow fiber (capillary) modules were first sold by Romicon in 1973, and spiral-wound modules, adapted to ultrafiltration applications by Abcor, became a commercial item by 1979–1980. Over the last 20 years, the ultrafiltration industry has grown steadily. The principal problem inhibiting wider application of the technology is membrane fouling. The problem is controlled, but not eliminated, by module

Membrane Technology and Applications R. W. Baker
 ISBN: 0-470-85445-6

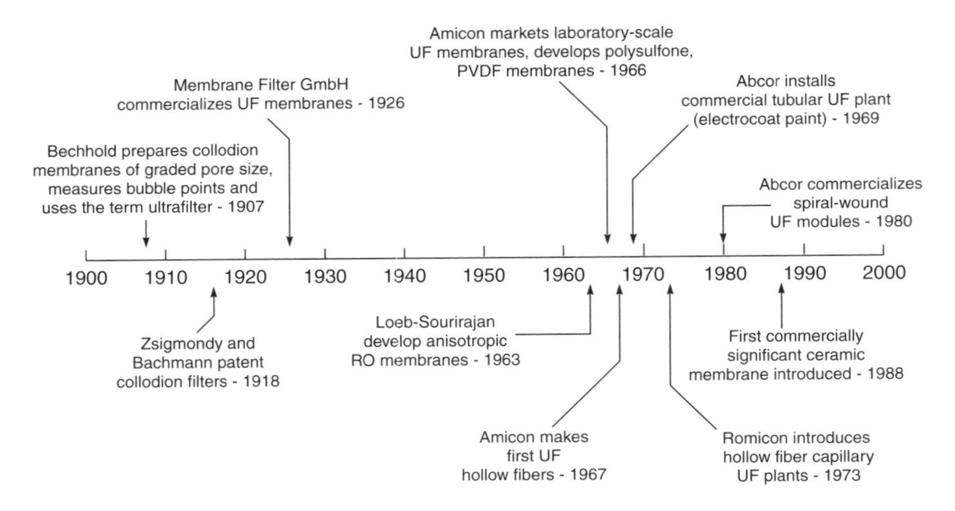

Figure 6.1 Milestones in the development of ultrafiltration

and system design and by regular membrane cleaning protocols. Development of membranes with surface properties designed to minimize fouling has also helped. Recently, several companies have developed ceramic-based ultrafiltration membranes. Although much more expensive than their polymeric equivalents, these have found a place in applications that require resistance to high temperatures or require regular cleaning with harsh solutions to control membrane fouling. Some of the milestones in the development of ultrafiltration membranes are charted in Figure 6.1.

Characterization of Ultrafiltration Membranes

Ultrafiltration membranes are usually anisotropic structures made by the Loeb–Sourirajan process. They have a finely porous surface layer or skin supported on a much more open microporous substrate. The finely porous surface layer performs the separation; the microporous substrate provides mechanical strength. The membranes discriminate between dissolved macromolecules of different sizes and are usually characterized by their molecular weight cut-off, a loosely defined term generally taken to mean the molecular weight of the globular protein molecule that is 90 % rejected by the membrane. Ultrafiltration and microfiltration are related processes—the distinction between the two lies in the pore size of the membrane. Microfiltration membranes have larger pores and are used to separate particles in the 0.1–10 μm range, whereas ultrafiltration is generally considered to be limited to membranes with pore diameters from 10 to 1000 Å.

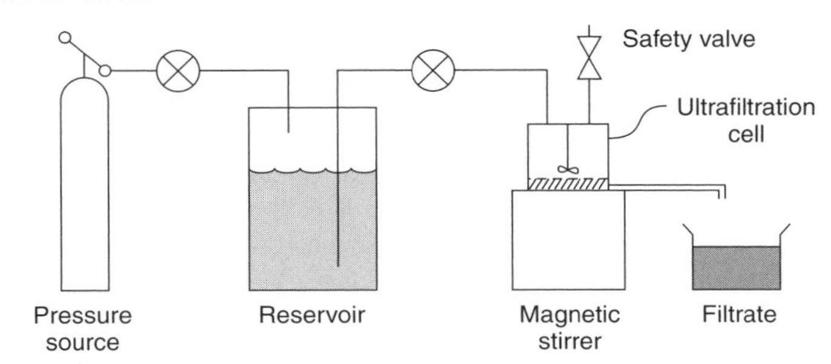

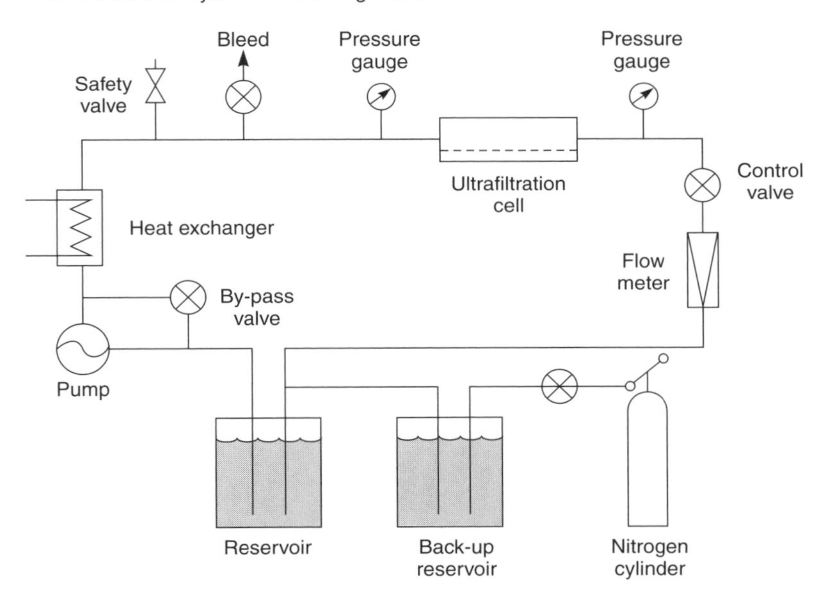

Figure 6.2 Laboratory ultrafiltration test systems

Laboratory-scale ultrafiltration experiments are performed with small, stirred batch cells or flow-through cells in a recirculation system. Diagrams of the two types of system are shown in Figure 6.2. Because ultrafiltration experiments are generally performed at pressures below 100 psi, plastic components can be used. Stirred batch cells are often used for quick experiments, but flow-through systems are preferred for systematic work. In flow-through systems, the feed solution can be more easily maintained at a constant composition, and the turbulence at the membrane surface required to control membrane fouling is high and easily reproducible. This allows reliable comparative measurements to be made.

The cut-off of ultrafiltration membranes is usually characterized by solute molecular weight, but several other factors affect permeation through these membranes. One important example is the shape of the molecule to be retained. When membrane retention measurements are performed with linear, water-soluble molecules such as polydextran, poly(ethylene glycol) or poly(vinyl pyrrolidone), the measured rejection is much lower than the rejection measured for proteins of the same molecular weight. It is believed that linear, water-soluble polymer molecules are able to snake through the membrane pores, as illustrated in Figure 6.3. Protein molecules, however, exist in solution as tightly wound globular coils held together by hydrogen bonds. These globular molecules cannot deform to pass through the membrane pores and are therefore rejected. Some results showing the rejection of different molecules for a polysulfone ultrafiltration membrane are listed in the table accompanying Figure 6.3 [8]. The membrane shows significant rejection to globular protein molecules as small as pepsin (MW 35 000) and cytochrome *c* (MW 13 000) but is completely permeable to a flexible linear polydextran, with an average molecular weight of more than 100 000.

The pH of the feed solution is another factor that affects permeation through ultrafiltration membranes, particularly with polyelectrolytes. For example,

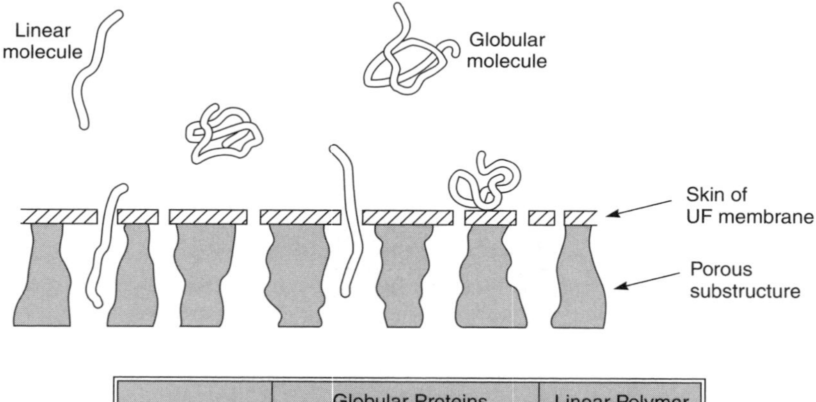

	Globular Proteins		Linear Polymer
Solute	Pepsin	Cytochrome *c*	Polydextran
MW (1000s) Rejection (%)	35 90	13 70	100 0

Figure 6.3 Ultrafiltration membranes are rated on the basis of nominal molecular weight cut-off, but the shape of the molecule to be retained has a major effect on retentivity. Linear molecules pass through a membrane, whereas globular molecules of the same molecular weight may be retained. The table shows typical results obtained with globular protein molecules and linear polydextran for the same polysulfone membrane [8]

poly(acrylic acid) is usually very well rejected by ultrafiltration membranes at pH 5 and above, but is completely permeable through the same membrane at pH 3 and below. This change in rejection behavior with pH is related to the change in configuration of the polyacid. In solutions at pH 5 and above, poly(acrylic acid) is ionized. In the ionized form, the negatively charged carboxyl groups along the polymer backbone repel each other; the polymer coil is then very extended and relatively inflexible. In this form, the molecule cannot readily permeate the small pores of an ultrafiltration membrane. At pH 3 and below, the carboxyl groups along the poly(acrylic acid) polymer backbone are all protonated. The resulting neutral molecule is much more flexible and can pass through the membrane pores.

Concentration Polarization and Membrane Fouling

A key factor determining the performance of ultrafiltration membranes is concentration polarization, which causes membrane fouling due to deposition of retained colloidal and macromolecular material on the membrane surface. A number of reviews have described the process in detail [9–13]. The pure water flux of ultrafiltration membranes is often very high—greater than 1 $cm^3/cm^2 \cdot min$ (350 $gal/ft^2 \cdot day$). However, when membranes are used to separate macromolecular or colloidal solutions, the flux falls within seconds, typically to 0.1 $cm^3/cm^2 \cdot min$. This immediate drop in flux is caused by the formation of a gel layer of retained solutes on the membrane surface due to concentration polarization. This gel layer forms a secondary barrier to flow through the membrane, as illustrated in Figure 6.4 and described in detail below. This first decline in flux is determined by the composition of the feed solution and its fluid hydrodynamics. Sometimes the resulting flux is constant for a prolonged period, and when the membrane is retested with pure water, its flux returns to the original value. More commonly, however, a further slow decline in flux occurs over a period of hours to weeks, depending on the feed solution. Most of this second decrease in flux is caused by slow consolidation of the secondary layer formed by concentration polarization on the membrane surface. Formation of this consolidated gel layer, called membrane fouling, is difficult to control. Control techniques include regular membrane cleaning, back flushing, or using membranes with surface characteristics that minimize adhesion. Operation of the membrane at the lowest practical operating pressure also delays consolidation of the gel layer.

A typical plot illustrating the slow decrease in flux that can result from consolidation of the secondary layer is shown in Figure 6.5 [14]. The pure water flux of these membranes is approximately 50 gal/min but, on contact with an electrocoat paint solution containing 10–20 % latex, the flux immediately falls to about 10–12 gal/min. This first drop in flux is due to the formation of the gel layer of latex particles on the membrane surface, as shown in Figure 6.4. Thereafter, the flux declines steadily over a 2-week period. This second drop in flux is caused by slow densification of the gel layer under the pressure of the

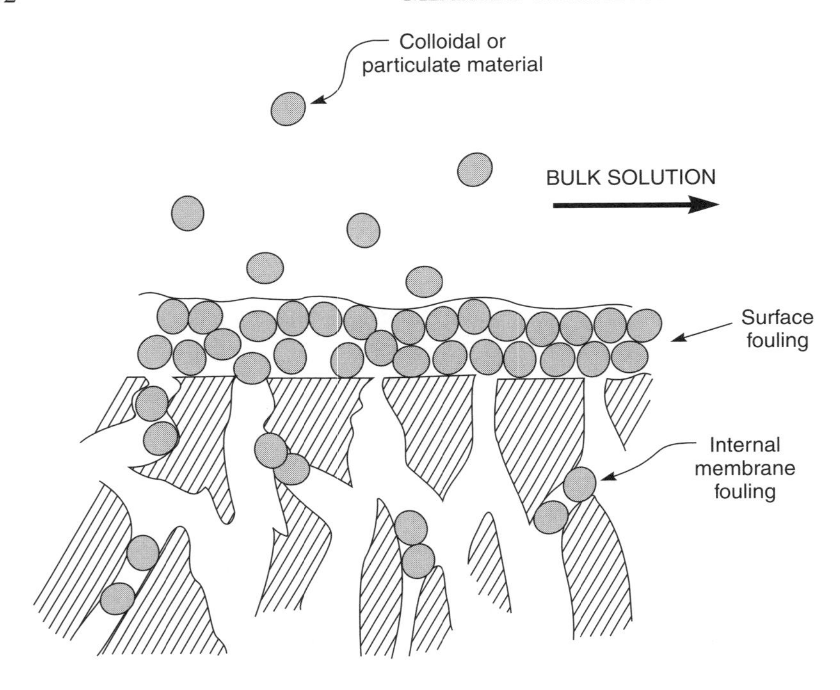

Figure 6.4 Schematic representation of fouling on an ultrafiltration membrane. *Surface fouling* is the deposition of solid material on the membrane that consolidates over time. This fouling layer can be controlled by high turbulence, regular cleaning and using hydrophilic or charged membranes to minimize adhesion to the membrane surface. Surface fouling is generally reversible. *Internal fouling* is caused by penetration of solid material into the membrane, which results in plugging of the pores. Internal membrane fouling is generally irreversible

system. In this particular example, the densified gel layer could be removed by periodic cleaning of the membrane. When the cleaned membrane is exposed to the latex solution again, the flux is restored to that of a fresh membrane.

If the regular cleaning cycle shown in Figure 6.5 is repeated many times, the membrane flux eventually does not return to the original value on cleaning. Part of this slow, permanent loss of flux is believed to be due to precipitates on the membrane surface that are not removed by the cleaning procedure. A further cause of the permanent flux loss is believed to be internal fouling of the membrane by material that penetrates the membrane pores and becomes lodged in the interior of the membrane, as illustrated in Figure 6.4. Ultrafiltration membranes are often used to separate colloids from water and microsolutes. In this case the tendency is to use relatively high-molecular-weight cut-off membranes, but the higher fluxes of these membranes can be transitory because they are more susceptible to internal fouling. A membrane with a lower molecular weight cut-off, even

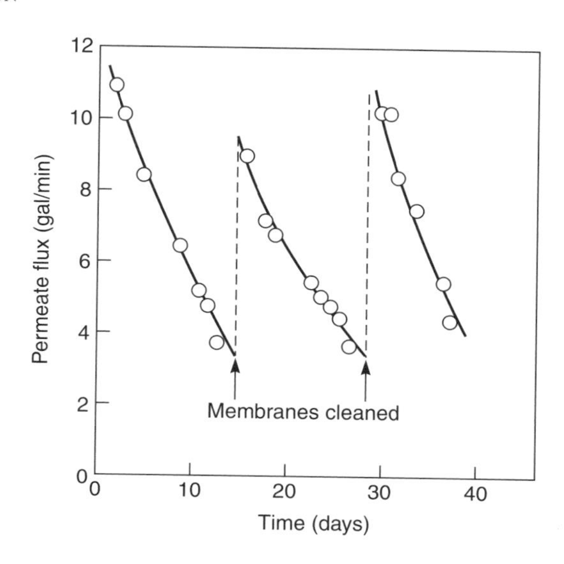

Figure 6.5 Ultrafiltration flux as a function of time of an electrocoat paint latex solution. Because of fouling, the flux declines over a period of days. Periodic cleaning is required to maintain high fluxes [14]. Reprinted from R. Walker, Recent Developments in Ultrafiltration of Electrocoat Paint, *Electrocoat* **82**, 16 (1982) with permission from Gardner Publications, Inc., Cincinnati, OH

though it may have a lower pure water flux, often provides a more sustained flux with the actual feed solutions because less internal fouling occurs.

As described above, the initial cause of membrane fouling is concentration polarization, which results in deposition of a layer of material on the membrane surface. The phenomenon of concentration polarization is described in detail in Chapter 4. In ultrafiltration, solvent and macromolecular or colloidal solutes are carried towards the membrane surface by the solution permeating the membrane. Solvent molecules permeate the membrane, but the larger solutes accumulate at the membrane surface. Because of their size, the rate at which the rejected solute molecules can diffuse from the membrane surface back to the bulk solution is relatively low. Thus their concentration at the membrane surface is typically 20–50 times higher than the feed solution concentration. These solutes become so concentrated at the membrane surface that a gel layer is formed and becomes a secondary barrier to flow through the membrane. The formation of this gel layer on the membrane surface is illustrated in Figure 6.6. The gel layer model was developed at the Amicon Corporation in the 1960s [8].

The formation of the gel layer is easily described mathematically. At any point within the boundary layer shown in Figure 6.6, the convective flux of solute to the membrane surface is given by the volume flux, J_v, of the solution through the membrane multiplied by the concentration of the solute, c_i. At steady state, this

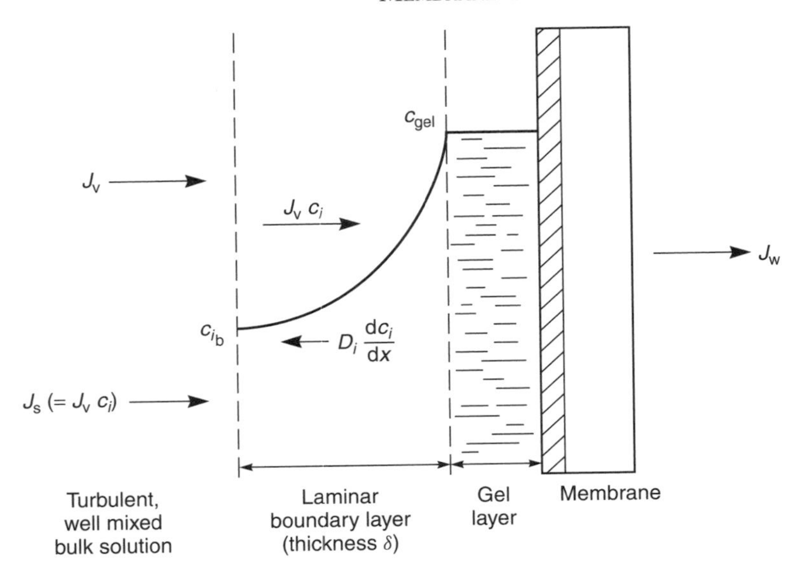

Figure 6.6 Illustration of the formation of a gel layer of colloidal material on the surface of an ultrafiltration membrane by concentration polarization

convective flux within the laminar boundary layer is balanced by the diffusive flux of retained solute in the opposite direction. This balance is expressed by the equation

$$J_v c_i = D_i \frac{\mathrm{d}c_i}{\mathrm{d}x} \tag{6.1}$$

where D_i is the diffusion coefficient of the macromolecule in the boundary layer. Once the gel layer has formed, the concentrations of solute at both surfaces of the boundary layer are fixed. At one surface the concentration is the feed solution concentration c_{i_b}; at the other surface it is the concentration at which the solute forms an insoluble gel (c_{gel}). Integration of Equation (6.1) over the boundary layer thickness (δ) then gives

$$\frac{c_{\text{gel}}}{c_{i_b}} = \exp\left(\frac{J_v \delta}{D_i}\right) \tag{6.2}$$

where c_{gel} is the concentration of retained solute at the membrane surface where the solute gels and c_{i_b} is the concentration in the bulk solution. In any particular ultrafiltration test, the terms c_{i_b}, c_{gel}, D_i and δ in Equation (6.2) are fixed because the solution and the operating conditions of the test are fixed. From Equation (6.2) this means that the volume flux J_v through the membrane is also fixed and quite independent of the intrinsic permeability of the membrane. In physical terms, this is because a membrane with a higher intrinsic permeability only causes a thicker gel layer to form on the surface of the membrane. This lowers the membrane flux until the rate at which solutes are brought toward the membrane surface and the rate at which they are removed are again balanced, as expressed in Equation (6.1).

The formation of a gel layer of colloidal material at the ultrafiltration membrane surface produces a limiting or plateau flux that cannot be exceeded at any particular operating condition. Once a gel layer has formed, increasing the applied

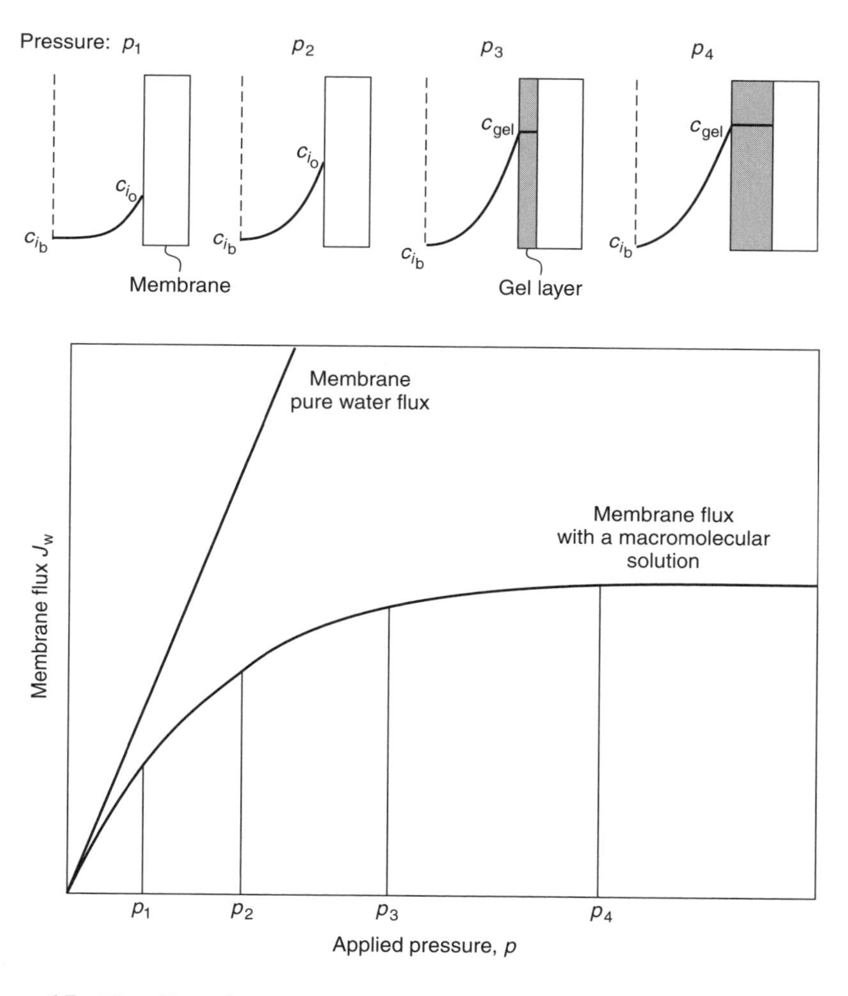

Figure 6.7 The effect of pressure on ultrafiltration membrane flux and the formation of a secondary gel layer. Ultrafiltration membranes are best operated at pressures between p_2 and p_3 at which the gel layer is thin. Operation at high pressures such as p_4 leads to formation of thick gel layers, which can consolidate over time, resulting in permanent fouling of the membrane

pressure does not increase the flux but merely increases the gel thickness. This is also shown in Equation (6.2), which contains no term for the applied pressure.

The effect of the gel layer on the flux through an ultrafiltration membrane at different feed pressures is illustrated in Figure 6.7. At a very low pressure p_1, the flux J_v is low, so the effect of concentration polarization is small, and a gel layer does not form on the membrane surface. The flux is close to the pure water flux of the membrane at the same pressure. As the applied pressure is increased to pressure p_2, the higher flux causes increased concentration polarization, and the concentration of retained material at the membrane surface increases. If the pressure is increased further to p_3, concentration polarization becomes enough for the retained solutes at the membrane surface to reach the gel concentration c_{gel} and form the secondary barrier layer. This is the limiting flux for the membrane. Further increases in pressure only increase the thickness of the gel layer, not the flux.

Experience has shown that the best long-term performance of an ultrafiltration membrane is obtained when the applied pressure is maintained at or just below the plateau pressure p_3 shown in Figure 6.7. Operating at higher pressures does not increase the membrane flux but does increase the thickness and density of retained material at the membrane surface layer. Over time, material on the membrane surface can become compacted or precipitate, forming a layer of deposited material that has a lower permeability; the flux then falls from the initial value.

A series of experimental results obtained with latex solutions illustrating the effect of concentration and pressure on flux are shown in Figure 6.8. The point at which the flux reaches a plateau value depends on the concentration of the latex in the solution: the more concentrated the solution, the lower the plateau flux. The exact relationship between the maximum flux and solute concentration can be obtained by rearranging Equation (6.2) to obtain

$$J_{max} = -\frac{D}{\delta}(\ln c_{i_b} - \ln c_{gel}) \tag{6.3}$$

where J_{max} is the plateau or limiting flux through the membrane.

Plots of the limiting flux J_{max} as a function of solution concentration for latex solution data are shown in Figure 6.9 for a series of latex solutions at various feed solution flow rates. A series of straight line plots is obtained, and these extrapolate to the gel concentration c_{gel} at zero flux. The slopes of the plots in Figure 6.9 are proportional to the term D/δ in Equation (6.3). The increase in flux resulting from an increase in the fluid recirculation rate is caused by the decrease in the boundary layer thickness δ.

Plots of maximum flux as a function of solute concentration for different solutes using the same membrane under the same conditions are shown in Figure 6.10 [15]. Protein or colloidal solutions, which easily form precipitated gels, have low fluxes and extrapolate to low gel concentrations. Particulate

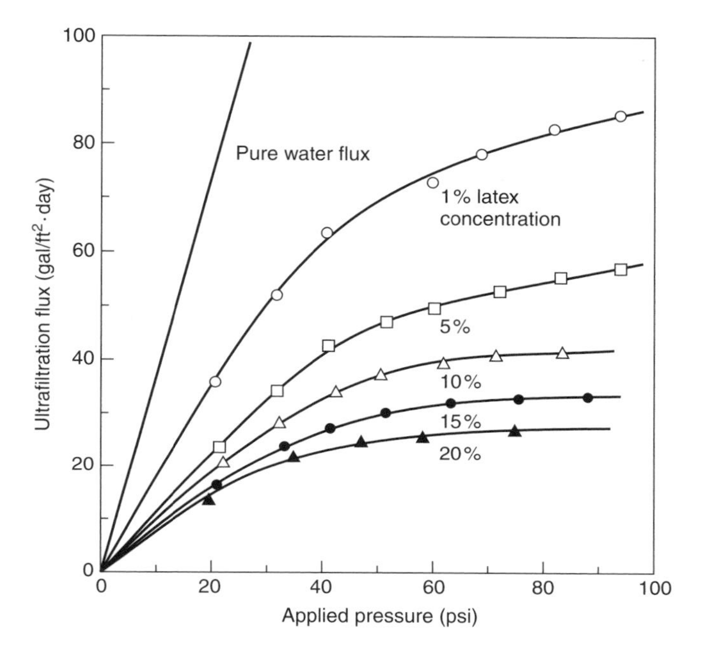

Figure 6.8 The effect of pressure on membrane flux for styrene–butadiene polymer latex solutions in a high-turbulence, thin-channel test cell [13]

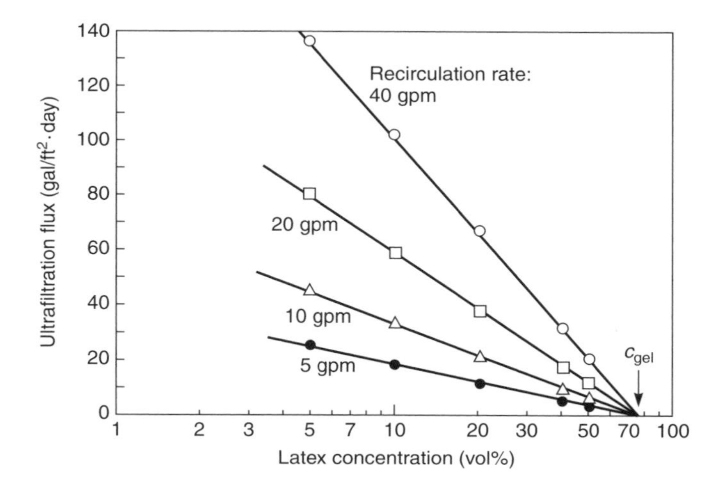

Figure 6.9 Ultrafiltration flux with a latex solution at an applied pressure of 60 psi (in the limiting flux region) as a function of feed solution latex concentration. These results were obtained in a high-turbulence, thin-channel cell. The solution recirculation rate is shown in the figure [13]

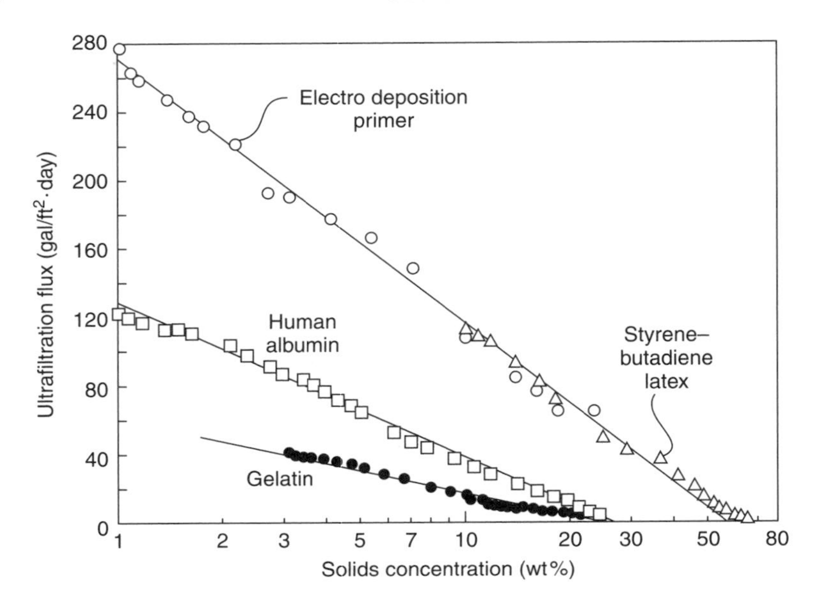

Figure 6.10 Effect of solute type and concentration on flux through the same type of ultrafiltration membrane operated under the same conditions [15]. Reproduced from M.C. Porter, Membrane Filtration, in *Handbook of Separation Techniques for Chemical Engineers*, P.A. Schweitzer (ed.), p. 2.39, Copyright © 1979, with permission of McGraw-Hill, New York, NY

suspensions, pigments, latex particles, and oil-in-water emulsions, which do not easily form gels, have higher fluxes at the same concentration and operating conditions and generally extrapolate to higher gel concentrations.

Studies of concentration polarization such as those illustrated in Figures 6.8–6.10 are usually performed during the first few hours of the membrane use. Compaction of the secondary membrane layer has then only just begun, and membrane fluxes are often high. Fluxes obtained in industrial processes, which must operate for days or weeks without cleaning, are usually much lower.

The gel layer model described above is very appealing and is widely used to rationalize the behavior of ultrafiltration membranes. Unfortunately a number of issues cannot be easily explained by this simple form of the model:

- The flux of many macromolecular colloidal and particulate solutions is too high (sometimes by an order of magnitude) to be rationalized by a reasonable value of the diffusion coefficient and the boundary layer thickness in Equation (6.2).
- In the plateau region of the flux–pressure curves of the type shown in Figure 6.8, different solutes should have fluxes proportional to the value

of their diffusion coefficients D in Equation (6.3). This is not the case, as shown in Figure 6.10. For example, latex and particulate solutes with very small diffusion coefficients typically have higher ultrafiltration limiting fluxes than protein solutions measured with the same membranes under the same conditions. This is the opposite of the expected behavior.

- Experiments with different ultrafiltration membranes and the same feed solution often yield very different ultrafiltration limiting fluxes. But according to the model shown in Figure 6.6 and represented by Equation (6.2), the ultrafiltration limiting flux is independent of the membrane type.

Contrary to normal experience that falling bread always lands jam-side down, the trend of these observations is that experiment produces a better result than theory predicts. For this reason the observations are lumped together and called the flux paradox [9]. The best working model seems to be that, in addition to simple diffusion, solute is also being removed from the membrane surface as undissolved gel particles by a scouring action of the feed fluid [16]. This explains why protein solutions that form tough adherent gels have lower fluxes under the same conditions than pigment and latex solutions that form looser gels. The model also explains why increasing the hydrophilicity of the membrane surface or changing the charge on the surface can produce higher limiting fluxes. Decreased adhesion between the gel and the membrane surface allows the flowing feed solution to remove gel particles more easily.

Figure 6.11 illustrates how turbulent eddies caused by the high velocity of the solution passing through the narrow channel of a spiral-wound module might remove gel particles from the membrane surface. Because of the high velocity of the feed solution and the feed spacer netting used in ultrafiltration modules, the feed liquid is normally very turbulent. Although a relatively laminar boundary layer may form next to the membrane surface, as described by the film model, periodic turbulent eddies may also occur. These eddies can dislodge gel from the membrane surface, carrying it away with the feed solution.

The most important effect of concentration polarization is to reduce the membrane flux, but it also affects the retention of macromolecules. Retention data obtained with dextran polysaccharides at various pressures are shown in Figure 6.12 [17]. Because these are stirred batch cell data, the effect of increased concentration polarization with increased applied pressure is particularly marked. A similar drop of retention with pressure is observed with flow-through cells, but the effect is less because concentration polarization is better controlled in such cells. With macromolecular solutions, the concentration of retained macromolecules at the membrane surface increases with increased pressure, so permeation of the macromolecules also increases, lowering rejection. The effect is particularly noticeable at low pressures, under which conditions increasing the applied pressure produces the largest increase in flux, and hence concentration polarization, at the membrane surface. At high pressure, the change in flux with

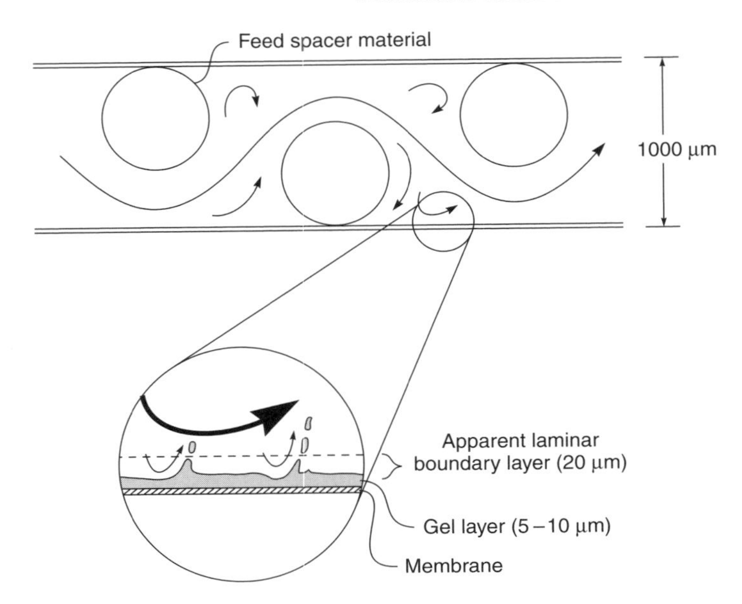

Figure 6.11 An illustration of the channel of a spiral-wound module showing how periodic turbulent eddies can dislodge deposited gel particles from the surface of ultrafiltration membranes

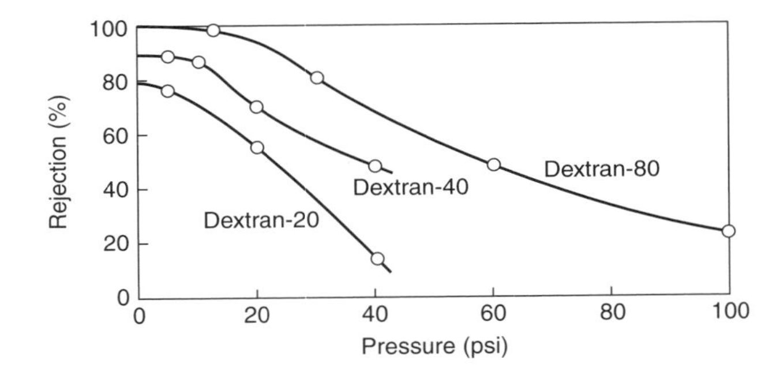

Figure 6.12 Rejection of 1 % dextran solutions as a function of pressure using Dextran 20 (MW 20 000), Dextran 40 (MW 40 000), and Dextran 80 (MW 80 000). Batch cell experiments performed at a constant stirring speed [17]

increased pressure is smaller, so the decrease in rejection by the membrane is less apparent.

Concentration polarization can also interfere with the ability of an ultrafiltration membrane to fractionate a mixture of dissolved macromolecules. Figure 6.13 [8] shows the results of experiments with a membrane with a molecular weight

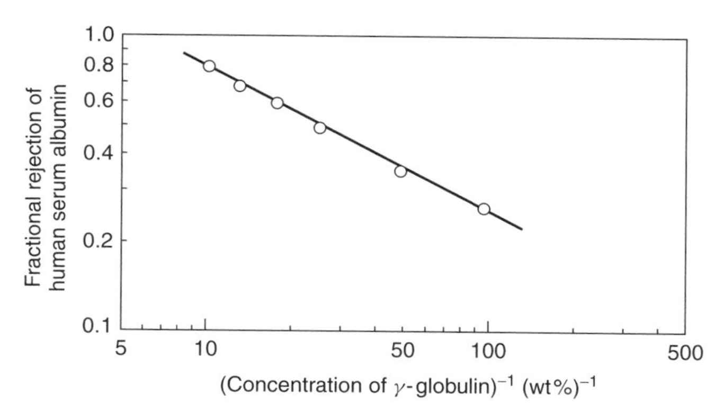

Figure 6.13 The retention of albumin (MW 65 000) in the presence of varying concentrations of γ-globulin (MW 156 000) by a membrane with a nominal molecular weight cut-off based on one-component protein solutions of MW 200 000. As the concentration of γ-globulin in the solution increases, the membrane water flux decreases, and the albumin rejection increases from 25 % at 0.01 wt% γ-globulin to 80 % rejection at 0.1 wt% γ-globulin [8]

cut-off of about 200 000 used to separate albumin (MW 65 000) from γ-globulin (MW 156 000). Tests with the pure components show that albumin passes through the membrane almost completely unhindered, but rejection of γ-globulin is significant. However, addition of even a small amount of γ-globulin to the albumin causes almost complete rejection of both components. The increased rejection is accompanied by a sharp decrease in membrane flux, suggesting that rejected globulin forms a secondary barrier layer. The secondary layer is eliminated only at very low γ-globulin concentrations, resulting in partial fractionation of the two proteins. Unfortunately, at such low dilutions the separation is no longer of commercial interest.

Because of the effect of the secondary layer on selectivity, ultrafiltration membranes are not commonly used to fractionate macromolecular mixtures. Most commercial ultrafiltration applications involve processes in which the membrane completely rejects all the dissolved macromolecular and colloidal material in the feed solution while completely passing water and dissolved microsolutes. Efficient fractionation by ultrafiltration is only possible if the species differ in molecular weight by a factor of 10 or more.

Membrane Cleaning

Several cleaning methods are used to remove the densified gel layer of retained material from the membrane surface. The easiest is to circulate an appropriate cleaning solution through the membrane modules for 1 or 2 h. The most common ultrafiltration fouling layers—organic polymer colloids and gelatinous materials—are best treated with alkaline solutions followed by hot detergent solutions. Enzymatic detergents are particularly effective when the fouling layer is a proteinaceous gel. Calcium, magnesium, and silica scales, often a problem with reverse osmosis membranes, are generally not a problem in ultrafiltration because these ions permeate the membrane (ultrafiltration of cheese whey, in which high calcium levels can lead to calcium scaling, is an exception). Because many feed waters contain small amounts of soluble ferrous iron salts, hydrated iron oxide scaling is a problem. In the ultrafiltration system these salts are oxidized to ferric iron by entrained air. Ferric iron is insoluble in water, so an insoluble iron hydroxide gel forms and accumulates on the membrane surface. Such deposits are usually removed with a citric or hydrochloric acid wash.

Regular cleaning is required to maintain the performance of all ultrafiltration membranes. The period of the cleaning cycle can vary from daily for food applications, such as ultrafiltration of whey, to once a month or more for ultrafiltration membranes used as polishing units in ultrapure water systems. A typical cleaning cycle is as follows:

1. Flush the system several times with hot water at the highest possible circulation rate.
2. Treat the system with an appropriate acid or alkali wash, depending on the nature of the layer.
3. Treat the system with a hot detergent solution.
4. Flush the system thoroughly with water to remove all traces of detergent; measure the pure water flux through the membrane modules under standard test conditions. Even after cleaning, some degree of permanent flux loss over time is expected. If the restoration of flux is less than expected, repeat steps 1–3.

Ultrafiltration systems should never be taken off line without thorough flushing and cleaning. Because membrane modules are normally stored wet, the final rinse solutions should contain a bacteriostat such as 0.5 % formaldehyde to inhibit bacterial growth.

In addition to regular cleaning with chemical solutions, mechanical cleaning of the membrane may be used, particularly if chemical cleaning does not restore the membrane flux. Early electrocoat paint systems used 1-in.-diameter tubular membrane modules. These tubes could be effectively cleaned by forcing sponge balls with a slightly larger diameter than the tube through the tube—the balls gently scraped the membrane surface, removing deposited material. Sponge-ball cleaning is an effective but relatively time-consuming process, so it is performed rather infrequently. However, automatic equipment for sponge-ball cleaning has been devised and has found a limited use.

Backflushing is another way of cleaning heavily fouled membranes. The method is widely used to clean capillary and ceramic membrane modules that can withstand a flow of solution from permeate to feed without damaging the membrane. Backflushing is not usually used for spiral-wound modules because the membranes are too easily damaged. In a backflushing procedure a slight over-pressure is applied to the permeate side of the membrane, forcing solution from the permeate side to the feed side of the membrane. The flow of solution lifts deposited materials from the surface. Backflushing must be done carefully to avoid membrane damage. Typical backflushing pressures are 5–15 psi.

One method of achieving a backflushing effect used with capillary ultrafiltration modules is initiated by closing the permeate port from the membrane module, as shown in Figure 6.14 [18]. In normal operation a pressure drop of 5–10 psi occurs between the feed and residue side of a membrane module. This pressure difference is required to drive the feed solution through the module. If the permeate port from the module is closed, the pressure on the permeate side of the membrane will increase to a pressure intermediate between those of the feed and residue streams. This produces a slight positive pressure difference at one end of the module and a slight negative pressure difference on the other end of the module, as shown in Figure 6.14(b). The pressure difference sets up a backflushing condition in which permeate-quality water that has permeated one-half of the module becomes a backflushing solution in the other half of the module. Deposited materials lifted from the membrane surface in the back-flushed area are swept away by the fast feed flow. If the direction of the feed flow is reversed, as shown in Figure 6.14(c), the other half of the module is then backflushed. This *in-situ* backflushing technique is used in capillary ultrafiltration modules in which the feed-to-residue pressure drop is quite large. An advantage of the procedure is that it can be done without stopping normal operation of the ultrafiltration system.

Because of the challenging environment in which ultrafiltration membranes are operated and the regular cleaning cycles, membrane lifetime is significantly shorter than that of reverse osmosis membranes. Ultrafiltration module lifetimes are rarely more than 2–3 years, and modules may be replaced annually in cheese whey or electrocoat paint applications. In contrast, reverse osmosis membranes are normally not cleaned more than once or twice per year and can last 4–5 years.

Membranes and Modules

Membrane Materials

Most of today's ultrafiltration membranes are made by variations of the Loeb–Sourirajan process. A limited number of materials are used, primarily polyacrylonitrile, poly(vinyl chloride)–polyacrylonitrile copolymers, polysulfone, poly(ether sulfone), poly(vinylidene fluoride), some aromatic polyamides, and cellulose acetate. In general, the more hydrophilic membranes are more fouling-resistant than the completely hydrophobic materials. For this reason water-soluble

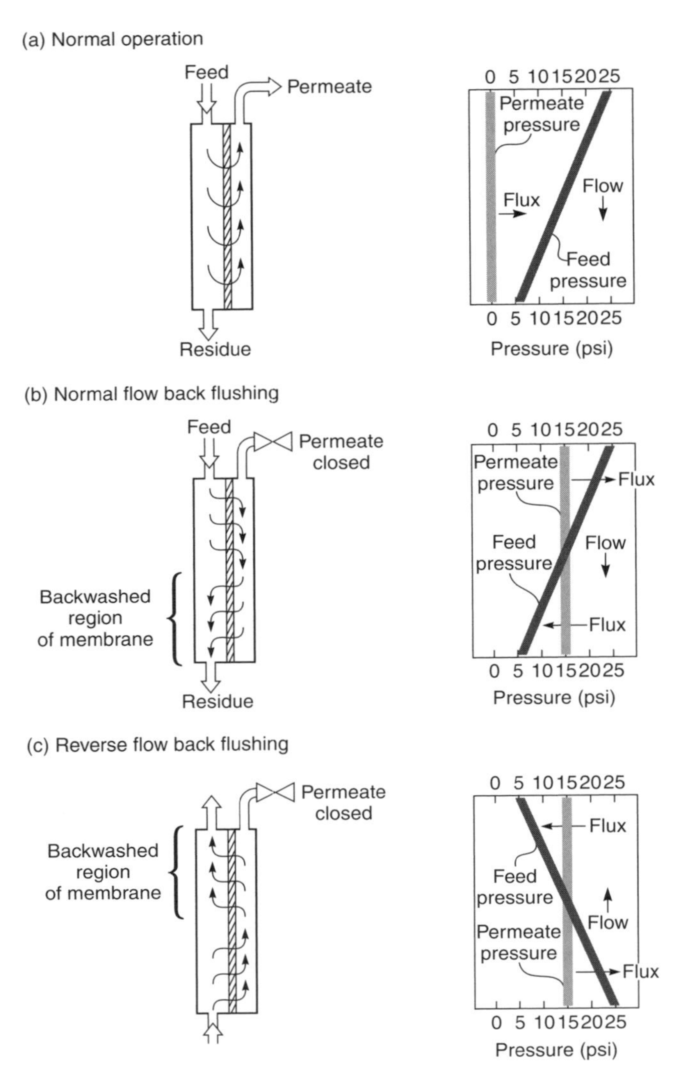

Figure 6.14 Backflushing of membrane modules by closing the permeate port. This technique is particularly applicable to capillary fiber modules

polymers such as poly(vinyl pyrrolidone) or poly(vinyl methyl ether) are often added to the membrane casting solutions used for hydrophobic polymers such as polysulfone or poly(vinylidene fluoride). During the membrane precipitation step, most of the water-soluble polymer is leached from the membrane, but enough remains to make the membrane surface hydrophilic.

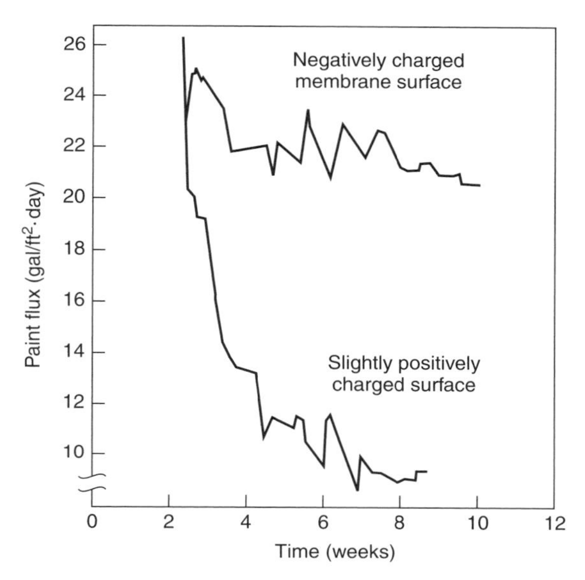

Figure 6.15 Effect of membrane surface charge on ultrafiltration flux decline. These membranes were used to ultrafilter cathodic electrocoat paint, which has a net negative charge. Electrostatic repulsion made the negatively charged membrane significantly more resistant to fouling than the similar positively charged membrane [13]

The charge on the membrane surface is important. Many colloidal materials have a slight negative charge from carboxyl, sulfonic or other acid groups. If the membrane surface also has a slight negative charge, adhesion of the colloidal gel layer to the membrane is reduced, which helps to maintain a high flux and inhibit membrane fouling. The effect of a slight positive charge on the membrane is the opposite. Charge and hydrophilic character can be the result of the chemical structure of the membrane material or can be applied to a preformed membrane surface by chemical grafting or surface treatment. The appropriate treatment depends on the application and the feed solution.

The importance of membrane surface characteristics on performance is illustrated by Figure 6.15. The feed solution in this example was a cathodic electrocoat paint solution in which the paint particulates had a net positive charge. As a result, membrane flux declined rapidly with the positively charged membranes but much more slowly with essentially identical membranes that had been treated to give the surface a net negative charge [13].

Ultrafiltration Modules

The need to control concentration polarization and membrane fouling dominates the design of ultrafiltration modules. The first commercially successful ultrafiltration systems were based on tubular and plate-and-frame modules. Over the years, many improvements have been made to these module designs, and they are still used for highly fouling solutions. The lower cost of spiral-wound and capillary modules has resulted in a gradual trend to replace tubular and plate-and-frame systems with these lower-cost modules. In relatively non-fouling applications, such as the use of ultrafiltration as part of a treatment train to produce ultrapure water, spiral-wound modules are universally used. Spiral-wound and capillary modules are also used in some food applications, such as ultrafiltration of cheese whey and clarification of apple juice.

Because of their large diameter, tubular ultrafiltration modules can be used to treat solutions that would rapidly foul other module types. In a number of demanding applications, such as treatment of electrocoat paint, concentration of latex solutions, or separation of oil–water emulsions, the fouling resistance and ease of cleaning of tubular modules outweighs their high cost, large volume, and high energy consumption. In a typical tubular module system, 5- to 8-ft-long tubes are manifolded in series. The feed solution is circulated through the module array at velocities of 2–6 m/s. This high solution velocity causes a pressure drop of 2–3 psi per module or 10–30 psi for a module bank. Because of the high circulation rate and the resulting pressure drop, large pumps are required, so tubular modules have the highest energy consumption of any module design. Most tubular ultrafiltration plants use 30–100 kWh of energy per 1000 gallons of permeate produced. At an electrical energy cost of US$0.06/kWh, this corresponds to an energy cost of US$2–6 per 1000 gallons of permeate, a major cost factor.

The diameter of the early tubular membrane modules was 1 in. Later, more energy-efficient, higher-membrane-area modules were produced by nesting four to six smaller-diameter tubes inside a single housing (see Chapter 3). Typical tubular module costs vary widely but are generally from US$200 to 500/$m^2$. Recently ceramic tubular modules have been introduced; these are more expensive, typically from US$1000 to 2000/$m^2$. This high cost limits their use to a few applications with extreme feed operating conditions.

Plate-and-frame units compete with tubular units in some applications. These modules are not quite as fouling resistant as tubular modules but are less expensive. Most consist of a flat membrane envelope with a rubber gasket around the outer edge. The membrane envelope, together with appropriate spacers, forms a plate that is contained in a stack of 20–30 plates. Typical feed channel heights are 0.5–1.0 mm, and the system operates in high-shear conditions. Plate-and-frame systems can be operated at higher pressures than tubular or capillary modules—operating pressures up to 150 psi are not uncommon. This can be an advantage in some applications. The compact design, small hold-up volume, and absence of stagnant areas also makes sterilization easy. For these reasons plate-and-frame units are used in several types of food industry operations, particularly

Figure 6.16 Horizontal DDS plate-and-frame ultrafiltration system. Courtesy of Alfa Laval Nakskov A/S, Naksvov, Denmark

in Europe where Rhône Poulanc and De Danske Sukkerfabrikker (DDS) [now Alfa Laval] pioneered these applications in the 1970s. A photograph of an Alfa Laval plate-and-frame system is shown in Figure 6.16.

Capillary hollow fiber modules were introduced by Romicon in the early 1970s. A typical capillary module contains 500–2000 fibers with a diameter of 0.5–1.0 mm housed in a 30-in.-long, 3-in.-diameter cartridge. Modules have a membrane area of 2–10 m^2. Feed solution is pumped down the bore of the fibers. Operating pressures are quite low, normally not more than 25 psi (to avoid breaking the fibers). This low operating pressure is a disadvantage in the treatment of some clean feed solutions for which high-pressure operation would be advantageous. The normal feed-to-residue pressure drop of a capillary module is 10–15 psi. Under these conditions, capillary modules achieve good throughputs with many solutions. High-temperature sanitary systems are available; this, combined with the small hold-up volume and clean flow path, has encouraged the use of these modules in biotechnology applications in which small volumes of expensive solutions are treated. A major advantage of capillary fiber systems is that the membrane can be cleaned easily by backflushing. With capillary modules it is important to avoid 'blinding' the fibers with particulates caught at the fiber entrance. Prefiltration to remove all particulates larger than one-tenth of the fiber's inside diameter is required to avoid blinding.

The use of spiral-wound modules in ultrafiltration applications has increased recently. This design was first developed for reverse osmosis modules in which the feed channel spacer is a fine window-screen material. In ultrafiltration a coarser feed spacer material is used, often as much as 45 mil thick. This coarse spacer prevents particulates from lodging in the spacer corners. However, prefiltration of the ultrafiltration feed down to 5–10 μm is still required for long-term operation. In the past, spiral-wound modules were limited to ultrafiltration of clean feed waters, such as preparation of ultrapure water for the electronics or pharmaceutical industries. Development of improved pretreatment and module spacer designs now allows these modules to be used for more highly fouling solutions such as cheese whey. In these food applications, the stagnant volume between the module insert and the module housing is a potentially unsterile area. To eliminate this dead space, the product seal is perforated to allow a small bypass flow to continuously flush this area.

In the last few years a number of companies, most notably New Logic International (Emeryville, CA), have introduced plate-and-frame modules in which the membrane plate is vibrated or rotated. Thus, control of concentration polarization at the membrane surface is by movement of the membrane rather than by movement of the feed solution [19]. Moving the membrane concentrates most of the turbulence right at the membrane surface, where it is most needed. These modules achieve very high turbulence at the membrane surface at a relatively low energy cost. The fluxes obtained are high and stable. Vibrating–rotating modules are considerably more expensive than cross-flow modules so the first applications have been with high-value, highly fouling feed solutions that are difficult to treat with standard modules.

System Design

Batch Systems

The simplest type of ultrafiltration system is a batch unit, shown in Figure 6.17. In such a unit, a limited volume of feed solution is circulated through a module at a high flow rate. The process continues until the required separation is achieved, after which the concentrate solution is drained from the feed tank, and the unit is ready to treat a second batch of solution. Batch processes are particularly suited to the small-scale operations common in the biotechnology and pharmaceutical industries. Such systems can be adapted to continuous use but this requires automatic controls, which are expensive and can be unreliable.

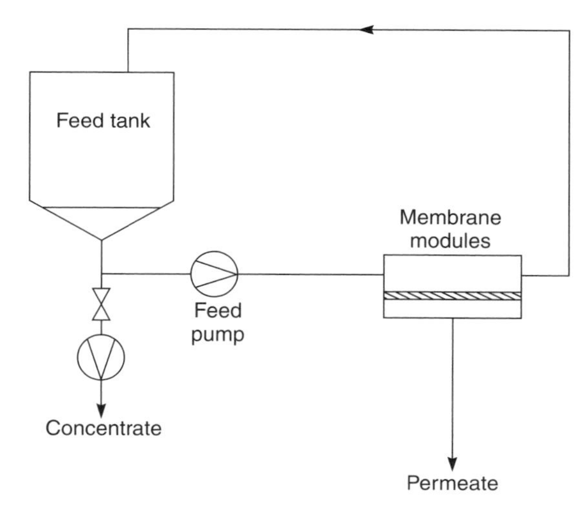

Figure 6.17 Flow schematic of a batch ultrafiltration process

The easiest way to calculate the performance of a batch system is to assume the membrane is completely retentive for the solute of interest. That is,

$$\mathbb{R} = \left(1 - \frac{c_p}{c_b}\right) = 1 \tag{6.4}$$

where c_p is the solute concentration in the permeate and c_b is the solute concentration in the feed. It follows that the increase in concentration of the solute in the feed tank from the initial concentration $c_b(o)$, to the concentration at time t, $c_b(t)$ is proportional to the volume of solution remaining in the feed tank, that is,

$$\frac{c_b(t)}{c_b(o)} = \frac{V_{(o)}}{V_{(t)}} \tag{6.5}$$

where the volume of solution removed in the permeate is $V_{(o)} - V_{(t)}$. If, as is often the case, the membrane is slightly permeable to the solute ($\mathbb{R} < 1$), the concentration ratio achieved can be written as

$$\ln\left[\frac{c_b(t)}{c_b(o)}\right] = \mathbb{R} \ln\left(\frac{V_{(o)}}{V_{(t)}}\right) \tag{6.6}$$

When the rejection coefficient equals one, Equation (6.6) reduces to Equation (6.5). A plot of the concentration ratio of retained solute as a function of the volume reduction for membranes with varying rejection coefficients is shown in Figure 6.18. This figure illustrates the effect of partially retentive membranes on loss of solute.

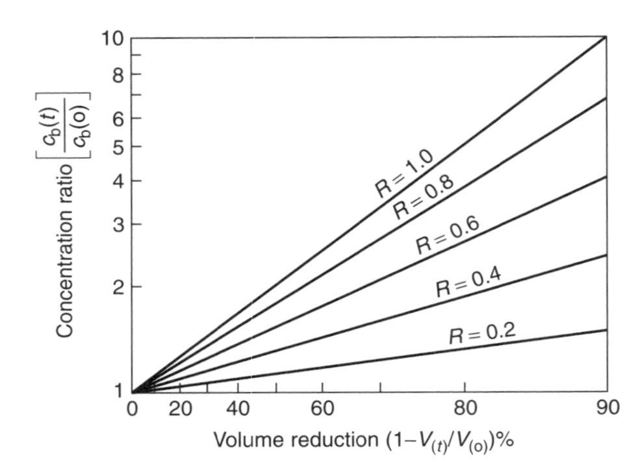

Figure 6.18 Increase in concentration of the retained feed solution as a function of volume reduction of the feed for membranes of different solute rejections. The difference between these lines and the $\mathbb{R} = 1$ line represents loss of solute through the membrane

Continuous Systems

Continuous ultrafiltration processes, in which modules are arranged in series to obtain the separation required in a single pass, are relatively common. This is because high feed solution flow rates are required to control concentration polarization; a single-pass process would not achieve the required removal under these conditions. Solution velocities in ultrafiltration modules are 5–10 times higher than in reverse osmosis. For these reasons, feed-and-bleed systems are commonly used in large ultrafiltration plants. Figure 6.19 shows one-, two- and three-stage feed-and-bleed systems. In these systems a large volume of solution is circulated continuously through a bank of membrane modules. Concurrently, a small volume of feed solution enters the recirculation loop just before the recirculation pump, and an equivalent volume of more concentrated solution is removed (or bled) from the recirculation loop just after the membrane module. The advantage of feed-and-bleed systems is that a high feed solution velocity through the modules is easily maintained independent of the volume of solution being treated. In most plants the flow rate of solution in the recirculation loop is 5–10 times the feed solution flow rate. This high circulation rate means that the concentration of retained material in the circulating solution is close to the concentration of the bleed solution and is significantly higher than the feed solution concentration. Because the flux of ultrafiltration membranes decreases with increasing concentration, more membrane area is required to produce the required separation than in a batch or a once-through continuous system operated at the same feed solution velocity.

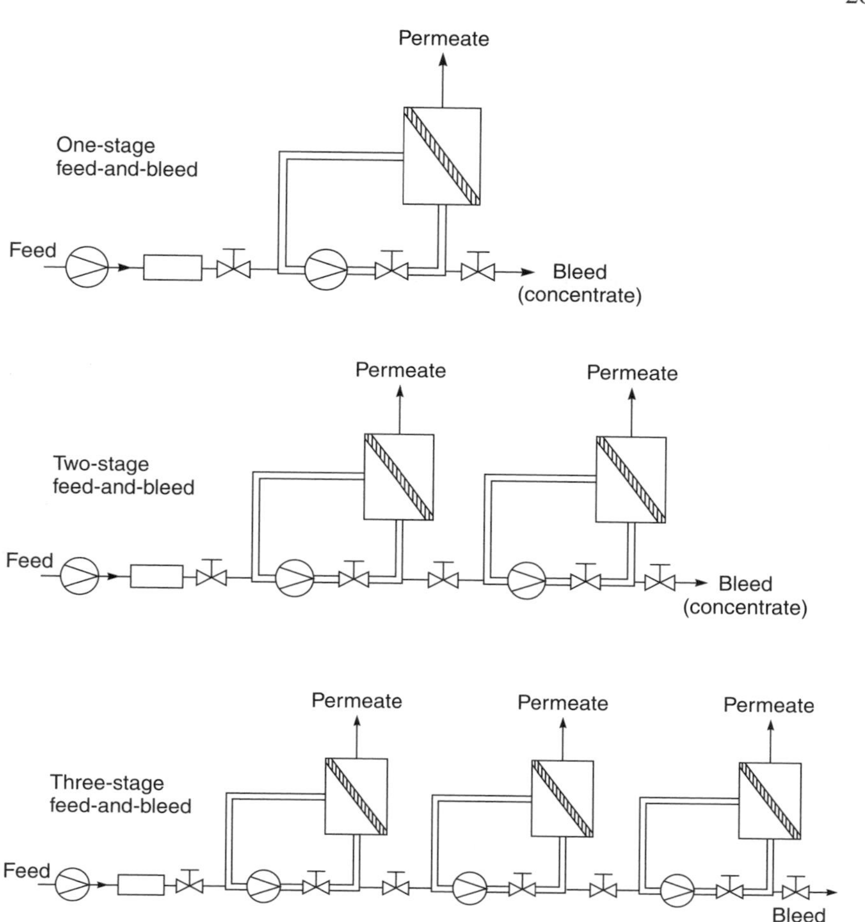

Figure 6.19 One-, two- and three-stage feed-and-bleed systems. In general, the most efficient design is achieved when all stages have approximately the same membrane area. As the number of stages is increased, the average concentration of the solution circulating through the membrane modules decreases, and the total membrane area of the system is significantly less than for a one-stage design

To overcome the inefficiency of one-stage feed-and-bleed designs, industrial systems are usually divided into multiple stages, as shown in Figure 6.19. By using multiple stages, the difference in concentration between the solution circulating in a stage and the feed solution entering the stage is minimized. The following numerical values illustrate this point. In this example, the membrane is assumed to be completely retentive and the goal is to concentrate the feed solution from 1 to 8 %. If this is done in a one-stage feed-and-bleed system, the average concentration of the solution passing through the modules is 8 %, and the flux is proportionately low. In a more efficient two-stage feed-and-bleed system, the first stage concentrates the solution from 1 to 3 %, and the second stage concentrates the solution from 3 to 8 %. Approximately three-quarters of the permeate is removed in the first stage, and the rest in the second stage. Because the modules in the first stage operate at a concentration of 3 % rather than 8 %, these modules have a higher membrane flux than in the one-stage unit. In fact, the membrane area of each stage is about equal although the volume of permeate produced by each stage is very different. The two-stage feed-and-bleed design has about 60 % of the area of the one-stage system. The three-stage system, which concentrates the solution in three equal-area stages—from 1 to 2 % in the first stage, from 2 to 4 % in the second stage, and from 4 to 8 % in the third stage—is even more efficient. In this case the total membrane area is about 40 % of the area of a one-stage system performing the same separation.

Because of the significantly lower membrane areas of multistage feed-and-bleed systems, most large plants have between three and five stages. The limit to the number of stages is reached when the reduction in membrane area does not offset the increase in complexity of the system. Also, because of the high fluid circulation rates involved in feed-and-bleed ultrafiltration plants, the cost of pumps can rise to 30 to 40 % of the total cost of the system. Electricity to power the pumps is a significant operating expense.

Applications

In the 1960s and early 1970s it was thought that ultrafiltration would be widely used to treat industrial wastewater. This application did not materialize. Ultrafiltration is far too expensive to be generally used for this application, however, it is used to treat small, concentrated waste streams from particular point sources before they are mixed with the general sewer stream. Ultrafiltration is also used if the value of the components to be separated is sufficient to offset the cost of the process. Examples exist in food processing, in which the ultrafiltered concentrate is used to produce a high-value product, or in the production of ultrapure water in the electronics industry.

The cost of ultrafiltration plants varies widely, depending on the size of the plant, the type of solution to be treated, and the separation to be performed. In general, ultrafiltration plants are much smaller than reverse osmosis systems. Typical flow rates are 10 000–100 000 gal/day, one-tenth that of the average reverse osmosis plant. Rogers [20] compiled the costs shown in Figure 6.20 that, adjusted for inflation, still seem reasonable. For typical plants treating 10 000–100 000 gal/day of feed solution, the capital cost is in the range US$2–5 gal/day capacity. The typical breakdown of these costs is shown in Table 6.1 [21]. Operating costs will normally be US$3–4/1000 gal/day capacity,

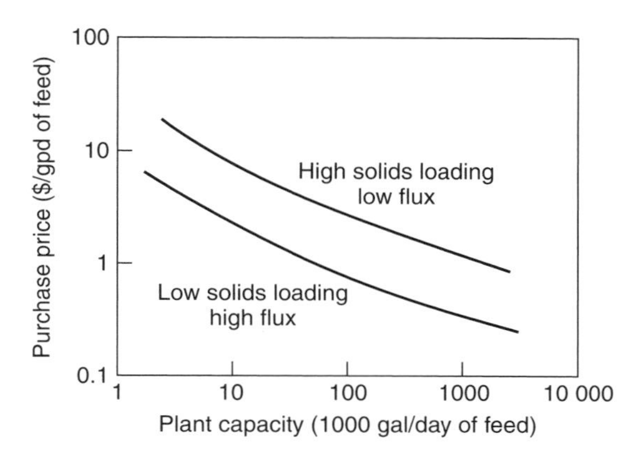

Figure 6.20 Purchase price in 2003 US dollars for ultrafiltration plants as a function of plant capacity. Data of Rogers corrected for inflation [20]. Reprinted from *Synthetic Membrane Processes*, A.N. Rogers, Economics of the Application of Membrane Processes, p. 454, G. Belfort (ed.), Copyright 1984, with permission from Elsevier

Table 6.1 Typical ultrafiltration capital and operating cost breakdown [21]

Capital costs	%
Pumps	30
Membrane modules	20
Module housings	10
Pipes, valves, frame	20
Controls/other	20
Total	100
Operating costs	
Membrane replacement	30–50
Cleaning costs	10–30
Energy	20–30
Labor	15
Total	100

with membrane module replacement costs about 30–50 %, and energy costs for the recirculation pumps 20–30 %, depending on the system design.

The current ultrafiltration market is approximately US$200 million/year but because the market is very fragmented, no individual segment is more than about US$10–30 million/year. Also, each of the diverse applications uses membranes, modules, and system designs tailored to the particular industry served. The result is little product standardization, many custom-built systems, and high costs compared to reverse osmosis. The first large successful application was the recovery of electrocoat paint in automobile plants. Later, a number of significant applications developed in the food industry [22,23], first in the production of cheese, then in the production of apple and other juices and, more recently, in the production of beer and wine. Industrial wastewater and process water treatment is a growing application, but high costs limit growth. Early plants were all tubular or plate-and-frame systems, but less expensive capillary and specially configured spiral-wound modules are now used more commonly. An overview of ultrafiltration applications is given in Cheryan and Alvarez's review article [23] and Cheryan's book [24].

Electrocoat Paint

In the 1960s, automobile companies began to use electrodeposition of paint on a large scale. The paint solution is an emulsion of charged paint particles. The metal piece to be coated is made into an electrode of opposite charge to the paint particles and is immersed in a large tank of the paint. When a voltage is applied between the metal part and the paint tank, the charged paint particles migrate under the influence of the voltage and are deposited on the metal surface, forming a coating over the entire wetted surface of the metal part. After electrodeposition, the piece is removed from the tank and rinsed to remove excess paint, after which the paint is cured in an oven.

The rinse water from the washing step rapidly becomes contaminated with excess paint, while the stability of the paint emulsion is gradually degraded by ionic impurities carried over from the cleaning operation before the paint tank. Both of these problems are solved by the ultrafiltration system shown in Figure 6.21. The ultrafiltration plant takes paint solution containing 15–20 % solids and produces a clean permeate containing the ionic impurities but no paint particles, which is sent to the counter-current rinsing operation, and a slightly concentrated paint to be returned to the paint tank. A portion of the ultrafiltration permeate is bled from the tank and replaced with water to maintain the ionic balance of the process.

Electrocoat paint is a challenging feed solution for an ultrafiltration process. The solids content of the solution is high, typically 15–20 wt%, so a gel layer easily forms on the membrane. The gel formation results in relatively low fluxes, generally 10–15 gal/ft^2 · day. However, the value of the paint recovered from the rinse water and elimination of other rinse-water cleanup steps made the ultrafiltration process an immediate success when introduced by Abcor. Tubular modules were used in the first plants [7] and are still installed in many electrocoat operations, although capillary and some spiral-wound modules are used in newer plants. The first electrocoat paint was anionic because the latex emulsion particles carried a negative charge. These emulsions were best treated with membranes having a slight negative charge to minimize fouling. Cationic latex paints carrying

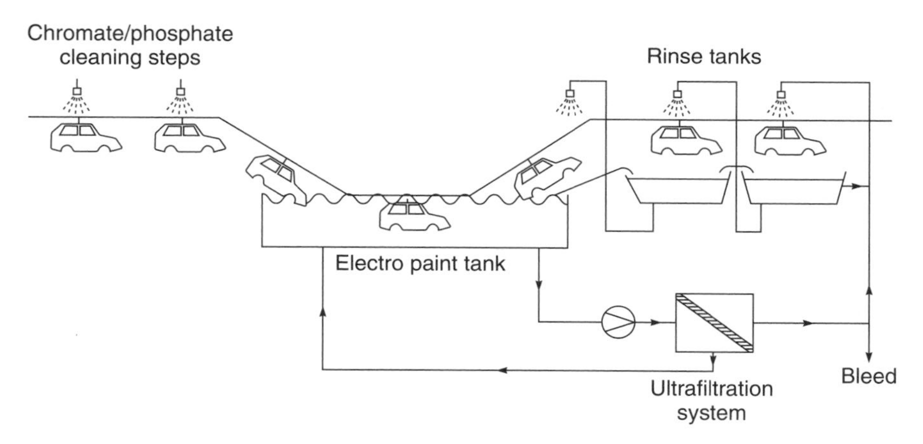

Figure 6.21 Flow schematic of an electrocoat paint ultrafiltration system. The ultrafiltration system removes ionic impurities from the paint tank carried over from the chromate/phosphate cleaning steps and provides clean rinse water for the counter-current rinsing operation

a positive charge were introduced in the late 1970s. Ultrafiltration of these paints required development of membranes carrying a slight positive charge.

Food Industry

Cheese Production

Ultrafiltration has found a major application in the production of cheese; the technology is now widely used throughout the dairy industry. During cheese production the milk is coagulated (or curdled) by precipitation of the milk proteins. The solid that forms (curd) is sent to the cheese fermentation plant. The supernatant liquor (whey) represents a disposal problem. The compositions of milk and whey are shown in Table 6.2. Whey contains most of the dissolved salts and sugars present in the original milk and about 25 % of the original protein. In the past, whey was often discharged to the sewer because its high salt and lactose content makes direct use as a food supplement difficult. Now about half of the whey produced in the United States is processed to obtain additional value and avoid a troublesome waste disposal problems. The traditional cheese production process and two newer processes using ultrafiltration membranes are shown in Figure 6.22.

The objective of the two membrane processes shown in Figure 6.22 is to increase the fraction of milk proteins used as cheese or some other useful product and to reduce the waste disposal problem represented by the whey. In the MMV

Table 6.2 Composition of milk and cheese whey

Component (wt%)	Milk	Whey
Total solids	12.3	7.0
Protein	3.3	0.9
Fat	3.7	0.7
Lactose/other carbohydrates	4.6	4.8
Ash	0.7	0.6

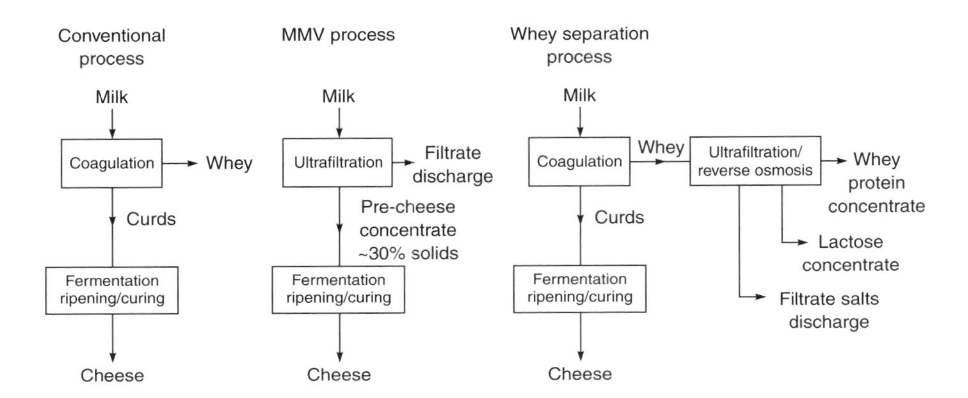

Figure 6.22 Simplified flow schematic showing the traditional cheese production method, and two new methods using ultrafiltration to increase the recycle of useful product

process, named after the developers Maubois, Mocquot and Vassal [25], whole or skimmed milk is concentrated three- to five-fold to produce a pre-cheese concentrate that can be used directly to produce soft cheeses and yogurt. Typically, the total solids level of the concentrate is about 30–35 %, containing 12–17 % protein. This protein concentration is sufficient for soft cheeses (Camembert, Mozzarella and Feta) but cannot be used directly to produce hard cheeses (Cheddar and Swiss), for which protein levels of 25 % are required. When ultrafiltration can be used, increased milk protein utilization increases cheese production by approximately 10 %, so the process is widely used.

The second whey separation process uses both ultrafiltration and reverse osmosis to obtain useful protein from the whey produced in the traditional cheese manufacturing process. A flow schematic of a combined ultrafiltration–reverse osmosis process is shown in Figure 6.23. The goal is to separate the whey into three streams, the most valuable of which is the concentrated protein fraction stripped of salts and lactose. Because raw whey has a high lactose concentration, before the whey protein can be used as a concentrate, the protein concentration must be increased to at least 60–70 % on a dry basis and the lactose content

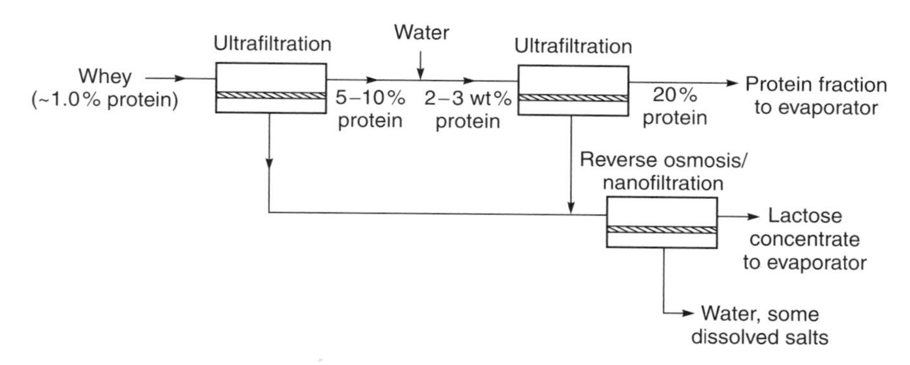

Figure 6.23 Simplified flow schematic of an ultrafiltration/reverse osmosis process to extract valuable components from cheese whey

reduced by 95 %. The objective of the ultrafiltration membrane step is to concentrate the protein as much as possible to minimize evaporator drying costs and to simultaneously remove the lactose. These two objectives are difficult to meet in a single ultrafiltration step because of the reduction in flux at the very high volume reduction required to achieve sufficient lactose removal. Therefore, whey plants commonly use an ultrafiltration step to achieve a 5- to 10-fold volume reduction and remove most of the lactose, after which the feed is diluted with water and reconcentrated in a second step which removes the remaining lactose. Most whey plants use spiral-wound ultrafiltration modules in multistage feed-and-bleed systems. Sanitary spiral-wound module designs are used to eliminate stagnant areas in the module housing, and the entire plant is sterilized daily with hot high- and low-pH cleaning solutions. This harsh cleaning treatment significantly reduces membrane lifetime.

Although whey protein products have several food uses, the lactose contained in the permeate is less valuable, and many plants discharge the permeate to a biological wastewater treatment plant. A few plants recover lactose as dry lactose sugar, as shown in Figure 6.23. Some plants also ferment the lactose concentrate to make ethanol. An introduction to membrane ultrafiltration in cheese production is given by Kosikowski [26].

Clarification of Fruit Juice

Apple, pear, orange and grape juices are all clarified by ultrafiltration. Ultrafiltration of apple juice is a particularly successful application. Approximately 200 plants have been installed, and almost all US apple juice is clarified by this method. In the traditional process, crude filtration was performed directly after crushing the fruit. Pectinase was added to hydrolyze pectin, which reduced the viscosity of the juice before it was passed through a series of decantation and

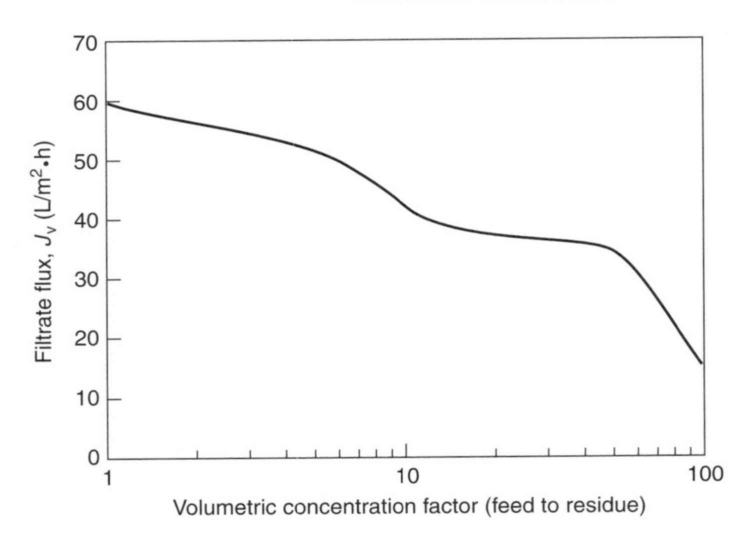

Figure 6.24 Ultrafiltration flux in apple juice clarification as a function of the volumetric feed-to-residue concentration factor. Tubular polysulfone membranes at 55 °C [27]. Reprinted from R.G. Blanck and W. Eykamp, Fruit Juice Ultrafiltration, in *Recent Advances in Separation Techniques-III*, N.N. Li (ed.), AIChE Symposium Series Number 250, 82 (1986). Reproduced by permission of the American Institute of Chemical Engineers. Copyright © 1986 AIChE. All rights reserved

diatomaceous filtration steps to yield clear juice with a typical yield of about 90 %. By replacing these final filtration steps with ultrafiltration, a very good-quality, almost-sterile product can be produced with a yield of almost 97 % [26,27].

Ultrafiltration membranes with a molecular weight cut-off of 10 000–50 000, packaged as tubular or capillary hollow fiber modules, are generally used. The initial feed solution is quite fluid, but in this application almost all of the feed solution is forced through the membrane, and overall concentration factors of 50 are normal. This means that the final residue solution is concentrated and viscous so the solution is usually filtered at 50–55 °C. Operation at this temperature also reduces bacterial growth. A flux versus concentration factor curve produced in this type of application is shown in Figure 6.24. As the concentration of the residue rises, the flux falls dramatically.

Oil–Water Emulsions

Oil–water emulsions are widely used in metal machining operations to provide lubrication and cooling. Although recycling of the fluids is widely practiced, spent waste streams are produced. Using ultrafiltration to recover the oil component and allow safe discharge of the water makes good economic sense, and this application covers a wide volume range. In large, automated machining operations such as automobile plants, steel rolling mills, and wire mills, a central ultrafiltration system may process up to 100 000 gal/day of waste emulsion. These are relatively sophisticated plants that operate continuously using several ultrafiltration feed-and-bleed stages in series. At the other end of the scale are very small systems dedicated to single machines, which process only a few gallons of emulsion per hour. The principal economic driver for users of small systems is the avoided cost of waste hauling. For larger systems the value of the recovered oil and associated chemicals can be important. In both cases, tubular or capillary hollow fiber modules are generally used because of the high fouling potential and very variable composition of emulsified oils. A flow diagram of an ultrafiltration system used to treat large machine oil emulsions is shown in Figure 6.25. The dilute, used emulsion is filtered to remove metal cuttings and is then circulated through a feed-and-bleed ultrafiltration system, producing a concentrated emulsion for reuse and a dilute filtrate that can be discharged or reused.

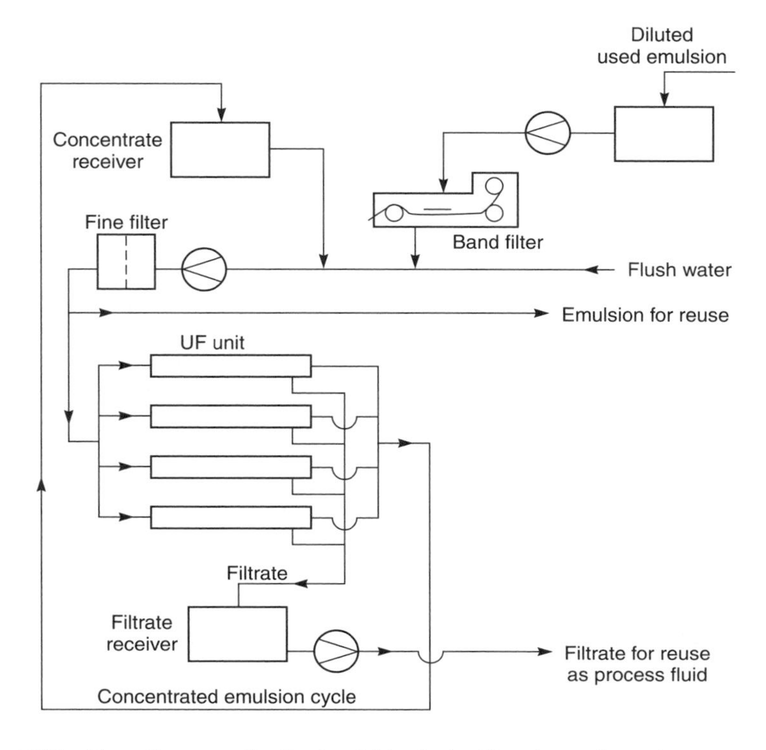

Figure 6.25 Flow diagram of a feed-and-bleed ultrafiltration unit used to concentrate a dilute oil emulsion

Process Water and Product Recycling

Ultrafiltration has been applied to a number of process and product recycling operations. Typical applications include cleaning and recycling hot water used in food processing applications, recovery of latex particles contained in wastewater produced in production of latex paints [28,29] and recovery of poly(vinyl alcohol) sizing agents used as process aids in synthetic fabric weaving operations [28]. The economic driving force for the applications can come from a number of sources:

- **Water recovery.** Depending on the plant's location, reduced municipal water costs can produce savings in the US$1.00–2.00/1000 gal range.
- **Heat recovery.** Many process streams are hot. Ultrafiltration usually works better with hot feeds so hot feed solutions are not a problem. If the hot, clean permeate can be recycled without cooling, the energy savings can be considerable. If the water is 50 °C above ambient temperature, the energy savings amount to about US$4.00/1000 gal.
- **Avoided water treatment costs.** These costs will vary over a wide range depending on the process. For a food processing plant they are likely to be relatively modest—perhaps only US$1/1000 gal or less—but treating latex emulsion plant effluents (called white water) can cost as much as US$10/1000 gal or more.
- **Product recovery value.** If the product concentrated by the ultrafiltration process can be recovered and reused in the plant, this is likely to be the most important credit.

A typical example of a process water and product recycling application, shown in Figure 6.26, is the recovery of poly(vinyl alcohol) sizing agent. In this application, all the above economic drivers contribute to the total plant economics. The feed stream is produced in fabric weaving when the fiber is dipped in a solution of poly(vinyl alcohol) to increase its strength. After weaving, the poly(vinyl alcohol) is removed in a desizing wash bath. The solution produced in this bath is hot (55 °C) and contains 0.5–1.0 % poly(vinyl alcohol). The purpose of the ultrafiltration unit is to concentrate the poly(vinyl alcohol) so it can be recycled to the sizing bath and to send the reclaimed, hot clean permeate stream back to the desizing step. After filtration the poly(vinyl alcohol) solution is relatively particulate-free and quite viscous, so spiral-wound modules are used to reduce costs. For very small plants with flows of less than 5 gal/min, batch systems are used. However, most plants are in the 10- to 100-gal/min range and are multistage feed-and-bleed systems, as shown in Figure 6.26. The environment is challenging for the membranes, which must be cleaned weekly with detergents to remove waxy deposits and with citric acid to remove iron scale. Even so, modules must be replaced every 12–18 months, representing a major operating cost.

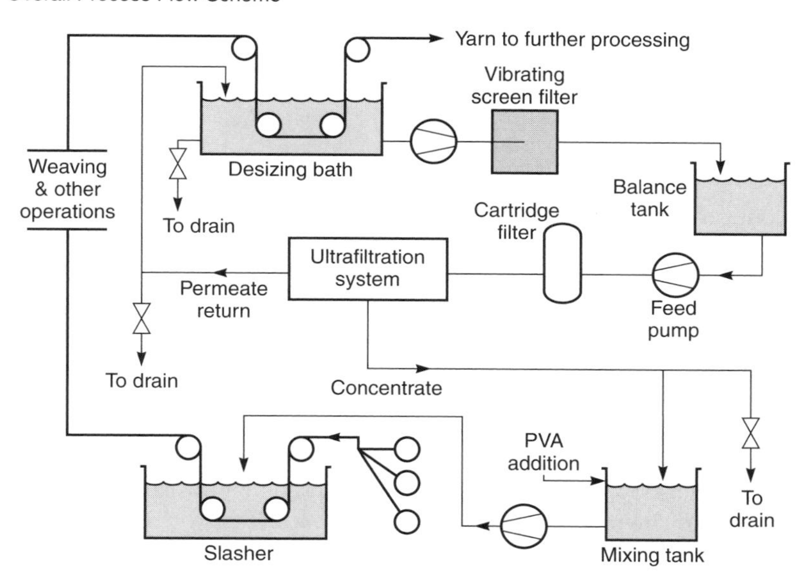

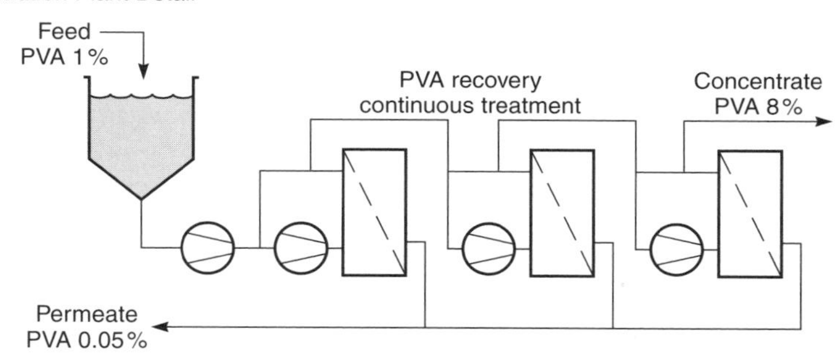

Figure 6.26 Flow schematic of a three-stage feed-and-bleed ultrafiltration system used to recover poly(vinyl alcohol) (PVA) sizing agents used in the production of cotton/synthetic blend fabrics [28]

Biotechnology

Many applications exist for ultrafiltration in the biotechnology industry. A typical application is the concentration and removal of products from fermentation operations used in enzyme production, cell harvesting, or virus production. Most of the systems are small; the volume processed is often only 100 to 1000 gal/day, but the value of products is often very high. Batch systems are commonly used.

Because of the glamour of biotechnology, these applications have received a disproportionate interest by academic researchers. However, this is not a major market for ultrafiltration equipment, and many of the plants use little more than bench-scale equipment.

Conclusions and Future Directions

The high cost per gallon of permeate produced limits the expansion of ultrafiltration into most large wastewater and industrial process stream applications. Costs are high because membrane fluxes are modest, large amounts of energy are used to circulate the feed solution to control fouling, membrane modules must be cleaned frequently, and membrane lifetimes are short. These are all different aspects of the same problem—membrane fouling.

Unfortunately, membrane fouling and gel layer formation are inherent features of ultrafiltration. Only limited progress in controlling these problems has been made in the last 20 years and, barring an unexpected breakthrough, progress is likely to remain slow. Development of inherently fouling-resistant membranes by changing the membrane surface absorption characteristics or charge is a promising approach. By reducing adhesion of the deposited gel layer to the surface, the scrubbing action of the feed solution can be enhanced. Another approach is to develop inherently more fouling-resistant modules. In principle, bore-side-feed capillary fiber modules offer high membrane areas, good flow distribution, and the potential for simple automatic flushing to clean the membrane. The capillary fibers used to date have generally been limited to relatively small diameters and low operating pressures. Development of economical ways to produce 2- to 3-mm-diameter capillary fiber modules, able to operate at 50–100 psi, could lead to lower energy consumption and higher, more stable membrane fluxes. Monolithic ceramic membrane modules have all of these features, but for these to be widely accepted, costs must be reduced by an order of magnitude from today's levels, that is, to less than US\$100–200/m^2. If this cost reduction were achieved, ceramics might replace polymeric membranes in many applications. Vibrating membrane modules have been introduced recently and, although costs are high, their performance is very good. Cost reductions could make this type of module more generally applicable in the future.

References

1. H. Bechhold, Kolloidstudien mit der Filtrationsmethode, *Z. Physik Chem.* **60**, 257 (1907).
2. R. Zsigmondy and W. Bachmann, Uber Neue Filter, *Z. Anorg. Chem.* **103**, 119 (1918).
3. J.D. Ferry, Ultrafilter Membranes and Ultrafiltration, *Chem. Rev.* **18**, 373 (1936).
4. W.J. Elford, Principles Governing the Preparation of Membranes Having Graded Porosities. The Properties of 'Gradocol' Membranes as Ultrafilters, *Trans. Faraday Soc.* **33**, 1094 (1937).

5. S. Loeb and S. Sourirajan, Sea Water Demineralization by Means of an Osmotic Membrane, in *Saline Water Conversion-II*, Advances in Chemistry Series Number 38, Washington, DC (1963).
6. A.S. Michaels, High Flow Membrane, US Patent 3,615,024 (October, 1971).
7. R.L. Goldsmith, R.P. deFilippi, S. Hossain and R.S. Timmins, Industrial Ultrafiltration, in *Membrane Processes in Industry and Biomedicine*, M. Bier (ed.), Plenum Press, New York, pp. 267–300 (1971).
8. R.W. Baker and H. Strathmann, Ultrafiltration of Macromolecular Solutions with High-Flux Membranes, *J. Appl. Polym. Sci.* **14**, 1197 (1970).
9. G. Belfort, R.H. Davis and A.L. Zydney, The Behavior of Suspensions and Macromolecular Solutions in Crossflow Microfiltration, *J. Membr. Sci.* **1**, 96 (1994).
10. M.C. Porter, Concentration Polarization with Membrane Ultrafiltration, *Ind. Eng. Chem. Prod. Res. Dev.* **11**, 234 (1972).
11. G. Jonsson and C.E. Boesen, Polarization Phenomena in Membrane Processes, in *Synthetic Membrane Processes*, G. Belfort (ed.), Academic Press, Orlando, FL, pp. 100–130 (1984).
12. C. Kleinstreuer and G. Belfort, Mathematical Modeling of Fluid Flow and Solute Distribution in Pressure-driven Membrane Modules, in *Synthetic Membrane Processes*, G. Belfort (ed.), Academic Press, Orlando, FL, pp. 131–190 (1984).
13. M.C. Porter, Ultrafiltration, in *Handbook of Industrial Membrane Technology*, M.C. Porter (ed.), Noyes Publication, Park Ridge, NJ, pp. 136–259 (1990).
14. R. Walker, Recent Developments in Ultrafiltration of Electrocoat Paint, *Electrocoat* **82**, 1 (1982).
15. M.C. Porter, Membrane Filtration, in *Handbook of Separation Techniques for Chemical Engineers*, P.A. Schweitzer (ed.), McGraw-Hill, New York, NY, pp. 2.3–2.103 (1979).
16. R.M. McDonogh, T. Gruber, N. Stroh, H. Bauser, E. Walitza, H. Chmiel and H. Strathmann, Criteria for Fouling Layer Disengagement During Filtration of Feed Containing a Wide Range of Solutes, *J. Membr. Sci.* **73**, 181 (1992).
17. R.W. Baker, Method of Fractionating Polymers by Ultrafiltration, *J. Appl. Polym. Sci.* **13**, 369 (1969).
18. B.R. Breslau, A.J. Testa, B.A. Milnes and G. Medjanis, Advances in Hollow Fiber Ultrafiltration Technology, in *Ultrafiltration Membranes and Applications*, A.R. Cooper (ed.), Plenum Press, New York, NY, pp. 109–128 (1980).
19. B. Culkin, A. Plotkin and M. Monroe, Solve Membrane Fouling Problems with High-shear Filtration, *Chem. Eng. Prog.* **94**, 29 (1998).
20. A.N. Rogers, Economics of the Application of Membrane Processes, in *Synthetic Membrane Processes*, G. Belfort (ed.), Academic Press, Orlando, FL, pp. 437–477 (1984).
21. W. Eykamp, Microfiltration and Ultrafiltration, in *Membrane Separation Technology: Principles and Applications*, R.D. Noble and S.A. Stern (eds), Elsevier Science, Amsterdam, pp. 1–40 (1995).
22. B.R. Breslau, P.H. Larsen, B.A. Milnes and S.L. Waugh, The Application of Ultrafiltration Technology in the Food Processing Industry, *The 1988 Sixth Annual Membrane Technology/Planning Conference*, Cambridge, MA (November, 1988).
23. M. Cheryan and F.R. Alvarez, Food and Beverage Industry Applications, in *Membrane Separation Technology: Principles and Applications*, R.D. Noble and S.A. Stern (eds), Elsevier Science, Amsterdam, pp. 415–460 (1995).
24. M. Cheryan, *Ultrafiltration and Microfiltration Handbook*, Technomic Publishing Co., Lancaster, PA (1998).
25. J.L. Maubois, G. Mocquot and L. Vassal, Preparation of Cheese Using Ultrafiltration, US Patent 4,205,080 (1980).

26. F.V. Kosikowski, Membrane Separations in Food Processing, in *Membrane Separations in Biotechnology*, W.C. McGregor (ed.), Marcel Dekker, New York, pp. 201–254 (1986).
27. R.G. Blanck and W. Eykamp, Fruit Juice Ultrafiltration, in *Recent Advances in Separation Techniques-III*, N.N. Li (ed.), AIChE Symposium Series Number 250, AIChE, New York, NY, p. 82 (1986).
28. L. Mir, W. Eykamp and R.L. Goldsmith, Current and Developing Applications for Ultrafiltration, *Indust. Water Eng.* May/June, 1 (1977).
29. B.R. Breslau and R.G. Buckley, The Ultrafiltration of 'Whitewater', An Application Whose Time Has Come!, *The 1992 Tenth Annual Membrane Technology/Separations Planning Conference*, Newton, MA (October, 1992).

7 MICROFILTRATION

Introduction and History

Microfiltration refers to filtration processes that use porous membranes to separate suspended particles with diameters between 0.1 and 10 μm. Thus, microfiltration membranes fall between ultrafiltration membranes and conventional filters. Like ultrafiltration, microfiltration has its modern origins in the development of collodion (nitrocellulose) membranes in the 1920s and 1930s. In 1926 Membranfilter GmbH was founded and began to produce collodion microfiltration membranes commercially. The market was very small, but by the 1940s other companies, including Sartorius and Schleicher and Schuell, were producing similar membrane filters.

The first large-scale application of microfiltration membranes was to culture microorganisms in drinking water; this remains a significant application. The test was developed in Germany during World War II, as a rapid method to monitor the water supply for contamination. The existing test required water samples to be cultured for at least 96 h. Mueller and others at Hamburg University devised a method in which a liter of water was filtered through a Sartorius microfiltration membrane. Any bacteria in the water were captured by the filter, and the membrane was then placed on a pad of gelled nutrient solution for 24 h. The nutrients diffused to the trapped bacteria on the membrane surface, allowing them to grow into colonies large enough to be easily counted under a microscope. After the war there was no US supplier of these membranes, so in 1947 the US Army sponsored a program by Goetz at CalTech to duplicate the Sartorius technology. The membranes developed there were made from a blend of cellulose acetate and nitrocellulose, and were formed by controlled precipitation with water from the vapor phase. This technology was passed to the Lowell Chemical Company, which in 1954 became the Millipore Corporation, producing the Goetz membranes on a commercial scale. Over the next 40 years Millipore became the largest microfiltration company. Membranes made from a number of noncellulosic materials, including poly(vinylidene fluoride), polyamides, polyolefins, and poly(tetrafluoroethylene), have been developed over the last 40 years by Millipore

Membrane Technology and Applications R. W. Baker
 ISBN: 0-470-85445-6

and others. Nonetheless, the cellulose acetate/cellulose nitrate blend membrane remains a widely used microfilter.

Until the mid-1960s, the use of microfiltration membranes was confined to laboratory or to very small-scale industrial applications. The introduction of pleated membrane cartridges by Gelman in the 1970s was an important step forward and made possible the use of microfiltration membranes in large-scale industrial applications. In the 1960s and 1970s, microfiltration became important in biological and pharmaceutical manufacturing, as did microfiltration of air and water in the production of microelectronics in the 1980s. The production of low-cost, single-use, disposable cartridges for pharmaceutical and electronics processes now represents a major part of the microfiltration industry. In most applications of microfiltration in these industries, trace amounts of particles are removed from already very clean solutions. The most widely used process design, illustrated in Figure 7.1(a), is dead-end or in-line filtration, in which the entire fluid flow is forced through the membrane under pressure. As particles accumulate on the membrane surface or in its interior, the pressure required to maintain the required flow increases, until at some point the membrane must be replaced. In

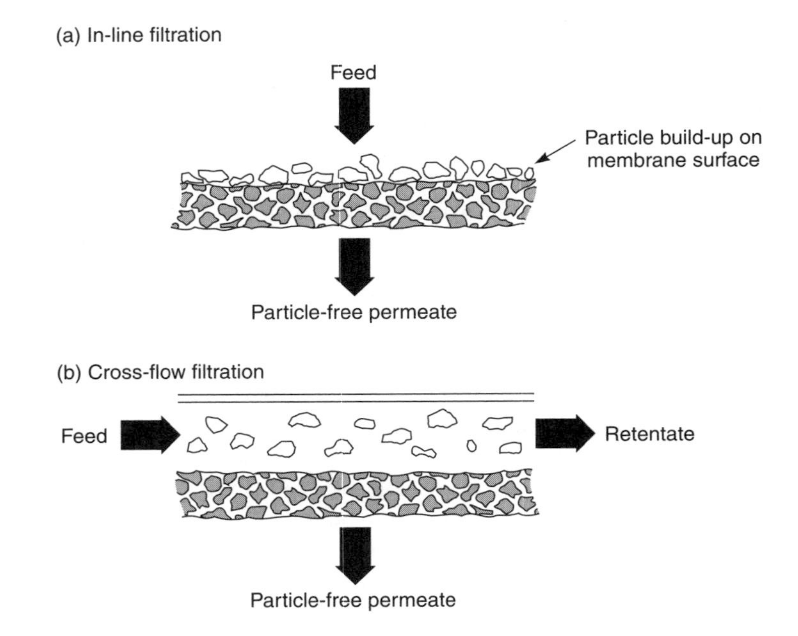

Figure 7.1 Schematic representation of (a) in-line and (b) cross-flow filtration with microfiltration membranes. The equipment used for in-line filtration is simple, but retained particles plug the membrane rapidly. The equipment required for cross-flow filtration is more complex, but the membrane lifetime is longer

the 1970s, an alternative process design known as cross-flow filtration, illustrated in Figure 7.1(b), began to be used.

In cross-flow systems, the feed solution is circulated across the surface of the filter, producing two streams: a clean particle-free permeate and a concentrated retentate containing the particles. The equipment required for cross-flow filtration is more complex, but the membrane lifetime is longer than with in-line filtration. The commercial availability of ceramic tubular cross-flow filters from Membralox (now a division of US Filter), starting in the mid-1980s, has increased the application of cross-flow filtration, particularly for solutions with high particle concentrations. Streams containing less than 0.1 % solids are almost always treated with in-line filters; streams containing 0.5 % solids are almost always treated with cross-flow filters. Between these two limits, both in-line and cross-flow systems can be used, depending on the particular characteristics of the application.

In the last few years, a third type of microfiltration operating system called semi-dead-end filtration has emerged. In these systems, the membrane unit is operated as a dead-end filter until the pressure required to maintain a useful flow across the filter reaches its maximum level. At this point, the filter is operated in cross-flow mode, while concurrently backflushing with air or permeate solution. After a short period of backflushing in cross-flow mode to remove material deposited on the membrane, the system is switched back to dead-end operation. This procedure is particularly applicable in microfiltration units used as final bacterial and virus filters for municipal water treatment plants. The feed water has a very low loading of material to be removed, so in-line operation can be used for a prolonged time before backflushing and cross-flow to remove the deposited solids is needed.

Beginning in 1990–1993, the first microfiltration/ultrafiltration systems began to be installed to treat municipal drinking water obtained from surface water. The US EPA and European regulators are implementing rules requiring this water to be treated to control giardia, coliform bacteria, and viruses. Large plants using back-flushable hollow fiber membrane modules are being built by a number of companies: US Filter (Memtec), Norit (X-Flow), Koch (Romicon), and Hydranautics.

Some of the important milestones in the development of microfiltration are charted in Figure 7.2.

Background

Types of Membrane

The two principal types of microfiltration membrane filter in use—depth filters and screen filters—are illustrated in Figure 7.3. Screen filters have small pores in their top surface that collect particles larger than the pore diameter on the surface of the membrane. Depth filters have relatively large pores on the top

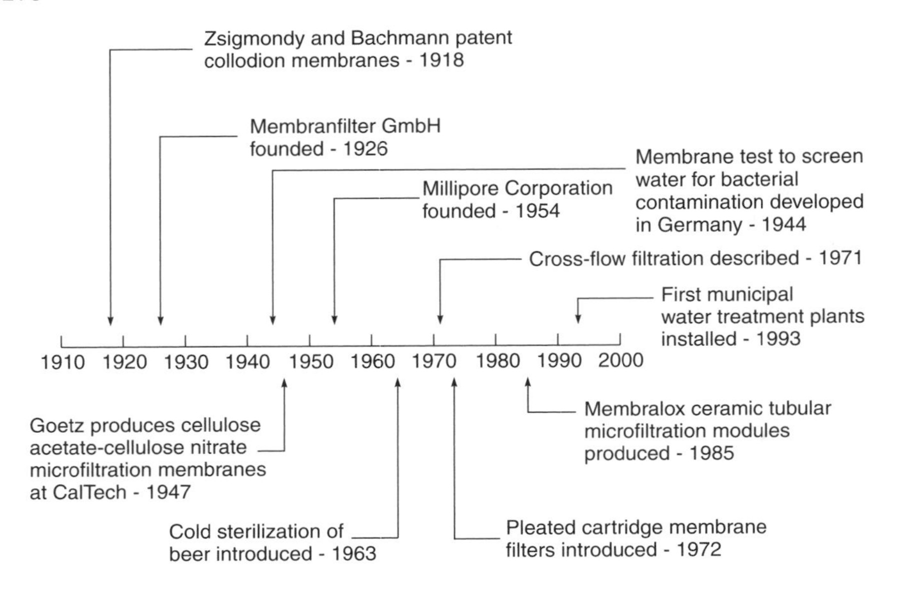

Figure 7.2 Milestones in the development of microfiltration

surface so particles pass to the interior of the membrane. The particles are then captured at constrictions in the membrane pores or by adsorption onto the pore walls. Screen filter membranes rapidly become plugged by the accumulation of retained particles at the top surface. Depth filters have a much larger surface area available for collection of the particles, providing a larger holding capacity before fouling. The mechanism of particle capture by these membranes is described in more detail in Chapter 2.

Depth membrane filters are usually preferred for in-line filtration. As particles are trapped within the membrane, the permeability falls, and the pressure required to maintain a useful filtrate flow increases until, at some point, the membrane must be replaced. The useful life of the membrane is proportional to the particle loading of the feed solution. A typical application of in-line depth microfiltration membranes is final polishing of ultrapure water just prior to use. Screen membrane filters are preferred for the cross-flow microfiltration systems shown in Figure 7.1(b). Because screen filters collect the retained particles on the surface of the membrane, the recirculating fluid helps to keep the filter clean.

Membrane Characterization

Microfiltration membranes are often used in applications for which penetration of even one particle or bacterium through the membrane can be critical. Therefore, membrane integrity, that is, the absence of membrane defects or oversized pores,

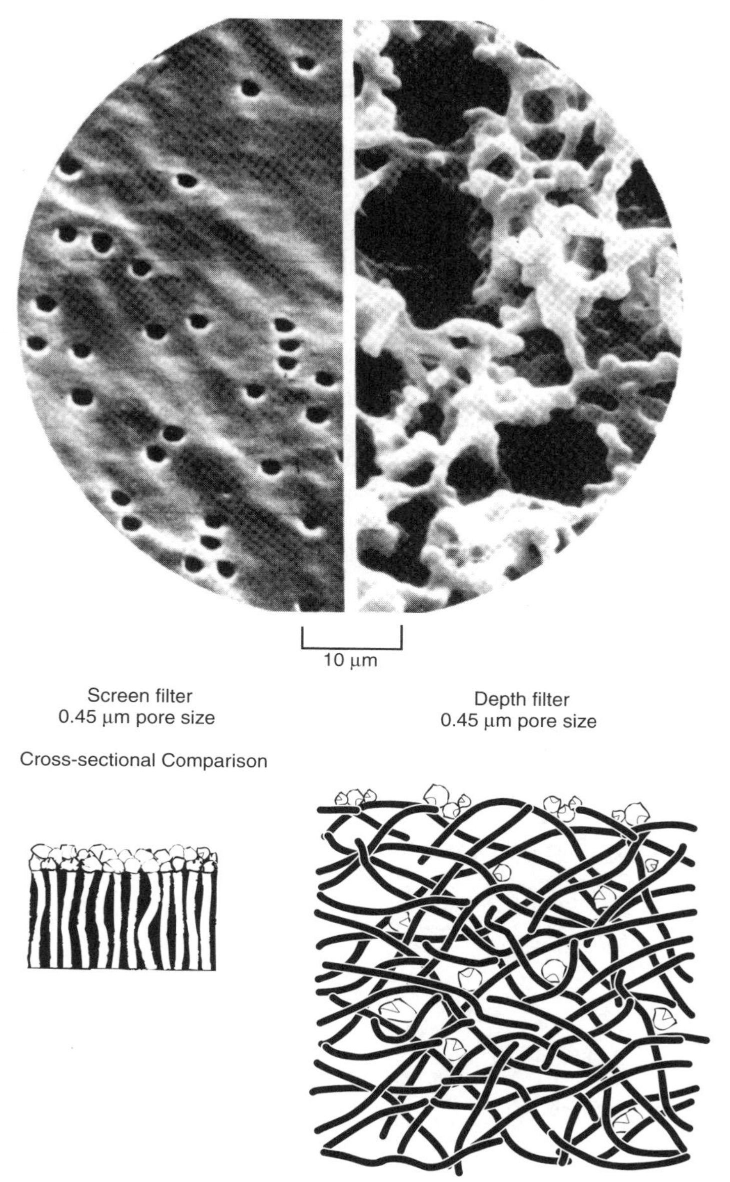

Figure 7.3 Surface scanning electron micrograph and schematic comparison of nominal 0.45-μm screen and depth filters. The screen filter pores are uniform and small and capture the retained particles on the membrane surface. The depth filter pores are almost 5–10 times larger than the screen filter equivalent. A few large particles are captured on the surface of the membrane, but most are captured by adsorption in the membrane interior

is extremely important. Several tests are used to characterize membrane pore size and pore size distribution.

Characterizing the pore size of microfiltration membranes is a problem for manufacturers. Most microfiltration membranes are depth filters, so electron micrographs usually show an image similar to that in Figure 7.3. The average pore diameter of these membranes appears to be about 5 μm, yet the membranes are complete filters for particles or bacteria of about 0.5-μm diameter. Therefore, most manufacturers characterize their membranes by the size of the bacteria that are completely filtered by the membrane. The ability of a membrane to filter bacteria from solutions depends on the pore size of the membrane, the size of the bacteria being filtered, and the number of organisms used to challenge the membrane. Some results of Elford [1] that illustrate these effects are shown in Figure 7.4. Elford found that membranes with relatively large pores could completely filter bacteria from the challenge solution to produce a sterile filtrate, providing the challenge concentration was low. If the organism concentration was increased, breakthrough of bacteria to the filtrate occurred. However, if the membrane pore size was small enough, a point was reached at which no breakthrough of bacteria to the filtrate occurred no matter how concentrated the challenge solution. This point is taken to be the pore size of the membrane.

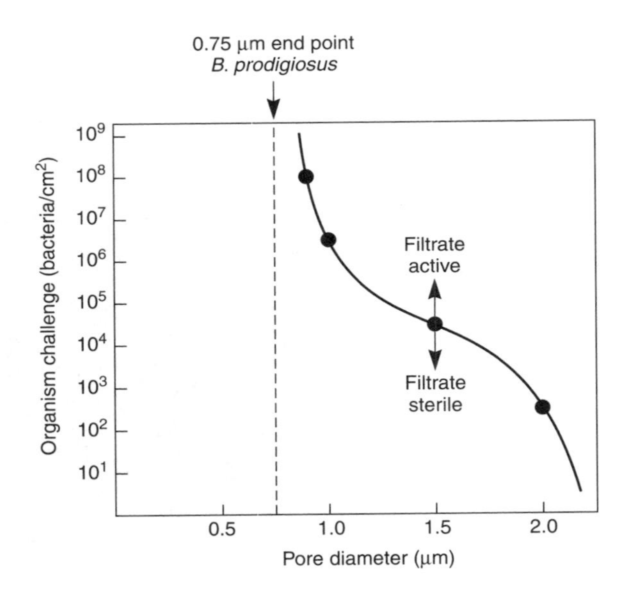

Figure 7.4 Membrane pore diameter from bubble point measurements versus *Bacillus prodigiosus* concentration [1]. Reprinted from W.J. Elford, The Principles of Ultrafiltration as Applied in Biological Studies, *Proc. R. Soc. London, Ser. B* **112**, 384 (1933) with permission from The Royal Society, London, UK

The industry has adopted two bacterial challenge tests to measure pore size and membrane integrity [2]. The tests are based on two bacteria: *Serrata marcescens*, originally thought to have a diameter of 0.45 μm, and *Pseudomonas diminuta*, originally thought to have a diameter of 0.22 μm. In fact, both organisms are ellipsoids with an aspect ratio of about 1.5:1. These tests have changed several times over the years, but by convention a membrane is designated 0.45-μm pore size if it is completely retentive when challenged with 10^7 *S. marcescens* organisms per cm^2 and 0.22-μm pore size if it is completely retentive when challenged with 10^7 *P. diminuta* organisms per cm^2. Most commercial microfiltration membranes are categorized as 0.22- or 0.45-μm-diameter pore size based on these tests. Membranes with larger or smaller pore sizes are classified by the penetration tests with latex particle or bubble point measurements described below, relative to these two primary standard measurements.

Currently, most bacterial challenge tests are performed with *P. diminuta*. This organism has an average size of 0.3–0.4 μm, although the size varies significantly with the culture conditions. In a rich culture medium, the cells can form much larger clumps. Thus, to obtain consistent results, the culture characteristics must be carefully monitored and control experiments performed with already qualified 0.45- and 0.22-μm filters to confirm that no clumping has occurred. The ASTM procedure is illustrated in Figure 7.5 [2]. Factors affecting this test are discussed in detail by Meltzer [3].

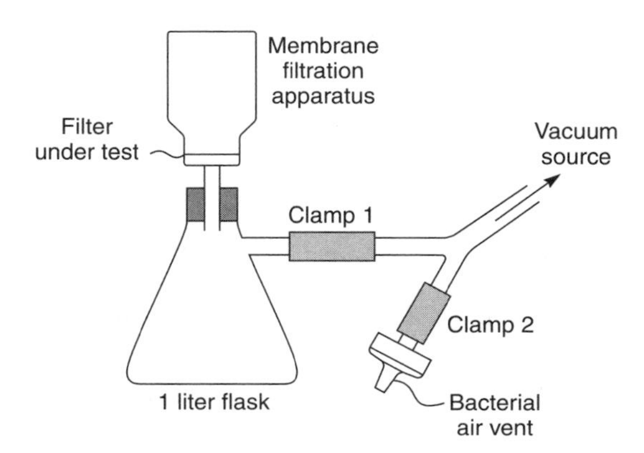

Figure 7.5 Apparatus for testing the microbial retention characteristics of membrane filters. The whole apparatus is sterilized, and initially the flask contains 140 mL of double-strength culture medium. The culture to be tested (100 mL) is passed through the filter with clamp 1 open and clamp 2 closed. The sides of the filter apparatus are washed with two 20 mL portions of sterile broth. Clamp 2 is then opened, the vacuum released, and clamp 1 closed. The filter apparatus is replaced by a sterile rubber stopper and the flask incubated. Absence of turbidity in the flask indicates that the filter has retained the test organism. From Brock [4]. Courtesy of Thomas D. Brock

The performance of membranes in bacterial challenge tests is often quantified by a log reduction value (LRV), defined as

$$LRV = \log_{10}\left(\frac{c_f}{c_p}\right) \tag{7.1}$$

where c_f is the concentration of bacteria in the challenge solution and c_p is the concentration in the permeate. It follows that at 99 % rejection, c_f/c_p is 100 and the LRV is 2; at 99.9 % rejection, the LRV is 3; and so on. In pharmaceutical and electronic applications, an LRV of 7 or 8 is usually required. In municipal water filtration, an LRV of 4 or 5 is the target.

Latex Challenge Tests

Bacterial challenge tests require careful, sterile laboratory techniques and an incubation period of several days before the results are available. For this reason, secondary tests based on filtration of suspensions of latex particles of precise diameters have been developed. In such a test, a monodisperse latex suspension with particle diameters from 0.1 to 10 μm is used. The test solution is filtered through the membrane, and the number of particles permeating the membrane is determined by filtering the permeate solution a second time with a tight membrane screen filter. The membrane screen filter captures the latex particles for easy counting. Although the latex challenge test has been used in fundamental studies of microfiltration membrane properties, it is not widely used by membrane producers. The bubble point test described below, backed by correlating the bubble point to the primary bacterial challenge test results, is more commonly used.

Bubble Point Test

The bubble point test is simple, quick and reliable and is by far the most widely used method of characterizing microfiltration membranes. The membrane is first wetted with a suitable liquid, usually water for hydrophilic membranes and methanol for hydrophobic membranes. The membrane is then placed in a holder with a layer of liquid on the top surface. Air is fed to the bottom of the membrane, and the pressure is slowly increased until the first continuous string of air bubbles at the membrane surface is observed. This pressure is called the *bubble point pressure* and is a characteristic measure of the diameter of the largest pore in the membrane. Obtaining reliable and consistent results with the bubble point test requires care. It is essential, for example, that the membrane be completely wetted with the test liquid; this may be difficult to determine. Because this test is so widely used by microfiltration membrane manufacturers, a great deal of work has been devoted to developing a reliable test procedure to address this and other issues. The use of this test is reviewed in Meltzer's book [3].

The bubble point pressure can be related to the membrane pore diameter, r, by the equation

$$\Delta p = \frac{2\gamma \cos\theta}{r} \quad (7.2)$$

where Δp is the bubble point pressure, γ is the fluid surface tension, and θ is the liquid–solid contact angle. For completely wetting solutions, θ is 0° so $\cos\theta$ equals 1. Properties of liquids commonly used in bubble point measurements are given in Table 7.1.

Microfiltration membranes are heterogeneous structures having a distribution of pore sizes. The effect of the applied gas pressure on the liquid in a bubble test is illustrated schematically in Figure 7.6. At pressures well below the bubble point, all pores are completely filled with liquid so gas can only pass through the membrane by diffusion through the liquid film. Just below the bubble point pressure, liquid begins to be forced out of the largest membrane pores. The diffusion rate then starts to increase until the liquid is completely forced out of the largest pore. Bubbles of gas then form on the membrane surface. As the gas pressure is increased further, liquid is forced out of more pores, and general convective flow of gas through the membrane takes place. This is sometimes called the 'foam all over pressure' and is a measure of the average pore size of the membrane.

The apparatus used to measure membrane bubble points is shown in its simplest form in Figure 7.7 [4]. Bubble point measurements are subjective, and different operators can obtain different results. Nonetheless the test is quick and simple and is widely used as a manufacturing quality control technique. Bubble point measurements are also used to measure the integrity of filters used in critical pharmaceutical or biological operations.

Bubble point measurements are most useful to characterize sheet stock or small membrane filters. The technique is more difficult to apply to formed membrane cartridges containing several square feet of membrane because diffusive flow of

Table 7.1 Properties of liquids commonly used in bubble point measurements. The conversion factor divided by the bubble pressure (in psi) gives the maximum pore size (in μm)

Wetting liquid	Surface tension (dyn/cm)	Conversion factor
Water	72	42
Kerosene	30	17
Isopropanol	21.3	12
Silicone fluid[a]	18.7	11
Fluorocarbon fluid[b]	16	9

[a] Dow Corning 200 fluid, 2.0 cSt.
[b] 3M Company, Fluorochemical FC-43.

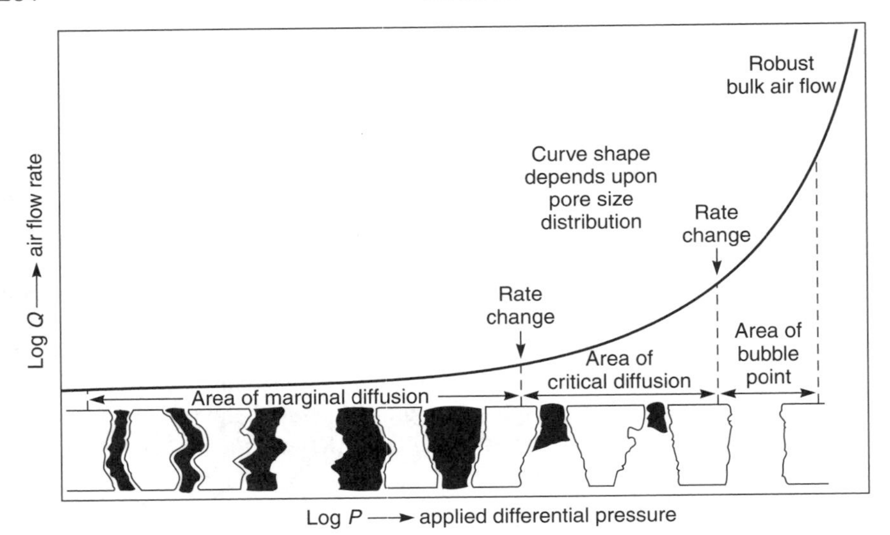

Figure 7.6 Schematic of the effect of applied gas pressure on gas flow through a wetted microporous membrane in a bubble pressure test (Meltzer) [3]. Reprinted from Meltzer [3] by courtesy of Marcel Dekker, Inc.

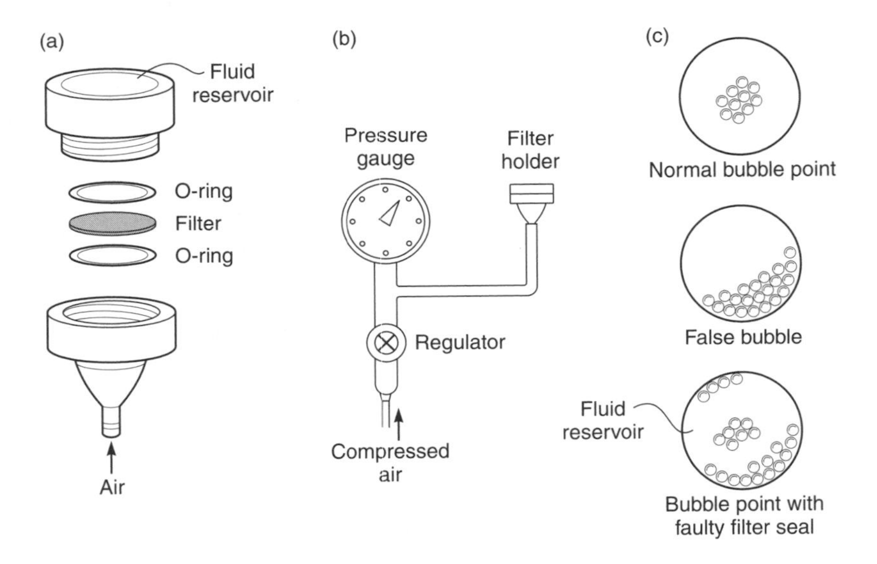

Figure 7.7 Bubble point measurements; (a) exploded view of filter holder; (b) test apparatus; and (c) typical bubble patterns produced. From Brock [4]. Courtesy of Thomas D. Brock

gas through the liquid film masks the bubble point. To test cartridges, the applied pressure is set at a few psi below the bubble point, typically at 80 % of the bubble point pressure. The diffusive flow of gas through the wetted cartridge filter is then measured [5]. This provides a good integrity test of large-area cartridge filters because even a small membrane defect increases gas flow significantly above the norm for defect-free cartridges.

Although bubble point measurements can be used to determine the pore diameter of membranes using Equation (7.1), the results must be treated with caution. Based on Equation (7.1), a 0.22-μm pore diameter membrane should have a bubble point of about 200 psig. In fact, based on the bacterial challenge test, a 0.22-μm pore diameter membrane has a bubble point pressure of 40–60 psig, depending on the membrane. That is, the bubble point test indicates that the membranes has a pore diameter of about 1 μm.

Figure 7.8 shows typical results comparing microbial challenge tests using 0.22-μm *P. diminuta* with membrane bubble points for a series of related membranes [6]. In these tests at a microbial reduction factor of 10^8–10^9, the membrane has a bubble point pressure of only 40 psig, far below the theoretical

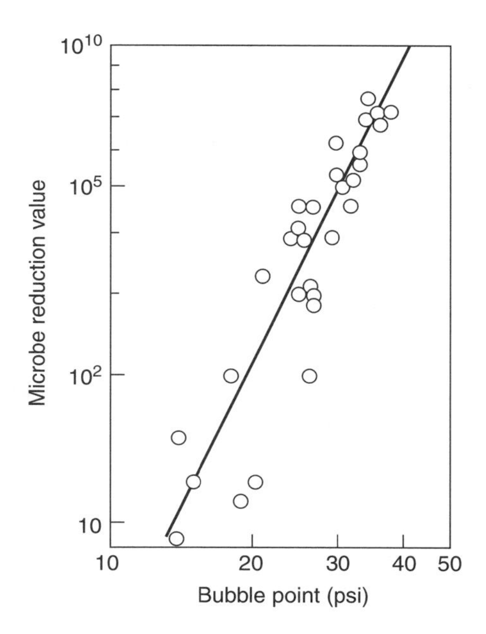

Figure 7.8 Correlation of *P. diminuta* microbial challenge and bubble point test data for a series of related membranes [6]. Reprinted from T.J. Leahy and M.J. Sullivan, Validation of Bacterial Retention Capabilities of Membrane Filters, *Pharm. Technol.* **2**, 65 (1978) with permission from Pharmaceutical Technology, Eugene, OR

value of 200 psig for a 0.22-μm pore diameter membrane. Such discrepancies are sometimes handled by a correction factor in Equation (7.1) to account for the shape of the membrane pores, but no reasonable shape factor can account for the four-fold discrepancy seen here. There are two possible reasons why the bubble point test overestimates the minimum pore size of the membrane. First, the test is a measure of the pore size of the membrane. However, a one-to-one relationship between the diameter of the bacteria able to penetrate the membrane and the pore diameter assumes that the only method of bacterial capture is direct filtration of the test organism somewhere in the membrane. If no organisms penetrate the membrane even at a high concentration, the conclusion is that no pores larger than the organism's diameter exist. However, this ignores other capture mechanisms, such as adsorption and electrostatic attraction, that can remove the organism even though the pore diameter is larger than the particle. As a result, although a small fraction of the membrane pores may be larger than 0.22 μm, leading to a low bubble point pressure, bacteria still cannot travel through these pores in a normal challenge test.

A second explanation, proposed by Williams and Meltzer [7], is illustrated in Figure 7.9. In liquid flow, all flow through the membrane is from the high-pressure (top) to the low-pressure (bottom) side of the membrane. In a bubble point test the membrane is filled with liquid, and gas is used to displace liquid from the large pores. The bubble point is reached when the first contiguous series of large pores through the membrane is formed. This path can be long and tortuous and may not follow the path taken by liquid flow.

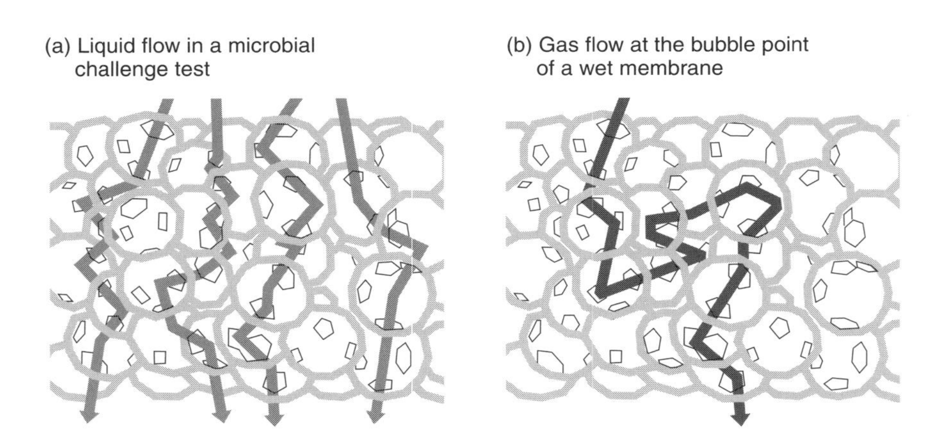

Figure 7.9 An illustration of the model of Williams and Meltzer [7] to explain the discrepancy between membrane pore diameter measurements based on the microbial challenge test and the bubble point test. Reprinted from R.E. Williams and T.H. Meltzer, Membrane Structure, the Bubble Point and Particle Retention, *Pharm. Technol.* **7** (5), 36 (1983) with permission from Pharmaceutical Technology, Eugene, OR

Microfiltration Membranes and Modules

The first major application of microfiltration membranes was for biological testing of water. This remains an important laboratory application in microbiology and biotechnology. For these applications the early cellulose acetate/cellulose nitrate phase separation membranes made by vapor-phase precipitation with water are still widely used. In the early 1960s and 1970s, a number of other membrane materials with improved mechanical properties and chemical stability were developed. These include polyacrylonitrile–poly(vinyl chloride) copolymers, poly(vinylidene fluoride), polysulfone, cellulose triacetate, and various nylons. Most cartridge filters use these membranes. More recently poly(tetrafluoroethylene) membranes have come into use.

In the early 1960s and 1970s, the in-line plate-and-frame module was the only available microfiltration module. These units contained between 1 and 20 separate membrane envelopes sealed by gaskets. In most operations all the membrane envelopes were changed after each use; the labor involved in disassembly and reassembly of the module was a significant drawback. Nonetheless these systems are still widely used to process small volumes of solution. A typical plate-and-frame filtration system is shown in Figure 7.10.

Figure 7.10 Sterile filtration of a small-volume pharmaceutical solution with a 142 mm plate-and-frame filter used as a prefilter in front of a small disposable cartridge final filter. From Gelman Science

More recently, a variety of cartridges that allow a much larger area of membrane to be incorporated into a disposable unit have become available. Disposable plate-and-frame cartridges have been produced, but by far the largest portion of the market is for pleated cartridges, first introduced in the early 1970s. A disposable cartridge filter of this type is shown in Figure 7.11. A typical cartridge is 10 in. long, has a diameter of 2–2.5 in., and contains about 3 ft^2 of membrane. Often the membrane consists of several layers: an outer prefilter facing the solution to be filtered, followed by a finer polishing membrane filter.

In these units, the membrane is pleated and then folded around the permeate core. The cartridge fits inside a specially designed housing into which the feed solution enters at a pressure of 10–120 psi. Pleated membrane cartridges, which are fabricated with high-speed automated equipment, are cheap, disposable, reliable, and hard to beat if the solution to be filtered has a relatively low particle level. Ideal applications are production of aseptic solutions in the pharmaceutical industry or ultrapure water for wafer manufacture in the electronics industry. The low particle load of these feed solutions allows small in-line cartridges to filter large volumes of solution before needing replacement. Manufacturers produce cartridge holders that allow a number of cartridges to be connected in series

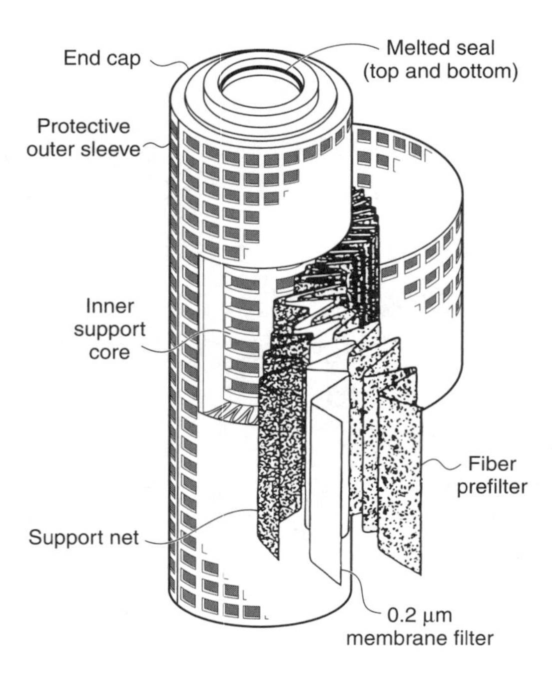

Figure 7.11 Cut-away view of a simple pleated cartridge filter. By folding the membrane a large surface area can be contacted with the feed solution producing a high particle loading capacity. (From Membrana product literature)

Figure 7.12 Standard-size disposable cartridges can be connected in series or parallel to handle large flows. This unit consists of nine cartridges arranged in a 3 × 3 array. (From Sartorius product literature)

or in parallel to handle large solution flows. A multicartridge unit is shown in Figure 7.12.

However, the short lifetime of in-line cartridge filters makes them unsuitable for microfiltration of highly contaminated feed streams. Cross-flow filtration, which overlaps significantly with ultrafiltration technology, described in Chapter 6, is used in such applications. In cross-flow filtration, long filter life is achieved by sweeping the majority of the retained particles from the membrane surface before they enter the membrane. Screen filters are preferred for this application, and an ultrafiltration membrane can be used. The design of such membranes and modules is covered under ultrafiltration (Chapter 6) and will not be repeated here.

Process Design

A typical in-line cartridge filtration application is illustrated in Figure 7.13. A pump forces liquid through the filter, and the pressure across the filter is measured by a pressure gauge. Initially, the pressure difference measured by the gauge is small, but as retained particles block the filter, the pressure difference increases until a predetermined limiting pressure is reached, and the filter is changed.

To extend its life, a microfiltration cartridge may contain two or more membrane filters in series, or as shown in Figure 7.13, a coarse prefilter cartridge

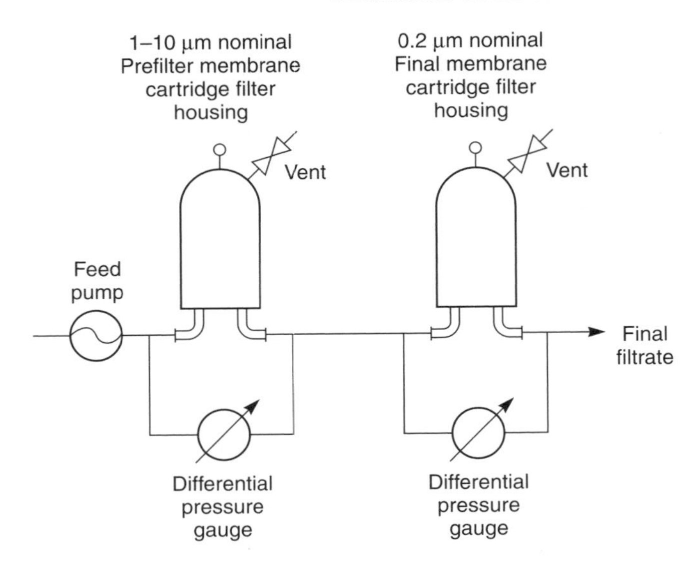

Figure 7.13 Typical in-line filtration operation using two cartridge filters in series. The prefilter removes all of the large particles and some of the smaller ones. The final polishing filter removes the remaining small particles

before the final polishing filter. The prefilter captures the largest particles, allowing smaller particles to pass and be captured by the following finely porous membrane. The use of a prefilter extends the life of the microfiltration cartridge significantly. Without a prefilter the fine microfiltration membrane would be rapidly blinded by accumulation of large particles on the membrane surface. The correct combination of prefilter and final membrane must be determined for each application. This can be done by placing the prefilter on top of the required final filter membrane in a small test cell, or better yet, with two test cells in series. With two test cells the pressure drop across each filter can be measured separately.

The objective of a prefilter is to extend the life of the final filter by removing the larger particles from the feed, allowing the final filter to remove the smaller particles. The results obtained with different prefilters are shown in Figure 7.14 [8]. Figure 7.14(a) shows the rate of pressure rise across the fine filter alone. The limited dirt-holding capacity of this filter means that it is rapidly plugged by a surface layer of large particles. Figure 7.14(b) shows the case when a too coarse prefilter is used. In this case, the pressure difference across the prefilter remains small, whereas the pressure difference across the final filter increases as rapidly as before because of plugging by particles passing the prefilter. Little improvement in performance is obtained. Figure 7.14(c) shows the case where the prefilter is too fine. This situation is the opposite of 7.14(a)—the pressure difference across

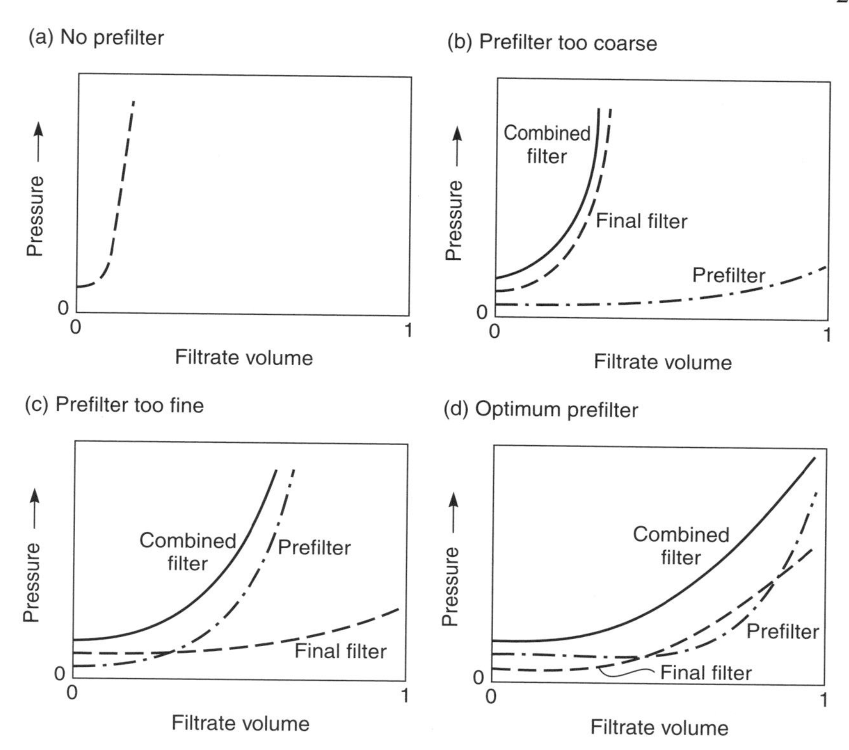

Figure 7.14 The pressure difference across the prefilter, the final filter and the combined filters for various combinations of prefilter and final filter. The optimum prefilter distributes the particle load evenly between the two filters so both filters reach their maximum particle load at the same time. This maximizes the useful life of the combination

the prefilter increases rapidly, and the lifetime of the combination filter is limited by this filter. Figure 7.14(d) shows the optimum combination in which the pressure difference is uniformly distributed across the prefilter and final filter. This condition maximizes the lifetime of the filter combination.

Recently, some membrane manufacturers have attempted to produce anisotropic microfiltration membranes in which the open microporous support is a built-in prefilter. Unlike most other applications of anisotropic membranes, these membranes are oriented with the coarse, relatively open pores facing the feed solution, and the most finely microporous layer is at the bottom of the membrane. The goal is to increase filter life by distributing the particle load more evenly across the filter than would be the case with an isotropic porous membrane.

Cartridge microfiltration is a stable area of membrane technology—few changes in cartridge design or use have occurred in the past 20 years. Most changes have focused on improving resistance to higher temperatures, solvents and extremes of pH, to allow application of these filters in more challenging environments.

Recent innovation in microfiltration has mainly concerned the development of cross-flow filtration technology and membranes. The design of these processing systems closely follows that of ultrafiltration described in Chapter 6. In cross-flow filtration, the membrane must retain particles at the membrane surface; therefore, only asymmetric membranes or screen filters with their smallest pores facing the feed solutions can be used. Ceramic filters of the type made by Membralox (now part of US Filter) and others are being used increasingly in this type of application. A ceramic microfiltration cross-flow filter is shown in Figure 7.15. Capillary hollow fiber membrane modules similar to those originally developed for ultrafiltration applications are also now being widely used for cross-flow microfiltration applications.

The key innovation that has led to the increased use of cross-flow microfiltration membrane modules in the last few years has been the development of back-pulsing or backflushing to control membrane fouling [9–11]. In this procedure, the water flux through the membrane is reversed to remove any particulate and fouling material that may have formed on the membrane surface. In microfiltration several types of backflushing can be used. Short, relatively frequent flow reversal lasting a few seconds and applied once every few minutes is called

Figure 7.15 Monolithic ceramic microfilter. The feed solution passes down the bores of the channels formed in a porous ceramic block. The channel walls are coated with a finely porous ceramic layer

back-pulsing. Longer flow reversal, lasting 1 or 2 min and applied once every 1 or 2 h, is called backflushing. The balance between the duration of back pulses and their frequency depends on the particular application.

Direct observations illustrating the efficiency of back-pulsing have been made by Mores and Davis [9] using a transparent test cell and cellulose acetate microfiltration membranes fouled with yeast cells. Figure 7.16(a) shows a photograph of the membrane surface after 2 h of operation with a yeast solution. The membrane surface is completely covered with yeast cells. Figure 7.16(b–d) shows the effect of back-pulsing for different times. Back-pulsing for 0.1 s removes about half the yeast, back-pulsing for 1 s removes about 90 %, and back-pulsing for 180 s removes all but a few yeast cells.

Microfiltration cross-flow systems are often operated at a constant applied transmembrane pressure in the same way as the reverse osmosis and ultrafiltration systems described in Chapters 5 and 6. However, microfiltration membranes tend to foul and lose flux much more quickly than ultrafiltration and reverse osmosis membranes. The rapid decline in flux makes it difficult to control system operation. For this reason, microfiltration systems are often operated as constant flux systems, and the transmembrane pressure across the membrane is slowly increased to maintain the flow as the membrane fouls. Most commonly the feed pressure is fixed at some high value and the permeate pressure

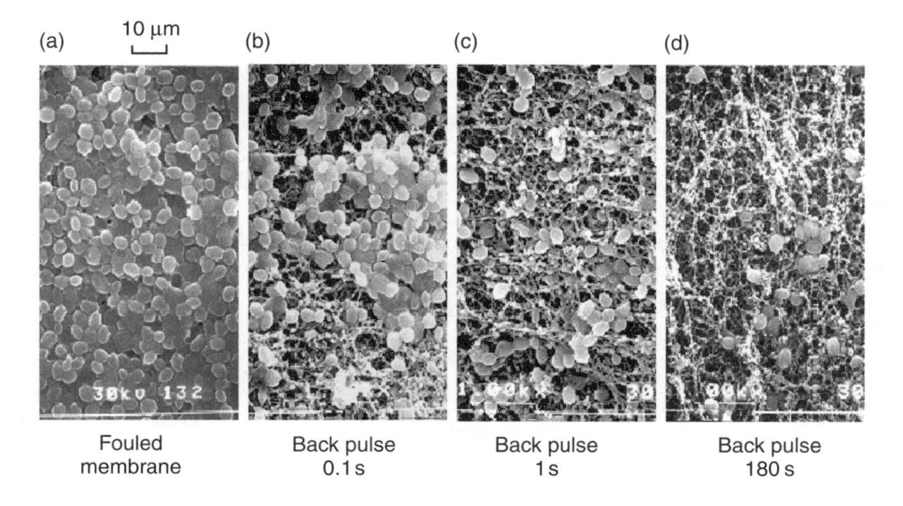

Figure 7.16 An illustration of the efficiency of back-pulsing in removing fouling materials from the surface of microfiltration membranes. Direct microscopic observations of Mores and Davis [9] of cellulose acetate membranes fouled with a 0.1 wt% yeast suspension. The membrane was backflushed with permeate solution at 3 psi for various times. Reprinted from *J. Membr. Sci.* **189**, W.D. Mores and R.H. Davis, Direct Visual Observation of Yeast Deposition and Removal During Microfiltration, p. 217, Copyright 2001, with permission from Elsevier

is set at a value just below the feed pressure. As the membrane is used, its permeability slowly decreases because of fouling. This decrease in permeability is compensated for by lowering the permeate pressure and so increasing the pressure driving force. When the permeate pressure reaches some predetermined value, the module is taken off-line and cleaned or backflushed to restore its permeability.

Some results illustrating the operation of microfiltration membranes using the constant-flux procedure are illustrated in Figure 7.17 [12]. The target flux of the module was set by choosing the initial pressure difference across the membrane. As the membrane fouled, the pressure difference increased until it reached the limiting pressure difference of 6–9 psi. In the example shown, modules set to produce a flux of 50 $L/m^2 \cdot h$ were completely fouled within 20 h. Reducing the flux to 40 $L/m^2 \cdot h$ increased the useful lifetime of the membrane three-fold; reducing the flux to 30 $L/m^2 \cdot h$ achieved a very long lifetime. Just as in ultrafiltration, the membranes perform best when operated at conditions where membrane fouling is controlled by the flow of liquid across the surface. Operating at high flux or high transmembrane pressure leads to deposition of a thick compacted fouling layer.

The advantages and disadvantages of in-line microfiltration and cross-flow filtration are compared in Table 7.2. In general, in-line filtration is preferred as a polishing operation for already clean solutions, for example, to sterilize water

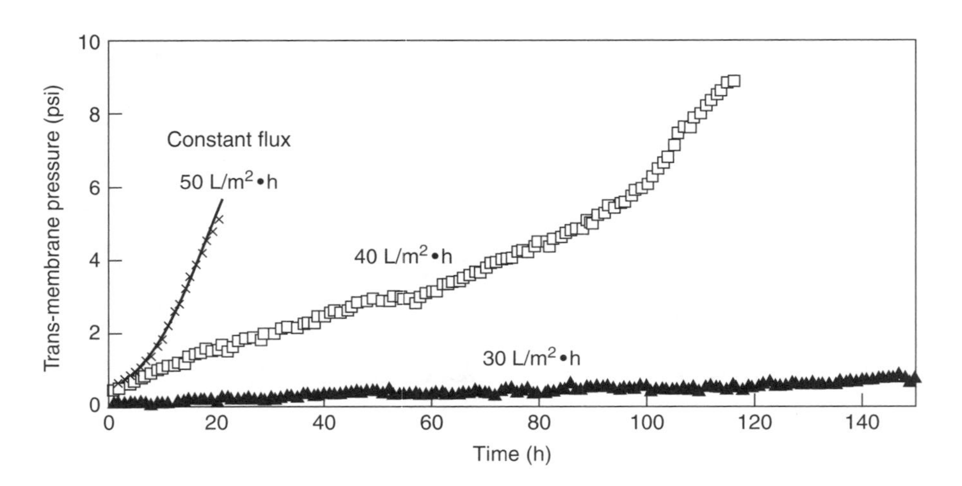

Figure 7.17 Experiments showing the rate of fouling of 0.22-μm microfiltration membranes used to treat dilute biomass solutions. The membranes were operated at the fluxes shown, by increasing transmembrane pressure over time to maintain this flux as the membranes fouled [12]. Reprinted from *J. Membr. Sci.* **209,** B.D. Cho and A.G. Fane, Fouling Transients in Nominally Sub-critical Flux Operation of a Membrane Bioreactor, p. 391, Copyright 2002, with permission from Elsevier

Table 7.2 Comparison of advantages and disadvantages of in-line and cross-flow microfiltration

In-line microfiltration	Cross-flow microfiltration
Low capital cost	High capital cost
High operating costs—membrane must be replaced after each use and disposal can be a problem	Operating costs modest—membranes have extended lifetimes if regularly cleaned
Operation is simple—no moving parts	Operation is complex—filters require regular cleaning
Best suited to dilute (low solid content) solutions. Membrane replacement costs increase with particle concentrations in the feed solution	Best suited to high solid content solutions. Costs are relatively independent of feed solution particle concentrations
Representative applications: Sterile filtration; Clarification/sterilization of beer and wine	Representative applications: Continuous culture/cell recycle; Filtration of oilfield produced water

in the pharmaceutical and electronics industries. Cross-flow filtration is more expensive than in-line filtration in this type of application but, if the water has a high particle content, cross-flow filtration is preferred.

Applications

The microfiltration market differs significantly from that of other membrane separation processes in that membrane lifetimes are often measured in hours. In a few completely passive applications, such as treating sterile air vents, membranes may last several months; in general the market is dominated by single-use cartridges designed to filter a relatively small mass of particles from a solution. The volume of solution that can be treated by a microfiltration membrane is directly proportional to the particle level in the water. As a rough rule of thumb, the particle-holding capacity of a cartridge filter in a noncritical use is between 100 and 300 g/m^2 of membrane area. Thus, the volume of fluid that can be treated may be quite large if the microfilter is a final safety filter for an electronics plant ultrapure water system, but much smaller if treating contaminated surface water or a food processing stream. The approximate volume of various solutions that can be filtered by a 5-μm filter before the filter is completely plugged is given in Table 7.3 [13].

Despite the limited volumes that can be treated before a filter must be replaced, microfiltration is economical because the cost of disposable cartridges is low. Currently, a 10-in.-long pleated cartridge costs between US$10 and US$20 and contains 0.3–0.5 m^2 of active membrane area. The low cost reflects the large numbers that are produced.

Table 7.3 Approximate volume of fluid that can be filtered by 1 m^2 of a 5-μm membrane before fouling [13]

Solution	Volume filtered (m^3/m^2)
Water from deep wells	1000
Solvents	500
Tap water	200
Wine	50
Pharmaceuticals for ampoules	50
20 % glucose solution	20
Vitamin solutions	10
Parenterals	10
Peanut oil	5
Fruit juice concentrate	2
Serum (7 % protein)	0.6

The primary market for the disposable cartridge is sterile filtration for the pharmaceutical industry and final point-of-use polishing of ultrapure water for the microelectronics industry. Both industries require very high-quality, particle-free water. The cost of microfiltration compared to the value of the products is small so these markets have driven the microfiltration industry for the past 15 years.

Sterile Filtration of Pharmaceuticals

Microfiltration is used widely in the pharmaceutical industry to produce injectable drug solutions. Regulating agencies require rigid adherence to standard preparation procedures to ensure a consistent, safe, sterile product. Microfiltration removes particles but, more importantly, all viable bacteria, so a 0.22-μm-rated filter is usually used. Because the cost of validating membrane suppliers is substantial, users usually develop long-term relationships with individual suppliers.

A microfilter for this industry is considered sterile if it achieves a log reduction factor of better than 7. This means that if 10^7 bacteria/cm^2 are placed on the filter, none appears in the filtrate. A direct relationship exists between the log reduction factor and the bubble point of a membrane.

Microfiltration cartridges produced for this market are often sterilized directly after manufacture and again just prior to use. Live steam, autoclaving at 120 °C, or ethylene oxide sterilization may be used, depending on the applications. A flow schematic of an ampoule-filling station (after material by Schleicher and Schuell) is shown in Figure 7.18.

In this process, feedwater is first treated by a deionization system consisting of reverse osmosis, followed by mixed bed ion exchange, and a final 5-μm filtration step. The requirements of water for injection are a good deal less stringent

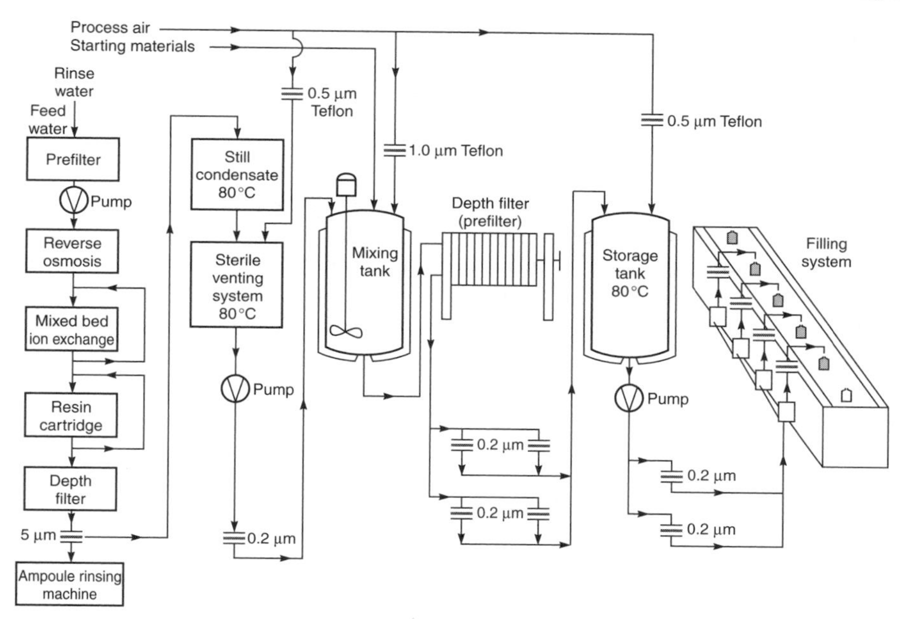

Figure 7.18 Flow diagram illustrating the use of microfiltration sterilization filters in a production line used to prepare ampoules of injectable drug solutions

than the requirements of ultrapure water for the electronics industry, so the water treatment system is relatively straightforward. The water is first sterilized with a 0.2-μm final filter before being mixed with the drug solution, then sent to a storage tank for the ampoule-filling station. Before use, the solution is filtered at least twice more with 0.2-μm filters to ensure sterility. Because pharmaceuticals are produced by a batch process, all filters are replaced at the end of each batch.

Sterilization of Wine and Beer

Cold sterilization of beer using microfiltration was introduced on a commercial scale in 1963. The process was not generally accepted at that time, but has recently become more common. Sterilization of beer and wine is much less stringent than pharmaceutical sterilization. The main objective is to remove yeast cells, which are quite large, so the product is clear and bright. Bacterial removal is also desirable; a 10^6 reduction in bacteria is equivalent to the best depth filters. The industry has found that 1-μm filters can remove essentially all the yeast as well as provide a 10^6 reduction in the common bacteria found in beer and wine. Because the cost structure of beer and wine production is very different from that of pharmaceuticals, the filtration system typically involves one or more prefilters to extend the life of the final polishing filter.

Microfiltration in the Electronics Industry

Microfilters are used in the electronics industry, principally as final point-of-use filters for ultrapure water. The water is already very pure and almost completely particle- and salt-free, so the only potential problem is contamination in the piping from the central water treatment plant to the device fabrication area. Although fine filters with 0.1 μm pore diameter or less may be used, lifetimes are relatively long.

The electronics industry also uses a variety of reactive gases and solvents which must be particle free. Teflon® microfilters are widely used to treat these materials.

Microfiltration for Drinking Water Treatment

Beginning in about 1990, the first microfiltration/ultrafiltration plants were installed to treat municipal surface water supplies [14,15]. The driver was implementation of an EPA surface water treatment rule requiring all utilities in the United

Figure 7.19 Photograph of a 25 million gal/day capillary hollow fiber module plant to produce potable water from a well, installed by Norit (X-Flow) in Keldgate, UK. Courtesy of Norit Membrane Technology BV

States to provide an LRV of 3 for *giardia* and an LRV of 4 for viruses. European regulators have adopted similar rules.

The plants installed have all been equipped with hollow fiber membrane modules. The feed water is generally fairly clean, so the modules are operated in a dead-end mode for 10–20 min and then backflushed with air or filtered water for 20–30 s. During backflushing, the modules are swept with water to remove the accumulated solids, after which the cycle is repeated. It is estimated that 40 000 water works in the United States are affected by the EPA ruling, so the potential market is very large. Many of these water works are small, but several large plants equipped with hundreds of modules have also been installed. A photograph of one such plant is shown in Figure 7.19. Similar plants are also being considered to prefilter and sterilize feed water for reverse osmosis desalination plants or for tertiary treatment and ultimate reuse of water from sewage treatment plants.

Conclusions and Future Directions

The main microfiltration market is for in-line disposable cartridge filters. These cartridges are sold into two growing modern industries—microelectronics and pharmaceuticals—so prospects for continued market growth of the industry are very good. In addition to these existing markets, significant potential markets exist for microfiltration in bacterial control of drinking water, tertiary treatment of sewage, and replacement of diatomaceous earth depth filters in the chemical processing and food industries. The particle load of all these waters is far higher than that presently treated by microfiltration and has required development of cross-flow filtration systems able to give filter lifetimes of months or even years. Such systems are now being installed in municipal water treatment plants. The units can be cleaned by backflushing and offer reliable performance. Municipal water treatment is likely to develop into a major future application of microfiltration technology.

References

1. W.J. Elford, The Principles of Ultrafiltration as Applied in Biological Studies, *Proc. R. Soc. London, Ser. B*, **112**, 384 (1933).
2. Determining Bacterial Retention of Membrane Filters Utilized for Liquid Filtration, *ASTM F838-83*, American Society for Testing and Materials, Philadelphia (1983).
3. T.H. Meltzer, *Filtration in the Pharmaceutical Industry*, Marcel Dekker, New York (1987).
4. T.D. Brock, *Membrane Filtration: A User's Guide and Reference*, Science Tech., Madison, WI (1983).
5. F. Hofmann, Integrity Testing of Microfiltration Membranes, *J. Parenteral Sci. Technol.* **38**, 148 (1984).
6. T.J. Leahy and M.J. Sullivan, Validation of Bacterial Retention Capabilities of Membrane Filters, *Pharm. Technol.* **2**, 65 (1978).

7. R.E. Williams and T.H. Meltzer, Membrane Structure, the Bubble Point and Particle Retention, *Pharm. Technol.* **7** (5), 36 (1983).
8. M.C. Porter, Microfiltration, in *Handbook of Industrial Membrane Technology*, M.C. Porter (ed.), Noyes Publications, Park Ridge, NJ, pp. 61–135 (1990).
9. W.D. Mores and R.H. Davis, Direct Visual Observation of Yeast Deposition and Removal During Microfiltration, *J. Membr. Sci.* **189**, 217 (2001).
10. R. Sondhi and R. Bhave, Role of Backpulsing in Fouling Minimization in Crossflow Filtration with Ceramic Membranes, *J. Membr. Sci.* **186**, 41 (2001).
11. P. Srijaroonrat, E. Julien and Y. Aurelle, Unstable Secondary Oil/Water Emulsion Treatment using Ultrafiltration: Fouling Control by Backflushing, *J. Membr. Sci.* **159**, 11 (1999).
12. B.D. Cho and A.G. Fane, Fouling Transients in Nominally Sub-critical Flux Operation of a Membrane Bioreactor, *J. Membr. Sci.* **209**, 391 (2002).
13. W. Hein, Mikrofiltration. Verfahren fur kritische Trenn-unde Reinigungsprobleme bei Flussigkeiter und Gasen, *Chem. Produkt.* November (1980).
14. R.A. Cross, Purification of Drinking Water with Ultrafiltration, *The 1993 Eleventh Annual Membrane Technology/Separations Planning Conference*, Newton, MA (October 1993).
15. M. Kolega, G.S. Grohmann, R.F. Chiew and A.W. Day, Disinfection and Clarification of Treated Sewage by Advanced Microfiltration, *Water Sci. Technol.* **23**, 1609 (1991).

8 GAS SEPARATION

Introduction and History

Gas separation has become a major industrial application of membrane technology only during the past 20 years, but the study of gas separation has a long history. Systematic studies began with Thomas Graham who, over a period of 20 years, measured the permeation rates of all the gases then known through every diaphragm available to him [1]. This was no small task because his experiments had to start with synthesis of the gas. Graham gave the first description of the solution-diffusion model, and his work on porous membranes led to Graham's law of diffusion. Through the remainder of the nineteenth and the early twentieth centuries, the ability of gases to permeate membranes selectively had no industrial or commercial use. The concept of the perfectly selective membrane was, however, used as a theoretical tool to develop physical and chemical theories, such as Maxwell's kinetic theory of gases.

From 1943 to 1945, Graham's law of diffusion was exploited for the first time, to separate $U^{235}F_6$ from $U^{238}F_6$ as part of the Manhattan project. Finely microporous metal membranes were used. The separation plant, constructed in Knoxville, Tennessee, represented the first large-scale use of gas separation membranes and remained the world's largest membrane separation plant for the next 40 years. However, this application was unique and so secret that it had essentially no impact on the long-term development of gas separation.

In the 1940s to 1950s, Barrer [2], van Amerongen [3], Stern [4], Meares [5] and others laid the foundation of the modern theories of gas permeation. The solution-diffusion model of gas permeation developed then is still the accepted model for gas transport through membranes. However, despite the availability of interesting polymer materials, membrane fabrication technology was not sufficiently advanced at that time to make useful gas separation membrane systems from these polymers.

The development of high-flux anisotropic membranes and large-surface-area membrane modules for reverse osmosis applications in the late 1960s and early 1970s provided the basis for modern membrane gas separation technology. The first company to establish a commercial presence was Monsanto, which

Membrane Technology and Applications R. W. Baker
© 2004 John Wiley & Sons, Ltd ISBN: 0-470-85445-6

launched its hydrogen-separating Prism® membrane in 1980 [6]. Monsanto had the advantage of being a large chemical company with ample opportunities to test pilot- and demonstration-scale systems in its own plants before launching the product. The economics were compelling, especially for the separation of hydrogen from ammonia-plant purge-gas streams. Within a few years, Prism systems were installed in many such plants [7].

Monsanto's success encouraged other companies to advance their own membrane technologies. By the mid-1980s, Cynara, Separex and Grace Membrane Systems were producing membrane plants to remove carbon dioxide from methane in natural gas. This application, although hindered by low natural gas prices in the 1990s, has grown significantly over the years. At about the same time, Dow launched Generon®, the first commercial membrane system for nitrogen separation from air. Initially, membrane-produced nitrogen was cost-competitive in only a few niche areas, but the development by Dow, Ube and Du Pont/Air Liquide of materials with improved selectivities has since made membrane separation much more competitive. This application of membranes has expanded very rapidly and is expected to capture more than one-half of the market for nitrogen separation systems within the next few years. To date, approximately 10 000 nitrogen systems

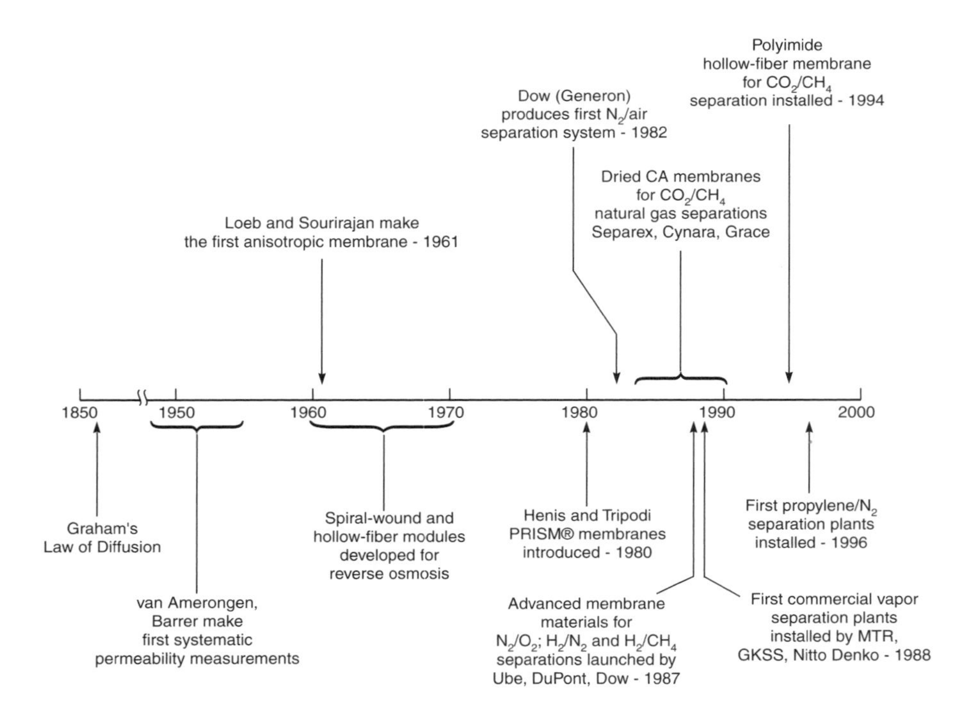

Figure 8.1 Milestones in the development of gas separation

have been installed worldwide. Gas separation membranes are also being used for a wide variety of other, smaller applications ranging from dehydration of air and natural gas to organic vapor removal from air and nitrogen streams. Application of the technology is expanding rapidly and further growth is likely to continue for the next 10 years or so. Figure 8.1 provides a summary of the development of gas separation technology.

Theoretical Background

Both porous and dense membranes can be used as selective gas separation barriers; Figure 8.2 illustrates the mechanism of gas permeation. Three types of porous membranes, differing in pore size, are shown. If the pores are relatively large—from 0.1 to 10 μm—gases permeate the membrane by convective flow, and no separation occurs. If the pores are smaller than 0.1 μm, then the pore diameter is the same size as or smaller than the mean free path of the gas molecules. Diffusion through such pores is governed by Knudsen diffusion, and the transport rate of any gas is inversely proportional to the square root of its molecular weight. This relationship is called Graham's law of diffusion. Finally, if the membrane pores are extremely small, of the order 5–20 Å, then gases are separated by molecular sieving. Transport through this type of membrane is complex and includes both diffusion in the gas phase and diffusion of adsorbed species on the surface of the pores (surface diffusion). These very small-pore membranes have not been used on a large scale, but ceramic and ultramicroporous

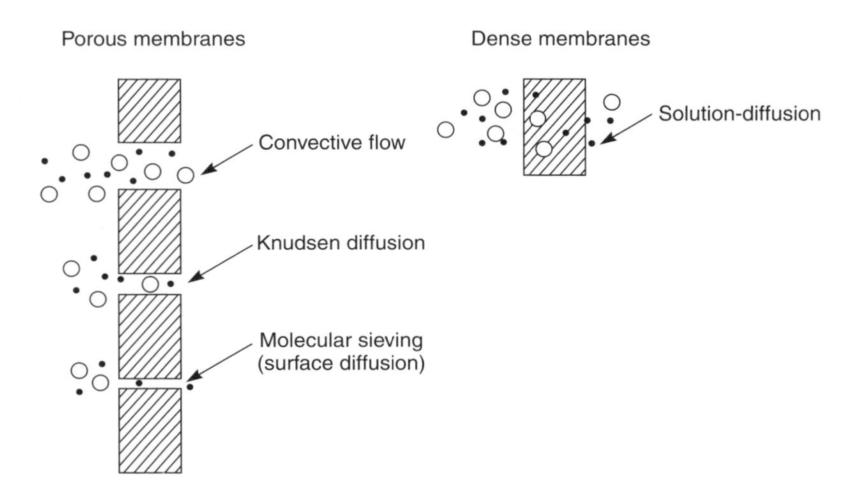

Figure 8.2 Mechanisms for permeation of gases through porous and dense gas separation membranes

glass membranes with extraordinarily high selectivities for similar molecules have been prepared in the laboratory.

Although these microporous membranes are topics of considerable research interest, all current commercial gas separations are based on the dense polymer membrane shown in Figure 8.2. Separation through dense polymer films occurs by a solution-diffusion mechanism.

In Chapter 2 [Equation (2.59)], it was shown that gas transport through dense polymer membranes is governed by the expression

$$J_i = \frac{D_i K_i^G (p_{i_o} - p_{i_\ell})}{\ell} \tag{8.1}$$

where J_i is the flux of component i (g/cm^2 · s), p_{i_o} and p_{i_ℓ} are the partial pressure of the component i on either side of the membrane, ℓ is the membrane thickness, D_i is the permeate diffusion coefficient, and K_i^G is the Henry's law sorption coefficient (g/cm^3 · pressure). In gas permeation it is much easier to measure the volume flux through the membrane than the mass flux and so Equation (8.1) is usually recast as

$$j_i = \frac{D_i K_i (p_{i_o} - p_{i_\ell})}{\ell} \tag{8.2}$$

where j_i is the volume (molar) flux expressed as [cm^3(STP) of component i]/cm^2 · s and K_i is a sorption coefficient with units [cm^3(STP) of component i/cm^3 of polymer]·pressure. The product $D_i K_i$ can be written as $\mathcal{P}_i$, which is called the membrane permeability, and is a measure of the membrane's ability to permeate gas.[1] A measure of the ability of a membrane to separate two gases, i and j, is the ratio of their permeabilities, α_{ij}, called the membrane selectivity

$$\alpha_{ij} = \frac{\mathcal{P}_i}{\mathcal{P}_j} \tag{8.3}$$

The relationship between polymer structure and membrane permeation was discussed in detail in Chapter 2 and is revisited only briefly here. Permeability can be expressed as the product $D_i K_i$ of two terms. The diffusion coefficient, D_i, reflects the mobility of the individual molecules in the membrane material; the gas sorption coefficient, K_i, reflects the number of molecules dissolved in the membrane material. Thus, Equation (8.3) can also be written as

$$\alpha_{ij} = \left[\frac{D_i}{D_j}\right]\left[\frac{K_i}{K_j}\right] \tag{8.4}$$

[1]The permeability of gases through membranes is most commonly measured in Barrer, defined as 10^{-10} cm^3(STP)/cm^2 · s · cmHg and named after R.M. Barrer, a pioneer in gas permeability measurements. The term $j_i/(p_{i_o} - p_{i_\ell})$, best called the pressure-normalized flux or permeance, is often measured in terms of gas permeation units (gpu), where 1 gpu is defined as 10^{-6} cm^3(STP)/cm^2 · s · cmHg. Occasional academic purists insist on writing permeability in terms of mol · m/m^2 · s · Pa (1 Barrer = 0.33×10^{-15} mol · m/m^2 · s · Pa), but fortunately this has not caught on.

The ratio D_i/D_j is the ratio of the diffusion coefficients of the two gases and can be viewed as the mobility selectivity, reflecting the different sizes of the two molecules. The ratio K_i/K_j is the ratio of the sorption coefficients of the two gases and can be viewed as the sorption or solubility selectivity, reflecting the relative condensabilities of the two gases. In all polymer materials, the diffusion coefficient decreases with increasing molecular size, because large molecules interact with more segments of the polymer chain than do small molecules. Hence, the mobility selectivity always favors the passage of small molecules over large

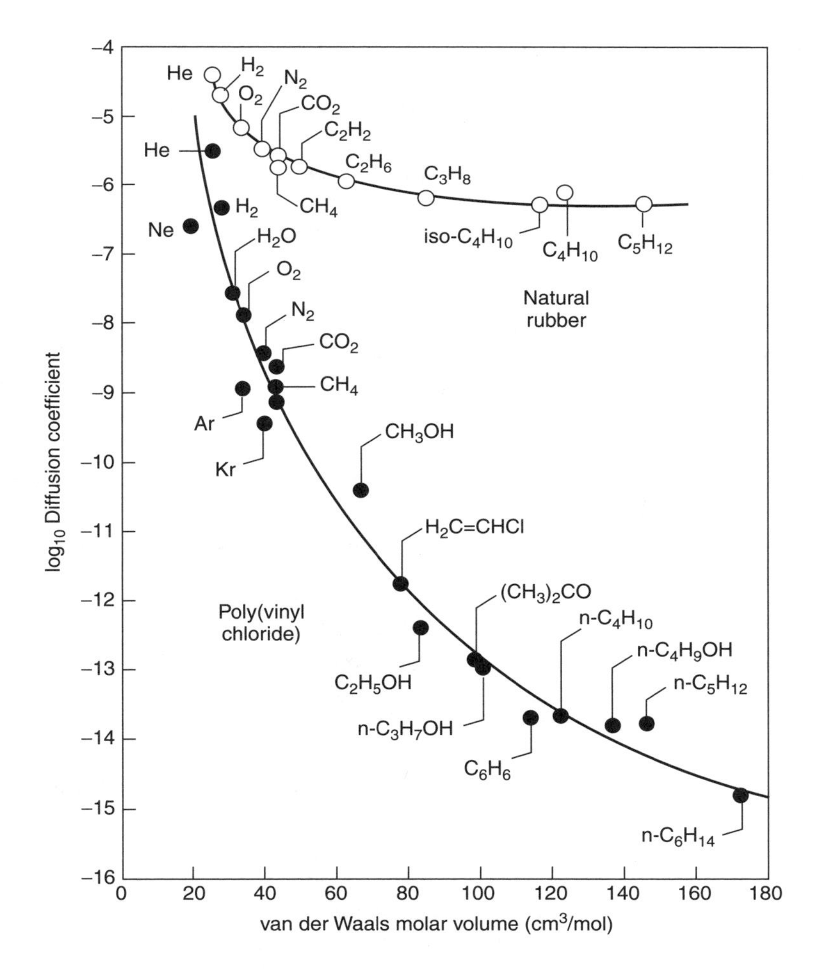

Figure 8.3 Diffusion coefficient as a function of molar volume for a variety of permeants in natural rubber and in poly(vinyl chloride), a glassy polymer. This type of plot was first drawn by Gruen [8], and has been used by many others since

ones. However, the magnitude of the mobility selectivity term depends greatly on whether the membrane material is above or below its glass transition temperature (T_g). If the material is below the glass transition temperature, the polymer chains are essentially fixed and do not rotate. The material is then called a glassy polymer and is tough and rigid. Above the glass transition temperature, the segments of the polymer chains have sufficient thermal energy to allow limited rotation around the chain backbone. This motion changes the mechanical properties of the polymer dramatically, and it becomes a rubber. The relative mobility of gases, as characterized by their diffusion coefficients, differs significantly in rubbers and glasses, as illustrated in Figure 8.3 [8]. Diffusion coefficients in glassy materials decrease much more rapidly with increasing permeate size than diffusion coefficients in rubbers. For example, the mobility selectivity of natural rubber for nitrogen over pentane is approximately 10. The mobility selectivity of poly(vinyl chloride), a rigid, glassy polymer, for nitrogen over pentane is more than 100 000.

The second factor affecting the overall membrane selectivity is the sorption or solubility selectivity. The sorption coefficient of gases and vapors, which is a measure of the energy required for the permeant to be sorbed by the polymer, increases with increasing condensability of the permeant. This dependence on condensability means that the sorption coefficient also increases with molecular diameter, because large molecules are normally more condensable than smaller ones. The gas sorption coefficient can, therefore, be plotted against boiling point or molar volume as shown in Figure 8.4 [9]. As the figure shows, sorption selectivity favors larger, more condensable molecules, such as hydrocarbon vapors, over permanent gases, such as oxygen and nitrogen. However, the difference between the sorption coefficients of permeants in rubbery and glassy polymers is far less marked than the difference in the diffusion coefficients.

It follows from the discussion above that the balance between the mobility selectivity term and the sorption selectivity term in Equation (8.4) [10] is different for glassy and rubbery polymers. This difference is illustrated by the data in Figure 8.5. In glassy polymers, the mobility term is usually dominant, permeability falls with increasing permeate size, and small molecules permeate preferentially. Therefore, when used to separate organic vapors from nitrogen, glassy membranes preferentially permeate nitrogen. In rubbery polymers, the sorption selectivity term is usually dominant, permeability increases with increasing permeate size, and larger molecules permeate preferentially. Therefore, when used to separate organic vapor from nitrogen, rubbery membranes preferentially permeate the organic vapor. The separation properties of polymer membranes for a number of the most important gas separation applications have been summarized by Robeson [11]. A review of structure/property relations has been give by Stern [12]. Properties of some representative and widely used membrane materials are summarized in Table 8.1.

Calculating the selectivity of a membrane using Equation (8.3) and using the permeabilities listed in Table 8.1 must be done with caution. Permeabilities in

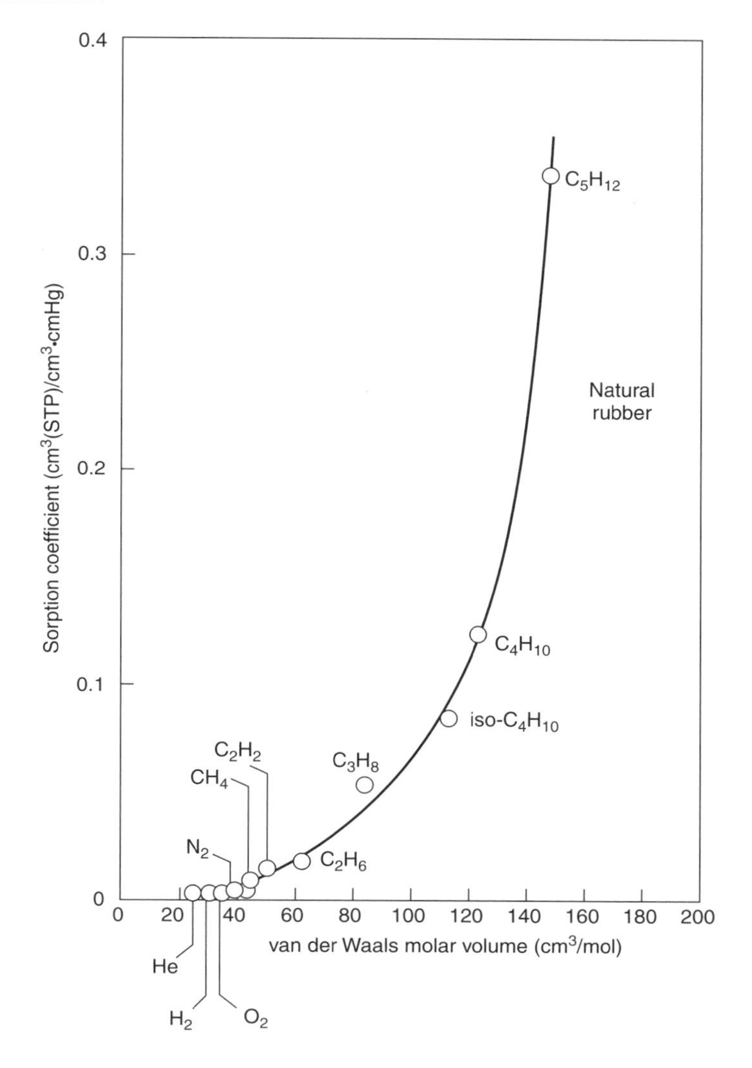

Figure 8.4 Gas sorption coefficient as a function of molar volume for natural rubber membranes. Larger permeants are more condensable and have higher sorption coefficients [9]

Table 8.1 are measured with pure gases; the selectivity obtained from the ratio of pure gas permeabilities gives the ideal membrane selectivity, an intrinsic property of the membrane material. However, practical gas separation processes are performed with gas mixtures. If the gases in a mixture do not interact strongly with the membrane material, the pure gas intrinsic selectivity and the mixed

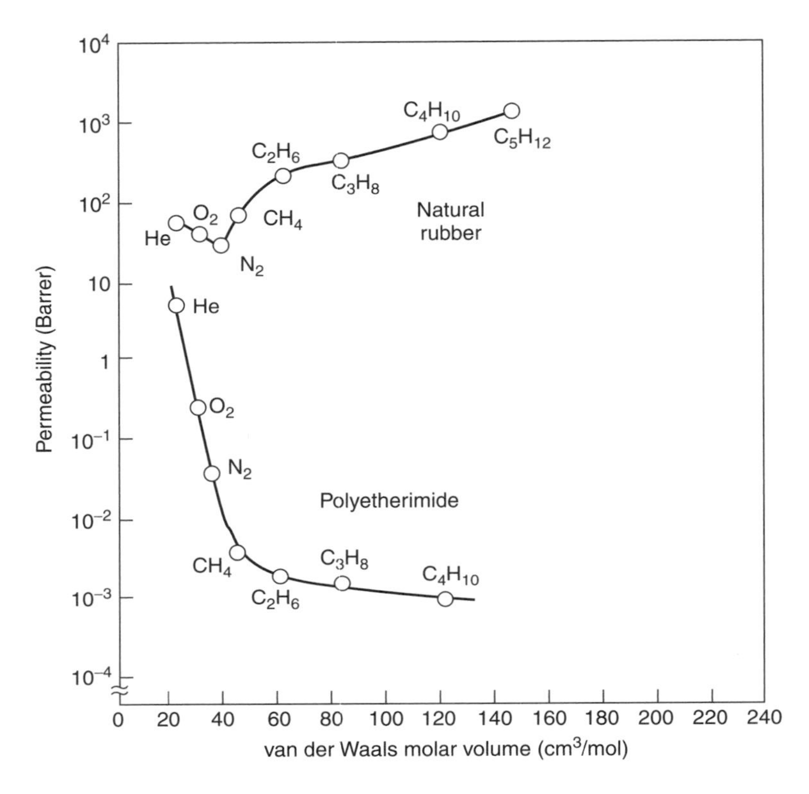

Figure 8.5 Permeability as a function of molar volume for a rubbery and a glassy polymer, illustrating the different balance between sorption and diffusion in these polymer types. The natural rubber membrane is highly permeable; permeability increases rapidly with increasing permeant size because sorption dominates. The glassy polyetherimide membrane is much less permeable; the permeability decreases with increasing permeant size because diffusion dominates [10]. Reprinted from R.D. Behling, K. Ohlrogge, K.-V. Peinemann and E. Kyburz, The Separation of Hydrocarbons from Waste Vapor Streams, in *Membrane Separations in Chemical Engineering*, A.E. Fouda, J.D. Hazlett, T. Matsuura and J. Johnson (eds), AIChE Symposium Series Number 272, Vol. 85, p. 68 (1989). Reproduced by permission of the American Institute of Chemical Engineers.

gas selectivity will be equal. This is usually the case for mixtures of oxygen and nitrogen, for example. In many other cases, such as a carbon dioxide and methane mixture, one of the components (carbon dioxide) is sufficiently sorbed by the membrane to affect the permeability of the other component (methane). The selectivity measured with a gas mixture may then be one-half or less of the selectivity calculated from pure gas measurements. Pure gas selectivities are much more commonly reported in the literature than gas mixture data because

Table 8.1 Permeabilities {Barrer [10^{-10} cm^3(STP) · cm/cm^2 · s · cmHg]} measured with pure gases, at the temperatures given, of widely used polymers

Gas	Rubbers		Glasses		
	Silicone rubber at 25 °C (T_g − 129 °C)	Natural rubber at 30 °C (T_g − 73 °C)	Cellulose acetate at 25 °C (T_g 40 − 124 °C)	Polysulfone at 35 °C (T_g 186 °C)	Polyimide (Ube Industries) at 60 °C (T_g > 250 °C)
H_2	550	41	24	14	50
He	300	31	33	13	40
O_2	500	23	1.6	1.4	3
N_2	250	9.4	0.33	0.25	0.6
CO_2	2700	153	10	5.6	13
CH_4	800	30	0.36	0.25	0.4
C_2H_6	2100	—	0.20	—	0.08
C_3H_8	3400	168	0.13	—	0.015
C_4H_{10}	7500	—	0.10	—	—

they are easier to measure. Neglecting the difference between these two values, however, has led a number of workers to seriously overestimate the ability of a membrane to separate a target gas mixture. Figure 8.6 [13] shows some data for the separation of methane and carbon dioxide with cellulose acetate membranes. The calculated pure gas selectivity is very good, but in gas mixtures enough carbon dioxide dissolves in the membrane to increase the methane permeability far above the pure gas methane permeability value. As a result the selectivities measured with gas mixtures are much lower than those calculated from pure gas data.

Membrane Materials and Structure

Metal Membranes

Although almost all industrial gas separation processes use polymeric membranes, interest in metal membranes continues, mostly for the high-temperature membrane reactor applications discussed in Chapter 13 and for the preparation of pure hydrogen for fuel cells. For completeness, the background to these membranes is described briefly here. The study of gas permeation through metals began with Graham's observation of hydrogen permeation through palladium. Pure palladium absorbs 600 times its volume of hydrogen at room temperature and is measurably permeable to the gas. Hydrogen permeates a number of other metals including tantalum, niobium, vanadium, nickel, iron, copper, cobalt and platinum [14]. In most cases, the metal membrane must be operated at high temperatures (>300 °C) to obtain useful permeation rates and to prevent embrittlement and cracking of

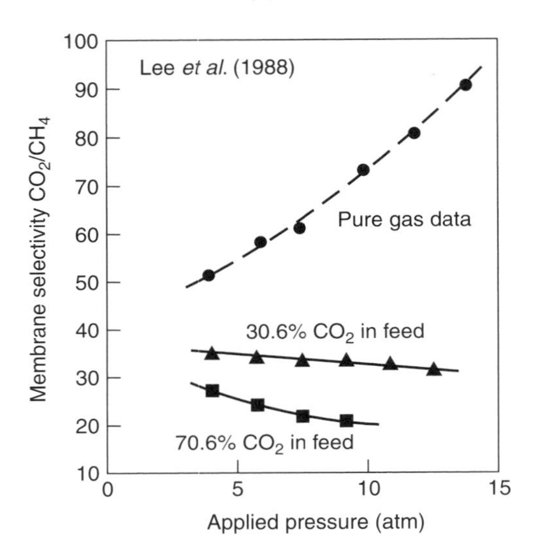

Figure 8.6 The difference between selectivities calculated from pure gas measurements and selectivities measured with gas mixtures can be large. Data of Lee *et al.* [13] for carbon dioxide/methane with cellulose acetate films. Reprinted from S.Y. Lee, B.S. Minhas and M.D. Donohue, Effect of Gas Composition and Pressure on Permeation through Cellulose Acetate Membranes, in *New Membrane Materials and Processes for Separation*, K.K. Sirkar and D.R. Lloyd (eds), AIChE Symposium Series Number 261, Vol. 84, p. 93 (1988). Reproduced with permission of the American Institute of Chemical Engineers. Copyright © 1988 AIChE. All rights reserved

the metal by sorbed hydrogen. Poisoning of the membrane surface by oxidation or sulfur deposition from trace amounts of hydrogen sulfide also occurs. A breakthrough in metal permeation studies occurred in the 1960s when Hunter at Johnson Matthey discovered that palladium/silver alloy membranes showed no hydrogen embrittlement even when used to permeate hydrogen at room temperature [15]. Although most work on gas permeation through membranes has focused on hydrogen, oxygen-permeable metal membranes are also known; however, the permeabilities are low.

Hydrogen-permeable metal membranes are extraordinarily selective, being extremely permeable to hydrogen but essentially impermeable to all other gases. The gas transport mechanism is the key to this high selectivity. Hydrogen permeation through a metal membrane is believed to follow the multistep process illustrated in Figure 8.7 [16]. Hydrogen molecules from the feed gas are sorbed on the membrane surface, where they dissociate into hydrogen atoms. Each individual hydrogen atom loses its electron to the metal lattice and diffuses through the lattice as an ion. Hydrogen atoms emerging at the permeate side of the membrane reassociate to form hydrogen molecules, then desorb, completing the permeation

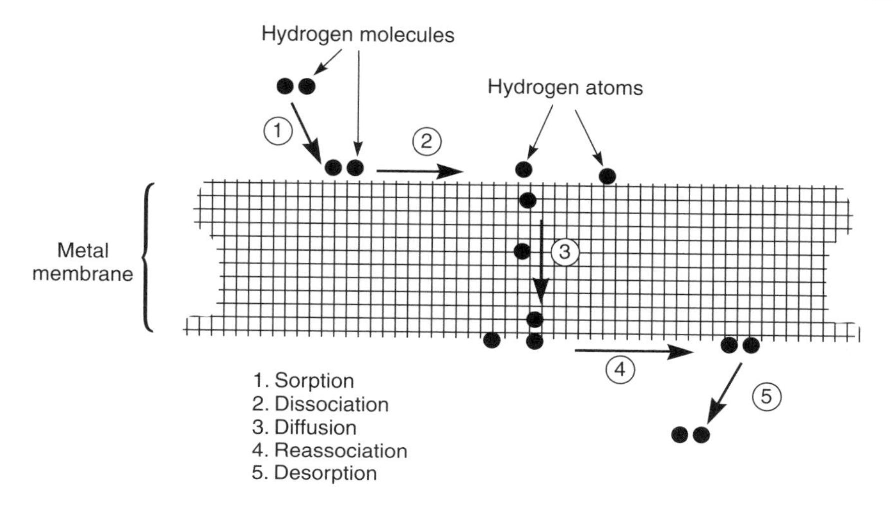

Figure 8.7 Mechanism of permeation of hydrogen through metal membranes

process. Only hydrogen is transported through the membrane by this mechanism; all other gases are excluded.

If the sorption and dissociation of hydrogen molecules is a rapid process, then the hydrogen atoms on the membrane surface are in equilibrium with the gas phase. The concentration, c, of hydrogen atoms on the metal surface is given by Sievert's law:

$$c = Kp^{1/2} \tag{8.5}$$

where K is Sievert's constant and p is the hydrogen pressure in the gas phase. At high temperatures (>300 °C), the surface sorption and dissociation processes are fast, and the rate-controlling step is diffusion of atomic hydrogen through the metal lattice. This is supported by the data of Holleck and others, who have observed that the hydrogen flux through the metal membrane is proportional to the difference of the square roots of the hydrogen pressures on either side of the membrane. At lower temperatures, however, the sorption and dissociation of hydrogen on the membrane surface become the rate-controlling steps, and the permeation characteristics of the membrane deviate from Sievert's law predictions.

Palladium-alloy membranes were studied extensively during the 1950s and 1960s, and this work led to the installation by Union Carbide of a full-scale demonstration plant to separate hydrogen from a refinery off-gas stream containing methane, ethane, carbon monoxide and hydrogen sulfide [17]. The plant could produce 99.9 % or better pure hydrogen in a single pass through the membrane. The plant operated with 25-μm-thick membranes, at a temperature of 370 °C and a feed pressure of 450 psi. The high cost of the membranes and the need to operate at high temperatures to obtain useful fluxes made the process uncompetitive with other hydrogen recovery technologies. In the 1970s and early 1980s, Johnson Matthey built a number of systems to produce on-site hydrogen by separation of hydrogen/carbon dioxide mixtures made by reforming methanol [18]. This was not a commercial success, but the company still produces small systems using palladium–silver alloy membranes to generate ultrapure hydrogen from 99.9 % hydrogen for the electronics industry.

Recently, attempts have been made to reduce the cost of palladium metal membranes by preparing composite membranes. In these membranes a thin selective palladium layer is deposited onto a microporous ceramic, polymer or base metal layer [19–21]. The palladium layer is applied by electrolysis coating, vacuum sputtering or chemical vapor deposition. This work is still at the bench scale.

Polymeric Membranes

Most gas separation processes require that the selective membrane layer be extremely thin to achieve economical fluxes. Typical membrane thicknesses are less than 0.5 μm and often less than 0.1 μm. Early gas separation membranes [22] were adapted from the cellulose acetate membranes produced for reverse osmosis by the Loeb–Sourirajan phase separation process. These membranes are produced by precipitation in water; the water must be removed before the membranes can be used to separate gases. However, the capillary forces generated as the liquid evaporates cause collapse of the finely microporous substrate of the cellulose acetate membrane, destroying its usefulness. This problem has been overcome by a solvent exchange process in which the water is first exchanged for an alcohol, then for hexane. The surface tension forces generated as liquid hexane is evaporated are much reduced, and a dry membrane is produced. Membranes produced by this method have been widely used by Grace (now GMS, a division of Kvaerner) and Separex (now a division of UOP) to separate carbon dioxide from methane in natural gas.

Experience has shown that gas separation membranes are far more sensitive to minor defects, such as pinholes in the selective membrane layer, than membranes used in reverse osmosis or ultrafiltration. Even a single small membrane defect can dramatically decrease the selectivity of gas separation membranes, especially with relatively selective membranes such as those used to separate hydrogen from nitrogen. For example, a good polymeric hydrogen/nitrogen separating membrane has a selectivity of more than 100. A small defect that allows as little as 1 % of the permeating gas to pass unseparated doubles the nitrogen flux and halves the membrane selectivity. The sensitivity of gas separation membranes to defects posed a serious problem to early developers. Generation of a few defects is very difficult to avoid during membrane preparation and module formation.

From 1978 to 1980, Henis and Tripodi [6,23], then at Monsanto, devised an ingenious solution to the membrane defect problem; their approach is illustrated in Figure 8.8. The Monsanto group made Loeb–Sourirajan hollow fiber membranes

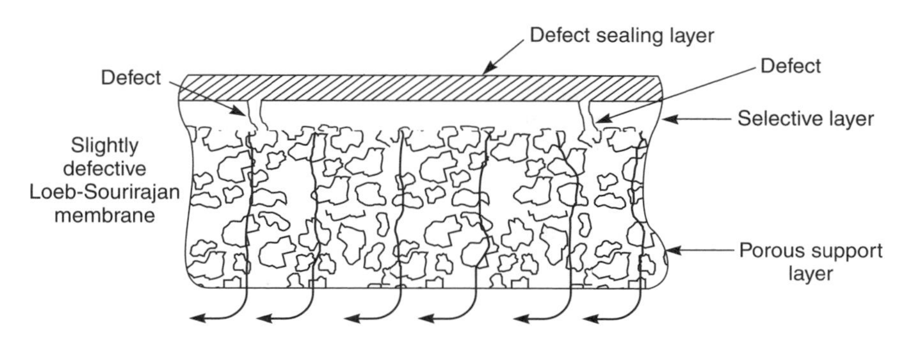

Figure 8.8 The technique devised by Henis and Tripodi [23] to seal defects in their selective polysulfone Loeb–Sourirajan membrane

(principally from polysulfone), then coated the membranes with a thin layer of silicone rubber. Silicone rubber is extremely permeable compared to polysulfone but has a much lower selectivity; thus, the silicone rubber coating did not significantly change the selectivity or flux through the defect-free portions of the polysulfone membrane. However, the coating plugged membrane defects in the polysulfone membrane and eliminated convective flow through these defects. The silicone rubber layer also protected the membrane during handling. The development of silicone rubber-coated anisotropic membranes was a critical step in the production by Monsanto of the first successful gas separation membrane for hydrogen/nitrogen separations.

Another type of gas separation membrane is the multilayer composite structure shown in Figure 8.9. In this membrane, a finely microporous support membrane is overcoated with a thin layer of the selective polymer, which is a different material from the support. Additional layers of very permeable materials such as silicone rubber may also be applied to protect the selective layer and to seal any defects. In general it has been difficult to make composite membranes with

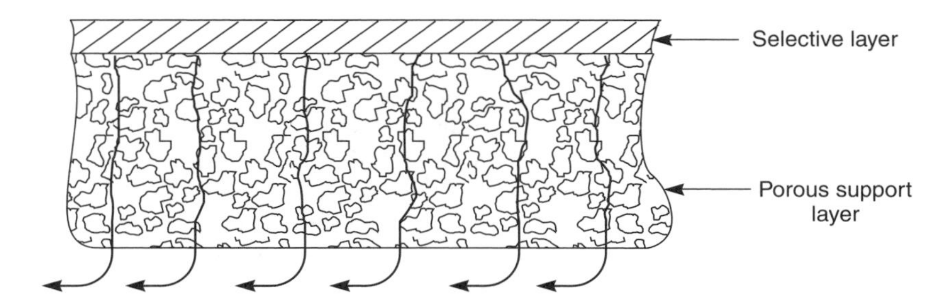

Figure 8.9 Two-layer composite membrane formed by coating a thin layer of a selective polymer on a microporous support that provides mechanical strength

glassy selective layers as thin and high-flux as good-quality Loeb–Sourirajan membranes. However, composites are the best way to form membranes from rubbery selective materials; the microporous support layer can be a tough glassy material to provide strength. Rubbery composite membranes of this type can withstand pressure differentials of 1500 psi or more.

Ceramic and Zeolite Membranes

During the last few years, ceramic- and zeolite-based membranes have begun to be used for a few commercial separations. These membranes are all multilayer composite structures formed by coating a thin selective ceramic or zeolite layer onto a microporous ceramic support. Ceramic membranes are prepared by the sol–gel technique described in Chapter 3; zeolite membranes are prepared by direct crystallization, in which the thin zeolite layer is crystallized at high pressure and temperature directly onto the microporous support [24,25].

Both Mitsui [26] and Sulzer [27] have commercialized these membranes for dehydration of alcohols by pervaporation or vapor/vapor permeation. The membranes are made in tubular form. Extraordinarily high selectivities have been reported for these membranes, and their ceramic nature allows operation at high temperatures, so fluxes are high. These advantages are, however, offset by the costs of the membrane modules, currently in excess of US$3000/m^2 of membrane.

Mixed-matrix Membranes

The ceramic and zeolite membranes described above have been shown to have exceptional selectivities for a number of important separations. However, the membranes are not easy to make and consequently are prohibitively expensive for many separations. One solution to this problem is to prepare membranes from materials consisting of zeolite particles dispersed in a polymer matrix. These membranes are expected to combine the selectivity of zeolite membranes with the low cost and ease of manufacture of polymer membranes. Such membranes are called mixed-matrix membranes.

Mixed-matrix membranes have been a subject of research interest for more than 15 years [28–33]. The concept is illustrated in Figure 8.10. At relatively low loadings of zeolite particles, permeation occurs by a combination of diffusion through the polymer phase and diffusion through the permeable zeolite particles. The relative permeation rates through the two phases are determined by their permeabilities. At low loadings of zeolite, the effect of the permeable zeolite particles on permeation can be expressed mathematically by the expression shown below, first developed by Maxwell in the 1870s [34].

$$P = P_c \left[\frac{P_d + 2P_c - 2\Phi(P_c - P_d)}{P_d + 2P_c + \Phi(P_c - P_d)} \right] \tag{8.6}$$

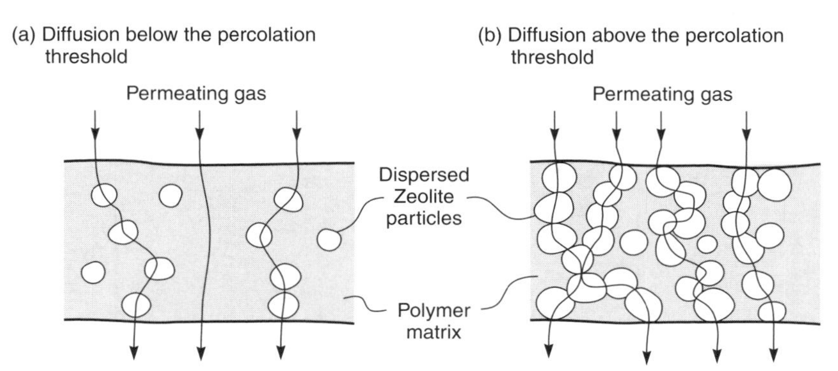

Figure 8.10 Gas permeation through mixed-matrix membranes containing different amounts of dispersed zeolite particles

where P is the overall permeability of the mixed-matrix material, Φ is the volume fraction of the dispersed zeolite phase, P_c is the permeability of the continuous polymer phase, and P_d is the permeability of the dispersed zeolite phase.

At low loadings of dispersed zeolite, individual particles can be considered to be well separated. At higher loadings, some small islands of interconnected particles form; at even higher loadings, these islands grow and connect to form extended pathways. At loadings above a certain critical value, continuous channels form within the membrane, and almost all the zeolite particles are connected to the channels. This is called the percolation threshold. At this particle loading, the Maxwell equation is no longer used to calculate the membrane permeability. The percolation threshold is believed to be achieved at particle loadings of about 30 vol %.

Figure 8.11, adapted from a plot by Robeson *et al.* [35], shows a calculated plot of permeation of a model gas through zeolite-filled polymer membranes in which the zeolite phase is 1000 times more permeable than the polymer phase. At low zeolite particle loadings, the average particle is only in contact with one or two other particles, and a modest increase in average permeability occurs following the Maxwell model. At particle loadings of 25–30 vol % the situation is different—most particles touch two or more particles, and most of the permeating gas can diffuse through interconnected zeolite channels. The percolation threshold has been reached, and the Maxwell model no longer applies. Gas permeation is then best described as permeation through two interpenetrating, continuous phases. At very high zeolite loadings, the mixed-matrix membrane may be best described as a continuous zeolite phase containing dispersed particles of polymer. The Maxwell model may then again apply, with the continuous and the dispersed phases in Equation (8.6) reversed.

The figure also shows that the highly permeable zeolite only has a large effect on polymer permeability when the percolation threshold is reached. That is, useful

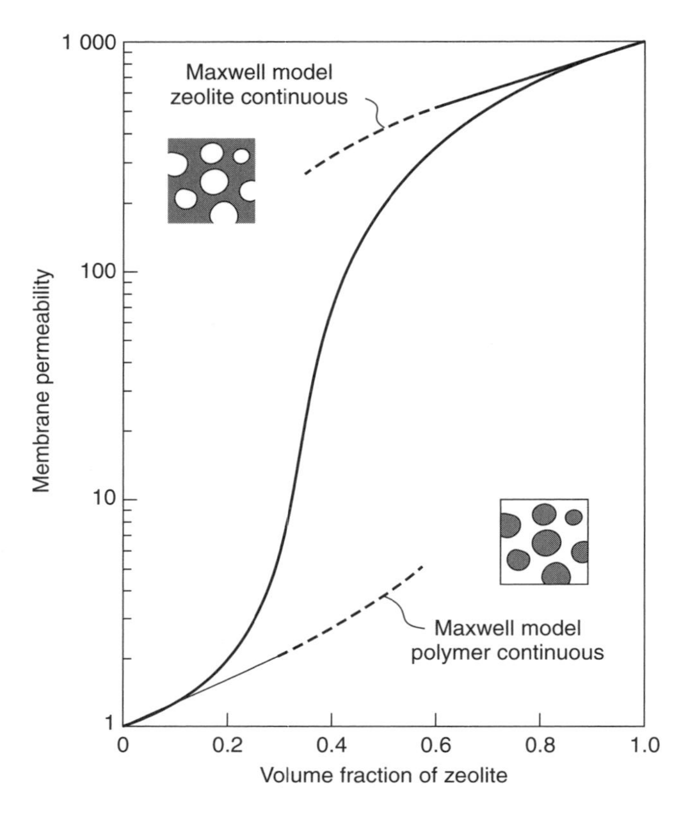

Figure 8.11 Change in membrane permeabilities for mixed-matrix membranes containing different volume fractions of zeolite. Adapted from Robeson *et al.* [35]

membranes must contain more than 30 vol % zeolite. This observation is borne out by the limited experimental data available.

Despite the great deal of recent research on mixed-matrix membranes, the results to date have been modest. Two general approaches have been used. The first, investigated by Koros [31–33] and Smolders [29], is to use the expected difference in the diffusion coefficients of gases in the zeolite particles. Koros, in particular, has focused on zeolites with small aperture sizes, for example, Zeolite 4A, with an effective aperture size of 3.8–4.0 Å, which has been used to separate oxygen (Lennard-Jones (LJ) diameter 3.47 Å) from nitrogen (LJ diameter 3.8 Å). The oxygen/nitrogen selectivity of the Zeolite 4A membrane has been calculated to be 37, with an oxygen permeability of 0.8 Barrer—an exceptional membrane. To maximize the effect of the zeolite in his mixed-matrix membrane, Koros used relatively low-permeability polymers, such as Matrimid® and other polyimides, or poly(vinyl acetate).

The second type of zeolite mixed-matrix membrane relies on relative sorption of different permeants to obtain an improved separation. For example, Smolders *et al.* [28] at the University of Twente, and Peinemann at GKSS, Geesthacht [30], showed that silicalite-silicone rubber mixed-matrix membranes had exceptional selectivities for the permeation of ethanol (kinetic diameter 4.5 Å) over water (kinetic diameter 2.6 Å). These zeolites separate by virtue of their higher sorption of ethanol compared to water on the hydrophobic silicalite surface. Differences in diffusion coefficients favor permeation of water, but this effect is overcome by the sorption effect. The net result is a more than seven-fold increase in the relative permeability of ethanol over water, compared to pure silicone rubber membranes. Because the aperture diameter of the silicalite particles is relatively large, permeabilities through the zeolite phase are also high, allowing rubbery, relatively high-permeability polymers to be used as the matrix phase.

Membrane Modules

Gas separation membranes are formed into spiral-wound or hollow fiber modules. Particulate matter, oil mist, and other potentially fouling materials can be completely and economically removed from gas streams by good-quality coalescing filters, so membrane fouling is generally more easily controlled in gas separation than with liquid separations. Therefore, the choice of module design is usually decided by cost and membrane flux. The hollow fiber membranes used in gas separation applications are often very fine, with lumen diameters of 50–200 μm. However, the pressure drop required on the lumen side of the membrane for these small-diameter fibers can become enough to seriously affect membrane performance. In the production of nitrogen from air, the membrane pressure-normalized fluxes are relatively low, from 1 to 2 gpu, and parasitic pressure drops are not a problem. However, in the separation of hydrogen from nitrogen or methane or carbon dioxide from natural gas, pressure-normalized fluxes are higher, and hollow fine fiber modules can develop excessive permeate-side pressure drops. The solution is to use capillary fibers or spiral-wound modules for this type of application. Nonetheless, these disadvantages of hollow fiber membranes may be partially offset by their lower cost per square meter of membrane. These factors are summarized for some important gas separation applications in Table 8.2.

Process Design

The three factors that determine the performance of a membrane gas separation system are illustrated in Figure 8.12. The role of membrane selectivity is obvious; not so obvious are the importance of the ratio of feed pressure (p_o) to permeate pressure (p_ℓ) across the membrane, usually called the pressure ratio, φ, and defined as

$$\varphi = \frac{p_o}{p_\ell} \tag{8.7}$$

Table 8.2 Module designs used for various gas separation applications

Application	Typical membrane material	Selectivity (α)	Average pressure-normalized flux [10^{-6} cm^3(STP)/$cm^2 \cdot s \cdot cmHg$]	Module design commonly used
O_2/N_2	Polyimide	6–7	1–2	Hollow fiber
H_2/N_2	Polysulfone	100	10–20	Hollow fiber
CO_2/CH_4	Cellulose acetate	15–20	2–5	Spiral or hollow fiber
VOC/N_2	Silicone rubber	10–30	100	Spiral
H_2O/Air	Polyimide	>200	5	Capillary —bore-side feed

and of the membrane stage-cut, θ, which is the fraction of the feed gas that permeates the membrane, defined as

$$\theta = \frac{\text{permeate flow}}{\text{feed flow}} \tag{8.8}$$

Pressure Ratio

The importance of pressure ratio in the separation of gas mixtures can be illustrated by considering the separation of a gas mixture with component concentrations of n_{i_o} and n_{j_o} at a feed pressure p_o. A flow of component i across the membrane can only occur if the partial pressure of i on the feed side of the membrane ($n_{i_o} p_o$) is greater than the partial pressure of i on the permeate side of the membrane ($n_{i_\ell} p_\ell$), that is,

$$n_{i_o} p_{i_o} > n_{i_\ell} p_\ell \tag{8.9}$$

It follows that the maximum separation achieved by the membrane can be expressed as

$$\frac{n_{i_\ell}}{n_{i_o}} \leq \frac{p_o}{p_\ell} \tag{8.10}$$

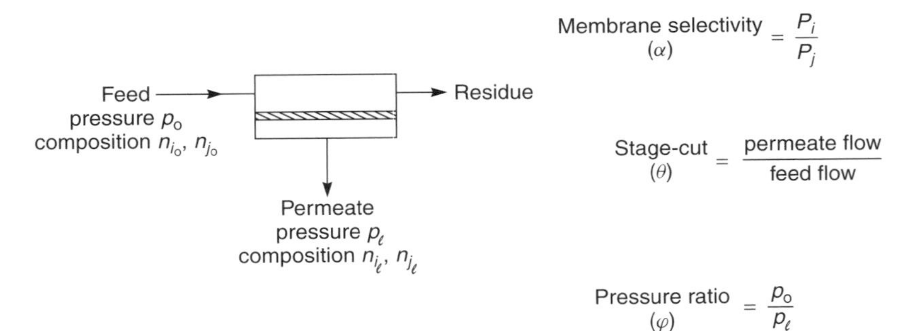

Figure 8.12 Parameters affecting the performance of membrane gas separation systems

That is, the separation achieved can never exceed the pressure ratio φ, no matter how selective the membrane:

$$\frac{n_{i_\ell}}{n_{i_o}} \leq \varphi \tag{8.11}$$

The relationship between pressure ratio and membrane selectivity can be derived from the Fick's law expression for the fluxes of components i and j

$$j_i = \frac{\mathcal{P}_i(p_{i_o} - p_{i_\ell})}{\ell} \tag{8.12}$$

and

$$j_j = \frac{\mathcal{P}_j(p_{j_o} - p_{j_\ell})}{\ell} \tag{8.13}$$

The total gas pressures on the feed and permeate side are the sum of the partial pressures. For the feed side

$$p_o = p_{i_o} + p_{j_o} \tag{8.14}$$

and for the permeate side

$$p_\ell = p_{i_\ell} + p_{j_\ell} \tag{8.15}$$

The volume fractions of components i and j on the feed and permeate side are also related to partial pressures. For the feed side

$$n_{i_o} = \frac{p_{i_o}}{p_o} \quad n_{j_o} = \frac{p_{j_o}}{p_o} \tag{8.16}$$

and for the permeate side

$$n_{i_\ell} = \frac{p_{i_\ell}}{p_\ell} \quad n_{j_\ell} = \frac{p_{j_\ell}}{p_\ell} \tag{8.17}$$

while from mass balance considerations

$$\frac{j_i}{j_j} = \frac{n_{i_\ell}}{n_{j_\ell}} = \frac{n_{i_\ell}}{1 - n_{i_\ell}} = \frac{1 - n_{j_\ell}}{n_{j_\ell}} \tag{8.18}$$

Combining Equations (8.14–8.18) yields an expression linking the concentration of component i on the feed and permeate sides of the membrane

$$n_{i_\ell} = \frac{\varphi}{2}\left[n_{i_o} + \frac{1}{\varphi} + \frac{1}{\alpha - 1} - \sqrt{\left(n_{i_o} + \frac{1}{\varphi} + \frac{1}{\alpha - 1}\right)^2 - \frac{4\alpha n_{i_o}}{(\alpha - 1)\varphi}}\right] \tag{8.19}$$

This somewhat complex expression breaks down into two limiting cases depending on the relative magnitudes of the pressure ratio and the membrane selectivity. First, if the membrane selectivity (α) is very much larger than the pressure ratio (φ), that is,

$$\alpha \gg \varphi \tag{8.20}$$

then Equation (8.20) becomes

$$n_{i_\ell} = n_{i_o}\varphi \tag{8.21}$$

This is called the pressure-ratio-limited region, in which the performance is determined only by the pressure ratio across the membrane and is independent of the membrane selectivity. If the membrane selectivity (α) is very much smaller than the pressure ratio (φ), that is,

$$\alpha \ll \varphi \tag{8.22}$$

then Equation (8.19) becomes

$$n_{i_\ell} = \frac{\alpha n_{i_o}}{1 - n_{i_o}(1 - \alpha)} \tag{8.23}$$

This is called the membrane-selectivity-limited region, in which the membrane performance is determined only by the membrane selectivity and is independent of the pressure ratio. There is, of course, an intermediate region between these two limiting cases, in which both the pressure ratio and the membrane selectivity affect the membrane system performance. These three regions are illustrated in Figure 8.13, in which the calculated permeate concentration (n_{i_ℓ}) is plotted versus pressure ratio (φ) for a membrane with a selectivity of 30 [36]. At a pressure ratio of 1, feed pressure equal to the permeate pressure, no separation is achieved by the membrane. As the difference between the feed and permeate pressure increases,

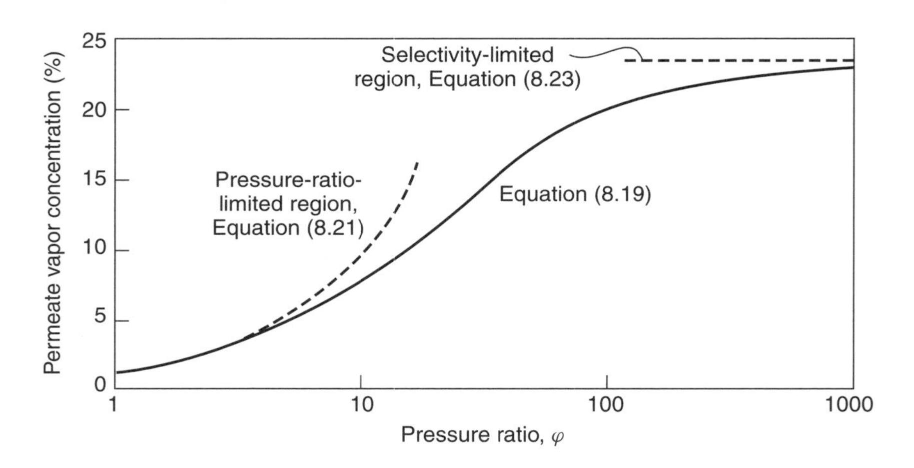

Figure 8.13 Calculated permeate vapor concentration for a vapor-permeable membrane with a vapor/nitrogen selectivity of 30 as a function of pressure ratio. The feed vapor concentration is 1 %. Below pressure ratios of about 10, separation is limited by the pressure ratio across the membrane. At pressure ratios above about 100, separation is limited by the membrane selectivity [36]

the concentration of the more permeable component in the permeate gas begins to increase, first according to Equation (8.21) and then, when the pressure ratio and membrane selectivity are comparable, according to Equation (8.19). At very high pressure ratios, that is, when the pressure ratio is four to five times higher than the membrane selectivity, the membrane enters the membrane-selectivity-controlled region. In this region the permeate concentration reaches the limiting value given by Equation (8.23).

The relationship between pressure ratio and selectivity is important because of the practical limitation to the pressure ratio achievable in gas separation systems. Compressing the feed stream to very high pressure or drawing a very hard vacuum on the permeate side of the membrane to achieve large pressure ratios both require large amounts of energy and expensive pumps. As a result, typical practical pressure ratios are in the range 5–20.

Because the attainable pressure ratio in most gas separation applications is limited, the benefit of very highly selective membranes is often less than might be expected. For example, as shown in Figure 8.14, if the pressure ratio is 20, then increasing the membrane selectivity from 10 to 20 will significantly improve system performance. However, a much smaller incremental improvement results from increasing the selectivity from 20 to 40. Increases in selectivity above 100 will produce negligible improvements. A selectivity of 100 is five times the pressure ratio of 20, placing the system in the pressure-ratio-limited region.

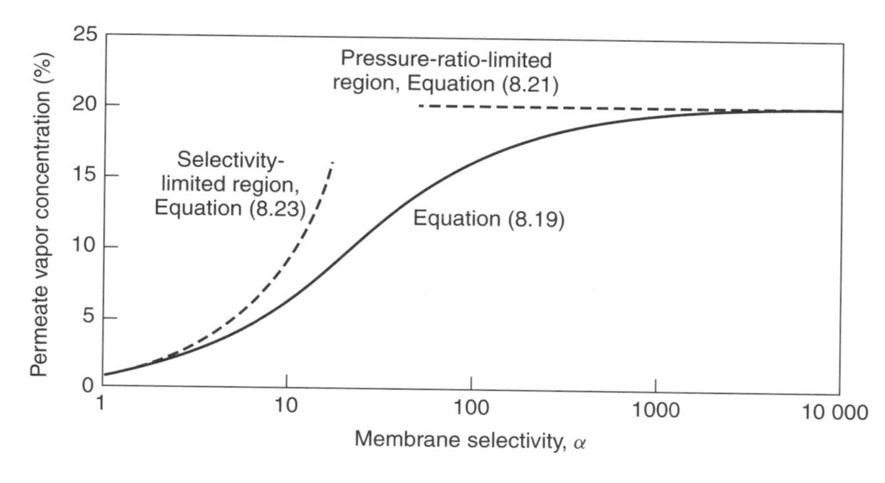

Figure 8.14 Calculated permeate vapor concentration as a function of selectivity. The feed vapor concentration is 1 %; the pressure ratio is fixed at 20. Below a vapor/nitrogen selectivity of about 10, separation is limited by the low membrane selectivity; at selectivities above about 100, separation is limited by the low pressure ratio across the membrane [36]

Stage-cut

Another factor that affects membrane system design is the degree of separation required. The usual target of a gas separation system is to produce a residue stream essentially stripped of the permeable component and a small, highly concentrated permeate stream. These two requirements cannot be met simultaneously; a trade-off must be made between removal from the feed gas and enrichment in the permeate. The system attribute that characterizes this trade-off is called the stage-cut. The effect of stage-cut on system performance is illustrated in Figure 8.15.

In the example calculation shown in Figure 8.15, the feed gas contains 50 % of a permeable gas (i) and 50 % of a relatively impermeable gas (j). Under the assumed operating conditions of this system (pressure ratio 20, membrane selectivity 20), it is possible at zero stage-cut to produce a permeate stream containing 94.8 % of component i. But the permeate stream is tiny and the residue stream is still very close to the feed gas concentration of 50 %. As the fraction of the feed gas permeating the membrane is increased by increasing the membrane area, the concentration of the permeable component in the residue and permeate streams falls. At a stage-cut of 25 %, the permeate gas concentration has fallen from 94.8 % (its maximum value) to 93.1 %. The residue stream concentration of permeable gas is then 35.5 %. Increasing the fraction of the feed gas that permeates the membrane to 50 % by adding more membrane area produces a residue

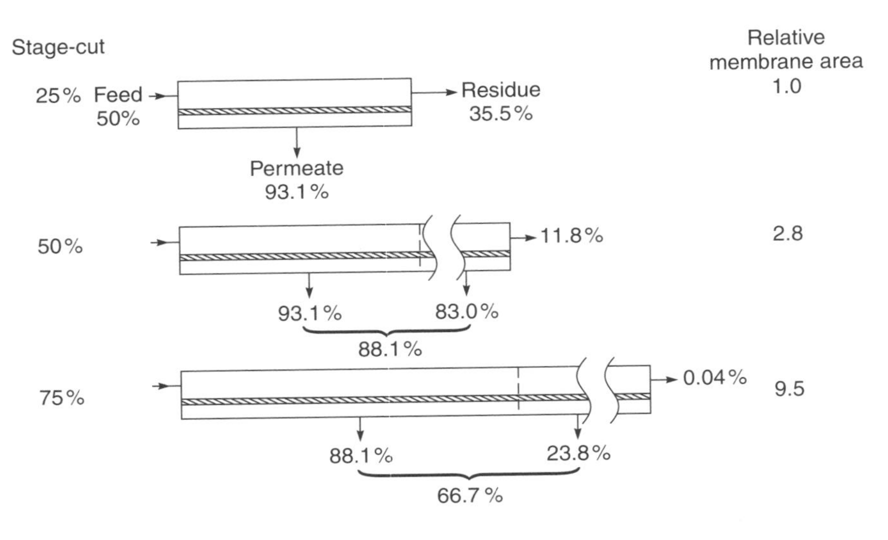

Figure 8.15 The effect of stage-cut on the separation of a 50/50 feed gas mixture (pressure ratio, 20; membrane selectivity, 20). At low stage-cuts a concentrated permeate product, but only modest removal from the residue, can be obtained. At high stage-cuts almost complete removal is obtained, but the permeate product is only slightly more enriched than the original feed

stream containing 11.8 % of the permeable gas. However, the gas permeating the added membrane area only contains 83.0 % of the permeable component, so the average concentration of permeable component in the permeate stream is reduced from 93.1 to 88.1 %. If the fraction of the feed gas that permeates the membrane is increased to 75 % by adding even more membrane area, the concentration of the permeable component in the residue stream is reduced to only 0.04 %. However, the gas permeating the added membrane area only contains 23.8 % of the permeable component, *less than the original feed gas*. The average concentration of the permeable component in the feed gas is, therefore, reduced to 66.7 %. This means that one-half of the less permeable component has been lost to the permeate stream.

The calculations shown in Figure 8.15 illustrate the trade-off between recovery and purity. A single-stage membrane process can be designed for either maximum recovery or maximum purity, but not both. The calculations also show that membranes can produce very pure residue gas streams enriched in the less permeable component, although at low recoveries. However, the enrichment of the more permeable component in the permeate can never be more than the membrane selectivity, so a membrane with low selectivity produces an only slightly enriched permeate. This is why membranes with an oxygen/nitrogen selectivity of 4–6 can produce very pure nitrogen (>99.5 %) from air on the residue side of the membrane, but the same membranes cannot produce better than 50–60 % oxygen on the permeate side. If the more permeable component must be pure, very selective membranes are required or multistage or recycle membrane systems must be used.

Finally, the calculations in Figure 8.15 show that increasing the stage-cut to produce a pure residue stream requires a disproportionate increase in membrane area. As the feed gas is stripped of the more permeable component, the average permeation rate through the membrane falls. In the example shown, this means that permeating the first 25 % of the feed gas requires a relative membrane area of 1, permeating the next 25 % requires a membrane area increment of 1.8, and permeating the next 25 % requires an increment of 6.7.

Multistep and Multistage System Designs

Because the membrane selectivity and pressure ratio achievable in a commercial membrane system are limited, a one-stage membrane system may not provide the separation desired. The problem is illustrated in Figure 8.16. The target of the process is 90 % removal of a volatile organic compound (VOC), which is the permeable component, from the feed gas, which contains 1 vol % of this component. This calculation and those that immediately follow assume a feed gas mixture VOC and nitrogen. Rubbery membranes such as silicone rubber permeate the VOC preferentially because of its greater condensability and hence solubility in the membrane. In this calculation, the pressure ratio is fixed at 20

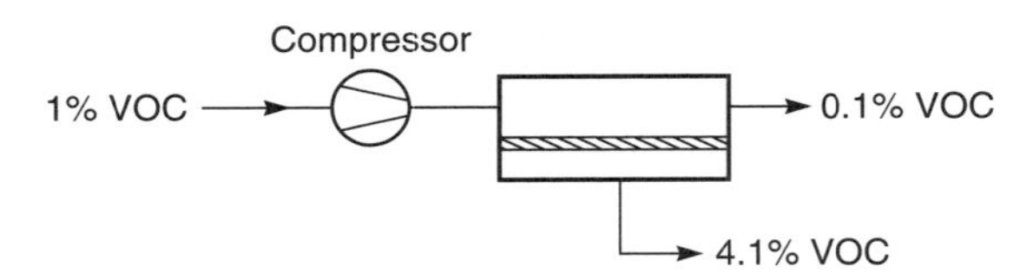

Figure 8.16 A one-stage vapor separation operation. The performance of this system was calculated from a crossflow model using a vapor/nitrogen selectivity of 20 and a pressure ratio of 20

by compressing the feed gas, and the permeate is maintained at atmospheric pressure. The membrane VOC/nitrogen selectivity is assumed to be 20.

Figure 8.16 shows that when 90 % of the VOC in the feed stream is removed, the permeate stream will contain approximately 4 % of the permeable component. In many cases, 90 % removal of VOC from the feed stream is insufficient to allow the residue gas to be discharged, and enrichment of the component in the permeate is insufficient also.

If the main problem is insufficient VOC removal from the feed stream, a two-step system as shown in Figure 8.17 can be used. In a two-step system, the residue stream from the first membrane unit is passed to a second unit, where the VOC concentration is reduced by a further factor of 10, from 0.1 to 0.01 %. Because the concentration of VOC in the feed to the second membrane unit is low, the permeate stream is relatively dilute and is recirculated to the feed stream.

A multistep design of this type can achieve almost complete removal of the permeable component from the feed stream to the membrane unit. However, greater removal of the permeable component is achieved at the expense of increases in membrane area and power consumption by the compressor. As a rule of thumb, the membrane area required to remove the last 9 % of a component from the feed equals the membrane area required to remove the first 90 %.

Sometimes, 90 % removal of the permeable component from the feed stream is acceptable for the discharge stream from the membrane unit, but a higher

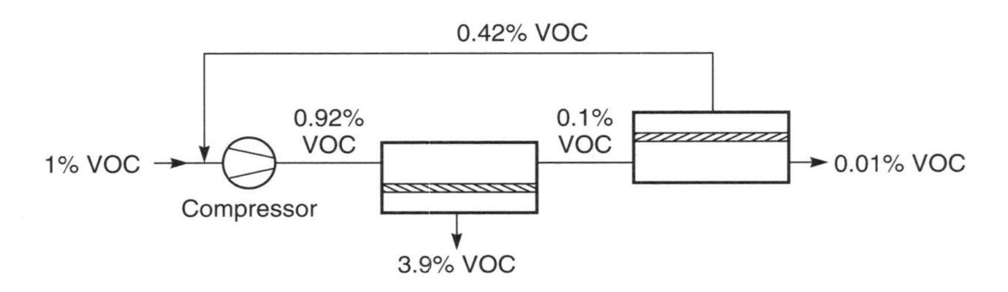

Figure 8.17 A two-step system to achieve 99 % vapor removal from the feed stream. Selectivity, 20; pressure ratio, 20

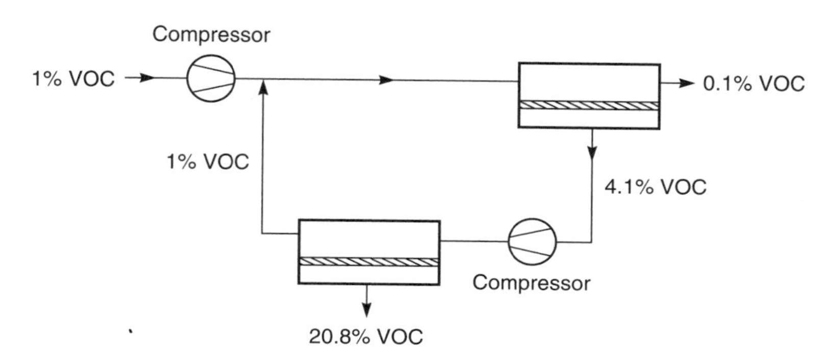

Figure 8.18 A two-stage system to produce a highly concentrated permeate stream. Selectivity, 20; pressure ratio, 20

concentration is needed to make the permeate gas usable. In this situation, a two-stage system of the type shown in Figure 8.18 is used. In a two-stage design, the permeate from the first membrane unit is recompressed and sent to a second membrane unit, where a further separation is performed. The final permeate is then twice enriched. In the most efficient two-stage design, the residue stream from the second stage is reduced to about the same concentration as the original feed gas, with which it is mixed. In the example shown in Figure 8.18, the permeate stream, concentrated a further five-fold, leaves the system at a concentration of 21 %. Because the volume of gas treated by the second-stage membrane unit is much smaller than in the first stage, the membrane area of the second stage is relatively small. Thus, incorporation of a second stage only increases the overall membrane area and power requirements by approximately 15–20 %.

Multistage/multistep combinations of two-step and two-stage processes can be designed but are seldom used in commercial systems—their complexity makes them uncompetitive with alternative separation technologies. More commonly some form of recycle design is used.

Recycle Designs

A simple recycle design, sometimes called a two-and-one-half-stage system, proposed by Wijmans [37] is shown in Figure 8.19. In this design, the permeate from the first membrane stage is recompressed and sent to a two-step second stage, where a portion of the gas permeates and is removed as enriched product. The remaining gas passes to another membrane stage, which brings the gas concentration close to the original feed value. The permeate from this stage is mixed with the first-stage permeate, forming a recycle loop. By controlling the relative size of the two second stages any desired concentration of the more permeable component can be achieved in the product. In the example shown, the permeable

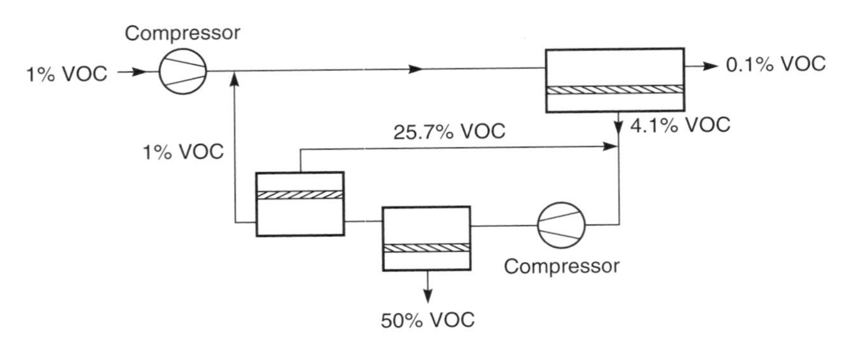

Figure 8.19 Two-and-one-half-stage system: by forming a recycle loop around the second stage, a small, very concentrated product stream is created. Selectivity, 20; pressure ratio, 20 [37]

component is concentrated to 50 % in the permeate. The increased performance is achieved at the expense of a slightly larger second-stage compressor and more membrane area. Normally, however, this design is preferable to a more complex three-stage system.

Figure 8.20 shows another type of recycle design in which a recycle loop increases the concentration of the permeable component to the point at which it can be removed by a second process, most commonly condensation [38]. The feed stream entering the recycle loop contains 1 % of the permeable component as in Figures 8.16–8.19. After compression to 20 atm, the feed gas passes through a condenser at 30 °C, but the VOC content is still below the condensation concentration at this temperature. The membrane unit separates the gas into a VOC-depleted residue stream and a vapor-enriched permeate stream, which is recirculated to the front of the compressor. Because the bulk of the vapor is recirculated, the concentration of vapor in the loop increases rapidly until the pressurized gas entering the condenser exceeds the vapor dew point of 6.1 %. At

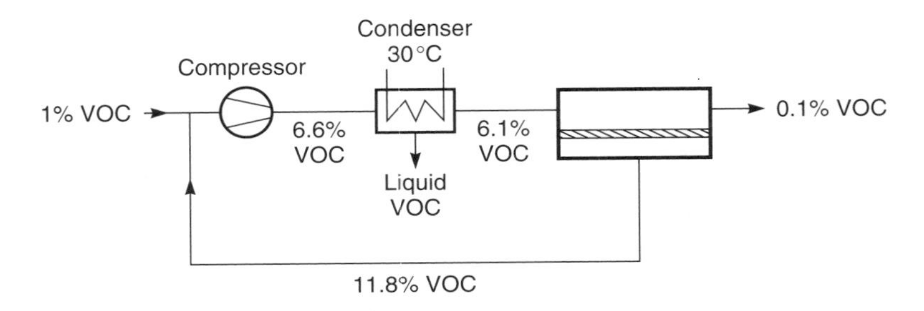

Figure 8.20 Recycle system design using one membrane stage, preceded by a compressor and condenser: feed stream, 1 % vapor in nitrogen; selectivity, 20; pressure ratio, 20

this point, the system is at steady state; the mass of VOC entering the recirculation loop is equal to the mass discharged in the residue stream plus the mass removed as liquid condensate.

Recycle designs of this type are limited to applications in which the components of the gas mixture, if sufficiently concentrated, can be separated from the gas by some other technique. With organic vapors, condensation is often possible; adsorption, chemical scrubbing or absorption can also be used. The process shown in Figure 8.20 is used to separate VOCs from nitrogen and air or to separate propane, butane, pentane and higher hydrocarbons from natural gas (methane).

Applications

The membrane gas separation industry is still growing and changing. Most of the large industrial gas companies now have membrane affiliates: Air Products (Permea), MG (Generon), Air Liquide (Medal) and Praxair (IMS). The affiliates focus mainly on producing membrane systems to separate nitrogen from air, but also produce some hydrogen separation systems. Another group of companies, UOP (Separex), Natco (Cynara), Kvaerner (GMS) and ABB Lummus Global (MTR), produces membrane systems for natural gas separations. A third group of smaller independents are focusing on the new applications, including vapor separation, air dehydration and oxygen enrichment. The final size and form of this industry are still unknown. The following section covers the major current applications. Overview articles on the main gas separation applications can be found in Paul and Yampol'skii [39], in Koros and Fleming [40] and elsewhere [41].

Hydrogen Separations

The first large-scale commercial application of membrane gas separation was the separation of hydrogen from nitrogen in ammonia purge gas streams. The process, launched in 1980 by Monsanto, was followed by a number of similar applications, such as hydrogen/methane separation in refinery off-gases and hydrogen/carbon monoxide adjustment in oxo-chemical synthesis plants [7]. Hydrogen is a small, noncondensable gas, which is highly permeable compared to all other gases. This is particularly true with the glassy polymers primarily used to make hydrogen-selective membranes; fluxes and selectivities of hydrogen through some of these materials are shown in Table 8.3. With fluxes and selectivities as high as these, it is easy to understand why hydrogen separation was the first gas separation process developed. Early hydrogen membrane gas separation plants used polysulfone or cellulose acetate membranes, but now a variety of specifically synthesized materials, such as polyimides (Ube, Praxair), polyaramide (Medal) or brominated polysulfone (Permea), are used.

Table 8.3 Hydrogen separation membranes

Membrane (developer)	Selectivity			Hydrogen pressure-normalized flux [10^{-6} $cm^3(STP)/cm^2 \cdot s \cdot cmHg$]
	H_2/CO	H_2/CH_4	H_2/N_2	
Polyaramide (Medal)	100	>200	>200	—
Polysulfone (Permea)	40	80	80	100
Cellulose acetate (Separex)	30–40	60–80	60–80	200
Polyimide (Ube)	50	100–200	100–200	80–200

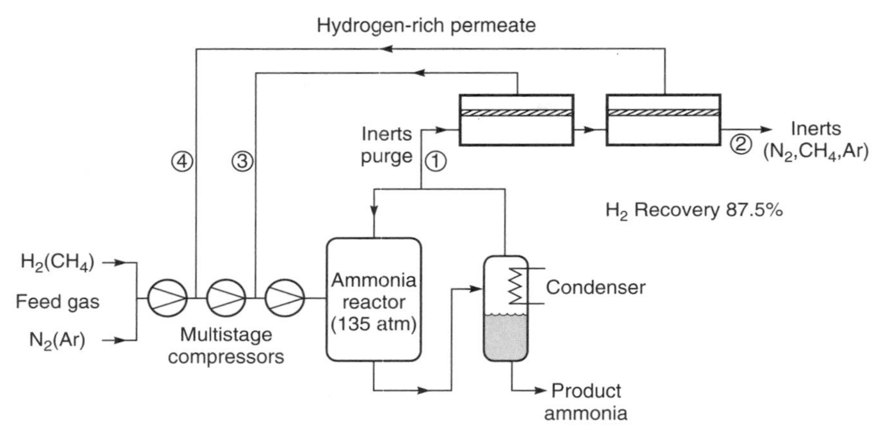

	Stream Composition (%)			
	Membrane Feed ①	Membrane Vent ②	High-Pressure Permeate ③	Low-Pressure Permeate ④
Hydrogen	62	21	87.3	84.8
Nitrogen	21	44	7.1	8.4
Methane	11	23	36	4.3
Argon	6	13	2.0	2.5
Pressure (atm)	135	132	70	28
Flow (scfm)	2000	740	830	430

Figure 8.21 Simplified flow schematic of the PRISM® membrane system to recover hydrogen from an ammonia reactor purge stream. A two-step membrane system is used to reduce permeate compression costs

A typical membrane system flow scheme for recovery of hydrogen from an ammonia plant purge gas stream is shown in Figure 8.21. A photograph of such a system is shown in Figure 8.22. During the production of ammonia from nitrogen and hydrogen, argon enters the high-pressure ammonia reactor as an impurity with the nitrogen stream and methane enters the reactor as an impurity with the hydrogen. Ammonia produced in the reactor is removed by condensation, so the argon and methane impurities accumulate until they represent as much as 15 % of the gas in the reactor. To control the concentration of these components, the reactor must be continuously purged. The hydrogen lost with this purge gas can represent 2–4 % of the total hydrogen consumed. These plants are very large, so recovery of the hydrogen for recycle to the ammonia reactor is economically worthwhile.

In the process shown in Figure 8.21, a two-step membrane design is used to reduce the cost of recompressing the hydrogen permeate stream to the very high

Figure 8.22 Photograph of an Air Products and Chemicals, Inc. PRISM® membrane system installed at an ammonia plant. The modules are mounted vertically

pressures of ammonia reactors. In the first step, the feed gas is maintained at the reactor pressure of 135 atm, and the permeate is maintained at 70 atm, giving a pressure ratio of 1.9. The hydrogen concentration in the feed to this first step is about 45 %, high enough that even at this low pressure ratio the permeate contains about 90 % hydrogen. However, by the time the feed gas hydrogen concentration has fallen to 30 %, the hydrogen concentration in the permeate is no longer high enough for recycle to the reactor. This remaining hydrogen is recovered in a second membrane step operated at a lower permeate pressure

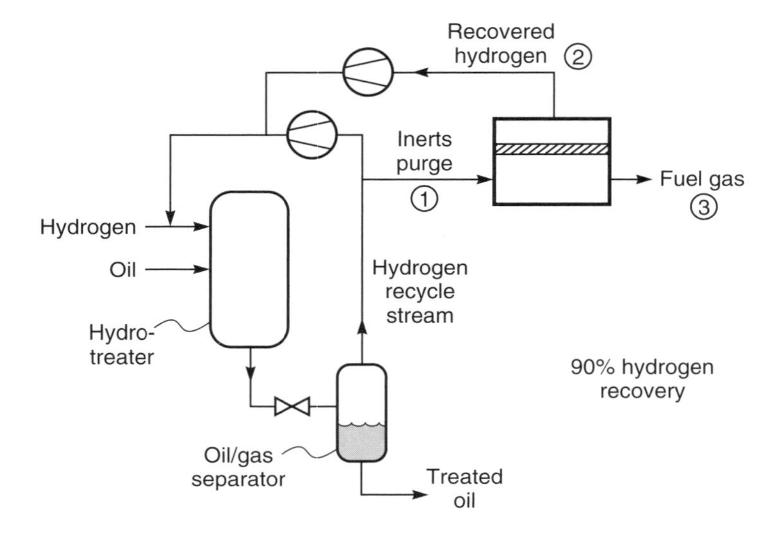

	Stream Composition		
	Untreated Purge ①	Recovered Hydrogen ②	Treated Purge ③
Hydrogen	82	96.5	34.8
Methane	12	2.6	43.3
Ethane	4.6	0.7	17.1
Propane	1.2	0.2	4.8
Pressure (psig)	1800	450	1450
Flow (MMscfd)	18.9	14.5	4.4

Figure 8.23 Hydrogen recovery from a hydrotreater used to lower the molecular weight of a refinery oil stream. Permea polysulfone membranes (PRISM®) are used [42]

of 28 atm and a pressure ratio of 4.7. The increased pressure ratio increases the hydrogen concentration in the permeate significantly. By dividing the process into two steps operating at different pressure ratios, maximum hydrogen recovery is achieved at minimum recompression costs.

A second major application of hydrogen-selective membranes is recovery of hydrogen from waste gases produced in various refinery operations [7,42,43]. A typical separation—treatment of the high-pressure purge gas from a hydrotreater—is shown in Figure 8.23. The hydrogen separation process is designed to recycle the hydrogen to the hydrotreater. As in the case of the ammonia plant, there is a trade-off between the concentration of hydrogen in the permeate and the permeate pressure and subsequent cost of recompression. In the example shown, a permeate of 96.5 % hydrogen is considered adequate at a pressure ratio of 3.9.

Another example of the use of highly hydrogen-selective membranes in the petrochemical industry is the separation of hydrogen from carbon monoxide/hydrogen mixtures to obtain the correct ratio of components for subsequent synthesis operations.

Oxygen/Nitrogen Separation

By far the largest gas separation process in current use is the production of nitrogen from air. The first membranes used for this process were based on poly(4-methyl-1-pentene) (TPX) and ethyl cellulose. These polymer materials have oxygen/nitrogen selectivities of 4; the economics of the process were marginal. The second-generation materials now used have selectivities of 6–7, providing very favorable economics, especially for small plants producing 5–500 scfm of nitrogen. In this range, membranes are the low-cost process, and most new small nitrogen plants use membrane systems.

Table 8.4 lists the permeabilities and selectivities of some of the materials that are used or have been used for this separation. There is a strong inverse relationship between flux and selectivity. Membranes with selectivities of 6–7 typically have 1 % of the permeability of membranes with selectivities of 2–3. This selectivity/permeability trade-off is very apparent in the plot of selectivity as a function of oxygen permeability shown in Figure 8.24, prepared by Robeson [11]. This plot shows data for a large number of membrane materials reported in the literature. A wide range of selectivity/permeability combinations are provided by different membrane materials; for gas separation applications only the most permeable polymers at a particular selectivity are of interest. The line linking these polymers is called the upper bound, beyond which no better material is currently known. The relative positions of the upper bound in 1991 and in 1980 show the progress that has been made in producing polymers specifically tailored for this separation. Development of better materials is a continuing research topic at the

Table 8.4 Permeabilities and selectivities of polymers of interest in air separation

Polymer	Oxygen permeability (Barrer)	Nitrogen permeability (Barrer)	Oxygen/ nitrogen selectivity
Poly(1-trimethylsilyl-1-propyne) (PTMSP)	7600	5400	1.4
Teflon AF 2400	1300	760	1.7
Silicone rubber	600	280	2.2
Poly(4-methyl-1-pentene) (TPX)	30	7.1	4.2
Poly(phenylene oxide) (PPO)	16.8	3.8	4.4
Ethyl cellulose	11.2	3.3	3.4
6FDA-DAF (polyimide)	7.9	1.3	6.2
Polysulfone	1.1	0.18	6.2
Polyaramide	3.1	0.46	6.8
Tetrabromo *bis* polycarbonate	1.4	0.18	7.5

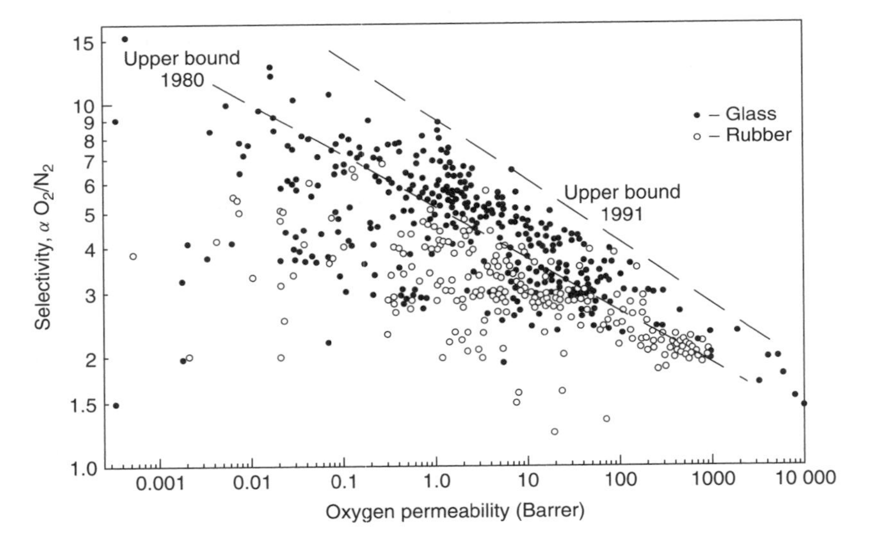

Figure 8.24 Oxygen/nitrogen selectivity as a function of oxygen permeability. This plot by Robeson [11] shows the wide range of combination of selectivity and permeability achieved by current materials. Reprinted from *J. Membr. Sci.* **62**, L.M. Robeson, Correlation of Separation Factor Versus Permeability for Polymeric Membranes, p. 165. Copyright 1991, with permission from Elsevier

major gas separation companies and in some universities, so further but slower movement of the upper bound may be seen in the future.

High oxygen/nitrogen selectivity is required for an economical nitrogen production process. The effect of improved membrane selectivities on the efficiency of nitrogen production from air is illustrated in Figure 8.25. This figure shows the

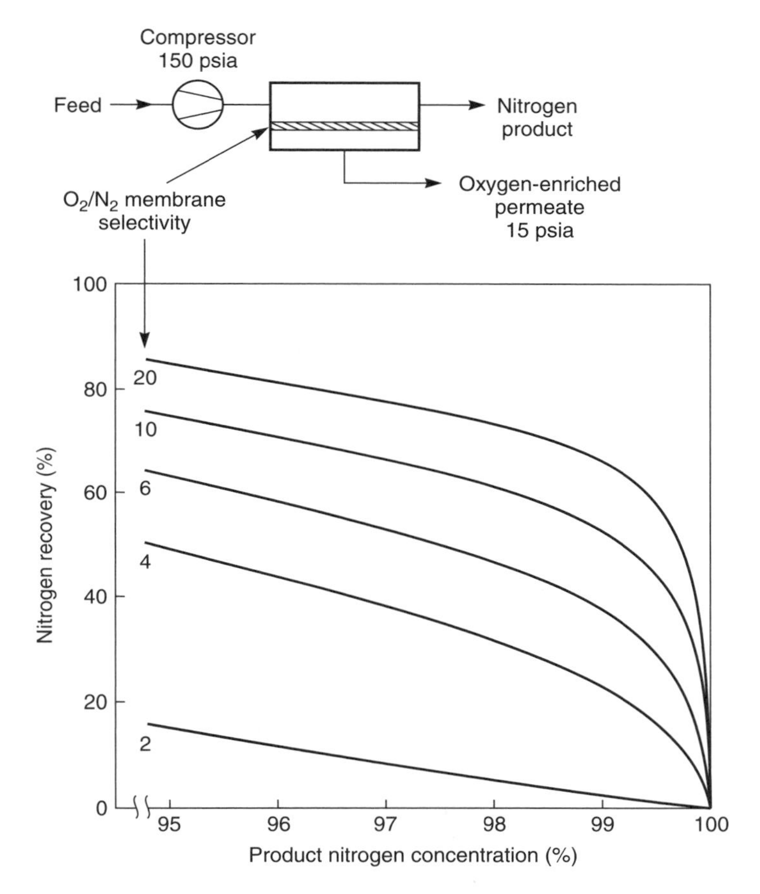

Figure 8.25 Nitrogen recovery as a function of product nitrogen concentration for membranes with selectivities between 2 and 20

trade-off between the fraction of nitrogen in the feed gas recovered as nitrogen product gas as a function of the nitrogen concentration in the product gas. All oxygen-selective membranes, even membranes with an oxygen/nitrogen selectivity as low as 2, can produce better than 99 % nitrogen, albeit at very low recoveries. The figure also shows the significant improvement in efficiency that results from an increase in oxygen/nitrogen selectivity from 2 to 20.

The first nitrogen production systems used membranes made from TPX with a selectivity of about 4. These membranes were incorporated in one-stage designs to produce 95 % nitrogen used to render flammable-liquid storage tanks inert. As the membranes improved, more complex process designs, of the type shown in Figure 8.26, were used to produce purer gas containing >99 % nitrogen. The

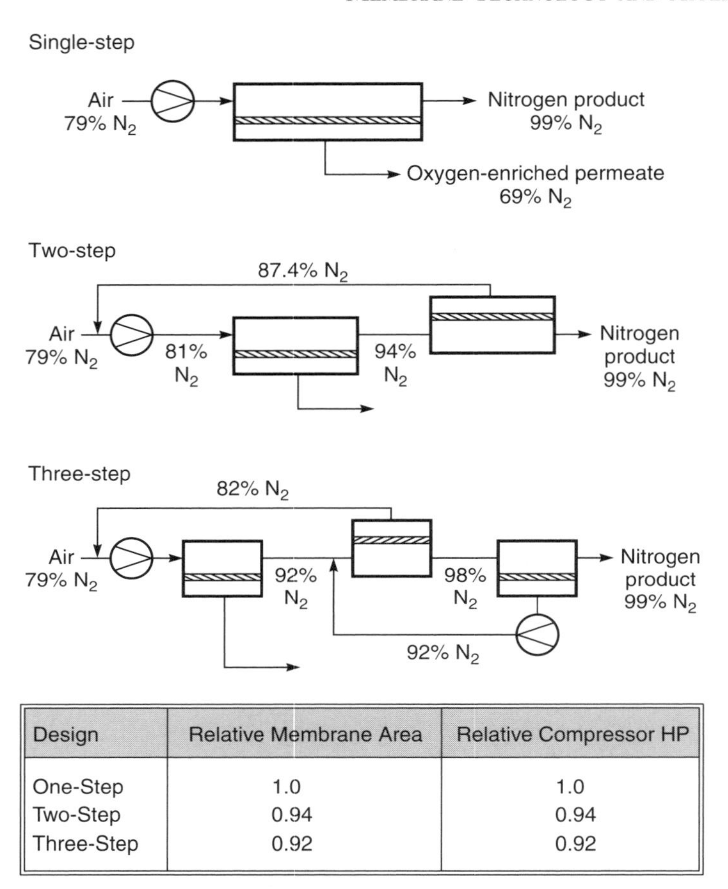

Design	Relative Membrane Area	Relative Compressor HP
One-Step	1.0	1.0
Two-Step	0.94	0.94
Three-Step	0.92	0.92

Figure 8.26 Single-, two- and three-step designs for nitrogen production from air

first improvement was the two-step process. As oxygen is removed from the air passing through the membrane modules, the concentration in the permeating gas falls. At some point the oxygen concentration in the permeate gas is less than the concentration in normal ambient feed air. Mixing this oxygen-depleted gas permeate with the incoming air then becomes worthwhile. The improvement will be most marked when the system is used to produce high-quality nitrogen containing less than 1 % oxygen. In the example shown in Figure 8.26, the second-step permeate gas contains 12.5 % oxygen, and recycling this gas to the incoming feed air reduces the membrane area and compressor load by about 6 %. This relatively small saving is worthwhile because it is achieved at essentially no cost by making a simple piping change to the system. In the two-step design, the 12.5 % oxygen permeate recycle stream is mixed with ambient air containing 21 % oxygen. A more efficient design would be to combine the recycle and feed gas where the feed gas has approximately the same concentration. This is the objective of the three-step process shown in Figure 8.26. This design saves a further 2 % in membrane area and some compressor power, but now two compressors are needed. Three-step processes are, therefore, generally limited to large systems in which the energy and membrane area savings compensate for the extra complexity and higher maintenance cost of a second compressor. A discussion of factors affecting the design of nitrogen plants is given by Prasad *et al.* [44,45].

Membrane nitrogen production systems are now very competitive with alternative technologies. The competitive range of the various methods of obtaining nitrogen is shown in Figure 8.27. Very small nitrogen users generally purchase gas cylinders or delivered liquid nitrogen, but once consumption exceeds 5000 scfd of nitrogen, membranes become the low-cost process. This is particularly true if the required nitrogen purity is between 95 % and 99 % nitrogen. Membrane systems can still be used if high quality nitrogen (up to 99.9 %) is required, but the cost of the system increases significantly. Very large nitrogen users—above 10 MMscfd of gas—generally use pipeline gas or on-site cryogenic systems. Pressure swing adsorption (PSA) systems are also sometimes used in the 1–10 MMscfd range.

A membrane process to separate nitrogen from air inevitably produces oxygen-enriched air as a by-product. Sometimes this by-product gas, containing about 35 % oxygen, can be used beneficially, but usually it is vented. A market for oxygen or oxygen-enriched air exists, but because oxygen is produced as the permeate gas stream it is much more difficult to produce high-purity oxygen than

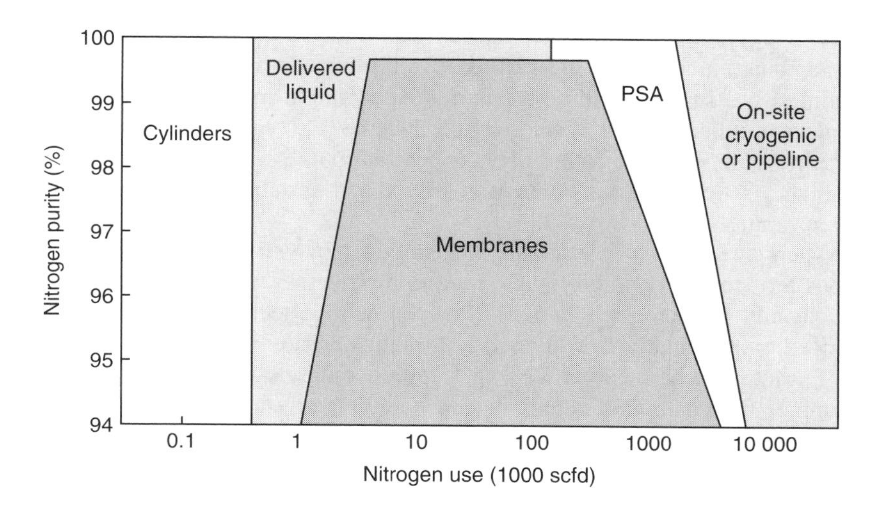

Figure 8.27 Approximate competitive range of current membrane nitrogen production systems. Many site-specific factors can affect the actual system selection

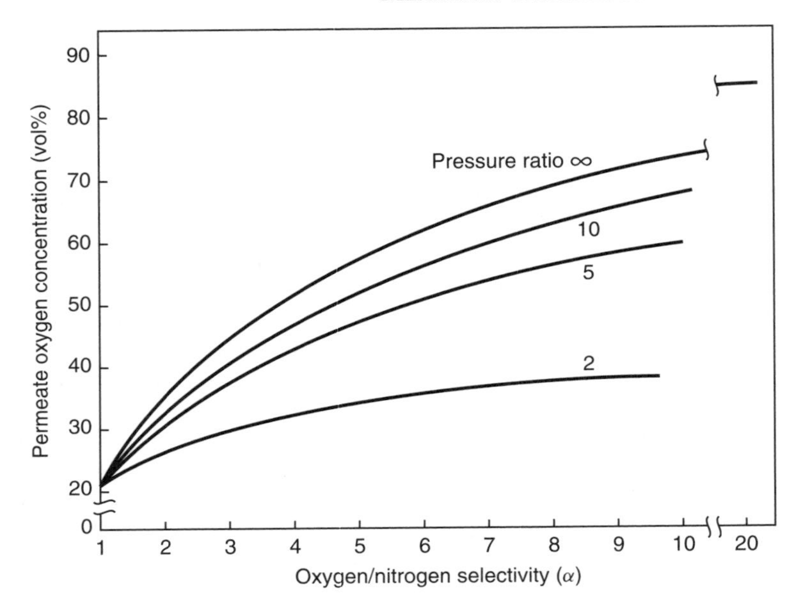

Figure 8.28 The maximum possible oxygen concentration in the permeate from a one-step membrane process with membranes of various selectivities (assumes zero stage-cut). Even the best current membrane materials, with a selectivity of 8, only produce 68 % oxygen in the permeate at an infinite pressure ratio

high-purity nitrogen with membrane systems. Figure 8.28 shows the maximum permeate oxygen concentration that can be produced by a one-step membrane process using membranes of various selectivities. Even at zero stage-cut and an infinite pressure ratio, the best currently available membrane, with an oxygen/nitrogen selectivity of 8, can only produce 68 % oxygen. At useful stage-cuts and achievable pressure ratios this concentration falls. These constraints limit membrane systems to the production of oxygen-enriched air in the 30–50 % oxygen range.

Oxygen-enriched air is used in the chemical industry, in refineries, and in various fermentation and biological digestion processes, but it must be produced very cheaply for these applications. The competitive technology is pure oxygen produced cryogenically then diluted with atmospheric air. The quantity of pure oxygen that must be blended with air to produce the desired oxygen enrichment determines the cost. This means that in membrane systems producing oxygen-enriched air, only the fraction of the oxygen above 21 % can be counted as a credit. This fraction is called the equivalent pure oxygen (EPO_2) basis.

A comparison of the cost of oxygen-enriched air produced by membranes and by cryogenic separation shows that current membranes are generally uncompetitive. The only exception is for very small users in isolated locations, where the logistics of transporting liquid oxygen to the site increase the oxygen cost to US$80–100/ton.

Development of better membranes for producing oxygen-enriched air has been, and continues to be, an area of research because of the potential application of the gas in combustion processes. When methane, oil, and other fuels are burned with air, a large amount of nitrogen passes as an inert diluent through the burners and is discarded as hot exhaust gas. If oxygen-enriched air were used, the energy lost with the hot exhaust gas would decrease considerably. Use of oxygen-enriched air also improves the efficiency of diesel engines [46]. The useful energy that can be extracted from the same amount of fuel increases significantly even if air is enriched only from 25 to 35 % oxygen. But to make this process worthwhile, the fuel savings achieved must offset the cost of the oxygen-enriched air used. Calculations show that the process would be cost-effective for some applications at an EPO_2 cost as high as US$60/ton and, for many applications, at an EPO_2 cost of US$30–40/ton. Bhide and Stern [47] have published an interesting

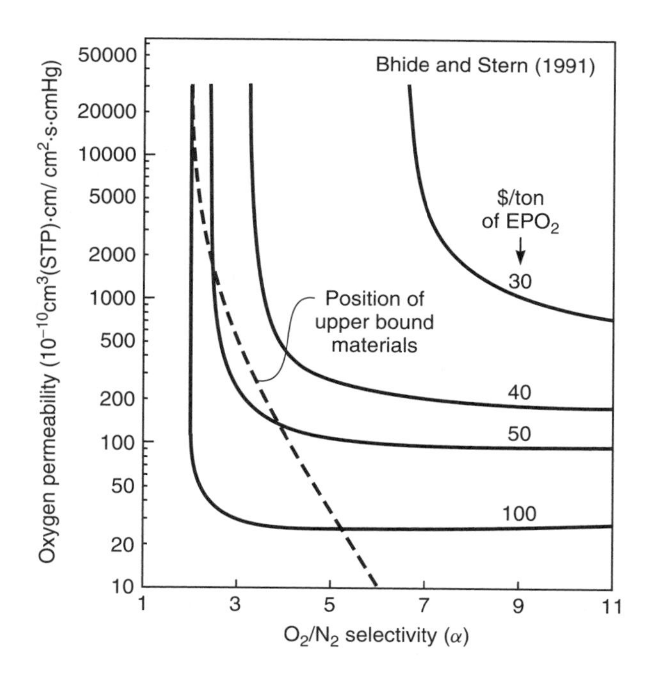

Figure 8.29 Cost of oxygen-enriched air produced by membrane separation on an EPO_2 basis as a function of the oxygen permeability and oxygen/nitrogen selectivity of the membrane. The performance of today's best membranes is represented by the upper bound performance line from Robeson's plot (Figure 8.24) [35,47]. Reprinted from *J. Membr. Sci.* **62**, B.O. Bhide and S.A. Stern, A New Evaluation of Membrane Processes for the Oxygen-enrichment of Air, p. 87. Copyright 1991, with permission from Elsevier

analysis of this problem, the results of which are shown in Figure 8.29. The figure shows the cost of oxygen-enriched air produced by a membrane process for membranes of various permeabilities and selectivities. The assumptions were optimistic–low-cost membrane modules (US$54/m^2) and membranes with extremely thin selective separating layers (1000 Å). Also shown in Figure 8.29 is the portion of the upper-bound curve obtained from the permeability/selectivity trade-off plot shown in Figure 8.24. As the figure shows, a number of materials at the upper-bound limit, with oxygen/nitrogen selectivities of 3–4 and permeabilities of 50–500, are within striking distance of the US$30–40/ton target. Production of these very high-performance membrane modules is at the outer limit of current technology but improvements in the technology could open up new, very large applications of membranes in the future.

Natural Gas Separations

US production of natural gas is about 20 trillion scf/year; total worldwide production is about 40 trillion scf/year. All of this gas requires some treatment, and approximately 20 % of the gas requires extensive treatment before it can be delivered to the pipeline. As a result, several billion dollars' worth of natural gas separation equipment is installed annually worldwide. The current membrane market share is about 2 %, essentially all for carbon dioxide removal. However, this fraction is expected to increase because applications of membranes to other separations in the natural gas processing industry are under development [48].

Raw natural gas varies substantially in composition from source to source. Methane is always the major component, typically 75–90 % of the total. Natural gas also contains significant amounts of ethane, some propane and butane, and 1–3 % of other higher hydrocarbons. In addition, the gas contains undesirable impurities: water, carbon dioxide, nitrogen and hydrogen sulfide. Although raw natural gas has a wide range of compositions, the composition of gas delivered to the pipeline is tightly controlled. Typical US natural gas specifications are shown in Table 8.5. The opportunity for membranes lies in the processing of gas to meet these specifications.

Table 8.5 Composition of natural gas required for delivery to the US national pipeline grid

Component	Specification
CO_2	<2 %
H_2O	<120 ppm
H_2S	<4 ppm
C_{3+}	950–1050 Btu/scf
Content	Dew point, −20 °C
Total inerts (N_2, CO_2, He, etc.)	<4 %

Natural gas is usually produced from the well and transported to the gas processing plant at high pressure, in the range 500–1500 psi. To minimize recompression costs, the membrane process must remove impurities from the gas into the permeate stream, leaving the methane, ethane, and other hydrocarbons in the high-pressure residue gas. This requirement determines the type of membranes that can be used for this separation. Figure 8.30 is a graphical representation of the factors of molecular size and condensability that affect selection of membranes for natural gas separations.

As Figure 8.30 shows, water is small and condensable; therefore, it is easily separated from methane by both rubbery and glassy polymer membranes. Both rubbery and glassy membranes can also separate carbon dioxide and hydrogen sulfide from natural gas. However, in practice carbon dioxide is best separated by glassy membranes (utilizing size selectivity) [49,50], whereas hydrogen sulfide, which is larger and more condensable than carbon dioxide, is best separated by rubbery membranes (utilizing sorption selectivity) [51,52]. Nitrogen can be separated from methane by glassy membranes, but the difference in size is small, so the separations achieved are small. Finally, propane and other hydrocarbons, because of their condensability, are best separated from methane with rubbery sorption-selective membranes. Table 8.6 shows typical membrane materials and the selectivities that can be obtained with good-quality membranes.

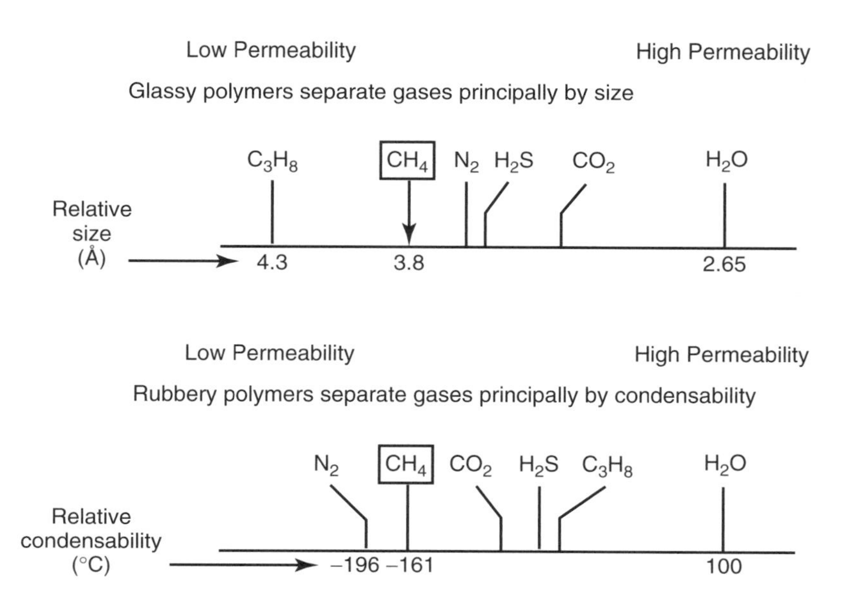

Figure 8.30 The relative size and condensability (boiling point) of the principal components of natural gas. Glassy membranes generally separate by differences in size; rubbery membranes separate by differences in condensability

Table 8.6 Membrane materials and selectivities for separation of impurities from natural gas

Component to be permeated	Category of preferred polymer material	Typical polymer used	Typical selectivity over methane
CO_2	Glass	Cellulose acetate, polyimide	10–20
H_2S	Rubber	Ether-amide block copolymer	20–40
N_2	Glass	Polyimide, perfluoro polymers	2–3
H_2O	Rubber or glass	Many	>200
Butane	Rubber	Silicone rubber	7–10

Carbon Dioxide Separation

Removal of carbon dioxide is the only membrane-based natural gas separation process currently practiced on a large scale—more than 200 plants have been installed, some very large. Most were installed by Grace (now Kvaerner-GMS), Separex (UOP) and Cynara and all use cellulose acetate membranes in hollow fiber or spiral-wound module form. More recently, hollow fiber polyaramide (Medal) membranes have been introduced because of their higher selectivity.

The designs of two typical carbon dioxide removal plants are illustrated in Figure 8.31. One-stage plants, which are simple, contain no rotating equipment, and require minimal maintenance, are preferred for small gas flows. In such plants methane loss to the permeate is often 10–15 %. If there is no fuel use for this gas, it must be flared, which represents a significant revenue loss. Nonetheless, for gas wells producing 1–2 MMscfd, one-stage membrane units with their low capital and operating costs may still be the optimum treatment method.

As the natural gas stream increases in size, the methane loss from a one-stage system becomes prohibitive. Often the permeate gas is recompressed and passed through a second membrane stage. This second stage reduces the methane loss to a few percent. However, the recompression cost is considerable, and the membrane system may no longer compete with amine absorption, the alternative technology. In general, membrane systems have proved to be most competitive for gas streams below 30 MMscfd containing high concentrations of carbon dioxide. Spillman [48] and McKee *et al.* [53] have reviewed the competitive position of membrane systems for this application. Currently the market for membrane carbon dioxide gas separation systems can be summarized as follows:

1. Very small systems (less than 5 MMscfd). At this flow rate, membrane units are very attractive. Often the permeate is flared or used as fuel, so the system is a simple bank of membrane modules.
2. Small systems (5–30 MMscfd). Two-stage membrane systems are used to reduce methane loss. In this gas flow range, amine and membrane systems

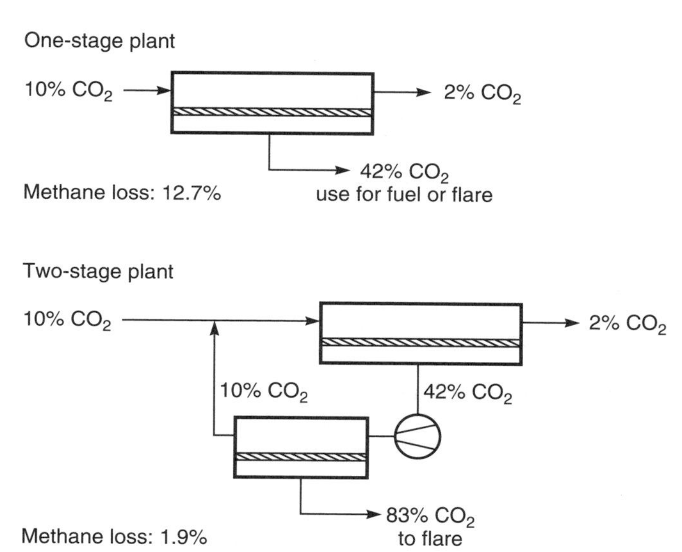

Figure 8.31 Flow scheme of one-stage and two-stage membrane separation plants to remove carbon dioxide from natural gas. Because the one-stage design has no moving parts, it is very competitive with other technologies especially if there is a use for the low-pressure permeate gas. Two-stage processes are more expensive because a large compressor is required to compress the permeate gas. However, the loss of methane with the fuel gas is much reduced

compete; the choice between the two technologies depends on site-specific factors.

3. Medium to large systems (greater than 30 MMscfd). In general, membrane systems are too expensive to compete head-to-head with amine plants. However, a number of large membrane systems have been installed on offshore platforms, at carbon dioxide flood operations, or where site-specific factors particularly favor membrane technology. As membranes improve, their market share is increasing.

In principle, the combination of membranes for bulk removal of the carbon dioxide with amine units as polishing systems offers a low-cost alternative to all-amine plants for many streams. However, this approach has not been generally used because the savings in capital cost are largely offset by the increased complexity of the plant, which now contains two separation processes. The one exception has been in carbon dioxide flood enhanced oil-recovery projects [49,54], in which carbon dioxide is injected into an oil formation to lower the viscosity of the oil. Water, oil and gas are removed from the formation; the carbon dioxide is separated from the gas produced and reinjected. In these projects,

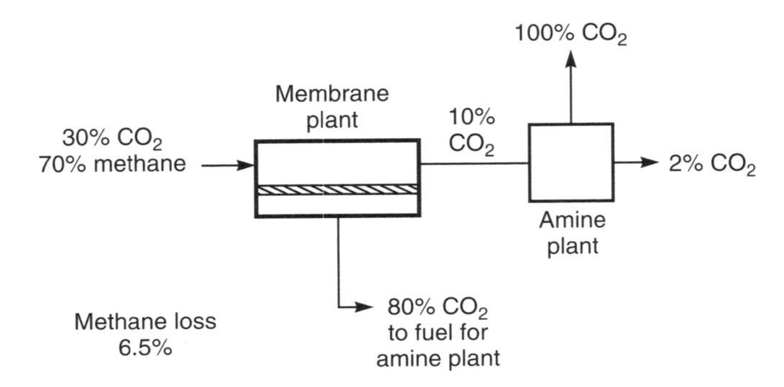

Figure 8.32 A typical membrane/amine plant for the treatment of associated natural gas produced in carbon dioxide/enhanced oil projects. The membrane permeate gas is often used as a fuel for the amine absorption plant

the composition and volume of the gas changes significantly over the lifetime of the project. The modular nature of membrane units allows easy retrofitting to an existing amine plant, allowing the performance of the plant to be adjusted to meet the changing separation needs. Also, the capital cost of the separation system can be spread more evenly over the project lifetime. An example of a membrane/amine plant design is shown in Figure 8.32. In this design, the membrane unit removes two-thirds of the carbon dioxide, and the amine plant removes the remainder. The combined plant is usually significantly less expensive than an all-amine or all-membrane plant.

Dehydration

All natural gas must be dried before entering the national distribution pipeline to control corrosion of the pipeline and to prevent formation of solid hydrocarbon/water hydrates that can choke valves. Currently glycol dehydrators are widely used; approximately 50 000 units are in service in the United States. However, glycol dehydrators are not well suited for use on small gas streams or on offshore platforms, increasingly common sources of natural gas. In addition, these units coextract benzene, a known carcinogen and trace contaminant in natural gas, and release the benzene to the atmosphere. The Environmental Protection Agency (EPA) has announced its intention to require benzene emission control systems to be fitted to large glycol units.

Membrane processes offer an alternative approach to natural gas dehydration and are being developed by a number of companies. Membranes with intrinsic selectivities for water from methane of more than 500 are easily obtained, but because of concentration polarization effects, actual selectivities are typically about 200. Two possible process designs are shown in Figure 8.33. In the first

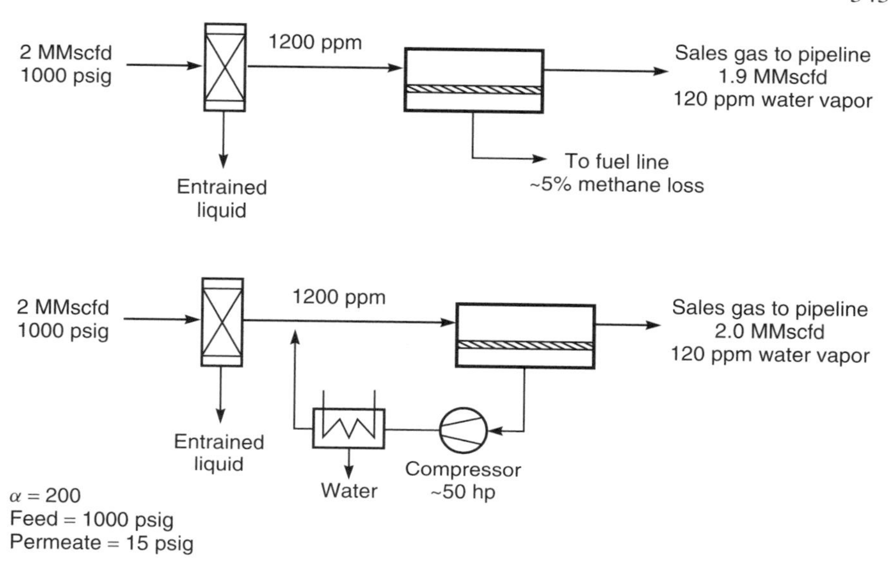

Figure 8.33 Dehydration of natural gas is easily performed by membranes but high cost may limit its scope to niche applications

design, a small one-stage system removes 90 % of the water in the feed gas, producing a low-pressure permeate gas representing 5–6 % of the initial gas flow. This gas contains the removed water. If the gas can be used as low-pressure fuel at the site, this design is economical and competitive with glycol dehydration. In the second design, the wet, low-pressure permeate gas is recompressed and cooled, so the water vapor condenses and is removed as liquid water. The natural gas that permeates the membrane is then recovered. However, if the permeate gas must be recompressed, as in the second design, the capital cost of the system approximately doubles, and membranes are then only competitive in special situations where glycol dehydration is not possible.

Dew Point Adjustment, C_{3+} Recovery

Natural gas usually contains varying amounts of ethane, propane, butane, and higher hydrocarbons. The gas is often close to its saturation point with respect to some of these hydrocarbons, which means liquids will condense from the gas at cold spots in the pipeline transmission system. To avoid the problems caused by condensation of liquids, the dew point of US natural gas is lowered to about −20 °C before delivery to the pipeline by removing portions of the propane and butane and higher hydrocarbons. For safety reasons the Btu rating of the pipeline gas is also usually controlled within a narrow range, typically

950–1050 Btu per cubic foot. Because the Btu values of ethane, propane and pentane are higher than that of methane, natural gas that contains significant amounts of these hydrocarbons may have an excessive Btu value, requiring their removal. Of equal importance, these higher hydrocarbons are generally more valuable as recovered liquids than their fuel value in the gas. For all of these reasons almost all natural gas is treated to control the C_{3+} hydrocarbon content.

The current technology used to separate the higher hydrocarbons from natural gas streams is condensation, shown schematically in Figure 8.34. The natural gas stream is cooled by refrigeration or expansion to between −20 °C and −40 °C. The condensed liquids, which include the higher hydrocarbons and water, are separated from the gas streams and subjected to fractional distillation to recover the individual components. Because refrigeration is capital-intensive and uses large amounts of energy, there is considerable interest in alternative techniques, such as membrane gas separation.

A typical flow diagram of a membrane system for C_{3+} liquids recovery is also shown in Figure 8.34. The natural gas is fed to modules containing a higher-hydrocarbon-selective membrane, which removes the higher hydrocarbons as the permeate stream. This stream is recompressed and cooled by a cold-water exchanger to condense higher hydrocarbons. The non-condensed bleed

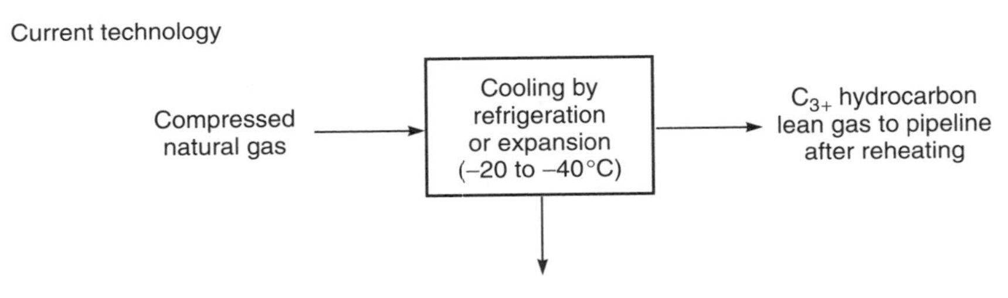

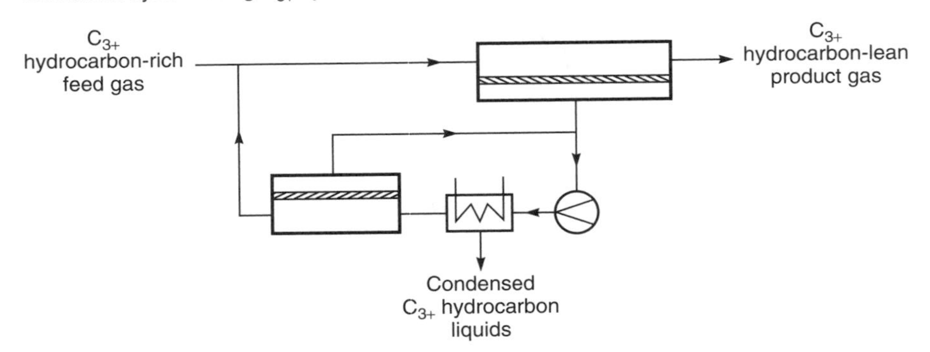

Figure 8.34 Recovery of C_{3+} hydrocarbons from natural gas

stream from the condenser to the inlet will normally still contain more heavy hydrocarbons than the raw gas, so prior to returning the gas to the feed stream, the condenser bleed stream is passed through a second set of membrane modules. The permeate streams from the two sets of modules are combined, creating a recirculation loop around the condenser, which continuously concentrates the higher hydrocarbons [37].

The competitiveness of membrane systems in this application is very sensitive to the selectivity of the membranes for propane, butane and higher hydrocarbons over methane. If the membranes are very selective (propane/methane selectivity of 5–7, butane/methane selectivity of 10–15), the permeate stream from the main set of modules will be small and concentrated, minimizing the cost of the recompressor. Currently, silicone rubber membranes are being considered for this application, but other, more selective materials have been reported [55].

Vapor/Gas Separations

In the separation of vapor/gas mixtures, rubbery polymers, such as silicone rubber, can be used to permeate the more condensable vapor, or glassy polymers can be used to permeate the smaller gas. Although glassy, gas-permeable membranes have been proposed for a few applications, most installed plants use vapor-permeable membranes, often in conjunction with a second process such as condensation [36,38] or absorption [56]. The first plants, installed in the early 1990s, were used to recover vapors from gasoline terminal vent gases or chlorofluorocarbon (CFC) vapors from the vents of industrial refrigeration plants. More recently, membranes have begun to be used to recover hydrocarbons and processing solvents from petrochemical plant purge gas. Some of these streams are quite large and discharge vapors with a recovery value of US$1–2 million/year.

One of the most successful petrochemical applications is treatment of resin degassing vent gas in polyolefin plants [57,58]. Olefin monomer, catalyst, solvents, and other co-reactants are fed at high pressure into the polymerization reactor. The polymer product (resin) is removed from the reactor and separated from excess monomer in a flash separation step. The recovered monomer is recycled to the reactor. Residual monomer is removed from the resin by stripping with nitrogen. The composition of this degassing vent stream varies greatly, but it usually contains 20–50 % of mixed hydrocarbon monomers in nitrogen. The monomer content represents about 1 % of the hydrocarbon feedstock entering the plant. This amount might seem small, but because polyolefin plants are large operations, the recovery value of the stream can be significant.

Several membrane designs can be used; two are shown in Figure 8.35 [59]. The two-step, two-stage system shown in Figure 8.35(a) achieves the target separation by linking several membrane units together, each unit performing a partial separation. The first unit removes a portion of the propylene from the feed gas to produce a concentrated permeate. The residue gas from this step is then sent to a

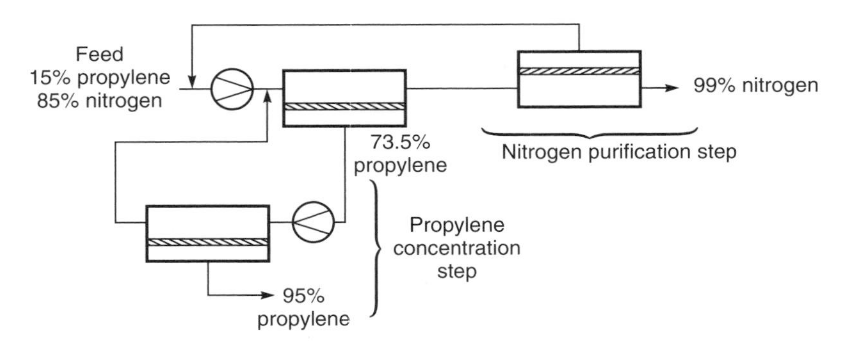

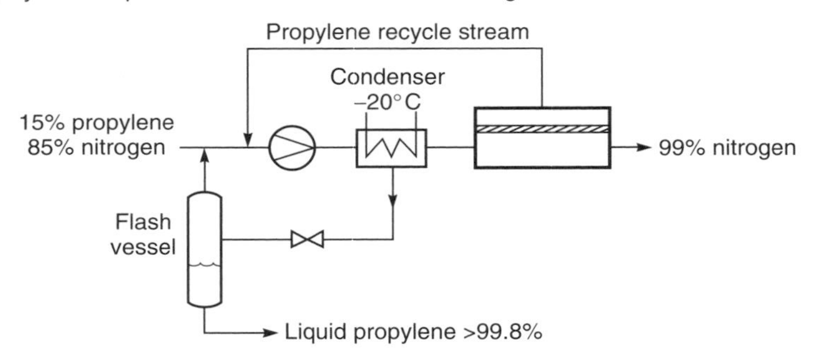

Figure 8.35 Vapor separation process designs able to achieve high vapor recovery and high-purity product streams

second membrane unit which produces a nitrogen gas stream containing less than 1 % propylene. The permeate from this second step is only partially enriched in propylene, so it is mixed with the incoming feed gas.

To increase the propylene concentration in the permeate gas from the first step, a second-stage membrane unit is used. The permeate gas from the first stage is compressed and then passed through the second stage to produce a final permeate stream containing 95 % propylene. The partially depleted second stage residue stream is recycled back to the feed. By linking together the three membrane separation units, the target recovery for both propylene and nitrogen can be achieved.

A number of multistep vapor separation systems of the type shown in Figure 8.35(a) have been installed. These systems have the advantage of operating at ambient temperatures and having as the main rotating equipment gas compressors with which petrochemical plant operators are very familiar.

However, if very good removal of the nitrogen from the recycle gas is required, the hybrid process combining condensation and membrane separation shown in Figure 8.35(b) is preferred. In this design, the compressed feed gas is sent to a condenser. On cooling the feed gas, a portion of the propylene content is removed as a condensed liquid. The remaining, uncondensed propylene is removed by the membrane separation system to produce a 99 % nitrogen stream. The permeate gas is recycled to the incoming feed gas from the purge bin.

Because the gas sent to the membrane stage is cooled, the solubility of propylene in the membrane is enhanced, and the selectivity of the membrane unit is increased. The propylene condensate contains some dissolved nitrogen so the liquid is flashed at low pressure to remove this gas, producing a better than 99.5 % pure hydrocarbon product. A photograph of a propylene/nitrogen vent gas treatment system is shown in Figure 8.36.

Vapor/Vapor Separations

A final group of separations likely to develop into a major application area for membranes is vapor/vapor separations, such as ethylene (bp −103.9 °C) from

Figure 8.36 Photograph of a membrane unit used to recover nitrogen and propylene from a polypropylene plant vent gas

ethane (bp −88.9 °C), propylene (bp −47.2 °C) from propane (bp −42.8 °C), and *n*-butane (bp −0.6 °C) from isobutane (bp −10 °C). These close-boiling mixtures are separated on a very large scale in the synthesis of ethylene and propylene, the two largest-volume organic chemical feedstocks, and in the synthesis of isobutane in refineries to produce high-octane gasoline. Because the mixtures are close-boiling, large towers and high reflux ratios are required to achieve good separations.

If membranes are to be used for these separations, highly selective materials must be developed. Several groups have measured the selectivities of polymeric membranes for ethylene/ethane and propylene/propane mixtures. Burns and Koros have reviewed these results [60]. The data should be treated with caution. Some authors report selectivities based on the ratio of the permeabilities of the pure gases; others use a hard vacuum or a sweep gas on the permeate side of the membrane. Both procedures produce unrealistically high selectivities. In an industrial plant, the feed gas will be at 100–150 psig and a temperature sufficient to maintain the gas in the vapor phase; the permeate gas will be at a pressure of 10–20 psig. Under these operating conditions, plasticization and loss of selectivity occur with even the most rigid polymer membranes, so selectivities are usually low. Because of these problems, this application might be one for which the benefits of ceramic or carbon fiber membranes can justify their high cost. Caro *et al.* have recently reviewed the ceramic membrane literature [24].

Dehydration of Air

The final application of gas separation membranes is dehydration of compressed air. The competitive processes are condensation or solid desiccants, both of which are established, low-cost technologies. Membranes with water/air selectivities of more than 500 are known, although actual selectivities obtained in membrane modules are less because of concentration polarization effects. Nonetheless, existing membranes are more than adequate for this separation. The problem inhibiting their application is the loss of compressed feed air through the membrane. Compressed air is typically supplied at about 7 atm (105 psi), so the pressure ratio across the membrane is about 7. Because air dehydration membranes have a selectivity of more than 200, these membranes are completely pressure-ratio-limited. Based on Equation (8.10), this means that the permeate gas cannot be more than seven times more concentrated than the feed. The result is that a significant fraction of the feed gas must permeate the membrane to carry away the permeate water vapor. Typically 15–20 % of the pressurized feed gas permeates the membrane, which affects the productivity of the compressor significantly. Counterflow-sweep designs of the type discussed in Chapter 4 are widely used to reduce permeant loss. For this reason, membrane air dehydration systems have not found a wide market except where the reliability and simplicity of the membrane design compared to adsorbents or cooling are particularly attractive.

Conclusions and Future Directions

The application of membranes to gas separation problems has grown rapidly since the installation of the first industrial plants in the early 1980s. The current status of membrane gas separation processes is summarized in Table 8.7, in which the processes are divided into four groups. The first group consists of the established processes: nitrogen production from air, hydrogen recovery and air drying. These processes represent more than 80 % of the current gas separation membrane market. All have been used on a large commercial scale for 10 years, and dramatic improvements in membrane selectivity, flux and process designs have been made during that time. For example, today's hollow fine fiber nitrogen production module generates more than 10 times the amount of nitrogen, with better quality and at a lower energy consumption, than the modules produced in the early 1980s. However, the technology has now reached a point at which, barring a completely unexpected breakthrough, further changes in productivity are likely to be the result of a number of small incremental changes.

Developing processes are the second group of applications. These include carbon dioxide separation from natural gas, organic vapor separation from air and nitrogen, and recovery of light hydrocarbons from refinery and petrochemical plant purge gases. All of these processes are performed on a commercial scale, and in total several hundred plants have been installed. Significant expansion in these applications, driven by the development of better membranes and process designs, is occurring. For example, carbon dioxide removal from natural gas has been practiced using cellulose acetate membranes for more than 15 years. Introduction of more selective and higher-flux membranes has begun and, in time, is likely to make membrane processes much more competitive with amine absorption. The application of silicone rubber vapor separation membranes in petrochemical and refinery applications is currently growing.

The 'to be developed' membrane processes represent the future expansion of gas separation technology. Natural gas treatment processes, including dehydration, natural gas liquids (C_{3+} hydrocarbons) recovery, and hydrogen sulfide removal, are currently being studied at the field testing and early commercial stage by several companies. The market is very large, but the fraction that membranes will ultimately capture is unknown. The production of oxygen-enriched air is another large potential application for membranes. The market size depends completely on the properties of the membranes that can be produced. Improvements of a factor of two in flux at current oxygen/nitrogen selectivities would probably produce a limited membrane market; improvements by a factor of five to ten would make the use of oxygen-enriched air in natural gas combustion processes attractive. In this case the market could be very large indeed. The final application listed in Table 8.7 is the separation of organic vapor mixtures using membranes in competition, or perhaps in combination, with distillation.

Table 8.7 Status of membrane gas separation processes

Process	Application	Comments
Established processes		
Oxygen/nitrogen	Nitrogen from air	Processes are all well developed. Only incremental improvements in performance expected
Hydrogen/methane; hydrogen/nitrogen; hydrogen/carbon monoxide	Hydrogen recovery; ammonia plants and refineries	
Water/air	Drying compressed air	
Developing processes		
VOC/air	Air pollution control applications	Several applications being developed. Significant growth expected as the process becomes accepted
Light hydrocarbons from nitrogen or hydrogen	Reactor purge gas, petrochemical process streams, refinery waste gas	Application is expanding rapidly
Carbon dioxide/methane	Carbon dioxide from natural gas	Many plants installed but better membranes are required to change market economics significantly
To-be-developed processes		
C_{3+} hydrocarbons/methane	NGL recovery from natural gas	Field trials and demonstration system tests under way. Potential market is large
Hydrogen sulfide, water/methane	Natural gas treatment	Niche applications, difficult for membranes to compete with existing technology
Oxygen/nitrogen	Oxygen enriched air	Requires better membranes to become commercial. Size of ultimate market will depend on properties of membranes developed. Could be very large
Organic vapor mixtures	Separation of organic mixtures in refineries and petrochemical plants	Requires better membranes and modules. Potential size of application is large

References

1. T. Graham, On the Absorption and Dialytic Separation of Gases by Colloid Septa, *Philos. Mag.* **32**, 401 (1866).
2. R.M. Barrer, *Diffusion In and Through Solids*, Cambridge University Press, London (1951).
3. G.J. van Amerongen, Influence of Structure of Elastomers on their Permeability to Gases, *J. Appl. Polym. Sci.* **5**, 307 (1950).
4. S.A. Stern, Industrial Applications of Membrane Processes: The Separation of Gas Mixtures, in *Membrane Processes for Industry*, Proceedings of the Symposium, Southern Research Institute, Birmingham, AL, pp. 196–217 (1966).
5. P. Meares, Diffusion of Gases Through Polyvinyl Acetate, *J. Am. Chem. Soc.* **76**, 3415 (1954).
6. J.M.S. Henis and M.K. Tripodi, A Novel Approach to Gas Separations Using Composite Hollow Fiber Membranes, *Sep. Sci. Technol.* **15**, 1059 (1980).
7. D.L. MacLean, W.A. Bollinger, D.E. King and R.S. Narayan, Gas Separation Design with Membranes, in *Recent Developments in Separation Science*, N.N. Li and J.M. Calo (eds), CRC Press, Boca Raton, FL, p. 9 (1986).
8. F. Gruen, Diffusionmessungen an Kautschuk (Diffusion in Rubber), *Experimenta* **3**, 490 (1947).
9. G.J. van Amerongen, The Permeability of Different Rubbers to Gases and Its Relation to Diffusivity and Solubility, *J. Appl. Phys.* **17**, 972 (1946).
10. R.D. Behling, K. Ohlrogge, K.V. Peinemann and E. Kyburz, The Separation of Hydrocarbons from Waste Vapor Streams, in *Membrane Separations in Chemical Engineering*, A.E. Fouda, J.D. Hazlett, T. Matsuura and J. Johnson (eds), AIChE Symposium Series Number 272, AIChE, New York, NY, Vol. 85, p. 68 (1989).
11. L.M. Robeson, Correlation of Separation Factor versus Permeability for Polymeric Membranes, *J. Membr. Sci.* **62**, 165 (1991).
12. S.A. Stern, Polymers for Gas Separation: The Next Decade, *J. Membr. Sci.* **94**, 1 (1994).
13. S.Y. Lee, B.S. Minhas and M.D. Donohue, Effect of Gas Composition and Pressure on Permeation through Cellulose Acetate Membranes, in *New Membrane Materials and Processes for Separation*, K.K. Sirkar and D.R. Lloyd (eds), AIChE Symposium Series Number 261, AIChE, New York, NY, Vol. 84, p. 93 (1988).
14. G. Alefeld and J. Völkl (eds), *Hydrogen in Metals I: Basic Properties*, Springer-Verlag, Germany (1978).
15. J.B. Hunter, Silver-Palladium Film for Separation and Purification of Hydrogen, US Patent 2,773,561 (December, 1956).
16. J.B. Hunter, Ultrapure Hydrogen by Diffusion through Palladium Alloys, *Disv. Pet. Chem. Prepr.* **8**, 4 (1963).
17. R.B. McBride and D.L. McKinley, A New Hydrogen Recovery Route, *Chem. Eng. Prog.* **61**, 81 (1965).
18. J.E. Philpott, Hydrogen Diffusion Technology, Commercial Applications of Palladium Membrane, *Platinum Metals Rev.* **29**, 12 (1985).
19. A.L. Athayde, R.W. Baker and P. Nguyen, Metal Composite Membranes for Hydrogen Separation, *J. Membr. Sci.* **94**, 299 (1994).
20. R.E. Buxbaum and A.B. Kinney, Hydrogen Transport Through Tubular Membranes of Palladium-coated Tantalum and Niobium, *Ind. Eng. Chem. Res.* **35**, 530 (1996).
21. D.J. Edlund and J. McCarthy, The Relationship Between Intermetallic Diffusion and Flux Decline in Composite-metal Membranes: Implications for Achieving Long Membrane Lifetime, *J. Membr. Sci.* **107**, 147 (1995).

22. U. Merten and P.K. Gantzel, Method and Apparatus for Gas Separation by Diffusion, US Patent 3,415,038 (December, 1968).
23. J.M.S. Henis and M.K. Tripodi, Multicomponent Membranes for Gas Separations, US Patent 4,230,436 (October, 1980).
24. J. Caro, M. Noack, P. Kolsch and R. Schäfer, Zeolite Membranes: State of Their Development and Perspective, *Microporous Mesoporous Mater.* **38**, 3 (2000).
25. J. Brinker, C.-Y. Tsai and Y. Lu, Inorganic Dual-Layer Microporous Supported Membranes, US Patent 6,536,604 (March 2003).
26. M. Kondo, M. Komori, H. Kita and K. Okamoto, Tubular-type Pervaporation Module with Zeolite NaA Membrane, *J. Membr. Sci.* **133**, 133 (1997).
27. N. Wynn, Pervaporation Comes of Age, *Chem Eng. Prog.* **97**, 66 (2001).
28. H.J.C. te Hennepe, D. Bargeman, M.H.V. Mulder and C.A. Smolders, Zeolite-filled Silicone Rubber Membranes Part I: Membrane Preparation and Pervaporation Results, *J. Membr. Sci.* **35**, 39 (1987).
29. J.-M. Duval, B. Folkers, M.H.V. Mulder, G. Desgrandchamps and C.A. Smolders, Adsorbent Filled Membranes for Gas Separation, *J. Membr. Sci.* **80**, 189 (1993).
30. M.-D. Jia, K.V. Peinemann and R.-D. Behling, Preparation and Characterization of Thin-film Zeolite-PDMS Composite Membranes, *J. Membr. Sci.* **73**, 119 (1992).
31. R. Mahajan and W.J. Koros, Factors Controlling Successful Formation of Mixed-matrix Gas Separation Materials, *Ind. Eng. Chem. Res.* **39**, 2692 (2000).
32. R. Mahajan and W.J Koros, Mixed-matrix Materials with Glassy Polymers Part 1, *Polym. Eng. Sci.* **42**, 1420 (2002).
33. R. Mahajan and W.J Koros, Mixed-matrix Materials with Glassy Polymers Part 2, *Polym. Eng. Sci.* **42**, 1432 (2002).
34. J.C. Maxwell, *Treatise on Electricity and Magnetism Vol. I*, Oxford University Press, London (1873).
35. L.M. Robeson, A. Noshay, M. Matzner and C.N. Merian, Physical Property Characteristics of Polysulfone/poly(dimethyl siloxane) Block Copolymers, *Angew. Makromol. Chem.* **29**, 47 (1973).
36. R.W. Baker and J.G. Wijmans, Membrane Separation of Organic Vapors from Gas Streams, in *Polymeric Gas Separation Membranes*, D.R. Paul and Y.P. Yampol'skii (eds), CRC Press, Boca Raton, FL, pp. 353–398 (1994).
37. R.W. Baker and J.G. Wijmans, Two-stage Membrane Process and Apparatus, US Patents 5,256,295 and 5,256,296 (October, 1993).
38. J.G. Wijmans, Process for Removing Condensable Components from Gas Streams, US Patent 5,199,962 (April, 1993) and 5,089,033 (February, 1992).
39. D.R. Paul and Y.P. Yampol'skii (eds), *Polymeric Gas Separation Membranes*, CRC Press, Boca Raton, FL (1994).
40. W.J. Koros and G.K. Fleming, Membrane Based Gas Separation, *J. Membr. Sci.* **83**, 1 (1993).
41. R.W. Baker, Future Directions of Membrane Gas Separation Technology, *Ind. Eng. Chem. Res.* **41**, 1393 (2002).
42. W.A. Bollinger, D.L. MacLean and R.S. Narayan, Separation Systems for Oil Refining and Production, *Chem. Eng. Prog.* **78**, 27 (1982).
43. J.M.S. Henis, Commercial and Practical Aspects of Gas Separation Membranes, in *Polymeric Gas Separation Membranes*, D.R. Paul and Y.P. Yampol'skii (eds), CRC Press, Boca Raton, FL, pp. 441–530 (1994).
44. R. Prasad, R.L. Shaner and K.J. Doshi, Comparison of Membranes with Other Gas Separation Technologies, in *Polymeric Gas Separation Membranes*, D.R. Paul and Y.P. Yampol'skii (eds), CRC Press, Boca Raton, FL, pp. 531–614 (1994).
45. R. Prasad, F. Notaro and D.R. Thompson, Evolution of Membranes in Commercial Air Separation, *J. Membr. Sci.* **94**, 225 (1994).

46. G.R. Rigby and H.C. Watson, Application of Membrane Gas Separation to Oxygen Enrichment of Diesel Engines, *J. Membr. Sci.* **87**, 159 (1994).
47. B.O. Bhide and S.A. Stern, A New Evaluation of Membrane Processes for the Oxygen-enrichment of Air, *J. Membr. Sci.* **62**, 87 (1991).
48. R.W. Spillman, Economics of Gas Separation by Membranes, *Chem. Eng. Prog.* **85**, 41 (1989).
49. D. Parro, Membrane Carbon Dioxide Separation, *Energy Prog.* **5**, 51 (1985).
50. W.J. Schell, C.G. Wensley, M.S.K. Chen, K.G. Venugopal, B.D. Miller and J.A. Stuart, Recent Advances in Cellulosic Membranes for Gas Separation and Pervaporation, *Gas Sep. Purif.* **3**, 162 (1989).
51. G. Chatterjee, A.A. Houde and S.A. Stern, Poly(ether methane) and Poly(ether urethane urea) Membranes with High H_2S/CH_4 Selectivity, *J. Membr. Sci.* **135**, 99 (1997).
52. K.A. Lokhandwala, R.W. Baker and K.D. Amo, Sour Gas Treatment Process, US Patent 5,407,467 (April, 1995).
53. R.L. McKee, M.K. Changela and G.J. Reading, Carbon Dioxide Removal: Membrane Plus Amine, *Hydrocarbon Process.* **70**, 63 (1991).
54. R.J. Hamaker, Evolution of a Gas Separation Membrane, 1983–1990, in *Effective Industrial Membrane Processes*, M.K. Turner (ed.), Elsevier, NY, pp. 337–344 (1991).
55. J. Schultz and K.-V. Peinemann, Membranes for Separation of Higher Hydrocarbons from Methane, *J. Membr. Sci.* **110**, 37 (1996).
56. K. Ohlrogge, J. Wind and R.D. Behling, Off Gas Purification by Means of Membrane Vapor Separation Systems, *Sep. Sci Technol.* **30**, 1625 (1995).
57. R.W. Baker and M.L. Jacobs, Improve Monomer Recovery from Polyolefin Resin Degassing, *Hydrocarbon Process.* **75**, 49 (1996).
58. R.W. Baker, K.A. Lokhandwala, M.L. Jacobs and D.E. Gottschlich, Recover Feedstock and Product from Reactor Vent Streams, *Chem. Eng. Prog.* **96**, 51 (2000).
59. R.W. Baker, J.G. Wijmans and J. Kaschemekat, The Design of Membrane Vapor-Gas Separation Systems, *J. Membr. Sci.* **151**, 55 (1998).
60. R.L. Burns and W.J. Koros, Defining the Challenges for C_3H_6/C_3H_8 Separation Using Polymeric Membranes, *J. Membr. Sci.* **211**, 299 (2003).

9 PERVAPORATION

Introduction and History

The pervaporation process to separate liquid mixtures is shown schematically in Figure 9.1. A feed liquid mixture contacts one side of a membrane; the permeate is removed as a vapor from the other side. Transport through the membrane is induced by the vapor pressure difference between the feed solution and the permeate vapor. This vapor pressure difference can be maintained in several ways. In the laboratory, a vacuum pump is usually used to draw a vacuum on the permeate side of the system. Industrially, the permeate vacuum is most economically generated by cooling the permeate vapor, causing it to condense; condensation spontaneously creates a partial vacuum.

The origins of pervaporation can be traced to the nineteenth century, but the word itself was coined by Kober in 1917 [1]. The process was first studied in a systematic fashion by Binning and co-workers at American Oil in the 1950s [2–5]. Binning was interested in applying the process to the separation of organic mixtures. Although this work was pursued at the laboratory and bench scales for a number of years and several patents were obtained, the process was not commercialized. Membrane technology at that time could not produce the high-performance membranes and modules required for a commercially competitive process. The process was picked up in the 1970s at Monsanto by Eli Perry and others. More than a dozen patents assigned to Monsanto issued from 1973 to 1980 cover a wide variety of pervaporation applications [6], but none of this work led to a commercial process. Academic research on pervaporation was also carried out by Aptel, Neel and others at the University of Toulouse [7,8]. By the 1980s, advances in membrane technology made it possible to prepare economically viable pervaporation systems.

Pervaporation systems are now commercially available for two applications. The first and most important is the removal of water from concentrated alcohol solutions. GFT, now owned by Sulzer, the leader in this field, installed the first pilot plant in 1982 [9]. The ethanol feed to the membrane contains about 10 % water. The pervaporation process removes the water as the permeate, producing a residue of pure ethanol containing less than 1 % water. All

Membrane Technology and Applications R. W. Baker
 ISBN: 0-470-85445-6

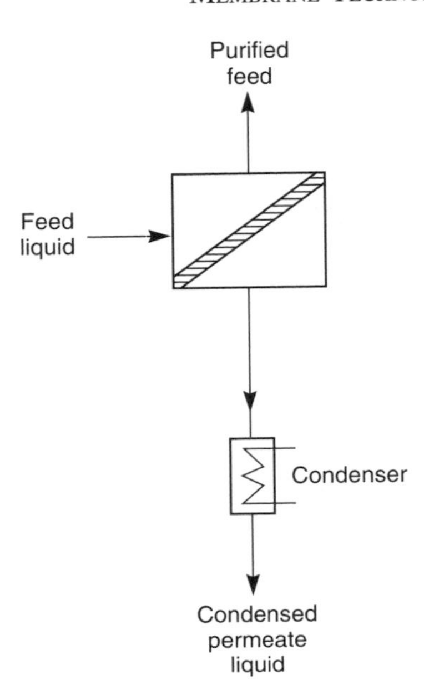

Figure 9.1 In the pervaporation process, a liquid mixture contacts the membrane, which preferentially permeates one of the liquid components as a vapor. The vapor enriched in the more permeable component is cooled and condensed, spontaneously generating a vacuum that drives the process

the problems of azeotropic distillation are avoided. More than 100 plants have since been installed by Sulzer (GFT) and its licensees for this application [10]. The current largest plant was installed at Bethenville, France in 1988; this unit contains 2400 m^2 of membranes and processes 5000 kg/h of ethanol. The second commercial application of pervaporation is the removal of small amounts of volatile organic compounds (VOCs) from contaminated water. This technology was developed by Membrane Technology and Research [11–13]; the first commercial plant was sold in 1996.

Both of the current commercial pervaporation processes concentrate on the separation of VOCs from contaminated water. This separation is relatively easy, because organic solvents and water have very different polarities and exhibit distinct membrane permeation properties. No commercial pervaporation systems have yet been developed for the separation of organic/organic mixtures. However, current membrane technology makes pervaporation for these applications possible, and the process is being actively developed by a number of companies. The first pilot-plant results for an organic–organic application, the separation of methanol from methyl *tert*-butyl ether/isobutene mixtures, was reported by Separex in 1988 [14,15]. This is a particularly favorable application

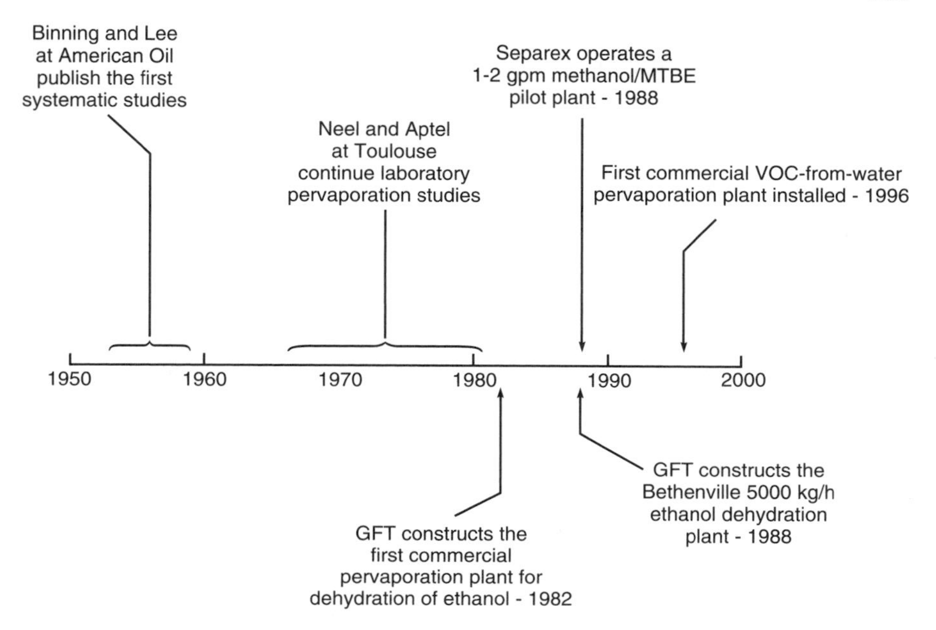

Figure 9.2 Milestones in the development of pervaporation

because available cellulose acetate membranes achieve a good separation. More recently, Exxon, now ExxonMobil, started a pervaporation pilot plant for the separation of aromatic/aliphatic mixtures, a separation problem in refineries; polyimide/polyurethane block copolymer membranes were used [16,17]. A time line illustrating some of the key milestones in the development of pervaporation is shown in Figure 9.2.

Theoretical Background

In Chapter 2 it was shown that the flux of a component i through a pervaporation membrane can be expressed in terms of the partial vapor pressures on either side of the membrane, p_{i_o} and p_{i_ℓ}, by the equation

$$J_i = \frac{P_i^G}{\ell}(p_{i_o} - p_{i_\ell}) \tag{9.1}$$

where J_i is the flux, ℓ is the membrane thickness and P_i^G is the gas separation permeability coefficient. A similar equation can be written for component j. The separation achieved by a pervaporation membrane is proportional to the fluxes J_i and J_j through the membrane.

Equation (9.1) is the preferred method of describing membrane performance because it separates the two contributions to the membrane flux: the membrane contribution, P_i^G/ℓ and the driving force contribution, $(p_{i_o} - p_{i_\ell})$. Normalizing membrane performance to a membrane permeability allows results obtained under different operating conditions to be compared with the effect of the operating condition removed. To calculate the membrane permeabilities using Equation (9.1), it is necessary to know the partial vapor pressure of the components on both sides of the membrane. The partial pressures on the permeate side of the membrane, p_{i_ℓ} and p_{j_ℓ}, are easily obtained from the total permeate pressure and the permeate composition. However, the partial vapor pressures of components i and j in the feed liquid are less accessible. In the past, such data for common, simple mixtures would have to be found in published tables or calculated from an appropriate equation of state. Now, commercial computer process simulation programs calculate partial pressures automatically for even complex mixtures with reasonable reliability. This makes determination of the feed liquid partial pressures a trivial exercise.

Having said this, the bulk of the pervaporation literature continues to report membrane performance in terms of the total flux through the membrane and a separation factor, β_{pervap}, defined for a two-component fluid as the ratio of the two components on the permeate side of the membrane divided by the ratio of the two components on the feed side of the membrane. The term β_{pervap} can be written in several ways.

$$\beta_{\text{pervap}} = \frac{c_{i_\ell}/c_{j_\ell}}{c_{i_o}/c_{j_o}} = \frac{n_{i_\ell}/n_{j_\ell}}{n_{i_o}/n_{j_o}} = \frac{p_{i_\ell}/p_{j_\ell}}{p_{i_o}/p_{j_o}} \tag{9.2}$$

where c_i and c_j are the concentrations, n_i and n_j are the mole fractions, and p_i and p_j are the vapor pressures of the two components i and j.

The separation factor, β_{pervap}, contains contributions from the intrinsic permeation properties of the membrane, the composition and temperature of the feed liquid, and the permeate pressure of the membrane. The contributions of these factors are best understood if the pervaporation process is divided into two steps, as shown in Figure 9.3 [18]. The first step is evaporation of the feed liquid to form a saturated vapor in contact with the membrane; the second step is diffusion of this vapor through the membrane to the low-pressure permeate side. This two-step description is only a conceptual representation; in pervaporation no vapor phase actually contacts the membrane surface. Nonetheless, the representation of the process shown in Figure 9.3 is thermodynamically completely equivalent to the actual pervaporation process shown in Figure 9.1.

In the process illustrated in Figure 9.3, the first step is evaporation from the feed liquid to form a saturated vapor phase in equilibrium with the liquid. This evaporation step produces a separation because of the different volatilities of the components of the feed liquid. The separation can be defined as β_{evap}, the ratio of the component concentrations in the feed vapor to their concentrations in the

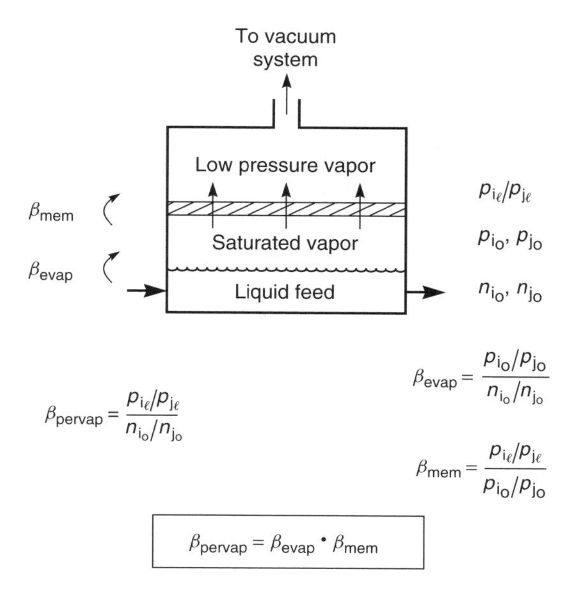

Figure 9.3 The pervaporation process shown in Figure 9.1 can be described by the thermodynamically equivalent process illustrated here. In this model the total pervaporation separation β_{pervap} is made up of an evaporation step followed by a membrane permeation step [18]

feed liquid:

$$\beta_{evap} = \frac{p_{i_o}/p_{j_o}}{n_{i_o}/n_{j_o}} \tag{9.3}$$

The second step in the process is permeation of components i and j through the membrane; this step is equivalent to conventional gas separation. The driving force for permeation is the difference in the vapor pressures of the components in the feed and permeate vapors. The separation achieved in this step, β_{mem}, can be defined as the ratio of the components in the permeate vapor to the ratio of the components in the feed vapor

$$\beta_{mem} = \frac{p_{i_\ell}/p_{j_\ell}}{p_{i_o}/p_{j_o}} \tag{9.4}$$

Equation (9.4) shows that the separation achieved in pervaporation is equal to the product of the separation achieved by evaporation of the liquid and the separation achieved by selective permeation through the membrane.[1]

$$\beta_{pervap} = \beta_{evap} \cdot \beta_{mem} \tag{9.5}$$

[1]Figure 9.3 illustrates the concept of permeation from a saturated vapor phase in equilibrium with the feed liquid as a tool to obtain Equation (9.5). A number of workers have experimentally compared vapor permeation and pervaporation separations and have sometimes shown that permeation from the

The first application of pervaporation was the removal of water from an azeotropic mixture of water and ethanol. By definition, the evaporative separation term β_{evap} for an azeotropic mixture is 1 because, at the azeotropic concentration, the vapor and the liquid phases have the same composition. Thus, the 200- to 500-fold separation achieved by pervaporation membranes in ethanol dehydration is due entirely to the selectivity of the membrane, which is much more permeable to water than to ethanol. This ability to achieve a large separation where distillation fails is why pervaporation is also being considered for the separation of aromatic/aliphatic mixtures in oil refinery applications. The evaporation separation term in these closely boiling mixtures is again close to 1, but a substantial separation is achieved due to the greater permeability of the membrane to the aromatic components.

The β_{pervap} term in Equation (9.2) and Equation (9.5) can be derived in terms of β_{evap}, membrane permeabilities, and membrane operating conditions using the standard solution-diffusion model from Chapter 2. The membrane fluxes can be written as

$$J_i = \frac{P_i^G(p_{i_o} - p_{i_\ell})}{\ell} \tag{9.6}$$

and

$$J_j = \frac{P_j^G(p_{j_o} - p_{j_\ell})}{\ell} \tag{9.7}$$

where J is the permeation flux through the membrane, P^G is the permeability coefficient of the vapors i and j, and ℓ is the thickness of the separating layer of the membrane. Dividing Equation (9.6) by Equation (9.7) gives

$$\frac{J_i}{J_j} = \frac{P_i^G}{P_j^G}\frac{(p_{i_o} - p_{i_\ell})}{(p_{j_o} - p_{j_\ell})} \tag{9.8}$$

The fluxes J_i and J_j in Equation (9.8) are weight fluxes (g/cm$^2 \cdot$ s); similarly the permeabilities P_i^G and P_j^G are weight-based (g $\cdot$ cm/cm$^2 \cdot$ s $\cdot$ cmHg). Equation (9.8) is more conveniently written in molar terms as

$$\frac{j_i}{j_j} = \frac{\mathcal{P}_i^G}{\mathcal{P}_j^G}\frac{(p_{i_o} - p_{i_\ell})}{(p_{j_o} - p_{j_\ell})} \tag{9.9}$$

where j_i and j_j are molar fluxes with unit mols/cm$^2 \cdot$ s or cm^3(STP)/cm$^2 \cdot$ s and $\mathcal{P}_i$ and $\mathcal{P}_j$ are molar permeabilities with units mol $\cdot$ cm/cm$^2 \cdot$ s $\cdot$ unit pressure

---

liquid is faster and less selective than permeation from the equilibrium vapor. This is an experimental artifact. In vapor permeation experiments the vapor in contact with the membrane is not completely saturated. This means that the activities of the feed components in vapor permeation experiments are less than their activity in pervaporation experiments. Because sorption by the membrane in this range is extremely sensitive to activity, the vapor permeation fluxes are lower than pervaporation fluxes. Kataoka *et al.* [19] have illustrated this point in a series of careful experiments.

or more conventionally $cm^3(STP) \cdot cm/cm^2 \cdot s \cdot cmHg$. The ratio of the molar membrane permeability coefficients $\mathcal{P}_i^G/\mathcal{P}_j^G$ is the conventional gas membrane selectivity, α_{mem} [see Equation (8.3)].

The ratio of the molar fluxes is also the same as the ratio of the permeate partial pressures

$$\frac{j_i}{j_j} = \frac{p_{i_\ell}}{p_{j_\ell}} \tag{9.10}$$

Combining Equations (9.4), (9.5), (9.9) and (9.10) yields

$$\beta_{pervap} = \frac{\beta_{evap}\alpha_{mem}(p_{i_o} - p_{i_\ell})}{(p_{j_o} - p_{j_\ell})(p_{i_o}/p_{j_o})} \tag{9.11}$$

Equation (9.11) identifies the three factors that determine the performance of a pervaporation system. The first factor, β_{evap}, is the vapor–liquid equilibrium, determined mainly by the feed liquid composition and temperature; the second is the membrane selectivity, α_{mem}, an intrinsic permeability property of the membrane material; and the third includes the feed and permeate vapor pressures, reflecting the effect of operating parameters on membrane performance. This equation is, in fact, the pervaporation equivalent of Equation (8.19) that describes gas separation in Chapter 8.

As in gas separation, the separation achieved by pervaporation is determined both by the membrane selectivity and by the membrane pressure ratio. The interaction of these two factors is expressed in Equation (9.11). Also, as in gas separation, there are two limiting cases in which one of the two factors dominates the separation achieved. The first limiting case is when the membrane selectivity is very large compared to the vapor pressure ratio between the feed liquid and the permeate vapor:

$$\alpha_{mem} \gg \frac{p_o}{p_\ell} \tag{9.12}$$

This means that for a membrane with infinite selectivity for component i, the permeate vapor pressure of component i will equal the feed partial vapor pressure of i. That is,

$$p_{i_\ell} = p_{i_o} \tag{9.13}$$

Equation (9.13) combined with Equation (9.4) gives

$$\beta_{mem} = \frac{p_{j_o}}{p_{j_\ell}} \tag{9.14}$$

which, combined with Equation (9.5), leads to the limiting case

$$\beta_{pervap} = \beta_{evap} \cdot \frac{p_{j_o}}{p_{j_\ell}} \text{ when } \alpha_{mem} \gg \frac{p_o}{p_\ell} \tag{9.15}$$

Similarly, in the case of a very large membrane selectivity in favor of component j

$$\beta_{\text{pervap}} = \beta_{\text{evap}} \frac{p_{i_o}}{p_{i_\ell}} \tag{9.16}$$

For the special case in which component i is the minor component in the feed liquid, p_{j_o} approaches p_o, p_{j_ℓ} approaches p_ℓ, and Equation (9.15) reverts to

$$\beta_{\text{pervap}} = \beta_{\text{evap}} \frac{p_o}{p_\ell} \tag{9.17}$$

where p_o/p_ℓ is the feed-to-permeate ratio of the total vapor pressures.

The second limiting case occurs when the vapor pressure ratio is very large compared to the membrane selectivity. This means that the permeate partial pressure is smaller than the feed partial vapor pressures, and p_{i_ℓ} and $p_{j_\ell} \to 0$. Equation (9.11) then becomes

$$\beta_{\text{pervap}} = \beta_{\text{evap}}\ \alpha_{\text{mem}} \text{ when } \alpha_{\text{mem}} \ll \frac{p_o}{p_\ell} \tag{9.18}$$

The relationship between the three separation factors, β_{pervap}, β_{evap} and β_{mem}, is illustrated in Figure 9.4. This type of plot was introduced by Shelden and Thompson [20] to illustrate the effect of permeate pressure on pervaporation separation

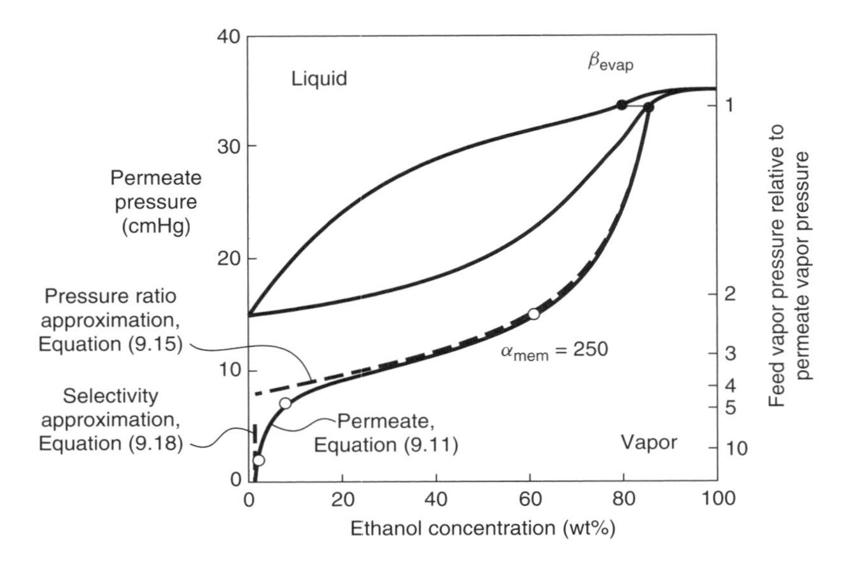

Figure 9.4 The effect of permeate pressure on the separation of ethanol/water mixtures with a poly(vinyl alcohol) membrane. The feed solution contains 20 wt% water and 80 wt% ethanol. The line drawn through the experimental data points is calculated from Equation (9.11)

and is a convenient method to represent the pervaporation process graphically. When the permeate pressure, $p_\ell = p_{i_\ell} + p_{j_\ell}$, approaches the feed vapor pressure, $p_o = p_{i_o} + p_{j_o}$, the vapor pressure ratio across the membrane shown on the right-hand axis of the figure approaches unity. The composition of the permeate vapor then approaches the composition obtained by simple evaporation of the feed liquid, shown by the point at which β_{pervap} equals β_{evap}. As the permeate pressure is decreased to below the feed vapor pressure, the vapor pressure ratio increases. The overall separation obtained, β_{pervap}, is then the product of the separation due to evaporation of the feed liquid, β_{evap}, and the separation due to permeation through the membrane, β_{mem}. The line labeled 'permeate' in Figure 9.4 can be calculated from Equation (9.11). Two limiting cases are also shown on the figure. The first limiting case, when the membrane selectivity α_{mem} is much larger than the pressure ratio p_o/p_ℓ, is calculated from Equation (9.15). The second limiting case, for the region in which the membrane selectivity is much smaller than the pressure ratio, is calculated from Equation (9.18). Figure 9.4 is the pervaporation equivalent of Figure 8.13 for gas separation discussed in Chapter 8.

The example shown in Figure 9.4 includes experimental data from GFT, which show the effect of permeate pressure on the pervaporation separation of an 80 wt% aqueous ethanol solution using a highly selective poly(vinyl alcohol) membrane. The line obtained from Equation (9.11) passes through all the data points when the water/ethanol membrane selectivity (α_{mem}) is assumed to be 250. Although this membrane is very selective, a good separation between the feed solution and the permeate vapor is only achieved at low permeate pressures when a high vapor pressure ratio exists across the membrane and the high intrinsic selectivity of the GFT membrane is utilized. For this reason, in practical applications of this membrane the feed solution is heated to about 120 °C (raising the feed vapor pressure to 2–4 atm), and the permeate vapor is condensed at −10 to −20 °C (to lower the vapor pressure of the permeate to about 1 cmHg). This combination achieves the vapor pressure ratio of more than 200 required for a good separation.

Membrane Materials and Modules

Membrane Materials

The selectivity (α_{mem}) of pervaporation membranes critically affects the overall separation obtained and depends on the membrane material. Therefore, membrane materials are tailored for particular separation problems. As with other solution-diffusion membranes, the permeability of a component is the product of the membrane sorption coefficient and the diffusion coefficient (mobility). The membrane selectivity term α_{mem} in Equation (9.11) can be written as

$$\alpha_{\text{mem}} = \frac{\mathcal{P}_i^G}{\mathcal{P}_j^G} = \left(\frac{D_i}{D_j}\right)\left(\frac{K_i}{K_j}\right) \tag{9.20}$$

(see Equation (8.3) in Chapter 8). This expression shows that membrane selectivity is the product of the mobility selectivity (D_i/D_j) of the membrane material, generally governed by the relative mobility of the permeants, and the solubility selectivity (K_i/K_j), generally governed by the chemistry of the membrane material. In gas permeation the total sorption of gases by the membrane material is usually low, often less than 1 wt%, so the membrane selectivity measured with gas mixtures is often close to the selectivity calculated from the ratio of the pure gas permeabilities. In pervaporation the membrane is in contact with the feed liquid, and typical sorptions are 2–20 wt%. Sorption of one of the components of the feed can then change the sorption and diffusion of the second component. As a rule of thumb, the total sorption of the feed liquid by the membrane material should be in the range 3–15 wt%. Below 3 wt% sorption, the membrane selectivity may be good, but the flux through the material will be too low. Above 15 wt% sorption, fluxes will be high, but the membrane selectivity will generally be low because the mobility selectivity will decrease as the material becomes more swollen and plasticized. The sorption selectivity will also tend towards unity.

By manipulating the chemistry of membrane materials, either sorption- or diffusion-selectivity-controlled membranes can be made. The range of results that can be obtained with different membranes with the same liquid mixture is illustrated in Figure 9.5 for the separation of acetone from water [21]. The figure shows the concentration of acetone in the permeate as a function of the concentration in the feed. The two membranes shown have dramatically different properties. The silicone rubber membrane, made from a hydrophobic rubbery material, preferentially sorbs acetone, the more hydrophobic organic compound. For rubbery materials the diffusion selectivity term, which would favor permeation of the smaller component (water), is small. Therefore, the silicone rubber membrane is sorption-selectivity-controlled and preferentially permeates acetone. In contrast, the poly(vinyl alcohol) membrane is made from a hydrophilic, rigid, crosslinked material. Because poly(vinyl alcohol) is hydrophilic, the sorption selectivity favors permeation of water, the more hydrophilic polar component. Also, because poly(vinyl alcohol) is glassy and crosslinked, the diffusion selectivity favoring the smaller water molecules over the larger acetone molecules is substantial [22]. As a result, poly(vinyl alcohol) membranes permeate water several hundred times faster than acetone.

In any membrane process, it is desirable for the minor components to permeate the membrane, so the acetone-selective silicone rubber membrane is best used to treat dilute acetone feed streams, concentrating most of the acetone in a small volume of permeate. The water-selective poly(vinyl alcohol) membrane is best used to treat concentrated acetone feed streams, concentrating most of the water in a small volume of permeate. Both membranes are more selective than distillation, which relies on the vapor–liquid equilibrium to achieve separation.

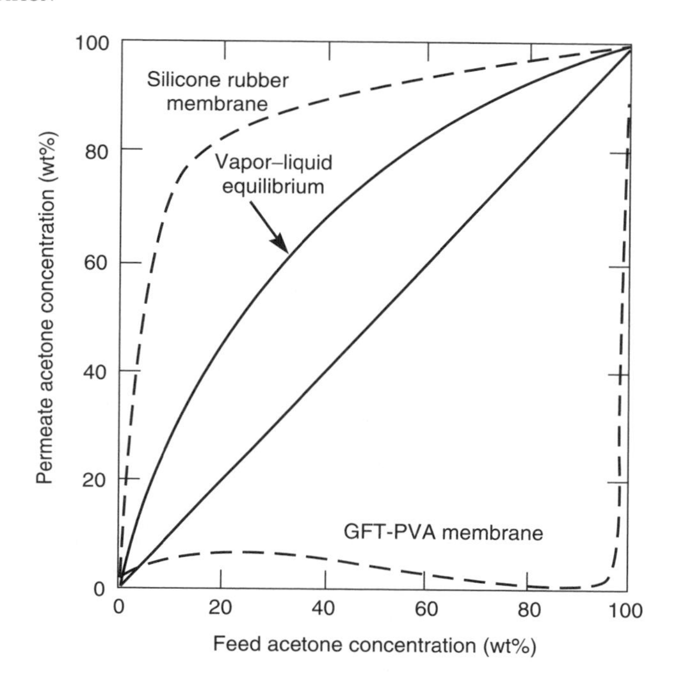

Figure 9.5 Pervaporation separation of acetone–water mixtures achieved with a water-selective membrane poly(vinyl alcohol) (PVA), and an acetone-selective membrane (silicone rubber) [21]. Reprinted from Hollein *et al.* [21], p. 1051 by courtesy of Marcel Dekker, Inc.

Most pervaporation membranes are composites formed by solution-coating the selective layer onto a microporous support. Some of the more commonly used membranes are listed in Table 9.1. For dehydration of organic solvents, such as ethanol, isopropanol and acetone, several excellent membrane materials are available. The technical and economic feasibility of these processes is controlled by membrane module and system engineering issues rather than by membrane flux and selectivity. Chemically crosslinked poly(vinyl alcohol), formed as a composite membrane by solution casting onto a polyacrylonitrile microporous support, was developed by GFT and has been used for a long time. The poly(vinyl alcohol) layer is crosslinked by heat or addition of crosslinking agents such as gluteraldehyde. This membrane has a water/alcohol selectivity (α_{mem}) of more than 200 [18] and can achieve extremely good separation of water from ethanol or isopropanol solutions. However, the membrane is swollen and even dissolved by hot acid or base solutions such as hot acetic acid or hot aniline. Membranes stable to such feed solutions can be prepared by plasma polymerization [23].

For separating VOCs from water, silicone rubber composite membranes are the state-of-the-art material. Silicone rubber is easy to fabricate, is mechanically and

Table 9.1 Widely used pervaporation membrane materials

Dehydration of organics	
Water/ethanol Water/isopropanol Water/glycol, etc.	Microporous polyacrylonitrile coated with a 5–20 μm layer of crosslinked poly(vinyl alcohol) is the most commonly used commercial material [10]. Chitosan [24] and polyelectrolyte membranes such as Nafion [25,26] have equivalent properties
VOC/water separation	
Toluene/water Trichloroethylene/water Methylene chloride/water	Membranes comprising silicone rubber coated onto polyimides, polyacrylonitrile or other microporous supports membranes are widely used [12,27]. Other rubbers such as ethylene-propylene terpolymers have been reported to have good properties also [28]. Polyamide-polyether block copolymers have also been used for pervaporation of some polar VOCs [29,30]
Organic/organic separation	
	The membrane used depends on the nature of the organics. Poly(vinyl alcohol) and cellulose acetate [14] have been used to separate alcohols from ethers. Polyurethane-polyimide block copolymers have been used for aromatic/aliphatic separations [17]

Table 9.2 Typical silicone rubber membrane module pervaporation separation factors (VOC removal from water)

Separation factor for VOC over water	Volatile organic compound (VOC)
200–1000	Benzene, toluene, ethyl benzene, xylenes, TCE, chloroform, vinyl chloride, ethylene dichloride, methylene chloride, perchlorofluorocarbons, hexane
20–200	Ethyl acetate, propanols, butanols, MEK, aniline, amyl alcohol
5–20	Methanol, ethanol, phenol, acetaldehyde
1–5	Acetic acid, ethylene glycol, DMF, DMAC

TCE, trichloroethylene; MEK, methyl ethyl ketone; DMF, dimethyl formamide; DMAC, dimethyl acetamide.

chemically strong, and has good separation factors for many common organic compounds, as shown in Table 9.2. These representative data were obtained with industrial-scale modules under normal operating conditions. The performance of silicone membranes in laboratory test cells operated under ideal conditions is usually better.

A number of academic studies have produced rubbery hydrophobic membrane materials with far higher selectivities than silicone rubber [27]. For example,

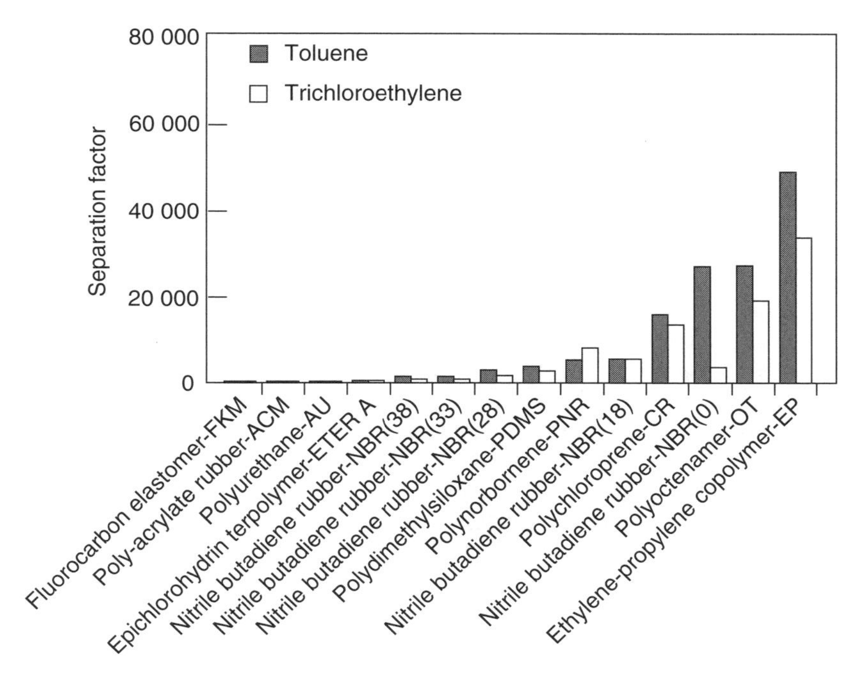

Figure 9.6 Comparative separation factors for toluene and trichloroethylene from water with various rubbery membranes [28]. These experiments were performed with thick films in laboratory test cells. In practice, separation factors obtained with membrane modules are far less because of concentration polarization effects. Reprinted from Nijhuis *et al.* [28], p. 248 with permission of Bakish Materials Corporation, Englewood, NJ

Figure 9.6 shows the separation factors measured by Nijhuis *et al.* [28] for various membranes with dilute toluene and trichloroethylene solutions. The separation factor of silicone rubber is in the 4000–5000 range, but other materials have separation factors as high as 40 000. However, in practice, an increase in membrane separation factor beyond about 1000 provides very little additional benefit. Once a separation factor of this magnitude is obtained, other factors, such as ease of manufacture, mechanical strength, chemical stability, and control of concentration polarization become more important. This is why silicone rubber remains prevalent, even though polymers with higher selectivities are known.

Membranes with improved separation factors would be useful for hydrophilic VOCs such as ethanol, methanol and phenol, for which the separation factor of silicone rubber is in the range 5–10. As yet, no good replacement for silicone rubber has been developed. The most promising results to date have been obtained with silicone rubber membranes containing dispersed zeolite particles [31]. Apparently, ethanol preferentially permeates the pores of the zeolite particles; membranes have been produced in the laboratory with ethanol/water selectivities of 40 or more. Membranes with these properties could be applied in fermentation processes and solvent recovery if they can be made on a large scale.

Polyamide-polyether block copolymers (Pebax®, Elf Atochem, Inc., Philadelphia, PA) have been used successfully with polar organics such as phenol and aniline [32–34]. The separation factors obtained with these organics are greater than 100, far higher than the separation factors obtained with silicone rubber. The improved selectivity reflects the greater sorption selectivity obtained with the polar organic in the relatively polar polyamide-polyether membrane. On the other hand, toluene separation factors obtained with polyamide-polyether membranes are below those measured with silicone rubber.

For the separation of organic/organic mixtures, current membranes are only moderately selective, generally because the differences in sorption between different organic molecules are small, and many membrane materials swell excessively in organic solvent mixtures, especially at high temperatures. One approach is to use rigid backbone polymers to control swelling, for example, the Matrimid® polyimides developed by Grace [35]. However, the permeability of these materials is often very low. Another approach, used by Exxon [17], is to use block copolymers consisting of rigid polyimide segments that provide a strong network and softer segments formed from more flexible polymers through which permeant transport occurs. The permeation properties of the polymer were varied by tailoring the size and chemistry of the two blocks. Notwithstanding this work, development of more selective membranes is required for application of pervaporation to other important organic/organic separations, such as separation of aromatics from aliphatics, olefins from paraffins, and branched hydrocarbons.

Membrane Modules

Pervaporation applications often involve hot feed solutions containing organic solvents. Such solutions can degrade the seals and plastic components of membrane modules. As a result, the first-generation commercial pervaporation modules used a stainless steel plate-and-frame design. Recently attempts have been made to switch to lower-cost module designs. Texaco, Membrane Technology and Research (MTR) and Separex (UOP) have all used spiral-wound modules for pervaporation, and Zenon developed hollow fiber modules. One particular issue affecting pervaporation module design is that the permeate side of the membrane often operates at a vacuum of less than 100 torr. The pressure drop required to draw the permeate vapor to the permeate condenser may then be a significant fraction of the permeate pressure. Efficient pervaporation modules must have short, porous permeate channels to minimize this permeate pressure drop.

Process Design

Transport through pervaporation membranes is produced by maintaining a vapor pressure gradient across the membrane. As in gas separation, the flux through the membrane is proportional to the vapor pressure difference [Equation (9.1)], but the separation obtained is determined by the membrane selectivity and the pressure ratio [Equation (9.11)]. Figure 9.7 illustrates a number of ways to achieve the required vapor pressure gradient.

In the laboratory, the low vapor pressure required on the permeate side of the membrane is often produced with a vacuum pump, as shown in Figure 9.7(a). In a commercial-scale system, however, the vacuum pump requirement would be impossibly large. In the early days of pervaporation research, the calculated vacuum pump size was sometimes used as proof that pervaporation would never be commercially viable. An attractive alternative to a vacuum pump, illustrated in Figure 9.7(b), is to cool the permeate vapor to condense the liquid; condensation of the liquid spontaneously generates the permeate side vacuum. The feed solution may also be heated to increase the vapor pressure driving force. In this process, sometimes called thermo-pervaporation, the driving force is the difference in vapor pressure between the hot feed solution and the cold permeate liquid at the temperature of the condenser. This type of design is preferred for commercial operations, because the cost of providing the required cooling and heating is much less than the cost of a vacuum pump, and the process is operationally more reliable.

A third possibility, illustrated in Figure 9.7(c), is to sweep the permeate side of the membrane with a counter-current flow of carrier gas. In the example shown, the carrier gas is cooled to condense and recover the permeate vapor, and the gas is recirculated. This mode of operation has little to offer compared to temperature-gradient-driven pervaporation, because both require cooling water for the condenser. However, if the permeate has no value and can be discarded without condensation (for example, in the pervaporative dehydration of an organic solvent with an extremely water-selective membrane), this is the preferred mode of operation. In this case, the permeate would contain only water plus a trace of organic solvent and could be discharged or incinerated at low cost. No permeate refrigeration is required [36].

An alternative carrier-gas system uses a condensable gas, such as steam, as the carrier sweep fluid. One variant of this system is illustrated in Figure 9.7(d). Low-grade steam is often available at low cost, and, if the permeate is immiscible with the condensed carrier, water, it can be recovered by decantation. The condensed water will contain some dissolved organic and can be recycled to the evaporator and then to the permeate side of the module. This operating mode is limited to water-immiscible permeates and to feed streams for which contamination of the feed liquid by water vapor permeating from the sweep gas is not a problem. This idea has been discovered, rediscovered, and patented a number of times, but never used commercially [37,38]. If the permeate is soluble in the condensable

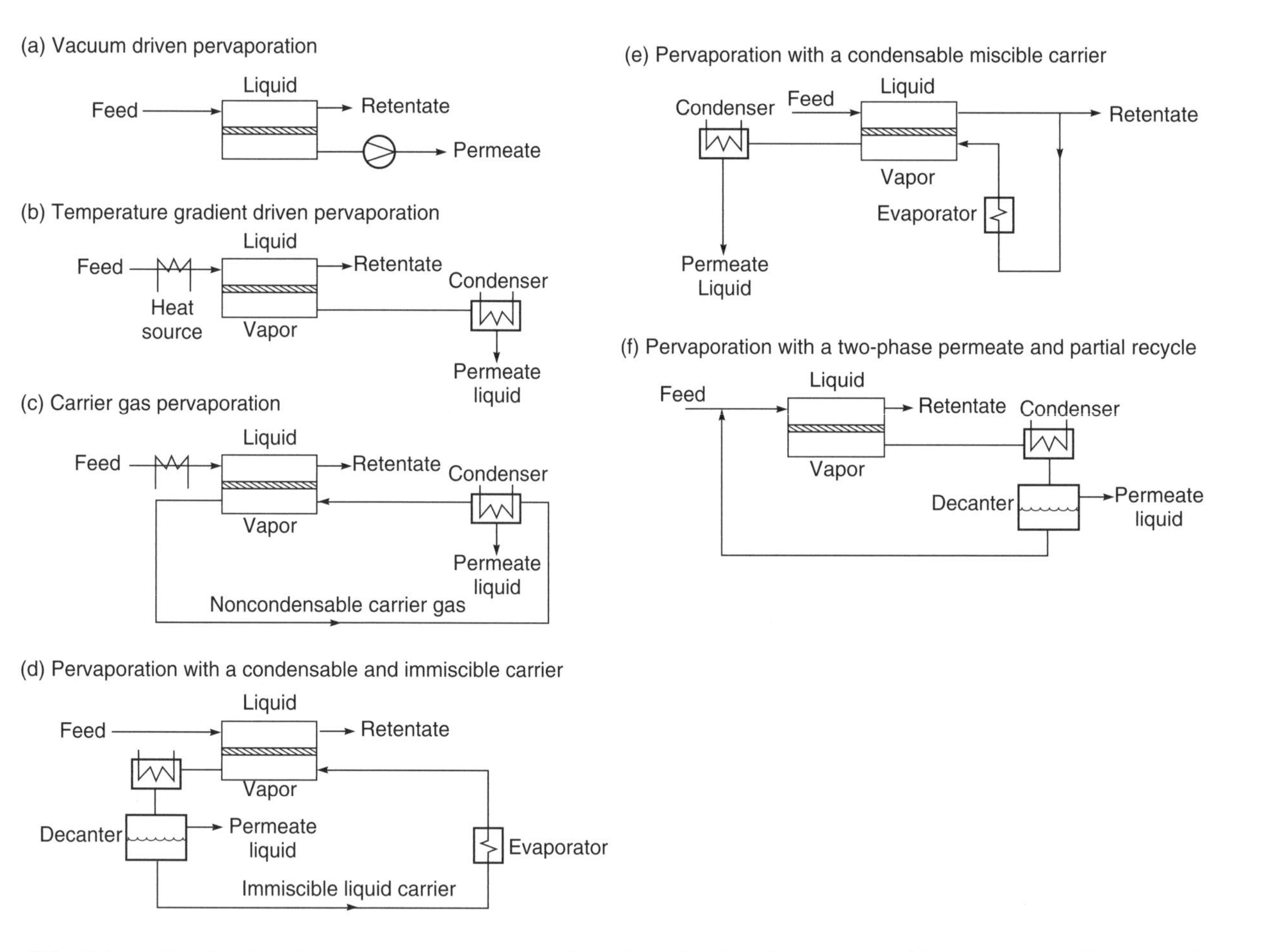

Figure 9.7 Schematics of potential pervaporation process configurations that have been suggested but not necessarily practiced

sweep generated, then the sweep gas is best obtained by evaporating a portion of the residue liquid as shown in Figure 9.7(e). The final pervaporation process, illustrated in Figure 9.7(f), is a system of particular interest for removing low concentrations of dissolved VOCs from water. The arrangement shown is used when the solubility of the permeating solvent in water is limited. In this case, the condensed permeate liquid separates into two phases: an organic phase, which can be treated for reuse, and an aqueous phase saturated with organic, which can be recycled to the feed stream for reprocessing.

In the process designs shown in Figures 9.1 and 9.7, the permeate vapor is condensed to yield a single liquid permeate condensate. A simple improvement

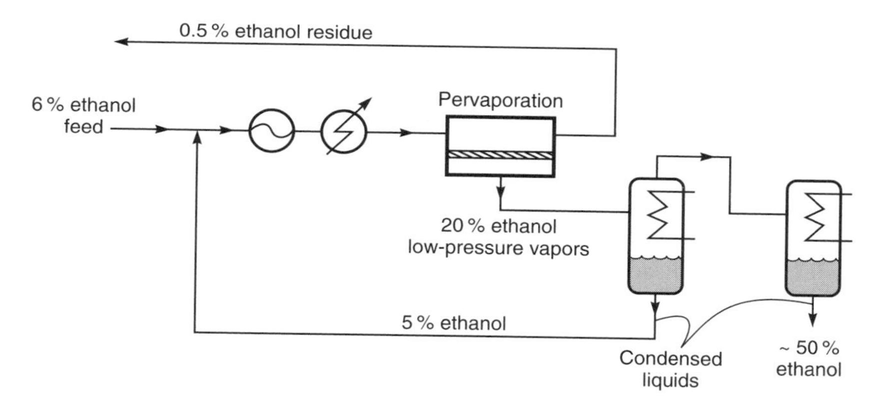

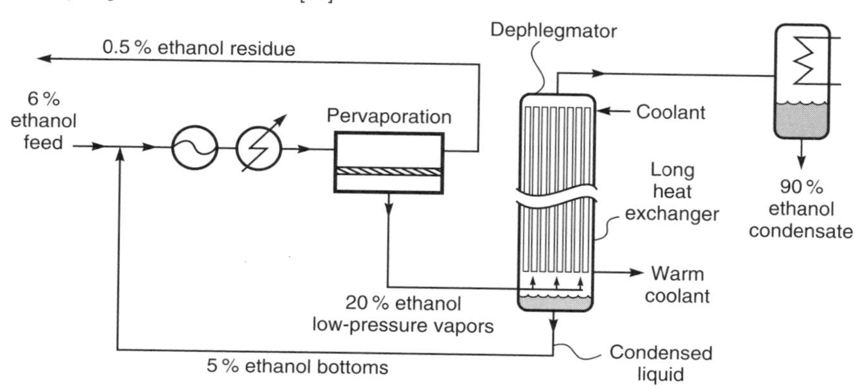

Figure 9.8 The use of permeate vapor fractional condensation systems to improve the separation achieved in pervaporation of dilute ethanol solutions: (a) Two-stage fractional condensation [40] and (b) Dephlegmator condensation [39]

to the pervaporation process is to use fractional condensation of the permeate vapor to achieve an improved separation. Two process designs are shown in Figure 9.8. In Figure 9.8(a), the permeate vapor is condensed in two condensers in series. In the example shown, the recovery of ethanol from water, the first (higher temperature) condenser produces a first condensate containing about 5 % ethanol that is recycled to the incoming feed. The second (lower temperature) condenser condenses the remaining vapor to produce an ethanol product stream containing about 50 % ethanol. This design has not been widely used because the increase in product quality usually does not compensate for the increased complexity of the process.

The condensation system shown in Figure 9.8(b) uses a dephlegmator to achieve the separation required [39]. A dephlegmator in its simplest form is a vertical heat exchanger. Warm, low-pressure permeate vapor from the pervaporation unit enters the dephlegmator at the bottom. As the vapor rises up the column, some condenses on the cold tube wall. The resultant liquid flows downward within the feed passage countercurrent to the rising feed vapor. Mass transfer between the liquid and vapor enriches the liquid in the less volatile components as the more volatile components are revaporized. As a result, several theoretical stages of separation are achieved. The degree of separation achieved can be impressive. In the example shown, the 20 wt% ethanol permeate vapor is separated into 5 wt% bottoms, which is recycled to the pervaporation unit, and a 90–95 wt% overhead ethanol product stream.

Applications

The three current applications of pervaporation are dehydration of solvents, water purification, and organic/organic separations as an alternative to distillation. Currently dehydration of solvents, in particular ethanol and isopropanol, is the only process installed on a large scale. However, as the technology develops, the other applications are expected to grow. Separation of organic mixtures, in particular, could become a major application. Each of these applications is described separately below.

Solvent Dehydration

Several hundred plants have been installed for the dehydration of ethanol by pervaporation. This is a particularly favorable application for pervaporation because ethanol forms an azeotrope with water at 95 % and a 99.5 % pure product is needed. Because the azeotrope forms at 95 % ethanol, simple distillation does not work. A comparison of the separation of ethanol and water obtained by various pervaporation membranes and the vapor–liquid equilibrium line that controls separation obtained by distillation is shown in Figure 9.9 [40]. The membranes

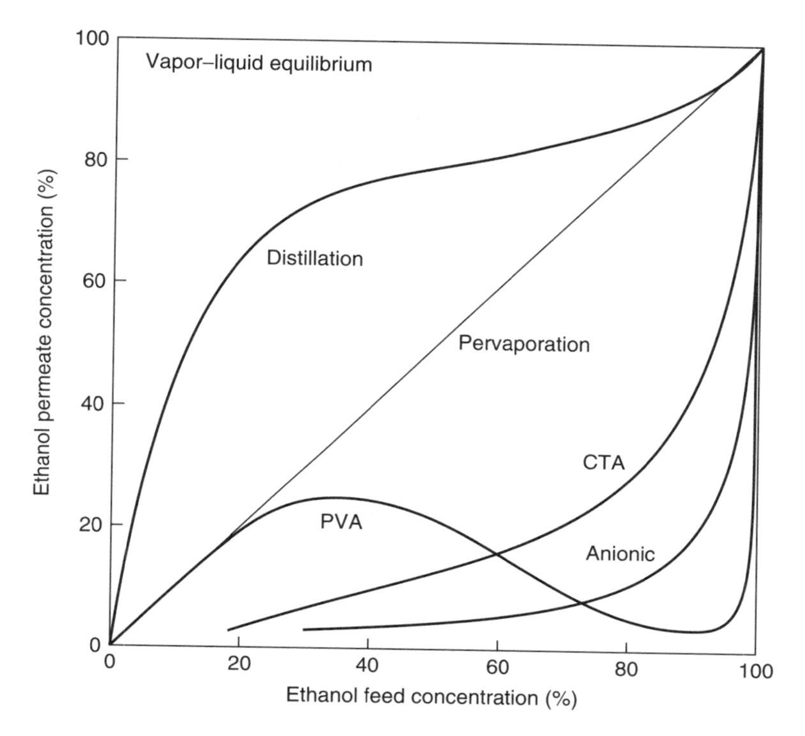

Figure 9.9 Comparison of separation of ethanol/water mixtures by distillation and by three pervaporation membranes: cellulose triacetate (CTA), an anionic polyelectrolyte membrane, and GFT's poly(vinyl alcohol) (PVA) membrane [40]

all achieve a good separation, but the GFT poly(vinyl alcohol) membrane performance is the best. Most pervaporation dehydration systems installed to date have been equipped with this membrane, although Mitsui is producing zeolite tubular modules [41,42]. Because an ethanol/water azeotrope forms at 95 % ethanol, the concentration of ethanol from fermentation feeds to high degrees of purity requires rectification with a benzene entrainer, some sort of molecular-sieve drying process, or a liquid–liquid extraction process. All of these processes are expensive. However, the availability of extremely water-selective pervaporation membranes allows pervaporation systems to produce almost pure ethanol (>99.9 % ethanol from a 90 % ethanol feed). The permeate stream contains approximately 50 % ethanol and can be returned to the distillation column.

A flow scheme for an integrated distillation–pervaporation plant operating on a 5 % ethanol feed from a fermentation mash is shown in Figure 9.10. The distillation column produces an ethanol stream containing 80–90 % ethanol, which is fed to the pervaporation system. To maximize the vapor pressure difference and the pressure ratio across the membrane, the pervaporation module usually

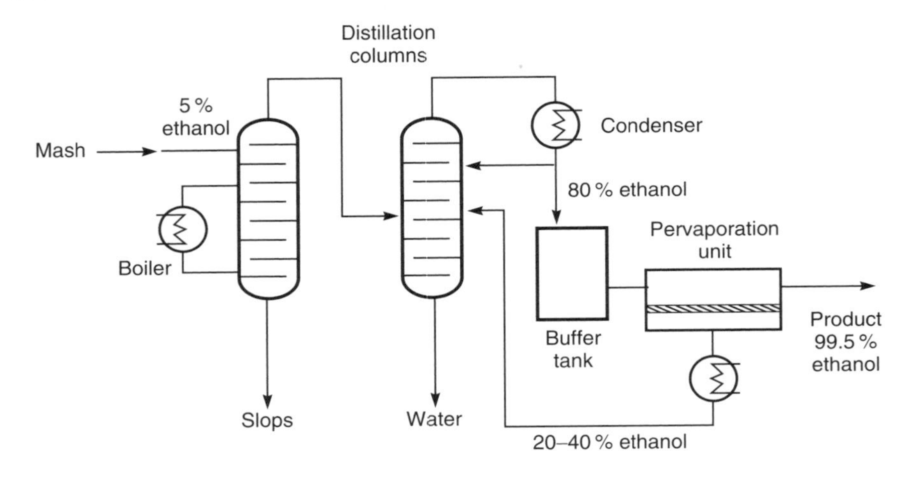

Figure 9.10 Integrated distillation–pervaporation plant for ethanol recovery from fermentors

operates in the temperature range 105–130 °C with a corresponding feed stream vapor pressure of 2–6 atm. Despite these harsh conditions, the membrane lifetime is good and manufacturers give qualified guarantees of up to 4 years.

Figure 9.10 shows a single-stage pervaporation unit. In practice, three to five pervaporation units are usually used in series, with additional heat supplied to the ethanol feed between each stage. This compensates for pervaporative cooling of the feed and maintains the feed temperature. The heat required is obtained by thermally integrating the pervaporation system with the condenser of the final distillation column. Therefore, most of the energy used in the process is low-grade heat. Generally, about 0.5 kg of steam is required for each kilogram of ethanol produced. The energy consumption of the pervaporation process is, therefore, about 500 Btu/L of product, less than 20 % of the energy used in azeotropic distillation, which is typically about 3000 Btu/L.

Reliable capital and operating cost comparisons between pervaporation and distillation are not available. Pervaporation is less capital and energy intensive than distillation or adsorption processes for small plants treating less than 5000 L/h of feed solution. However, because of the modular nature of the process, the costs of pervaporation are not as sensitive to economies of scale as are the costs of distillation and adsorption processes. Distillation costs, on the other hand, scale at a rate proportional to 0.6–0.7 times the power consumption. Thus, distillation remains the most economical process for large plants. The cross-over point at which distillation becomes preferable to pervaporation from an energy and economic point of view currently appears to be 5000 L/h processing capacity. Bergdorf has made an analysis of the comparative costs of pervaporation, distillation and other processes [43].

Because most of the installed pervaporation alcohol dehydration plants are relatively small, in the 500–5000 L/h range, the membrane module cost is generally only 15–40 % of the total plant cost [44] even when relatively high-cost stainless steel plate-and-frame modules of the type originally developed by GFT are used. Cost savings could undoubtedly be achieved by using more economical spiral-wound or capillary fiber modules, but Sulzer (GFT) apparently does not regard these savings sufficient to cover the significant development costs involved in producing such modules able to operate at 100 °C with hot ethanol solutions. Photographs of the Sulzer (GFT) plate-and-frame module and of an ethanol dehydration system are shown in Figure 9.11 [44].

There is an increasing trend to replace liquid pervaporation with vapor permeation in some dehydration applications, particularly dehydration of ethanol and isopropanol. The main disadvantage of vapor permeation is that energy is used to evaporate the liquid, only a portion of which can be recovered when the vapor streams are ultimately recovered. On the other hand, by evaporating the liquid feed, any dissolved salts and solid contaminants are left in the evaporator, giving a purer product, important in the recovery of isopropanol, for example, in the electronics industry. More importantly, pervaporation requires the feed liquid to be repeatedly reheated to supply the latent heat of evaporation removed by the permeating vapor. The need for interstage reheating complicates the system design and leads to lower average fluxes, as the example calculations of Sander [45] show in Figure 9.12. In liquid pervaporation, the feed stream must be reheated five times as the water concentration drops from 6 to 1 % and the average temperature of the fluid is at about 95 °C. The vapor feed stream, however, requires no reheating and remains at about the initial feed temperature. High

Figure 9.11 Photograph of a 50-m^2 GFT plate-and-frame module and an ethanol dehydration system fitted with this type of module. The module is contained in the large vacuum chamber on the left-hand side of the pervaporation system [44]

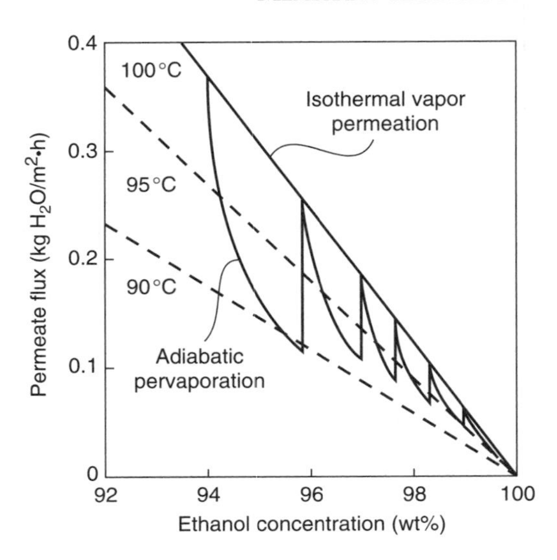

Figure 9.12 Isothermal vapor permeation and multistage pervaporation with intermediate heating. GFT poly(vinyl alcohol) membranes [45]

feed temperatures are needed to produce a high vapor pressure driving force to improve membrane separation performance and flux. The improvement in membrane flux achieved by increasing the average feed temperature as little as 5 or 10 °C is significant.

Most of the early solvent dehydration systems were installed for ethanol dehydration. More recently pervaporation has been applied to dehydration of other solvents, particularly isopropanol used as a cleaning solvent. Dehydration of other solvents, including glycols, acetone and methylene chloride, has been considered. Schematics of pervaporation processes for these separations are shown in Figure 9.13.

Dewatering of glycol is a difficult separation by distillation alone, so a hybrid process of the type shown in Figure 9.13(a) has been proposed. The product of the distillation step is approximately 90 % glycol/10 % water. This mixture is then sent to a pervaporation unit to remove most of the water as a dischargeable product. The glycol concentrate produced by the pervaporation unit contains 1–2 wt% water and can be sent to an optional adsorption dryer if further dehydration is required.

Figure 9.13(b) shows the use of pervaporation to dry a chlorinated solvent, in this case water-saturated ethylene dichloride containing 2000 ppm water. A poly(vinyl alcohol) dehydration membrane can easily produce a residue containing less than 10 ppm water and a permeate containing about 50 wt% water. On condensation the permeate vapor separates into two phases, a very small water

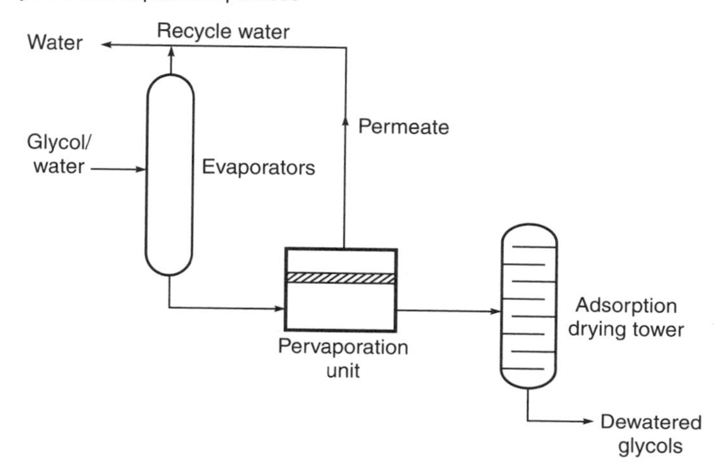

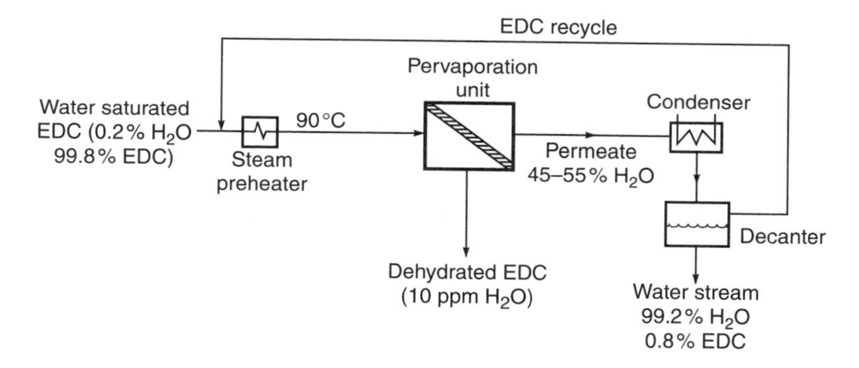

Figure 9.13 Other solvent dehydration processes under investigation

phase that can be discharged and an ethylene dichloride phase that can be recycled to the incoming feed.

A final interesting application of dehydration membranes is to shift the equilibrium of chemical reactions. For example, esterification reactions of the type

$$\text{acid} + \text{alcohol} \rightleftharpoons \text{ester} + \text{water}$$

are usually performed in batch reactors, and the degree of conversion is limited by buildup of water in the reactor. By continuously removing the water, the equilibrium reaction can be forced to the right. In principle, almost complete conversion can be achieved. This process was first suggested by Jennings and

Binning in the 1960s [46]. A number of groups have since studied this type of process, and at least one commercial plant has been installed [47–49].

Separation of Dissolved Organics from Water

A number of applications exist for pervaporation to remove or recover VOCs from water. If the aqueous stream is very dilute, pollution control is the principal economic driving force. However, if the stream contains more than 1–2 % VOC, recovery for eventual reuse can enhance the process economics.

Several types of membrane have been used to separate VOCs from water and are discussed in the literature [11,28]. Usually the membranes are made from rubbery polymers such as silicone rubber, polybutadiene, natural rubber, and polyamide-polyether copolymers. Rubbery pervaporation membranes are remarkably effective at separating hydrophobic organic solutes from dilute aqueous solutions. The concentration of VOCs such as toluene or trichloroethylene (TCE) in the condensed permeate is typically more than 1000 times that in the feed solution. For example, a feed solution containing 100 ppm of such VOCs yields a permeate vapor containing 10–20 % VOC. This concentration is well above the saturation limit, so condensation produces a two-phase permeate. This permeate comprises an essentially pure condensed organic phase and an aqueous phase containing a small amount of VOC that can be recycled to the aqueous feed. The flow scheme for this process is shown in Figure 9.7(f). The separations achieved with moderately hydrophobic VOCs, such as ethyl acetate, methylene chloride and butanol, are still impressive, typically providing at least 100-fold enrichment in the permeate. However, the separation factors obtained with hydrophilic solvents, such as methanol, acetic acid and ethylene glycol, are usually modest, at 5 or below [8].

Some data showing measured pervaporation separation factors for dilute aqueous VOC solutions are shown in Figure 9.14, in which the total separation factor, β_{pervap}, is plotted against the theoretical evaporative separation factor, β_{evap}, obtained from the equation of state. Two sets of data, both obtained with silicone rubber membranes, are shown. One set was obtained with thick membranes in laboratory test cells under very well stirred conditions [33] that largely eliminate concentration polarization. The other set was obtained with high-flux membranes in spiral-wound modules [12]. The difference between the curves is due to the concentration polarization effects discussed in Chapter 4. With VOCs such as acetone, methyl ethyl ketone (MEK) and ethyl acetate, the difference between separation factors measured in the laboratory test cells and in spiral-wound modules is relatively small. The difference becomes very large for more hydrophobic VOCs with high separation factors. Concentration polarization effects reduce the separation factor for VOCs such as toluene or TCE 5- to 10-fold.

The data in Figure 9.14 also allow determination of the relative contributions of the evaporative separation term β_{evap} and the membrane selectivity term β_{mem}

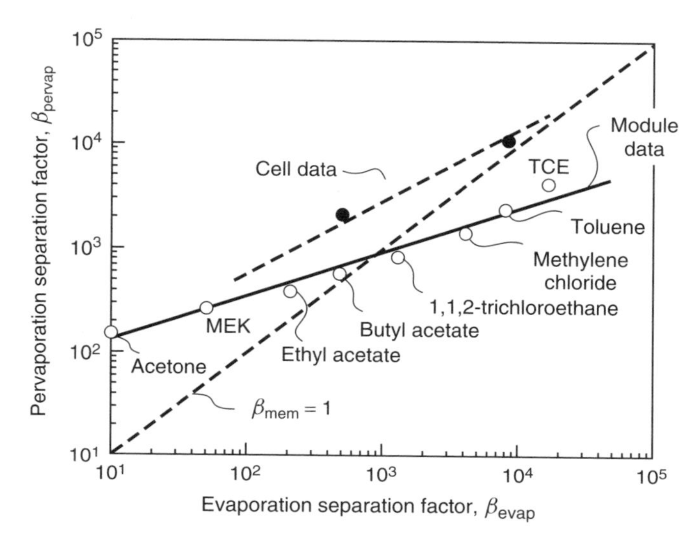

Figure 9.14 Pervaporation separation factor, β_{pervap}, as a function of the VOC evaporation separation factor, β_{evap}. Data obtained with laboratory-scale spiral-wound modules containing a composite silicone rubber membrane and in laboratory cells with thick membranes

to the total separation achieved by pervaporation β_{pervap} [Equation (9.5)]. Earlier it was shown that membranes used to dehydrate ethanol achieved almost all of the total pervaporation separation as a result of a high membrane selectivity term, in the 100–500 range. With these membranes the evaporative separation term was usually close to 1. In the case of the separation of VOCs from water, the relative contribution of evaporation and membrane permeation to the separation is quite different. For example, MEK has a pervaporation separation factor of approximately 280. In this case, the evaporation contribution β_{evap} is 40; therefore, from Equation (9.5), the membrane contribution β_{mem} is 7. For more hydrophobic VOCs, the total separation factor increases, because the evaporative separation term is larger. For example, the separation factor β_{pervap} for toluene measured in cell experiments is an impressive 10 000, but most of the separation is due to the evaporation step β_{evap}, which is 8000. The membrane contribution β_{mem} is only 1.2, and the approximate selectivity of the membrane falls to 0.3 when concentration polarization effects are taken into account.

Concentration polarization plays a dominant role in the selection of membrane materials, operating conditions, and system design in the pervaporation of VOCs from water. Selection of the appropriate membrane thickness and permeate pressure is discussed in detail elsewhere [50]. In general, concentration polarization effects are not a major problem for VOCs with separation factors less than 100–200. With solutions containing such VOCs, very high feed velocities through

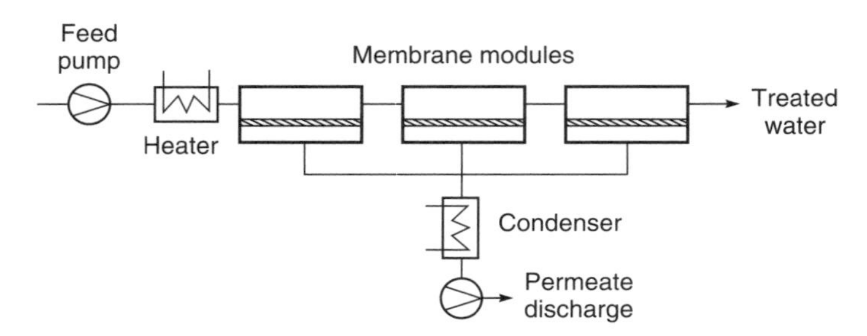

Figure 9.15 Once-through pervaporation system design. This design is most suitable for removal of VOCs with modest separation factors for which concentration polarization is not a problem

the membrane modules are not needed to control concentration polarization, so a once-through process design as illustrated in Figure 9.15 can be used. In the once-through design, the membrane modules are arranged in series, and the feed solution only passes through the modules once. The velocity of the solution in the modules is determined by the number of modules in series and the feed flow rate. With VOCs having modest separation factors, such a system can provide both adequate fluid velocities to control concentration polarization and sufficient residence time within the module to remove the required amount of VOC from the feed. With VOCs having large separation factors, such as toluene and TCE, it is difficult to balance the fluid velocity required to control concentration polarization with the residence time required to achieve the target VOC removal in a single pass.

For treating water containing VOCs with separation factors of more than 500, for which concentration polarization is a serious problem, feed-and-bleed systems similar to those described in the chapter on ultrafiltration can be used. For small feed volumes a batch process as illustrated in Figure 9.16 is more suitable. In a batch system, feed solution is accumulated in a surge tank. A portion of this solution is then transferred to the feed tank and circulated at high velocity through the pervaporation modules until the VOC concentration reaches the desired level. At this time, the treated water is removed from the feed tank, the tank is loaded with a new batch of untreated solution, and the cycle is repeated.

Applications for VOC-from-water pervaporation systems include treatment of contaminated wastewaters and process streams in the chemical industry, removal of small amounts of VOCs from contaminated groundwater, and the recovery of volatile flavor and aroma elements from streams produced in the processing of fruits and vegetables. A number of factors enter into the selection of pervaporation over a competing technology:

- *VOC type*. Pervaporation is best applied to recovery of VOCs with medium to high volatility, for which separation factors of 50 or more can be achieved.

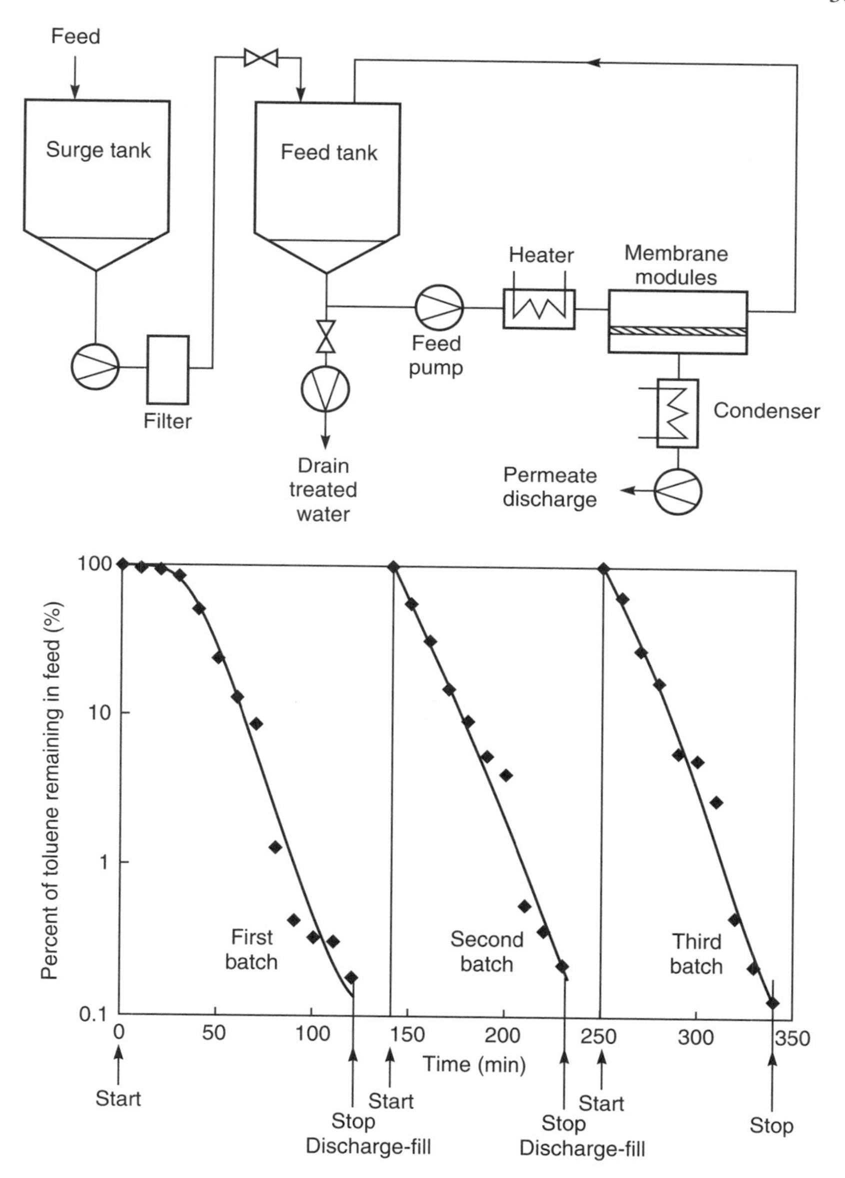

Figure 9.16 Flow diagram and typical performance for a 50-gal cyclic batch pervaporation system. The treatment time for the first 50-gal batch was set at 120 min because the unit was cold; thereafter, the cycle time was set at 90 min. The system achieved 99.8 % removal of toluene from the feed water [13]

These VOCs are normally more hydrophobic than acetone and have a Henry's law coefficient of greater than about 2 atm/mol fraction.

- *VOC concentration.* In general, the optimum VOC concentration range for pervaporation is 200–50 000 ppm (5 wt%). In this range, the conventional technology is steam stripping. Generally streams containing more than 5 wt% organic are better treated by distillation or, if the organic has no recovery value, simply incinerated. If the water contains less than 100 ppm VOC, recovery of the VOC is not an objective, so a destructive technology such as UV oxidation is used, or the stream is treated by air stripping followed by carbon adsorption to remove the VOC from the effluent air stream.
- *Stream flow rate.* The costs of pervaporation, like other membrane processes, increase linearly with increasing system size, whereas processes such as steam stripping scale to the 0.6–0.7 power. This makes pervaporation most competitive for small- to medium-sized streams. For streams containing highly volatile VOCs with Henry's law coefficients greater than 100 atm/mol fraction, pervaporation will generally be limited to streams smaller than 100 gal/min. For moderately volatile VOCs (0.5–100 atm/mol fraction), pervaporation is preferred for streams smaller than 10–20 gal/min. Steam stripping is preferred for very large streams or for those containing VOCs with a poor pervaporation VOC/water separation factor.
- *VOC thermal stability.* Separation of VOCs from water by pervaporation generally requires heating the feed water to only 50–70 °C. This is significantly lower than the temperatures involved in distillation or steam stripping, a considerable advantage if the VOCs are valuable, thermally labile compounds. This feature is important in applications such as flavor and aroma recovery in the food industry.

Commercial development of pervaporation for VOC removal/recovery has been slower than many predicted; only a few plants have been installed. The first significant applications are likely to be in the food industry, processing aqueous condensate streams generated in the production of concentrated orange juice, tomato paste, apple juice and the like. These condensates contain a complex mixture of alcohols, esters and ketones that are the flavor elements of the juice. Steam distillation could be used to recover these elements, but the high temperatures involved would damage the product. Pervaporation recovers essentially all of these components, producing a concentrated, high-value oil without exposing the flavor elements to high temperatures [51,52]. Figure 9.17 shows gas chromatography (GC) traces of the feed and permeate streams produced by pervaporation of an orange juice evaporator condensate stream.

Another potential pervaporation application is removing small amounts of VOCs from industrial wastewaters, allowing the water to be discharged to the sewer and concentrating the VOCs in a small-volume stream that can be sent to a hazardous waste treater [13]. Without a treatment system such as pervaporation,

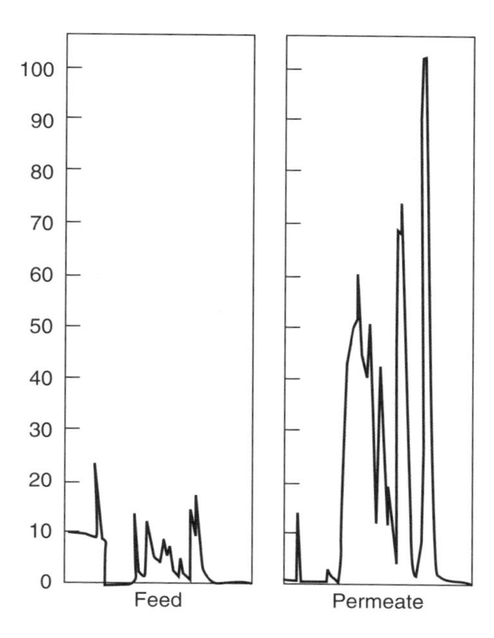

Figure 9.17 GC traces showing recovery of flavor and aroma components from orange juice evaporator condensate

the entire waste stream would have to be trucked off site. The avoided trucking cost can be considerable; a photograph of a batch pervaporation system installed for this purpose is shown in Figure 9.18.

Separation of Organic Mixtures

The third application area for pervaporation is the separation of organic/organic mixtures. The competitive technology is generally distillation, a well-established and familiar technology. However, a number of azeotropic and close-boiling organic mixtures cannot be efficiently separated by distillation; pervaporation can be used to separate these mixtures, often as a combination membrane-distillation process. Lipnizki *et al.* have recently reviewed the most important applications [53].

The degree of separation of a binary mixture is a function of the relative volatility of the components, the membrane selectivity, and the operating conditions. For azeotropic or close-boiling mixtures, the relative volatility is close to 1, so separation by simple distillation is not viable. However, if the membrane permeation selectivity is much greater than 1, a significant separation is possible using pervaporation. An example of such a separation is given in Figure 9.19, which shows a plot of the pervaporation separation of benzene/cyclohexane mixtures

Figure 9.18 Photograph of a 300–500 gal/day pervaporation system installed to treat wastewater contaminated with methylene chloride. This system has been operating at Applied Biosystems in Redwood City, California since 1995 [13]

using a 20-μm-thick crosslinked cellulose acetate-poly(styrene phosphate) blend membrane [54]. The vapor–liquid equilibrium for the mixture is also shown; the benzene/cyclohexane mixture forms an azeotrope at approximately 50 % benzene. A typical distillation stage could not separate a feed stream of this composition. However, pervaporation treatment of this mixture produces a vapor permeate containing more than 95 % benzene. This example illustrates the advantages of pervaporation over simple distillation for separating azeotropes and close-boiling mixtures.

It would be unusual for a pervaporation process to perform an entire organic/organic separation. Rather, pervaporation will be most efficient when combined with distillation in a hybrid process [55]. The two main applications of pervaporation–distillation hybrid processes are likely to be in breaking azeotropes and in removing a single-component, high-purity side stream from a multicomponent distillation separation. Figure 9.20 shows some potential pervaporation–distillation combinations. In Figure 9.20(a) pervaporation is combined with distillation to break an azeotrope that is concentrated in one component (>90 %). This approach is used in the production of high-purity ethanol. The ethanol/water azeotrope from the top of the distillation column is fed to a pervaporation unit where the water is removed as the permeate and returned to the column as a reflux.

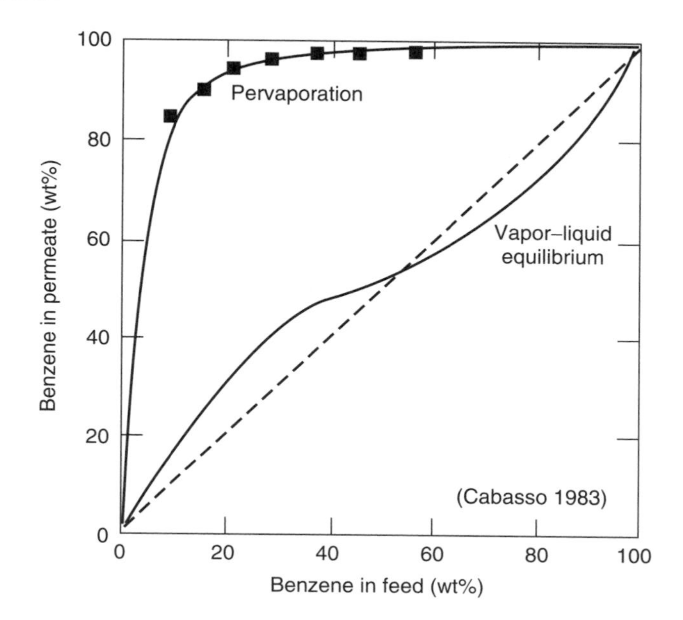

Figure 9.19 Fraction of benzene in permeate as a function of feed mixture composition for pervaporation at the reflux temperature of a binary benzene/cyclohexane mixture. A 20-μm-thick crosslinked blend membrane of cellulose acetate and poly(styrene phosphate) was used [54]. Reprinted with permission from I. Cabasso, Organic Liquid Mixtures Separation by Selective Polymer Membranes, *Ind. Eng. Chem. Prod. Res. Dev.* **22**, 313. Copyright 1983 American Chemical Society

Figure 9.20(b) illustrates the use of pervaporation with two distillation columns to break a binary azeotrope such as benzene/cyclohexane. The feed is supplied at the azeotropic composition and is split into two streams by the pervaporation unit. The residue stream, rich in cyclohexane, is fed to a distillation column that produces a pure bottom product and an azeotropic top stream, which is recycled to the pervaporation unit. Similarly, the other distillation column treats the benzene-rich stream to produce a pure benzene product and an azeotropic mixture that is returned to the pervaporation unit.

Pervaporation can also be used to unload a distillation column, thereby reducing energy consumption and operating cost and increasing throughput. The example shown in Figure 9.20(c) is for the recovery of pure methanol by pervaporation of a side stream from a column separating a methanol/isobutene/methyl tertiary butyl ether (MTBE) feed mixture [14,15].

The principal problem hindering the development of commercial systems for organic/organic separations is the lack of membranes and modules able to withstand long-term exposure to organic compounds at the elevated temperatures required for pervaporation. Membrane and module stability problems are not

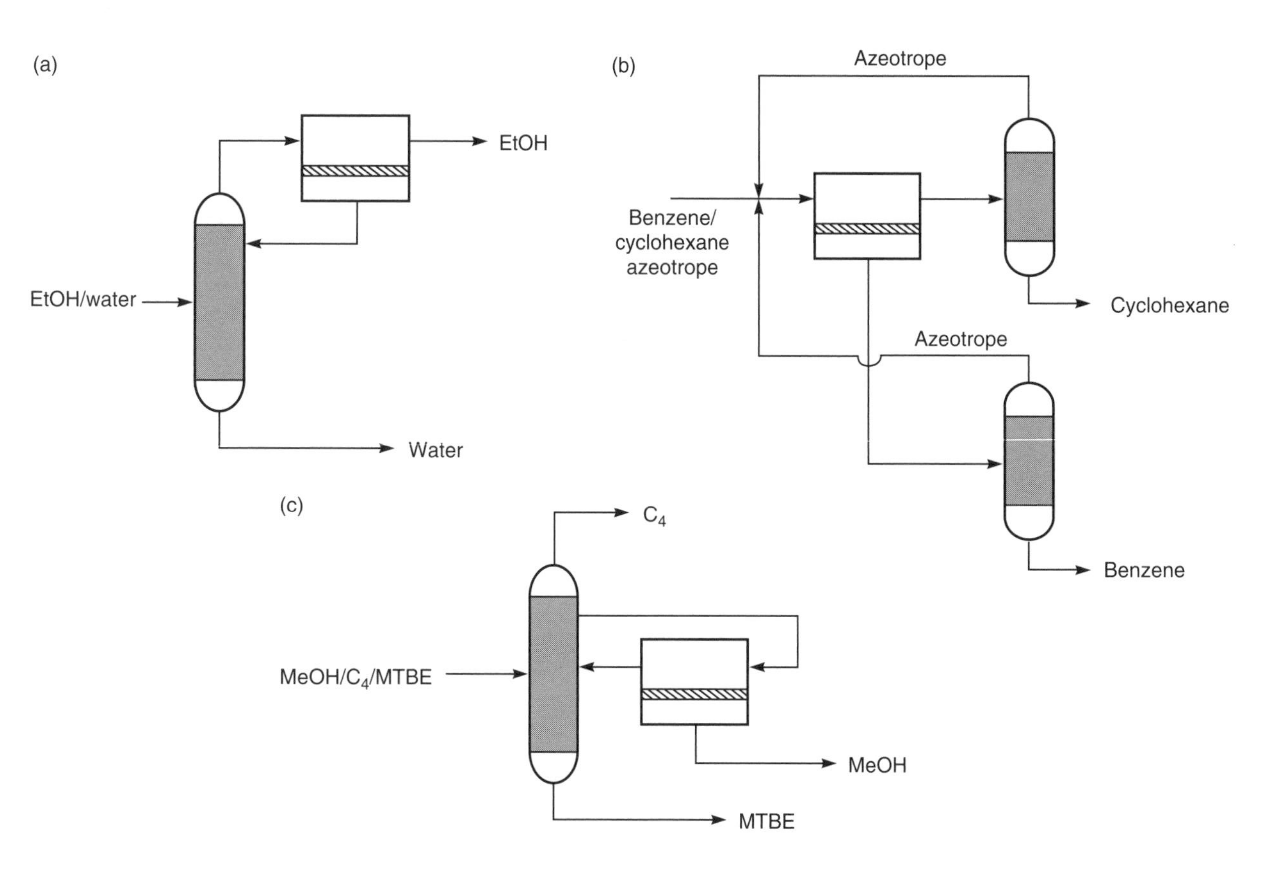

Figure 9.20 Potential configurations for pervaporation–distillation hybrid processes

MTBE Reaction Chemistry

$$CH_3OH + C(CH_3)=CH_2 \rightleftharpoons CH_2OH-C(CH_3)_2-CH_3$$

Methanol Isobutene MTBE

Pervaporation Process before Debutanizer

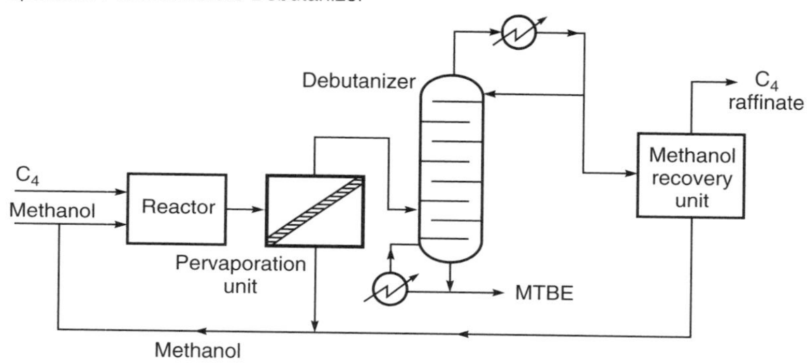

Pervaporation on Debutanizer Sidedraw

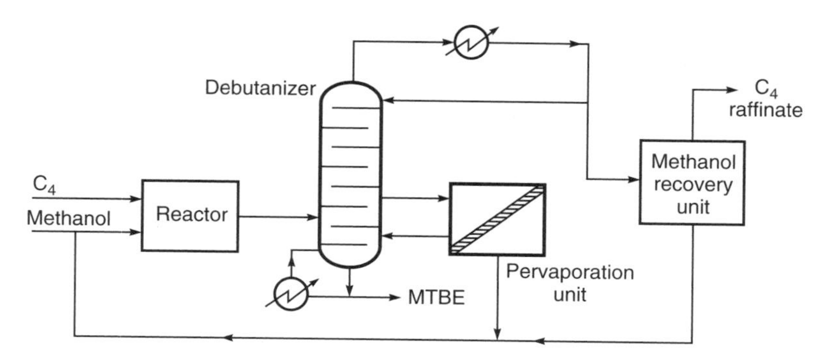

Figure 9.21 Methods of integrating pervaporation membranes in the recovery of methanol from the MTBE production process [15]. Courtesy of Air Products and Chemicals, Inc., Allentown, PA

insurmountable, however, as shown by the successful demonstration of a pervaporation process for the separation of methanol from an isobutene/MTBE mixture. This mixture is generated during the production of MTBE; the product from the reactor is an alcohol/ether/hydrocarbon mixture in which the alcohol/ether and the alcohol/hydrocarbon both form azeotropes. The product stream is treated by pervaporation to yield a methanol-enriched permeate, which can be recycled to the etherification process. Two alternative ways of integrating the pervaporation process into a complete reaction scheme are illustrated in Figure 9.21.

The process shown in Figure 9.21 was first developed by Separex, using cellulose acetate membranes. The separation factor for methanol from MTBE is high (>1000) because the membrane material, cellulose acetate, is relatively glassy and hydrophilic. Thus, both the mobility selectivity term and the sorption term in Equation (9.5) significantly favor permeation of the smaller molecule, methanol, because methanol is more polar than MTBE or isobutene, the other feed components. These membranes are reported to work well for feed methanol concentrations up to 6 %. Above this concentration, the membrane is plasticized, and selectivity is lost. More recently, Sulzer (GFT) has also studied this separation using their plasma-polymerized membrane [56].

Another application that has developed to the pilot scale is the separation of aromatic/aliphatic mixtures in refining crude oils into transportation fuels [16,17,57]. For hydrocarbons with approximately the same boiling point range, the permeability is generally in the order aromatics > unsaturated hydrocarbons > saturated hydrocarbons. For aliphatic hydrocarbons in approximately the same boiling point range, the order of permeabilities is straight chain > cyclic chain > branched chain. The goal in these processes is to perform a bulk separation of the hydrocarbon mixture by pervaporation; therefore, the membrane must be highly permeable and selective. The general approach [17,58] is to prepare segmented block copolymers consisting of hard segments not swollen by the hydrocarbon oil, to control the swelling of the soft segments through which the oil would permeate. Crosslinking was also used to control swelling of the membrane materials, polyester-polyimide and polyurea-polyurethane polymers.

Recently Sulzer, working with Grace Davison [35,59] and using polyimide, polysiloxane or polyurea urethane membranes, and ExxonMobil [60], using Nafion® or cellulose triacetate membranes, have described processes to separate sulfur compounds from various refinery streams.

Conclusions and Future Directions

During the 10 years after GFT installed the first commercial pervaporation plant in 1982, there was a surge of interest in all types of pervaporation. Much of this interest has now ebbed, and the number of companies involved in developing pervaporation has decreased considerably. The oil companies Texaco, British Petroleum and Exxon, all of which had large research groups working on pervaporation problems in the 1980s, with the exception mentioned above, seem to have abandoned this approach. The key problem seems to be economic. Pervaporation is a competitive technology on a small scale, but current membranes and modules are insufficiently selective and economical to compete with distillation, steam stripping, or solvent extraction in larger plants. As a result, most of the systems sold are small, and the total value of new systems installed annually is probably less than US$5 million/year. This market is likely to expand over the next few years, particularly if recovery of high-value aroma and flavor elements in food processing operations meets the developers' expectations.

References

1. P.A. Kober, Pervaporation, Perstillation, and Percrystallization, *J. Am. Chem. Soc.* **39**, 944 (1917).
2. R.C. Binning and J.M. Stuckey, Method of Separating Hydrocarbons Using Ethyl Cellulose Selective Membrane, US Patent 2,958,657 (November, 1960).
3. R.C. Binning, R.J. Lee, J.F. Jennings and E.C. Martin, Separation of Liquid Mixtures by Permeation, *Ind. Eng. Chem.* **53**, 45 (1961).
4. R.C. Binning and W.F. Johnston, Jr, Aromatic Separation Process, US Patent 2,970,106 (January, 1961).
5. R.C. Binning, J.F. Jennings and E.C. Martin, Process for Removing Water from Organic Chemicals, US Patent 3,035,060 (May, 1962).
6. W.F. Strazik and E. Perry, Process for the Separation of Styrene from Ethyl Benzene, US Patent 3,776,970. Also US Patents 4,067,805 and 4,218,312 and many others.
7. P. Aptel, N. Challard, J. Cuny and J. Neel, Application of the Pervaporation Process to Separate Azeotropic Mixtures, *J. Membr. Sci.* **1**, 271 (1976).
8. J. Neel, Q.T. Nguyen, R. Clement and L. Le Blanc, Fractionation of a Binary Liquid Mixture by Continuous Pervaporation, *J. Membr. Sci.* **15**, 43 (1983).
9. A.H. Ballweg, H.E.A. Brüschke, W.H. Schneider, G.F. Tüsel and K.W. Böddeker, Pervaporation Membranes, in *Proceedings of Fifth International Alcohol Fuel Technology Symposium*, Auckland, New Zealand, pp. 97–106 (May, 1982).
10. H.E.A. Brüschke, State of Art of Pervaporation, in *Proceedings of Third International Conference on Pervaporation Processes in the Chemical Industry*, R. Bakish (ed.), Bakish Materials Corp., Englewood, NJ, pp. 2–11 (1988).
11. I. Blume, J.G. Wijmans and R.W. Baker, The Separation of Dissolved Organics from Water by Pervaporation, *J. Membr. Sci.* **49**, 253 (1990).
12. A.L. Athayde, R.W. Baker, R. Daniels, M.H. Le and J.H. Ly, Pervaporation for Wastewater Treatment, *CHEMTECH* **27**, 34 (1997).
13. G. Cox and R.W. Baker, Pervaporation for the Treatment of Small Volume VOC-contaminated Waste Water Streams, *Indust. Wastewater* **6**, 35 (1998).
14. M.S.K. Chen, R.M. Eng, J.L. Glazer and C.G. Wensley, Pervaporation Process for Separating Alcohols from Ethers, US Patent 4,774,365 (September, 1988).
15. M.S.K. Chen, G.S. Markiewicz and K.G. Venugopal, Development of Membrane Pervaporation TRIM™ Process for Methanol Recovery from CH_3OH/MTBE/C_4 Mixtures, in *Membrane Separations in Chemical Engineering*, AIChE Symposium Series Number 272, A.E. Fouda, J.D. Hazlett, T. Matsuura and J. Johnson (eds), AIChE, New York, NY, p. 85 (1989).
16. R.C. Schucker, Separation of Organic Liquids by Perstraction, in *Proceedings of Seventh International Conference on Pervaporation Processes in the Chemical Industry*, R. Bakish (ed.), Bakish Materials Corp., Englewood, NJ, pp. 321–332 (1995).
17. R.C. Schucker, Highly Aromatic Polyurea/Urethane Membranes and Their Use of the Separation of Aromatics from Non-aromatics, US Patent 5,063,186, 5,055,632 and 4,983,338 and many others.
18. J.G. Wijmans and R.W. Baker, A Simple Predictive Treatment of the Permeation Process in Pervaporation, *J. Membr. Sci.* **79**, 101 (1993).

19. T. Kataoka, T. Tsuru, S.-I. Nakao and S. Kimura, Membrane Transport Properties of Pervaporation and Vapor Permeation in Ethanol–Water System using Polyacryonitrile and Cellulose Acetate Membranes, *J. Chem. Eng. Jpn* **24**, 334 (1991).
20. R.A. Shelden and E.V. Thompson, Dependence of Diffusive Permeation Rates on Upstream and Downstream Pressures, *J. Membr. Sci.* **4**, 115 (1978).
21. M.E. Hollein, M. Hammond and C.S. Slater, Concentration of Dilute Acetone–Water Solutions Using Pervaporation, *Sep. Sci. Technol.* **28**, 1043 (1993).
22. S. Yamada, T. Nakagawa and T. Abo, Pervaporation of Water–Ethanol with PVA-Fluoropore Composite Membranes, in *Proceedings of Fourth International Conference on Pervaporation Processes in the Chemical Industry*, R. Bakish (ed.), Bakish Materials Corp., Englewood, NJ, pp. 64–74 (1989).
23. G. Ellinghorst, H. Steinhauser and A. Hubner, Improvement of Pervaporation Plant by Choice of PVA or Plasma Polymerized Membranes, in *Proceedings of Sixth International Conference on Pervaporation Processes in the Chemical Industry*, R. Bakish (ed.), Bakish Materials Corp., Englewood, NJ, pp. 484–493 (1992).
24. K. Watanabe and S. Kyo, Pervaporation Performance of Hollow-fiber Chitosan-Polyacrylonitrile Composite Membrane in Dehydration of Ethanol, *J. Chem. Eng. Jpn* **25**, 17 (1992).
25. A. Wenzlaff, K.W. Böddeker and K. Hattenbach, Pervaporation of Water–Ethanol Through Ion Exchange Membranes, *J. Membr. Sci.* **22**, 333 (1985).
26. I. Cabasso and Z.-Z. Liu, The Permselectivity of Ion-exchange Membranes for Non-electrolyte Liquid Mixtures, *J. Membr. Sci.* **24**, 101 (1985).
27. M. Bennett, B.J. Brisdon, R. England and R.W. Feld, Performance of PDMS and Organofunctionalized PDMS Membranes for the Pervaporation Recovery of Organics from Aqueous Streams, *J. Membr. Sci.* **137**, 63 (1997).
28. H.H. Nijhuis, M.V.H. Mulder and C.A. Smolders, Selection of Elastomeric Membranes for the Removal of Volatile Organic Components from Water, in *Proceedings of Third International Conference on Pervaporation Processes in the Chemical Industry*, R. Bakish (ed.), Bakish Materials Corp., Englewood, NJ, pp. 239–251 (1988).
29. G. Bengtson and K.W. Böddeker, Pervaporation of Low Volatiles from Water, in *Proceedings of Third International Conference on Pervaporation Processes in the Chemical Industry*, R. Bakish (ed.), Bakish Materials Corp., Englewood, NJ, pp. 439–448.
30. I. Blume and I. Pinnau, Composite Membrane, Method of Preparation and Use, US Patent 4,963,165 (October, 1990).
31. H.J.C. te Hennepe, D. Bargeman, M.H.V. Mulder and C.A. Smolders, Zeolite-filled Silicone Rubber Membranes, *J. Membr. Sci.* **35**, 39 (1987).
32. K.W. Böddeker and G. Bengtson, Pervaporation of Low Volatility Aromatics from Water, *J. Membr. Sci.* **53**, 143 (1990).
33. K.W. Böddeker and G. Bengtson, Selective Pervaporation of Organics from Water, in *Pervaporation Membrane Separation Processes*, R.Y.M. Huang (ed.), Elsevier, Amsterdam, pp. 437–460 (1991).
34. K. Meckl and R.N. Lichtenthaler, Hybrid-processes Including Pervaporation for the Removal of Organic Compounds from Process and Waste Water, in *Proceedings of Sixth International Conference on Pervaporation Processes in the Chemical Industry*, R. Bakish (ed.), Bakish Materials Corp., Englewood, NJ (1992).
35. L.S. White, R.F. Wormsbecher and M. Lesmann, Membrane Separation for Sulfer Reduction, US Patent Application US 2002/0153284 A1 (October, 2002).
36. S. Yuan and H.G. Schwartzberg, Mass Transfer Resistance in Cross Membrane Evaporation into Air, *Recent Advances in Separation Science*, AIChE Symposium Series Number 68, AIChE, New York, NY, Vol. 120, p. 41 (1972).
37. A.E. Robertson, Separation of Hydrocarbons, US Patent 2,475,990 (July, 1949).

38. D.T. Friesen, D.D. Newbold, S.B. McCray and R.J. Ray, Pervaporation by Countercurrent Condensable Sweep, US Patent 5,464,540 (November, 1995).
39. L.M. Vane, F.R. Alvarez, A.P. Mairal and R.W. Baker, Separation of Vapor-phase Alcohol/Water Mixtures Via Fractional Condensation Using a Pilot-scale Dephlegmator: Enhancement of the Pervaporation Process Separation Factor, *Ind. Eng. Chem. Res.* (in press).
40. J. Kaschemekat, B. Barbknecht and K.W. Böddeker, Konzentrierung von Ethanol durch Pervaporation, *Chem. -Ing. -Tech.* **58**, 740 (1986).
41. M. Kondo, M. Komori, H. Kita and K.-I. Okamoto, Tubular-type Pervaporation Module with NaA Zeolite Membrane, *J. Membr. Sci.* **133**, 133 (1997).
42. D. Shah, K. Kissick, A. Ghorpade, R. Hannah and D. Bhattacharyya, Pervaporation of Alcohol–Water and Dimethylformamide–Water Mixtures using Hydrophilic Zeolite NaA Membranes: Mechanisms and Results, *J. Membr. Sci.* **179**, 185 (2000).
43. J. Bergdorf, Case Study of Solvent Dehydration in Hybrid Processes With and Without Pervaporation, in *Proceedings of Fifth International Conference on Pervaporation Processes in the Chemical Industry*, R. Bakish (ed.), Bakish Materials Corp., Englewood, NJ, pp. 362–382 (1991).
44. R. Abouchar and H. Brüschke, Long-Term Experience with Industrial Pervaporation Plants, in *Proceedings of Sixth International Conference on Pervaporation Processes for the Chemical Industry*, R. Bakish (ed.), Bakish Materials Corp., Englewood, NJ, pp. 494–502 (1992).
45. U.H.F. Sander, Development of Vapor Permeation for Industrial Applications, in *Pervaporation Membrane Separation Processes*, R.Y.M. Huang (ed.), Elsevier, Amsterdam, pp. 509–534 (1991).
46. J.F. Jennings and R.C. Binning, Organic Chemical Reactions Involving Liberation of Water, US Patent 2,956,070 (October, 1960).
47. A. Dams and J. Krug, Pervaporation Aided Esterification—Alternatives in Plant Extension for an Existing Chemical Process, in *Proceedings of Fifth International Conference on Pervaporation Processes in the Chemical Industry*, R. Bakish (ed.), Bakish Materials Corp., Englewood, NJ, pp. 338–348 (1991).
48. H.E.A. Brüschke, G. Ellinghorst and W.H. Schneider, Optimization of a Coupled Reaction—Pervaporation Process, in *Proceedings of Seventh International Conference on Pervaporation Processes in the Chemical Industry*, R. Bakish (ed.), Bakish Materials Corp., Englewood, NJ, pp. 310–320 (1995).
49. Y. Zhu, R.G. Minet and T.T. Tsotsis, A Continuous Pervaporation Membrane Reactor for the Study of Esterification Reactions Using a Composite Polymeric/Ceramic Membrane, *Chem. Eng. Sci.* **51**, 4103 (1996).
50. R.W. Baker, J.G. Wijmans, A.L. Athayde, R. Daniels, J.H. Ly and M. Le, Separation of Volatile Organic Compounds from Water by Pervaporation, *J. Membr. Sci.* **137**, 159 (1998).
51. N. Rajagopalan and M. Cheryan, Pervaporation of Grape Juice Aroma, *J. Membr. Sci.* **104**, 243 (1995).
52. H.O.E. Karlsson and G. Trägårdth, Applications of Pervaporation in Food Processing, *Trends Food Sci. Technol.* **7**, 78 (1996).
53. F. Lipnizki, R.F. Feld and P.-K. Ten, Pervaporation-based Hybrid Processes: A Review of Process Design Applications and Economics, *J. Membr. Sci.* **153**, 183 (1999).
54. I. Cabasso, Organic Liquid Mixtures Separation by Selective Polymer Membranes, *Ind. Eng. Chem. Prod. Res. Dev.* **22**, 313 (1983).
55. W. Stephan, R.D. Nobel and C.A. Koval, Design Methodology for a Membrane/Distillation Hybrid Process, *J. Membr. Sci.* **99**, 259 (1995).

56. C. Streicher, P. Kremer, V. Tomas, A. Hubner and G. Ellinghorst, Development of New Pervaporation Membranes, Systems and Processes to Separate Alcohols/Ethers/ Hydrocarbons Mixtures, in *Proceedings of Seventh International Conference on Pervaporation Processes in the Chemical Industry*, R. Bakish (ed.), Bakish Materials Corp., Englewood, NJ, pp. 297–309 (1995).
57. S. Matsui and D.R. Paul, Pervaporation Separation of Aromatic/Aliphatic Hydrocarbons by a Series of Ionically Crosslinked Poly(n-alkyl acrylate) Membranes, *J. Membr. Sci.* **213**, 67 (2003).
58. N. Tanihara, N. Umeo, T. Kawabata, K. Tanaka, H. Kita and K. Okamoto, Pervaporation of Organic Liquids Through Poly(etherimide) Segmented Copolymer Membranes, *J. Membr. Sci.* **104**, 181 (1995).
59. J. Balko, G. Bourdillon and N. Wynn, Membrane Separation for Producing Gasoline, *Petrol. Q.* **8**, 17 (2003).
60. B. Minhas, R.R. Chuba and R.J. Saxton, Membrane Process for Separating Sulfur Compounds from FCC Light Naphtha, US Patent Application US 2002/0111524 A1 (August, 2002).

10 ION EXCHANGE MEMBRANE PROCESSES – ELECTRODIALYSIS

Introduction and History

Ion exchange membranes are used in a number of separation processes, the most important of which is electrodialysis. In ion exchange membranes, charged groups are attached to the polymer backbone of the membrane material. These fixed charge groups partially or completely exclude ions of the same charge from the membrane. This means that an anionic membrane with fixed positive groups excludes positive ions but is freely permeable to negatively charged ions. Similarly a cationic membrane with fixed negative groups excludes negative ions but is freely permeable to positively charged ions, as illustrated in Figure 10.1.

In an electrodialysis system, anionic and cationic membranes are formed into a multicell arrangement built on the plate-and-frame principle to form up to 100 cell pairs in a stack. The cation and anion exchange membranes are arranged in an alternating pattern between the anode and cathode. Each set of anion and cation membranes forms a cell pair. Salt solution is pumped through the cells while an electrical potential is maintained across the electrodes. The positively charged cations in the solution migrate toward the cathode and the negatively charged anions migrate toward the anode. Cations easily pass through the negatively charged cation exchange membrane but are retained by the positively charged anion exchange membrane. Similarly, anions pass through the anion exchange membrane but are retained by the cation exchange membrane. The overall result of the process is that one cell of the pair becomes depleted of ions while the adjacent cell becomes enriched in ions. The process, which is widely used to remove dissolved ions from water, is illustrated in Figure 10.2.

Experiments with ion exchange membranes were described as early as 1890 by Ostwald [1]. Work by Donnan [2] a few years later led to development of the concept of membrane potential and the phenomenon of Donnan exclusion. These early charged membranes were made from natural materials or chemically

Membrane Technology and Applications R. W. Baker
 ISBN: 0-470-85445-6

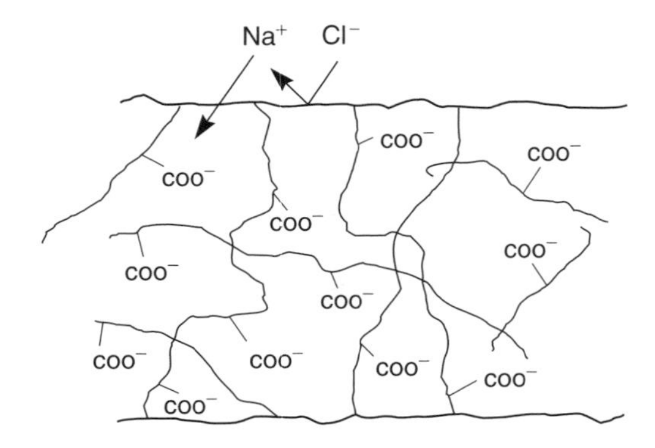

Figure 10.1 This cationic membrane with fixed carboxylic acid groups is permeable to cations such as sodium but is impermeable to anions such as chloride

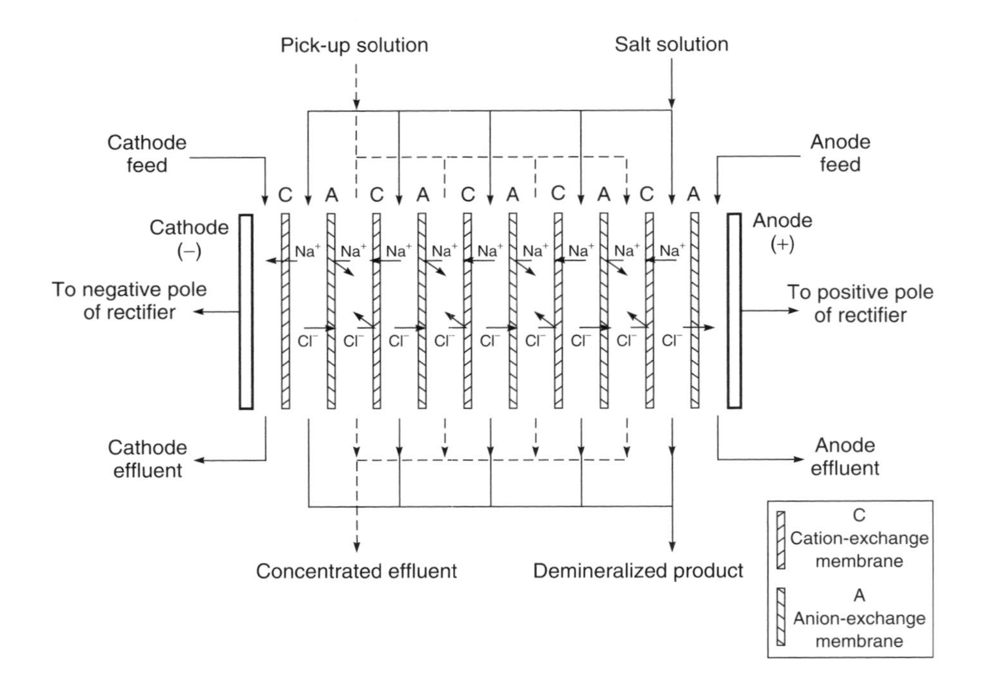

Figure 10.2 Schematic diagram of a plate-and-frame electrodialysis stack. Alternating cation- and anion-permeable membranes are arranged in a stack of up to 100 cell pairs

treated collodion membranes–their mechanical and chemical properties were very poor. Nonetheless, as early as 1939, Manegold and Kalauch [3] suggested the application of selective anionic and cationic exchange membranes to separate ions from water, and within another year Meyer and Strauss [4] described the concept of a multicell arrangement between a single pair of electrodes. The advances in polymer chemistry during and immediately after the Second World War led to the production of much better ion exchange membranes by Kressman [5], Murphy *et al.* [6] and Juda and McRae [7] at Ionics. With the development of these membranes, electrodialysis became a practical process. Ionics was the principal early developer and installed the first successful plant in 1952; by 1956 eight plants had been installed.

In the United States, electrodialysis was developed primarily for desalination of water, with Ionics being the industry leader. In Japan, Asahi Glass, Asahi Chemical (a different company), and Tokuyama Soda developed the process to concentrate seawater [8]. This application of electrodialysis is confined to Japan, which has no domestic salt sources. Electrodialysis membranes concentrate the salt in seawater to about 18–20 % solids, after which the brine is further concentrated by evaporation and the salt recovered by crystallization.

All of the electrodialysis plants installed in the 1950s through the 1960s were operated unidirectionally, that is, the polarity of the two electrodes, and hence the position of the dilute and concentrated cells in the stack, were fixed. In this mode of operation, formation of scale on the membrane surface by precipitation of colloids and insoluble salts was often a severe problem. To prevent scale, pH adjustment and addition of antiscaling chemicals to the feed water was required, together with regular membrane cleaning using detergents and descaling chemicals. Nevertheless, scaling and membrane fouling remained major problems, affecting plant on-stream time and widespread acceptance of the process. In the early 1970s, a breakthrough in system design, known as electrodialysis polarity reversal, was made by Ionics [9]. In these systems the polarity of the DC power applied to the membrane electrodes is reversed two to four times per hour. When the electrode polarity is reversed, the desalted water and brine chambers are also reversed by automatic valves that control the flows in the stack. By switching cells and reversing current direction, freshly precipitated scale is flushed from the membrane before it can solidify. The direction of movement of colloidal particulates drawn to the membrane by the flow of current is also reversed, so colloids do not form a film on the membrane. Electrodialysis plants using the reverse polarity technique have been operating since 1970 and have proved more reliable than their fixed polarity predecessors.

Electrodialysis is now a mature technology, with Ionics remaining the worldwide industry leader except in Japan. Desalting of brackish water and the production of boiler feed water and industrial process water were the main applications until the 1990s, but electrodialysis has since lost market share due to stiff competition from improved reverse osmosis membranes. Beginning in the 1990s, electrodeionization, a combination process using electrodialysis and ion exchange, began to be used to achieve very good salt removal in ultrapure water plants. This is now a major use of electrodialysis. Other important applications are control of ionic impurities from industrial effluent streams, water softening and desalting certain foods, particularly milk whey [10,11]. Over the last 20 years a number of other uses of ion exchange membranes have been found. Perhaps the most important is the development by Asahi, Dow and DuPont of perfluoro-based ion exchange membranes with exceptional chemical stability for membrane chlor-alkali cells [12]. More than 1 million square meters of these membranes have been installed. Ion exchange membranes are also finding an increasing market in electrolysis processes of all types. One application that has received a great deal of attention is the use of bipolar membranes to produce acids and alkalis by electrolysis of salts. Bipolar membranes are laminates of anionic and cationic membranes. The first practical bipolar membranes were developed by K.J. Liu and others at Allied Chemicals in about 1977 [13]; they were later employed in Allied's Aquatech acid/base production process [14]. A final, growing use of ion exchange membranes is in advanced fuel cells and battery systems in which the membranes regulate ion transport from various compartments in the cells [15]. A time line illustrating the major milestones in the development of ion exchange membranes is shown in Figure 10.3.

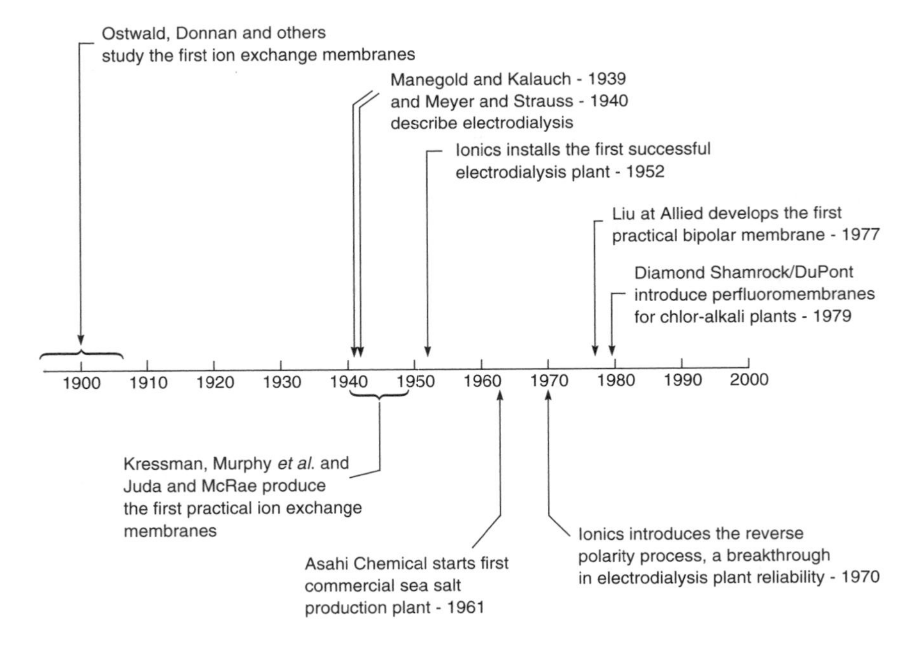

Figure 10.3 Milestones in the development of ion exchange membrane processes

Theoretical Background

Transport Through Ion Exchange Membranes

In electrodialysis and the other separation processes using ion exchange membranes, transport of components generally occurs under the driving forces of both concentration and electric potential (voltage) gradients. However, because the two types of ion present, anions and cations, move in opposite directions under an electric potential gradient, ion exchange membrane processes are often more easily treated in terms of the amount of charge transported than the amount of material transported. Consider, for example, a simple univalent–univalent electrolyte such as sodium chloride, which can be considered to be completely ionized in dilute solutions. The concentration of sodium cations is then c^+, and the concentration of chloride anions is c^-. The velocity of the cations in an externally applied field of strength, E, is u (cm/s), and the velocity of the anions measured in the same direction is $-v$ (cm/s). Each cation carries the protonic charge $+e$ and each anion the electronic charge of $-e$, so the total amount of charge transported per second across a plane of 1 cm^2 area is

$$\frac{I}{F} = c^+(u)(+e) + c^-(-v)(-e) = ce(u+v) \tag{10.1}$$

where I is the current and F is the Faraday constant to convert transport of electric charge to a current flow in amps. This equation links the electric current with the transport of ions.

It has been found that the fractions of the current carried by the anions and cations do not necessarily have to be equal. The fraction of the total current carried by any particular ion is known as the transport number of that ion. Thus, the transport number for the cations is t^+ and the transport number for the anions is t^-. It follows that

$$t^+ + t^- = 1 \tag{10.2}$$

Combining Equations (10.1) and (10.2), the transport number of the cations in the univalent–univalent electrolyte described above is given as

$$t^+ = \frac{c^+ue}{ce(u+v)} = \frac{u}{u+v} \tag{10.3}$$

and similarly for the anion

$$t^- = \frac{v}{u+v} \tag{10.4}$$

Transport numbers for different ions, even in aqueous solutions, can vary over a wide range, reflecting the different sizes of the ions. Ions with the same charge as the fixed charge groups in an ion exchange membrane are excluded from the membrane and, therefore, carry a very small fraction of the current through the membrane. In these membranes the transport number of the excluded ions is very small, normally between 0 and 0.05. Counter ions with a charge opposite to the fixed charged groups permeate the membrane freely and carry almost all of the current through the membrane. The transport numbers of these ions are between 0.95 and 1.0. This difference in transport number, a measure of relative permeability, allows separations to be achieved with ion exchange membranes.

Equation (10.1) shows that, as in other transport processes, the flux of the permeating component is the product of a mobility term (u or v) and a concentration term (c^+ or c^-). In ion exchange transport processes, most of the separation is achieved by manipulating the concentration terms. When the membrane carries fixed charges, the counter ions of the same charge will tend to be excluded from the membrane. As a result, the concentration of ions of the same charge is reduced, while the concentration of ions of opposite charge is elevated. This makes the membrane selective for ions of the opposite charge.

The ability of ion exchange membranes to discriminate between oppositely charged ions was put on a mathematical basis by Donnan in 1911 [2]. Figure 10.4 shows the distribution of ions between a salt solution and an ion exchange membrane containing fixed negative charges, R^-.

The equilibrium between the ions in the membrane (m) and the surrounding solution (s) can be expressed as

$$c^+{}_{(m)} \cdot c^-{}_{(m)} = kc^+{}_{(s)} \cdot c^-{}_{(s)} \tag{10.5}$$

where k is an equilibrium constant. Charge balance considerations lead to the expression

$$c^+{}_{(m)} = c^-{}_{(m)} + c_{R^-{}_{(m)}} \tag{10.6}$$

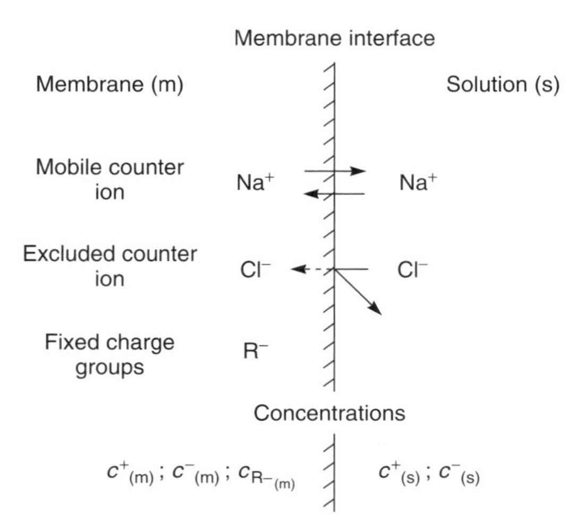

Figure 10.4 An illustration of the distribution of ions between a cationic membrane with fixed negative ions and the surrounding salt solution

For a fully dissolved salt, such as sodium chloride, the total molar concentration of the salt $c_{(s)}$ is equal to the concentration of each of the ions, so

$$c_{(s)} = c^{+}{}_{(s)} = c^{-}{}_{(s)} \tag{10.7}$$

Combining these three equations and rearranging gives the expression

$$\frac{c^{+}{}_{(m)}}{c^{-}{}_{(m)}} = \frac{[c^{-}{}_{(m)} + c_{R^{-}{}_{(m)}}]^2}{k[c_{(s)}]^2} \tag{10.8}$$

Because the membrane is cationic (fixed negative charges), the concentration of negative counter-ions in the membrane will be small compared to the concentration of fixed charges, that is,

$$c_{R^{-}{}_{(m)}} \gg c^{-}{}_{(m)} \tag{10.9}$$

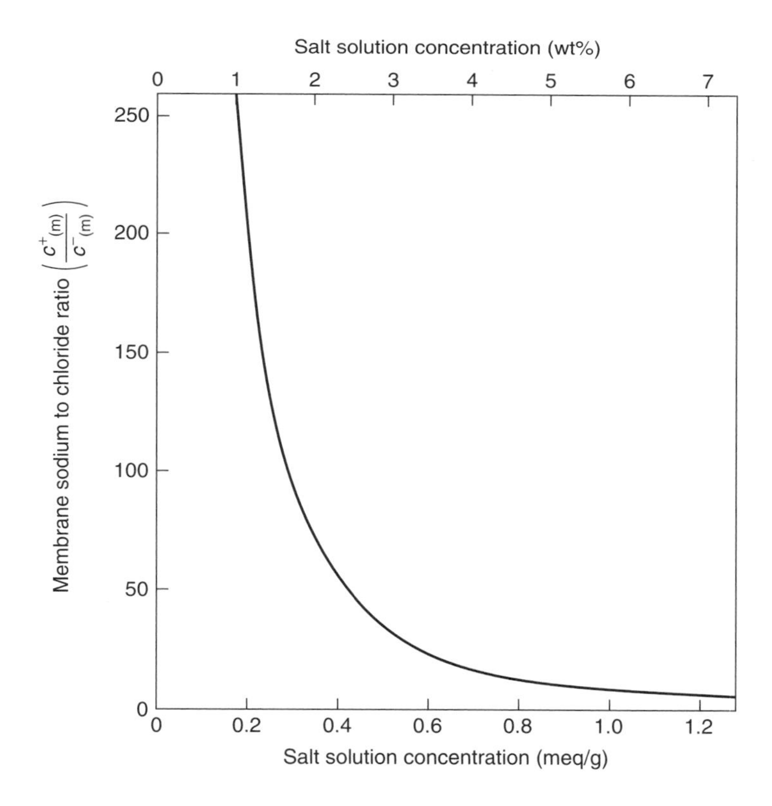

Figure 10.5 The sodium-to-chloride ion concentration ratio inside a negatively charged ion exchange membrane containing a concentration of fixed negative groups of 3 meq/g as a function of salt concentration. At salt concentrations in the surrounding solution of less than about 1 wt% sodium chloride (0.2 meq/g), chloride ions are almost completely excluded from the membrane

so it can be assumed that

$$c^{-}{}_{(m)} + c_{R^{-}{}_{(m)}} \approx c_{R^{-}{}_{(m)}} \tag{10.10}$$

Equation (10.8) can then be written

$$\frac{c^{+}{}_{(m)}}{c^{-}{}_{(m)}} = \frac{1}{k}\left(\frac{c_{R^{-}(m)}}{c_{(s)}}\right)^2 \tag{10.11}$$

This expression shows that the ratio of sodium to chloride ions in the membrane ($c^{+}{}_{(m)}/c^{-}{}_{(m)}$) is proportional to the square of the ratio of the fixed charge groups in the membrane to the salt concentration in the surrounding solution ($c_{R-(m)}/c_{(s)}$). In the commonly used ion exchange membranes, the fixed ion concentration in the membrane is very high, typically at least 3–4 milliequivalents per gram (meq/g). Figure 10.5 shows a plot of the sodium-to-chloride concentration ratio in a cationic membrane calculated using Equation (10.11) The ion exchange membrane is assumed to have a fixed negative charge concentration of 3 meq/g. The plot shows that, at salt solution concentrations of less than 0.2 meq/g (~1 wt% sodium chloride), chloride ions are almost completely excluded from the ion exchange membrane. This means that in this concentration range the transport number for sodium is close to one and for chloride is close to zero. Only at high salt concentrations–above about 0.6 meq/g (3 wt% sodium chloride)–does the ratio of sodium to chloride ions in the membrane fall below 30, and the membrane becomes measurably permeable to chloride ions.

Chemistry of Ion Exchange Membranes

A wide variety of ion exchange membrane chemistries has been developed. Typically each electrodialysis system manufacturer produces its own membrane tailored for the specific applications and equipment used. An additional complication is that many of these developments are kept as trade secrets or are only described in the patent literature. Korngold [16] gives a description of ion exchange membrane manufacture.

Current ion exchange membranes contain a high concentration of fixed ionic groups, typically 3–4 meq/g or more. When placed in water, these ionic groups tend to absorb water; charge repulsion of the ionic groups can then cause the membrane to swell excessively. This is why most ion exchange membranes are highly crosslinked to limit swelling. However, high crosslinking densities make polymers brittle, so the membranes are usually stored and handled wet to allow absorbed water to plasticize the membrane. Most ion exchange membranes are produced as homogenous films 50–200 μm thick. Typically the membrane is reinforced by casting onto a net or fabric to maintain the shape and to minimize swelling.

Ion exchange membranes fall into two broad categories: homogeneous and heterogeneous. In homogeneous membranes, the charged groups are uniformly distributed through the membrane matrix. These membranes swell relatively uniformly when exposed to water, the extent of swelling being controlled by their crosslinking density. In heterogeneous membranes, the ion exchange groups are contained in small domains distributed throughout an inert support matrix, which provides mechanical strength. Heterogeneous membranes can be made, for example, by dispersing finely ground ion exchange particles in a polymer support matrix. Because of the difference in the degree of swelling between the ion exchange portion and the inert portion of heterogeneous membranes, mechanical failure, leading to leaks at the boundary between the two domains, can be a problem.

Homogeneous Membranes

A number of early homogeneous membranes were made by simple condensation reactions of suitable monomers, such as phenol–formaldehyde condensation reactions of the type:

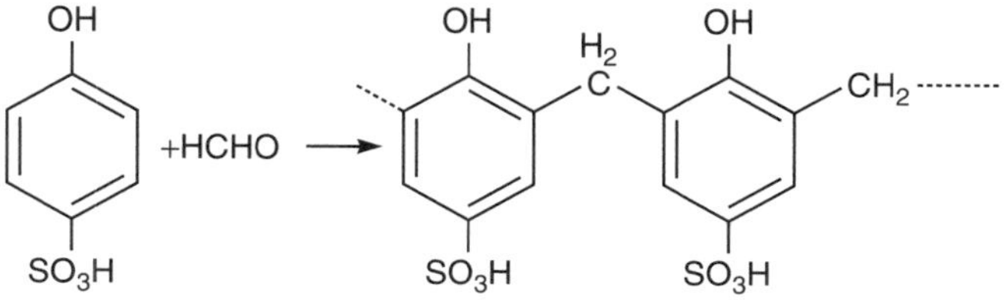

OH, SO_3H +HCHO → OH, H_2C, OH, CH_2, SO_3H, SO_3H

The mechanical stability and ion exchange capacity of these condensation resins were modest. A better approach is to prepare a suitable crosslinked base membrane, which can then be converted to a charged form in a subsequent reaction. Ionics is believed to use this type of membrane in many of their systems. In a typical preparation procedure, a 60:40 mixture of styrene and divinyl benzene is cast onto a fabric web, sandwiched between two plates and heated in an oven to form the membrane matrix. The membrane is then sulfonated with 98 % sulfuric acid or a concentrated sulfur trioxide solution. The degree of swelling in the final membrane is controlled by varying the divinyl benzene concentration in the initial mix to control crosslinking density. The degree of sulfonation can also be varied. The chemistry of the process is:

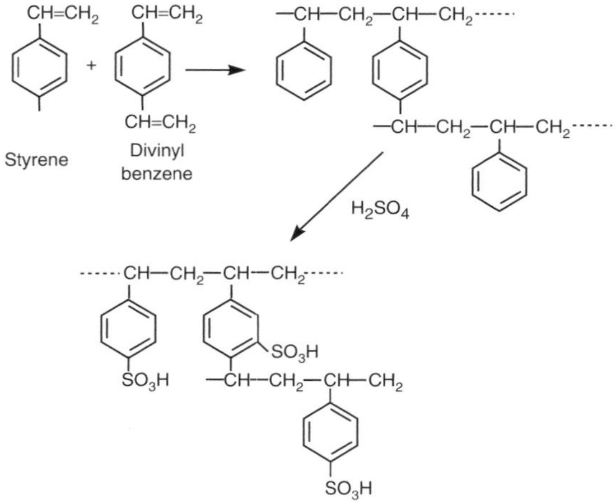

Anion exchange membranes can be made from the same crosslinked polystyrene membrane base by post-treatment with monochloromethyl ether and aluminum chloride to introduce chloromethyl groups into the benzene ring, followed by formation of quaternary amines with trimethyl amine:

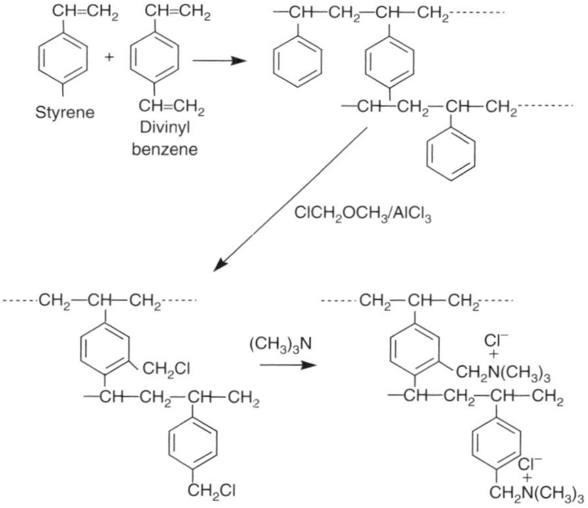

A particularly important category of ion exchange polymers is the perfluorocarbon type made by DuPont under the trade name Nafion® [17,18]. The base polymer is made by polymerization of a sulfinol fluoride vinyl ether with tetrafluoroethylene. The copolymer formed is extruded as a film about 120 μm thick, after which the sulfinol fluoride groups are hydrolyzed to form sulfonic acid groups:

$$
\begin{array}{l}
-(CF_2CF_2)_n - CFCF_2 - \\
\qquad\qquad\qquad\quad | \\
\qquad\qquad\qquad\quad (OCF_2CF-)_m OCF_2CF_2SO_3H \\
\qquad\qquad\qquad\qquad\qquad | \\
\qquad\qquad\qquad\qquad\quad CF_3 \qquad\qquad m = 1\text{–}3
\end{array}
$$

Asahi Chemical [8] and Tokuyama Soda [19] have developed similar chemistries in which the $-CF_2SO_2F$ groups are replaced by carboxylic acid groups. In these perfluoro polymers, the backbone is extremely hydrophobic whereas the charged acid groups are strongly polar. Because the polymers are not crosslinked, some phase separation into different domains takes place. The hydrophobic perfluoro-polymer domains provide a nonswelling matrix, ensuring the integrity of the membrane. The ionic hydrophilic domains absorb water and form as small clusters distributed throughout the perfluoro-polymer matrix. This configuration, illustrated in Figure 10.6, minimizes both the hydrophobic interaction of ions and water with the backbone and the electrostatic repulsion of close sulfonate groups. These perfluorocarbon membranes are completely inert to concentrated sodium hydroxide solutions and have been widely used in membrane electrochemical cells in the chlor-alkali industry.

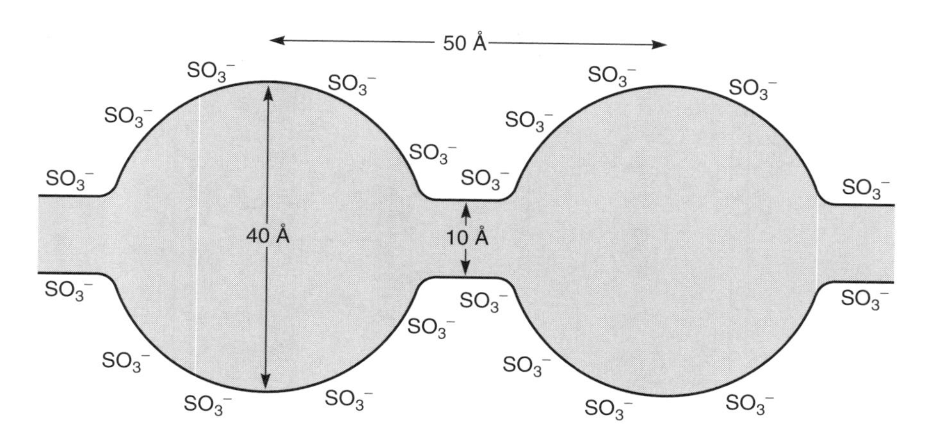

Figure 10.6 Schematic of the cluster model used to describe the distribution of sulfonate groups in perfluorocarbon-type cation exchange membranes such as Nafion [18]

Heterogeneous Membranes

Heterogeneous membranes have been produced by a number of Japanese manufacturers. The simplest form has very finely powdered cation or anion exchange particles uniformly dispersed in polypropylene. A film of the material is then extruded to form the membrane. The mechanical properties of these membranes are often poor because of swelling of the relatively large—10–20 μm diameter—ion exchange particles. A much finer heterogeneous dispersion of ion exchange particles, and consequently a more stable membrane, can be made with a poly(vinyl chloride) (PVC) plastisol. A plastisol of approximately equal parts PVC, styrene monomer and crosslinking agent in a dioctyl phthalate plasticizing solvent is prepared. The mixture is then cast and polymerized as a film. The PVC and polystyrene polymers form an interconnected domain structure. The styrene groups are then sulfonated by treatment with concentrated sulfuric acid or sulfur trioxide to form a very finely dispersed but heterogeneous structure of sulfonated polystyrene in a PVC matrix, which provides toughness and strength.

Transport in Electrodialysis Membranes

Concentration Polarization and Limiting Current Density

Transport of ions in an electrodialysis cell, in which the salt solutions in the chambers formed between the ion exchange membranes are very well stirred, is shown in Figure 10.7. In this example, chloride ions migrating to the left easily

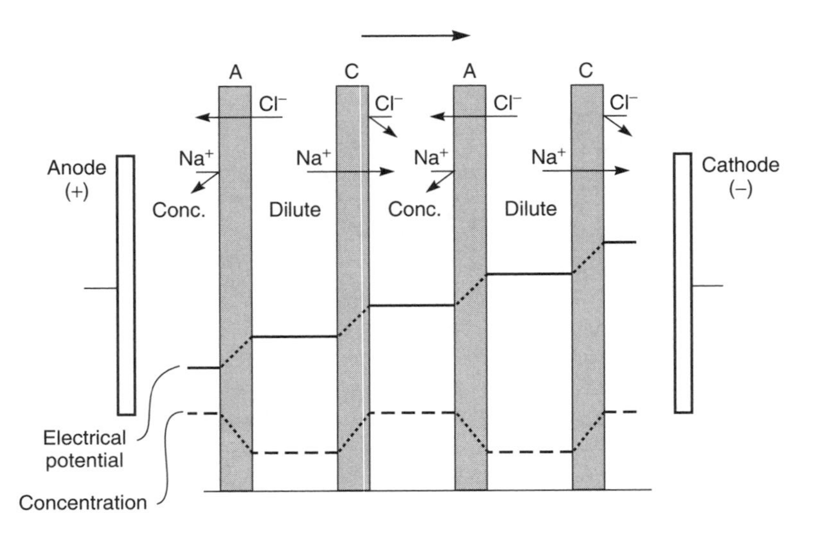

Figure 10.7 Schematic of the concentration and potential gradients in a well-stirred electrodialysis cell

permeate the anionic membranes containing fixed positive groups and are stopped by the cationic membranes containing fixed negative groups. Similarly, sodium ions migrating to the right permeate the cationic membranes but are stopped by the anionic membranes. The overall result is increased salt concentration in alternating compartments while the other compartments are simultaneously depleted of salt. The drawing shown implies that the voltage potential drop caused by the electrical resistance of the apparatus takes place entirely across the ion exchange membrane. This is the case for a very well-stirred cell, in which the solutions in the compartments are completely turbulent. In a well-stirred cell the flux of ions across the membranes and hence the productivity of the electrodialysis system can be increased without limit by increasing the current across the stack. In practice, however, the resistance of the membrane is often small in proportion to the resistance of the water-filled compartments, particularly in the dilute compartment where the concentration of ions carrying the current is low. In this compartment the formation of ion-depleted regions next to the membrane places an additional limit on the current and hence the flux of ions through the membranes. Ion transport through this ion-depleted aqueous boundary layer generally controls electrodialysis system performance.

Concentration polarization controls the performance of practical electrodialysis systems. Because ions selectively permeate the membrane, the concentration of some of the ions in the solution immediately adjacent to the membrane surface becomes significantly depleted compared to the bulk solution concentration. As the voltage across the stack is increased to increase the flux of ions through the membrane, the solution next to the membrane surface becomes increasingly depleted of the permeating ions. Depletion of the salt at the membrane surface means that an increasing fraction of the voltage drop is dissipated in transporting ions across the boundary layer rather than through the membrane. Therefore the energy consumption per unit of salt transported increases significantly. A point can be reached at which the ion concentration at the membrane surface is zero. This represents the maximum transport rate of ions through the boundary layer. The current through the membrane at this point is called the limiting current density, that is, current per unit area of membrane (mA/cm^2). Once the limiting current density is reached, any further increase in voltage difference across the membrane will not increase ion transport or current through the membrane. Normally the extra power is dissipated by side reactions, such as dissociation of the water in the cell into ions, and by other effects. Concentration polarization can be partially controlled by circulating the salt solutions at high flow rates through the cell chambers. But even when very turbulent flow is maintained in the cells, significant concentration polarization occurs.

The formation of concentration gradients caused by the flow of ions through a single cationic membrane is shown in Figure 10.8. As in the treatment of concentration polarization in other membrane processes, the resistance of the aqueous solution is modeled as a thin boundary layer of unstirred solution separating the

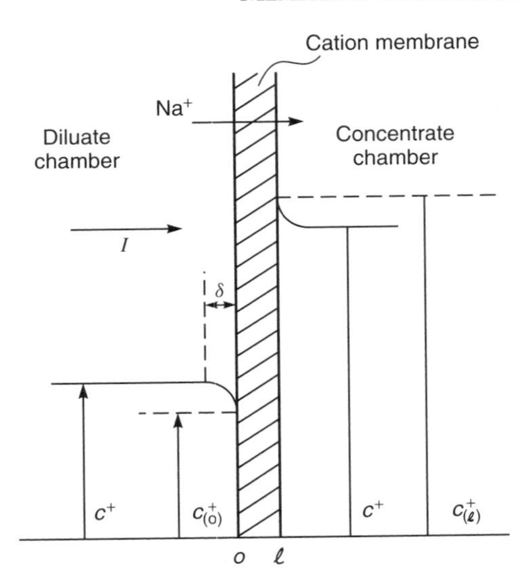

Figure 10.8 Schematic of the concentration gradients adjacent to a single cationic membrane in an electrodialysis stack. The effects of boundary layers that form on each side of the membrane on sodium ion concentrations are shown

membrane surface from the well-stirred bulk solution. In electrodialysis the thickness (δ) of this unstirred layer is generally 20–50 μm. Concentration gradients form in this layer because only one of the ionic species is transported through the membrane. This species is depleted in the boundary layer on the feed side and enriched in the boundary layer on the permeate side.

Figure 10.8 shows the concentration gradient of univalent sodium ions next to a cationic membrane. Exactly equivalent gradients of anions, such as chloride ions, form adjacent to the anionic membranes in the stack. The ion gradient formed on the left, dilute side of the membrane can be described by Fick's law. Thus the rate of diffusion of cations to the surface is given by:

$$J^+ = D^+ \frac{(c^+ - c_{(o)}{}^+)}{\delta} \tag{10.12}$$

where D^+ is the diffusion coefficient of the cation in water, c^+ is the bulk concentration of the cation in the solution, and $c^+{}_{(o)}$ is the concentration of the cation in the solution adjacent to the membrane surface (o).

The rate at which the cations approach the membrane by electrolyte transport is t^+I/F. It follows that the total flux of sodium ions to the membrane surface (J^+) is the sum of these two terms

$$J^+ = \frac{D^+(c^+ - c_{(o)}{}^+)}{\delta} + \frac{t^+I}{F} \tag{10.13}$$

Transport through the membrane is also the sum of two terms, one due to the voltage difference, the other due to the diffusion caused by the difference in ion concentrations on each side of the membrane. Thus, the ion flux through the membrane can be written

$$J^+ = \frac{t_{(m)}{}^+ I}{F} + \frac{P^+(c_{(o)}{}^+ - c_{(\ell)}{}^+)}{\ell} \tag{10.14}$$

where P^+ is the permeability of the sodium ions in a membrane of thickness ℓ. The quantity $P^+(c_{(o)}{}^+ - c^+_{(\ell)})/\ell$ is much smaller than transport due to the voltage gradient, so Equations (10.13) and (10.14) can be combined and simplified to

$$\frac{D^+(c^+ - c_{(o)}{}^+)}{\delta} + \frac{t^+ I}{F} = \frac{t_{(m)}{}^+ I}{F} \tag{10.15}$$

For a selective cationic ion exchange membrane for which $t_{(m)}{}^+ \approx 1$, Equation (10.15) can be further simplified to

$$I = \frac{F}{1 - t^+} \cdot \frac{D^+}{\delta}(c^+ - c_{(o)}{}^+) \tag{10.16}$$

This important equation has a limiting value when the concentration of the ion at the membrane surface is zero ($c_{(o)}{}^+ \approx 0$). At this point the current reaches its maximum value; the limiting current is given by the equation

$$I_{\text{lim}} = \frac{D^+ F c^+}{\delta(1 - t^+)} \tag{10.17}$$

This limiting current, I_{lim}, is the maximum current that can be employed in an electrodialysis process. If the potential required to produce this current is exceeded, the extra current will be carried by other processes, first by transport of anions through the cationic membrane and, at higher potentials, by hydrogen and hydroxyl ions formed by dissociation of water. Both of these undesirable processes consume power without producing any separation. This decreases the current efficiency of the process, that is, the separation achieved per unit of power consumed. A more detailed discussion of the effect of the limiting current density on electrodialysis performance is given by Krol *et al.* [20].

The limiting current can be determined experimentally by plotting the electrical resistance across the membrane stack against the reciprocal electric current. This is called a Cowan–Brown plot after its original developers [21]; Figure 10.9 shows an example for a laboratory cell [22]. At a reciprocal current of 0.1/A, the resistance has a minimum value. When the limiting current is exceeded, the excess current is not used to transport ions. Instead the current causes water to dissociate into protons and hydroxyl ions. The pH of the solutions in the cell chambers then begins to change, reflecting this water splitting. This change in pH, also shown in Figure 10.9, can be used to determine the value of the

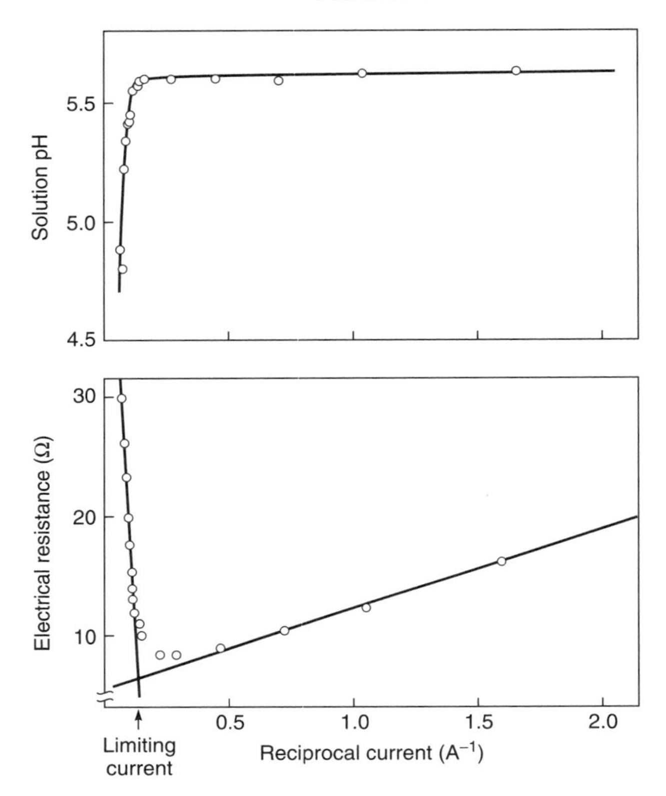

Figure 10.9 Cowan–Brown plots showing how the limiting current density can be determined by measuring the stack resistance or the pH of the dilute solution as a function of current [22]. Redrawn from R. Rautenbach and R. Albrecht, *Membrane Processes*, Copyright © 1989. This material is used by permission of John Wiley & Sons, Ltd

limiting current density. In industrial-scale electrodialysis systems, determining the limiting current is not so easy. In large membrane stacks the boundary layer thickness will vary from place to place across the membrane surface. The limiting current, where the boundary layer is relatively thick because of poor fluid flow distribution, will be lower than where the boundary layer is thinner. Thus, the measured limiting current may be only an approximate value. In practice, systems are operated at currents substantially below the limiting value.

The limiting current density for an electrodialysis system operated at the same feed solution flow rate is a function of the feed solution salt concentration, as shown in Equation (10.17). As the salt concentration in the solution increases, more ions are available to transport current in the boundary layer, so

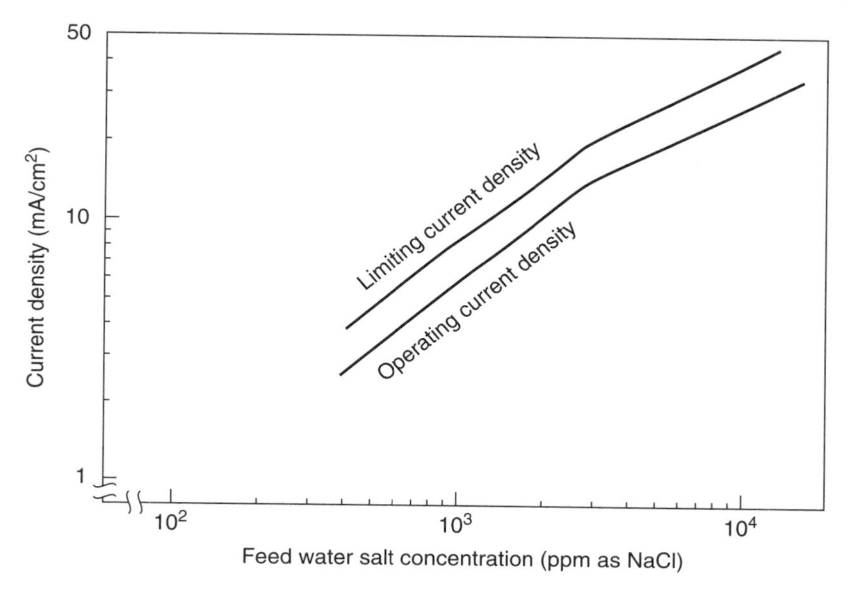

Figure 10.10 Limiting current density and operating current density as a function of feed water salt concentration. The change in slope of the curves at about 3000 ppm salt reflects the change in the activity coefficient of the ions at high salt concentrations [23]

the limiting current density also increases. For this reason large electrodialysis systems with several electrodialysis stacks in series will operate with different current densities in each stack, reflecting the change in the feed water concentration as salt is removed. The normal range of limiting current densities and actual operating current densities used in an electrodialysis system is shown in Figure 10.10[23]. These values may change depending on the particular cell design employed.

Current Efficiency and Power Consumption

A key factor determining the overall efficiency of an electrodialysis process is the energy consumed to perform the separation. Energy consumption E in kilowatts, is linked to the current I through the stack and the resistance R of the stack by the expression

$$E = I^2R \tag{10.18}$$

The theoretical electric current I_{theor} required to perform the separation is directly proportional to the number of charges transported across the ion exchange membrane and is given by the expression

$$I_{\text{theor}} = z\Delta CFQ \tag{10.19}$$

where Q is the feed flow rate, ΔC is the difference in molar concentration between the feed and the dilute solutions, z is the valence of the salt and F is the Faraday constant. Thus the theoretical power consumption E_{theor} to achieve a given separation is given by substituting Equation (10.19) into Equation (10.18) to give:

$$E_{\text{theor}} = IRz\Delta CQF \tag{10.20}$$

or

$$E_{\text{theor}} = Vz\Delta CQF \tag{10.21}$$

where V is the theoretical voltage drop across the stack. In the absence of concentration polarization and any resistance losses in the membrane or solution compartments, the energy required to achieve a separation and a flow of ions out of the concentrated feed solution into the dilute solution for any cell pair is as shown in Figure 10.11.

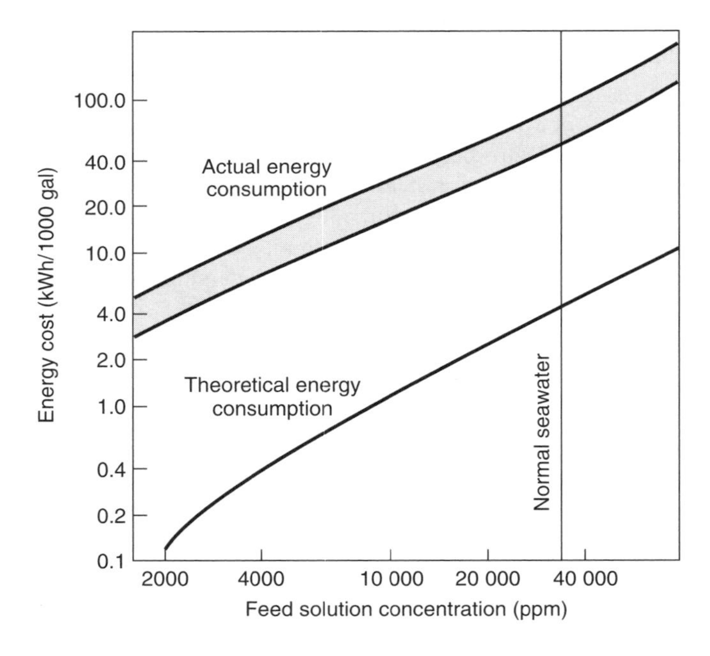

Figure 10.11 Comparison of the theoretical energy consumption and the actual energy consumption of electrodialysis desalination systems. Most of the difference results from concentration polarization effects [24]

The actual voltage drop and hence the energy consumed are higher than the theoretical value for two reasons [24]. First, as shown in Figure 10.8, the concentrations of ions in the solutions adjacent to the membrane surfaces are significantly lower than the bulk solution values. That is, the actual voltage drop used in Equation (10.21) is several times larger than the voltage drop in the absence of polarization. The result is to increase the actual energy consumption five to ten times above the theoretical minimum value. In commercial electrodialysis plants, concentration polarization is controlled by circulating the solutions through the stack at a high rate. Various feed spacer designs are used to maximize turbulence in the cells. Because electric power is used to power the feed and product solution circulation pumps, a trade-off exists between the power saved because of the increased efficiency of the electrodialysis stack and the power consumed by the pumps. In current electrodialysis systems, the circulation pumps consume approximately one-quarter to one-half of the total power. Even under these conditions concentration polarization is not fully controlled and actual energy consumption is substantially higher than the theoretical value.

Most inefficiencies in electrodialysis systems are related to the difficulty in controlling concentration polarization. The second cause is current utilization losses, arising from the following factors [10]:

1. Ion exchange membranes are not completely semipermeable; some leakage of co-ions of the same charge as the membrane can occur. This effect is generally negligible at low feed solution concentrations, but can be serious with concentrated solutions, such as the seawater treated in Japan.
2. Ions permeating the membrane carry solvating water molecules in their hydration shell. Also, osmotic transport of water from the dilute to the concentrated chambers can occur.
3. A portion of the electric current can be carried by the stack manifold, bypassing the membrane cell. Modern electrodialysis stack designs generally make losses due to this effect negligible.

System Design

An electrodialysis plant consists of several elements:

- a feed pretreatment system;
- the membrane stack;
- the power supply and process control unit;
- the solution pumping system.

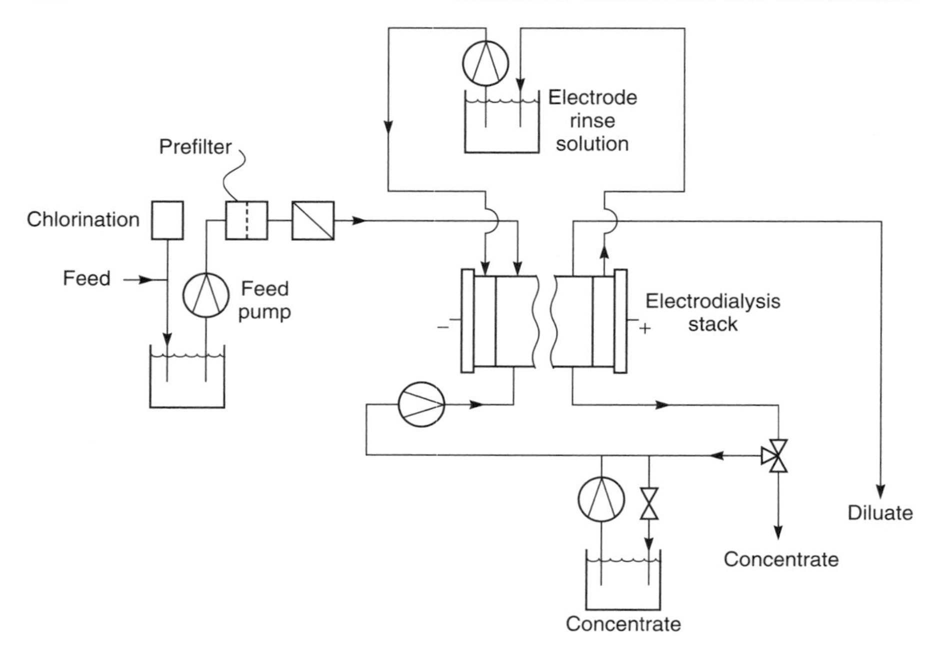

Figure 10.12 Flow diagram of a typical electrodialysis plant [10]

Many plants use a single electrodialysis stack, as shown in Figure 10.12. Manifolding may be used to allow the feed and brine solutions to pass through several cell pairs, but the entire procedure is performed in the single stack.

In large systems, using several electrodialysis stacks in series to perform the same overall separation is more efficient [25]. The current density of the first stack is higher than the current density of the last stack, which is operating on a more dilute feed solution. As in the single-stack system, the feed solution may pass through several cell pairs in each stack. Because concentration polarization becomes more important as the solution becomes more dilute, the solution velocity increases in the stacks processing the most dilute solution. The velocity is controlled by the number of cell pairs through which the solution passes in each stack. The number of cell pairs decreases from the first to the last stack; this is known as the taper of the system. The flow scheme of a three-stage design is shown in Figure 10.13.

Feed Pretreatment. The type and complexity of the feed pretreatment system depends on the content of the water to be treated. As in reverse osmosis, most feed water is sterilized by chlorination to prevent bacterial growth on the membrane.

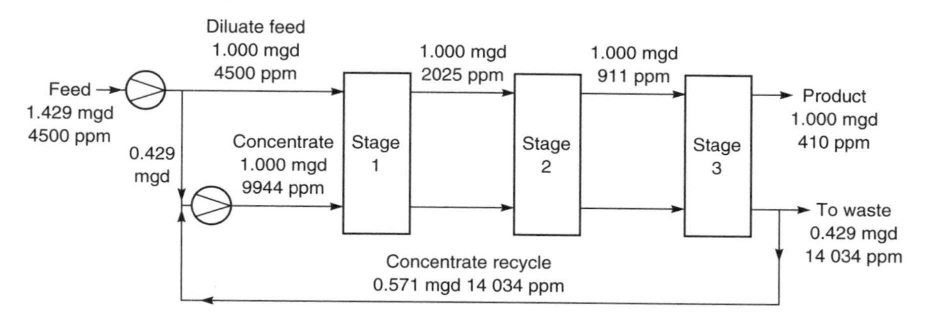

Figure 10.13 Flow scheme of a three-stage electrodialysis plant [25]. Reprinted from A.N. Rogers, Design and Operation of Desalting Systems Based on Membrane Processes, in *Synthetic Membrane Processes*, G. Belfort (ed.), Academic Press, Copyright 1977, with permission from Elsevier

Scaling on the membrane surface by precipitation of sparingly soluble salts such as calcium sulfate is usually controlled by adding precipitation inhibitors such as sodium hexametaphosphate. The pH may also be adjusted to maintain salts in their soluble range. Large, charged organic molecules or colloids such as humic acid are particularly troublesome impurities, because they are drawn by their charge to the membrane surface but are too large to permeate. They then accumulate at the dilute solution side of the membrane and precipitate, causing an increase in membrane resistance. Filtration of the feed water may control these components, and operation in the polarity reversal mode is often effective.

Membrane Stack. After the pretreatment step, the feed water is pumped through the electrodialysis stack. This stack normally contains 100–200 membrane cell pairs each with a membrane area between 1 and 2 m^2. Plastic mesh spacers form the channels through which the feed and concentrate solutions flow. Most manufacturers use one of the spacer designs shown in Figure 10.14. In the tortuous path cell design of Figure 10.14(a), a solid spacer grid forms a long open channel through which the feed solution flows at relatively high velocity. The channel is not held open by netting, so the membranes must be thick and sturdy to prevent collapse of the channels. In the sheet flow design of Figure 10.14(b), the gap between the membrane leaves is maintained by a polyolefin mesh spacer. The spacer is made as thin as possible without producing an excessive pressure drop.

Two membranes and two gasket spacers form a single cell pair. Holes in the gasket spacers are aligned with holes in the membrane sheet to form the manifold

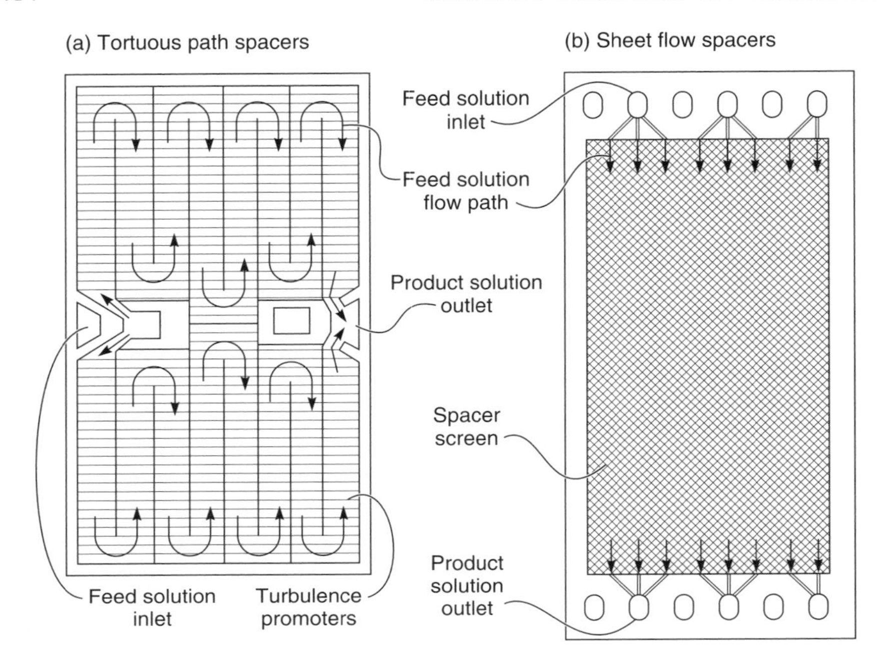

Figure 10.14 The two main types of feed solution flow distribution spacers used in electrodialysis [10]

channels through which the dilute and concentrated solutions are introduced into each cell. The end plate of the stack is a rigid plastic frame containing the electrode compartment. The entire arrangement is compressed together with bolts between the two end flow plates. The perimeter gaskets of the gasket spacers are tightly pressed into the membranes to form the cells. A large electrodialysis stack has several hundred meters of fluid seals around each cell. Early units often developed small leaks over time, causing unsightly salt deposits on the outside of the stacks. These problems have now been largely solved. In principle, an electrodialysis stack can be disassembled and the membranes cleaned or replaced on-site. In practice, this operation is performed infrequently and almost never in the field.

Power Supply and Process Control Unit. Electrodialysis systems use large amounts of direct current power; the rectifier required to convert AC to DC and to control the operation of the system represents a significant portion of a plant's capital cost. A typical voltage drop across a single cell pair is in the range 1–2 V and the normal current flow is 40 mA/cm^2. For a 200-cell-pair stack containing 1 m^2 of membrane, the total voltage is about 200–400 V and the current about

400 A per stack. This is a considerable amount of electric power, and care must be used to ensure safe operation.

Solution Pumping System. A surprisingly large fraction of the total power used in electrodialysis systems is consumed by the water pumps required to circulate feed and concentrate solutions through the stacks. This fraction increases as the average salt concentration of the feed decreases and can become dominant in electrodialysis of low-concentration solutions (less than 500 ppm salt). The pressure drop per stack varies from 15 to 30 psi for sheet flow cells to as much as 70–90 psi for tortuous path cells. Depending on the separation required, the fluid will be pumped through two to four cells in series, requiring interstage pumps for each stack.

Applications

Brackish Water Desalination

Brackish water desalination is the largest application of electrodialysis. The competitive technologies are ion exchange for very dilute saline solutions, below 500 ppm, and reverse osmosis for concentrations above 2000 ppm. In the 500–2000 ppm range electrodialysis is often the low-cost process. One advantage of electrodialysis applied to brackish water desalination is that a large fraction, typically 80–95 % of the brackish feed, is recovered as product water. However, these high recoveries mean that the concentrated brine stream produced is five to twenty times more concentrated than the feed. The degree of water recovery is limited by precipitation of insoluble salts in the brine.

Since the first plants were produced in the early 1950s, several thousand brackish water electrodialysis plants have been installed around the world. Modern plants are generally fully automated and require only periodic operator attention. This has encouraged production of many small trailer-mounted plants. However, a number of large plants with production rates of 10 million gal/day or more have also been installed.

The power consumption of an electrodialysis plant is directly proportional to the salt concentration of the feed water, varying from 4 kWh/1000 gal for 1000 ppm feed water to 10–15 kWh/1000 gal for 5000 ppm feed water. About one-quarter to one-third of this power is used to drive the feed water circulation pumps.

Salt Recovery from Seawater

The second major application of electrodialysis is the production of table salt by concentration of seawater [8]. This process is only practiced in Japan, which has no other domestic salt supply. The process is heavily subsidized by the

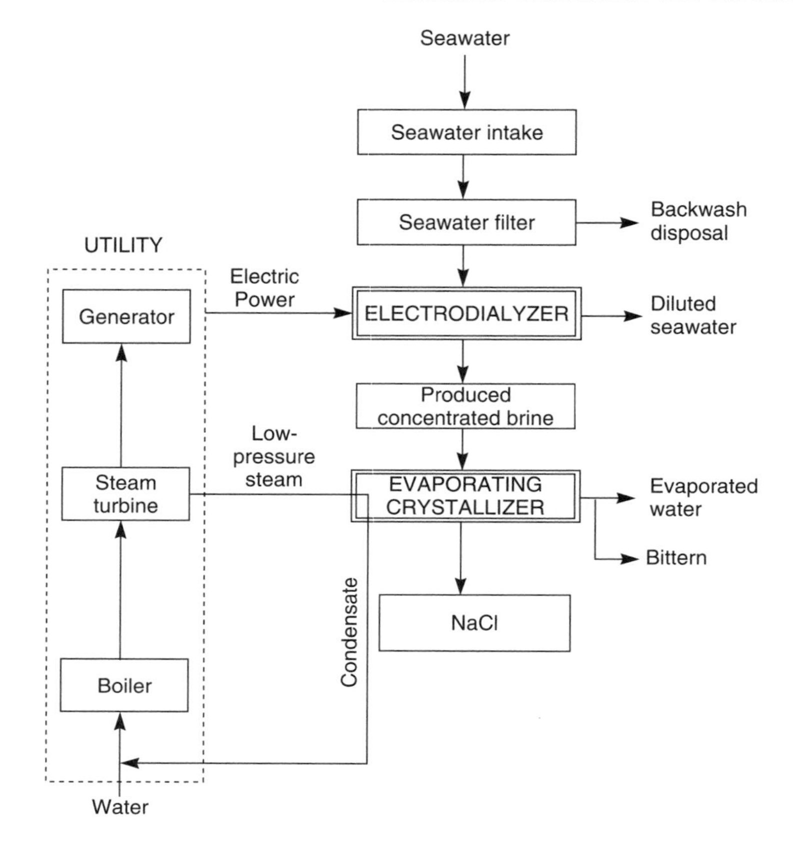

Figure 10.15 Flow scheme of the electrodialysis unit used in a seawater salt concentration plant [8]

government, and total production is approximately 1.2 million tons/year of salt. In total, these plants use more than 500 000 m^2 of membrane.

A flow scheme of one such seawater salt production plant is shown in Figure 10.15. A cogeneration unit produces the power required for the electrodialysis operation, which concentrates the salt in sea water to about 18–20 wt%. The waste stream from the power plant is then used to further concentrate the salt by evaporation.

Seawater contains relatively high concentrations of sulfate (SO_4^{2-}), calcium (Ca^{2+}), magnesium (Mg^{2+}) and other multivalent ions that can precipitate in the concentrated salt compartments of the plant and cause severe scaling. This problem has been solved by applying a thin polyelectrolyte layer of opposite charge to the ion exchange membrane on the surface facing the seawater solution. A cross-section of a coated anionic membrane is shown in Figure 10.16.

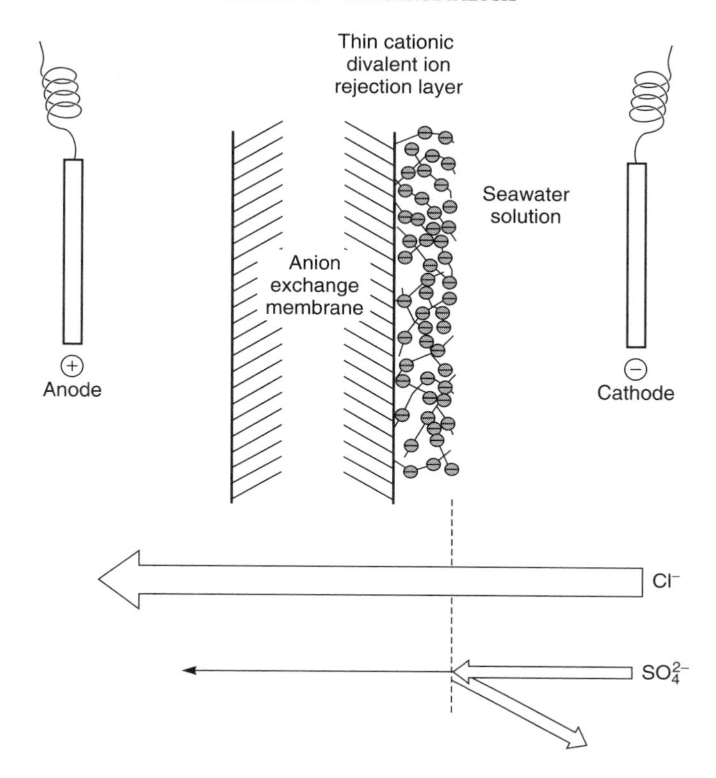

Figure 10.16 Polyelectrolyte-coated ion exchange membranes used to separate multivalent and monovalent ions in seawater salt concentration plants [8]

Because the Donnan exclusion effect is much stronger for multivalent ions than for univalent ions, the polyelectrolyte layer rejects multivalent ions but allows the univalent ions to pass relatively unhindered.

Other Electrodialysis Separation Applications

The two water desalination applications described above represent the majority of the market for electrodialysis separation systems. A small application exists in softening water, and recently a market has grown in the food industry to desalt whey and to remove tannic acid from wine and citric acid from fruit juice. A number of other applications exist in wastewater treatment, particularly regeneration of waste acids used in metal pickling operations and removal of heavy metals from electroplating rinse waters [11]. These applications rely on the ability of electrodialysis membranes to separate electrolytes from nonelectrolytes and to separate multivalent from univalent ions.

The arrangement of membranes in these systems depends on the application. Figure 10.17(a) shows a stack comprising only cation exchange membranes to soften water, whereas Figure 10.17(b) shows an all-anion exchange membrane stack to deacidify juice [26]. In the water-softening application, the objective is to exchange divalent cations such as calcium and magnesium for sodium ions. In the juice deacidification process, the all-anion stack is used to exchange citrate ions for hydroxyl ions. These are both ion exchange processes, and the salt concentration of the feed solution remains unchanged.

Continuous Electrodeionization and Ultrapure Water

Electrodeionization systems were first suggested to remove small amounts of radioactive elements from contaminated waters [27], but the principal current application is the preparation of ultrapure water for the electronics and pharmaceutical industries [28]. The process is sometimes used as a polishing step after the water has been pretreated with a reverse osmosis unit.

In the production of ultrapure water for the electronics industry, salt concentrations must be reduced to the ppb range. This is a problem with conventional electrodialysis units because the low conductivity of very dilute feed water streams generally limits the process to producing water in the 10 ppm range. This limitation can be overcome by filling the dilute chambers of the electrodialysis stack with fine mixed-bed ion exchange beads as shown in Figure 10.18. The ions enter the chamber, partition into the ion exchange resin beads and are concentrated many times. As a result ion and current flow occur through the resin bed, and the resistance of the cell is much lower than for a normal cell operating on the same very dilute feed. An additional benefit is that, towards the bottom of the bed where the ion concentration is in the ppb range, a certain amount of water splitting occurs. This produces hydrogen and hydroxyl ions that also migrate to the membrane surface through the ion exchange beads. The presence of these ions maintains a high pH in the anion exchange beads and a low pH in the cation exchange beads. These extreme pHs enhance the ionization and removal of weakly ionized species such as carbon dioxide and silica that would otherwise be difficult to remove. Such modified electrodialysis systems can reduce most ionizable solutes to below ppb levels.

Bipolar Membranes

Bipolar membranes consist of an anionic and a cationic membrane laminated together [13]. When placed between two electrodes, as shown in Figure 10.19, the interface between the anionic and cationic membranes becomes depleted of ions. The only way a current can then be carried is by the water splitting reaction, which liberates hydrogen ions that migrate to the cathode and hydroxyl ions that

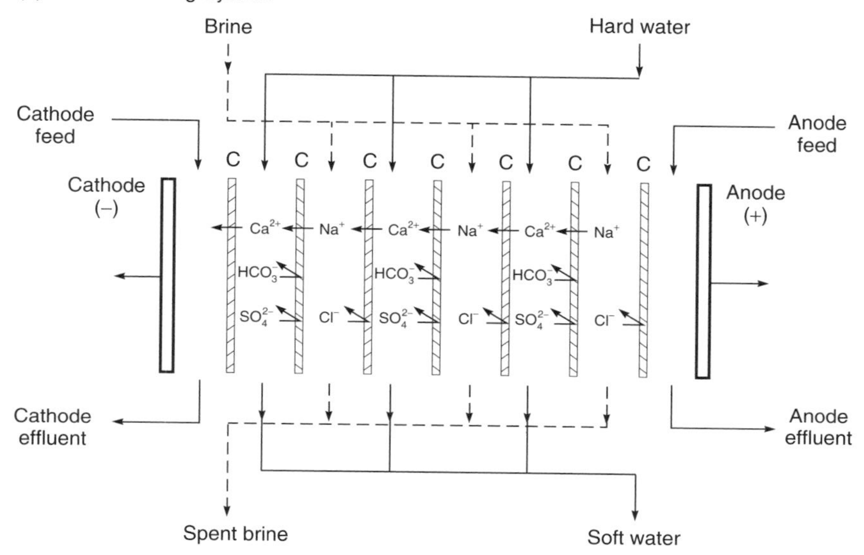

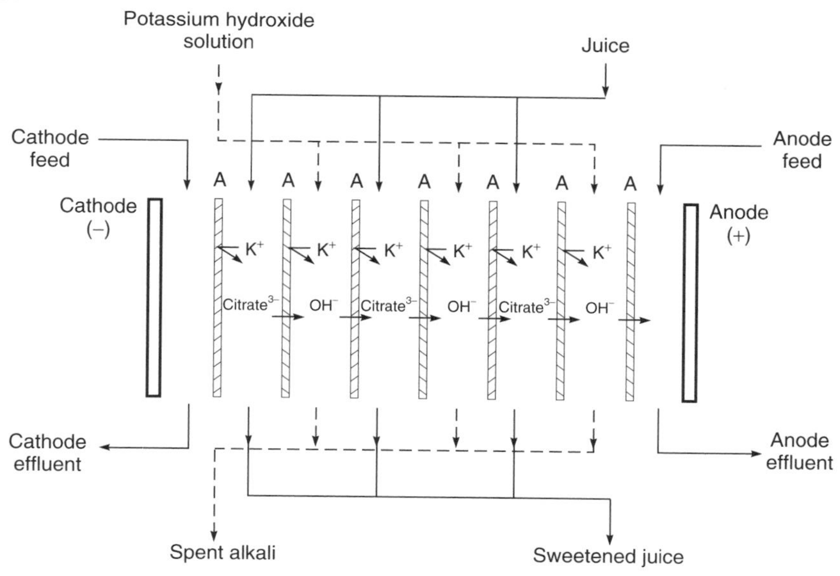

Figure 10.17 Flow schematic of electrodialysis systems used to exchange target ions in the feed solution. (a) An all-cation exchange membrane stack to exchange sodium ions for calcium ions in water softening. (b) An all-anion exchange membrane stack to exchange hydroxyl ions for citrate ions in deacidification of fruit juice

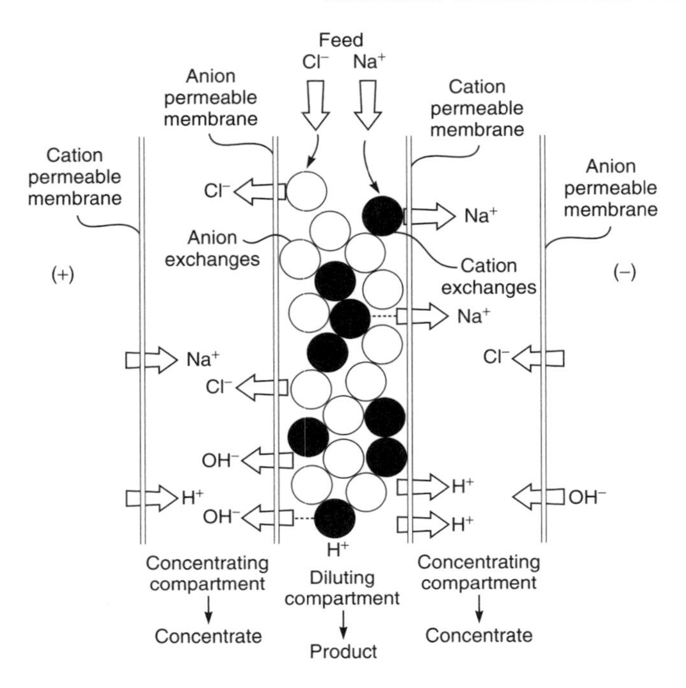

Figure 10.18 Schematic of the electrodeionization process using mixed-bed ion exchange resin to increase the conduction of the dilute compartments of the electrodialysis stack

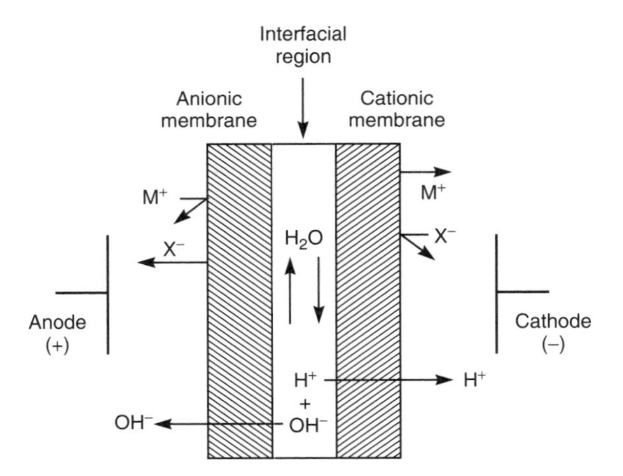

Figure 10.19 Schematic of a single bipolar membrane (not to scale) showing generation of hydroxyl and hydrogen ions by water splitting in the interior of the membrane. Electrolysis takes place in the thin interfacial region between the anodic and cathodic membranes

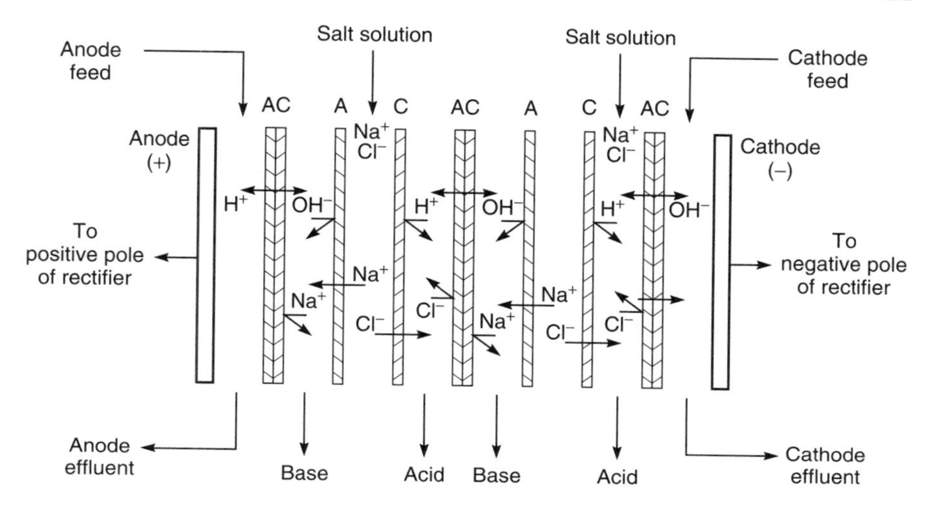

Figure 10.20 Schematic of a bipolar membrane process to split sodium chloride into sodium hydroxide and hydrochloric acid

migrate to the anode. The mechanism of water splitting in these membranes has been discussed in detail by Strathmann *et al.* [29]. The phenomenon can be utilized in an electrodialysis stack composed of a number of sets of three-chamber cells between two electrodes, as shown in Figure 10.20. Salt solution flows into the middle chamber; cations migrate to the chamber on the left and anions to the chamber on the right. Electrical neutrality is maintained in these chambers by hydroxyl and hydrogen ions provided by water splitting in the bipolar membranes that bound each set of three chambers [30].

Several other arrangements of bipolar membranes can achieve the same overall result, namely, dividing a neutral salt into the conjugate acid and base. The process is limited to the generation of relatively dilute acid and base solutions. Also, the product acid and base are contaminated with 2–4 % salt. Nevertheless the process is significantly more energy efficient than the conventional electrolysis process because no gases are involved. Total current efficiency is about 80 % and the system can often be integrated into the process generating the feed salt solution. A process utilizing bipolar membranes was first reported by Liu in 1977 [13]. Aquatech, originally a division of Allied Chemicals and now part of Graver Water, has pursued commercialization of the process for almost 20 years, but only a handful of plants have been installed. The principal problem appears to have been membrane instability, but these problems may now have been solved. Prospects for the process appear to be improving. A recent review of bipolar membrane technology has been produced by Kemperman [31].

Three other processes using ion exchange membranes (Donnan dialysis, diffusion dialysis and piezodialysis) are covered in Chapter 13.

Conclusions and Future Directions

Electrodialysis is by far the largest use of ion exchange membranes, principally to desalt brackish water or (in Japan) to produce concentrated brine. These two processes are both well established, and major technical innovations that will change the competitive position of the industry do not appear likely. Some new applications of electrodialysis exist in the treatment of industrial process streams, food processing and wastewater treatment systems but the total market is small. Long-term major applications for ion exchange membranes may be in the non-separation areas such as fuel cells, electrochemical reactions and production of acids and alkalis with bipolar membranes.

References

1. W. Ostwald, Elektrische Eigenschaften Halbdurchlässiger Scheidewande, *Z. Physik. Chem.* **6**, 71 (1890).
2. F.G. Donnan, Theory of Membrane Equilibria and Membrane Potentials in the Presence of Non-dialyzing Electrolytes, *Z. Elektrochem.* **17**, 572 (1911).
3. E. Manegold and C. Kalauch, Uber Kapillarsysteme, XXII Die Wirksamkeit Verschiedener Reinigungsmethoden (Filtration, Dialyse, Electrolyse und Ihre Kombinationen), *Kolloid Z.* **86**, 93 (1939).
4. K.H. Meyer and W. Strauss, *Helv. Chem. Acta* **23**, 795 (1940).
5. T.R.E. Kressman, Ion Exchange Resin Membranes and Impregnated Filter Papers, *Nature* **165**, 568 (1950).
6. E.A. Murphy, F.J. Paton and J. Ansel, US Patent 2,331,494 (1943).
7. W. Juda and W.A. McRae, Coherent Ion-exchange Gels and Membranes, *J. Am. Chem. Soc.* **72,** 1044 (1950).
8. M. Seko, H. Miyauchi and J. Omura, Ion Exchange Membrane Application for Electrodialysis, Electroreduction, and Electrohydrodimerisation, in *Ion Exchange Membranes*, D.S. Flett (ed.), Ellis Horwood Ltd, Chichester, pp. 121–136 (1983).
9. L.R. Siwak, Here's How Electrodialysis Reverses... and Why EDR Works, *Desalination Water Reuse* **2**, 16 (1992).
10. H. Strathmann, Electrodialysis, in *Membrane Separation Systems*, R.W. Baker, E.L. Cussler, W. Eykamp, W.J. Koros, R.L. Riley and H. Strathmann (eds), Noyes Data Corp., Park Ridge, NJ, pp. 396–448 (1991).
11. H. Strathmann, Electrodialysis and Its Application in the Chemical Process Industry, *Sep. Purif. Methods* **14**, 41 (1985).
12. M.P. Grotheer, Electrochemical Processing (Inorganic), in *Kirk Othmer Encyclopedia of Chemical Technology*, 4th Edn, Wiley Interscience, Wiley, New York, NY, Vol. 9, p. 124 (1992).
13. K.-J. Liu, F.P. Chlanda and K.J. Nagasubramanian, Use of Biopolar Membranes for Generation of Acid and Base: An Engineering and Economical Analysis, *J. Membr. Sci.* **2**, 109 (1977).
14. K. Nagasubramanian, F.P. Chlanda and K.-J. Liu, Bipolar Membrane Technology: An Engineering and Economic Analysis, *Recent Advances in Separation Tech-II*, AIChE Symposium Series Number 1192, AIChE, New York, NY, Vol. 76, p. 97 (1980).
15. R. Yeo, Application of Perfluorosulfonated Polymer Membranes in Fuel Cells, Electrolysis, and Load Leveling Devices, in *Perfluorinated Ionomer Membranes*, A. Eisenberg and H.L. Yeager (eds), ACS Symposium Series Number 180, American Chemical Society, Washington, DC, pp. 453–473 (1982).

16. E. Korngold, Electrodialysis—Membranes and Mass Transport, in *Synthetic Membrane Processes*, G. Belfort (ed.), Academic Press, Orlando, FL, pp. 191–220 (1984).
17. A. Eisenberg and H.L. Yeager (eds.), Perfluorinated Ionomer Membranes, ACS Symposium Series Number 180, American Chemical Society, Washington, DC (1982).
18. T.D. Gierke, Ionic Clustering in Nafion Perfluorosulfonic Acid Membranes and Its Relationship to Hydroxyl Rejection and Chlor-Alkali Current Efficiency, Paper presented at the Electrochemical Society Fall Meeting, Atlanta, GA (1977).
19. T. Sata, K. Motani and Y. Ohaski, Perfluorinated Ion Exchange Membrane, Neosepta-F and Its Properties, in *Ion Exchange Membrane*, D.S. Flett (ed.), Ellis Horwood Ltd, Chichester, pp. 137–150 (1983).
20. J.J. Krol, M. Wessling and H. Strathmann, Concentration Polarization with Monopolar Ion Exchange Membranes: Current–Voltage Curves and Water Dissociation, *J. Membr. Sci.* **162**, 145 (1999)
21. D.A. Cowan and J.H. Brown, Effects of Turbulence on Limiting Current in Electrodialysis Cells, *Ind. Eng. Chem.* **51**, 1445 (1959).
22. R. Rautenbach and R. Albrecht, *Membrane Processes*, John Wiley & Sons, Ltd, Chichester (1989).
23. M. Hamada, Brackish Water Desalination by Electrodialysis, *Desalination Water Reuse* **2**, 8 (1992).
24. L.H. Shaffer and M.S. Mintz, Electrodialysis, in *Principles of Desalination*, K.S. Spiegler (ed.), Academic Press, New York, pp. 200–289 (1966).
25. A.N. Rogers, Design and Operation of Desalting Systems Based on Membrane Processes, in *Synthetic Membrane Processes*, G. Belfort (ed.), Academic Press, Orlando, FL, pp. 437–476 (1984).
26. J.A. Zang, Sweetening Citrus Juice in Membrane Process for Industry, *Proc. Southern Res. Inst. Conf.* 35 (1966).
27. W.R. Walters D.W. Weisner and L.J. Marek, Concentration of Radioactive Aqueous Wastes: Electromigration Through Ion-exchange Membranes, *Ind. Eng. Chem.* **47**, 61 (1955).
28. G.C. Ganzi, A.D. Jha, F. DiMascio and J.H. Wood, Electrodeionization: Theory and Practice of Continuous Electrodeionization, *Ultrapure Water* **14**, 64 (1997).
29. H. Strathmann, J.J. Krol, H.-J. Rapp and G. Eigenberger, Limiting Current Density and Water Dissociation in Bipolar Membranes, *J. Membr. Sci.* **125**, 123 (1997).
30. C. Carmen, Bipolar Membrane Pilot Performance in Sodium Chloride Salt Splitting, *Desalination Water Reuse* **4**, 46 (1994).
31. A.J.B. Kemperman (ed.), *Handbook on Bipolar Membrane Technology*, Twente University Press, Twente (2000).

11 CARRIER FACILITATED TRANSPORT

Introduction and History

Carrier facilitated transport membranes incorporate a reactive carrier in the membrane. The carrier reacts with and helps to transport one of the components of the feed across the membrane. Much of the work on carrier facilitated transport has employed liquid membranes containing a dissolved carrier agent held by capillary action in the pores of a microporous film.

The types of transport that can occur in a liquid membrane are illustrated in Figure 11.1 Passive diffusion down a concentration gradient is the most familiar—this process is usually relatively slow and nonselective. In facilitated transport, the liquid membrane phase contains a carrier agent that chemically combines with the permeant to be transported. In the example shown, the carrier is hemoglobin, which transports oxygen. On the upstream side of the membrane, hemoglobin reacts with oxygen to form oxyhemoglobin, which then diffuses to the downstream membrane interface. There, the reaction is reversed: oxygen is liberated to the permeate gas and hemoglobin is re-formed. The hemoglobin then diffuses back to the feed side of the membrane to pick up more oxygen. In this way, hemoglobin acts as a shuttle to selectively transport oxygen through the membrane. Other gases that do not react with hemoglobin, such as nitrogen, are left behind.

Coupled transport resembles facilitated transport in that a carrier agent is incorporated in the membrane. However, in coupled transport, the carrier agent couples the flow of two species. Because of this coupling, one of the species can be moved against its concentration gradient, provided the concentration gradient of the second coupled species is sufficiently large. In the example shown in Figure 11.1, the carrier is an oxime that forms an organic-soluble complex with copper ions. The reaction is reversed by hydrogen ions. On the feed side of the membrane two oxime carrier molecules pick up a copper ion, liberating two hydrogen ions to the feed solution. The copper–oxime complex then diffuses to the downstream membrane interface, where the reaction is reversed because of

Membrane Technology and Applications R. W. Baker
 ISBN: 0-470-85445-6

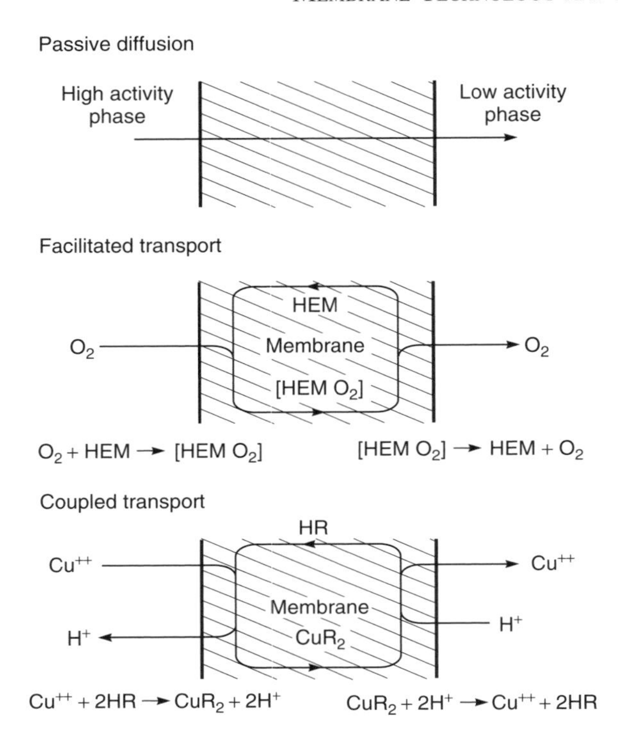

Figure 11.1 Schematic examples of passive diffusion, facilitated transport and coupled transport. The facilitated transport example shows permeation of oxygen across a membrane using hemoglobin as the carrier agent. The coupled transport example shows permeation of copper and hydrogen ions across a membrane using a reactive oxime as the carrier agent

the higher concentration of hydrogen ions in the permeate solution. The copper ion is liberated to the permeate solution, and two hydrogen ions are picked up. The re-formed oxime molecules diffuse back to the feed side of the membrane. Because carrier facilitated transport has so often involved liquid membranes, the process is sometimes called liquid membrane transport, but this is a misnomer, because solid membranes containing carriers dispersed or dissolved in the polymer matrix are being used increasingly.

Coupled transport was the first carrier facilitated process developed, originating in early biological experiments involving natural carriers contained in cell walls. As early as 1890, Pfeffer postulated that the transport in these membranes involved carriers. Perhaps the first coupled transport experiment was performed by Osterhout, who studied the transport of ammonia across algae cell walls [1]. A biological explanation of the coupled transport mechanism in liquid membranes is shown in Figure 11.2 [2].

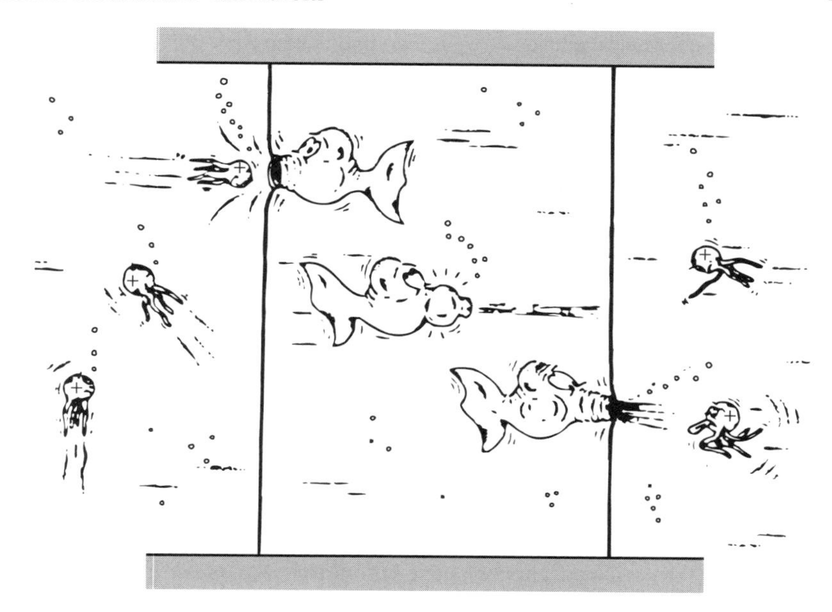

Figure 11.2 Gliozzi's biological model of coupled transport [2]

By the 1950s, the carrier concept was well established, and workers began to develop synthetic analogs of the natural systems. For example, in the mid-1960s, Shean and Sollner [3] studied a number of coupled transport systems using inverted U-tube membranes. At the same time, Bloch and Vofsi published the first of several papers in which coupled transport was applied to hydrometallurgical separations, namely the separation of uranium using phosphate ester carriers [4–6]. Because phosphate esters also plasticize poly(vinyl chloride) (PVC), Bloch and Vofsi prepared immobilized liquid films by dissolving the esters in a PVC matrix. The PVC/ester film, containing 60 wt% ester, was cast onto a paper support. Bloch and others actively pursued this work until the early 1970s. At that time, interest in this approach lagged, apparently because the fluxes obtained did not make the process competitive with conventional separation techniques.

Following the work of Bloch and Vofsi, other methods of producing immobilized liquid films were introduced. In one approach, the liquid carrier phase was held by capillarity within the pores of a microporous substrate, as shown in Figure 11.3(a). This approach was first used by Miyauchi [7] and by Largman and Sifniades and others [8,9]. The principal objective of this early work was to recover copper, uranium and other metals from hydrometallurgical solutions. Despite considerable effort on the laboratory scale, the first pilot plant was not installed until 1983 [10]. The main problem was instability of the liquid carrier phase held in the microporous membrane support.

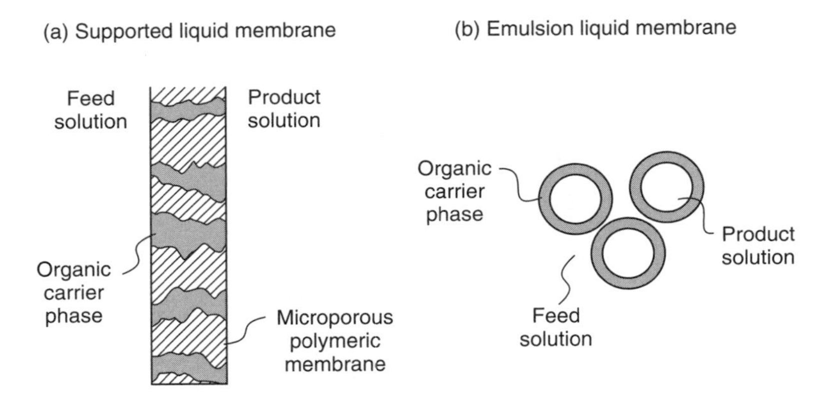

Figure 11.3 Methods of forming liquid membranes

Another type of immobilized liquid carrier is the emulsion or 'bubble' membrane. This technique employs a surfactant-stabilized emulsion as shown in Figure 11.3(b). The organic carrier phase forms the wall of an emulsion droplet separating the aqueous feed from the aqueous product solution. Metal ions are concentrated in the interior of the droplets. When sufficient metal has been extracted, the emulsion droplets are separated from the feed, and the emulsion is broken to liberate a concentrated product solution and an organic carrier phase. The carrier phase is decanted from the product solution and recycled to make more emulsion droplets. One technical problem is the stability of the liquid membrane. Ideally, the emulsion membrane would be completely stable during the extraction step to prevent the two aqueous phases mixing, but would be completely broken and easily separated in the stripping step. Achieving this level of control over emulsion stability has proved difficult. The technique of emulsion membranes was invented and popularized by Li and his co-workers at Exxon, starting in the late 1960s and continuing for more than 20 years [11–14]. The first use of these membranes was as passive devices to extract phenol from water. In 1971–1973 Cussler used this technique with carriers to selectively transport metal ions [15,16]. The Exxon group's work led to the installation of a pilot plant in 1979 [17]. The process is still not commercial, although a number of pilot plants have been installed, principally on hydrometallurgical feed streams [18].

More recently the use of membrane contactors to solve the stability problem of liquid membranes has been proposed [19–21]. The concept is illustrated in Figure 11.4. Two membrane contactors are used, one to separate the organic carrier phase from the feed and the other to separate the organic carrier phase from the permeate. In the first contactor metal ions in the feed solution diffuse across the microporous membrane and react with the carrier, liberating hydrogen counter ions. The organic carrier solution is then pumped from the first to the second membrane contactor, where the reaction is reversed. The metal ions are

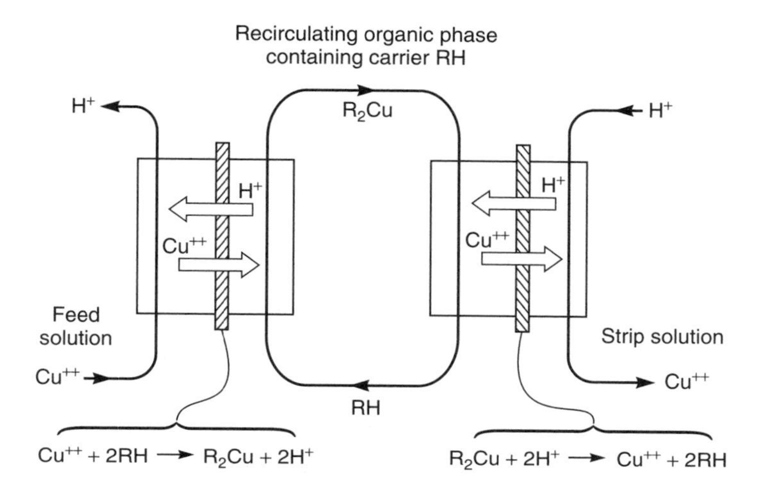

Figure 11.4 Use of two contactors in a liquid membrane process

liberated to the permeate solution, and hydrogen ions are picked up. The reformed carrier solution is then pumped back to the first membrane contactor. Sirkar [19,20] has used this system to separate metal ions. A similar process was developed to the large demonstration plant scale by Davis *et al.* at British Petroleum for the separation of ethylene/ethane mixtures, using a silver nitrate solution as the carrier phase [21].

Carrier facilitated transport processes often achieve spectacular separations between closely related species because of the selectivity of the carriers. However, no coupled transport process has advanced to the commercial stage despite a steady stream of papers in the academic literature. The instability of the membranes is a major technical hurdle, but another issue has been the marginal improvements in economics offered by coupled transport processes over conventional technology such as solvent extraction or ion exchange. Major breakthroughs in performance are required to make coupled transport technology commercially competitive.

Facilitated transport membranes are also a long way from the commercial stage and are plagued by many difficult technical problems. However, the economic rationale for developing facilitated transport membranes is at least clear. Practical facilitated transport membranes, able to separate gas mixtures for which polymeric membranes have limited selectivity, would be adopted. Target applications meeting this criterion are the separation of oxygen and nitrogen and the separation of paraffin/olefin mixtures. The selectivities of current polymeric membranes are modest in both of these separations. Scholander [22] reported the first work on facilitated transport in 1960—he studied the transport of oxygen through aqueous hemoglobin solutions. In the late 1960s through the early 1980s a great

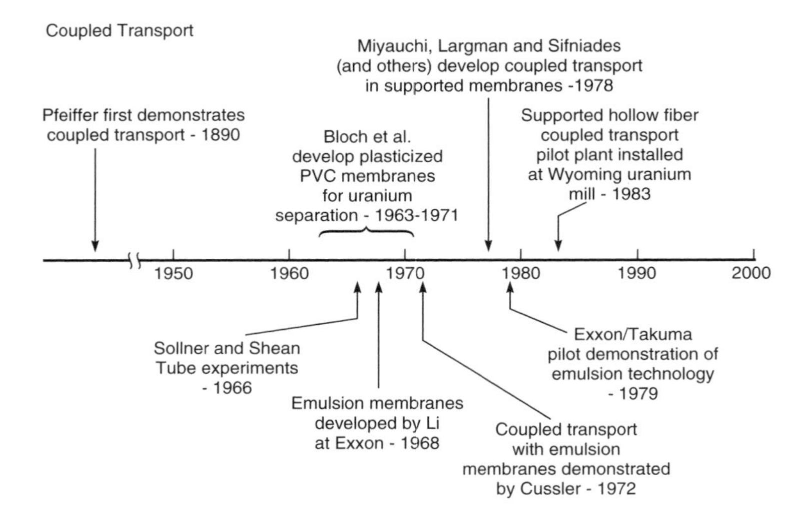

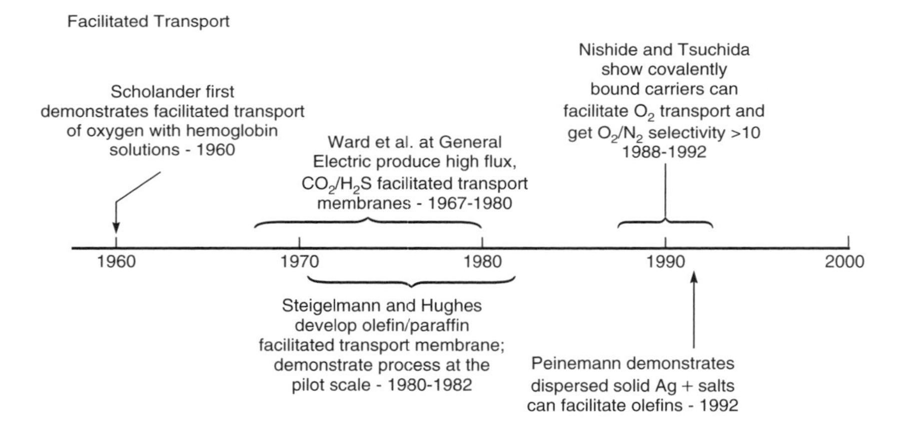

Figure 11.5 Milestones in the development of carrier facilitated transport

deal of work was performed by Ward and others at General Electric [23–26] and Steigelmann and Hughes [27] at Standard Oil. Ward's work focused on carbon dioxide and hydrogen sulfide separation, and some remarkable selectivities were obtained. However, the problems of membrane stability and scale-up were never solved. This group eventually switched to the development of passive polymeric gas separation membranes. Steigelmann and Hughes at Standard Oil concentrated most of their efforts on propylene/propane and ethylene/ethane separation, using concentrated silver salt solutions as carriers. Propylene/propane selectivities of several hundred were obtained, and the process was developed to the pilot plant stage. The principal problem was instability of the silver–olefin complex, which led to a decline in membrane flux and stability over 10–20 days. Although the membrane could be regenerated periodically, this was impractical in an industrial plant.

Following the development of good quality polymeric gas separation membranes in the early 1980s, industrial interest in facilitated transport waned. However, in the last few years, a number of workers have shown that facilitated transport membranes can be made by dispersing or complexing the carrier into a solid polymeric film. Such membranes are more stable than immobilized liquid film membranes, and formation of these membranes into thin, high-flux membranes by conventional techniques should be possible. Nishide, Tsuchida and others in Japan, working with immobilized oxygen carriers [28–30] and Peinemann in Germany [31] and Ho [32] and Pinnau [33] in the US, working with silver salts for olefin separation, have reported promising results. Apparently, the carrier mechanism in these membranes involves the permeant gas molecule hopping from active site to active site.

A milestone chart showing the historical development of carrier facilitated transport membranes is given in Figure 11.5. Because of the differences between coupled and facilitated transport applications these processes are described separately. Reviews of carrier facilitated transport have been given by Ho *et al.* [18], Cussler, Noble and Way [34–37,39], Laciak [38] and Figoli *et al.* [40].

Coupled Transport

Background

Carrier facilitated transport involves a combination of chemical reaction and diffusion. One way to model the process is to calculate the equilibrium between the various species in the membrane phase and to link them by the appropriate rate expressions to the species in adjacent feed and permeate solutions. An expression for the concentration gradient of each species across the membrane is then calculated and can be solved to give the membrane flux in terms of the diffusion coefficients, the distribution coefficients, and the rate constants for all the species involved in the process [41,42]. Unfortunately, the resulting expressions are too complex to be widely used.

An alternative approach is to make the simplification that the rate of chemical reaction is fast compared to the rate of diffusion; that is, the membrane diffusion is rate controlling. This approximation is a good one for most coupled transport processes and can be easily verified by showing that flux is inversely proportional to membrane thickness. If interfacial reaction rates were rate controlling, the flux would be constant and independent of membrane thickness. Making the assumption that chemical equilibrium is reached at the membrane interfaces allows the coupled transport process to be modeled easily [9]. The process is

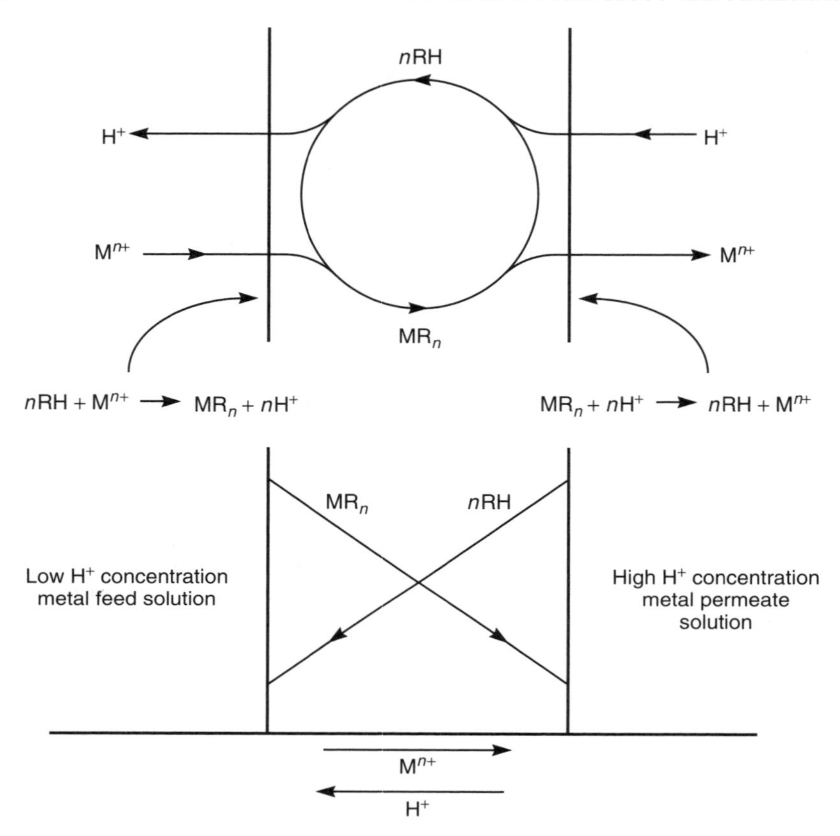

Figure 11.6 An illustration of the carrier agent concentration gradients that form in coupled transport membranes

shown schematically in Figure 11.6, in which the reaction of the carrier (RH) with the metal (M^{n+}) and hydrogen ion (H^+) is given as

$$nRH + M^{n+} \rightleftharpoons MR_n + nH^+ \tag{11.1}$$

This reaction is characterized by an equilibrium constant

$$K = \frac{[MR_n][H]^n}{[RH]^n[M]} \tag{11.2}$$

where the terms in square brackets represent the molar concentrations of the particular chemical species. The equilibrium equation can be written for the organic phase or the aqueous phase. As in earlier chapters the subscripts o and ℓ represent the position of the feed and permeate interfaces of the membrane. Thus the term $[\mathrm{MR}_n]_o$ represents the molar concentration of component MR in the aqueous solution at the feed/membrane interface. The subscript m is used to represent the membrane phase. Thus, the term $[\mathrm{MR}_n]_{o(m)}$ is the molar concentration of component MR_n in the membrane at the feed interface (point o).

Only $[\mathrm{MR}_n]$ and [RH] are measurable in the organic phase, where [H] and [M] are negligibly small. Similarly, only [H] and [M] are measurable in the aqueous phase, where $[\mathrm{MR}_n]$ and [RH] are negligibly small. Equation (11.2) can, therefore, be written for the feed solution interface as

$$K' = \frac{[\mathrm{MR}_n]_{o(m)}[\mathrm{H}]_o^n}{[\mathrm{RH}]_{o(m)}^n[\mathrm{M}]_o} = \frac{k_m}{k_a} \cdot K \tag{11.3}$$

where k_m and k_a are the partition coefficients of M and H between the aqueous and organic phases. This form of Equation (11.2) is preferred because all the quantities are easily accessible experimentally. For example, $[\mathrm{MR}_n]_{o(m)}/[\mathrm{M}]_o$ is easily recognizable as the distribution coefficient of metal between the organic and aqueous phases.

The same equilibrium applies at the permeate-solution interface, and Equation (11.3) can be recast to

$$K' = \frac{[\mathrm{MR}_n]_{\ell(m)}[\mathrm{H}]_\ell^n}{[\mathrm{RH}]_{\ell(m)}^n[\mathrm{M}]_\ell} \tag{11.4}$$

Consider now the situation when a counter ion concentration gradient that exactly balances the metal ion concentration gradient is established, so no flux of either ion across the membrane occurs. Under this condition, $[\mathrm{MR}_n]_{o(m)} = [\mathrm{MR}_n]_{\ell(m)}$ and $[\mathrm{RH}]_{o(m)}^n = [\mathrm{RH}]_{\ell(m)}^n$, producing the expression

$$\frac{[\mathrm{M}]_o}{[\mathrm{M}]_\ell} = \left(\frac{[\mathrm{H}]_o}{[\mathrm{H}]_\ell}\right)^n \tag{11.5}$$

Thus, the maximum concentration factor of metal ion that can be established across the membrane varies with the counter ion (hydrogen ion) concentration ratio (in the same direction) raised to the nth power.

This development, of course, says nothing about the metal ion flux across the membrane under non-equilibrium conditions; this is described by Fick's law. At steady state, the flux j_{MR_n}, in mol/cm$^2 \cdot$ s, of metal complex MR_n across the liquid membrane is given by

$$j_{\mathrm{MR}_n} = \frac{D_{\mathrm{MR}_n}([\mathrm{MR}_n]_{o(m)} - [\mathrm{MR}_n]_{\ell(m)})}{\ell} \tag{11.6}$$

where D_{MR_n} is the mean diffusion coefficient of the complex in the membrane of thickness ℓ. To put Equation (11.6) into a more useful form, the terms in $[\mathrm{MR}_n]$ are eliminated by introduction of Equation (11.3). This results in a complex expression involving the desired quantities [M] and [H], but also involving [RH]. However, mass balance provides the following relationship

$$n[\mathrm{MR}_n]_{(m)} + [\mathrm{RH}]_{(m)} = [\mathrm{R}]_{(m)\mathrm{tot}} \tag{11.7}$$

where $[\mathrm{R}]_{(m)\mathrm{tot}}$ is the total concentration of R in the membrane.

Substitution of Equations (11.3) and (11.4) into (11.6) gives an expression for the metal ion flux in terms of only constants and the concentrations of metal and counter ion in the aqueous solutions on the two sides of the membrane [9]. The solution is simple only for $n = 1$, in which case

$$j_{\mathrm{MR}_n} = \frac{D_{\mathrm{MR}_n}[\mathrm{R}]_{(m)\mathrm{tot}}}{\ell}\left[\left(\frac{1}{[\mathrm{H}]_o/[\mathrm{M}]_o K' + 1}\right) - \left(\frac{1}{[\mathrm{H}]_\ell/[\mathrm{M}]_\ell K' + 1}\right)\right] \tag{11.8}$$

This equation shows the coupling effect between the metal ion [M] and the hydrogen ion [H] because both appear in the concentration term of the Fick's law expression linked by the equilibrium reaction constant K'. Thus, there will be a positive 'uphill' flux of metal ion from the downstream to the upstream solution (that is, in the direction $\ell \rightarrow o$) as long as

$$\frac{[\mathrm{M}]_o}{[\mathrm{H}]_o} > \frac{[\mathrm{M}]_\ell}{[\mathrm{H}]_\ell} \tag{11.9}$$

When the inequality is opposite, the metal ion flux is in the conventional or 'downhill' direction. The maximum concentration factor, that is, the point at which metal ion flux ceases, can be determined in terms of the hydrogen ion concentration in the two aqueous phases

$$\frac{[\mathrm{M}]_o}{[\mathrm{M}]_\ell} = \frac{[\mathrm{H}]_o}{[\mathrm{H}]_\ell} \tag{11.10}$$

This expression is identical to Equation (11.5) for the case of a monovalent metal ion.

Characteristics of Coupled Transport Membranes

Concentration Effects

Equations (11.1)–(11.10) provide a basis for rationalizing the principal features of coupled transport membranes. It follows from Equation (11.8) that coupled transport membranes can move metal ions from a dilute to a concentrated solution against the metal ion concentration gradient, provided the gradient in the second coupled ion concentration is sufficient. A typical experimental result demonstrating this unique feature of coupled transport is shown in Figure 11.7. The process is counter-transport of copper driven by hydrogen ions, as described in Equation (11.1). In this particular experiment, a pH difference of 1.5 units is

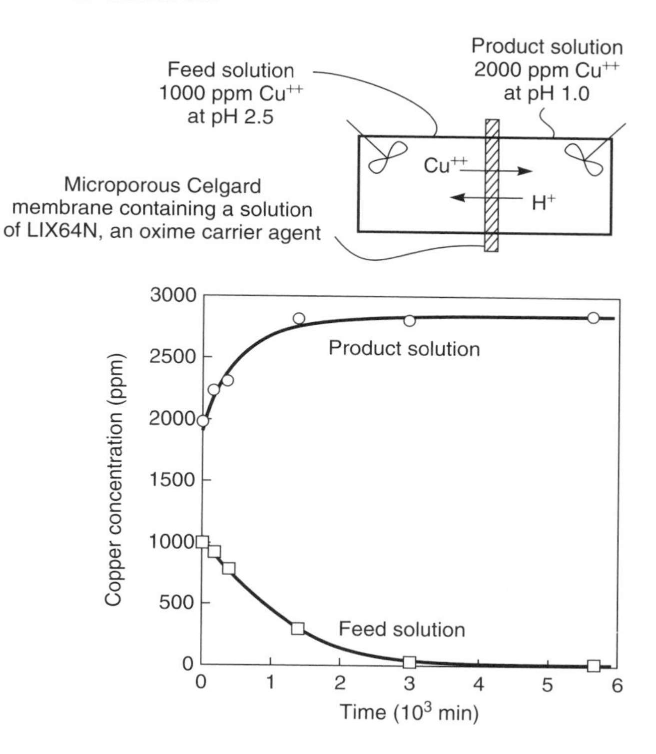

Figure 11.7 Demonstration of coupled transport. In a two-compartment cell, copper flows from the dilute (feed) solution into the concentrated (product) solution, driven by a gradient in hydrogen ion concentration [9]. Membrane, microporous Celgard 2400/LIX 64N; feed, pH 2.5; product, pH 1.0

maintained across the membrane. The initial product solution copper concentration is higher than the feed solution concentration. Nonetheless, copper diffuses against its concentration gradient from the feed to the product side of the membrane. The ratio of the counter hydrogen ions between the solutions on either side of the membrane is about 32 to 1 which, according to the appropriate form of Equation (11.5), should give a copper concentration ratio of

$$\frac{[\mathrm{Cu}^{2+}]_\ell}{[\mathrm{Cu}^{2+}]_o} = \left(\frac{[\mathrm{H}^+]_\ell}{[\mathrm{H}^+]_o}\right)^2 = (32)^2 \approx 1000 \qquad (11.11)$$

In the experiment shown in Figure 11.7, this means that the feed solution copper concentration should drop to just a few ppm, and this is the case.

A more convenient method of measuring the copper concentration factor is to maintain the product solution at some high copper concentration and to allow the feed solution copper concentration to reach a measurable steady-state value.

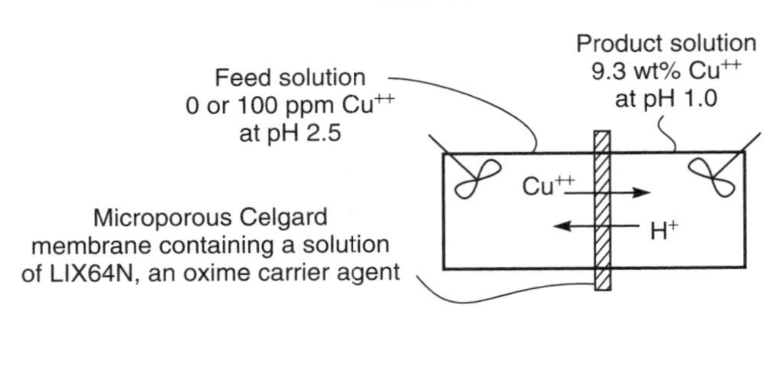

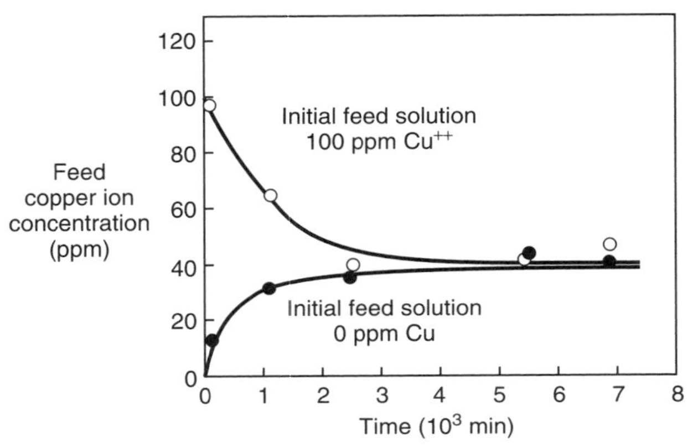

Figure 11.8 Experiments to demonstrate the maximum achievable concentration factor. Membrane, microporous Celgard 2400/LIX 64N; feed, pH 2.5, copper ion concentration, 0 or 100 ppm; product, pH 1.0, 9.3 wt% copper [9]. The concentration in the feed solution moves to a plateau value of 40 ppm at which the copper concentration gradient across the membrane is balanced by the hydrogen ion gradient in the other direction

Figure 11.8 shows the feed copper concentration in such an experiment, in which the steady-state feed solution concentration was about 40 ppm. The feed solution was allowed to approach steady state from both directions, that is, with initial copper concentrations higher and lower than the predicted value for the given pH gradient. As Figure 11.8 shows, regardless of the starting point, the copper concentration factors measured by this method are in reasonable agreement with the predictions of Equation (11.11).

Feed and Product Metal Ion Concentration Effects

A second characteristic of coupled transport membranes is that the membrane flux usually increases with increasing metal concentration in the feed solution, but is usually independent of the metal concentration in the product solution. This behavior follows from the flux Equations (11.6) and (11.8). In typical coupled transport experiments, the concentration of the driving ion (H^+) in the product solution is very high. For example, in coupled transport of copper, the driving ions are hydrogen ions, and 100 g/L sulfuric acid is often used as the product solution. As a result, on the product side of the membrane the carrier is in the protonated form, the term $[MR_n]_{\ell(m)}$ is very small compared to $[MR_n]_{o(m)}$, and Equation (11.8) reduces to

$$j_{MR_n} = \frac{D_{MR_n}[R]_{(m)tot}}{\ell} \cdot \frac{1}{[H]_o/[M]_o K' + 1} \tag{11.12}$$

The permeate solution metal ion concentration, $[M]_\ell$, does not appear in the flux equation, which means that the membrane metal ion flux is independent of the concentration of metal on the permeate side. However, the flux does depend on the concentration of metal ions, $[M]_o$, on the feed solution side. At low values of $[M]_o$, the flux will increase linearly with $[M]_o$, but at higher concentrations the flux reaches a plateau value as the term $[H]_o/[M]_o K'$ becomes small compared to 1. At this point all of the available carrier molecules are complexed and no further increase in transport rate across the membrane is possible. The form of this dependence is illustrated for the feed and product solution metal ion concentrations in Figure 11.9.

pH and Metal Ion Effects

It follows from flux Equation (11.12) that the concentration of the counter hydrogen ion and the equilibrium coefficient K' for a particular metal ion will affect the metal ion flux. The effect of these factors can best be understood by looking at curves of metal ion extraction versus pH. Examples are shown in Figure 11.10 for copper and other metals using the carrier LIX 64N [43]. The counter ion

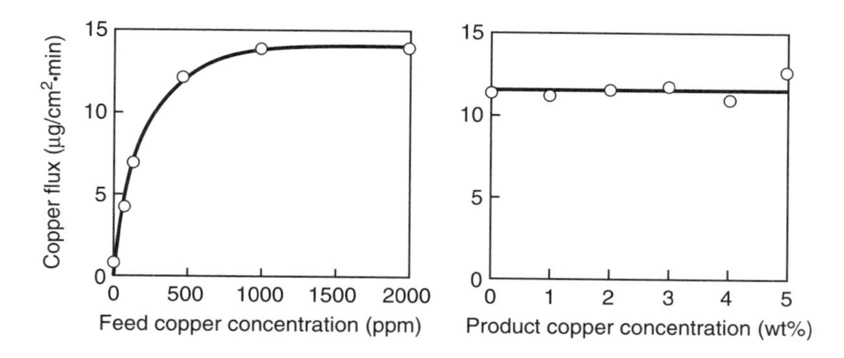

Figure 11.9 Effect of metal concentration in the feed and product solution on flux. Membrane, microporous Celgard 2400/30 % Kelex 100 in Kermac 470B; feed, pH 2.5; product, 100 g/L H_2SO_4 [9]

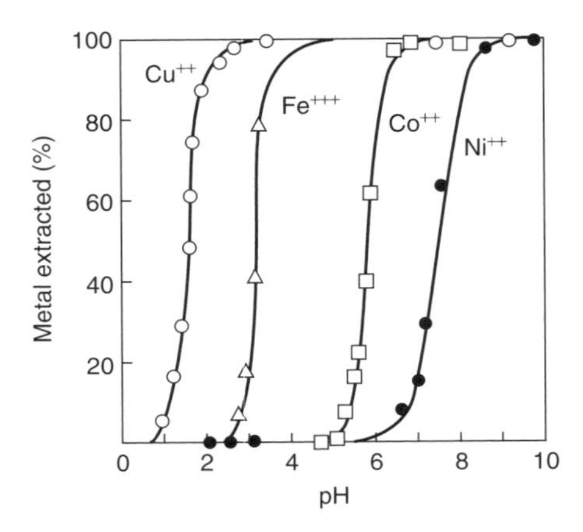

Figure 11.10 Metal extraction curves for four metal ions by LIX 64N. The aqueous phase initially contained 1000 ppm metal as the sulfate salt [43]

is hydrogen and the metal ions are extracted by reactions of the type shown in Equation (11.1).

The pH at which metal ions are extracted depends on the distribution coefficient for the particular metal and complexing agent. As a result, the pH at which the metal ions are extracted varies, as shown by the results in Figure 11.10. This behavior allows one metal to be separated from another. For example, consider the separation of copper and iron with LIX 64N. As Figure 11.10 shows, LIX 64N extracts copper at pH 1.5–2.0, but iron is not extracted until above pH 2.5. The separations obtained when 0.2 % solutions of copper and iron are tested with a LIX 64N membrane at various pHs are shown in Figure 11.11. The copper flux is approximately 100 times higher than the iron flux at a feed pH of 2.5.

Carrier Agent

In the examples given in Figures 11.9–11.11 to illustrate coupled transport, the two oxime carriers used for copper were LIX 64N and Kelex 100, which have the structures:

C_9H_{19} C N OH OH

LIX 64N

$C_{12}H_{25}$ OH N

Kelex 100

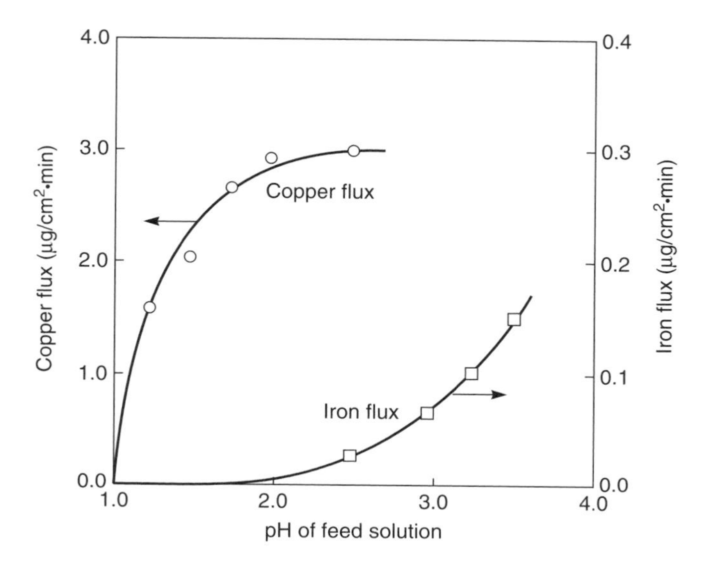

Figure 11.11 Copper and iron fluxes as a function of feed pH [9]. Membrane, Celgard 2400/LIX 64N; feed, 0.2 % metal; product, pH 1.0

However, a large number of complexing agents of all kinds with chemistries designed for specific metal ions have been reported in the literature. The tertiary amine Alamine 336 is widely used to transport anions such as $UO_2(SO_4)^{4-}$ and $Cr_2O_7^{2-}$ [44–46]. The macrocyclic crown ether family has also been used to transport alkali and rare earth metals [47,48]:

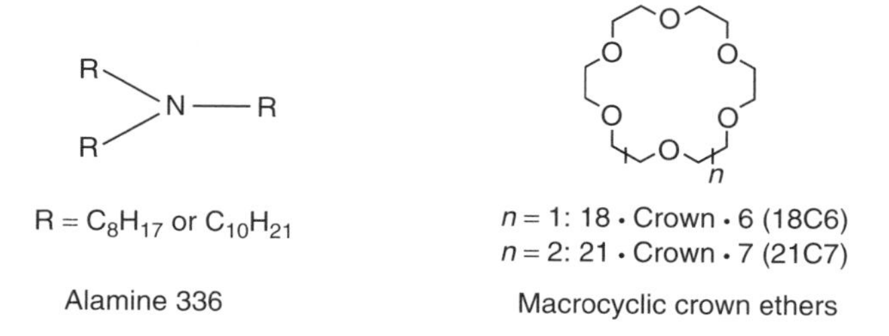

Coupled Transport Membranes

Supported Liquid Membranes

In supported liquid membranes, a microporous support impregnated with the liquid complexing agent separates the feed and product solutions. In coupled transport, the fluid on both sides of the membrane must be circulated to avoid concentration polarization, which is much more significant on the feed side than on the permeate side. In the laboratory, concentration polarization is easily avoided by using flat sheet membranes in a simple permeation cell with stirred solutions on both sides of the membrane. On a larger scale, hollow-fiber systems with the feed solution circulated down the bore of the fibers have been the most common form of membrane.

Large-scale processes require many modules to remove most of the metal from a continuous feed stream. In general, a multistage system operating in a feed-and-bleed mode is the most efficient design; a schematic representation of a three-stage system is shown in Figure 11.12 [49]. A fixed feed volume circulates through each module at a high rate to control concentration polarization; this flow is indicated by the solid lines in the figure. Feed solution is continuously introduced into the circulating volume of the first stage and is bled off at the

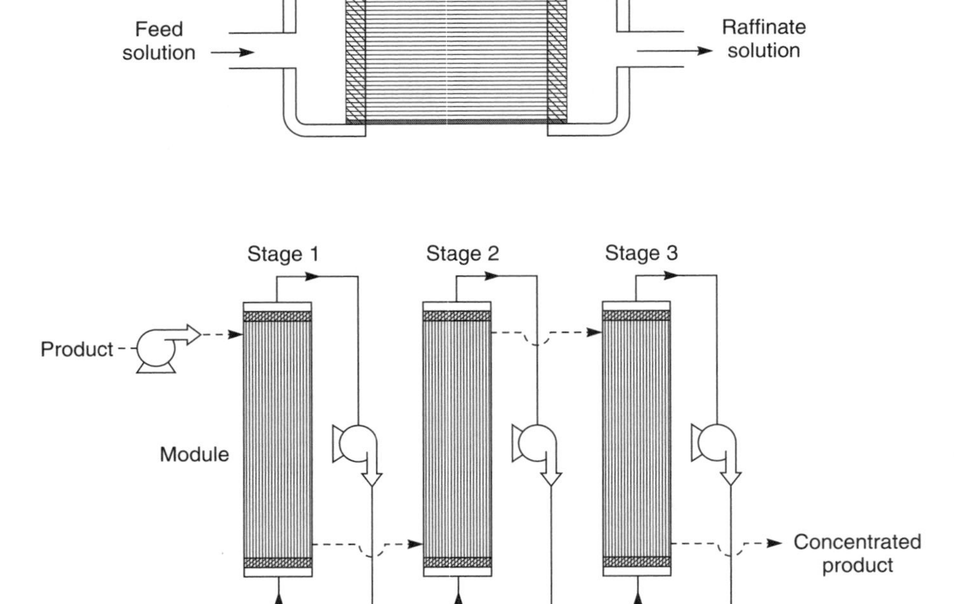

Figure 11.12 Schematic of a three-stage feed-and-bleed hollow fiber coupled transport concentrator [49]

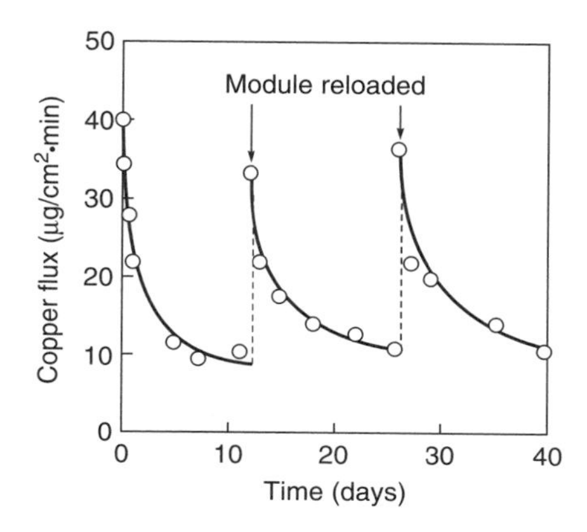

Figure 11.13 The effect of replenishing a hollow fiber coupled transport module with fresh complexing agent. Membrane, polysulfone, hollow fiber/Kelex 100; feed, 0.2 % copper, pH 2.5; product, 2 % copper, 100 g/L H_2SO_4 [49]

same rate. The bleed from the first stage constitutes the feed for the second, and the bleed from the second stage constitutes the feed for the third. In operation, the concentration of metal in the feed solution decreases as it flows from stage 1 to stage 3, with the final raffinate concentration depending on the feed-and-bleed flow rate. The product solution flows in series through the stages. The advantage of this multistage design over a single-stage system is that only the final stage operates on feed solution depleted of metal.

Liquid membranes supported by hollow fibers are relatively easy to make and operate, and the membrane fluxes can be high. However, membrane stability is a problem. The variation in coupled transport flux during long-term tests is illustrated in Figure 11.13 [49]. The detailed mechanism for this flux instability is not completely established but appears to be related to loss of the organic complexing agent phase from the support membrane [49–51]. Although membrane fluxes could be restored to their original values by reloading the membrane with fresh complexing agent for copper-coupled transport, this is not practical in a commercial system.

Emulsion Liquid Membranes

A form of liquid membrane that received a great deal of attention in the 1970s and 1980s was the bubble or emulsion membrane, first developed by Li at Exxon [11–13]. Figure 11.14 is a schematic illustration of an emulsion liquid membrane process, which comprises four main operations. First, fresh product solution is emulsified in the liquid organic membrane phase. This water/oil emulsion then enters a large mixer vessel, where it is again emulsified to form a water/oil/water emulsion. Metal ions in the feed solution permeate by coupled

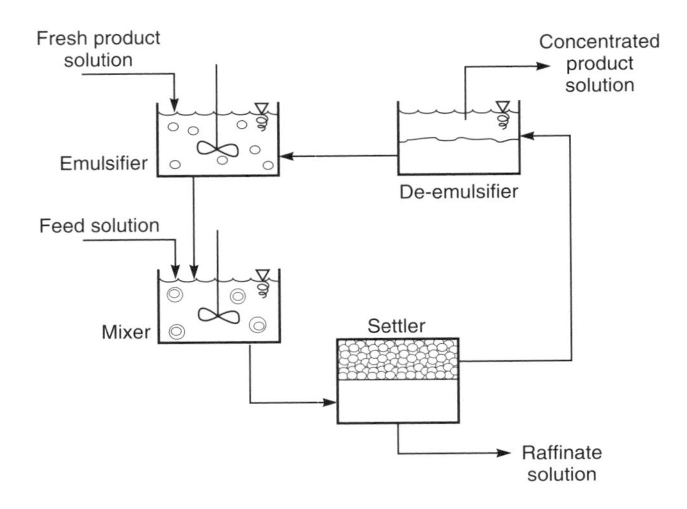

Figure 11.14 Flow diagram of a liquid emulsion membrane process

transport through the walls of the emulsion to the product solution. The mixture then passes to a settler tank where the oil droplets separate the metal-depleted raffinate solution. A single mixer/settler step is shown in Figure 11.14, but in practice a series of mixer/settlers may be used to extract the metal completely. The emulsion concentrate then passes to a de-emulsifier where the organic and concentrated product solutions are separated. The regenerated organic solution is recycled to the first emulsifier.

The optimum operating conditions for this type of process vary a great deal. The first water/oil emulsion is typically an approximately 50/50 mixture, which is then mixed with the aqueous feed solution phase at a ratio of 1 part emulsion phase to 5–20 parts feed solution phase. Typical extraction curves for copper using LIX 64N are shown in Figure 11.15. The extraction rate generally follows a first order expression [52]. The slope of the curve in Figure 11.15 is proportional to the loading of complexing agent in the organic phase and the rate of agitation in the mixer vessel.

Figure 11.15 also illustrates one of the problems of emulsion membrane systems, namely, degradation of the emulsion on prolonged contact with the feed solution and high-speed mixing of the product and feed solutions. Prolonged stirring of the emulsion with the feed solution causes the copper concentration to rise as some of the emulsion droplets break. Careful tailoring of the stirring rate and surfactant composition is required to minimize premature emulsion breakdown [52,53].

Although emulsion degradation must be avoided in the mixer and settler tanks, complete and rapid breakdown is required in the de-emulsifier in which the

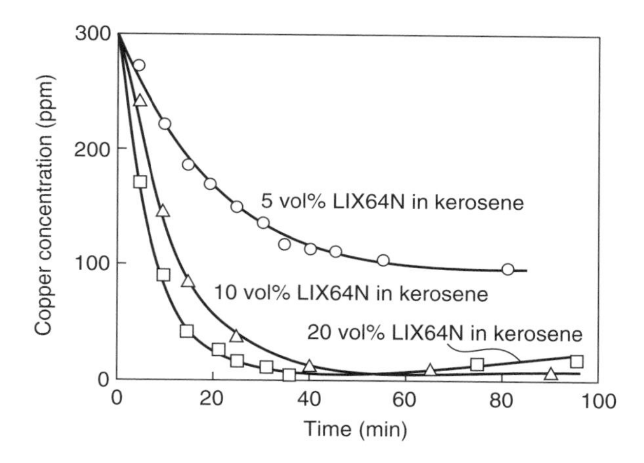

Figure 11.15 Copper extraction by a liquid emulsion membrane process [52]. Feed, 200 ml, pH 2.0, 300 ppm Cu^{2+}; membrane, 15 mL LIX 64N in kerosene, 3 % Span 80; stripping solution, 15 mL H_2SO_4. Reprinted from *J. Membr. Sci.* **6**, W. Völkel, W. Halwachs and K. Schügerl, Copper Extraction by Means of a Liquid Surfactant Membrane Process, p. 19, Copyright 1980, with permission from Elsevier

product solution is separated from the organic complexing agent. Currently, electrostatic coalescers seem to be the best method of breaking these emulsions. Even then, some of the organic phase is lost with the feed raffinate.

Applications

The best application of coupled transport is removal and recovery of metals from large, dilute feed solutions such as contaminated ground water or dilute hydrometallurgical process streams. Treatment of such streams by chemical precipitation, conventional solvent extraction with liquid ion exchange reagents, or extraction with ion exchange resins is often uneconomical. The ability of coupled transport to treat large-volume, dilute streams with relatively small amounts of the expensive carrier agent is an advantage.

An application that has received a good deal of attention is the recovery of copper from dilute hydrometallurgical process streams. Such streams are produced by extraction of low-grade copper ores with dilute sulfuric acid. Typically, the leach stream contains 500–5000 ppm copper and various amounts of other metal ions, principally iron. Currently, copper is removed from these streams by precipitation with iron or by solvent extraction. A scheme for recovering the copper by coupled transport is shown in Figure 11.16. The dilute copper solution from the dump forms the feed solution; concentrated sulfuric acid from the electrowinning operation forms the product solution. Copper from the feed solution permeates the membrane, producing a raffinate solution containing 50–100 ppm copper,

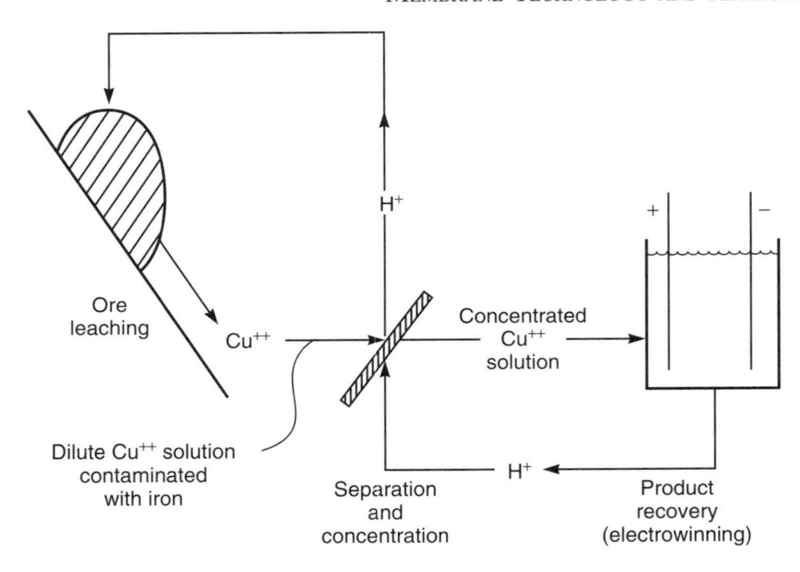

Figure 11.16 Schematic of copper recovery by coupled transport from dump leach streams. The concentrated copper solution produced by coupled transport separation of the dump leach liquid is sent to an electrolysis cell where copper sulfate is electrolyzed to copper metal and sulfuric acid

which is returned to the dump. The product solution, which contains 2–5 % copper, is sent to the electrowinning tankhouse. Many papers have described this application of coupled transport with supported [9,49] and emulsion [13,52] membranes. Membrane stability was a problem and, although the economics appeared promising, the advantage was insufficient to encourage adoption of this new process.

Facilitated Transport

Background

The transport equations used for facilitated transport parallel those derived for coupled transport [27]. The major difference is that only one species is transported across the membrane by the carrier. The carrier-species equilibrium in the membrane is

$$R + A \rightleftharpoons RA \tag{11.13}$$

where R is the carrier, A is the permeant transported by the carrier and RA is the permeant-carrier complex. Examples of reactions used in facilitated transport processes are shown in Table 11.1.

Table 11.1 Facilitated transport carrier reactions

CO_2	$CO_2 + H_2O + Na_2CO_3 \rightleftharpoons 2NaHCO_3$
O_2	$O_2 +$ CoSchiffs base $\rightleftharpoons$ CoSchiffs base(O_2)
SO_2	$SO_2 + H_2O + Na_2SO_3 \rightleftharpoons 2NaHSO_3$
H_2S	$H_2S + Na_2CO_3 \rightleftharpoons NaHS + NaHCO_3$
CO	$CO + CuCl_2 \rightleftharpoons CuCl_2(CO)$
C_2H_4	$C_2H_4 + AgNO_3 \rightleftharpoons AgNO_3(C_2H_4)$

As with coupled transport, two assumptions are made to simplify the treatment: first, that the rate of chemical reaction is fast compared to the rate of diffusion across the membrane, and second, that the amount of material transported by carrier facilitated transport is much larger than that transported by normal passive diffusion, which is ignored. The facilitated transport process can then be represented schematically as shown in Figure 11.17.

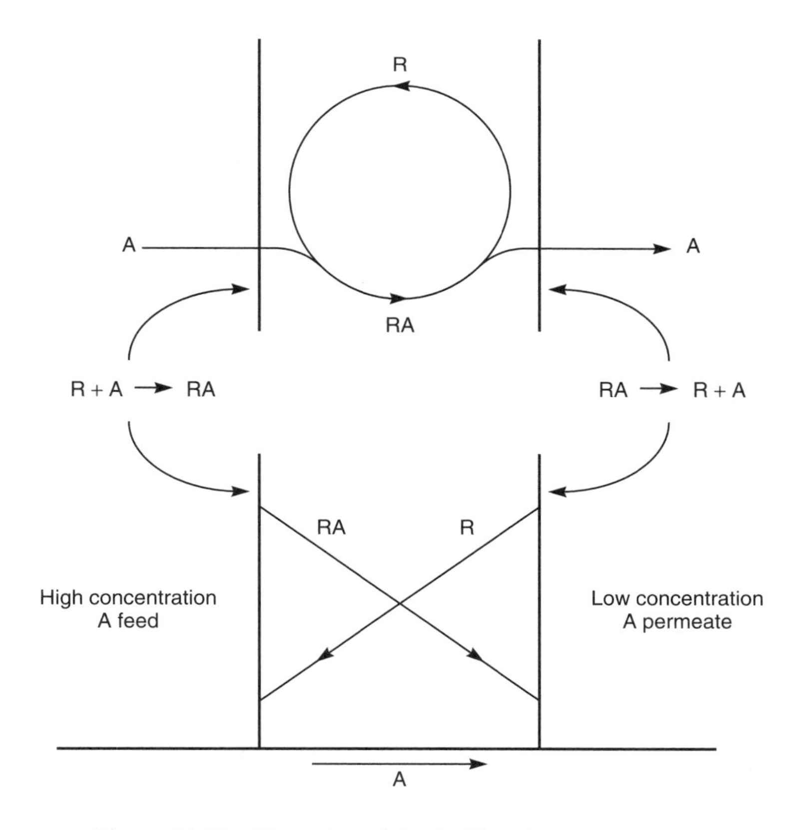

Figure 11.17 Illustration of the facilitated transport process

The carrier–permeate reaction within the membrane is described by the equilibrium constant

$$K = \frac{[\mathrm{RA}]_{(m)}}{[\mathrm{R}]_{(m)}[\mathrm{A}]_{(m)}} \tag{11.14}$$

The concentration of permeant, $[\mathrm{A}]_{(m)}$, within the membrane phase can be linked to the concentration (pressure) of permeant A in the adjacent gas phase, [A], by the Henry's law expression

$$[\mathrm{A}]_{(m)} = k[\mathrm{A}] \tag{11.15}$$

Hence Equation (11.14) can be written

$$\frac{[\mathrm{RA}]_{(m)}}{[\mathrm{R}]_{(m)}[\mathrm{A}]} = K \cdot k = K' \tag{11.16}$$

The components $[\mathrm{R}]_{(m)}$ and $[\mathrm{RA}]_{(m)}$ can be linked by a simple mass balance expression to the total concentration of carrier $[\mathrm{R}]_{(m)\mathrm{tot}}$ within the membrane phase, so Equation (11.16) can be rearranged to

$$[\mathrm{RA}]_{(m)} = \frac{[\mathrm{R}]_{(\mathrm{m})_{\mathrm{tot}}}}{1 + 1/[\mathrm{A}]K'} \tag{11.17}$$

Equation (11.17) shows the fraction of the carrier that reacts to form a carrier complex. At very large values of the term $[A]K'$, all the carrier is complexed, and $[\mathrm{RA}]_{(m)} \rightarrow [\mathrm{R}]_{(m)\mathrm{tot}}$. At low values, $[\mathrm{A}]K' \rightarrow 0$, none of the carrier is complexed ($[\mathrm{RA}]_{(m)} \rightarrow 0$). Equation (11.17) allows the concentration of the carrier–permeant complex at each side of the membrane to be calculated in terms of the equilibrium constant between the carrier and the permeant, and the concentration (pressure) of the permeant in the adjacent feed and permeant fluids. This allows transport through the membrane to be calculated using Fick's law. The flux, j_{RA}, of RA through the membrane is given by

$$j_{\mathrm{RA}} = \frac{D_{\mathrm{RA}}([\mathrm{RA}]_{o(m)} - [\mathrm{RA}]_{\ell(m)})}{\ell} \tag{11.18}$$

Substituting Equation (11.17) into Equation (11.18) yields

$$j_{\mathrm{RA}} = \frac{D_{\mathrm{RA}}[\mathrm{RA}]_{(m)\mathrm{tot}}}{\ell}\left[\frac{1}{1 + 1/[\mathrm{A}]_o K'} - \frac{1}{1 + 1/[\mathrm{A}]_\ell K'}\right] \tag{11.19}$$

To illustrate the dependence of the membrane flux on the equilibrium constant K' and the pressure gradient across the membrane, the flux, j_{RA}, when the permeant pressure is close to zero, that is, $[\mathrm{A}]_\ell \approx 0$, can be written as

$$j_{\mathrm{RA}} = \frac{D_{\mathrm{RA}}[\mathrm{R}]_{(m)\mathrm{tot}}}{\ell}\left(\frac{1}{1 + 1/[\mathrm{A}]_o K'}\right) \tag{11.20}$$

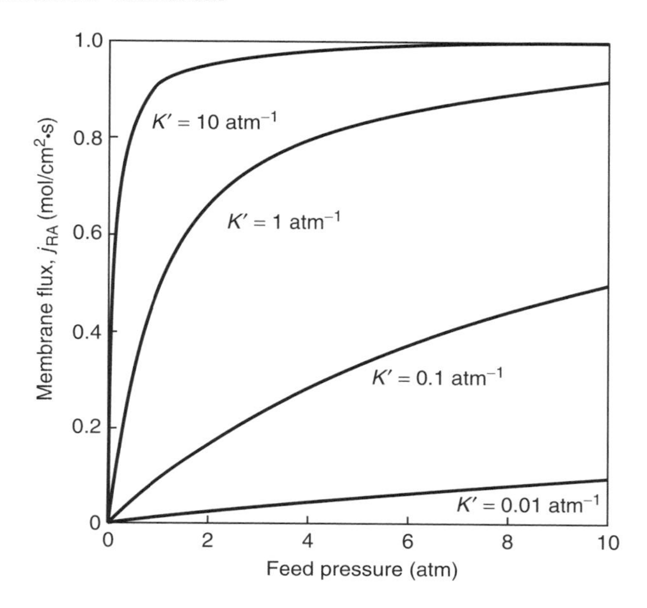

Figure 11.18 Flux through a facilitated transport membrane calculated using Equation (11.20) ($[A]_\ell \approx 0$ and $D_{RA}[R]_{(m)tot}/\ell \approx 1$)

This expression is plotted in Figure 11.18 as flux as a function of feed pressure for different values of the equilibrium constant, K'. In this example, at an equilibrium constant K' of 0.01 atm^{-1}, very little of carrier R reacts with permeant A even at a feed pressure of 10 atm, so the flux is low. As the equilibrium constant increases, the fraction of carrier reacting with permeant at the feed side of the membrane increases, so the flux increases. This result would suggest that, to achieve the maximum flux, a carrier with the highest possible equilibrium constant should be used. For example, the calculations shown in Figure 11.18 indicate a carrier with an equilibrium constant of 10 atm^{-1} or more.

The calculations shown in Figure 11.18 assume that a hard vacuum is maintained on the permeate side of the membrane. The operating and capital costs of vacuum and compression equipment prohibit these conditions in practical systems. More realistically, a carrier facilitated process would be operated either with a compressed gas feed and atmospheric pressure on the permeate side of the membrane, or with an ambient-pressure feed gas and a vacuum of about 0.1 atm on the permeate side. By substitution of specific values for the feed and permeate pressures into Equation (11.19), the optimum values of the equilibrium constant can be calculated. A plot illustrating this calculation for compression and vacuum operation is shown in Figure 11.19.

Under the assumptions of this calculation, the optimum equilibrium constant is 0.3 atm^{-1} for compression operation (feed pressure, 10 atm; permeate pressure,

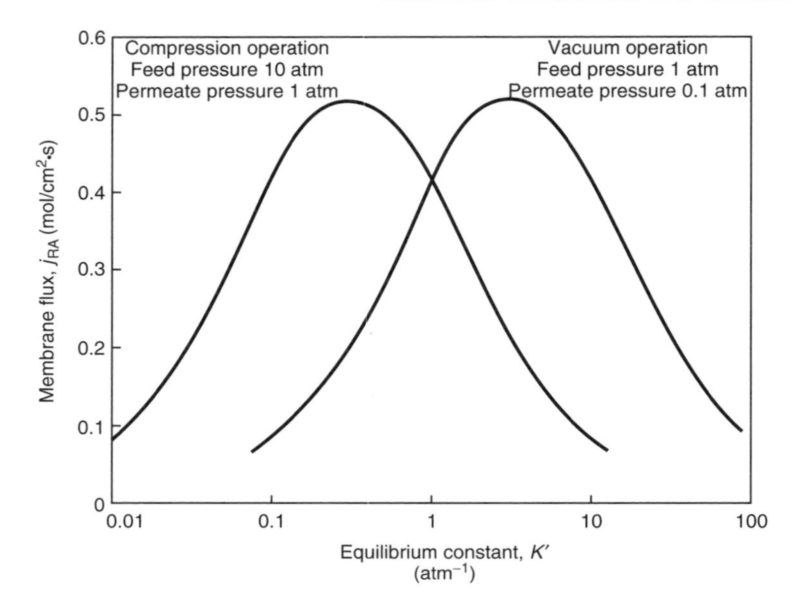

Figure 11.19 Illustration of the effect of feed and permeate pressure on the optimum carrier equilibrium constant. $D_{RA}[R]_{(m)tot}/\ell \approx 1$; vacuum operation, feed pressure 1 atm, permeate pressure 0.1 atm; compression operation, feed pressure 10 atm, permeate pressure 1 atm

1 atm) and 3 atm^{-1} for vacuum operation (feed pressure, 1 atm; permeate pressure, 0.1 atm). The results show that rather precise control of the equilibrium constant is required to achieve a useful facilitated transport process. In this example calculation, although carriers with equilibrium constants less than 0.3 atm^{-1} or greater than 3 atm^{-1} can transport the permeant across the membrane, obtaining the maximum flux for the process would require operation at feed and permeate pressures likely to make the process uneconomical.

Process Designs

Until quite recently, most of the facilitated transport results reported in the literature were obtained with supported liquid membranes held by capillarity in microporous films. The instability of these membranes has inhibited commercial application of the process. Three factors contribute to this instability and the consequent loss of membrane performance over time:

1. Evaporation of the solvent used to prepare the liquid membrane, leading to membrane failure.

2. Expulsion of liquid held by capillarity within the microporous membrane pores. The membrane must always be operated well below the average bubble point of the membrane, because liquid expulsion from even a few larger-than-average pores can cause unacceptable leakage of gas.
3. Degradation of the carrier agent by the permeant gas or by minor components such as water, carbon dioxide or hydrogen sulfide in the feed gas.

Significant progress has been made in alleviating the first two physical causes of membrane instability. The magnitude of the long-term chemical stability problem depends on the process. It is a major issue for carriers used to transport oxygen and olefins, but for carriers used to transport carbon dioxide, chemical stability is a lesser problem.

Several techniques can minimize the pressure difference across supported liquid membranes to improve membrane stability. In the laboratory, flow of an inert sweep gas such as helium on the permeate side can be used to maintain low partial pressure of the permeating component, while the hydrostatic pressure is about equal to that of the feed. A variation on this approach proposed by Ward is to use a condensable sweep gas such as steam. The permeate-steam mixture is cooled and condensed, separating the permeate gas from the condensed water, which is then sent to a boiler to regenerate the steam [25,26]. A simplified flow scheme of this process is shown in Figure 11.20(a). An alternative approach is to sweep the permeate side of the membrane with an absorbent liquid in which the permeate gas dissolves. Hughes *et al.* [27], for example, used liquid hexane to sweep the permeate side of their propylene/propane separating membrane, as illustrated in Figure 11.20(b). The hexane/propylene mixture leaving the membrane permeator is sent to a small distillation column to recover the hexane liquid and a concentrated propylene gas stream.

Both techniques shown in Figure 11.20 increase the complexity of the separation process significantly, and neither has advanced to a commercial process. The focus of much of the recent work on facilitated transport has been to produce membranes that are inherently stable and can be used in conventional gas separation systems. Laciak has recently reviewed this work [38].

One approach used with ionic carriers is to impregnate ion exchange membranes with the carrier feed solution. Ion exchange sites in the membrane are ion-paired to the facilitated transport carrier [54–56]. The membrane is swollen with a solvent, usually water but sometimes glycerol, so that the carrier ions have some mobility. These membranes are, in effect, swollen polymeric gels, so the problem of carrier fluid displacement from the membrane pores if the bubble pressure is exceeded does not occur. Evaporation of the solvent remains a problem, and addition of solvent vapor to the feed gas is generally required.

Another method of solving the solvent evaporation problem was devised by Pez, Laciak and others [38,57–60] at Air Products, using carriers in the form

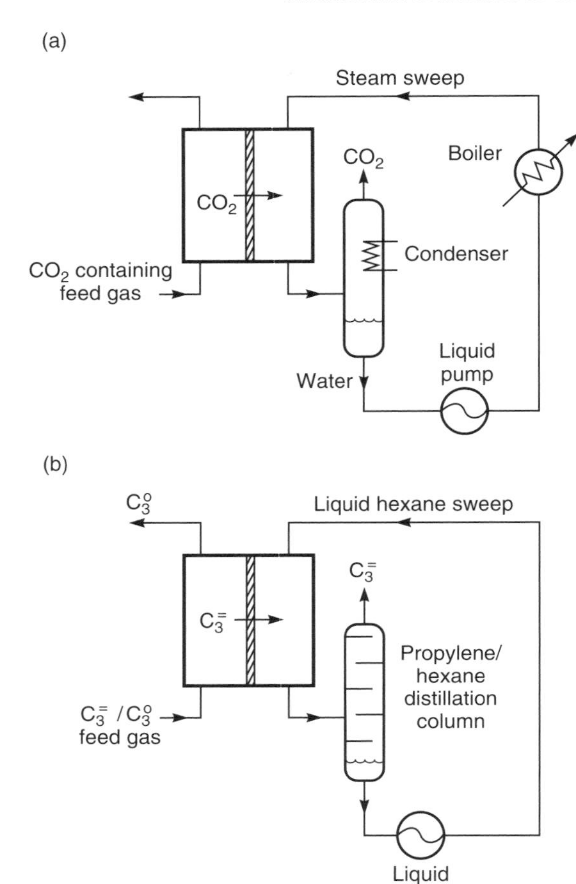

Figure 11.20 Flow schematics of (a) a steam sweep configuration used in the facilitated transport of carbon dioxide [25,26] and (b) a liquid hexane sweep used in the transport of propylene [27]

of organic salts that become liquids (molten salts) at ambient temperatures. Examples of such salts are:

- triethyl ammonium chlorocuprate $(C_2H_5)_3 \cdot NHCuCl_2$, a carrier for carbon monoxide;
- tetrahexyl ammonium benzoate $(C_6H_{13})_4 \cdot N^+C_6H_5CO_2{}^-$, a carrier for carbon dioxide;
- tetrahexyl ammonium fluoride tetrahydrate $(C_6H_{13})_4 \cdot NF \cdot 4H_2O$, a carrier for carbon dioxide.

Under the membrane test conditions, these carrier salts are liquids with essentially no vapor pressure, so the solvent evaporation problem is eliminated.

Yet another approach to stabilizing facilitated transport membranes is to form multilayer structures in which the supported liquid-selective membrane is encapsulated between thin layers of very permeable but nonselective dense polymer layers. The coating layers must be very permeable to avoid reducing the gas flux through the membrane; materials such as silicone rubber or poly(trimethylsiloxane) are usually used [26].

Despite many years of effort, none of these methods of stabilizing liquid membranes has had real success. For this reason, a number of workers are trying to develop solid carrier facilitated membranes. Two approaches are being tried. One is to covalently link the carrier complex to the matrix polymer. So far the improvement in selectivity obtained by this approach has been modest. This may be due to the difficulty of obtaining high loadings of the carrier in the polymer matrix. The second approach, which has been more successful, is to produce facilitated transport membranes in which the polymer matrix acts as a partial solvent for the carrier. For example, poly(ethylene oxide) or ethylene oxide copolymers can dissolve covalent salts such as silver tetrafluoroborate ($AgBF_4$), a facilitated transport carrier for olefins [31–33,61–63]. Significant facilitation of some gases has been achieved with these membranes.

Examples of the best results obtained are shown in Figure 11.21 [33,61]. The composite membranes with which these data were obtained were formed by casting a solution of 80 wt% silver tetrafluoroborate in a propylene oxide copolymer matrix onto a microporous support. When subjected to a 40-day test with a gas

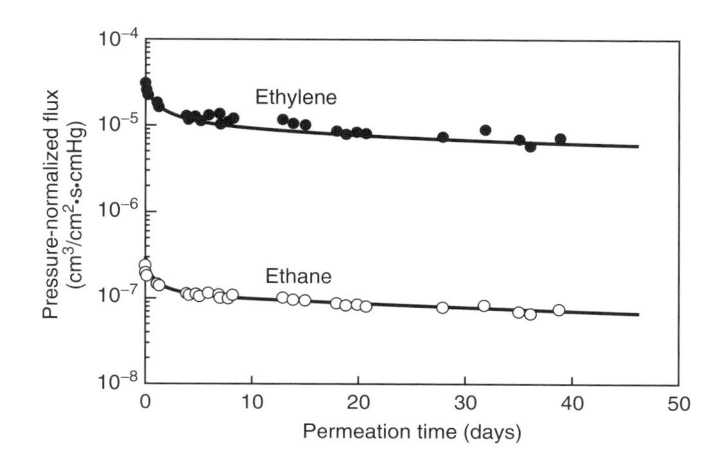

Figure 11.21 Long-term performance of a composite solid polymer electrolyte membrane consisting of 80 wt% $AgBF_4$ dissolved in a propylene oxide copolymer matrix. Feed gas, 70 vol% ethylene/30 vol% ethane at 50 psig; permeate pressure, atmospheric [33,61]

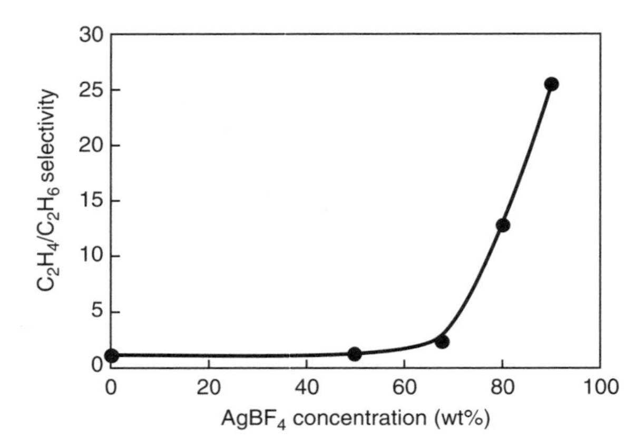

Figure 11.22 Mixed-gas ethylene/ethane selectivity of a solid polymer electrolyte membrane as a function of $AgBF_4$ concentration in the polyamide-polyether matrix [62]

mixture of 70 vol% ethylene/30 vol% ethane, the membrane produced a permeate containing 98.7–99.4 % ethylene over the entire test period. Conventional wisdom suggests that these immobilized carriers would lose their facilitation ability, but this does not appear to be the case. The exact transport mechanism is not clear but appears to involve the permeable gas molecules moving between fixed sites through the polymer [28–30,34]. These solid matrix membranes often show clear evidence of a percolation threshold. At low carrier loadings, little or no facilitation is observed until, at a certain critical loading, facilitation occurs, and thereafter increases rapidly [62,64]. Some results illustrating this effect are shown in Figure 11.22. At loadings below 70 wt% $AgBF_4$, essentially no facilitation is seen; at loadings greater than this threshold value, facilitation occurs. It is believed that the percolation threshold level is the point at which carrier sites are close enough that the permeating complex molecule can hop from carrier site to carrier site through the membrane.

Applications

Over the last 30 years a variety of facilitated transport carriers have been studied for a number of important separation problems, reviewed briefly below.

Carbon Dioxide/Hydrogen Sulfide Separation

From the late 1960s to the early 1980s, Ward and others at General Electric studied facilitated transport membranes, particularly for separation of the acid gases carbon dioxide and hydrogen sulfide from methane and hydrogen [23–26].

This work was finally abandoned after the development of selective polymeric membranes in the 1980s.

Although many carriers are available for carbon dioxide and hydrogen sulfide transport, one of the most studied chemistries uses aqueous carbonate/bicarbonate solutions. Four principal reactions occur in the film

$$CO_2 + H_2O \rightleftharpoons H^+ + HCO_3^- \quad (11.21)$$

$$CO_2 + OH^- \rightleftharpoons HCO_3^- \quad (11.22)$$

$$HCO_3^- \rightleftharpoons H^+ + CO_3^{2-} \quad (11.23)$$

$$H^+ + OH^- \rightleftharpoons H_2O \quad (11.24)$$

Equations (11.21) and (11.22) are measurably slow reactions; Reactions (11.23) and (11.24) are essentially instantaneous. All four reactions determine the equilibrium concentrations, but the process can be illustrated in simple form by Figure 11.23 [25].

At the feed side of the membrane, carbon dioxide dissolves in the aqueous carbonate/bicarbonate solution and reacts with water and carbonate ions according to Equations (11.21) and (11.23).

$$\begin{array}{l} CO_2 + H_2O \rightleftharpoons H^+ + HCO_3^- \\ H^+ + CO_3^{2-} \rightleftharpoons HCO_3^- \\ \hline CO_2 + H_2O + CO_3^{2-} \rightleftharpoons 2HCO_3^- \end{array} \quad (11.25)$$

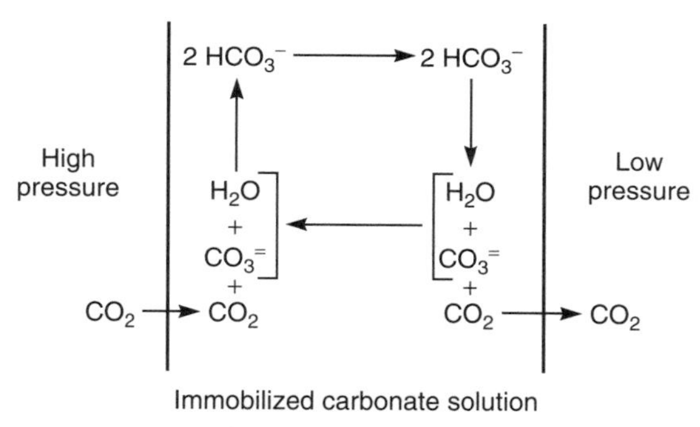

Figure 11.23 Facilitated transport of carbon dioxide through an immobilized carbonate/bicarbonate solution [25]. Reprinted with permission from S.G. Kimura, S.L. Matson and W.J. Ward III, Industrial Applications of Facilitated Transport, in *Recent Developments in Separation Science*, N.N. Li, J.S. Dranoff, J.S. Schultz and P. Somasundaran (eds) (1979). Copyright CRC Press, Boca Raton, FL

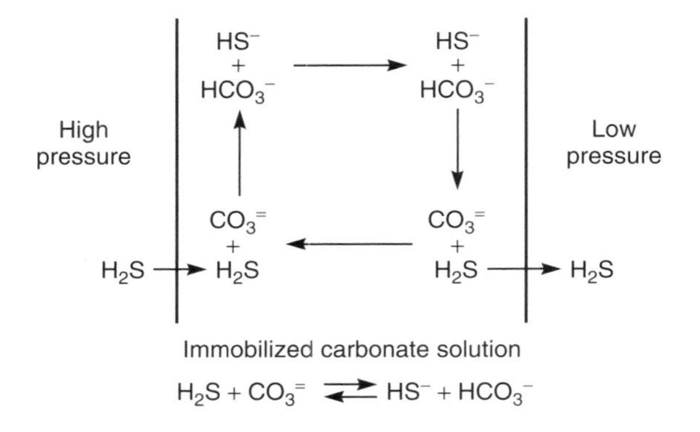

Figure 11.24 Facilitated transport of hydrogen sulfide through an immobilized carbonate/bicarbonate solution [26]. Reprinted with permission from S.L. Matson, C.S. Herrick and W.J. Ward III, Progress on the Selective Removal of H_2S from Gasified Coal Using an Immobilized Liquid Membrane, *Ind. Eng. Chem., Prod. Res. Dev.* **16**, 370. Copyright 1977, American Chemical Society and American Pharmaceutical Association

At the permeate side of the membrane the reaction is reversed, and bicarbonate ions form carbon dioxide, water, and carbonate ions.

Coupled transport of hydrogen sulfide through the same carbonate/bicarbonate membrane is shown in Figure 11.24 [26]. The overall reaction is simple

$$H_2S + CO_3{}^{2-} \rightleftharpoons HS^- + HCO_3{}^- \tag{11.26}$$

but, again, a number of reactions occur simultaneously to establish the equilibrium concentrations.

Because some of the reactions involved in establishing equilibrium at the membrane surface are slow compared to diffusion, the calculated concentration gradients formed in the liquid membrane do not have a simple form. The equations for partial reaction rate control have been derived by Ward and Robb [23].

The transport rates of carbon dioxide and hydrogen sulfide through these carbonate membranes can be significantly increased by adding catalysts to increase the rates of the slow reactions of Equations (11.21) and (11.22). A variety of materials can be used, but the anions of the weak acids such as arsenite, selenite and hypochlorite have been found to be the most effective. Small concentrations of these components increase permeation rates three- to five-fold.

Membranes selective to carbon dioxide and hydrogen sulfide have been considered for removal of these gases from natural gas and various synthetic gas streams. Again, the main problem has been instability of available supported liquid membranes under the typical pressure gradients of several hundred psi. Because the membranes are generally more permeable to hydrogen sulfide than to carbon dioxide, their use to selectively remove hydrogen sulfide from streams contaminated with both gases has also been studied.

Olefin Separation

Concurrently with the work on carbon dioxide and hydrogen sulfide at General Electric, Steigelmann and Hughes [27] and others at Standard Oil were developing facilitated transport membranes for olefin separations. The principal target was the separation of ethylene/ethane and propylene/propane mixtures. Both separations are performed on a massive scale by distillation, but the relative volatilities of the olefins and paraffins are so small that large columns with up to 200 trays are required. In the facilitated transport process, concentrated aqueous silver salt solutions, held in microporous cellulose acetate flat sheets or hollow fibers, were used as the carrier.

Silver ions react readily with olefins, forming a silver–olefin complex according to the reaction:

$$Ag^+ + \text{olefin} \rightleftharpoons Ag^+(\text{olefin}) \quad (11.27)$$

Hughes and Steigelmann used silver nitrate solutions mainly because of the low cost and relatively good stability compared to other silver salts. Silver tetrafluoroborate ($AgBF_4$) has been used by others. The absorption isotherm obtained with a 4 M silver nitrate solution equilibrated with ethylene is shown in Figure 11.25 [26].

The propylene isotherm is reported to be very similar. Based on these data, silver salt membranes are best used with pressurized ethylene feed streams; pressures of 3–6 atm are generally used. The Standard Oil work was continued for a number of years and was taken to the pilot-plant stage using hollow fiber

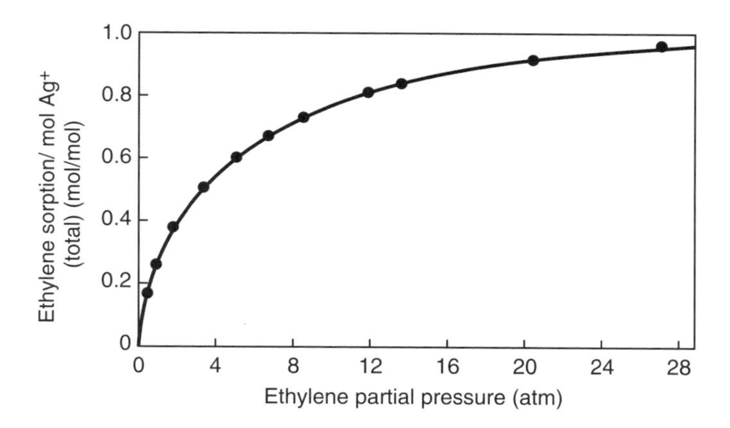

Figure 11.25 Solubility of ethylene in a 4 M silver nitrate solution [26]. Reprinted with permission from S.L. Matson, C.S. Herrick and W.J. Ward III, Progress on the Selective Removal of H_2S from Gasified Coal Using an Immobilized Liquid Membrane, *Ind. Eng. Chem. Prod. Res. Dev.* **16**, 370. Copyright 1977, American Chemical Society and American Pharmaceutical Association

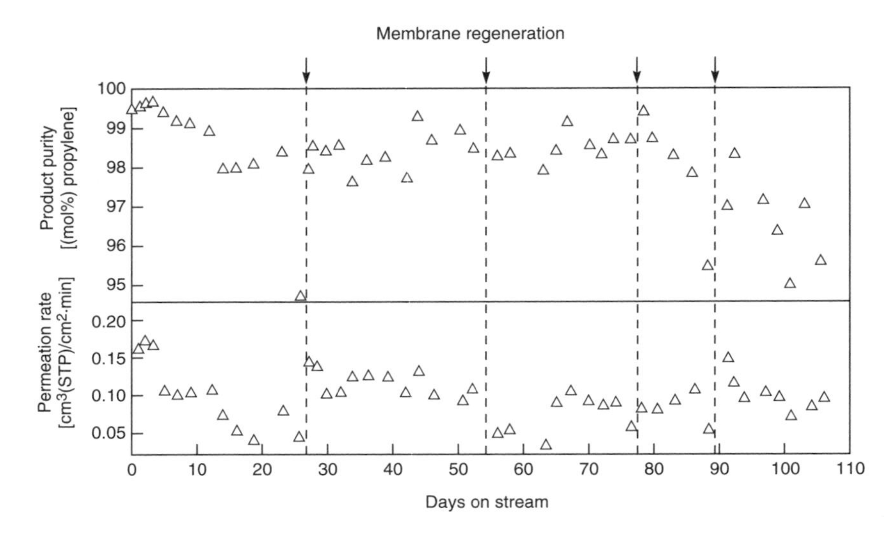

Figure 11.26 Performance of a 37 m^2 hollow fiber silver-nitrate-impregnated facilitated transport membrane for the separation of propylene/propane mixtures. The feed pressure was 5–13 atm; the permeate was a hexane liquid sweep stream. The vertical dotted lines show when the membrane was regenerated with fresh silver nitrate solution [27]. Reprinted with permission from R.D. Hughes, J.A. Mahoney and E.F. Steigelmann, Olefin Separation by Facilitated Transport Membranes, in *Recent Developments in Separation Science*, N.N. Li and J.M. Calo (eds) (1986). Copyright CRC Press, Boca Raton, FL

modules containing almost 40 m^2 membrane area. Some typical data are shown in Figure 11.26 [27]. In these experiments, the feed pressure was maintained at 5–13 atm with liquid hexane circulated on the permeate side of the fibers. In laboratory tests propylene/propane selectivities of more than 100 were obtained routinely; in the large pilot system the initial selectivity was not quite as high but was still very good. Unfortunately, the selectivity and flux deteriorated over a period of a few weeks, partly due to loss of water from the fibers, which could not be prevented even when the feed gas was humidified. Periodic regeneration by pumping fresh silver nitrate solution through the fibers partially restored their properties. However, this technique is not practical in an industrial plant. These instability problems caused Standard Oil to halt the program, which remains the largest facilitated transport trial to date.

The best hope for olefin/paraffin facilitated membrane separations seems to be the solid polymer electrolyte membranes discussed earlier, the results of which are shown in Figures 11.21 and 11.22. If stable membranes with these properties can be produced on an industrial scale, significant applications could develop in treating gases from steam crackers that manufacture ethylene and from polyolefin plants.

Oxygen/Nitrogen Separations

The first demonstration of facilitated transport of oxygen was performed by Scholander [22] using thin films of cellulose acetate impregnated with aqueous hemoglobin solutions. Later Bassett and Schultz [65] demonstrated the process with cobalt dihistidine, a synthetic carrier. The enhancements obtained in these

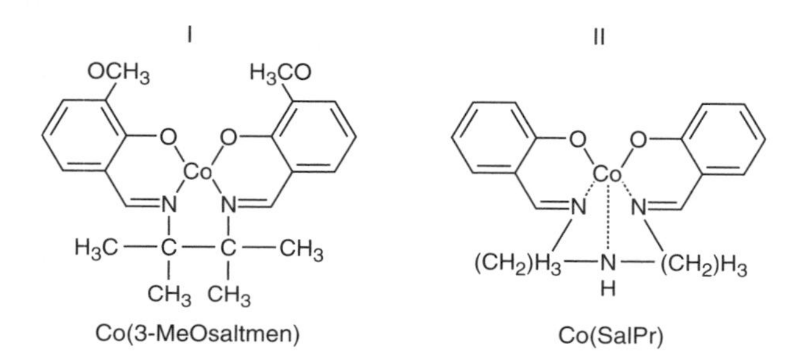

Figure 11.27 Examples of cobalt-based facilitated transport oxygen carriers [66]

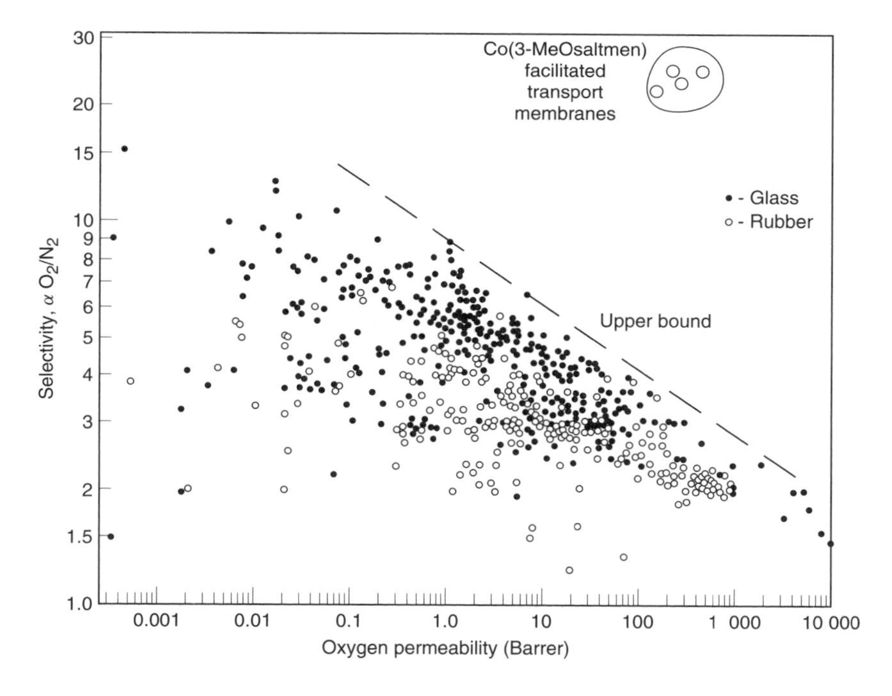

Figure 11.28 The oxygen/nitrogen selectivity plotted against oxygen permeability for polymeric membranes [68] and Co(3-MeOsaltmen)-based facilitated transport membranes [66]

experiments were low, but Johnson and others [66,67] demonstrated very large enhancements using a series of cobalt-based metal chelate carriers. The chemical structures of two typical cobalt Schiffs-base carriers of the type used in this study and in most later work are shown in Figure 11.27. All of these compounds have a central cobalt (II) ion with four coordinating atoms in a planar array. The oxygen molecule coordinates with the cobalt ion from one side of the plane while another coordinating atom, usually a nitrogen group, acts as an electron-donating axial base. In compound I, referred to as Co(3-MeOsaltmen), the coordinating base is usually an imidazole or pyridine group, which must be present for oxygen complexation to occur. In compound II, referred to as Co(SalPr), the coordinating base group is provided by a donor nitrogen atom that is part of the structure.

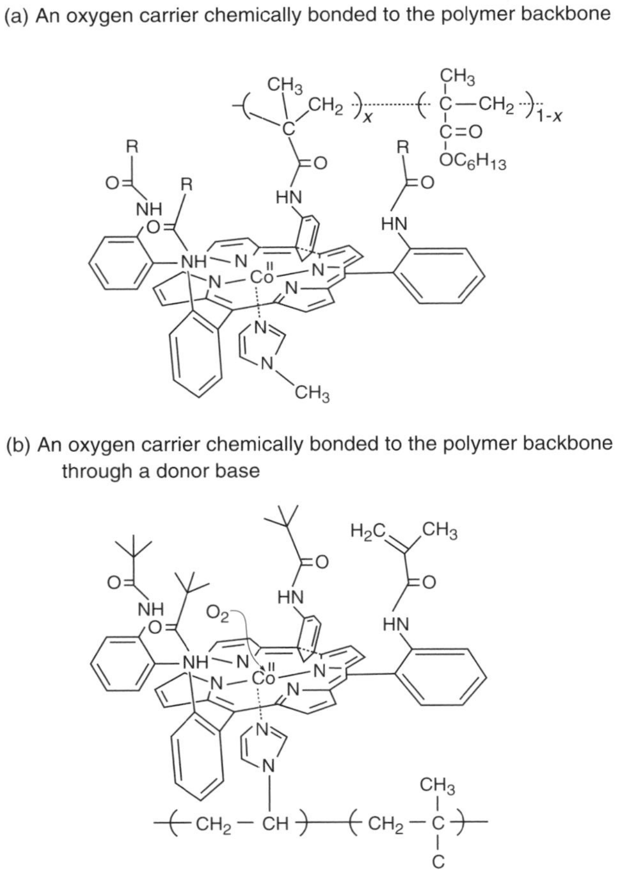

(a) An oxygen carrier chemically bonded to the polymer backbone

(b) An oxygen carrier chemically bonded to the polymer backbone through a donor base

Figure 11.29 Methods of forming bound oxygen carriers

Using carriers of this type, high degrees of facilitation can be achieved. Some data from Johnson's work plotted on the Robeson plot [68] for the polymeric oxygen/nitrogen separating materials described in Chapter 8 are shown in Figure 11.28. This figure shows the promise of facilitated transport membranes and why, even after many failures, interest in this topic has not waned. If stable, thin membranes with these permeabilities and selectivities could be made, major reductions in the cost of membrane-produced oxygen and nitrogen—the second and third largest volume industrial chemicals—would result.

One promising approach to facilitated transport pioneered by Nishide and coworkers at Wasada University is to chemically bind the oxygen carrier to the polymer backbone, which is then used to form a dense polymer film containing no solvent [28]. In some examples, the carrier species is covalently bonded to the polymer matrix as shown in Figure 11.29(a). In other cases, the polymer matrix contains base liquids which complex with the carrier molecule through the base group as shown in Figure 11.29(b). Because these films contain no liquid solvent, they are inherently more stable than liquid membranes and also could be formed into thin films of the selective material in composite membrane form. So far the selectivities and fluxes of these membranes have been moderate.

Conclusions and Future Directions

Carrier facilitated transport membranes have been the subject of serious study for more than 30 years, but no commercial process has resulted. These membranes are a popular topic with academic researchers, because spectacular separations can be achieved with simple laboratory equipment. Unfortunately, converting these laboratory results into practical processes requires the solution of a number of intractable technological problems.

Coupled transport with supported and emulsion liquid membranes has made very little real progress towards commercialization in the last 15 years. In addition, it is now apparent that only a few important separation problems exist for which coupled transport offers clear technical and economic advantages over conventional technology. Unless some completely unexpected breakthrough occurs, it is difficult to imagine that coupled transport will be used on a significant commercial scale within the next 10–20 years. The future prospects for coupled transport are, therefore, dim.

The prospects for facilitated transport membranes for gas separation are better because these membranes offer clear potential economic and technical advantages for a number of important separation problems. Nevertheless, the technical problems that must be solved to develop these membranes to an industrial scale are daunting. Industrial processes require high-performance membranes able to operate reliably without replacement for at least one and preferably several years. No current facilitated transport membrane approaches this target, although some of the solid polymer electrolyte and bound-carrier membranes show promise.

Development of industrial-scale facilitated transport membranes and systems requires access to membrane technology not generally available in universities, and a commitment to a long-term development program that few companies are willing to undertake.

References

1. W.J.V. Osterhout, How do Electrolytes Enter the Cells? *Proc. Natl. Acad. Sci.* **21**, 125 (1935).
2. A. Gliozzi, Carriers and Channels in Artificial and Biological Membranes, in *Bioenergetics and Thermodynamics*, A. Braibanti (ed.), D. Reidel Publishing Co., New York, pp. 299–353 (1980).
3. G.M. Shean and K. Sollner, Carrier Mechanisms in the Movement of Ions Across Porous and Liquid Ion Exchanger Membranes, *Ann. N.Y. Acad. Sci.* **137**, 759 (1966).
4. R. Bloch, O. Kedem and D. Vofsi, Ion Specific Polymer Membrane, *Nature* **199**, 802 (1963).
5. R. Bloch, Hydrometallurgical Separations by Solvent Membranes, in *Proceedings of Membrane Science and Technology*, J.E. Flinn (ed.), Columbus Laboratories of Battelle Memorial Institute, Plenum Press, pp. 171–187 (1970).
6. J. Jagur-Grodzinski, S. Marian and D. Vofsi, The Mechanism of a Selective Permeation of Ions Through 'Solvent Polymer Membranes', *Sep. Sci.* **8**, 33 (1973).
7. T. Miyauchi, Liquid–Liquid Extraction Process of Metals, US Patent 4,051,230 (September, 1977).
8. T. Largman and S. Sifniades, Recovery of Copper (II) from Aqueous Solutions by Means of Supported Liquid Membranes, *Hydrometall.* **3**, 153 (1978).
9. R.W. Baker, M.E. Tuttle, D.J. Kelly and H.K. Lonsdale, Coupled Transport Membranes, I: Copper Separations, *J. Membr. Sci.* **2**, 213 (1977).
10. W.C. Babcock, R.W. Baker, D.J. Kelly and E.D. LaChapelle, Coupled Transport Membranes for Uranium Recovery, in *Proceedings of ISEC'80, University of Liege*, Liege, Belgium (1980).
11. N.N. Li, Permeation Through Liquid Surfactant Membranes, *AIChE J.* **17**, 459 (1971).
12. N.N. Li, Facilitated Transport Through Liquid Membranes—An Extended Abstract, *J. Membr. Sci.* **3**, 265 (1978).
13. N.N. Li, R.P. Cahn, D. Naden and R.W.M. Lai, Liquid Membrane Processes for Copper Extraction, *Hydrometall.* **9**, 277 (1983).
14. E.S. Matulevicius and N.N. Li, Facilitated Transport Through Liquid Membranes, *Sep. Purif. Meth.* **4**, 73 (1975).
15. E.L. Cussler, Membranes Which Pump, *AIChE J.* **17**, 1300 (1971).
16. C.F. Reusch and E.L. Cussler, Selective Membrane Transport, *AIChE J.* **19**, 736 (1973).
17. H.C. Hayworth, A Case in Technology Transfer, *CHEMTECH* **11**, 342 (1981).
18. W.S.W. Ho, N.N. Li, Z. Gu, R.J. Marr and J. Drexler, Emulsion Liquid Membranes, in *Membrane Handbook*, W.S.W. Ho and K.K. Sirkar (eds), Van Nostrand Reinhold, New York, pp. 595–724 (1992).
19. A.R. Sangupta, R. Basin and K.K. Sirkar, Separation of Solutes from Aqueous Solutions by Contained Liquid Membranes, *AIChE J.* **34**, 1698 (1988).
20. S. Majumdar and K.K. Sirkar, Hollow-fiber Contained Liquid Membrane, in *Membrane Handbook*, W.S.W. Ho and K.K. Sirkar (eds), Van Nostrand Reinhold, New York (1992).
21. J.C. Davis, R.J. Valus, R. Eshraghi and A.E. Velikoff, Facilitated Transport Membrane Hybrid Systems for Olefin Purification, *Sep. Sci. Technol.* **28**, 463 (1993).

22. P.F. Scholander, Oxygen Transport Through Hemoglobin Solutions, *Science* **131**, 585 (1960).
23. W.J. Ward III and W.L. Robb, Carbon Dioxide–Oxygen Separation: Facilitated Transport of Carbon Dioxide Across a Liquid Film, *Science* **156**, 1481 (1967).
24. W.J. Ward III, Analytical and Experimental Studies of Facilitated Transport, *AIChE J.* **16**, 405 (1970).
25. S.G. Kimura, S.L. Matson and W.J. Ward III, Industrial Applications of Facilitated Transport, in *Recent Developments in Separation Science*, N.N. Li, J.S. Dranoff, J.S. Schultz and P. Somasundaran (eds), CRC Press, Boca Raton, FL, pp. 11–26 (1979).
26. S.L. Matson, C.S. Herrick and W.J. Ward III, Progress on the Selective Removal of H_2S from Gasified Coal Using an Immobilized Liquid Membrane, *Ind. Eng. Chem. Prod. Res. Dev.* **16**, 370 (1977).
27. R.D. Hughes, J.A. Mahoney and E.F. Steigelmann, Olefin Separation by Facilitated Transport Membranes, in *Recent Developments in Separation Science*, N.N. Li and J.M. Calo (eds), CRC Press, Boca Raton, FL, pp. 173–196 (1986).
28. H. Nishide, H. Kawakami, T. Suzuki, Y. Azechi, Y. Soejima and E. Tsuchida, Effect of Polymer Matrix on the Oxygen Diffusion via a Cobalt Porphyrin Fixed in a Membrane, *Macromolecules* **24**, 6306 (1991).
29. T. Suzuki, H. Yasuda, H. Nishide, X.-S. Chen and E. Tsuchida, Electrochemical Measurement of Facilitated Oxygen Transport Through a Polymer Membrane Containing Cobaltporphyrin as a Fixed Carrier, *J. Membr. Sci.* **112**, 155 (1996).
30. H. Nishide, H. Kawakami, S.Y. Sasame, K. Ishiwata and E. Tsuchida, Facilitated Transport of Molecular Oxygen in Cobaltporphyrin/Poly(1-trimethylsilyl-1-propyne) Membrane, *J. Polym. Sci., Part A: Polym. Chem.* **30**, 77 (1992).
31. K.-V. Peinemann, S.K. Shukla and M. Schossig, Preparation and Properties of Highly Selective Inorganic/Organic Blend Membranes for Separation of Reactive Gases, *The 1990 International Congress on Membranes and Membrane Processes*, North American Membrane Society, Chicago, pp. 792–794 (1990).
32. W.S. Ho, Polymeric Membrane and Processes for Separation of Aliphatically Unsaturated Hydrocarbons, US Patents 5,062,866 (November, 1991) and 5,015,268 (May, 1991).
33. I. Pinnau, L.G. Toy and C.G. Casillas, Olefin Separation Membrane and Processes, US Patent 5,670,051 (September, 1997).
34. E.L. Cussler, Facilitated Transport, in *Membrane Separation Systems*, R.W. Baker, E.L. Cussler, W. Eykamp, W.J. Koros, R.L. Riley and H. Strathmann (eds), Noyes Data Corp., Park Ridge, NJ, pp. 242–275 (1991).
35. J.D. Way and R.D. Noble, Facilitated Transport, in *Membrane Handbook*, W.S.W. Ho and K.K. Sirkar (eds), Van Nostrand Reinhold, New York, pp. 833–866 (1992).
36. E.L. Cussler, *Diffusion: Mass Transfer in Fluid Systems*, Cambridge University Press, New York (1984).
37. R.D. Noble and J.D. Way, *Liquid Membranes*, ACS Symposium Series Number 347, American Chemical Society, Washington, DC (1987).
38. D.V. Laciak, Development of Active-transport Membrane Devices, *Department of Energy Technical Report*, NTIS DE94015947 (July, 1994).
39. E.L. Cussler, Facilitated and Active Transport, in *Polymeric Gas Separation Membranes*, D.R. Paul and Y.P. Yampol'skii (eds), CRC Press, Boca Raton, FL, pp. 273–300 (1994).
40. A. Figoli, W.F.C. Sager and M.H.V. Mulder, Facilitated Oxygen Transport in Liquid Membranes: Review and New Concepts, *J. Membr. Sci.* **181**, 97 (2001).
41. K.A. Smith, J.H. Meldon and C.K. Colton, An Analysis of Carrier-facilitated Transport, *AIChE J.* **19**, 102 (1973).

42. J.S. Schultz, J.D. Goddard and S.R. Suchdeo, Facilitated Transport via Carrier-mediated Diffusion in Membranes, *AIChE J.* **20**, 417 (1974).
43. R.W. Baker and I. Blume, Coupled Transport Membranes, in *Handbook of Industrial Membrane Technology*, M.C. Porter (ed.), Noyes Publications, Park Ridge, NJ, pp. 511–558 (1990).
44. W.C. Babcock, R.W. Baker, M.G. Conrod and K.L. Smith, Coupled Transport Membranes for Removal of Chromium from Electroplating Rinse Solutions, in *Chemistry in Water Reuse*, Ann Arbor Science Publishers, Ann Arbor, MI (1981).
45. W.C. Babcock, R.W. Baker, E.D. LaChapelle and K.L. Smith, Coupled Transport Membranes, II: The Mechanism of Uranium Transport with a Tertiary Amine, *J. Membr. Sci.* **7**, 17 (1980).
46. W.C. Babcock, R.W. Baker, E.D. LaChapelle and K.L. Smith, Coupled Transport Membranes, III: The Rate-limiting Step in Uranium Transport with a Tertiary Amine, *J. Membr. Sci.* **7**, 89 (1980).
47. J.D. Lamb, P.R. Brown, J.J. Christensen, J.S. Bradshaw, D.G. Garrick and R.M. Izatt, Cation Transport at 25 °C from Binary Na^+-Mn^+, Cs^+-Mn^+ and Sr^{2+}-Mn^+ Nitrate Mixtures in a H_2O–$CHCl_3$–H_2O Liquid Membrane System Containing a Series of Macrocyclic Carriers, *J. Membr. Sci.* **13**, 89 (1983).
48. R.M. Izatt, R.M. Haws, J.D. Lamb, D.V. Dearden, P.R. Brown, D.W. McBride, Jr and J.J. Christensen, Facilitated Transport from Ternary Cation Mixtures Through Water–Chloroform–Water Membrane Systems Containing Macrocyclic Ligands, *J. Membr. Sci.* **20**, 273 (1984).
49. W.C. Babcock, R.W. Baker, D.J. Kelly, E.D. LaChapelle and H.K. Lonsdale, *Coupled Transport Membranes for Metal Separations. Final Report, Phase IV*, US Bureau of Mines Technical Report, Springfield, VA (1979).
50. A.M. Neplenbroek, D. Bargeman and C.A. Smolders, Mechanism of Supported Liquid Membrane Degradation, *J. Membr. Sci.* **67**, 121, 133 (1992).
51. F.F. Zha, A.G. Fane and C.J.D. Fell, Instability Mechanisms of Supported Liquid Membranes in Phenol Transport Processes, *J. Membr. Sci.* **107**, 59 (1995).
52. W. Völkel, W. Halwachs and K. Schügerl, Copper Extraction by Means of a Liquid Surfactant Membrane Process, *J. Membr. Sci.* **6**, 19 (1980).
53. W. Völkel, W. Poppe, W. Halwachs and K. Schügerl, Extraction of Free Phenols from Blood by a Liquid Membrane Enzyme Reactor, *J. Membr. Sci.* **11**, 333 (1982).
54. O.H. LeBlanc, W.J. Ward, S.L. Matson and S.G. Kimura, Facilitated Transport in Ion Exchange Membranes, *J. Membr. Sci.* **6**, 339 (1980).
55. J.D. Way, R.D. Noble, D.L. Reed and G.M. Ginley, Facilitated Transport of CO_2 in Ion Exchange Membranes, *AIChE J.* **33**, 480 (1987).
56. R. Rabago, D.L. Bryant, C.A. Koval and R.D. Noble, Evidence for Parallel Pathways in the Facilitated Transport of Alkenes through Ag^+-exchanged Nafion Films, *Ind. Eng. Chem. Res.* **35**, 1090 (1996).
57. G.P. Pez and D.V. Laciak, Ammonia Separation Using Semipermeable Membranes, US Patent 4,762,535 (August, 1988).
58. G.P. Pez, R.T. Carlin, D.V. Laciak and J.C. Sorensen, Method for Gas Separation, US Patent 4,761,164 (August, 1988).
59. R. Quinn, J.B. Appleby and G.P. Pez, New Facilitated Transport Membranes for the Separation of Carbon Dioxide from Hydrogen and Methane, *J. Membr. Sci.* **104**, 139 (1995).
60. D.V. Laciak, R. Quinn, G.P. Pez, J.B. Appleby and P. Puri, Selective Permeation of Ammonia and Carbon Dioxide by Novel Membranes, *Sep. Sci. Technol.* **26**, 1295 (1990).
61. I. Pinnau and L.G. Toy, Solid Polymer Electrolyte Composite Membranes for Olefin/Paraffin Separation, *J. Membr. Sci.* **184**, 39 (2001).

62. A. Morisato, Z. He, I. Pinnau and T.C. Merkel, Transport Properties of PA12-PTMO/ $AgBF_4$ Solid Polymer Electrolyte Membranes for Olefin/Paraffin Separation, *Desalination* **145**, 347 (2002).
63. S.U. Hong, C.K. Kim and Y.S. Kang, Measurement and Analysis of Propylene Solubility in Polymer Electrolytes Containing Silver Salts, *Macromolecules* **33**, 7918 (2000).
64. K.M. White, B.D. Smith, P.J. Duggan, S.L. Sheahan and E.M. Tyndall, Mechanism of Facilitated Saccharide Transport through Plasticized Cellulose Triacetate Membranes, *J. Membr. Sci.* **194**, 165 (2001).
65. R.J. Bassett and J.S. Schultz, Nonequilibrium Facilitated Diffusion of Oxygen Through Membranes of Aqueous Cobalt Dihistidine, *Biochim. Biophys. Acta* **211**, 194 (1970).
66. B.M. Johnson, R.W. Baker, S.L. Matson, K.L. Smith, I.C. Roman, M.E. Tuttle and H.K. Lonsdale, Liquid Membranes for the Production of Oxygen-enriched Air–II. Facilitated Transport Membranes, *J. Membr. Sci.* **31**, 31 (1987).
67. R.W. Baker, I.C. Roman and H.K. Lonsdale, Liquid Membranes for the Production of Oxygen-Enriched Air–I. Introduction and Passive Liquid Membranes, *J. Membr. Sci.* **31**, 15 (1987).
68. L.M. Robeson, Correlation of Separation Factor Versus Permeability for Polymeric Membranes, *J. Membr. Sci.* **62**, 165 (1991).

12 MEDICAL APPLICATIONS OF MEMBRANES

Introduction

In this chapter, the use of membranes in medical devices is reviewed briefly. In terms of total membrane area produced, medical applications are at least equivalent to all industrial membrane applications combined. In terms of dollar value of the products, the market is far larger. In spite of this, little communication between these two membrane areas has occurred over the years. Medical and industrial membrane developers each have their own journals, societies and meetings, and rarely look over the fence to see what the other is doing. This book cannot reverse 50 years of history, but every industrial membrane technologist should at least be aware of the main features of medical applications of membranes. Therefore, in this chapter, the three most important applications—hemodialysis (the artificial kidney), blood oxygenation (the artificial lung) and controlled release pharmaceuticals—are briefly reviewed.

Hemodialysis

The kidney is a key component of the body's waste disposal and acid–base regulation mechanisms. Each year approximately one person in ten thousand suffers irreversible kidney failure. Before 1960, this condition was universally fatal [1] but now a number of treatment methods can maintain these patients. Of these, hemodialysis is by far the most important, and approximately 800 000 patients worldwide benefit from the process. Each patient is dialyzed approximately three times per week with a dialyzer containing about 1 m^2 of membrane area. Economies of scale allow these devices to be produced for about US$15 each; the devices are generally discarded after one or two uses. As a result the market for dialyzers alone is about US$1.3 billion [2,3].

The operation of the human kidney simulated by hemodialyzers is illustrated in Figure 12.1. The process begins in the glomerulus, a network of tiny capillaries surrounding spaces called Bowman's capsules. Blood flowing through these

Membrane Technology and Applications R. W. Baker
 ISBN: 0-470-85445-6

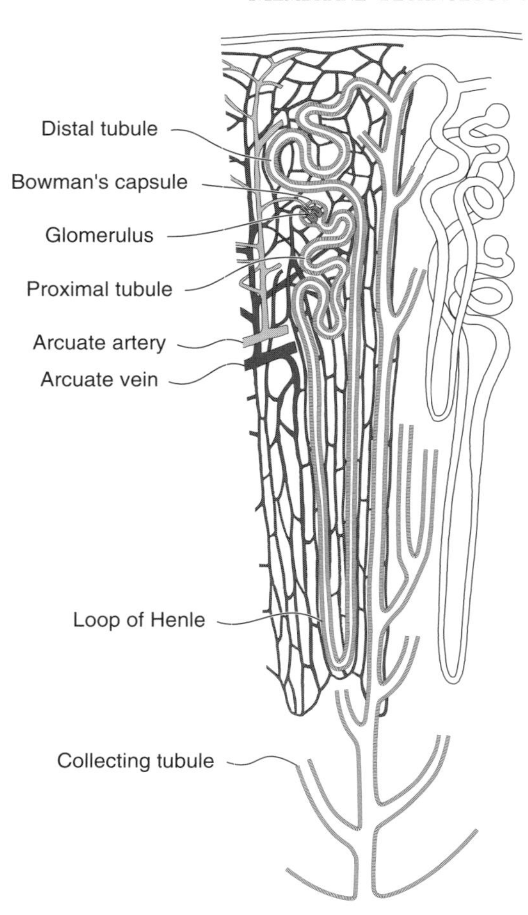

Figure 12.1 Schematic of a single nephron, the functional unit of the kidney. Microsolutes are filtered from blood cells in Bowman's capsules. As the filtrate passes towards the collection tubule most of the microsolutes and water are reabsorbed by a type of facilitated transport process. The fluid finally entering the collecting tubule contains the nitrogenous wastes from the body and is excreted as urine. There are about 1 million nephrons in the normal kidney [1]

capillaries is at a higher pressure than the fluid in Bowman's capsules, and the walls of the capillaries are finely microporous. As a result, water, salts and other microsolutes in the blood are ultrafiltered into the capsule while blood cells stay behind. Each Bowman's capsule is connected by a relatively long, thin duct to the collecting tubule, ultimately forming urine, which is sent via the urethra to the bladder. The average kidney has approximately 1 million tubules and many Bowman's capsules are connected to each tubule. As the fluid that permeates into Bowman's capsules from the blood travels down the collection duct to the central tubule, more than 99 % of the water and almost all of the salts, sugars and proteins are reabsorbed into the blood by a process similar to facilitated transport. The remaining concentrated fluid ultimately forms urine and is rich in urea and creatinine. This is the principal method by which these nitrogen-containing metabolites are discharged from the body. The acid–base balance of the body is also controlled by the bicarbonate level of urine, and many drugs and toxins are excreted from the body this way.

The first successful hemodialyzer was constructed by Kolf and Berk in The Netherlands in 1945 [4,5]. Kolf's device used dialysis to remove urea and other waste products directly from blood. A flat cellophane (cellulose) tube formed the dialysis membrane; the tube was wound around a rotating drum immersed in a bath of saline. As blood was pumped through the tube, urea and other low molecular weight metabolites diffused across the membrane to the dialysate down a concentration gradient. The cellophane tubing did not allow diffusion of larger components in the blood such as proteins or blood cells. By maintaining the salt, potassium and calcium levels in the dialysate solution at the same levels as in the blood, loss of these components from the blood was prevented.

Kolf's early devices were used for patients who had suffered acute kidney failure as a result of trauma or poisoning and needed dialysis only a few times. Such emergency treatment was the main application of hemodialysis until the early 1960s, because patients suffering from chronic kidney disease require dialysis two to three times per week for several years, which was not practical with these early devices. However, application of hemodialysis to this class of patient was made possible by improvements in the dialyzer design in the 1960s. The development of a plastic shunt that could be permanently fitted to the patient to allow easy access to their blood supply was also important. This shunt, developed by Scribner *et al.* [6], allowed dialysis without the need for surgery to connect the patient's blood vessels to the dialysis machine for each treatment.

Kolf's first tubular dialyzer, shown in Figure 12.2, required several liters of blood to prime the system, a major operational problem. In the 1950s, tubular dialyzers were replaced with coil (spiral) devices, also developed by Kolf and coworkers. This coil system was the basis for the first disposable dialyzer produced commercially in the early 1960s. The blood volume required to prime the device was still excessive, however, and during the 1960s the plate-and-frame and hollow fiber devices shown in Figure 12.3 were developed. In the US in 1975, about 65 % of all dialyzers were coil, 20 % hollow fiber systems and 15 % plate-and-frame. Within 10 years the coil dialyzer had essentially disappeared, and the market was divided two-thirds hollow fibers and one-third plate-and-frame. By 1996, hollow fiber dialyzers had more than 95 % of the market.

Hollow fiber dialyzers typically contain 1–2 m^2 of membrane in the form of fibers 0.1–0.2 mm in diameter. A typical dialyzer module may contain several thousand fibers housed in a 2-in.-diameter tube, 1–2 ft long. Approximately

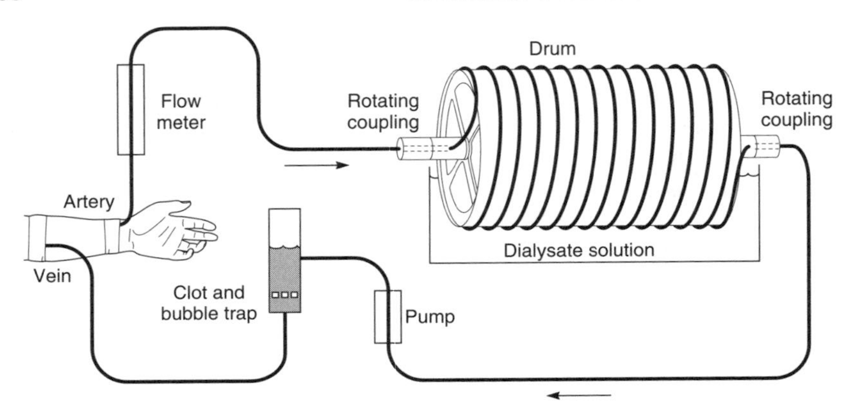

Figure 12.2 Schematic of an early tubular hemodialyzer based on the design of Kolf's original device. The device required several liters of blood to fill the tubing and minor surgery to connect to the patient. Nonetheless, it saved the lives of patients suffering from acute kidney failure [1]

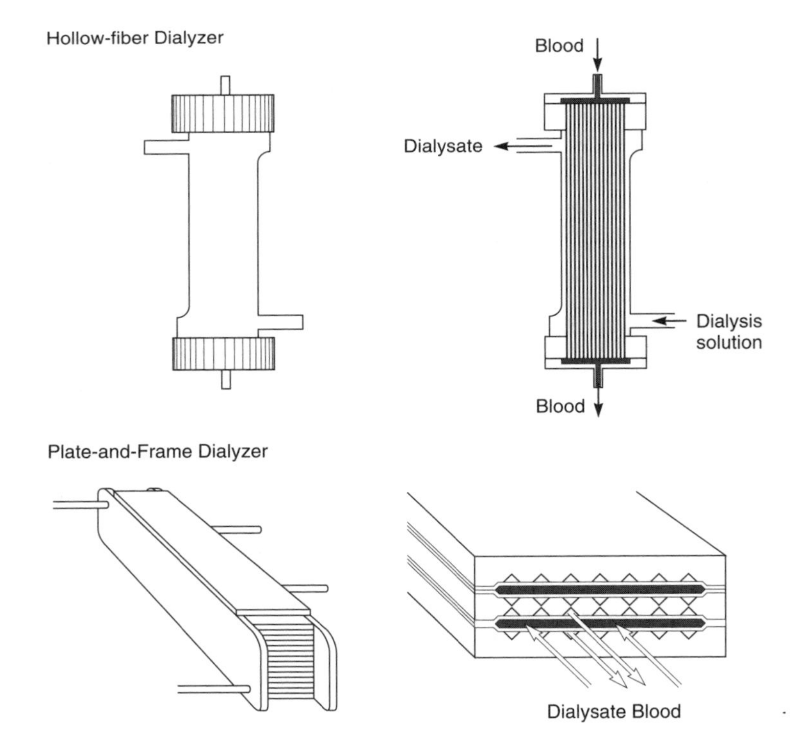

Figure 12.3 Schematic of hollow fiber and plate-and-frame dialyzers

100 million hemodialysis procedures are performed annually worldwide. Because hollow fiber dialyzers are produced in such large numbers, prices are very low. Today a 1–2 m^2 hollow fiber dialyzer sells for about US$15, which is well below the module costs of any other membrane technology. These low costs have been achieved by the use of high speed machines able to spin several hundred fibers simultaneously around the clock. The entire spinning, cutting, module potting and testing process is automated.

In a hollow fiber dialyzer the blood flows down the bore of the fiber, providing good fluid flow hydrodynamics. An advantage of the hollow fiber design is that only 60–100 mL of blood is required to fill the dialyzer. At the end of a dialysis procedure hollow fiber dialyzers can also be easily drained, flushed with sterilizing agent, and reused. Dialyzer reuse is widely practiced, in part for economic reasons, but also because the biocompatibility of the membrane appears to improve after exposure to blood.

The regenerated cellulose membranes used in Kolf's first dialyzer are still in use in some dialyzers. Cellulose membranes are isotropic hydrogels generally about 10 μm thick and, although very water swollen, they have a high wet strength. The hydraulic permeability of cellulose is relatively low, and the membrane has a molecular weight cut-off of about 2000 dalton. The permeability of cellulose hydrogel membranes compared to the calculated permeability of an aqueous film of equal thickness is shown in Figure 12.4.

Although cellulose has been used successfully in hemodialyzers for many years, there is some concern about the ability of the free hydroxyl groups on the

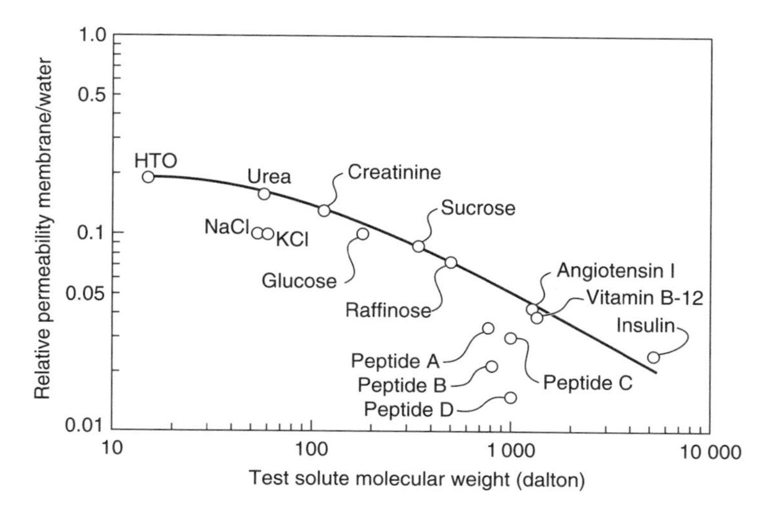

Figure 12.4 Solute permeability relative to the permeability of a film of water for various solutes in a regenerated cellulose membrane (Cuprophan 150). This type of membrane is still widely used in hemodialysis devices

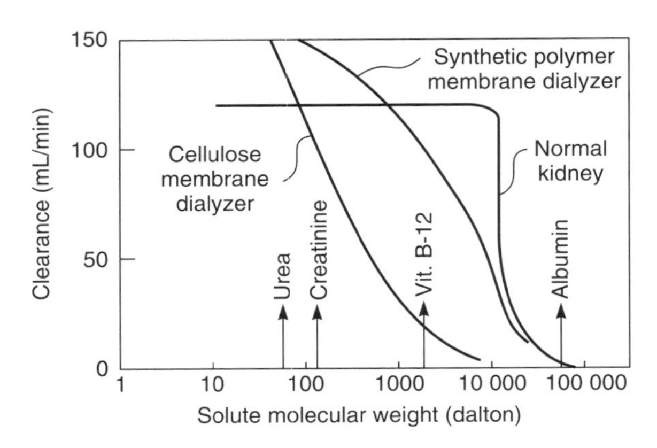

Figure 12.5 Clearance, a measure of membrane permeability, as a function of molecular weight for hemodialyzers and the normal kidney [7]

membrane surface to activate the blood clotting process. When cellulose-based dialyzers are reused, the membrane's blood compatibility improves because a coating of protein has formed on the membrane surface. Recently, synthetic polymers have begun to replace cellulose. These membrane materials are substituted cellulose derivatives, specifically cellulose acetate or polymers such as polyacrylonitrile, polysulfone, polycarbonate, polyamide and poly(methyl methacrylate). These synthetic fiber membranes are generally microporous with a finely microporous skin layer on the inside, blood-contacting surface of the fiber. The hydraulic permeability of these fibers is up to 10 times that of cellulose membranes, and they can be tailored to achieve a range of molecular weight cut-offs using different preparation procedures. The blood compatibility of the synthetic polymer membranes is good, and these membranes are likely to largely replace unsubstituted cellulose membranes over the next few years.

One attractive feature of some of the new synthetic polymer membranes is their ability to remove some of the middle molecular weight metabolites in blood. This improvement in performance is illustrated by Figure 12.5. Cellulose membranes efficiently remove the major metabolites, urea and creatinine, from blood, but metabolites with molecular weights between 1000 and 10 000 are removed poorly. Patients on long-term dialysis are believed to accumulate these metabolites, which are associated with a number of health issues. The new synthetic polymer membranes appear to simulate the function of the normal kidney more closely.

Blood Oxygenators

Blood oxygenators are used during surgery when the patient's lungs cannot function normally. Pioneering work on these devices was carried out in the 1930s and 1940s by J.H. Gibbon [8,9], leading to the first successful open heart surgery on a human patient in 1953. Gibbon's heart–lung machine used a small tower filled with stainless steel screens to contact blood with counter-flowing oxygen. Direct oxygenation of the blood was used in all such devices until the early 1980s. Screen oxygenators of the type devised by Gibbon were first replaced with a disk oxygenator, which consisted of 20–100 rotating disks in a closed cylinder containing 1–2 L of blood. Later, bubble oxygenators were developed, in which blood was oxygenated in a packed plastic tower through which blood flowed. Because these direct-contact oxygenators required rather large volumes of blood to prime the device and, more importantly, damaged some of the blood components, they were used in only a few thousand operations per year in the 1980s. The introduction of indirect-contact membrane oxygenators resulted in significantly less blood damage and lower blood priming volumes. The devices were rapidly accepted, and the total number of procedures performed following their introduction expanded rapidly. The first membrane oxygenators were introduced in 1980, and by 1985 represented more than half of the oxygenators in use. This percentage had risen to 70 % by 1990; now, only membrane oxygenators are used. Over the same period, the number of procedures using blood oxygenators has risen to approximately 1 million per year worldwide. Each device costs around US$500–600, so the total annual market is about US$500 million.

The function of a membrane blood oxygenator is shown schematically in Figure 12.6. In the human lung, the total exchange membrane area between

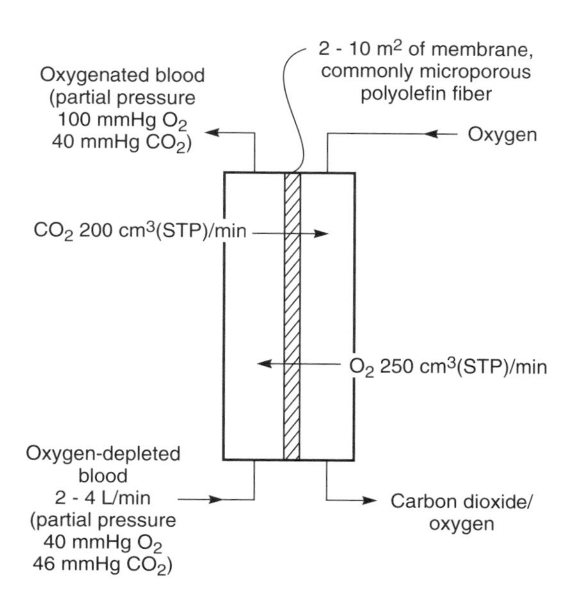

Figure 12.6 Flow schematic of a membrane blood oxygenator

the blood capillaries and the air drawn in and out is about 80 m^2. The human lung membrane is estimated to be about 1 μm thick, and the total exchange capacity of the lung is far larger than is normally required. This allows people with impaired lung capacity to lead relatively normal lives. A successful heart–lung machine must deliver about 250 cm^3(STP)/ min oxygen and remove about 200 cm^3(STP)/ min carbon dioxide [10]. Because of the limited solubility of these gases in the blood, relatively large blood flows through the device are required, typically 2–4 L/min, which is approximately 10 times the blood flow through a kidney dialyzer. The first membrane oxygenators used silicone rubber membranes, but now microporous polyolefin fibers are used. To maintain good mass transfer with minimal pressure drop through the device, blood is generally circulated on the outside of the fibers.

Controlled Drug Delivery

In controlled drug delivery systems a membrane is used to moderate the rate of delivery of drug to the body. In some devices the membrane controls permeation of the drug from a reservoir to achieve the drug delivery rate required. Other devices use the osmotic pressure produced by diffusion of water across a membrane to power miniature pumps. In yet other devices the drug is impregnated into the membrane material, which then slowly dissolves or degrades in the body. Drug delivery is then controlled by a combination of diffusion and biodegradation.

The objective of all of these devices is to deliver a drug to the body at a rate predetermined by the design of the device and independent of the changing environment of the body. In conventional medications, only the total mass of drug delivered to a patient is controlled. In controlled drug delivery, both the mass and the rate at which the drug is delivered can be controlled, providing three important therapeutic benefits:

1. The drug is metered to the body slowly over a long period; therefore, the problem of overdosing and underdosing associated with conventional periodic medication is avoided.
2. The drug is given locally, ideally to the affected organ directly, rather than systemically as an injection or tablet. Localized delivery results in high concentrations of the drug at the site of action, but low concentrations and hence fewer side effects elsewhere.
3. As a consequence of metered, localized drug delivery, controlled release devices generally equal or improve the therapeutic effects of conventional medications, while using a fraction of the drug. Thus, the problems of drug-related side effects are correspondingly lower.

The concept of controlled delivery is not limited to drugs. Similar principles are used to control the delivery of agrochemicals, fertilizers and pesticides, for example, and in many household products. However, most of the technology development in the past 30 years has focused on drug delivery; only this aspect of the topic is covered here.

The origins of controlled release drug delivery can be traced to the 1950s. Rose and Nelson [11], for example, described the first miniature osmotic pump in 1955. A key early publication was the paper of Folkman and Long [12] in 1964 describing the use of silicone rubber membranes to control the release of anesthetics and cardiovascular drugs. Concurrent discoveries in the field of hormone regulation of female fertility quickly led to the development of controlled release systems to release steroids for contraception [13–15]. The founding of Alza Corporation by Alex Zaffaroni in the late 1960s gave the entire technology a decisive thrust. Alza was dedicated to developing novel controlled release drug delivery systems [16]. The products developed by Alza during the subsequent 25 years stimulated the entire pharmaceutical industry. The first pharmaceutical product in which the drug registration document specified both the total amount of drug in the device and the delivery rate was an Alza product, the Ocusert®, launched in 1974. This device, shown in Figure 12.7, consisted of a three-layer laminate with the drug sandwiched between two rate-controlling polymer membranes. The device is an ellipse about 1 mm thick and 1 cm in diameter. The device is placed in a cul de sac of the eye where it delivers the drug (pilocarpine) at a constant rate for 7 days, after which it is removed and replaced. The Ocusert was a technical tour de force, although only a limited marketing success. Alza later developed a number of more widely used products, including multilayer transdermal patches designed to deliver drugs through the skin [17]. The drugs include scopolamine (for motion sickness), nitroglycerine (for angina), estradiol (for hormone replacement), and nicotine (for control of smoking addiction). Many imitators have followed Alza's success, and more than 20 transdermal products delivering a variety of drugs are now available.

Membrane Diffusion-controlled Systems

In membrane diffusion-controlled systems, a drug is released from a device by permeation from its interior reservoir to the surrounding medium. The rate of diffusion of the drug through the membrane governs its rate of release. The reservoir device illustrated in Figure 12.8 is the simplest diffusion-controlled system. An inert membrane encloses the drug to be released; the drug diffuses through the membrane at a finite, controllable rate. If the concentration (or thermodynamic activity) of the material in equilibrium with the inner surface of the enclosing membrane is constant then the concentration gradient, the driving force for

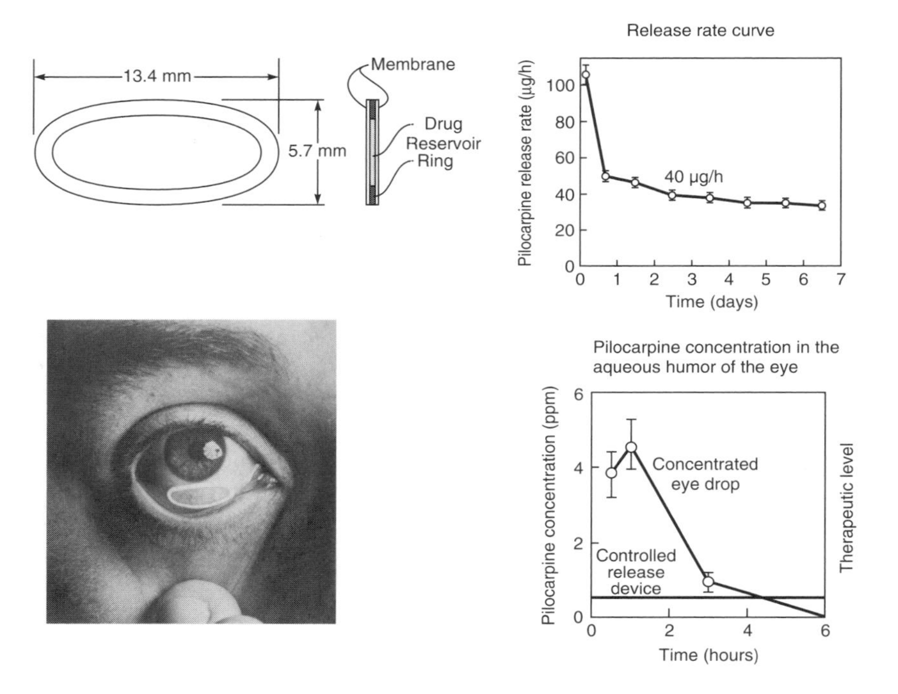

Figure 12.7 The Ocusert pilocarpine system is a thin multilayer membrane device. The central sandwich consists of a core containing the drug pilocarpine. The device is placed in the eye, where it releases the drug at a continuous rate for 7 days. Devices with release rates of 20 or 40 μg/h are used. Controlled release of the drug eliminates the over- and under-dosing observed with conventional eyedrop formulations, which must be delivered every 4–6 h to maintain therapeutic levels of the drug in the eye tissue [18]

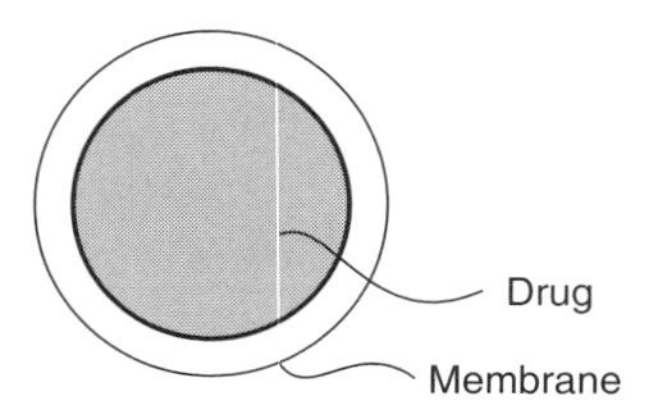

Figure 12.8 Reservoir device

diffusional release of the drug, is constant. This occurs when the inner reservoir contains a saturated solution of the material, providing a constant release rate for as long as excess solid is maintained in the solution. This is called zero-order release. If, however, the active drug within the device is initially present as an unsaturated solution, its concentration falls as it is released. The release rate declines exponentially, producing a first-order release profile.

For a device containing a saturated solution of drug, and excess solid drug, Fick's law

$$J = -DK\frac{dc_s}{dx} \tag{12.1}$$

can be restated for a slab or sandwich geometry as

$$\frac{dM_t}{dt} = \frac{AJ}{l} = \frac{ADKc_s}{l} \tag{12.2}$$

where M_t is the mass of drug released at any time t, and hence dM_t/dt is the steady-state release rate at time t; A is the total surface area of the device (edge effects being ignored); c_s is the saturation solubility of the drug in the reservoir layer; and J is the membrane-limiting flux.

The Ocusert system illustrated in Figure 12.7 is one example of a diffusion-controlled reservoir device. Another is the steroid-releasing intrauterine device (IUD) shown in Figure 12.9. Inert IUDs of various shapes were widely used for

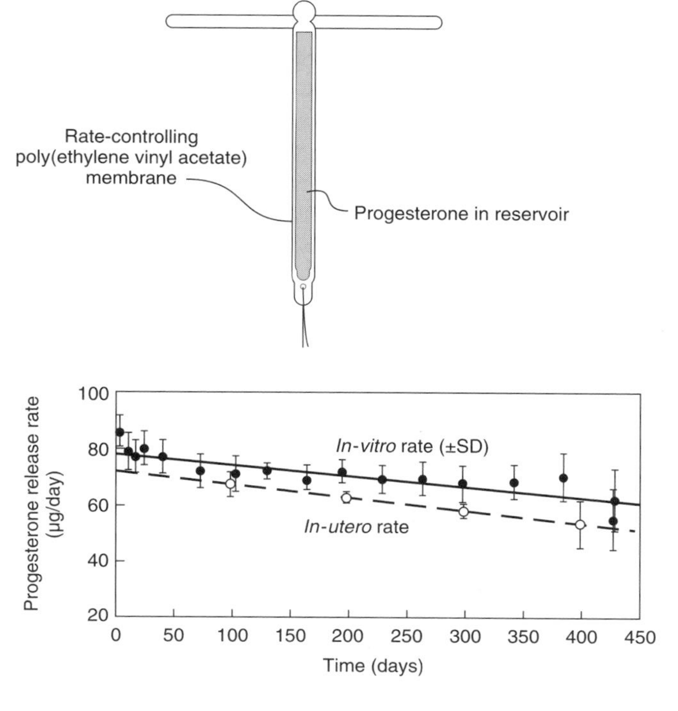

Figure 12.9 Progestasert® intrauterine device (IUD) designed to deliver progesterone for contraception at 65 μg/day for 1 year [19]

birth control in the 1950s and 1960s. The contraceptive effect of these IUDs was based on physical irritation of the uterus. Thus, the devices resulting in the lowest pregnancy rate were often associated with unacceptable levels of pain and bleeding, whereas more comfortable devices were associated with unacceptably high pregnancy rates. Researchers tried a large number of different IUD shapes in an attempt to produce a device that combined a low pregnancy rate with minimal pain and bleeding, but without real success.

Steroid-releasing IUDs, in which the contraceptive effect of the device comes largely from the steroid, offer a solution to the discomfort caused by inert IUDs. Such devices can use IUDs with a low pain and bleeding level as a platform for the steroid-releasing system. Scommegna *et al.* performed the first clinical trials to test this concept [15]. The commercial embodiment of these ideas is shown in Figure 12.9, together with the drug release rate curve [19]. Inspection of this curve shows an initial high drug release during the first 30–40 days, representing drug that has migrated into the polymer during storage of the device and which is released as an initial burst. Thereafter, the device maintains an almost constant drug release rate until it is exhausted at about 400 days. Later versions of this device contained enough drug to last 2 years or more.

The second common category of diffusion-controlled devices is the monolithic system, in which the agent to be released is dispersed uniformly throughout the rate-controlling polymer medium, as shown in Figure 12.10. The release profile is then determined by the loading of dispersed agent, the nature of the components, and the geometry of the device. Thin spots, pinholes, and other similar defects, which can be problems with reservoir systems, do not substantially alter the release rate from monolithic devices. This, together with the ease with which dispersions can be compounded (by milling and extruding, for example), results in low production costs. These advantages often outweigh the less desirable feature of the declining release rate with time characteristic of these systems.

There are two principal types of monolithic device. If the active agent is dissolved in the polymer medium, the device is called a *monolithic solution*. Examples of this type of device are pesticide-containing cat and dog collars to control ticks and fleas. Such devices are often used when the active agent is a liquid; some polymers [for example, poly(vinyl chloride)] can easily sorb up to 20 % or more of these liquids. However, if the solubility of the active agent in the

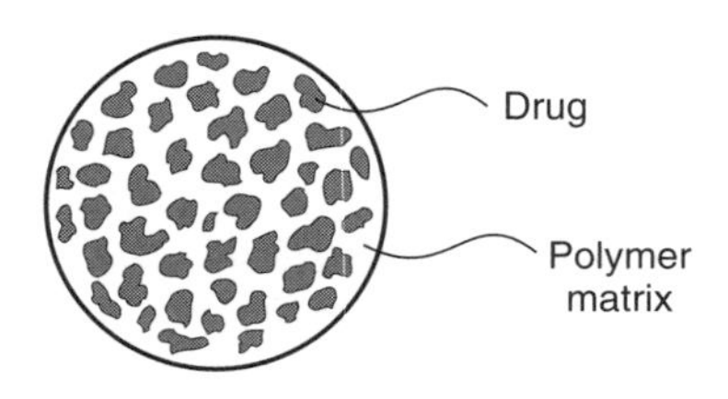

Figure 12.10 Monolithic device

polymer medium is more limited, then only a portion of the agent is dissolved and the remainder is dispersed as small particles throughout the polymer. A device of this type is called a *monolithic dispersion.*

The kinetics of release from a monolithic solution system have been derived for a number of geometries by Crank [20]. For a slab geometry, the release kinetics can be expressed by either of two series, both given here for completeness

$$\frac{M_t}{M_0} = 1 - \sum_{n=0}^{\infty} \frac{8\exp[-D(2n+1)^2\pi^2 t/l^2]}{(2n+1)^2\pi^2} \tag{12.3}$$

or

$$\frac{M_t}{M_0} = 4\left(\frac{Dt}{l^2}\right)^{1/2}\left[\pi^{-1/2} + \sum_{n=0}^{\infty}(-1)^n \mathrm{ierfc}\left(\frac{nl}{2\sqrt{Dt}}\right)\right] \tag{12.4}$$

where M_0 is the total amount of drug sorbed, M_t is the amount desorbed at time t and l is the thickness of the device.

Fortunately, these complex expressions reduce to two approximations, reliable to better than 1 %, valid for different parts of the desorption curve. The early time approximation, which holds for the initial portion of the curve, derived from Equation (12.4), is

$$\frac{M_t}{M_0} = 4\left(\frac{Dt}{\pi l^2}\right)^{1/2} \quad \text{for } 0 \le \frac{M_t}{M_0} \le 0.6 \tag{12.5}$$

The late time approximation, which holds for the final portion of the desorption curve, derived from Equation (12.3), is

$$\frac{M_t}{M_0} = 1 - \frac{8}{\pi^2}\exp\left(\frac{-\pi^2 Dt}{l^2}\right) \quad \text{for } 0.4 \le \frac{M_t}{M_0} \le 1.0 \tag{12.6}$$

These approximations are plotted in Figure 12.11, which illustrates their different regions of validity.

In general, the rate of release at any particular time is of more interest than the accumulated total release. This rate is easily obtained by differentiating Equations (12.5) and (12.6) to give

$$\frac{\mathrm{d}M_t}{\mathrm{d}t} = 2M_0\left(\frac{D}{\pi l^2 t}\right)^{1/2} \tag{12.7}$$

for the early time approximation and

$$\frac{\mathrm{d}M_t}{\mathrm{d}t} = \frac{8DM_0}{l^2}\exp\left(\frac{-\pi^2 Dt}{l^2}\right) \tag{12.8}$$

for the late time approximation. These two approximations are plotted against time in Figure 12.12. Again, for simplicity, M_0 and D/l^2 have been set at unity.

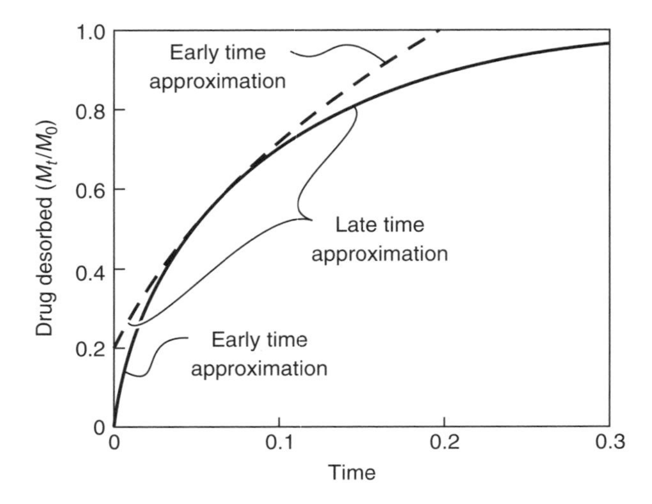

Figure 12.11 The fraction of drug desorbed from a slab as a function of time using the early time and late time approximations. The solid line shows the portion of the curve over which the approximations are valid ($D/l^2 = 1$) [21]

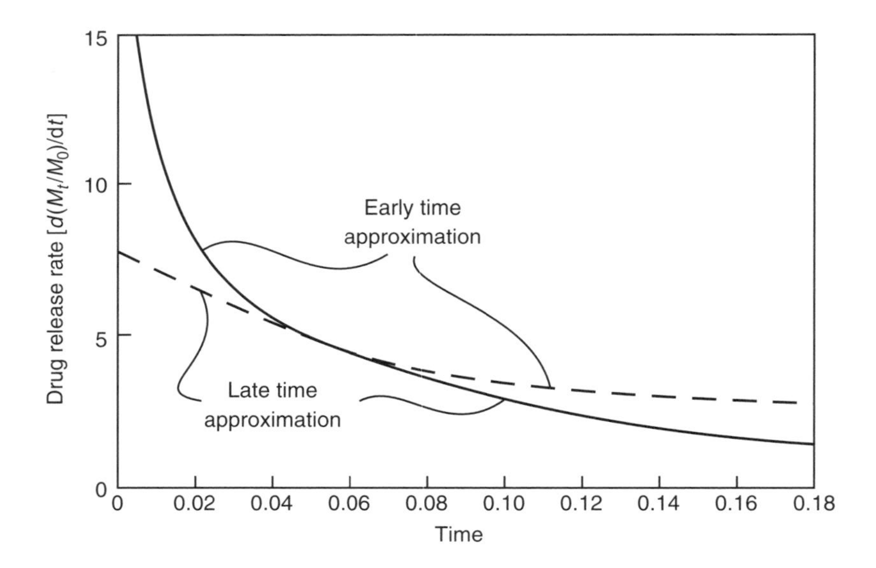

Figure 12.12 The release rate of drug initially dissolved in a slab as a function of time, using the early time and late time approximations. The solid line shows the portion of the curve over which the approximations are valid ($D/l^2 = 1$)

The release rate falls off in proportion to $t^{-1/2}$ until 60 % of the agent has been desorbed; thereafter, it decays exponentially.

The expressions (12.5)–(12.8) are also convenient ways of measuring diffusion coefficients in polymers. A permeant is contacted with a film of material of known geometry until equilibrium is reached. The film is then removed from the permeant solution, washed free of contaminants, and the rate of release of the permeant is measured. From the release curves, the diffusion coefficient and permeant sorption can be obtained.

A monolithic dispersion system consists of a dispersion of solid active drug in a rate-limiting polymer matrix. As with monolithic solution systems, the release rate varies with the geometry of the device; it also varies with drug loading. The starting point for release of drug from these systems can be described by a simple model due to Higuchi [22] and is shown schematically in Figure 12.13.

Higuchi's model assumes that solid drug in the surface layer of the device dissolves in the polymer matrix and diffuses from the device first. When the surface layer becomes exhausted of drug, the next layer begins to be depleted. Thus, the interface between the region containing dispersed drug and the region containing only dissolved drug moves into the interior as a front. The validity of Higuchi's model has been demonstrated experimentally numerous times by comparing the predicted release rate calculated from the model with the actual release rate. In addition, the movement of a dissolving front can be monitored directly by sectioning and examining monolithic devices that have been releasing agent for various lengths of time [23]. The proof of Higuchi is straightforward

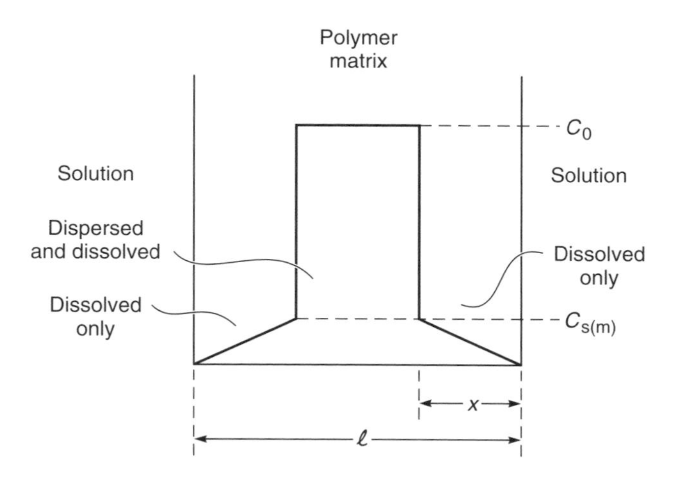

Figure 12.13 Schematic representation of a cross-section through a polymer matrix initially containing dispersed solid drug. The interface between the region containing dispersed drug and the region containing only dissolved drug has moved a distance x from the surface [22]

and leads to the equation

$$\begin{aligned} M_t &= A[DKtc_s(2c_0 - c_s)]^{1/2} \\ &\simeq A(2DKtc_sc_0)^{1/2} \qquad \text{for } c_0 \gg c_s \end{aligned} \tag{12.9}$$

The release rate at any time is then given by

$$\begin{aligned} \frac{\mathrm{d}M_t}{\mathrm{d}t} &= \frac{A}{2}\left[\frac{DKc_2}{t}(2c_0 - c_s)\right]^{1/2} \\ &\simeq \frac{A}{2}\left(\frac{2DKc_sc_0}{t}\right)^{1/2} \qquad \text{for } c_0 \gg c_s \end{aligned} \tag{12.10}$$

The Higuchi model is an approximate solution in that it assumes a 'pseudosteady state', in which the concentration profile from the dispersed drug front to the outer surface is linear. Paul and McSpadden [24] have shown that the correct expression can be written as:

$$M_t = A[2DKtc_s(c_0 - Kc_s)]^{1/2} \tag{12.11}$$

which is almost identical to Equation (12.9), and reduces to it when $c_0 \gg c_s$. Clearly the release rate is proportional to the square root of the loading; thus, it can be easily varied by incorporating more or less agent. Furthermore, although the release rate is by no means constant, the range of variation is narrower than would be the case if the agent were merely dissolved, rather than dispersed, in the matrix. An example of the release rate of the drug from an ethylene-vinyl acetate slab containing dispersed antibiotic chloramphenicol is shown in Figure 12.14. The drug release rate decreases in proportion to the square root of time in accordance with Equation (12.10).

Biodegradable Systems

The diffusion-controlled devices outlined so far are permanent, in that the membrane or matrix of the device remains implanted after its delivery role is completed. In some applications, particularly in the medical field, this is undesirable; such applications require a device that degrades during or subsequent to its delivery role.

Many polymer-based devices that slowly biodegrade when implanted in the body have been developed; the most important are based on polylactic acid, polyglycolic acid and their copolymers. In principle, the release of an active agent can be programmed by dispersing the material within such polymers, with erosion of the polymer effecting release of the agent [25,26]. One class of biodegradable polymers is *surface eroding*; the surface area of such polymers decreases with time as the conventionally cylindrical- or spherical-shaped device erodes. This

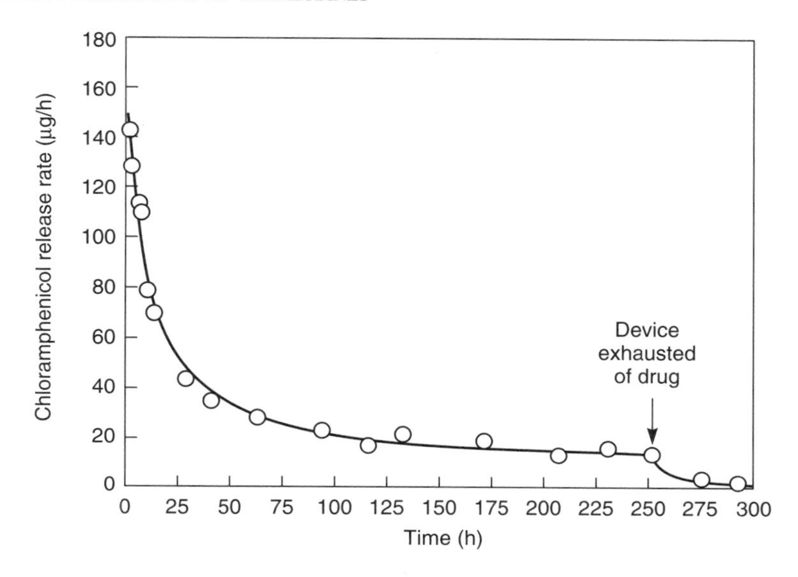

Figure 12.14 Release of the antibiotic drug chloramphenicol dispersed in a matrix of poly(ethylene-vinyl acetate). The solid line is calculated from Equation (12.10) [21]

results in a decreasing release rate unless the geometry of the device is appropriately manipulated or the device is designed to contain a higher concentration of the agent in the interior than in the surface layers. In a more common class of biodegradable polymer, the initial period of degradation occurs very slowly, after which the degradation rate increases rapidly. The bulk of the polymer then erodes over a comparatively short period. In the initial period of exposure to the body, the polymer chains are being cleaved but the molecular weight is still high, so the polymer's mechanical properties are not seriously affected. As chain cleavage continues, a point is reached at which the polymer fragments become swollen or soluble in water. At this point the polymer begins to dissolve. This type of polymer can be used to make reservoir or monolithic diffusion-controlled systems that degrade after their delivery role is over. A final category of polymer has the active agent covalently attached by a labile bond to the backbone of a matrix polymer. When placed at the site of action the labile bonds slowly degrade, releasing the active agent and forming a soluble polymer. The methods by which these concepts can be formulated into actual practical systems are illustrated in Figure 12.15.

Osmotic Systems

Osmotic effects are often a problem in diffusion-controlled systems because imbibition of water swells the device or dilutes the drug. However, several devices

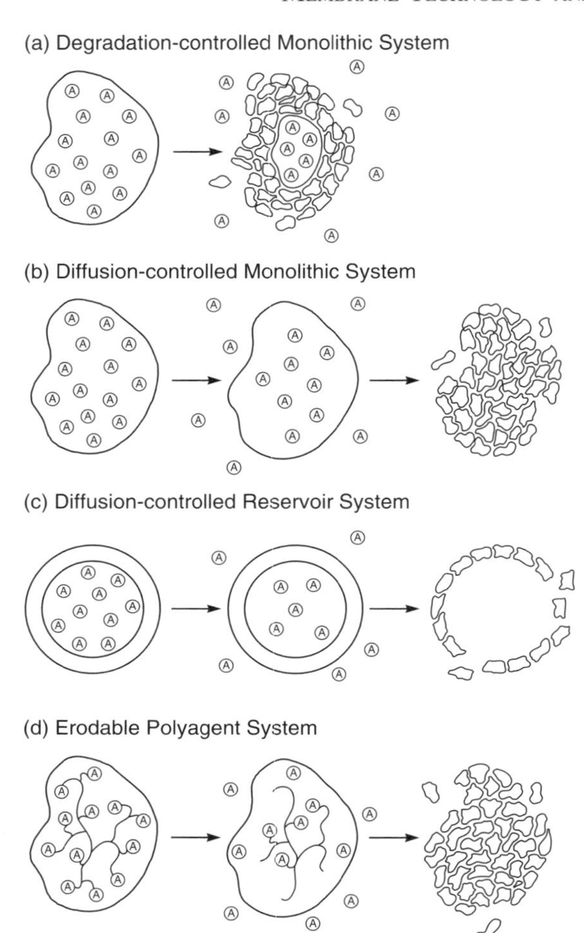

Figure 12.15 Methods of using biodegradable polymers in controlled release implantable devices to release the active agent, A

have been developed that actually use osmotic effects to control the release of drugs. These devices, called osmotic pumps, use the osmotic pressure developed by diffusion of water across a semipermeable membrane into a salt solution to push a solution of the active agent from the device. Osmotic pumps of various designs are applied widely in the pharmaceutical area, particularly in oral tablet formulations [27].

The forerunner of modern osmotic devices was the Rose–Nelson pump. Rose and Nelson were two Australian physiologists interested in the delivery of drugs to the gut of sheep and cattle [11]. Their pump, illustrated in Figure 12.16,

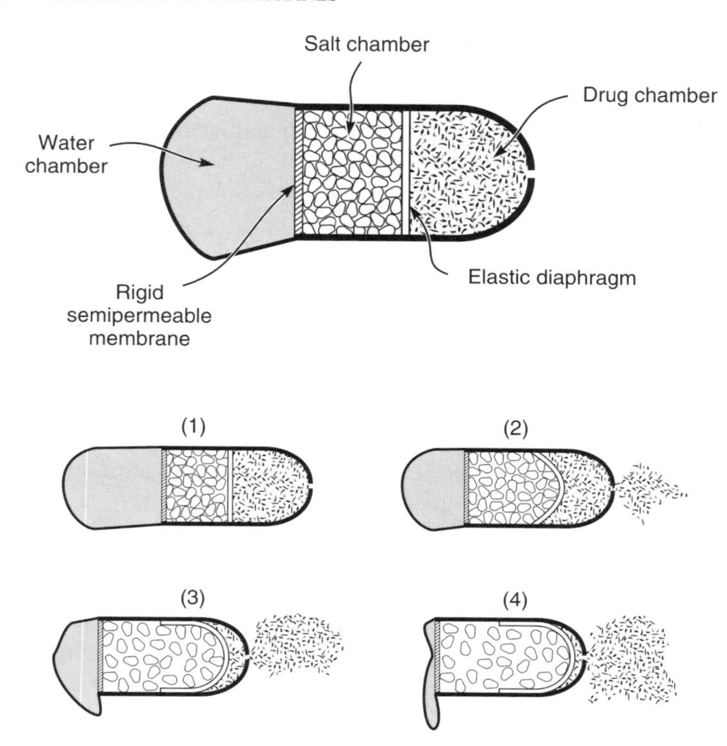

Figure 12.16 Principle of the three-chamber Rose–Nelson osmotic pump first described in 1955 [11]

consists of three chambers: a drug chamber, a salt chamber containing excess solid salt and a water chamber. The salt and water chambers are separated by a rigid semipermeable membrane. The difference in osmotic pressure across the membrane moves water from the water chamber into the salt chamber. The volume of the salt chamber increases because of this water flow, which distends the latex diaphragm separating the salt and drug chambers, thereby pumping drug out of the device.

The pumping rate of the Rose–Nelson pump is given by the equation

$$\frac{\mathrm{d}M_t}{\mathrm{d}t} = \frac{\mathrm{d}V}{\mathrm{d}t}c \tag{12.12}$$

where $\mathrm{d}M_t/\mathrm{d}t$ is the drug release rate, $\mathrm{d}V/\mathrm{d}t$ is the volume flow of water into the salt chamber and c is the concentration of drug in the drug chamber. The osmotic water flow across a membrane is given by the equation

$$\frac{\mathrm{d}V}{\mathrm{d}t} = \frac{A\theta\,\Delta\pi}{l} \tag{12.13}$$

where dV/dt is a water flow across the membrane of area A, thickness l, and osmotic permeability θ ($cm^3 \cdot cm/cm^2 \cdot h \cdot atm$), and $\Delta\pi$ is the osmotic pressure difference between the solutions on either side of the membrane. This equation is only strictly true for completely selective membranes—that is, membranes permeable to water but completely impermeable to the osmotic agent. However, this is a good approximation for most membranes. Substituting Equation (12.13) for the flux across the membrane gives

$$\frac{dM_t}{dt} = \frac{A\theta\Delta\pi c}{l} \quad (12.14)$$

The osmotic pressure of the saturated salt solution is high, on the order of tens of atmospheres, and the small pressure required to pump the suspension of active agent is insignificant in comparison. Therefore, the rate of water permeation across the semipermeable membrane remains constant as long as sufficient solid salt is present in the salt chamber to maintain a saturated solution and hence a constant osmotic pressure driving force.

The Higuchi–Leeper pump designs represent the first of a series of simplifications of the Rose–Nelson pump made by Alza Corporation beginning in the early 1970s. An example of one of these designs [28] is shown in Figure 12.17. The

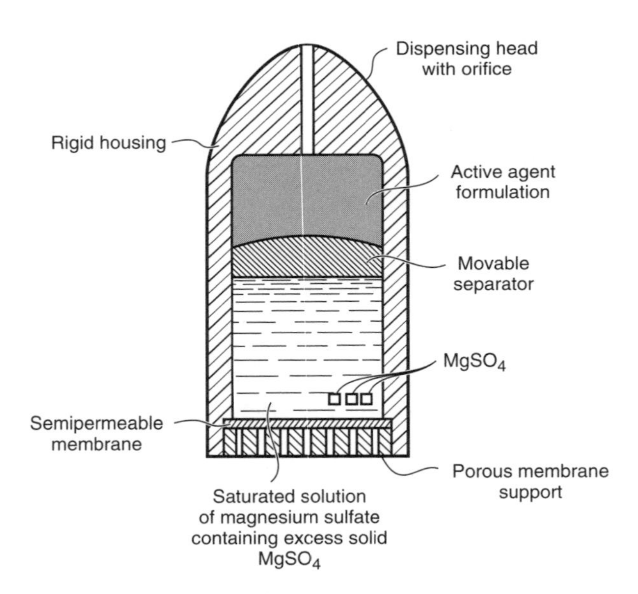

Figure 12.17 The Higuchi–Leeper osmotic pump design [28]. This device has no water chamber and can be stored in a sealed foil pouch indefinitely. However, once removed from the pouch and placed in an aqueous environment, for example, by an animal swallowing the device, the pumping action begins. The active agent is pumped at a constant rate according to Equation (12.14)

Higuchi–Leeper pump has no water chamber; the device is activated by water imbibed from the surrounding environment. This means that the drug-laden pump can be prepared and then stored for weeks or months prior to use. The pump is only activated when it is swallowed or implanted in the body. Higuchi–Leeper pumps contain a rigid housing, and the semipermeable membrane is supported on a perforated frame. This type of pump usually has a salt chamber containing a fluid solution with excess solid salt. The target application of this device was the delivery of antibiotics and growth hormones to animals because repeated delivery of oral medications to animals is difficult. The problem is solved by these devices, which are designed to be swallowed by the animal and then to reside in the rumen, delivering a full course of medication over a period of days to weeks.

In the early 1970s, Higuchi and Theeuwes developed another, even simpler variant of the Rose–Nelson pump [29,30]. One such device is illustrated in Figure 12.18. As with the Higuchi–Leeper pump, water to activate the osmotic action of the pump comes from the surrounding environment. The Higuchi–Theeuwes device, however, has no rigid housing—the membrane acts as the outer casing of the pump. This membrane is quite sturdy and is strong enough to withstand the pumping pressure developed inside the device. The device is loaded with the desired drug prior to use. When the device is placed in an aqueous environment, release of the drug follows a time course set by the salt used in the salt chamber and the permeability of the outer membrane casing.

The principal application of these small osmotic pumps has been as implantable controlled release delivery systems in experimental studies on the effect of continuous administration of drugs. The devices are made with volumes of 0.2–2 mL. Figure 12.18 shows one such device being implanted in a laboratory rat. The delivery pattern obtained with the device is constant and independent of the site of implantation, as shown by the data in Figure 12.19.

The development that made osmotic delivery a major method of achieving controlled drug release was the invention of the elementary osmotic pump by Theeuwes in 1974 [31]. The concept behind this invention is illustrated in Figure 12.20. The device is a further simplification of the Higuchi–Theeuwes pump, and eliminates the separate salt chamber by using the drug itself as the osmotic agent. The device is formed by compressing a drug having a suitable osmotic pressure into a tablet using a tableting machine. The tablet is then coated with a semipermeable membrane, usually cellulose acetate, and a small hole is drilled through the membrane coating. When the tablet is placed in an aqueous environment, the osmotic pressure of the soluble drug inside the tablet draws water through the semipermeable coating, forming a saturated aqueous solution inside the device. The membrane does not expand, so the increase in volume caused by the imbibition of water raises the hydrostatic pressure inside the tablet slightly. This pressure is relieved by a flow of saturated drug solution out of the device through the small orifice. Thus, the tablet acts as a small pump, in which water is drawn osmotically into the tablet through the membrane wall and then

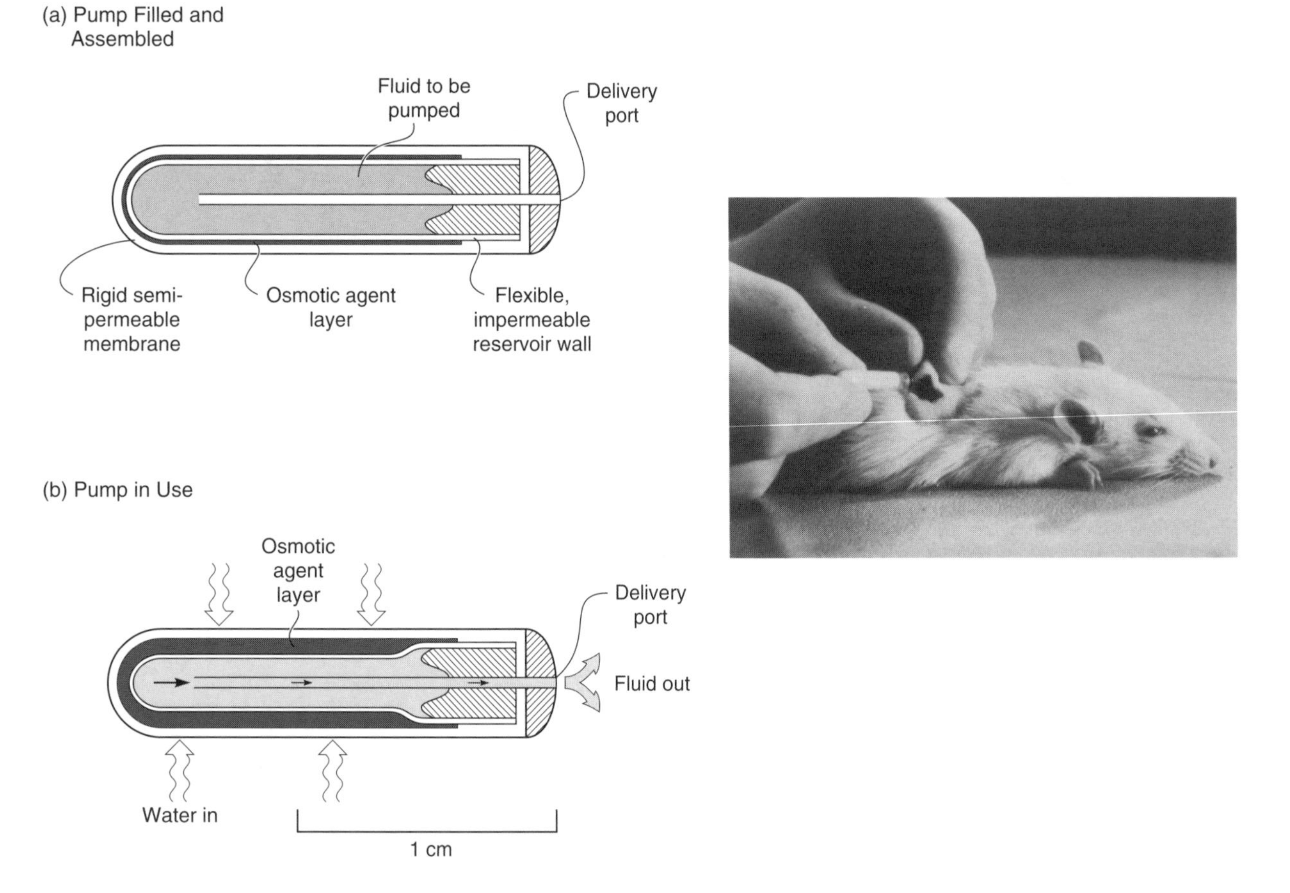

Figure 12.18 The Higuchi–Theeuwes osmotic pump has been widely used in drug delivery tests in laboratory animals. The device is small enough to be implanted under the skin of a rat and delivers up to 1000 μL of drug solution over a 3- to 4-day period [29,30]

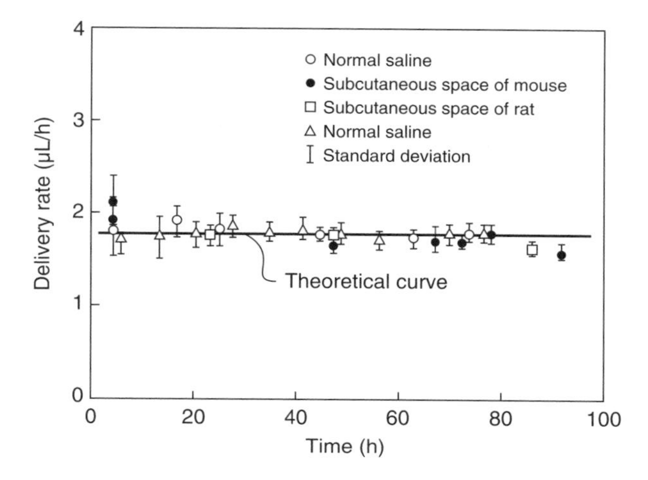

Figure 12.19 Drug delivery curves obtained with an implantable osmotic pump [30]. Reprinted from F. Theeuwes and S.I. Yum, Principles of the Design and Operation of Generic Osmotic Pumps for the Delivery of Semisolid or Liquid Drug Formulations, *Ann. Biomed. Eng.* **4**, 343, 1976, with permission of Biomedical Engineering Society

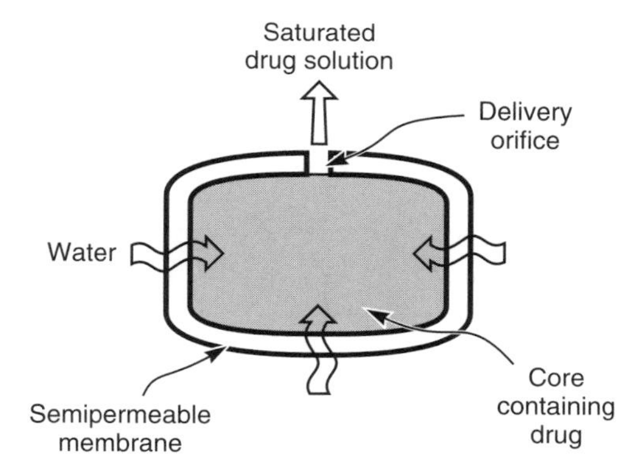

Figure 12.20 The Theeuwes elementary osmotic pump [31]. Reprinted with permission from F. Theeuwes, Elementary Osmotic Pump, *J. Pharm. Sci.* **64**, 1987. Copyright 1975, American Chemical Society and American Pharmaceutical Association

leaves as a saturated drug solution through the orifice. This process continues at a constant rate until all the solid drug inside the tablet has been dissolved and only a solution-filled shell remains. This residual dissolved drug continues to be delivered, but at a declining rate, until the osmotic pressures inside and outside the tablet are equal. The driving force that draws water into the device is the difference in osmotic pressure between the outside environment and a saturated

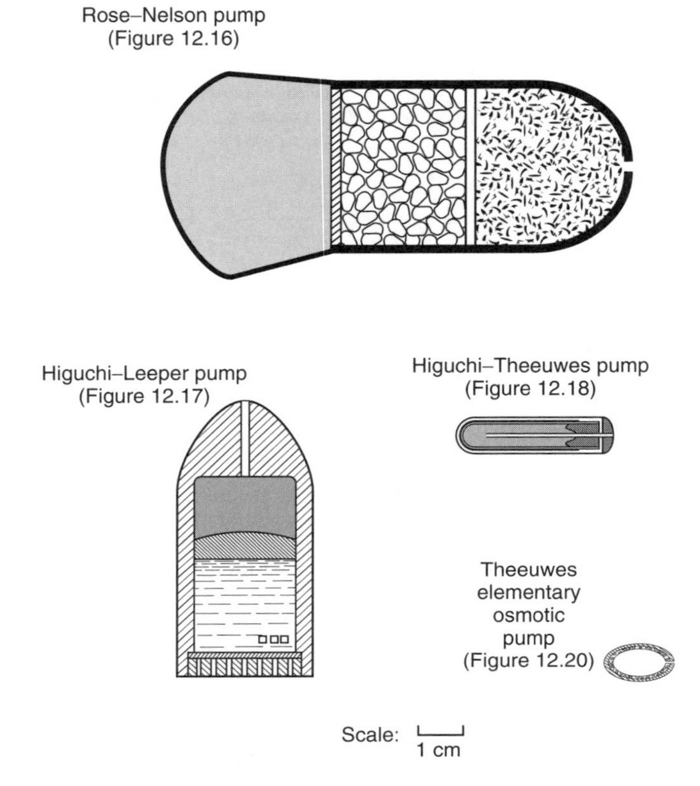

	Rose–Nelson pump	Higuchi–Leeper pump	Higuchi–Theeuwes pump	Theeuwes elementary osmotic pump
Approximate volume (cm^3)	80	35	3	< 1
Components	Rigid housing Water chamber Salt chamber Drug chamber Elastic diaphragm Membrane	Rigid housing Salt chamber Drug chamber Elastic diaphragm Membrane	 Salt chamber Drug chamber Elastic diaphragm Membrane	 Drug chamber Membrane
Number of components	6	5	4	2

Figure 12.21 The main types of osmotic pump drawn to scale

drug solution. Therefore, the osmotic pressure of the dissolved drug solution has to be relatively high to overcome the osmotic pressure of the body, but for drugs with solubilities greater than 5–10 wt% these devices function very well. Later variations on the simple osmotic tablet design have been made to overcome the solubility limitation. The elementary osmotic pump was developed by Alza under the name OROS®, and is commercially available for a number of drugs.

The four types of osmotic pump described above are interesting examples of how true innovation is sometimes achieved by leaving things out. The first osmotic pump produced by Rose and Nelson contained six critical components, had a volume of 80 cm^3, and was little more than a research tool. In the early 1980s, Felix Theeuwes and others progressively simplified and refined the concept, leading in the end to the elementary osmotic pump, a device that looks almost trivially simple. It has been described as a tablet with a hole, but is, in fact, a truly elegant invention having a volume of less than 1 cm^3, containing only two components, achieving almost constant drug delivery, and allowing manufacture on an enormous scale at minimal cost. Figure 12.21 shows examples of the four main types of osmotic pumps taken from the patent drawings. The pumps are drawn to scale to illustrate the progression that occurred as the design was simplified.

References

1. J.P. Merrill, The Artificial Kidney, *Sci. Am.* **205**, 56 (1961).
2. M.J. Lysaght, D.R. Boggs and M.H. Taimisto, Membranes in Artificial Organs, in *Synthetic Membranes*, M.B. Chenoweth (ed.), Hardwood Academic Publishers, Chur, Switzerland, pp. 100–117 (1986).
3. J.C. Van Stowe, Hemodialysis Apparatus, in *Handbook of Dialysis*, 2nd Edn, J.T. Daugirdas and T.S. Ing (eds), Lippincott-Raven Publishers, New York, NY, pp. 30–49 (1994).
4. W.J. Kolf and H.T. Berk, The Artificial Kidney: A Dialyzer with a Great Area, *Acta Med. Scan* **117**, 121 (1944).
5. W.J. Kolf, *New Ways of Treating Uremia*, J. and A. Churchill, London (1947).
6. W. Quinton, D. Dilpard and B.H. Scribner, Cannulation of Blood Vessels for Prolonged Hemodialysis, *Trans. Am. Soc. Artif. Inter. Organs* **6**, 104 (1960).
7. M.L. Keen and F.A. Gotch, Dialyzers and Delivery Systems in *Introduction to Dialysis*, M.C. Cogan, P. Schoenfeld and F.A. Gotch (eds) Churchill Livingston, New York, NY, pp. 1–7 (1991).
8. J.H. Gibbon, Jr, Artificial Maintenance of Circulation During Experimental Occlusions of Pulmonary Artery, *Arch. Surg. Chicago* **34**, 1105 (1937).
9. J.H. Gibbon, Jr, Application of a Mechanical Heart and Lung Apparatus to Cardiac Surgery, *Minn. Med.* **37**, 171 (1954).
10. P.M. Galletti and G.A. Brecher, *Heart–Lung Bypass*, Grun and Stratton, New York, NY (1962).
11. S. Rose and J.F. Nelson, A Continuous Long-term Injector, *Aust. J. Exp. Biol.* **33**, 415 (1955).
12. J.M. Folkman and D.M. Long, The Use of Silicone Rubber as a Carrier for Prolonged Drug Therapy, *J. Surg. Res.* **4**, 139 (1964).

13. P.J. Dziuk and B. Cook, Passage of Steroids Through Silicone Rubber, *Endocrinology* **78**, 208 (1966).
14. H.B. Croxatto, S. Diaz, R. Vera, M. Etchart and P. Atria, Fertility Control in Women with a Progestagen Released in Microquantities from Subcutaneous Capsules, *Am. J. Obstet. Gynecol.* **105**, 1135 (1969).
15. A. Scommegna, G.N. Pandya, M. Christ, A.W. Lee and M.R. Cohen, Intrauterine Administration of Progesterone by a Slow Releasing Device, *Fert. Steril.* **21**, 201 (1970).
16. A. Zaffaroni, Applications of Polymers in Rate-controlled Drug Delivery, *Polym. Sci. Tech.* **14**, 293 (1981).
17. J. Shaw, Development of Transdermal Therapeutic Systems, *Drug Dev. Ind. Pharm.* **9**, 579 (1983).
18. *The Ocusert (pilocarpine) Pilo-20/Pilo-40 Ocular Therapeutic System*, Alza Corporation, Palo Alto, CA (1974).
19. B.B. Pharriss, R. Erickson, J. Bashaw, S. Hoff, V.A. Place and A. Zaffaroni, Progestasert: A Uterine Therapeutic System for Long-term Contraception, *Fert. Steril.* **25**, 915 (1974).
20. J. Crank, *The Mathematics of Diffusion*, Oxford University Press, London (1956).
21. R.W. Baker, *Controlled Release of Biologically Active Agents*, Wiley, New York, NY (1987).
22. T. Higuchi, Rate of Release of Medicaments from Ointment Bases Containing Drugs in Suspension, *J. Pharm. Sci.* **50**, 874 (1961).
23. T.J. Roseman and W.I. Higuchi, Release of Medroxprogesterone Acetate from a Silicone Polymer, *J. Pharm. Sci.* **59**, 353 (1970).
24. D.R. Paul and S.K. McSpadden, Diffusional Release of a Solute from a Polymer Matrix, *J. Membr. Sci.* **1**, 33 (1976).
25. J. Heller, Controlled Release of Biologically Active Compounds from Bioerodible Polymers, *Biomaterials* **1**, 51 (1980).
26. C.G. Pitt and A. Schindler, The Design of Controlled Drug Delivery Systems Based on Biodegradable Polymers, in *Biodegradables and Delivery Systems for Contraception*, E.S.E. Hafez and W.A.A. van Os (eds), MTP Press, Lancaster, pp. 17–46 (1980).
27. G. Santus and R.W. Baker, Osmotic Drug Delivery: A Review of the Patent Literature, *J. Controlled Release* **35**, 1 (1995).
28. T. Higuchi and H.M. Leeper, Improved Osmotic Dispenser Employing Magnesium Sulfate and Magnesium Chloride, US Patent 3,760,804 (September, 1973).
29. F. Theeuwes and T. Higuchi, Osmotic Dispensing Agent for Releasing Beneficial Agent, US Patent 3,845,770 (November, 1974).
30. F. Theeuwes and S.I. Yum, Principles of the Design and Operation of Generic Osmotic Pumps for the Delivery of Semisolid or Liquid Drug Formulations, *Ann. Biomed. Eng.* **4**, 343 (1976).
31. F. Theeuwes, Elementary Osmotic Pump, *J. Pharm. Sci.* **64**, 1987 (1975).

13 OTHER MEMBRANE PROCESSES

Introduction

Any book must leave something out, and this one has left out a good deal; it does not cover membranes used in packaging materials, sensors, ion-selective electrodes, fuel cells, battery separators, electrophoresis and thermal diffusion. In this final chapter, five processes that come under the general title of 'other' are covered briefly.

Dialysis

Dialysis was the first membrane process to be used on an industrial scale, with the development of the Cerini dialyzer in Italy [1]. The production of rayon from cellulose expanded rapidly in the 1930s, resulting in a need for technology to recover sodium hydroxide from the large volumes of hemicellulose-sodium hydroxide by-product solutions. The hemicellulose was of little value, but the 17–18 wt% sodium hydroxide, if separated, could be reused directly in the process. Hemicellulose has a much higher molecular weight than sodium hydroxide, so parchmentized woven fabric or impregnated cotton cloth made an adequate dialysis membrane. The Cerini dialyzer, illustrated in Figure 13.1, consisted of a large tank containing 50 membrane bags. Feed liquid passed through the tank while the dialysate solution passed countercurrently through each bag in parallel. The product dialysate solution typically contained 7.5–9.5 % sodium hydroxide and was essentially free of hemicellulose. About 90 % of the sodium hydroxide in the original feed solution was recovered. The economics of the process were very good, and the Cerini dialyzer was widely adopted. Later, improved membranes and improved dialyzer designs, mostly of the plate-and-frame type, were produced. A description of these early industrial dialyzers is given in Tuwiner's book [2].

Dialysis was also used in the laboratory in the 1950s and 1960s, mainly to purify biological solutions or to fractionate macromolecules. A drawing of the

Membrane Technology and Applications R. W. Baker
 ISBN: 0-470-85445-6

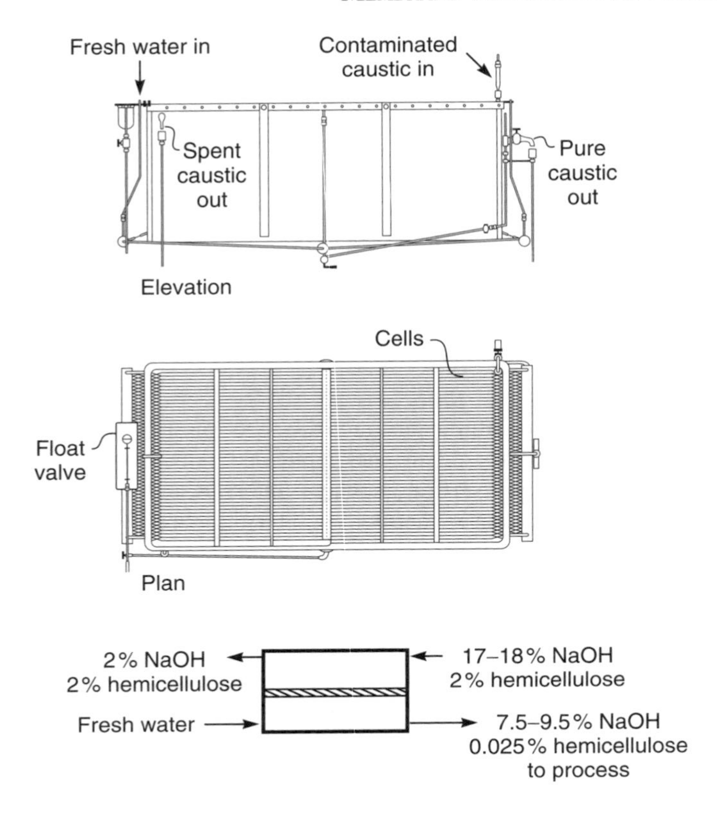

Figure 13.1 Elevation, plan drawing and flow scheme of the Cerini dialyzer, the first successful industrial dialyzer used to recover sodium hydroxide from waste streams resulting from the production of rayon [2]

laboratory dialyzer used by Craig and described in a series of papers in the 1960s is shown in Figure 13.2 [3,4]. Until ultrafiltration membranes became available in the late 1960s, this device was the only way to separate many large-volume biological solutions.

Now the major application of dialysis is the artificial kidney and, as described in Chapter 12, more than 100 million of these devices are used annually. Apart from this one important application, dialysis has essentially been abandoned as a separation technique, because it relies on diffusion, which is inherently unselective and slow, to achieve a separation. Thus, most potential dialysis separations are better handled by ultrafiltration or electrodialysis, in both of which an outside force and more selective membranes provide better, faster separations. The only three exceptions—Donnan dialysis, diffusion dialysis and piezodialysis—are described in the following sections.

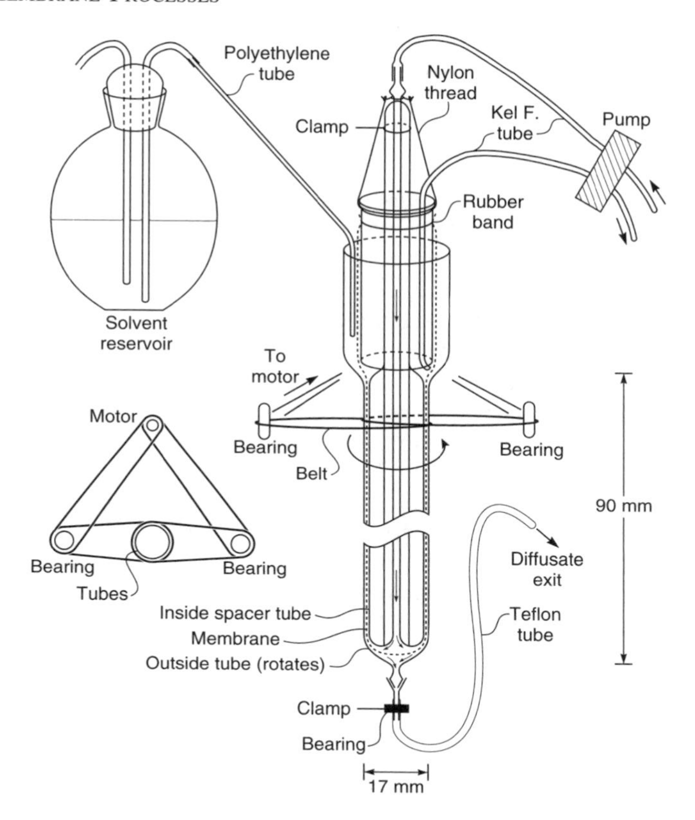

Figure 13.2 Schematic drawing of laboratory dialyzer developed by Craig [4] to separate low-molecular-weight impurities from biological solutions. This was the best method of performing this separation until ultrafiltration membranes became available in the late 1960s. The feed solution was circulated through the inside of the membrane tube; solvent solution was circulated on the outside. Boundary layer formation was overcome by rotating the outer shell with a small motor

Donnan Dialysis and Diffusion Dialysis

One dialysis process for which the membrane does have sufficient selectivity to achieve useful separations is Donnan dialysis. If salt solutions are separated by a membrane permeable only to ions of one charge, such as a cation exchange membrane containing fixed negatively charged groups, then distribution of two different cations M^+ and N^+ across the membrane can be expressed by the Donnan expression

$$\frac{[\mathrm{M}]_o}{[\mathrm{M}]_\ell} = \frac{[\mathrm{N}]_o}{[\mathrm{N}]_\ell} \tag{13.1}$$

where $[M]_o$ and $[N]_o$ are the concentrations of the two ions in the feed solution, and $[M]_\ell$ and $[N]_\ell$ are the concentrations of the two ions in the product solution. The derivation of the expression is given in Chapter 10. This equation has the same form as Equation (11.10) derived for coupled transport in Chapter 11, in which a cation-permeable, anion-impermeable membrane separates the two solutions. The difference between coupled transport and Donnan dialysis lies in how the membrane performs the separation.

Donnan dialysis was first described as a separation technique in 1967 by Wallace [5,6], who was interested in concentrating small amounts of radioactive metal ions. He used cation exchange membranes to treat a large volume of nearly neutral feed solution containing small amounts of metal salts such as uranyl nitrate $UO_2(NO_3)_2$. A small volume of 2 M nitric acid was used as the receiving solution. Because the membrane contained fixed negative charges, negative ions from the surrounding solutions were essentially excluded from the membrane, and only hydrogen ions (H^+) and uranyl ions (UO_2^{++}) could permeate the membrane. The very large difference in hydrogen ion concentration across the membrane meant that a large driving force was generated for hydrogen ions to diffuse to the dilute feed solution. To maintain electrical neutrality, an equal number of uranyl ions had to diffuse to the receiving solution. Wallace's apparatus and the results of one of his experiments are shown in Figure 13.3. In this experiment, 98 % of the

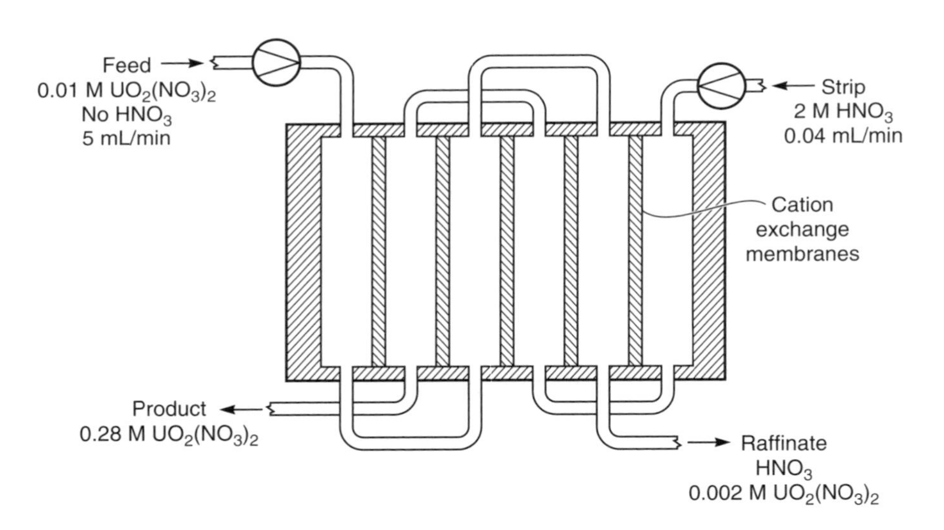

Figure 13.3 Illustration of a Donnan dialysis experiment to separate and concentrate uranyl nitrate, $UO_2(NO_3)_2$, after Wallace [5]

uranyl ions were stripped from the feed solution and concentrated 28-fold in the product nitric acid strip solution.

Like coupled transport, Donnan dialysis can concentrate metal ions many fold. The process is usually driven by an appropriate pH gradient. Because the membranes are normal cation or anion exchange membranes, the stability problems that plague the liquid membranes used in coupled transport are avoided. On the other hand, coupled transport uses carriers selective for one particular ion, excluding others. This property allows coupled transport membranes to selectively transport one particular ion across the membrane, both concentrating and separating the target ion from similar ions in the feed solution. Donnan dialysis membranes are essentially nonselective—all ions of the same charge in the feed solution are transported to the product solution at about the same rate.

Donnan dialysis can be made more selective if a complexing agent specific to one of the ions being transported across the membrane is added to the strip solution. For example, Huang *et al.* [7] used cation exchange membranes driven by sodium ions to transport copper and nickel ions across the membrane. Addition of complexing agents specific for nickel ions, such as oxalic acid or glycine, to the strip solution increased the selectivity of the membrane for nickel over copper dramatically. By removing nickel ions from the receiving solution, the complexing agent maintained a high driving force for nickel transport even when the copper ion concentration had reached equilibrium.

Although Donnan dialysis membranes can perform interesting separations, these membranes are a solution to few industrially important applications. Consequently, Donnan dialysis remains a solution in search of a problem.

A related dialysis process, diffusion dialysis, has found an application, mostly to recover acids from spent metal pickling agents such as sulfuric acid, hydrochloric acid or nitric–hydrofluoric acids [8]. These pickling chemicals remove scale from metal parts and become contaminated over time with iron, chromium, copper, nickel, zinc and other heavy metals. Acid recovery by electrodialysis is possible but diffusion-dialysis—a completely passive process—is often preferred because of its simplicity. The process utilizes the difference in permeability of hydrogen ions and multivalent metal ions through anion exchange membranes. A flow schematic is shown in Figure 13.4. The feed solution, containing heavy metal salts and acid, flows countercurrent to water, from which it is separated by an anion exchange membrane. The membrane, which is freely permeable to anions, also preferentially permeates hydrogen ions over heavy-metal cations. As a result, the acids in the feed solution, sulfuric acid in the example of Figure 13.4, are removed from the spent liquor and metal ions remain behind. Recovery of 70–80 % acid, contaminated with only a few percent of the metal ions, is possible.

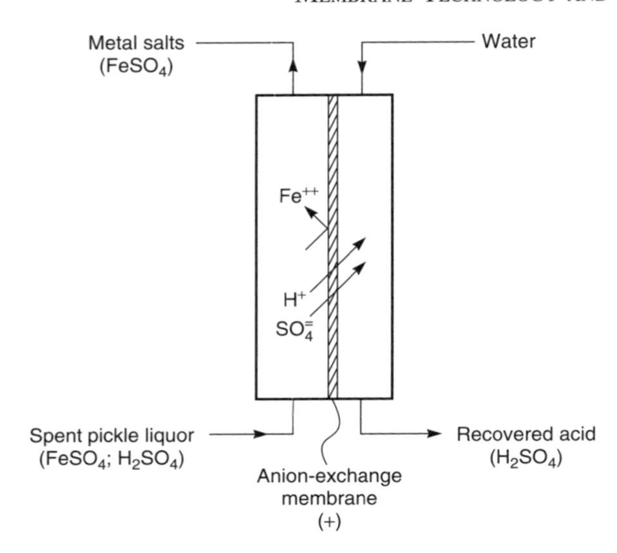

Figure 13.4 Schematic of a diffusion dialysis process to separate acids from heavy metal/acid mixtures

Charge Mosaic Membranes and Piezodialysis

Donnan dialysis, described in the previous section, is a type of ion exchange process. Ions of the same charge are redistributed across the membrane, but no net flow of salt from one side of the membrane to the other occurs. This is because ion exchange membranes are quite impermeable to salts. Although counter-ions to the fixed charge groups in the membrane can easily permeate the membrane, ions with the same charge as the fixed charge groups are excluded and do not permeate. Sollner [9] proposed that, if ion exchange membranes consisting of separated small domains of anionic and cationic membranes could be made, they would be permeable to both anions and cations. These membranes are now called charge mosaic membranes; the concept is illustrated in Figure 13.5. Cations permeate the cationic membrane domain; anions permeate the anionic domain.

Charge mosaic membranes can preferentially permeate salts from water. This is because the principle of electroneutrality requires that the counter-ion concentration inside the ion exchange regions be at least as great as the fixed charge density. Because the fixed charge density of ion exchange membranes is typically greater than 1 M, dilute counter-ions present in the feed solution are concentrated 10- to 100-fold in the membrane phase. The large concentration gradient that forms in the membrane leads to high ion permeabilities. Water and neutral solutes are not concentrated in the membrane and permeate at low rates. When used as dialysis membranes, therefore, these charged mosaic membranes are permeable to salts but relatively impermeable to non-ionized solutes.

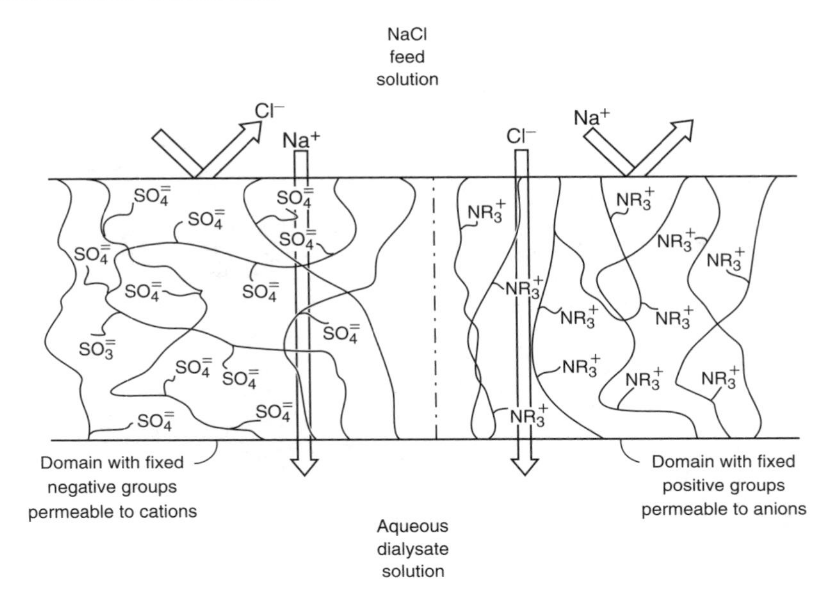

Figure 13.5 Charge mosaic membranes, consisting of finely dispersed domains containing fixed negatively and fixed positively charged groups, are salt permeable [15]

For charge mosaic membranes to work most efficiently, the cationic and anionic domains in the membrane must be close together to minimize charge separation effects [10–12]. The first charge mosaic membranes were made by distributing very small ion exchange beads in an impermeable support matrix of silicone rubber [13,14]. A second approach, used by Platt and Schindler [15], was to use the mutual incompatibility of most polymers that occurs when a solution containing a mixture of two different polymers is evaporated. Figure 13.6 shows a photomicrograph of a film cast from poly(styrene-*co*-butadiene) and poly(2-vinyl pyridine-*co*-butadiene). The *co*-butadiene fraction makes these two polymers mutually soluble in tetrahydrofuran but, on evaporation of the solvent, a two-phase-domain structure extending completely through the membrane layer forms. Once formed, the poly(2-vinyl pyridine-*co*-butadiene) portion of the membrane is quaternarized to form fixed positive groups, and the poly(styrene-*co*-butadiene) portion of the membrane is sulfonated to form fixed negative groups.

Miyaki and Fujimoto and co-workers [16,17] have obtained an even finer distribution of fixed charge groups by casting films from multicomponent block copolymers such as poly(isoprene-*b*-styrene-*b*-butadiene-*b*-(4-vinyl benzyl)dimethylamine- *b*-isoprene). These films show a very regular domain structure with a 200–500 Å spacing. After casting the polymer film, the (4-vinyl benzyl) dimethylamine blocks were quaternarized with methyl iodide vapor, and the styrene blocks were sulfonated with chlorosulfuric acid.

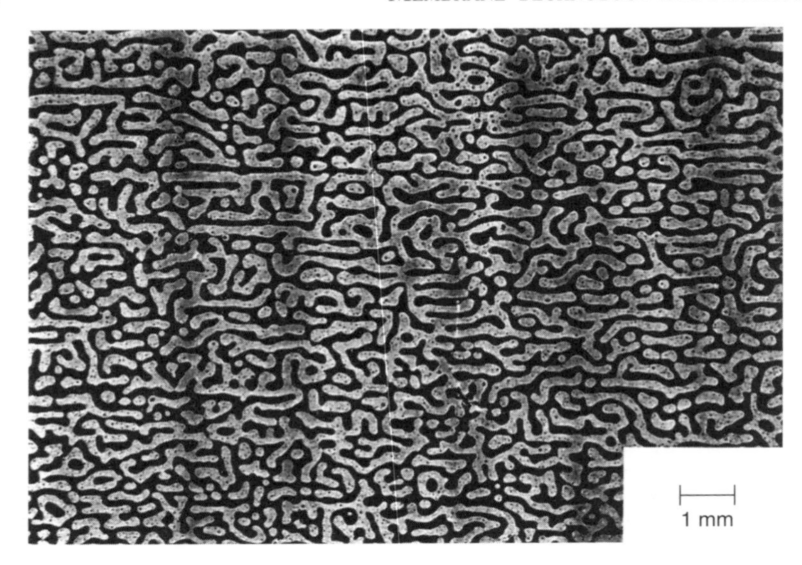

Figure 13.6 Film cast from a 1:2 mixture of poly(styrene-*co*-butadiene) and poly(2-vinyl pyridine-*co*-butadiene) with about 15 mol% butadiene content (10 wt% solution of the copolymers in tetrahydrofuran). Dark areas, poly(styrene-*co*-butadiene); light areas, poly (2-vinyl pyridine-*co*-butadiene) [15]. Courtesy of Dr A. Schindler

Table 13.1 Solute flux measured in well-stirred dialysis cells at 25 °C using 0.1 M feed solutions [17]. Reprinted with permission from K. Hirahara, S.-I. Takahashi, M. Iwata, T. Fujimoto and Y. Miyaki, Artificial Membranes from Multiblock Copolymers (5), *Ind. Eng. Chem. Prod. Res. Dev.* **305**, 25. Copyright 1986, American Chemical Society and American Pharmaceutical Association

Solute	Flux (10^{-8} mol/cm^2 · s)
Sodium chloride	7.5
Potassium chloride	9
Hydrochloric acid	18
Sodium hydroxide	10
Glucose	0.08
Sucrose	0.04

Using the block copolymer membranes described above, significant selectivities for electrolytes over non-electrolytes have been observed. Some data reported by Hirahara *et al.* [17] are shown in Table 13.1. The ionizable electrolytes were 100 times more permeable than non-ionized solutes such as glucose and sucrose, suggesting a number of potential applications in which deionization of mixed solutions is desirable. The permeabilities of salts in these membranes are also orders of magnitude higher than values measured for normal ion exchange membranes. In principle then, these membranes can be used in deionization processes, for example, to remove salts from sucrose solutions in the sugar industry.

A second potential application is pressure-driven desalination. When a pressure difference is applied across the membrane, the concentrated ionic groups in the ion exchange domains are swept through the membrane, producing a salt-enriched permeate on the low-pressure side. This process, usually called piezodialysis, has a number of conceptual advantages over the alternative, conventional reverse osmosis, because the minor component (salt), not the major component (water), permeates the membrane.

Piezodialysis has been the subject of sporadic research for a number of years but so far has met with little success. It was originally hoped that the flow of water and salt through charge mosaic membranes would be strongly coupled. If this were the case, the 100-fold enrichment of ions within the charged regions of the membrane would provide substantial enrichment of salt in the permeate solution. In practice, the enrichment obtained is relatively small, and the salt fluxes are low even at high pressures. The salt enrichment also decreases substantially as the salt concentration in the feed increases, limiting the potential applications of the process to desalination of low concentration solutions. Some results of piezodialysis experiments with block copolymer membranes and a potassium chloride solution are shown in Figure 13.7.

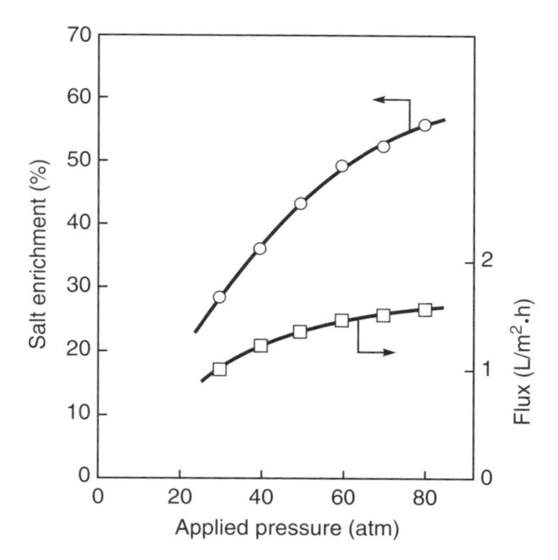

Figure 13.7 Piezodialysis of 0.02 M potassium chloride solution with block copolymer charge mosaic membranes [14]. Enrichment is calculated using the expression:

$$\text{enrichment} = 100\left(\frac{\text{concentration permeate}}{\text{concentration feed}} - 1\right)$$

Membrane Contactors and Membrane Distillation

In the membrane processes discussed elsewhere in this book, the membrane acts as a selective barrier, allowing relatively free passage of one component while retaining another. In membrane contactors, the membrane functions as an interface between two phases but does not control the rate of passage of permeants across the membrane. The use of a membrane as a contactor in a process to deoxygenate water is shown in Figure 13.8. Typically the membrane is a microporous hollow fiber that separates oxygen-containing water from a nitrogen sweep gas. Dissolved oxygen in the water diffuses to the nitrogen sweep gas. Even though the dissolved oxygen concentration in the water is very low, its equilibrium concentration in the gas phase in contact with the water is several thousand times higher. This means that oxygen permeation through the membrane down the concentration gradient to the nitrogen sweep gas is high. The function of the membrane in this application is to provide a high surface area for contact between the water and the nitrogen sweep gas. The relative permeabilities of oxygen and water vapor through the membrane are not a factor; exactly the same separation could be achieved by running the water and nitrogen countercurrent to each other in a packed tower. However, as shown later, membrane contactors can offer useful advantages over packed towers.

Membrane contactors are typically shell-and-tube devices containing microporous capillary hollow fiber membranes. The membrane pores are sufficiently small that capillary forces prevent direct mixing of the phases on either side of the membrane. The membrane contactor shown in Figure 13.8 separates a liquid and a gas phase: this is a liquid/gas contactor [18]. Membrane contactors

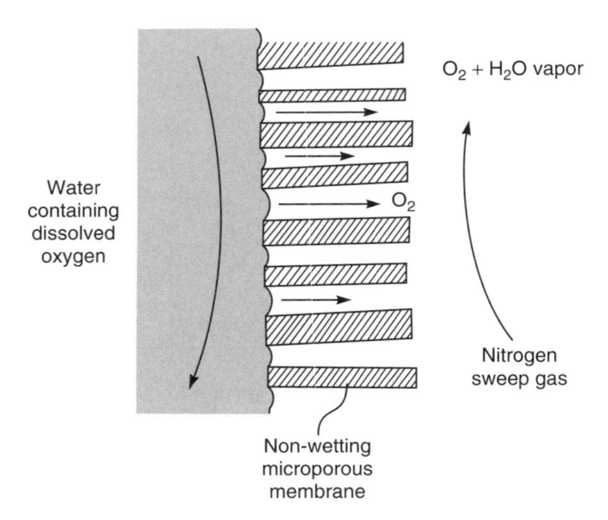

Figure 13.8 Application of a membrane contactor to remove dissolved oxygen from water

can also be used to separate two immiscible liquids (liquid/liquid contactors) or two miscible liquids (usually called membrane distillation). Contactors can also be used to selectively absorb one component from a gas mixture into a liquid (gas/liquid contactors). The various types of membrane contactors that have been used are illustrated in Figure 13.9.

Contactors have a number of advantages compared to simple liquid/gas absorber/strippers or liquid/liquid extractors. Perhaps the most important advantage is high surface area per volume. The contact area of membrane contactors compared to traditional contactor columns is shown in Table 13.2. Membrane contactors provide 10-fold higher contactor areas than equivalent-sized towers. This makes membrane

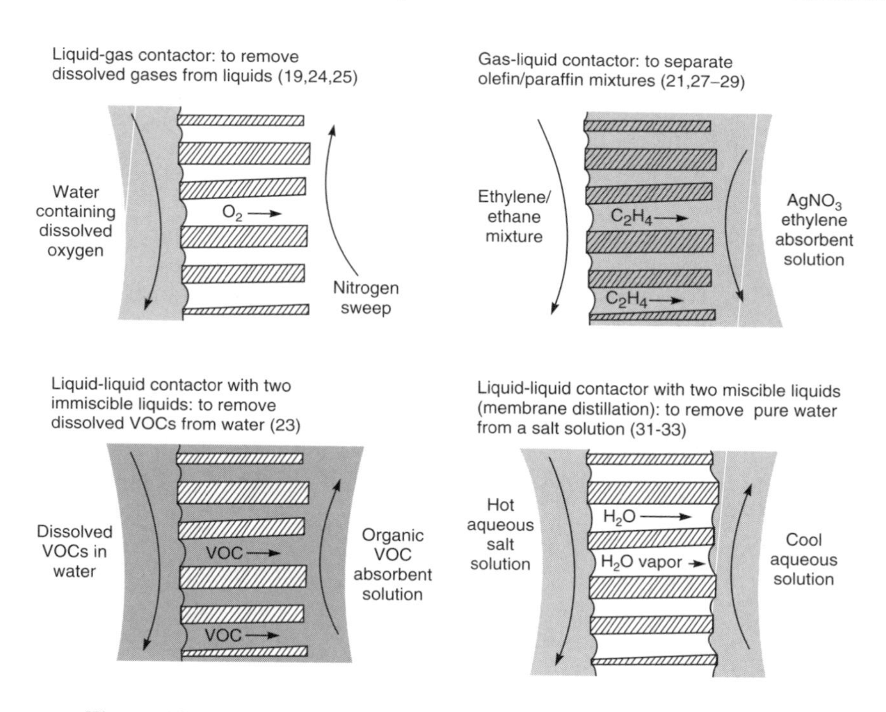

Figure 13.9 Examples of membrane contactors and their applications

Table 13.2 The specific surface areas of different contactors [20]

Contactor	Surface area per volume (cm^2/cm^3)
Free dispersion columns	0.03–0.3
Packed/trayed columns	0.3–3
Mechanically agitated columns	2–5
Membranes	10–50

contactors small and light, sometimes an important advantage. Blood oxygenators were not widely used for open heart surgery until the membrane blood oxygenator was developed, reducing the volume of blood required to operate the device to a manageable level. Similarly, the principal motivation to develop membrane contactors for offshore dehydration and carbon dioxide removal from natural gas is the reduction in weight and footprint possible. Kvaerner have shown that membrane contactors for this service have one-fourth of the footprint and one-tenth of the weight of conventional absorber/strippers [19].

A second advantage of membrane contactors is the physical separation of the counter-flowing phases by the membrane. The membrane area between the two phases is independent of their relative flow rates, so large flow ratio differences can be used without producing channeling or flooding or poor phase contact, and maximum advantage can be taken of the ability of counter-flow to separate and concentrate the components crossing the membrane. Small volumes of high cost extractants can be used to treat large volumes of low value feed. Separation of the two phases also eliminates entrainment of one phase by the other, as well as foaming. Finally, unlike traditional contactors, fluids of equal density can be used for the two phases.

The main disadvantages of contactors are related to the nature of the membrane interface. The membrane acts as an additional barrier to transport between the two phases that can slow the rate of separation. Over time, the membranes can foul, reducing the permeation rate further, or develop leaks, allowing direct mixing of the two phases. Finally, the polymeric membranes are necessarily thin (to maximize their permeation rate) and consequently cannot withstand large pressure differences across the membrane or exposure to harsh solvents and chemicals. In many industrial settings, this lack of robustness prohibits the use of membrane contactors.

Despite these caveats, the use of membrane contactors is growing rapidly. Positive reviews are given by Reed *et al.* [20], Qi and Cussler [21], Gabelman and Hwang [22] and Prasad and Sirkar [23].

Table 13.3 shows the dimensions of a series of industrial hollow fiber contactors produced by Hoechst Celanese under the trade name Liqui-Cel®. The

Table 13.3 Details of Liqui-Cel® hollow fiber membrane contactor modules

Module dimensions		Number of fibers (×1000)	Membrane area (m^2)	Area/unit volume (cm^2/cm^3)
Diameter (cm)	Length (cm)			
8	28	8	1.4	29
10	71	45	19	36
25	71	300	130	39

contact area per unit volume (cm^2/cm^3) is between 25 and 40. This high surface-to-volume ratio is achieved by making the fluid space between the membranes small—in the case of Liqui-Cel devices, between 200 and 400 μm. This means that the fluids passed through these devices must be particulate-free to avoid rapid plugging with retained particulates.

Applications of Membrane Contactors

Gas Exchange Contactors

Delivery or removal of gases from liquids is currently the largest commercial application of membrane contactors. One example is blood oxygenators, described in Chapter 12. Industrial applications of similar devices include deoxygenation of ultrapure water for the electronics industry or boiler feed water [24] and the adjustment of carbonation levels in beverages. [25] The performance of an industrial-scale oxygen removal system is shown in Figure 13.10. This unit consists of a 10-in.-diameter, rather short, capillary device containing about 135 m^2 of contactor area. The aqueous phase is circulated on the outer, shell-side of the fibers to avoid the excessive pressure required to circulate fluid at a high velocity down the fiber bore. The major resistance to mass transfer is in the liquid boundary on the outside of the fiber, so a baffled hollow fiber membrane module design [26] is used to cause radial flow of the fluid across the membrane from a central fluid distribution tube. Nitrogen sweep gas flows down the inside of the fibers. This design produces good turbulent mixing in the contactor at moderate pressure drops.

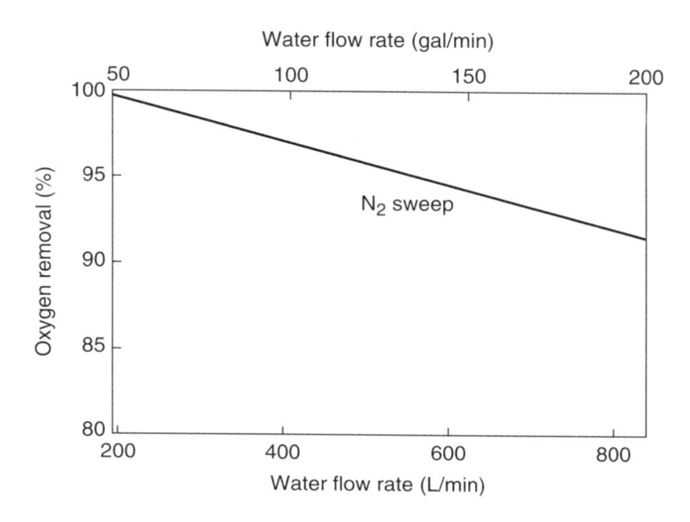

Figure 13.10 Oxygen removal from water with a 10-in.-diameter membrane contactor (135 m^2 membrane area) [24]

In Europe, the TNO [27] and Kvaerner [19] are both developing contactors to remove water and carbon dioxide from natural gas. Glycol or amines are used as the absorbent fluid. The goal is to reduce the size and weight of the unit to allow use on offshore platforms, so oftentimes only the absorber, the largest piece of equipment in a traditional absorber/stripper, is replaced with a membrane contactor. Kvaerner has taken this technology to the demonstration phase and commercial units are expected to be introduced soon.

Another type of gas exchange process, developed to the pilot plant stage, is separation of gaseous olefin/paraffin mixtures by absorption of the olefin into silver nitrate solution. This process is related to the separation of olefin/paraffin mixtures by facilitated transport membranes described in Chapter 11. A membrane contactor provides a gas–liquid interface for gas absorption to take place; a flow schematic of the process is shown in Figure 13.11 [28,29]. The olefin/paraffin gas mixture is circulated on the outside of a hollow fiber membrane contactor, while a 1–5 M silver nitrate solution is circulated countercurrently down the fiber bores. Hydrophilic hollow fiber membranes, which are wetted by the aqueous silver nitrate solution, are used.

The olefin fraction of the feed gas crosses the membrane and reacts reversibly with silver ions to form a soluble silver–olefin complex

$$Ag^{+} + \text{olefin} \rightleftharpoons Ag^{+}(\text{olefin}) \tag{13.2}$$

The olefin-laden silver solution is then pumped to a flash tank, where the pressure is lowered and the temperature raised sufficiently to reverse the complexation reaction and liberate pure ethylene. The regenerated silver nitrate solution is returned to the contactor. In this process, the high cost of the silver nitrate carrier

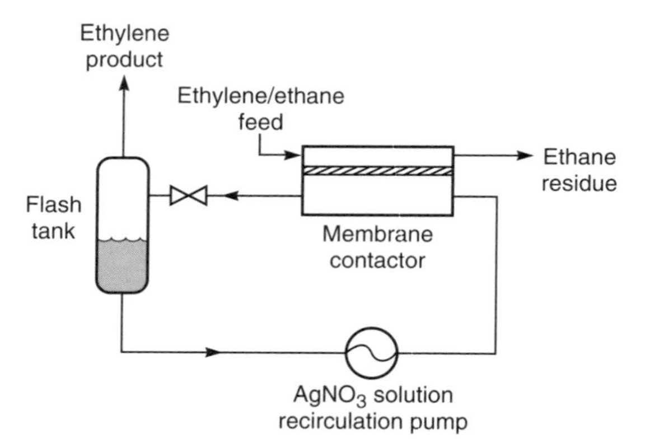

Figure 13.11 Flow schematic of the membrane contactor process developed by British Petroleum to separate ethylene/ethane mixtures by absorption into silver nitrate solution [28,29]

must be balanced against the cost of the membrane contactor. If the silver solution is circulated through the contactor at a very high rate, high fluxes are obtained, but the silver utilization calculated from the silver ion amount complexed in the contactor is low compared to the maximum possible complexation achievable under the condition of the test.

Absorption of olefin from olefin/paraffin mixtures has been scaled up to the pilot plant scale, and a number of successful trials were performed in the early 1990s. Separation factors of 200 or more were obtained, producing 99.7 % pure ethylene. However, slow degradation of the silver nitrate solution is a problem, and a portion of the recirculating degraded silver nitrate solution must be bled off and replaced with fresh solution continuously. Boundary layer problems on the liquid side of the membrane are also a serious issue in these devices [21].

To reduce the relatively large volume of silver nitrate solution held in the flash tank portion of the plant shown in Figure 13.11, Bessarabov *et al.* [30] have proposed using two membrane contactors in series, as shown in Figure 13.12. One contactor functions as an absorber, the other as a stripper. The first contactor removes ethylene from the pressurized feed gas into cold silver nitrate solution. The solution is then warmed and pumped to the second contactor where ethylene is desorbed from the silver nitrate solution into a low-pressure product ethylene gas stream. The regenerated silver nitrate solution is cooled and returned to the first contactor.

Bessarabov's devices use composite membranes consisting of a thin silicone rubber polymer layer coated onto a microporous poly(vinylidene fluoride) support layer. These membranes have high fluxes and minimal selectivities for the hydrocarbon gases, but the dense silicone layer provides a more positive barrier to bleed-through of liquid than do capillary effects with simple microporous membranes.

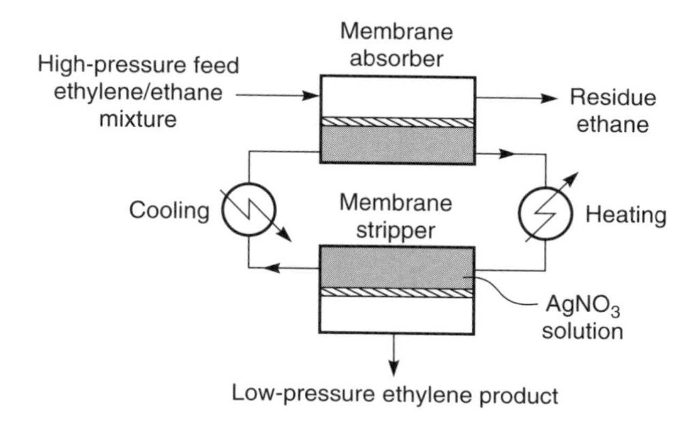

Figure 13.12 Flow schematic of process using two membrane contactors for the separation of ethylene/ethane mixtures proposed by Bessarabov *et al.* [30]

Liquid/Liquid Membrane Contactors (Membrane Distillation)

The most important example of liquid/liquid membrane contactors is membrane distillation, shown schematically in Figure 13.13. In this process, a warm, salt-containing solution is maintained on one side of the membrane and a cool pure distillate on the other. The hydrophobic microporous membrane is not wetted by either solution and forms a vapor gap between the two solutions. Because the solutions are at different temperatures, their vapor pressures are different; as a result, water vapor flows across the membrane. The water vapor flux is proportional to the vapor pressure difference between the warm feed and the cold permeate. Because of the exponential rise in vapor pressure with temperature, the flux increases dramatically as the temperature difference across the membrane is increased. Dissolved salts in the feed solution decrease the vapor pressure driving force, but this effect is small unless the salt concentration is very high. Some typical results illustrating the dependence of flux on the temperature and vapor pressure difference across a membrane are shown in Figure 13.14.

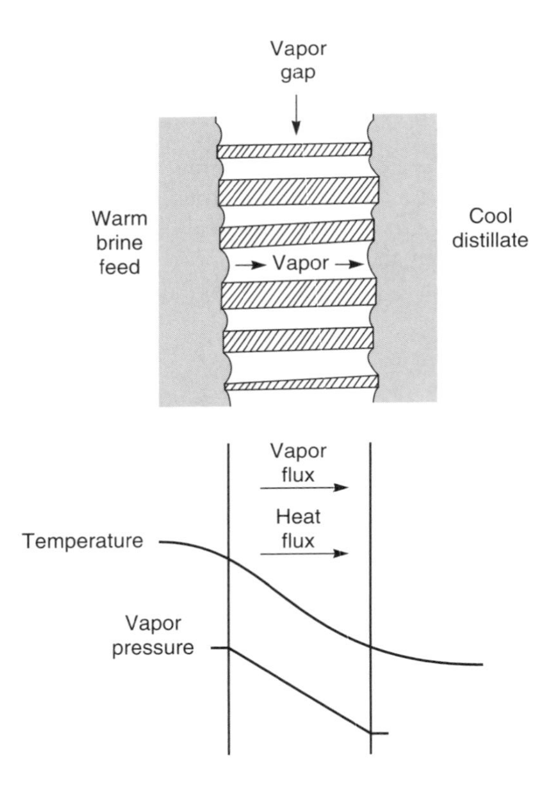

Figure 13.13 A schematic illustration of the membrane distillation process showing temperature and water vapor pressure gradients that drive the process

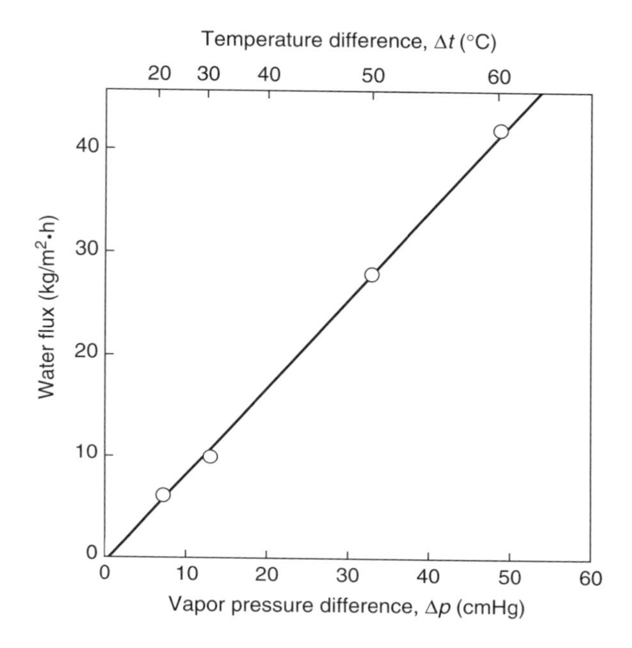

Figure 13.14 Water flux across a microporous membrane as a function of temperature and vapor pressure difference (distillate temperature, 18–38 °C; feed solution temperature, 50–90 °C). Taken from the data of Schneider *et al.* [31]

Membrane distillation offers a number of advantages over alternative pressure-driven processes such as reverse osmosis. Because the process is driven by temperature gradients, low-grade waste heat can be used and expensive high-pressure pumps are not required. Membrane fluxes are comparable to reverse osmosis fluxes, so membrane areas are not excessive. Finally, the process is still effective with slightly reduced fluxes even for very concentrated solutions. This is an advantage over reverse osmosis, in which the feed solution osmotic pressure places a practical limit on the concentration of a salt in the feed solution to be processed.

The principal application proposed for the technique is the separation of water from salt solutions. In the 1980s Enka, a division of Akzo, developed membrane distillation to the commercial scale using microporous polypropylene capillary membrane modules. The design and performance of their process are shown in Figure 13.15 [31]. The condensed distillate produced is almost salt free, whereas the salt concentration on the warm brine side of the membrane increases from 0.05 % to more than 20 %. At very high salt concentrations the flux across the membrane decreases slightly. Essentially all the power to drive the process is provided as low-grade heat. Despite the technical success of the device, a significant market did not develop. For large applications such as seawater desalination,

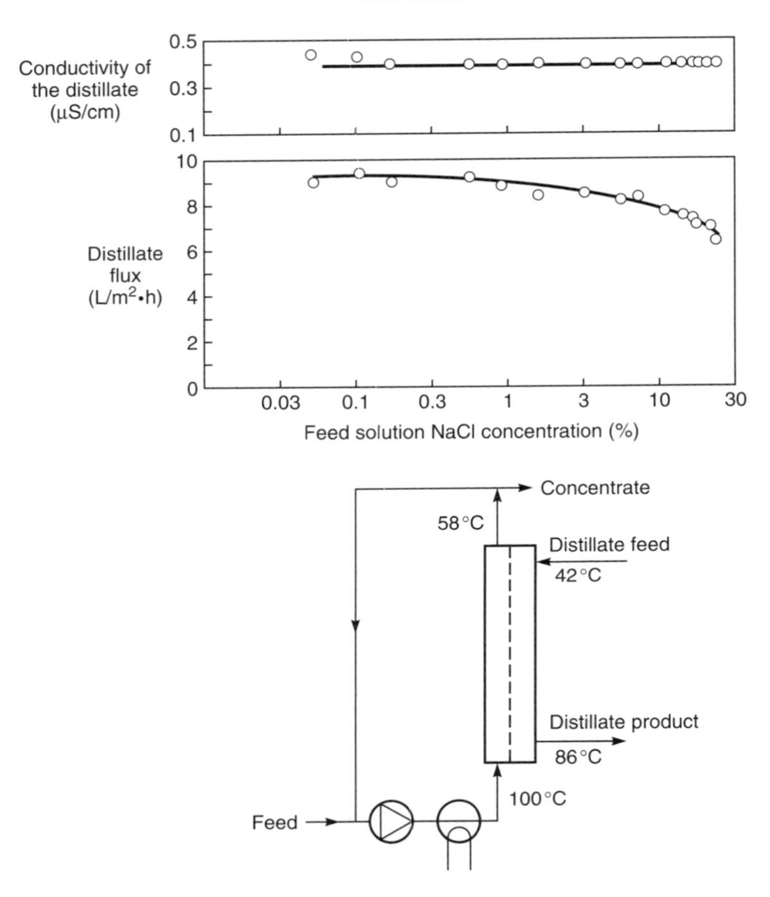

Figure 13.15 Flow scheme and performance data for a membrane distillation process for the production of water from salt solutions [31]. Feed salt solution is heated to 100 °C and passed counter-current to cool distillate that enters at 42 °C. The distillate product is almost salt-free as shown by its low conductivity. The distillate flux is almost constant up to salt concentrations as high as 20 % NaCl. Reprinted from *J. Membr. Sci.* **39**, K. Schneider, W. Hölz, R. Wollbeck and S. Ripperger, Membranes and Modules for Transmembrane Distillation, p. 25. Copyright 1988, with permission from Elsevier

for which the potential energy savings were significant, the capillary membrane contactor modules were too expensive compared to low-cost, reliable reverse osmosis modules. For smaller applications on chemical process streams, the energy savings were not important, so cost and reliability compared to simple evaporation were an issue. Currently there are no commercial producers of membrane distillation systems, although the process is still the subject of academic interest [32,33].

Membrane Reactors

By the early 1980s membrane technology had developed to the point at which a number of industrial groups began to consider using membranes to control the products of chemical reactions. Two properties of membranes are used; the first is the membrane as a contactor, as illustrated in Figure 13.16(a). The membrane separates the reaction medium in one chamber from a second chamber containing a catalyst, enzymes or a cell culture. This type of application has a long history in fermentation processes involving so-called bioreactors [34]. More recently, membrane reactors are being developed for conventional chemical

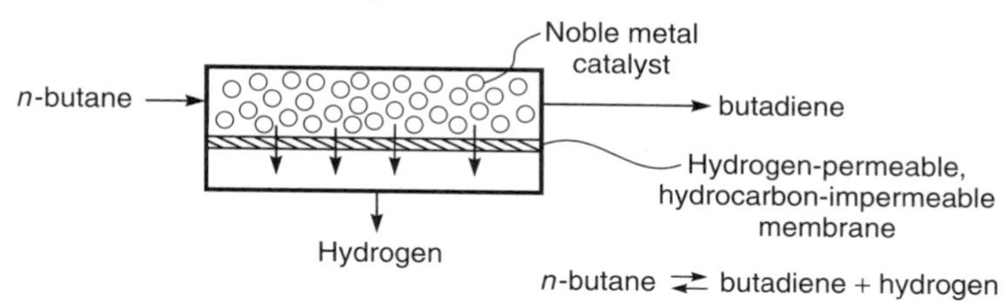

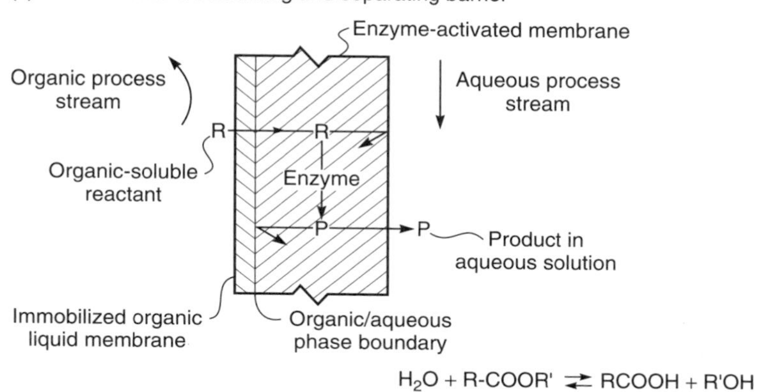

Figure 13.16 Examples of the three types of membrane reactor

separations [35]. As the reaction medium flows through the first chamber, membrane reactants diffuse through the membrane, react in the second chamber, and then diffuse back out to be collected as a product stream. The membrane provides a large exchange area between the catalytic material and the reaction medium but performs no separation function. In the example shown, the reactant is pectin, a high molecular weight polysaccharide present in citrus juice that causes an undesirable haze in the juice. Degradation of the pectin to galacturonic acid by the enzyme pectinase eliminates the haze [36].

The second type of membrane reactor, illustrated in Figure 13.16(b), uses the separative properties of a membrane. In this example, the membrane shifts the equilibrium of a chemical reaction by selectively removing one of the components of the reaction. The example illustrated is the important dehydrogenation reaction converting *n*-butane to butadiene and hydrogen

$$C_4H_{10} \rightleftharpoons C_4H_6 + 2H_2 \tag{13.3}$$

Removing hydrogen from the reaction chamber by permeation through the membrane causes the chemical equilibrium to shift to the right, and the conversion of butane to butadiene increases [37].

A third type of membrane reactor combines the functions of contactor and separator. An example of this combination membrane reactor is shown in Figure 13.16(c), in which the membrane is a multilayer composite. The layer facing the organic feed stream is an immobilized organic liquid membrane; the layer facing the aqueous product solution contains an enzyme catalyst for the de-esterification reaction

$$H_2O + R - COOR' \rightleftharpoons RCOOH + R'OH \tag{13.4}$$

Organic-soluble ester is brought to the reactor with the organic feed solution and freely permeates the immobilized organic liquid membrane to reach the catalyst enzyme. The ester is then hydrolyzed. The alcohol and acid products of hydrolysis are much more polar than the ester and, as such, are water soluble but relatively organic insoluble. These products diffuse to the aqueous permeate solution. The membrane both provides an active site for the reaction and separates the products of reaction from the feed [38].

In the membrane reactor shown in Figure 13.16(c), the chemical reaction and the separation step use the same membrane. However, in some processes it is desirable to separate reaction and separation into two distinct operations. If the net result of the process is to change the products of the chemical reaction, the process is still classified under the broad heading of membrane reactor. Two examples in which chemical reaction and separation are physically separated are shown in Figure 13.17. Figure 13.17(a) shows the use of a pervaporation membrane to shift the equilibrium of the de-esterification reaction [39,40]. A portion of the organic solution in the esterification reactor is continuously circulated past the

(a) Removal of the water of reaction from batch esterification processes to drive the reaction to completion

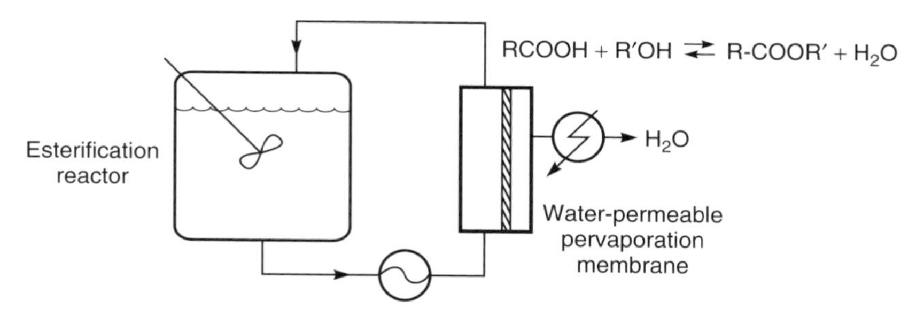

(b) Removal of hydrogen in the dehydrogenation of *n*-butane

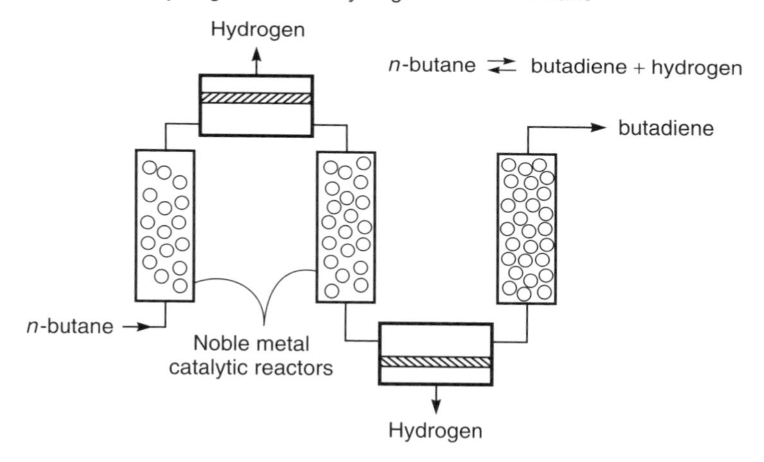

Figure 13.17 Examples of membrane reactors used to change the products of chemical reaction in which the membrane separation step is physically separated from the chemical reaction step

surface of a water-permeable membrane. Water produced in the esterification reaction is removed through the membrane. By removing the water, the reaction can be driven to completion.

Figure 13.17(b) shows the use of a hydrogen-permeable membrane to shift the equilibrium of the *n*-butane dehydrogenation reaction. The catalytic reactor is divided into steps, and the hydrogen-permeable membrane placed between each step. Because the hydrogen is removed from the reactor in two discrete steps, some inefficiency results, but separating the membrane separation step from the catalytic reactor allows the gas to be cooled before being sent to the membrane separator. Polymeric membranes can then be used for the gas separation operation [37]. Such membranes can remove hydrogen very efficiently from the butane–butadiene/hydrogen mixture but cannot be used at the 400–500 °C operating temperature of the catalytic reactor.

Applications of Membrane Reactors

Membrane reactors are being considered for many processes, and some are already being used on an industrial scale. A detailed description of this work is beyond the scope of this book; the three main application categories are described briefly below.

Cell Culture and Fermentation Processes

The traditional, and still the most common, fermentation process involves the addition of microbial cell cultures to the reaction medium in a batch reactor. This type of batch process is inherently slow, and microbial cells are lost with each batch of product. Recently there has been a great deal of interest in developing continuous fermentation processes using membrane bioreactors [34,41,42]. Much of this work has concentrated on fermentation of ethanol or acetone/butanol from low-grade food processing waste such as cheese whey, using a recycle membrane reactor design as shown in Figure 13.18. The principal advantages of the reactor are its continuous operation, the high cell densities that are maintained, and the lack of build-up of reaction products that inhibit the reaction.

Another type of microbiological reactor is the hollow fiber membrane bioreactor shown in Figure 13.19. In this device, the microbial cells are trapped on

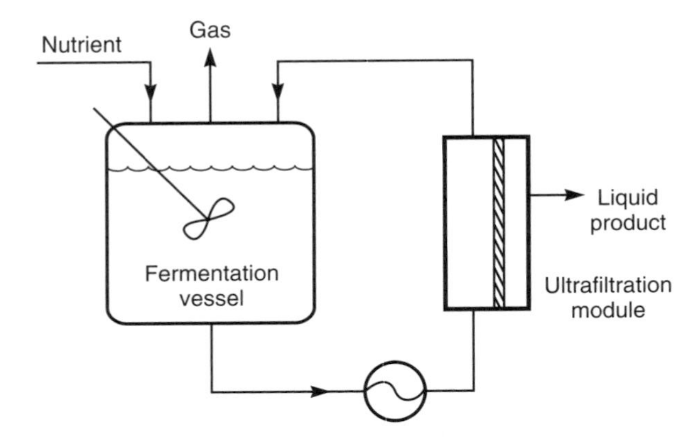

Figure 13.18 Continuous recycle fermentor membrane reactor. An ultrafiltration module removes the liquid products of fermentation as a clean product. This system is being developed for production of ethanol, acetone and butanol by fermentation of food processing waste streams

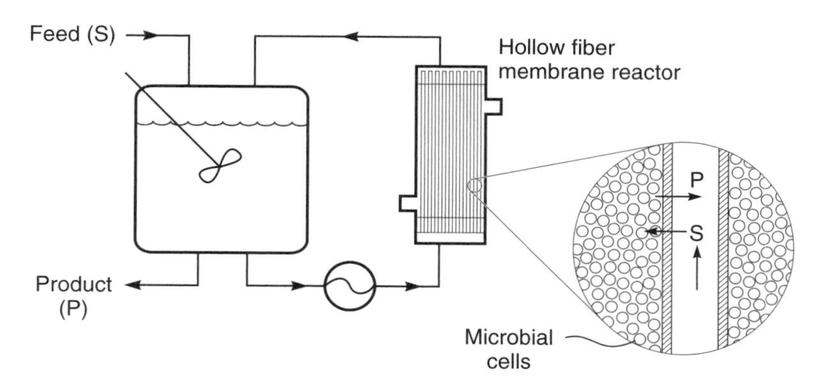

Figure 13.19 A hollow fiber membrane reactor. Nutrients (S) diffuse to the microbial cells on the shell side of the reactor and undergo reaction to form products (P) such as monoclonal antibodies [31]. Reprinted from *J. Membr. Sci.* **39**, K. Schneider, W. Hölz, R. Wollbeck and S. Ripperger, Membranes and Modules for Transmembrane Distillation, p. 38. Copyright 1988, with permission from Elsevier

the shell side of a capillary hollow fiber module. The feed solution, containing substrate and the products of microbial reaction, is circulated down the bore of the fibers [43,44]. This device has proved particularly useful in producing protein antibodies by genetically engineered mammalian cells. By manipulating the molecular weight cut-off of the fiber, the flux of molecules of different molecular weights across the filter can be controlled. Very high cell densities can be achieved in these hollow fiber cartridges, which have been used to produce monoclonal antibodies.

Light Hydrocarbon Gas-phase Catalytic Reactions

Several important refinery and chemical feedstock reactions appear to be good candidates for membrane reactor systems; some such reactions are listed in Table 13.4. Because of the high temperatures involved, developing the appropriate

Table 13.4 Petrochemical reactions being considered as applications for membrane reactors

$C_4H_{10} \rightleftharpoons C_4H_8 + H_2$
$C_6H_{11}CH_3 \rightleftharpoons C_6H_5CH_3 + 3H_2$
$C_6H_{12} \rightleftharpoons C_6H_6 + 3H_2$
$H_2S \rightleftharpoons H_2 + S$
$2CH_4 + O_2 \rightleftharpoons C_2H_4 + 2H_2O$
$2CH_4 + O_2 \rightleftharpoons 2CO + 4H_2$

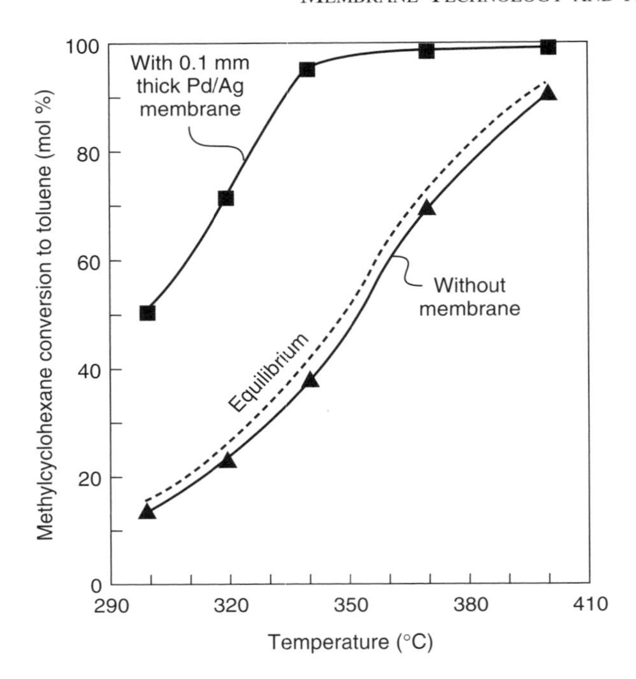

Figure 13.20 Methylcyclohexane conversion to toluene as a function of reactor temperature in a membrane and a nonmembrane reactor [45]. Reprinted with permission from J.K. Ali and D.W.T. Rippin, Comparing Mono and Bimetallic Noble Metal Catalysts in a Catalytic Membrane Reactor for Methyl-cyclohexane Dehydrogenation, *Ind. Eng. Chem. Res.* **34**, 722. Copyright 1995, American Chemical Society and American Pharmaceutical Association

selective membranes is difficult, and this type of membrane reactor has not moved beyond the laboratory stage.

The first four reactions listed in Table 13.4 are dehydrogenation reactions in which one of the reaction products is hydrogen. By removing hydrogen, the reaction equilibrium can be driven to completion, increasing the degree of conversion of the dehydrogenated product significantly. An example of the improvement in conversion that is possible is shown in Figure 13.20 [45]. In this figure, the fractional conversion of methylcyclohexane to toluene in a simple tube reactor is compared to that in a reactor with hydrogen-permeable, palladium-silver-alloy walls. Without the membrane, the degree of conversion is limited to the equilibrium value of the reaction. By removing the hydrogen, higher degrees of conversion can be achieved. Figure 13.20 also illustrates the problem that has inhibited widespread use of membrane reactors—the high temperature of the reactions. The reactions listed in Table 13.4 are all normally performed at 300–500 °C. These temperatures are far above the normal operating range of polymeric membranes, so hydrogen-permeable metal membranes, microporous carbon membranes, or ceramic membranes must be used. A review of membranes used in this application is given by Hsieh [46]. Currently metal or ceramic membranes are too expensive and too unreliable to be used in a commercial process. Rezac *et al.* [37,47] have tried to circumvent the problem by using reactors in series and cooling the gas between the reactor stages to the point that a high-temperature, polyimide-based polymer can be used to remove the hydrogen. Improved conversions are obtained by this process, and in the target dehydrogenation reaction studied, dehydrogenation of butane to butene and butadiene, good conversions can be obtained at relatively low reactor temperatures. Low-temperature reactor operations have a number of process benefits, so this approach may be developed further.

Another membrane application that could become a business is the use of ion-conducting membranes in membrane reactors. In the past 3 years, more than 70 US patents have appeared on this topic, as well as many papers. The overall concept is to use ceramic membranes that conduct oxygen or hydrogen ions at high temperatures. Materials that can conduct both ions and electrons are called mixed-conducting matrices. Several important early papers describing these materials were published by Teraoka *et al.* in the 1980s [48,49]. Various complex metal oxide compositions, including some better known for their properties as superconductors, have mixed-conducting properties; recent efforts in the field focus on these materials. Examples are perovskites having the structure $La_xA_{1-x}Co_yFe_{1-y}O_{3-z}$, where A is barium, strontium, or calcium; x and y are 0–1; and the value of z makes the overall material charge neutral. Passage of oxygen ions and electrons is related to the defect structure of these materials; at temperatures of 800–1000 °C, disks of these materials have shown extraordinary permeabilities to oxygen. Similar mixed-oxide membranes can also conduct protons [50].

Two large consortia, one headed by Air Products [51] and the other by Praxair/BP [52], are developing the membranes. At the appropriate high operating temperatures, the membranes are perfectly selective for oxygen over nitrogen, and oxygen permeabilities of 10 000 Barrer can be obtained. This means that, if the membrane thickness can be reduced to 1 μm, pressure-normalized fluxes of $10\,000 \times 10^6$ $cm^3(STP)/cm^2 \cdot s \cdot cmHg$ are possible. On this basis, for a plant to produce 1 MMscfh of oxygen, about 4000 m^2 of membrane tube area will be required—a large, but not inconceivably large, membrane area.

The most practical application, and the principal driving force behind the development of these membranes, is membrane reactor processes, such as the production of synthesis gas (syngas) by partial oxidation of methane or the oligomerization of methane to produce ethylene. Both processes are illustrated in Figure 13.21. In syngas production, oxygen ions diffusing through the membrane react with methane to form carbon monoxide and hydrogen. This gas can then be used without further separation to form methanol or other petrochemicals.

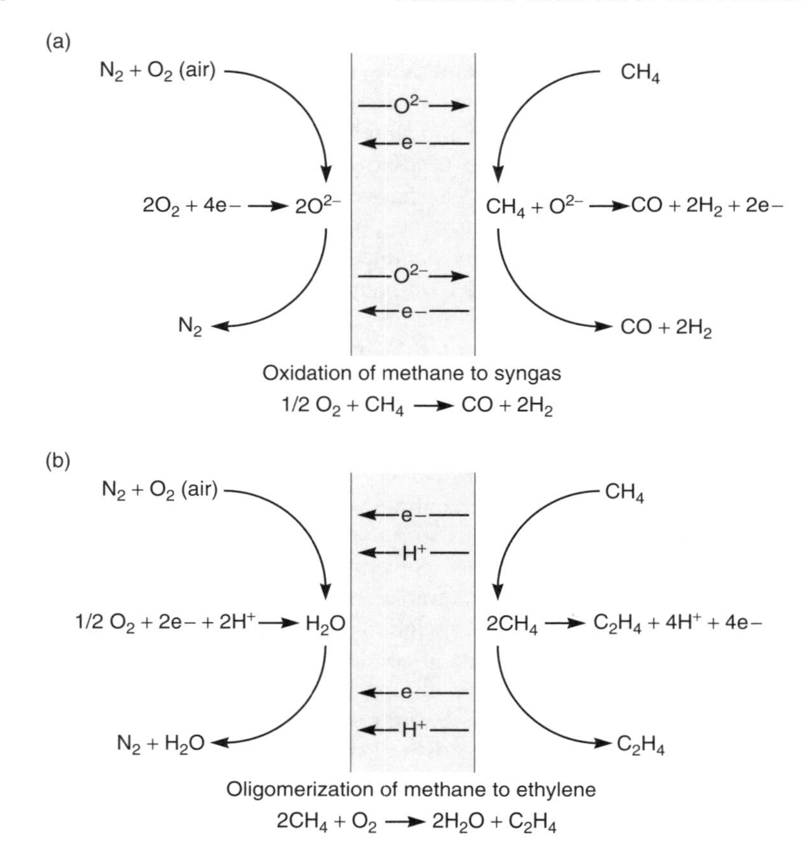

Figure 13.21 Use of ion-conducting ceramic membranes in a membrane reactor to produce (a) syngas ($CO + H_2$) and (b) ethylene

In ethylene production, methane is catalytically reacted to produce ethylene and hydrogen. The hydrogen permeates the membrane and then reacts with the oxygen in air to produce water. This second reaction produces the energy necessary to heat the process. Many large natural gas fields are not currently exploited because they are too far from the end-users. If this technology were developed, it would allow these fields to be developed, and so change the basis of the chemical and refining industries.

The membrane areas needed in these plants are not huge, but the technical challenges are substantial. Defect-free, anisotropic composite ceramic membranes that are 1–5 μm thick, able to operate continuously at 800–1000 °C, nonpoisoning, nonfouling and low-cost are required—not impossible, but difficult. Conceptual designs of the type of reactor required are beginning to appear.

The magnitude of the engineering task involved is indicated by the assumptions for the calculations [53]: a sealed vessel, containing 1000 1-in. diameter tubes, each 31 ft long, coated inside with perovskite membrane, in which the tubes are 1.5 in. apart, with a lower preheated section of 6 ft, a central reaction section of 18 ft, and an upper cooling section of 7 ft. The construction of such a vessel is neither simple nor cheap.

Chiral Drug Separation

Many drugs are produced as racemic mixtures of two mirror-image isomers. Often only one of these enantiomers has a beneficial pharmaceutical effect and the second enantiomer is much less active or, even worse, produces toxic side effects. For this reason many drugs must be resolved into their component enantiomers before being used. A number of techniques are available, but most are complex and costly. Resolution of racemic mixtures using stereoselective enzymatic reactions in a membrane bioreactor was pioneered by Sepracor and has been applied on an industrial scale for a number of important drugs [54,55]. Several ingenious process schemes have been proposed, one of which is illustrated in Figure 13.22.

The process shown in Figure 13.22 uses the stereospecific, enzymatically catalyzed hydrolysis of the ethyl ester of naproxen to the free acid to perform a chiral separation:

$$\underset{\text{Naproxen ethyl ester}}{CH_3O{-}C_{10}H_6{-}CH(CH_3)COOC_2H_5} + H_2O \xrightarrow{\text{Lipase enzyme}} \underset{\text{Naproxen}}{CH_3O{-}C_{10}H_6{-}CH(CH_3)COOH} + C_2H_5OH$$

The lipase enzyme stereospecifically hydrolyzes the (+) isomer of naproxen ester. The enzyme is immobilized in the wall of an inside-skinned hollow fiber membrane. The racemic *d* and *l* naproxen ester mixture, dissolved in methyl isobutyl ketone, is introduced on the shell side of the fiber and an aqueous buffer solution is circulated through the fiber lumen. The lipase enzyme hydrolyzes the *d* form of naproxen ester, forming ethanol and naproxen *d*. Naproxen *d* is a carboxylic acid soluble in aqueous buffer but insoluble in methyl isobutyl ketone. Consequently naproxen *d* is removed from the reactor with the buffer solution. The naproxen *l* ester remains in the methyl isobutyl ketone solution. This technique achieves an essentially complete separation of the *d* and *l* forms. In a clever final step

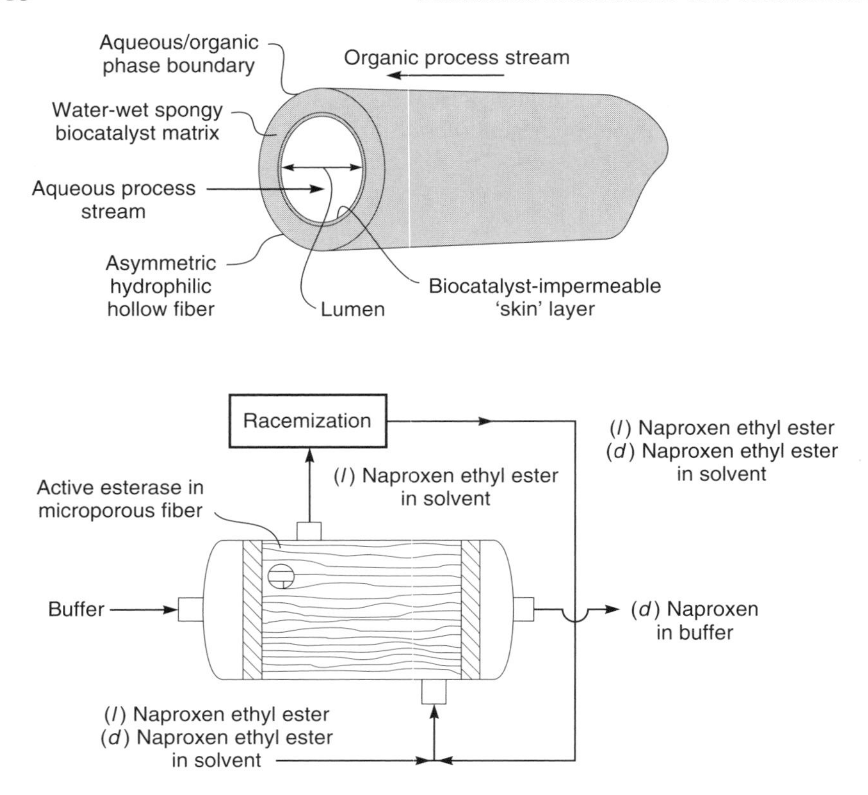

Figure 13.22 Application of a membrane bioreactor to separate chiral drug mixtures

the naproxen *l* ester is racemized to a *d* and *l* racemic mixture and recirculated to the membrane reactor. In this way all of the original mixture is eventually converted to the pure, pharmaceutically active *d* form. This technology has been applied to the chiral resolution of a number of drugs. A full-scale plant for the chiral separation of diltiazem intermediates, developed by Tanabe/Sepracor engineers, contains 1440 m^2 of hollow fiber membrane and produces 75 tons/year of resolved diltiazem intermediate [54].

Conclusions and Future Directions

Of the processes described in this chapter, membrane contactors and membrane reactors have the greatest potential to develop into large-scale commercial processes. Both technologies are already used on a small scale in niche applications, and both are being developed for much larger and more important processes. Membrane contactors are currently most widely used to deaerate liquids, but the best long-term application may be in the natural gas industry to replace amine absorber-strippers to remove carbon dioxide and hydrogen sulfide. Similarly, membrane reactors are currently used only in a few specialized biotech applications. However, long-term, this type of device will be used in the petrochemical industry, with very different membranes, for dehydrogenation processes or the partial oxidation of methane. Such applications could be much more important, but the development of suitable membranes poses a number of very challenging technical problems.

References

1. L. Cerini, Apparatus for the Purification of Impure Solutions of Caustic Soda and the Like on Osmotic Principals, US Patent 1,719,754 (July, 1929) and US Patent 1,815,761 (July, 1929).
2. S.B. Tuwiner, *Diffusion and Membrane Technology*, Reinhold Publishing Company, New York, NY (1962).
3. H.-C. Chen, C.H. O'Neal and L.C. Craig, Rapid Dialysis for Aminoacylation Assay of tRNA, *Anal. Chem.* **43**, 1017 (1971).
4. L.C. Craig and K. Stewart, Thin Film Counter-current Dialysis, *Science* **144**, 1093 (1964).
5. R.M. Wallace, Concentration and Separation of Ions by Donnan Membrane Equilibrium, *Ind. Eng. Chem. Process Des. Dev.* **6**, 423 (1967).
6. R.M. Wallace, Concentration of Ions Using Selective Membranes, US Patent 3,454, 490 (July, 1969).
7. T.-C. Huang, Y.-L. Lin and C.-Y. Chen, Selective Separation of Nickel and Copper from Complexing Solution by Cation-exchange Membrane, *J. Membr. Sci.* **37**, 131 (1988).
8. S.J. Oh, S.-H. Moon and T. Davis, Effects of Metal Ions on Diffusion Dialysis of Inorganic Acids, *J. Membr. Sci.* **169**, 95 (2000).
9. K. Sollner, Über Mosaik Membranen, *Biochem Z.* **244**, 370 (1932).
10. F.B. Leitz, *Piezodialysis in Membrane Separation Processes*, P. Meares (ed.), Elsevier Santi Publishing Company, Amsterdam, pp. 261–294 (1976).
11. C.R. Gardner, J.N. Weinstein and S.R. Caplan, Transport Properties of Charge-mosaic Membranes, *Desalination* **12**, 19 (1973).
12. U. Merten, Desalination by Pressure Osmosis, *Desalination* **297**, 1 (1966).
13. J.N. Weinstein and S.R. Caplan, Charge-mosaic Membranes: Dialytic Separation of Electrolytes from Nonelectrolytes and Amino Acids, *Science* **296**, 169 (1970).
14. S.R. Caplan, Transport in Natural and Synthetic Membranes in Membrane Processes, in *Industry and Biomedicine*, M. Bier (ed.), Plenum Press, New York, pp. 1–22 (1971).
15. K.L. Platt and A. Schindler, Ionic Membranes for Water Desalination, *Makromol. Chem.* **19**, 135 (1971).
16. H. Itou, M. Toda, K. Ohkoshi, M. Iwata, T. Fujimoto, Y. Miyaki and T. Kataoka, Artificial Membranes from Multiblock Copolymers (6), *Ind. Eng. Chem.* **983**, 27 (1988).
17. K. Hirahara, S.-I. Takahashi, M. Iwata, T. Fujimoto and Y. Miyaki, Artificial Membranes from Multiblock Copolymers (5), *Ind. Eng. Chem. Prod. Res. Dev.* **305**, 25 (1986).
18. N.A. D'Elia, L. Dahuron and E.L. Cussler, Liquid–Liquid Extraction with Microporous Hollow Fibers, *J. Membr. Sci.* **29**, 309 (1986).

19. O. Falk-Pedersen and H. Dannstrom, Method for Removing Carbon Dioxide from Gases, US Patent 6,228,145 (May, 2001).
20. B.W. Reed, M.J. Semmens and E.L. Cussler, Membrane Contactors, in *Membrane Separations Technology: Principles and Applications*, R.D. Noble and S.A. Stern (eds), Elsevier Science, Amsterdam, pp. 467–498 (1995).
21. Z. Qi and E.L. Cussler, Microporous Hollow Fiber for Gas Absorption, *J. Membr. Sci.* **23**, 321, 333 (1985).
22. A. Gabelman and S.-T. Hwang, Hollow Fiber Membrane Contactors, *J. Membr. Sci.* **159**, 61 (1999).
23. R. Prasad and K.K. Sirkar, Membrane Based Solvent Extraction, in *Membrane Handbook*, W.S.W. Ho and K.K. Sirkar (eds), Van Nostrand-Reinhold, New York, pp. 727–763 (1992).
24. F. Wiesler and R. Sodaro, Deaeration: Degasification of Water Using Novel Membrane Technology, *Ultrapure Water* **35**, 53 (1996).
25. J.K.R. Page and D.G. Kalhod, Control of Dissolved Gases in Liquids, US Patent 5,565,149 (October, 1996); K. Honda and M. Yamashita, Method for Deaerating Liquid Products, US Patent 5,522,917 (June, 1996).
26. R. Prasad, C.J. Runkle and H.F. Shuey, Spiral-wound Hollow-fiber Membrane Fabric Cartridges and Modules Having Flow Directing Baffles, US Patent 5,353,361 (October, 1994).
27. A.E. Jansen, P.H.M. Feron, J.H. Hanemaaijer and P. Huisjes, Apparatus and Method for Performing Membrane Gas/Liquid Absorption at Elevated Pressure, US Patent 6,355,092 (March 2002).
28. D.T. Tsou, M.W. Blachman and J.G. Davis, Silver-facilitated Olefin/Paraffin Separation in a Liquid Membrane Contactor System, *Ind. Eng. Chem. Res.* **33**, 3209 (1994).
29. R.J. Valus, R. Eshraghi, A.E. Velikoff and J.C. Davis, High Pressure Facilitated Membranes for Selective Separation and Process for the Use Thereof, US Patent 5,057,641 (October, 1991).
30. D.G. Bessarabov, R.D. Sanderson, E.P. Jacobs and I.N. Beckman, High Efficiency Separation of an Ethylene/Ethane Mixture, *Ind. Eng. Chem. Res.* **34**, 1769 (1995).
31. K. Schneider, W. Hölz, R. Wollbeck and S. Ripperger, Membranes and Modules for Transmembrane Distillation, *J. Membr. Sci.* **39**, 25 (1988).
32. R.W. Schofield, A.G. Fane and C.J.D. Fell, Gas and Vapor Transport Through Microporous Membranes, *J. Membr. Sci.* **53**, 173 (1990).
33. K.W. Lawson and D.R. Lloyd, Membrane Distillation, *J. Membr. Sci.* **124**, 25 (1997).
34. M. Cheryan, *Ultrafiltration and Microfiltration Handbook*, Technomic Publishing Co., Lancaster, PA, p. 445 (1998).
35. M.F. Kemmere and J.T.F. Keurentjes, Industrial Membrane Reactors, in *Membrane Technology in the Chemical Industry*, S.P. Nunes and K.-V. Peinemann (eds), Wiley-VCH, Weinheim, Germany, pp. 191–221 (2001).
36. X. Shao, Y. Fend, S. Hu and R. Govind, Pectin Degradation in a Spiral Membrane Reactor, in *Membrane Reactor Technology*, R. Govind and N. Itoh (eds), AIChE Symposium Series Number 268, AIChE, New York, NY, Vol. 85, pp. 85–92 (1989).
37. M.E. Rezac, W.J. Koros and S.J. Miller, Membrane-assisted Dehydrogenation of *n*-Butane: Influence of Membrane Properties on System Performance, *J. Membr. Sci.* **93**, 193 (1994).
38. S.L. Matson and J.A. Quinn, Membrane Reactors, in *Membrane Handbook*, W.S.W. Ho and K.K. Sirkar (eds), Van Nostrand Reinhold, New York, pp. 809–832 (1992).
39. H.H. Nijhuis, A. Kempermann, J.T.P. Derksen and F.P. Cuperus, Pervaporation Controlled Biocatalytic Esterification Reaction, in *Proc. Sixth International Conference on Pervaporation Process in the Chemical Industry*, Ottawa, Canada, R. Bakish (ed.), Bakish Materials Corp., Englewood, NJ, pp. 368–379 (1992).

40. A. Dams and J. King, Pervaporation Aided Esterification—Alternative in Plant Extension for an Existing Chemical Process, in *Proc. Fifth International Conference on Pervaporation Process in the Chemical Industry*, Heidelburg, Germany, R. Bakish (ed.), Bakish Materials Corp., Englewood, NJ, pp. 338–348 (1991).
41. M. Cheryan and M.A. Mehaia, Membrane Bioreactors, in *Membrane Separations in Biotechnology*, W.C. McGregor (ed.), Marcel Dekker, New York, pp. 255–302 (1986).
42. M. Cheryan and M.A. Mehaia, Membrane Bioreactors for High-performance Fermentations, in *Reverse Osmosis and Ultrafiltration*, S. Sourirajan and T. Matsura (eds), ACS Symposium Series Number 281, American Chemical Society, Washington, DC, pp. 231–246 (1985).
43. R.A. Kanazek, P.M. Gullino, P.O. Kohler and R.C. Dedrick, Cell Culture on Artificial Capillaries, An Approach to Tissue Growth, *In Vitro Science* **178**, 65 (1972).
44. W.S. Hu and T.C. Dodge, Cultivation of Mammalian Cells in Bioreactors, *Biotechnol. Prog.* **1**, 209 (1985).
45. J.K. Ali and D.W.T. Rippin, Comparing Mono and Bimetallic Noble Metal Catalysts in a Catalytic Membrane Reactor for Methyl-cyclohexane Dehydrogenation, *Ind. Eng. Chem. Res.* **34**, 722 (1995).
46. H.P. Hsieh, Inorganic Membrane Reactors: A Review, in *Membrane Reactor Technology*, R. Govind and N. Itoh (eds), AIChE Symposium Series Number 268, AIChE, New York, NY, Vol. 85, pp. 53–67 (1989).
47. S.J. Miller, M.E. Rezac and W.J. Koros, Dehydrogenation Using Dehydrogenation Catalyst and Polymer-porous Solid Composite Membrane, US Patent 5,430,218 (July, 1995).
48. Y. Teraoka, H. Zhang, S. Furukawa and N. Yamazoe, Oxygen Permeation Through Perovskite-type Oxides, *Chem. Lett.* 1743 (1985).
49. Y. Teraoka, T. Nobunaga and N. Yamazoe, The Effect of Cation Substitution on the Oxygen Semipermeability of Perovskite-type Oxides, *Chem. Lett.* 503 (1988).
50. Y.S. Lin, Microporous and Dense Inorganic Membranes: Current Status and Prospective, *Sep. Purif. Tech.* **25**, 39 (2001).
51. R.M. Thorogood, S. Srinvasan, T.F. Yee and M.P. Drake, Composite Mixed Conductor Membrane for Producing Oxygen, US Patent 5,240,480 (August, 1993); US Patents 5,240,473 (August, 1993); 5,261,932 (November, 1993); 5,269,822 (December, 1993); and 5,516,359 (May, 1996).
52. T.J. Mazanec, T.L. Cable, J.G. Frye and W.R. Kliwer, Solid-component Membranes, US Patents 5,591,315 (January, 1997); 5,306,411 (April, 1994); 5,648,304 (July, 1997).
53. C.F. Gottzmann, R. Prasad, J.M. Schwartz, V.E. Bergsten, J.E. White, T.J. Mazanec, T.L. Cable and J.C. Fagley, Tube and Shell Reactor with Oxygen Selective Ion Transport Ceramic Reaction Tubes, US Patent 6,139,810 (October, 2000).
54. J.L. Lopez and S.L. Matson, A Multiphase/Extractive Enzyme Membrane Reactor for Production of Diltiazem Chiral Intermediate, *J. Membr. Sci.* **125**, 189 (1997).
55. S.L. Matson, Method for Resolution of Stereoisomers in Multiphase and Extractive Membrane Reactors, US Patent 4,800,162 (January, 1989).

APPENDIX

Table A1 Constants

Mathematical
$e = 2.71828\ldots$
$\ln 10 = 2.30259\ldots$
$\pi = 3.14159\ldots$
Gas law constant, R
1.987 cal/g-mol · K
82.05 cm^3 · atm/g-mol · K
8.314×10^7 g · cm^2/s^2 · g-mol · K
8.314×10^3 kg · m^2/s^2 · kg-mol · K
Standard acceleration of gravity
980.665 cm/s^2
32.1740 ft/s^2
Avogadro's number
6.023×10^{23} molecules/g-mol
Faraday's constant, $\mathcal{F}$
9.652×10^4 abs-coulombs/g-equivalent
STP (standard temperature and pressure)
273.15 K and 1 atm pressure
Volume of 1 mol of ideal gas at STP = 22.41 L

Membrane Technology and Applications R. W. Baker
 ISBN: 0-470-85445-6

Table A2 Conversion factors for weight and volume

Given a quantity in these units	Multiply by	To convert quantity to these units
Pounds	453.59	Grams
Kilograms	2.2046	Pounds
Ton, short (US)	2000	Pounds
Ton, long (UK)	2240	Pounds
Ton, metric	1000	Kilograms
Gallons (US)	3.7853	Liters
Gallons (US)	231.00	Cubic inches
Gallons (US)	0.13368	Cubic feet
Cubic feet	28.316	Liters
Cubic meters	264.17	Gallons (US)

Table A3 Conversion factors—other

Given a quantity in these units	Multiply by	To convert quantity to these units
Inches	2.54	Centimeters
Meters	39.37	Inches
Mils	25.4	Microns
Square meters	10.764	Square feet
Dynes	1	$g \cdot cm/s$
Centipoises	10^{-3}	$kg/m \cdot s$

Table A4 Conversion factors for pressure

Given a quantity in these units	Multiply by value below to convert to corresponding units					
	Atmosphere (atm)	mmHg (torr)	lb/in.^{2} (psi)	kg/cm^2	kiloPascal (kPa)	Bar
Atmosphere (atm)	1	760	14.696	1.0332	101.325	1.01325
mmHg (torr)	1.3158×10^{-3}	1	1.9337×10^{-2}	1.3595×10^{-3}	0.13332	1.3332×10^{-3}
lb/in.^{2} (psi)	6.8046×10^{-2}	51.715	1	7.0305×10^{-2}	6.8948	6.8948×10^{-2}
kg/cm^2	0.96787	735.58	14.224	1	98.069	0.98069
kiloPascal (kPa)	9.8692×10^{-3}	7.5008	0.14504	10.197×10^{-3}	1	0.01
Bar	0.98692	750.08	14.504	1.0197	100	1

Table A5 Conversion factors for energy

Given a quantity in these units	Multiply by value below to convert to corresponding units					
	$g \cdot cm^2/s^2$ (ergs)	$kg \cdot m^2/s^2$ (joules)	Calories (cal)	British thermal units (Btu)	Horsepower · hour (hp · h)	Kilowatt hour (kWh)
$g \cdot cm^2/s^2$ (ergs)	1	10^{-7}	2.3901×10^{-8}	9.4783×10^{-11}	3.7251×10^{-14}	2.7778×10^{-14}
$kg \cdot m^2/s^2$ (joules)	10^7	1	2.3901×10^{-1}	9.4783×10^{-4}	3.7251×10^{-7}	2.7778×10^{-7}
Calories (cal)	4.1840×10^7	4.1840	1	3.9657×10^{-3}	1.5586×10^{-6}	1.1622×10^{-6}
British thermal units (Btu)	1.0550×10^{10}	1.0550×10^3	2.5216×10^2	1	3.9301×10^{-4}	2.9307×10^{-4}
Horsepower · hour (hp · h)	2.6845×10^{13}	2.6845×10^6	6.4162×10^5	2.5445×10^3	1	7.4570×10^{-1}
Kilowatt hour (kWh)	3.6000×10^{13}	3.6000×10^6	8.6042×10^5	8.4122×10^3	1.3410	1

Table A6 Conversion factors for liquid flux

Given a quantity in these units	Multiply by value below to convert to corresponding units					
	$L/m^2 \cdot h$ (Lmh)	gal (US)/ $ft^2 \cdot day$ (gfd)	$cm^3/cm^2 \cdot s$	$cm^3/cm^2 \cdot min$	$m^3/m^2 \cdot day$	$L/m^2 \cdot day$
$L/m^2 \cdot h$ (Lmh)	1	0.59	2.78×10^{-5}	1.6667×10^{-3}	2.40×10^{-2}	24.0
gal (US)/$ft^2 \cdot day$ (gfd)	1.70	1	4.72×10^{-5}	2.832×10^{-3}	4.07×10^{-2}	40.73
$cm^3/cm^2 \cdot s$	3.60×10^4	2.12×10^4	1	60	864	8.64×10^5
$cm^3/cm^2 \cdot min$	600	353	0.1667	1	14.4	1.44×10^3
$m^3/m^2 \cdot day$	41.67	24.55	1.16×10^{-3}	6.944×10^{-2}	1	10^3
$L/m^2 \cdot day$	4.17×10^{-2}	2.46×10^{-2}	1.16×10^{-6}	6.944×10^{-4}	1×10^{-3}	1

Table A7 Conversion factors for liquid permeance

Given a quantity in these units	Multiply by value below to convert to corresponding units				
	$L/m^2 \cdot h \cdot bar$	$L/m^2 \cdot h \cdot MPa$	$L/m^2 \cdot s \cdot Pa$	gal (US)/$ft^2 \cdot day \cdot psi$	$m^3/m^2 \cdot day \cdot bar$
$L/m^2 \cdot h \cdot bar$	1	10	2.777×10^{-9}	4.064×10^{-2}	2.4×10^{-2}
$L/m^2 \cdot h \cdot MPa$	0.1	1	2.777×10^{-10}	4.064×10^{-3}	2.4×10^{-3}
$L/m^2 \cdot s \cdot Pa$	3.601×10^8	3.601×10^9	1	1.463×10^7	8.642×10^6
gal (US)/$ft^2 \cdot day \cdot psi$	24.61	2.461×10^2	6.837×10^{-8}	1	0.5906
$m^3/m^2 \cdot day \cdot bar$	41.67	4.167×10^2	1.157×10^{-7}	1.693	1

Table A8 Conversion factors for gas flux

Given a quantity in these units	Multiply by value below to convert to corresponding units						
	$cm^3(STP)/cm^2 \cdot s$	$L(STP)/cm^2 \cdot s$	$m^3(STP)/m^2 \cdot day$	$ft^3(STP)/ft^2 \cdot h$	$mol/m^2 \cdot s$	$mol/cm^2 \cdot s$	$\mu mol/m^2 \cdot s$
$cm^3(STP)/cm^2 \cdot s$	1	1×10^{-3}	864	118.1	0.4462	0.4462×10^{-4}	0.4462×10^{6}
$L(STP)/cm^2 \cdot s$	1000	1	8.64×10^{5}	1.181×10^{5}	0.4462×10^{3}	0.4462×10^{-1}	0.4462×10^{9}
$m^3(STP)/m^2 \cdot day$	1.157×10^{-3}	1.157×10^{-6}	1	0.1366	0.5162×10^{-3}	0.5162×10^{-7}	0.5162×10^{3}
$ft^3(STP)/ft^2 \cdot h$	8.467×10^{-3}	8.467×10^{-6}	7.315	1	3.778×10^{-3}	3.778×10^{-7}	3.778×10^{3}
$mol/m^2 \cdot s$	2.241	2.241×10^{-3}	1.936×10^{3}	2.647×10^{2}	1	1×10^{-4}	1×10^{6}
$mol/cm^2 \cdot s$	2.241×10^{4}	2.241×10	1.936×10^{7}	2.647×10^{6}	1×10^{4}	1	1×10^{10}
$\mu mol/m^2 \cdot s$	2.241×10^{-6}	2.241×10^{-9}	1.936×10^{-3}	2.647×10^{-4}	1×10^{-6}	1×10^{-10}	1

Table A9 Conversion factors for gas permeance

Given a quantity in these units	Multiply by value below to convert to corresponding units					
	1×10^{-6} $cm^3(STP)/cm^2 \cdot s \cdot cmHg$ (gpu)	$L(STP)/m^2 \cdot h \cdot bar$	$cm^3(STP)/cm^2 \cdot min \cdot bar$	$scf/m^2 \cdot min \cdot bar$	$mol/m^2 \cdot s \cdot Pa$	$mol/m^2 \cdot h \cdot bar$
1×10^{-6} $cm^3(STP)/cm^2 \cdot s \cdot cmHg$ (gpu)	1	2.7	4.5×10^{-3}	1.589×10^{-3}	3.348×10^{-10}	0.1205
$L(STP)/m^2 \cdot h \cdot bar$	0.3703	1	1.645×10^{-3}	0.5808×10^{-3}	1.224×10^{-10}	4.404×10^{-2}
$cm^3(STP)/cm^2 \cdot min \cdot bar$	0.2222×10^{3}	0.6079×10^{3}	1	0.3530	0.7439×10^{-7}	26.78
$scf/m^2 \cdot min \cdot bar$	6.293×10^{2}	1.722×10^{3}	2.832	1	2.107×10^{-7}	75.83
$mol/m^2 \cdot s \cdot Pa$	0.2987×10^{10}	0.8172×10^{10}	1.344×10^{7}	0.4746×10^{7}	1	3.599×10^{8}
$mol/m^2 \cdot h \cdot bar$	8.299	22.71	37.35×10^{-3}	13.19×10^{-3}	27.87×10^{-10}	1

A 1-μm-thick membrane having a permeability of 1 Barrer has a permeance of 1 gpu.

Table A10 Conversion factors for gas permeability

Given a quantity in these units	Multiply by value below to convert to corresponding units				
	1×10^{-10} $cm^3(STP) \cdot cm/cm^2 \cdot s \cdot cmHg$ (Barrer)	$cm^3(STP) \cdot cm/cm^2 \cdot s \cdot bar$	$cm^3(STP) \cdot cm/cm^2 \cdot s \cdot Pa$	$L(STP) \cdot m/m^2 \cdot s \cdot Pa$	$mol \cdot m/m^2 \cdot s \cdot Pa$
1×10^{-10} $cm^3(STP) \cdot cm/cm^2 \cdot s \cdot cmHg$ (Barrer)	1	7.5×10^{-9}	7.5×10^{-14}	7.5×10^{-15}	3.347×10^{-16}
$cm^3(STP) \cdot cm/cm^2 \cdot s \cdot bar$	1.333×10^{8}	1	1×10^{-5}	1×10^{-6}	4.462×10^{-8}
$cm^3(STP) \cdot cm/cm^2 \cdot s \cdot Pa$	1.333×10^{13}	1×10^{5}	1	0.1	4.462×10^{-3}
$L(STP) \cdot m/m^2 \cdot s \cdot Pa$	1.333×10^{14}	1×10^{6}	10	1	4.462×10^{-2}
$mol \cdot m/m^2 \cdot s \cdot Pa$	2.989×10^{15}	2.242×10^{7}	2.242×10^{2}	22.42	1

Table A11 Vapor pressure of water and ice

Temp. (°C)	Pressure (mmHg)	Temp. (°C)	Pressure (mmHg)	Temp. (°C)	Pressure (mmHg)	Temp. (°C)	Pressure (mmHg)	Temp. (°C)	Pressure (mmHg)
−30	0.285	0	4.58	30	31.8	60	149	90	526
−29	0.317	1	4.93	31	33.7	61	156	91	546
−28	0.351	2	5.29	32	35.7	62	164	92	567
−27	0.389	3	5.69	33	37.7	63	171	93	589
−26	0.430	4	6.10	34	39.9	64	179	94	611
−25	0.476	5	6.54	35	41.2	65	188	95	634
−24	0.526	6	7.01	36	44.6	66	196	96	658
−23	0.580	7	7.51	37	47.1	67	205	97	682
−22	0.640	8	8.05	38	49.7	68	214	98	707
−21	0.705	9	8.61	39	52.4	69	224	99	733
−20	0.776	10	9.21	40	55.3	70	234	100	760
−19	0.854	11	9.84	41	58.3	71	244	101	788
−18	0.939	12	10.5	42	61.5	72	254	102	816
−17	1.03	13	11.2	43	64.8	73	266	103	845
−16	1.13	14	12.0	44	68.3	74	277	104	875
−15	1.24	15	12.8	45	71.9	75	289	105	906
−14	1.36	16	13.6	46	75.7	76	301	106	938
−13	1.49	17	14.5	47	79.6	77	314	107	971
−12	1.63	18	15.5	48	83.7	78	327	108	1004
−11	1.79	19	16.5	49	88.0	79	341	109	1039
−10	1.95	20	17.5	50	92.5	80	355	110	1075
−9	2.13	21	18.7	51	97.2	81	370	111	1111
−8	2.33	22	19.8	52	102	82	385	112	1149
−7	2.54	23	21.1	53	107	83	401	113	1187
−6	2.77	24	22.4	54	113	84	417	114	1228
−5	3.01	25	23.8	55	118	85	434	115	1267
−4	3.28	26	25.2	56	124	86	451	116	1310
−3	3.57	27	26.7	57	130	87	469	117	1353
−2	3.88	28	28.3	58	136	88	487	118	1397
−1	4.22	29	30.0	59	143	89	506	119	1443
0	4.58	30	31.8	60	149	90	526	120	1489

Table A12 Composition of air

Component	Concentration (vol%)	Concentration (wt%)
Nitrogen	78.09	75.52
Oxygen	20.95	23.15
Argon	0.933	1.28
Carbon dioxide	0.030	0.046
Neon	0.0018	0.0012
Helium	0.0005	0.00007
Krypton	0.0001	0.0003
Hydrogen	0.0005	0.00003
Xenon	0.000003	0.00004

Table A13 Typical osmotic pressures at 25 °C

Compound	Concentration (mg/L)	Concentration (mol/L)	Osmotic pressure (psi)
NaCl	35 000	0.60	398
Seawater	32 000	—	340
NaCl	2000	0.0342	22.8
Brackish water	2000–5000	—	15–40
$NaHCO_3$	1000	0.0119	12.8
Na_2SO_4	1000	0.00705	6.0
$MgSO_4$	1000	0.00831	3.6
$MgCl_2$	1000	0.0105	9.7
$CaCl_2$	1000	0.009	8.3
Sucrose	1000	0.00292	1.05
Dextrose	1000	0.0055	2.0

Table A14 Mean free path of gases (25 °C)

Gas	λ (Å)
Argon	1017
Hydrogen	1775
Helium	2809
Nitrogen	947
Neon	2005
Oxygen	1039
UF_6	279

Table A15 Estimated diameter of common gas molecules

Gas molecule	Kinetic diameter (Å)	Lennard–Jones diameter (Å)
Helium	2.60	2.55
Neon	2.75	2.82
Hydrogen	2.89	2.83
Nitrous oxide	3.17	3.49
Carbon dioxide	3.30	3.94
Acetylene	3.30	4.03
Argon	3.40	3.54
Oxygen	3.46	3.47
Nitrogen	3.64	3.80
Carbon monoxide	3.76	3.69
Methane	3.80	3.76
Ethylene	3.90	4.16
Propane	4.30	5.12
Propylene	4.50	4.68

Gas diameters can be determined as kinetic diameter based on molecular sieve measurements or estimated as Lennard–Jones diameters based on viscosity measurements. The absolute magnitude of the estimated diameters is not important, but the ratio of diameters can give a good estimate of the relative diffusion coefficients of different gas pairs [see Equation (8.4)]. On this basis the kinetic diameters do a better job of predicting the relative diffusion coefficients of carbon dioxide/methane (always greater than 1 and often as high as 5–10 in glassy polymers). However, the Lennard–Jones diameter does a better job of predicting the relative diffusion coefficients of propylene/propane (always greater than 1 and often as high as 5 in glassy polymers).

Table A16 Experimental diffusion coefficient of water in organic liquids at 20–25 °C at infinite dilution

Liquid	Temperature (°C)	Viscosity	$cm^2/s \times 10^5$
Methanol	20	—	2.2
Ethanol	25	1.15	1.2
1-Propanol	20	—	0.5
2-Propanol	20	—	0.5
1-Butanol	25	2.60	0.56
Isobutanol	20	—	0.36
Benzyl alcohol	20	6.5	0.37
Ethylene glycol	25	—	0.24
Triethylene glycol	30	30	0.19
Propane-1,2-diol	20	56	0.075
2-Ethylhexane-1,3-diol	20	320	0.019
Glycerol	20	1500	0.008
Acetone	25	0.33	4.6
Furfuraldehyde	20	1.64	0.90
Ethyl acetate	20	0.47	3.20
Aniline	20	4.4	0.70
n-Hexadecane	20	3.45	3.8
n-Butyl acetate	25	0.67	2.9
n-Butyric acid	25	1.41	0.79
Toluene	25	0.55	6.2
Methylene chloride	25	0.41	6.5
1,1,1-Trichloroethylene	25	0.78	4.6
Trichloroethylene	25	0.55	8.8
1,1,2,2-Tetrachloroethane	25	1.63	3.8
2-Bromo-2-chloro-1,1, 1-trifluoroethane	25	0.61	8.9
Nitrobenzene	25	1.84	2.8
Pyridine	25	0.88	2.7

Source: F.P. Lees and P. Sarram, *J. Chem. Eng. Data* **16**, 41 (1971).

Table A17 Diffusion coefficient of salts in water at 25 °C at infinite dilution

Salt	Diffusion coefficient ($cm^2/s \times 10^5$)
NH_4Cl	1.99
$BaCl_2$	1.39
$CaCl_2$	1.34
$Ca(NO_3)_2$	1.10
$CuSO_4$	0.63
LiCl	1.37
$LiNO_3$	1.34
$MgCl_2$	1.25
$Mg(NO_3)_2$	1.60
$MgSO_4$	0.85
KCl	1.99
KNO_3	1.89
K_2SO_4	1.95
Glycerol	0.94
NaCl	0.61
$NaNO_3$	1.57
Na_2SO_4	1.23
Sucrose	0.52
Urea	1.38

Source: Data correlated by Sourirajan from various sources in *Reverse Osmosis*, Academic Press, New York (1970).

Table A18 Interdiffusion of gases and vapors into air at 20 °C

Gas or vapor	Diffusion coefficient (cm^2/s)
O_2-Air	0.18
CO_2-Air	0.14
H_2-Air	0.61
H_2O-Air	0.22
n-Propyl alcohol	0.085
Ethyl acetate	0.072
Toluene	0.071
n-Octane	0.051

Source: Selected values from *International Critical Tables*, W.P. Boynton and W.H. Brattain.

Table A19 Interdiffusion of vapors into air, carbon dioxide or hydrogen

Gas/vapor	Diffusion coefficient (cm^2/s)		
	Air	CO_2	H_2
Oxygen	0.18	0.14	0.70
Water	0.22	0.14	0.75
Ethyl acetate	0.072	0.049	0.27
n-Propyl alcohol	0.085	0.058	0.32
Propyl butyrate	0.053	0.036	0.21

INDEX

Membrane Technology and Applications R. W. Baker

ISBN: 0-470-85445-6

D1337182

Henning Mankell

BEFORE THE FROST

Translated from the Swedish by
Ebba Segerberg

THE HARVILL PRESS
LONDON

Published by The Harvill Press, 2004
By arrangement with Leonhardt & Høier Literary Agency, Copenhagen

4 6 8 10 9 7 5

First published with the title *Innan frosten* by Leopard Förlag, Stockholm, 2002

First published in Great Britain in 2004 by
The Harvill Press
Random House
20 Vauxhall Bridge Road
London SW1V 2SA

Random House Australia (Pty) Limited
20 Alfred Street, Milsons Point, Sydney,
New South Wales 2061, Australia

Random House New Zealand Limited
18 Poland Road, Glenfield,
Auckland 10, New Zealand

Random House South Africa (Pty) Limited
Endulini, 5A Jubliee Road, Parktown 2193, South Africa

The Random House Group Limited Reg. No. 954009
www.randomhouse.co.uk/harvill

A CIP catalogue record for this book is available from the British Library

ISBN 1 84343 113 0 (hardback)
ISBN 1 84343 114 9 (paperback)

Map drawn by Reg Piggott

Papers used by Random House are natural, recyclable products made from wood grown in sustainable forests; the manufacturing processes conform to the environmental regulations of the country of origin

Typeset by Palimpsest Book Production Limited,
Polmont, Stirlingshire
Printed and bound in Great Britain by
Clays Ltd, St Ives plc

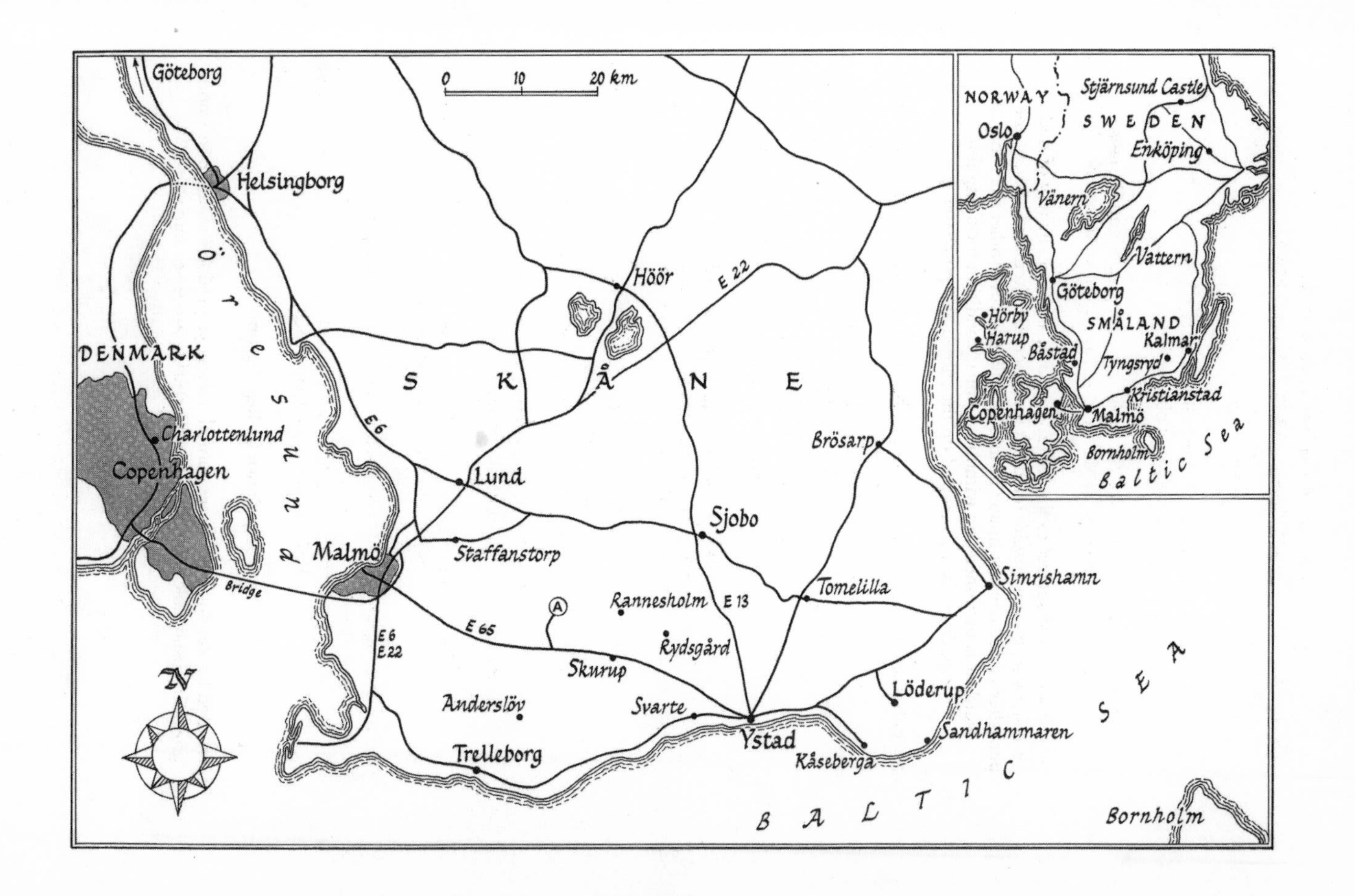

Göteborg
Helsingborg
0 10 20 km
Höör
E 22
DENMARK
Öresund
SKÅNE
Charlottenlund
Copenhagen
E6
Lund
Sjobo
Malmö
Staffanstorp
Brösarp
Bridge
Rannesholm
E 13
Tomelilla
Simrishamn
E 65
E6
E22
Rydsgård
Skurup
Svarte
Andersslöv
Trelleborg
Ystad
Löderup
Sandhammaren
Kåseberga
BALTIC SEA
Bornholm
N
NORWAY
Stjärnsund Castle
SWEDEN
Oslo
Enköping
Vänern
Vattern
Göteborg
Hörby
Harup
SMÅLAND
Båstad
Kalmar
Tyngsryd
Kristianstad
Copenhagen
Malmö
Bornholm
Baltic Sea

PROLOGUE

Jonestown, November 1978

His thoughts were like a shower of red-hot glowing needles in his head, an almost unbearable pain. He did his utmost to remain calm, to think clearly. The worst thing was fear. The fear that Jim would unleash his dogs and hunt him down, like the terrified beast of prey he had become. Jim's dogs: they were what he was most afraid of. All through that long night of November 18, when he had run until he was exhausted and had hidden among the decomposing roots of a fallen tree, he imagined that he could hear them closing in.

Jim never lets anyone escape, he thought. He seemed to me to be filled by an endless and divine source of love, but the man I have followed has turned out to be someone quite different. Unnoticed by us, he changed places with his shadow or with the Devil, whom he was always warning us about. The Devil of selfishness, who keeps us from serving God with obedience and submission. What appeared to me to be love turned into hate. I should have seen this earlier. Jim himself warned us about it time and again. He gave us the truth, but not all at once. It came slowly, a creeping realisation. But neither I nor anyone else wanted to hear it: the truth buried between the lines. It was my fault, I didn't want to see it. In his sermons and in all his teachings he did not only talk of the spiritual preparations we needed to undergo to ready ourselves for the Day of Judgment. He was also always telling us that we had to be ready to die.

He arrested the train of his thoughts and listened. Wasn't that the dogs barking? But no, it was only a sound inside him, generated by his fear. He went back in his confused and terrified mind to the apocalyptic events in Jonestown. He needed to understand what had happened.

Jim was their leader, shepherd, pastor. They had followed him in the exodus from California when they could no longer tolerate the persecution from the media and the state authorities. In Guyana, they were going to realise their dreams of a life of peaceful coexistence with nature and each other in God. And at first they had experienced something very close to that. But then it changed. Could they have been as threatened in Guyana as in California? Would they be safe anywhere? Perhaps only in death would they find the kind of protection they needed to construct the community they strove for. "I have seen far in my mind," Jim said. "I have seen much further than before. The Day of Judgment is near at hand and if we are not to perish in that terrible maelstrom we have to be ready to die. Only through physical death will we survive."

Suicide was the only answer. When Jim stood in the pulpit and mentioned it for the first time there was nothing frightening about his words. Initially parents were to give drinks laced with cyanide to their children; cyanide which Jim had stockpiled in plastic containers in a locked room at the back of his house. Then the grown-ups would take the poison. Those who were overcome with doubt in the final instance would be assisted by Jim and his closest associates. If they ran out of poison, they had guns. Jim would make sure that everybody was taken care of before he put the muzzle to his own head.

He lay under the tree, panting in the tropical heat. His ears strained to catch any sound of Jim's dogs, those large, red-eyed monsters that had inspired fear in all of them. Jim had told them that everyone in his congregation, everyone who had chosen to follow his path and come to Guyana, had no choice but to continue on the path laid out by God. The path which James Warren Jones had decided was the right one.

It had sounded so comforting. No-one else would have been able to make words like death, suicide, cyanide and weapons sound so beautiful and soothing.

He shivered. Jim has walked around and inspected the dead, he thought. He knows I am missing and he's going to send the dogs after me. The thought clawed its way out of his mind: the dead. Tears began to run down his face. For the first time, he took in the enormity of what had happened: Maria and the girl were dead, everyone was dead. But he did not want to believe it. Maria and he had talked about this in the small hours: Jim was no longer the same man they had once been drawn to, the one who had promised them salvation and a meaningful life if they joined the People's Temple. It was Maria who put her finger on it. "Jim's eyes have changed," she said. "He doesn't see us now. He looks past us and his eyes are cold, as if he wants nothing to do with any of us any more."

They spoke of running away together, but every morning they agreed that they could not abandon the path they had chosen. Jim would become his old self again. He was suffering some sort of crisis and it would soon be over; he was stronger than all of them. And without him they would never have had this brief experience of what seemed to them like heaven on earth.

There was one memory which stood out. It was from that time when the drugs, alcohol and guilt about leaving his little daughter had brought him close to ending it. He wanted to throw himself in front of a truck or train and then it would be over and no-one would miss him. During one of those last meandering walks through town, when he was saying goodbye to all the people who didn't care one way or another whether he lived or died, he happened to pass by the People's Temple. "It was God's plan," Jim said later. "He had already decided that you would be among the chosen, one of the few to experience His mercy." He didn't know what had made him walk up those steps and go into the building that looked nothing like a church. He still didn't know what it was, even now when he lay among the roots of a tree, waiting for Jim's dogs to track him down and tear him limb from limb.

He knew he should be making good his escape, but he did not leave his hiding place. He had abandoned one child already; he was not going to abandon another. Maria and the girl were still back there with the others.

What had really happened? They had got up as usual in the morning and gathered outside Jim's door. It stayed shut, as it so often had in the last days. They had therefore prayed without him, the 912 adults and the 320 children. Then they had left for their various jobs. He would never have survived had he not been one of a team given the task of finding two runaway cows. When he said goodbye to Maria and his daughter, he had no inkling of the terror to come. It was only when he and the other men reached the far side of the ravine that he understood that something was terribly wrong.

They had stopped dead in their tracks at the first sound of gunfire. And perhaps they heard human screams mingled with the chatter of the birds. They had looked at each other and then run back down towards the colony. He had become separated from the other men on the way back – possibly they had decided to flee rather than return. When he emerged from the shady forest and climbed the fence to the fruit orchards, everything was silent. Too silent. No-one was there, picking fruit. No-one at all was to be seen. He ran towards the houses, sure that something disastrous had occurred. Jim must have come out of his house this day with hate, not love, blazing from his eyes.

He had a cramp in his side and slowly shifted position, straining not to make any noise. What conclusion had he come to? As he ran through the fruit orchards, he tried to do what Jim had always taught them: to put his life in God's hands. He prayed as he ran. *Please, God, whatever may have happened, let Maria and the child be safe*. But God had chosen not to hear him.

In his desperation he started to believe that the shots he had heard from the ravine were the sounds of God and Jim taking aim at each other. When he came rushing into the dusty main street of Jonestown he half expected to catch the two of them at their duel. But God was nowhere to be seen. Jim Jones was there, the dogs barked like crazy in their cages and there were bodies everywhere. He could see at once that they were all dead. It was as if they had been struck down by a giant fist from the sky. Jim Jones and the six brothers who were his personal assistants and bodyguards had gone round and shot the children trying to crawl away from their parents' corpses. He ran among the dead, looking for Maria and the child, but without success.

It was when he shouted Maria's name out loud that he heard Jim calling him. He turned and saw his pastor aim a pistol at him. They were about 20 metres apart and between them, stretched out on the brown-burned ground, were the bodies of his friends, contorted as in their death throes. Jim pulled the trigger, but the shot missed. He ran before Jim had the chance to shoot again. He heard many shots fired and he heard Jim roar in rage, but he had not been hit and he made his stumbling way across the bodies and kept running until it was dark. He didn't know if he was the only survivor. Where were Maria and the girl? Why was he alone safe? Could one person escape the Day of Judgment? He didn't know, he only knew that this was no dream. This was all too real.

At dawn, heat began to rise like steam from the trees. That was when finally he realised that no dogs were coming. He crawled out from under the tree roots, shook his aching limbs and stood up. He started back towards the colony. He was dead tired and extremely thirsty. Everything was still very quiet. The dogs are dead, he thought. Jim must have meant it literally when he said no-one would escape judgment. Not even the dogs. He climbed over the fence and started running. The first of the dead he saw were those who had tried to escape. They had been shot in the back.

Then he stopped by the corpse of a familiar-looking man. Shaking, he bent and turned the body face up. It was Jim. His gaze has finally softened, he thought. And he's looking me straight in the eyes. He had a sudden impulse to strike him, to kick him in the face. But he quelled this violent urge and stood up. He was the only living soul among all these dead, and he could not rest until he found Maria and the girl.

Maria had tried to run; she had fallen forward when they shot her in the back. The girl was in her arms. He knelt beside them and cried. Now there's nothing left for me, he thought. Jim has turned our paradise into a hell.

He stayed with them until helicopters started circling over the area. He reminded himself of something Jim had told them shortly after they first came to Guyana, when life was still good. "The truth about a person can just as well be determined with the nose as with your eyes and ears," he had said. "The Devil hides inside people and the

Devil smells of sulphur. Whenever you catch a whiff of sulphur, raise the Cross for protection."

He didn't know what the future held, if anything. He didn't want to think about it. He wondered if he would ever be able to fill the void that God and Jim Jones had left behind.

PART I

The Eel Hunt

CHAPTER 1

The wind picked up shortly after 9.00 on the evening of August 21, 2001. In a valley to the south of the Rommele Hills, small waves were rippling across the surface of Marebo Lake. The man waiting in the shadows beside the water stretched out his hand to discover the direction of the wind. Virtually due south, he found to his satisfaction. He had chosen the right spot to put out food to attract the creatures he would soon be sacrificing.

He sat on the rock where he had spread out a sweater against the chill. It was a new moon and no light penetrated the thick layer of clouds. Dark enough for catching eels. That's what my Swedish playmate used to say when I was growing up. The eels start their migration in August. That's when they bump into the fishermen's traps and wander the length of the trap. And then the trap slams shut.

His ears, always alert, picked up the sound of a car passing some distance away. Apart from that there was nothing. He took out his torch and directed the beam over the shoreline and water. He could tell that they were approaching. He spotted at least two white patches against the dark water. Soon there would be more.

He switched off the light and tested his mind – exactingly trained – by thinking of the time. Three minutes past nine, he thought. Then he raised his wrist and checked the display. Three minutes past nine – he was right, of course. In another 30 minutes it would all be over. He had learned that humans were not alone in their need for regularity. Wild creatures could even be taught to respect time. It had taken him three months of patience and deliberation to prepare for tonight's sacrifice. He had made himself their friend.

He switched on the torch again. There were more white patches, and they were coming nearer to the shore. Briefly he lit up the tempting meal of broken bread crusts that he had set out on the ground, as well as the two petrol containers. He switched off the light and waited.

When the time came, he did exactly as he had planned. The swans had reached the shore and were pecking at the pieces of bread he had put out for them, oblivious of his presence or by now simply used to him. He set the torch aside and put on his night-vision goggles. There were six swans, three couples. Two were lying down while the rest were cleaning their feathers or still searching for bread.

Now. He got up, took a can in each hand and splashed the swans with petrol. Before they had a chance to fly away, he spread what remained in each of the cans and set light to a clump of dried grass among the swans. The burning petrol caught one swan and then all of them. In their agony, their wings on fire, they tried to fly away over the lake, but one by one plunged into the water like fireballs. He tried to fix the sight and sound of them in his memory; both the burning, screeching birds in the air and the image of hissing, smoking wings as they crashed into the lake. Their dying screams sound like broken trumpets, he thought. That's how I will remember them.

The whole thing was over in less than a minute. He was very pleased. It had gone according to plan, an auspicious beginning for what was to come.

He tossed the petrol cans into the water, tucked his jumper into the backpack and shone the torch around the place to be sure he had left nothing behind. When he was convinced he had remembered everything, he took a mobile phone from his coat pocket. He had bought the phone in Copenhagen a few days before.

When someone answered, he asked to be connected to the police. The conversation was brief. Then he threw the phone into the lake, put on his backpack and walked away into the night.

The wind was blowing from the east now and was growing stronger.

CHAPTER 2

It was the end of August and Linda Caroline Wallander wondered if there were any traits that she and her father had in common which yet remained to be discovered, even though she was almost 30 years old and ought to know who she was by this time. She had asked her father, had even tried to press him on it, but he seemed genuinely puzzled by her questions and brushed them aside, saying that she more resembled her grandfather. These "who-am-I-like?" conversations, as she called them, sometimes ended in fierce arguments. They kindled quickly, but they also died away almost at once. She forgot about most of them and supposed that he did too.

There had been one argument this summer which she had not been able to forget. It had been nothing really. They had been discussing their differing memories of a holiday they took to the island of Bornholm when she was little. For Linda there was more than this episode at stake; it was as if through reclaiming this memory she was on the verge of gaining access to a much larger part of her early life. She had been six, maybe seven years old, and both Mona and her father had been there. The idiotic argument had begun over whether or not it had been windy that day. Her father claimed she had been seasick and had thrown up all over his jacket, but Linda remembered the sea as blue and perfectly calm. They had only ever taken this one trip to Bornholm so it couldn't have been a case of their having mixed up several trips. Her mother had never liked boat journeys and her father was surprised she had agreed to this one holiday to Bornholm.

That evening, after the argument had ended, Linda had had trouble falling asleep. She was due to start working at the Ystad police station in two months. She had graduated from the police training college in Stockholm and would have much rather started working right away, but here she was with nothing to do all summer and her father couldn't keep her company since he had used up most of his holiday allowance in May. That was when he thought he had bought a house and would need extra time for moving. He had the house under contract. It was in Svarte, just south of the main road, right next to the sea. But the vendor changed her mind at the last minute. Perhaps because she couldn't stand the thought of entrusting her carefully tended roses and rhododendron bushes to a man who talked only about where he was going to put the kennel – when he finally bought a dog. She broke the contract and her father's agent suggested he ask for compensation, but he chose not to. The whole episode was already over in his mind.

He hunted for another house that cold and windy summer, but either they were too expensive or just not the house he had been dreaming of all those years in the flat on Mariagatan. He stayed on in the flat and asked himself if he was ever really going to move. When Linda graduated from the police training college, he drove up to Stockholm and helped her move her things to Ystad. She had arranged to rent a flat starting in September. Until then she could have her old room back.

They got on each other's nerves almost immediately. Linda was impatient to start working and accused her father of not pulling strings hard enough at the station to get her a temporary position. He said he had taken the matter up with Chief Lisa Holgersson. She would have welcomed the extra manpower, but there was nothing in the budget for additional staff. Linda would not be able to start until September 10, however much they might have wanted her to start sooner.

Linda spent the interval getting to know again two old school friends. One day she ran into Zeba, or "Zebra" as they used to call her. She had dyed her black hair red and also cut it short so Linda had not recognised her at first. Zeba's family came from Iran, and she and Linda had been in the same class until secondary school. When they bumped into each other on the street this July, Zeba had been pushing a toddler in a pushchair. They had gone to a café and had a coffee.

Zeba told her that she had trained as a barmaid, but her pregnancy had put a stop to her work plans. The father was Marcus. Linda remembered him, Marcus who loved exotic fruit and who had started his own plant nursery in Ystad at the age of 19. The relationship had soon ended, but the child remained a fact. Zeba and Linda chatted for a long time, until the toddler started screaming so loudly and insistently that they had to leave. But they had kept in touch since that chance meeting, and Linda noticed that she felt less impatient with the hiatus in her life whenever she managed to build these bridges between her present and the past that she had known in Ystad.

As she was going home to Mariagatan after her meeting with Zeba, it started to rain. She took cover in a shopping centre and – while she was waiting for the weather to clear up – she looked up Anna Westin's number in the directory. She felt a jolt inside when she found it. She and Anna had had no contact for ten years. The close friendship of their childhood had ended abruptly when they both fell in love with the same boy. Afterwards, when the feelings of infatuation were long gone, they had tried to resuscitate the friendship, but it had never been the same. Linda hadn't even thought much about Anna the last couple of years. But seeing Zeba again reminded her of her old friend and she was happy to discover that Anna still lived in Ystad.

Linda called her that evening and a few days later they met. Over the summer they would see each other several times a week, sometimes all three of them, but more often just Anna and Linda. Anna lived on her own as best as she could on her student budget. She was studying medicine.

Linda thought she was almost more shy now than when they were growing up. Anna's father had left home when she was five or six years old and they never once heard from him again. Anna's mother lived out in the country in Löderup, not far from where Linda's grandfather had lived and painted his favourite, unchanging motifs. Anna was apparently pleased that Linda had reestablished contact, but Linda soon realised that she had to tread carefully. There was something vulnerable, almost secretive about Anna and she would not let Linda come too close.

Still, being with her old friends helped to make Linda's summer go by, even though she was counting the days until she was allowed to pick up her uniform from fru Lundberg in the stockroom.

Her father worked flat out all summer, dealing with bank and post-office robberies in the Ystad area. From time to time Linda would hear about one case, which sounded like a series of well-planned attacks. Once her father had gone to bed, Linda would often sneak a look at his notebook and the case file he brought home. But whenever she asked him about the case directly he would avoid answering. She wasn't a police officer yet. Her questions would have to go unanswered until September.

The days went by. In the middle of one afternoon in August her father came home and said that the estate agent had called about a property near Mossbystrand. Would she like to come and see it with him? She called to postpone a rendezvous she had arranged with Zeba, then they got into her father's Peugeot and drove west. The sea was grey. Autumn was in the offing.

CHAPTER 3

The windows were boarded up, one of the drainpipes stuck out at an angle from the gutter, and several roof shingles were missing. The house stood on a hill with a sweeping view of the ocean, but there was something bleak and dismal about it. This is not a place where my father could find peace, Linda thought. Here he'll be at the mercy of his inner demons. But what are they, anyway? She began to list the chief sources of concern in his life, ordering them in her mind: first there was loneliness, then the creeping tendency to obesity and the stiffness in his joints. And beyond these? She put the question aside for the moment and joined her father as he inspected the outside of the house. The wind blew slowly, almost thoughtfully, in some nearby beech trees. The sea lay far below them. Linda squinted and spotted a ship on the horizon.

Kurt Wallander looked at his daughter.

"You look like me when you squint like that," he said.

"Only then?"

They kept walking and behind the house came across the rotting skeleton of a leather sofa. A field vole jumped from the rusting springs. Wallander looked around and shook his head.

"Remind me why I want to move to the country."

"I have no idea. Why *do* you want to move to the country?"

"I've always dreamed of being able to roll out of bed and walk outside to take my morning piss, if you'll pardon my language."

She looked at him with amusement. "Is that it?"

"Do I need a better reason than that? Come on, let's go."

"Let's walk round the house one more time."

This time she looked more closely at the place, as if she were the prospective buyer and her father the agent. She sniffed around like an animal.

"How much?"

"Four hundred thousand."

She raised her eyebrows.

"That's what it says," he said.

"Do you have that much money?"

"No, but the bank has pre-approved my loan. I'm a trusted customer, a policeman who has always been as good as his word. I think I'm even disappointed I don't like this place. An abandoned house is as depressing as a lonely person."

They drove away. Linda read a sign by the side of the road: Mossbystrand. He glanced at her.

"Do you want to go there?"

"Yes. If you have time."

This was where she had first told him of her decision to become a police officer. She was done with her vague plan to refinish furniture or to work in the theatre, as well as with her backpacking trips around the world. It was a long time since she had broken up with her first love, a boy from Kenya who had studied medicine in Lund. He had finally gone back to Kenya and she had stayed put. Linda had looked to her mother, Mona, to provide her with clues about how to live her own life, but all she saw in her mother was a woman who left everything half done. Mona had wanted two children and had only had one. She had thought that Kurt Wallander would be the great and only passion of her life, but she had divorced him and married a golf-playing retired banker from Malmö.

Eventually Linda had started looking more closely at her father, the detective chief inspector, the man who has always forgetting to pick her up at the airport when she came to visit. The one who never had time for her. She came to see that in spite of everything, now that her grandfather was dead, he was the one she was closest to. One morning just after she had woken up she realised that what she most wanted was to do what he did, to be a police officer. She had kept her thoughts to herself for a year and talked about it only with her boyfriend. When she was certain of it, she broke up with her boyfriend, flew down to Skåne, took her father to this beach and told him her news. He asked for a minute to digest what she had told him, which had made her suddenly unsure of herself. She had been convinced he would be happy. Watching his broad back, and his thinning hair blowing in the wind she had prepared for a fight, but when he turned and smiled at her, she knew.

They walked down to the beach. Linda poked her foot into some horse prints in the sand. Wallander looked at a gull hanging almost motionless above the sea.

"What are you thinking now?" she said.

"You mean about the house?"

"I mean about the fact that I'll soon be wearing a police uniform."

"It's hard for me even to imagine it. It will probably be upsetting for me, though I don't feel that way now."

"Why upsetting?"

"I know what lies in store for you. It's not hard to put the uniform on, but then to walk out in public is another thing. You'll notice that everyone looks at you. You become the police officer, the one who is supposed to jump in and take care of any trouble. I know what that feels like."

"I'm not afraid."

"I'm not talking about fear. I'm talking about the fact that from the first day you put on the uniform it will be, inescapably, in your life."

"How do you think I'll do?"

"You did well at the training college. You'll do well here. It's all up to you when it comes down to it."

They strolled along the beach. She told him she was going to go to Stockholm for a few days. Her graduating class was having a final party, a cadet ball, before everyone departed across the country to their new posts.

"We never had anything like that," Wallander said. "I didn't get much of an education, either. I still wonder how they sorted the applicants when I was young. I think they were interested in raw strength. You had to have some intelligence, of course. I do remember that I had quite a few beers with a friend after I graduated. Not in a bar, but at his place on South Förstadsgatan in Malmö."

He shook his head. Linda couldn't tell whether the memory amused or pained him.

"I was still living at home," he said. "I thought Dad was going to keel over when I came home in my uniform."

"How come he hated it so much – you becoming a police officer?"

"I think I only worked it out after he died. He tricked me."

Linda stopped. "Tricked you?"

He looked at her, smiling.

"Well, what I think now is that it was actually fine with him that I chose to be a policeman, but instead of telling me straight out, it amused him to keep me on my toes. And he certainly managed that, as you remember."

"You really believe that?"

"No-one knew him better than I did. I think I'm right. He was a scoundrel through and through. Wonderful, but a scoundrel. The only father I had."

They walked back to the car. The clouds were breaking up and it was getting warmer. Wallander looked at his watch as they were leaving.

"Are you in a hurry?" he said.

"I'm in a hurry to start working, that's all. Why do you ask?"

"There's something I should look into. I'll tell you about it as we go."

They turned on to the road to Trelleborg and turned off by Charlottenlund castle.

"I wanted to drive past since we were in the neighbourhood."

"Drive past what?"

"Marebo Manor, or – more precisely – Marebo Lake."

The road was narrow and windy. Wallander told her about it in a somewhat disjointed and confusing way. She wondered if his written reports were as disorganised as the summary she was getting.

Yesterday evening a man had called the Ystad police. He had given neither a name nor a location and spoke with an odd accent. He had said that burning swans were flying over Marebo Lake. When the officer on duty had asked him for more details, the man hung up. The conversation was duly logged, but no-one had followed it up because there had been a serious assault case in Svarte that evening, as well as two break-ins in central Ystad. The officer in charge had decided that it was most probably a hoax call, or possibly a hallucination, but when

Wallander heard of it from Martinsson he decided it was so bizarre that there might be some truth in it.

"Setting light to swans? Who would do a thing like that?"

"A sadist. Someone who hates birds."

"Do you honestly think it happened?"

Wallander turned on to a road signposted to Marebo Lake and took his time before answering.

"Didn't they teach you that at the training college? That policemen don't *think* anything, they only want to know. But they have to remain open to every possibility, however improbable. Which would include something like a report about swans on fire. So, yes, it could be true."

Linda didn't ask any more questions. They left the car park and headed down to the lake. Linda trailed behind her father and felt as if she was already wearing a uniform.

They walked round the lake but found no trace of a dead swan. Nor did they see that their progress was all the way observed through the lens of a telescope.

CHAPTER 4

A few days later Linda flew to Stockholm. Zeba had helped her make a dress for the ball. It was light blue and cut low across her chest and back. The class organisers had hired a big room on Hornsgatan. All 68 of them were there, even the prodigal son of the year, who had not managed to hide his drinking problem. No-one knew who had blown the whistle on him, so in a way they all felt responsible. Linda thought he was like their ghost; he would always be out there in the autumn darkness with a deep-seated longing to be forgiven and welcomed into their circle again.

On this occasion, their last chance to take their leave of each other and their teachers, Linda drank far too much wine. She wasn't a novice drinker by any means, but she could usually pace herself. This evening she knew she was drinking too much. She felt more impatient than ever to start working as she talked with student colleagues who had already been able to take the plunge. Her best friend from the training college, Mattias Olsson, had chosen not to go home to Sundsvall but to take a job in Norrköping. He had already distinguished himself by felling a weight-lifting idiot who had run amok from the effects of all the steroids in his body.

There was dancing, and speeches, and occasionally a comic song satirising the teachers. Linda's dress received many compliments. It would have been an altogether enjoyable evening had there not been a television set in the kitchen.

Someone heard on the late-night news that a police officer had been shot on the outskirts of Enköping. This news quickly spread among the dancing, intoxicated cadets and their teachers. The music was turned off and the television set brought out from the kitchen. Afterwards, Linda thought it was as if everyone had been kicked in the stomach. The party was over. They sat there in their long gowns and dark suits and watched footage of the crime scene as well as images of the officer who had been murdered. It had been a cold-blooded killing as he and his partner tried to question the driver of a stolen vehicle. Two men had jumped from the vehicle and opened fire on the policemen with automatic weapons. No warning shot had been fired. Their intention had clearly been to kill.

Linda was on her way to her Aunt Kristina's flat when she stopped at Mariatorget and called her father. It was after 3.00 and she could tell that he was barely awake. For some reason that made her furious. How could he sleep when a colleague had just been killed? That was also what she said to him.

"My not sleeping won't help anybody. Where are you?"

"On my way to Kristina's."

"The party went on until now? What time is it?"

"Three. It ended when we heard the news."

She heard him breathing heavily, as if his body had still not decided to become fully awake.

"What's that noise in the background?"

"Traffic. I'm looking for a taxi."

"Who's with you?"

"No-one."

"Are you crazy? You can't run around alone in Stockholm at this hour!"

"I'm fine, I'm not a child. Sorry I woke you." She hung up on him. This happens way too often, she thought. He has no idea how infuriating he is.

She flagged down a taxi and was driven to Gärdet where Kristina, her husband and 18-year-old son lived. Kristina had made up the sofa bed in the living room for her. Light from the street lamps penetrated the curtains. There was a photograph of Linda with her father and mother on the bookcase. She remembered when the picture was taken; she was 14, it was sometime in the spring and they had driven out to her grandfather's house in Löderup. Her father had won the camera in some office raffle and then, when they were about to take a family picture, her grandfather had suddenly baulked and locked himself in his studio. Her father had been extremely put out and Mona had sulked. Linda was the one who tried to convince her grandfather to come out and be in the picture.

"I won't have my picture taken with those two people and their fake smiles when I know they're about to leave each other," he said.

She could remember to this day how that had hurt. She knew how insensitive he could sometimes be, and the words still felt like a slap in the face. When she had collected herself she asked him if it was true what he had said, if he knew something she didn't.

"It won't help matters if you keep turning a blind eye," he said. "Go on. You're supposed to be in that picture. Maybe I'm wrong about all this."

Her grandfather was often wrong, but not in this case. And he had refused to be in that picture, which they took with the self-timer on the camera. The following year – the last year her parents lived together – the tensions in their home only grew.

That was the year she had tried to commit suicide. Twice. The first time, when she had slit her wrists, it was her father who found her. She remembered how frightened he had looked. But the doctors must have reassured him, since he and her mother said very little about it. Most of what they communicated was through looks and eloquent silences. But it propelled her parents into the last series of violent disputes which finally persuaded Mona to pack her bags and leave.

Linda had often thought it remarkable that she hadn't felt responsible for the break-up, but in fact she felt she had done them a favour, helping to catapult them out of a marriage that had ended in all but name a long time ago.

He didn't even know about the second time.

That was the biggest secret she kept from her father. Sometimes she supposed he must have heard about it, but in the end she was sure that he had never found out. The second time she tried to kill herself, it was for real.

She was 16, and had gone to stay with her mother in Malmö. It was a time of crushing defeats, the kind only a teenager can experience. She hated herself and her body, shunning the image she saw in the mirror while strangely enough also welcoming the changes she was undergoing. The depression hit her out of nowhere, beginning as a set of symptoms too vague to take seriously. Suddenly it was a fact and her mother had had absolutely no inkling of what was going on. What had most shaken Linda was that Mona had said no when she pleaded to be allowed to move to Malmö. It wasn't that there was anything wrong with her father, just that she wanted to get out of Ystad. But Mona had been unusually cool.

Linda had left the flat in a rage. It had been a day in early spring when there was still snow lying here and there. The wind blowing in from the sound had a sharp bite. She wandered along the city streets, not noticing where she was going. When she looked up she was on a bridge over the highway. Without really knowing why, she had climbed on to the railing and stood there, swaying. She looked down at the cars rushing past below, their sharp lights slicing the gloom. She wasn't aware of how long she stood there. She felt no fear or self-pity, she simply waited for the cold and the fatigue spreading in her limbs to get her finally to step out into the void.

Suddenly there was someone by her side, speaking in careful, soothing tones. It was a woman with a round, childish face, perhaps not so much older than Linda. She was wearing a police uniform and behind her there were two patrol cars with flashing lights. Only the officer with the childish face approached her. Linda sensed the presence of others further back, but they had clearly delegated the responsibility of talking that crazy teenager out of jumping to this one woman. She told Linda her name was Annika, that she wanted her to come down, that jumping wouldn't solve anything. Linda started defending herself – how could Annika possibly understand anything about her problems? But Annika hadn't backed down, she had stayed calm, as if she had infinite patience. When Linda finally did climb down from the railing and start crying, from a sense of disappointment that was actually relief, Annika had started crying too. They hugged each other and stood there for a long time. Linda told her that she didn't want her father to hear about it. Nor her mother for that matter, but especially not her father. Annika had promised to keep it under wraps and she had been true to her word. Linda had many times thought of calling the Malmö police station to thank her, but she never got further than lifting the receiver; she had never dialled the number.

* * *

She put the photograph back on the bookcase, thought briefly about the police officer who had been killed, and got ready for bed. She was woken in the morning by Kristina, who was getting ready for work. Kristina was her brother's opposite in almost every particular: tall, thin, with a pointed face and a shrill voice that Linda's father made fun of behind her back. But Linda loved her aunt: there was something refreshingly uncomplicated about her, and in this way too she was her brother's opposite. From his perspective, life was nothing but a heap of dense problems, the ones in his private life, unsolvable, all to be addressed with the force and fury of a ravenous bear in his work.

Linda took the bus to the airport shortly before 9.00 in the hope of getting a flight to Malmö. All the papers' headlines were of the murdered officer. She finally got a plane at noon for Sturup. Her father came to pick her up.

"Did you have a good time?" he said.

"What do you think?"

"How could I know? I wasn't there."

"But we talked last night – remember?"

"Obviously I remember. You were very unpleasant."

"I was tired and upset. A policeman was murdered. It destroyed the whole party atmosphere. No-one was in the mood after that."

He grunted, but he didn't say anything. He let her off when they got to Mariagatan.

"Have you found out anything more about this sadist?" she said.

He didn't understand what she was referring to.

"The bird hater. The burning swans."

"Probably just a hoax call. Quite a few people live by the lake. Someone would have seen something if it had been true."

Wallander drove back to the station and Linda went up to the flat. Her father had left a note by the phone. It was a message from Anna: *Important. Call soon.* Then her father had scribbled something she couldn't read. She called him at work.

“Why didn’t you tell me Anna called?”

“I forgot.”

“What have you written here? I can’t read your writing.”

“She sounded worried about something.”

“How do you mean?”

“Just that. She sounded worried. You’d better call her.”

Linda called, but Anna’s line was busy. When she tried again there was no answer. At 7.00, after she and her father had eaten, she put on her coat and walked over to Anna’s place. As soon as Anna opened the door, Linda could see what her father had meant. Anna’s expression was changed; her eyes darted anxiously; she pulled Linda into the flat and closed the door. It was as if she were in a hurry to keep the outside world at bay.

CHAPTER 5

Linda was reminded of Anna's mother, Henrietta. She was a thin woman with a bird-like, nervous way of moving and Linda had always been a little afraid of her.

Linda remembered the first time she had played at Anna's house. She may have been eight. Anna was in another class at school and they had never been sure what had drawn them to each other. It's as though there's an unseeable force that brings people together, she thought. At least, that's the way it was with us. We were inseparable, until we fell in love with the same boy.

Anna's father had only ever been present in pale photographs. Henrietta had carefully wiped away all traces of him, as if she were telling her daughter that there was no possibility of his return. The few photographs Anna had were hidden away in a drawer, under some socks and underwear. In the pictures he had long hair, glasses and a reluctant stance, as if he hadn't really wanted to pose for the camera. Anna had shown her the pictures as a mark of deepest confidence. When they became friends her father had already been gone for two years. Anna quietly rebelled against her mother's determination to keep the flat free of every trace of him. One time Henrietta had gathered up what remained of his clothes and stuffed them in a rubbish bag in the basement. Anna had gone down at night and rescued a shirt and some shoes, which she hid under her bed. For Linda this mysterious father had been a figure of adventure. She had often wished that she and Anna could trade places, that she could exchange her quarrelling parents for this man who had vanished one day like a grey wisp of smoke against a blue sky.

They sat on the sofa and Anna leaned back so that half of her face was in shadow.

"How was the ball?"

I know what she's doing, Linda thought. Whenever Anna has anything important to talk about she never comes out and says it. It takes time.

"We heard about the murdered police officer in the middle of it and that pretty much ended it right there. But my dress was a success. How is Henrietta?"

"Fine." Anna shook her head at her own words. "Fine – I don't know why I always say that. She's actually worse than ever. For the past two years she's been composing a requiem for herself. She calls it 'The Unnamed Mass' and she's thrown the thing into the fire at least twice. She managed to salvage most of the papers both times, but her self-esteem is about as low as it would be for a person with only one tooth left."

"What kind of music does she write?"

"I hardly know. She's tried to hum it for me – the few times she's been convinced that what she was working on had any merit. But it doesn't sound anything like a melody to me. It's the sort of music that sounds more like screams, that pokes and hits you. I can't imagine why anyone would ever listen to something like that. At the same time, I can't help but admire that she hasn't given up. I've tried to persuade her to do other things in life. She's not fifty yet. But each time she's reacted like a scalded cat. I wonder if she isn't a little mad."

Anna interrupted herself at this point as if she were afraid of having said too much. Linda waited for her to go on.

"Have you ever had the feeling you were going crazy?"

"Every day of the year."

Anna frowned. "No, not like that. I'm serious."

Linda was ashamed of her light-hearted comment.

"It happened to me once. You know all about that."

"When you slit your wrists . . . and when you tried to jump off the overpass. That's despair, Linda, it's not the same thing. Everyone has to face up to despair at least once in their life. It's a rite of passage. If you never find yourself raging at the sea or the moon or your parents you never really have the opportunity to grow up. The King and Queen of Contentment are damned in their own way. They have let their souls become numbed. Those of us who want to stay alive have to stay in touch with our sorrow and grief."

Linda had always envied Anna's fanciful way of expressing herself. I would have had to sit and write it all down if I was to come up with anything like that, she thought. The King and Queen of Contentment, forsooth.

"In that case, I suppose I've never really been afraid of losing my mind," she said lightly.

Anna got up and walked to the window. After a while she returned to the sofa. We're much more like our parents than we realise, Linda thought. I've seen Henrietta move in just that way when she's anxious: get up, walk around and then sit down again.

"I thought I saw my father yesterday," Anna said. "On a street in Malmö."

Linda raised her eyebrows. "Your father? You saw him on a street?"

"Yes."

Linda thought about it. "But you've never even seen him – not really, I mean. You were so young when he left."

"I have pictures of him."

Linda did the calculation. "It's been twenty-five years since he left."

"Twenty-four."

"OK, twenty-four. How much do you think a person changes in that many years? You can't know."

"It was him. My mother told me about his gaze. I'm sure it was him. It must have been him."

"I didn't even know you were in Malmö yesterday. I thought you were going to Lund, to study or whatever it is you do there."

Anna looked at her appraisingly. "You don't believe me."

"You don't believe it yourself."

"It was my dad." She took a deep breath. "You're right. I had been in Lund. When I got to Malmö, I had to change trains, there was a problem on the line. The train was cancelled and I had two hours to kill until the next one. It put me in a foul mood, since I hate waiting. I walked into town without any clear idea of what I was going to do, just to use up some of the unwanted, irritating time. Somewhere along the way I walked into a shop and bought a pair of socks I didn't even need. As I was walking past the St Jörgen Hotel a woman had fallen in the street. I didn't walk up close – I can't stand the sight of blood. Her skirt was bunched up and I remember thinking, why had no-one pulled it down for her? I was sure she was dead. A knot of onlookers had gathered, as if she were some creature washed up on the beach. I walked away, through the Triangle and into the big hotel there, to take their glass lift up to the roof. That's something I always do when I'm in Malmö. It's like taking a glass balloon up into the sky. But I wasn't allowed – now you can only operate the lift with your room key. That was a blow. It felt as if someone had taken away a toy. I sat in one of the plush armchairs in the lobby and looked out of the window. I was planning to stay there until it was time to walk back to the station.

"That's when I saw him. He was in the street. Gusts of wind were making the window pane rattle. I looked up and there he was on the pavement, looking at me. Our eyes met and we stared at each other for about five seconds. Then he looked away and walked. I was so shocked it didn't even cross my mind to follow him. To be perfectly honest, I didn't believe I really had seen him. Maybe I was hallucinating or it was a trick of the light. Sometimes you see someone and you think it's a person from your past, but it's just a stranger. When at last I did run out and look for him, he was gone. I felt a bit like an animal stalking its prey as I walked back to the train station – I tried to sniff out where he could be. I was so excited – upset, actually – that I hunted through the inner city and missed my train. He was nowhere to be found. But I was sure that it was him. He looked just as he did in the picture I have. And my mother once said he had a habit of first looking up before he said anything. I saw him make that exact gesture when he was standing on the other side of the window. When he left all those years ago he had long hair and thick, black-rimmed glasses; he doesn't look like that now. His hair is much shorter and his glasses are the kind without a frame round the lenses.

"I called you because I needed to talk to someone about it. I thought I would go nuts otherwise. It *was* him, it *was* my father. And it wasn't just that I recognised him; he had stopped on the pavement outside because *he* had recognised *me*."

Anna spoke with total conviction. Linda tried to remember what she had learned about eye-witness accounts – about the rate of accuracy in their reconstruction of events and the potential for embellishment. She also thought about what they had been taught about giving descriptions at the training college, and the computer exercises they had done. One assignment consisted of aging their own face by 20 years. Linda had seen how she started looking more and more like her father, even a little like her grandfather. Our ancestors survive somewhere in our faces, she thought. If you look like your mother as a child, you end up as your father when you age. When you no longer recognise your face, it's because an unknown ancestor has taken up residence for a while.

Linda found it hard to accept that Anna had indeed seen her father. He would hardly have recognised the grown woman his little girl had become, unless he had been discreetly following her development all these years. Linda ransacked her memory for what she knew about the mysterious Erik Westin. Anna's parents had both been young when she was born. They had both grown up in big cities but been beguiled by the green wave and had ended up in a collective in the isolated countryside of Småland. Linda had a vague memory that Westin was a handyman, that he specialised in making orthopaedic sandals. She had also heard Henrietta describe him as an impossible, hashish-smoking loser whose sole object in life was to do as little as possible and who had no sense what it meant to take responsibility for a child. And what had made him leave? He had left no letter, nor any signs of extensive preparation. The police had looked for him at first, but there had never been any indication of a crime and eventually they shelved the case.

Westin's disappearance must, nonetheless, have been carefully choreographed. He had taken his passport and what little money he had – most of it left over from selling their car, which had actually belonged to Henrietta. She was the only one with an income at the time, working as a night warden at a hospital.

Westin was there one day and gone the next. He had left without warning on trips before, so Henrietta waited two weeks before contacting the police. Linda recalled that her father had been involved in the investigation. Westin had no record – no previous arrests or convictions, nor any history of mental illness. A few months before he disappeared, he had had a complete physical check-up and been given a clean bill of health, apart from a mild anaemia.

Linda knew from her studies of police statistics that most missing persons turned up again. Of those who didn't, the majority were suicides. Only a few were the victims of crime, buried in unknown places, or decomposing at the bottom of a lake with weights attached to them.

"Have you told your mother?"

"Not yet."

"Why not?"

"I don't know. I think I'm still in shock."

"I'd say that you're not a hundred per cent convinced it was him."

Anna looked at her with pleading eyes. "I know it was him. If it wasn't, my brain has suffered a major short-circuit. That's why I asked you if you've ever been afraid you were going crazy."

"But why would he come back now, after twenty-four years? Why would he turn up in Malmö and look at you through a hotel window – how did he know you were there?"

"I've no idea."

Anna got up, made the same journey to the window and back.

"Sometimes I wonder if he really disappeared at all. Maybe he just chose to make himself invisible."

"But why would he have done that?"

"Because he wasn't up to it, to life. I don't mean just his responsibilities for me or my mum. He was probably looking for something more. That search drove him away from us. Or perhaps he was only trying to get away from himself. There are people who dream about being like a snake; about shedding their outer skin from time to time. But maybe he's been here all along, much closer than I realised."

"You wanted me to listen to you and then tell you what I thought. You say you're sure it was him, but I can't accept it. It's too much like a childhood fantasy, that he would suddenly reappear in your life. I'm sorry, but twenty-four years is too long."

"I know it was him. It was my dad. I'm not wrong about this."

They had reached an impasse. Linda sensed that Anna wanted to be left alone now, just as earlier she had craved company.

"At the same, I think you should tell your mum," Linda said, as she got up to leave. "Tell her that you saw him, or someone you thought was him."

"You're not going to believe me."

"What I believe is neither here nor there. You're the only one who knows what you saw. But you have to admit it sounds far-fetched. I'm not saying that I think you're making it up, obviously, you have no reason to. I'm just saying it's way against the odds for someone who's been gone for as long as your dad has to turn up again, that's all. Sleep on it and we'll talk about it tomorrow. I can come around five. Does that work for you?"

"I know it was him."

Linda frowned. There was something in Anna's voice that sounded shrill and hollow. Is she making it up after all? Linda thought. There's something that doesn't ring true in this. But why would she lie to me?

Linda walked home through the deserted streets. Outside the cinema on Stora Östergatan some teenagers were standing, chatting in a group. She wondered if they could see her invisible uniform.

CHAPTER 6

The next day Anna Westin disappeared without trace. Linda knew that something was wrong the moment she rang her doorbell at 5.00 and no-one answered. She rang a few more times and shouted Anna's name through the letter box, but she wasn't there. After 30 minutes, Linda took out a set of pass keys that one of her fellow students had given her. He had bought a collection of them on a trip to the United States and given them out to all his friends. In secret they had then spent a number of hours learning to use them. Linda was by now adept at opening any standard lock.

She picked Anna's lock without difficulty and walked in. She went quietly through the empty, well-kept rooms. Nothing looked amiss, the washing-up was done and the dish towels hanging neatly on the rack. Anna was orderly by nature. The fact that she wasn't here meant that something had happened. But what? Linda sat down on the sofa in the same spot as the previous night. Anna believes she saw her father, she thought. And now she disappears herself. These two events are obviously connected – but how? Linda sat for a long time, ostensibly trying to think it through, but in reality she was simply waiting for Anna to come back.

The day had started early for Linda. She had left for the police station at 7.30 to meet Martinsson, one of Inspector Wallander's oldest colleagues, who had been assigned to be her supervisor. They were not going to be working together as Linda – along with the other recruits – would be assigned to a patrol car with experienced colleagues. But Martinsson was the senior officer she would turn to with questions should the need arise. Linda remembered him from when she was little. Martinsson had been a young man then. According to her father,

Martinsson had often thought of quitting. Wallander had managed to talk him out of it on at least three separate occasions over the last ten years.

Linda had asked her father if he had had anything to do with Lisa Holgersson's decision to assign Martinsson as her mentor, but he vehemently denied any involvement. His intention was to stay as far removed from matters concerning her work as possible, he said. Linda had accepted this declaration with a pinch of salt. If there was one thing she feared, it was precisely that he would interfere with her work. That was the reason she had hesitated so long before deciding to apply for work in Ystad. She had thought about working in other areas of the country, but this was where she had ended up. In retrospect, it seemed fated.

Martinsson met her at reception and took her to his office. He had a picture of his smiling wife and their two children on his desk. Linda wondered whose picture she was going to have on hers. That decision was part of the everyday reality waiting for her. Martinsson started by talking about the two officers she was to be working with.

"They're both fine officers," he said. "Ekman can give the impression of being a bit tired and worn out, but no-one has a better grasp of police work than he does. Sundin is his polar opposite and focuses on the little things. He still tickets people for jumping a red light. But he knows what it means to be a policeman. You'll be in good hands."

"What do they say about working with a woman?"

"If they have anything to say about that, ignore them. It's not how it was even ten years ago."

"And my father?"

"What about him?"

"What do they say about my being his daughter?"

Martinsson waited a moment before answering. "There are probably one or two people in the station who would be happy to see you fall on your face. But you must have known that."

Then they talked at length about the state of affairs in the Ystad police district. The "state of affairs" was something Linda had heard about at home from early childhood, as she played under the table with the sounds of clinking glass above and the voice of her father and a colleague discussing the latest difficulties. She had never heard of a positive "state of affairs", there was invariably something to lament. A shipment of sub-standard uniforms, detrimental changes in patrol cars or radio systems, a rise in crime statistics, poor recruitment levels and so on. In fact, this continuing discussion of the "state of affairs", of how this era was different from the era before, seemed to be central to life on the force. But it's not an art they taught us at the training college, Linda thought. I know a lot about how to break up a fight in the main square and very little about pronouncing judgment on the general "state of affairs".

They went to the canteen for a cup of coffee. Martinsson's assessment of the present situation was concise: there were too few officers working in the field.

"Crime has never paid as well as it does today, it seems. I've been researching this. To find an equivalent level of what we may call successful crime you have to look as far back as the fifteenth century, before Gustav Vasa pulled the nation together. In that time, the time of small city states, there was widespread lawlessness and criminality just as there is today. We're not upholding law and order now, we're only in the business of trying to keep the growth of lawlessness from getting any more rapid."

Martinsson escorted her to the reception area.

"I hope all this talk hasn't depressed you. We certainly don't need more demoralised police officers – in order to be a good police officer you'll need all the courage and faith you can muster. A cheerful disposition also helps."

"Like my dad?"

Martinsson smiled.

"Kurt Wallander is very good at his job," he said. "You know that. But he's not renowned for being the life of the party around here, as I'm sure you've already worked out."

Before she left, he asked her about her reaction to the murdered officer. She told him about the cadet ball, the television in the kitchen and its effect on the festivities.

"It's always a blow," Martinsson said. "It affects all of us, as if we're suddenly surrounded by invisible guns out there, all aimed at our heads. Whenever a colleague is killed, many of us think about leaving the force, but in the end most people choose to stay. I'm one of them."

Linda left the station and walked to the apartment building in the eastern part of Ystad where Zeba lived. She thought about what Martinsson had told her – and what he had left out. That was something her father had drilled into her: always listen for what is not being said. It could prove to be the most important thing. But she didn't find anything like that in her conversation with Martinsson. He strikes me as the simple and straightforward type, she thought. Not someone who tries to read people's invisible signals.

She stayed with Zeba only briefly because her son had a stomach ache and cried the whole time. They decided to meet over the weekend when they would be able to have some peace and quiet to discuss the cadet ball and the success of Zeba's handiwork.

But August 27 did not go down in Linda's memory as the day on which she had her meeting with Martinsson, her mentor. It was filed away as the day that Anna vanished. After Linda had let herself into the flat by picking the lock, she sat on the sofa and tried to recall Anna's voice telling her about the man who had met her gaze through the hotel window and who bore a striking resemblance to her father. Anna had insisted that it was her father, but what was it she had really been trying to say? That was like Anna – she sounded convincing even when she was claiming as facts things that were imagined or invented. But she was never late for an appointment nor would she forget a date with a friend.

Linda walked once more through the flat, stopping in front of the bookcase by Anna's desk in the dining room. The books were mostly novels, she determined as she read the titles, and one or two travel guides. Not a single medical textbook. Linda frowned. The only thing even close to a course book was a volume on common ailments, the kind of encyclopaedia a lay person would have. There's something missing, she thought. This doesn't look like the bookcase of a medical student.

She proceeded to the kitchen and took note of the contents of the refrigerator. There were the usual food items, a sense of future use suggested by an unopened carton of milk with a sell-by date of September 2. Linda went back to the living room and pondered the question of why a medical student would have no textbooks in her bookcase. Did she keep them elsewhere? That made no sense, she lived in Ystad, and she had often told Linda she did most of her studying here.

Linda waited. At 7 p.m. she called her father, who answered with his mouth full.

"I thought you were coming home for dinner tonight," he grumbled.

Linda hesitated before replying. She was torn between wanting to tell him about Anna, and saying nothing.

"Something came up."

"What is it?"

"Something personal."

Her father growled on the other end.

"I had a meeting with Martinsson today."

"I know."

"How do you know?"

"He told me – only that you met, nothing else. Don't worry."

Linda went back to the sofa. At 8.00 she called Zeba and asked if she knew where Anna could be, but Zeba hadn't heard from her in several days. At 9.00, after she had helped herself to food from Anna's small pantry and fridge, she dialled Henrietta's number. The phone rang several times before she answered. Linda approached the subject as carefully as possible so as not to alarm the already fragile woman. Did she know if Anna was in Lund? Had she planned a trip to Malmö or Copenhagen? Linda asked the most harmless questions she could think of.

"I haven't talked to her since Thursday."

That's four days ago, Linda thought. That means Anna never told her about the man she saw through the window.

"Why do you need to know where she is?"

"I called her and there was no answer."

She sensed a tinge of anxiety on the other end.

"But you don't call me every time Anna doesn't answer her phone."

Linda was prepared for this question. "I had a sudden impulse to ask her over for dinner tonight. That was all." Linda steered the conversation to her own life. "Have you heard I'm going to start working here in Ystad?"

"Yes, Anna told me. But neither of us understands why you would want to be a policewoman."

"If I'd gone on learning how to refinish furniture as planned, I'd always have tacks in my mouth. A life in law enforcement just seemed more entertaining."

A clock struck somewhere in Anna's flat and Linda quickly ended the conversation. Then she thought it all through again. Anna wasn't a risk-taker. In contrast to Zeba and herself, Anna hated rollercoasters, was suspicious of strangers, and would never get into a taxi without looking the driver in the eye first. The simplest explanation was that she was still disconcerted by what she thought she had seen. She must have gone back to Malmö to look for the man she believed to be her father. This is the first time she's ever stood me up, Linda thought. But this is also the first time she's been convinced she saw her father.

Linda stayed in the flat until midnight.

There was no innocent explanation for Anna's absence. Something must have happened. But what?

CHAPTER 7

When Linda came back home shortly after midnight her father was asleep on the sofa. He woke at the sound of the closing door. Linda eyed the curve of his belly with disapproval.

"You're getting fatter," she said. "One day you're just going to go pop. Not like an old troll who wanders out into the sunshine, but like a balloon when it gets too full."

He pulled his dressing gown protectively across his chest.

"I do the best I can."

"No, you don't."

He sat up heavily. "I'm too tired to have this conversation," he said. "When you walked in I was in the middle of a beautiful dream. Do you remember Baiba?"

"The one from Latvia? Are you still in touch with each other?"

"About once a year, no more. She's found someone else, a German engineer who works at the municipal waterworks in Riga. She sounds very much in love when she talks about him, the wonderful Herman from Lubeck. I'm surprised it doesn't drive me insane with jealousy."

"You were dreaming about her?"

He smiled. "We had a child in the dream," he said. "A little boy building castles in the sand. An orchestra was playing in the distance and Baiba and I just stood there watching him. In my dream I thought, 'This is no dream, this is real', and I was incredibly happy."

"And you complain about having too many nightmares."

He wasn't listening.

"The door opened – that was you, of course – a car door. It was summer and very warm. The whole world was full of light like an over-exposed picture. Everyone's face was white and without shadows. It was beautiful. We were about to drive away when I woke up."

"I'm sorry."

He shrugged. "It was just a dream."

Linda wanted to tell him about Anna but her father lumbered out into the kitchen and drank some water from the tap. Linda followed him and when he was done he stood up and looked at her, smoothing his hair down at the back.

"You were out late. It's none of my business, I know, but I have an idea that you want me to ask you about it."

So Linda told him. He leaned against the refrigerator with his arms crossed. This is how I remember him from my childhood, she thought. This is how he always listened to me, like a giant. I used to think my father was as big as a mountain. Daddy Mountain.

He shook his head when she had finished. "That's not how it happens, not how people disappear."

"It's not like her. I've known her since I was seven. She's never been late for anything."

"However idiotic it sounds, sometime has to be the first. Let's say she was preoccupied by the fact that she thought she had seen her father. It's not unlikely – as you suggested yourself – that she went back to look for him."

Linda nodded. He was right. There was no reason to assume anything had happened.

Wallander sat down on the sofa.

"You'll learn that all events have their own logic. People kill each other, lie, break into houses, commit robberies, and sometimes they simply disappear. If you winch yourself far enough down the well – that's how I often think of my investigations – you'll find the explanation. It turns out that it was highly probable that such and such a person disappeared, that another robbed a bank. I'm not saying the unexpected never happens, but people are invariably wrong when they say, 'I never would have believed that about her'. Think it over and scrape the outermost layer of paint away, you'll find other colours underneath, other answers."

He yawned and let his hands fall to the table.

"Time for bed."

"No, let's stay here a few more minutes."

He looked at her intently. "You still think something has happened to your friend?"

"No, I'm sure you're right."

They sat quietly at the kitchen table. A gust of wind set a branch scraping against the window.

"I've been dreaming a lot recently," he said. "Maybe because you're always waking me up in the middle of the night. That means I remember my dreams. Yesterday I had the strangest dream. I was walking around a cemetery. Suddenly I found myself in front of a row of headstones where I started recognising the names. Stefan Fredman's name was among them."

Linda shivered. "I remember that case. Didn't he break into this flat?"

"I think so, but we were never able to prove it. He never told us."

"You went to his funeral. What happened?"

"He was sent to a psychiatric institution. One day he put on his war paint, climbed up to the roof and threw himself off."

"How old was he?"

"Eighteen or nineteen."

The branch scratched the window again.

"Who were the others? I mean on the headstones."

"A woman called Yvonne Ander. I even think the date on the stone was right, though it happened a long time ago."

"What did she do?"

"Do you remember the time when Ann-Britt Höglund was shot?"

"How could I forget? You left for Denmark after that happened and almost drank yourself to death."

"That's an extreme way of putting it."

"On the contrary, I think that's hitting the nail on the head. Anyway, I don't remember Yvonne Ander."

"She specialised in killing rapists, wife-beaters, men who had been abusive to women."

"That rings a bell."

"We found her in the end. Everyone thought she was a monster. But I thought she was one of the sanest people I had ever met."

"Is that one of the dangers of the profession?"

"What?"

"Do policemen fall in love with the female criminals they're hunting?"

He waved her insinuations away.

"Don't be stupid. I talked to her after she was brought in. She wrote me a letter before she committed suicide. What she told me was that she thought the justice system was like a fish net where the holes were too big. We don't catch, or choose not to catch, the killers who really deserve to be caught."

"Who was she referring to specifically? The police?"

He shook his head.

"I don't know. Everyone. The laws we live by are supposed to reflect the opinions of society at large. But Yvonne Ander had a point. I'll never forget her."

"How long ago was that?"

"Five, six years."

The phone rang.

Wallander jumped and he and Linda exchanged glances. It was 2 a.m. Wallander stretched his hand out for the kitchen phone. Linda worried for a moment that it was one of her friends, someone who didn't know she was staying with her father. She tried to interpret who it could be from her father's terse questions. She decided it had to be from the station. Perhaps Martinsson, or even Höglund. Something had happened near Rydsgård. Wallander signalled for her to get him something to write with and she handed him the pencil and pad of paper lying on the windowsill. He made some notes with the phone tucked into the crook of his neck. She peered over his shoulder. *Rydsgård, t. off > Charlottenlund Vik's farm.* That was close to the house they had looked at on the hill, the one her father wasn't going to buy. He wrote something else: *burned calf. Åkerblom.* Then a phone number. He hung up. Linda sat back down across from him.

"A burned calf? What's happened?"

"That's what I want to know."

He got up.

"I have to go out there."

"What about me?"

He hesitated. "You can come along if you like."

"You were there for the start of this thing," he said as they got in the car. "You might as well come along for the rest."

"The start of what?"

"The report about burning swans."

"It's happened again?"

"Yes and no. Some bastard let a calf out of the barn, sprayed it with petrol and set light to it. The farmer was the one who called the station. A patrol car was dispatched, but I'd left instructions to be contacted if anything along these lines happened again. It sounds like a sadistic pervert."

Linda knew there was more. "You're not telling me what you really think."

"No, I'm not."

He broke off the conversation. Linda wondered why he had let her come along.

They turned off the highway and drove through the deserted village of Rydsgård, then south towards the sea. A patrol car was waiting at the entrance of the farm. Together the two cars made their way towards the main buildings at Vik's farm.

"Who am I?" Linda asked quickly.

"My daughter. No-one will care. As long as you don't start pretending to be anything else – like a police officer, for instance."

They got out. The two officers from the other car came over and said hello. One was called Wahlberg, the other Ekman. Wahlberg had a heavy cold and Linda wished she didn't have to shake his hand. Ekman smiled and leaned towards her as if he were shortsighted.

"I thought you were starting in a couple of weeks."

"She's just keeping me company," Wallander said. "What's happened out here?"

They walked down behind the farmhouse to a new-looking barn. The farmer was kneeling next to the burned animal. He was a young man close to Linda's age. Farmers should be old, she thought. In my world there's no place for a farmer my own age.

Wallander stretched out his hand and introduced himself.

"Tomas Åkerblom," the farmer said.

"This is my daughter. She happened to be with me."

As Åkerblom looked over at Linda, a light from the barn illuminated his face. She saw that his eyes were wet with tears.

"Who would do a thing like this?" he said in a shaky voice.

He stepped aside to let them see, as if displaying a macabre installation. Linda had already picked up the smell of burned flesh. Now she saw the blackened body of the calf lying on its side in front of her. The eye socket nearer to her was completely charred. Smoke was still rising from the singed skin. The fumes were starting to make her nauseated and she took a step back. Wallander looked at her. She shook her head to indicate that she wasn't about to faint. He nodded and looked around at the others.

"Tell me what happened," he said.

Åkerblom started talking. He still sounded on the verge of tears.

"I had just gone to bed when I heard a sound. At first I thought I must have cried out in my sleep – that happens sometimes when I have a bad dream. Then I realised it came from the barn. The animals were braying and one of them sounded bad. I pulled the curtains away and saw fire. It was Apple – I couldn't identify him immediately of course, just that it was one of the calves. He ran straight into the wall of the barn. His whole body and head were in flames. I couldn't really take it in. I pulled on a pair of old boots and ran down. He had collapsed by the time I reached him; his legs were twitching. I grabbed an old piece of tarp and tried to put out the rest of the fire, but he was dead already. It was horrible. I remember thinking, 'This isn't happening, this isn't happening.' Who would do a thing like this?"

"Did you see anything else?" Wallander said.

"No, just what I told you."

"You said, 'Who would do a thing like this?' – why? Is there no way it could have been an accident?"

"You think a calf poured petrol over his own head and struck a match? How likely is that?"

"Let's assume it was a deliberate attack. Did you see anyone when you pulled the curtain away from the window?"

Åkerblom thought hard before answering. Linda tried to anticipate her father's next question.

"I only saw the burning animal."

"What kind of person do you imagine did this?"

"An insane . . . a fucking lunatic."

Wallander nodded. "That's all for now," he said. "Leave the animal as it is. Someone will be sent in the morning to take pictures, and of the area."

They walked to their cars.

"What kind of crazy bastard lunatic . . . ?" Åkerblom was muttering.

Wallander didn't answer him. Linda saw how tired he was. His forehead was deeply furrowed and he looked old. He's worried, she realised. First, the report about the swans, and then a calf named Apple is burned alive.

It was as if he read her mind. Wallander let his hand rest on the car door handle and turned to Åkerblom.

"Apple," he said. "That's an unusual name for an animal."

"I played table tennis when I was younger. I often name my animals after great Swedish champions. I have an ox called Waldner."

Linda could see that her father was smiling. She knew he appreciated originality.

They drove back to Ystad.

"What do you think this is about?" Linda asked.

"The best-case scenario is a pervert who gets a kick out of hurting animals."

"That's the best case?"

He hesitated. "The worst case would be someone who won't stop at animals," he said.

CHAPTER 8

When Linda woke up, she was alone in the flat. It was 7.30. The sound of her father shutting the front door must have woken her. He does that on purpose, she thought, and stretched out in bed. He doesn't like me sleeping in.

She got up and opened the window. It was a clear day and the weather seemed to be set fair. She thought about the events of the night; the still-smoking carcass and her father looking suddenly old and worn out. It's the anxiety, she thought. He can hide a lot from me, but not his anxiety.

She ate breakfast and put on her clothes from yesterday, then tried two other outfits before she could make up her mind. She called Anna. The answering machine cut in after five rings and she asked Anna to pick up if she was there. No reply. Linda walked into the hall and looked at herself in the mirror. Was she still worried about Anna? No, she said to herself, I'm not worried. Anna has her reasons. She's most likely chasing down that man she saw on the street, a man she thinks is her long-lost father.

Linda went out for a walk and picked up a newspaper from a bench. She turned to the motoring section and looked at the ads for used cars. There was a Saab for 19,000 kronor. Her father had already promised to chip in 10,000 and she knew she needed a car. But a Saab for 19,000? How long would it last?

She tucked the newspaper into her pocket and walked to Anna's flat. No-one answered the door. After letting herself in she was at once struck by the feeling that someone had been there since she left the place the night before. She froze and looked all around the hall – the coats hanging in their place, the shoes all in a row. Was anything different? She couldn't put her finger on it, but she was convinced there was something.

She continued into the living room and sat on the sofa. Dad would tell me to look for the impressions people have left behind in this room, she thought, of themselves and their dramatic interactions. But I see nothing, only that Anna isn't here.

After combing through the flat twice she convinced herself that Anna had not been home during the night. Nor anyone else. All she recognised were tiny, near-invisible traces she herself had left behind.

She went into Anna's bedroom and sat at the little desk. She hesitated at first, but her curiosity got the better of her. She knew that Anna kept a journal, had done so since she was a child. Linda remembered an incident from school where Anna had been sitting in a corner writing in her journal. A boy pulled it out of her hands as a joke, but she had been so furious she bit him on the shoulder and everyone had known to leave the journal well alone after that.

Linda pulled out the desk drawers. They were full of old diaries, well thumbed, crammed with writing. The dates were written on the spines. Up until Anna was 16 they were all red. Thereafter they were black.

Linda closed the drawers and riffled through some papers lying on the desk. She found the journal Anna was currently keeping. I'm only going to look at the last entry, she thought, telling herself it was justified since she was motivated only by concern for Anna's well-being. She opened the book at the last entry, for the day they were supposed to have met. She bent over the page; Anna's handwriting was cramped, as if she were trying to hide the words. Linda read the short text twice, first without understanding it, then with a sense of bafflement. What Anna had written was nonsense: *myth, fear, myth, fear*. Was it code?

Linda immediately broke her promise only to look at the last entry. She turned over the page and there she found a regular entry. Anna had written: *The Saxhausen textbook is a pedagogical disaster. Completely impossible to read or understand. How can textbooks like this be allowed? Future doctors will be scared off and turn to research, where there is also more money*. Further on: *Had low fever this morning. Weather clear but windy* – that much was true. Linda flipped the page to the last line and read it again. She tried to imagine that she was Anna writing the words.

There were never changes in the text, no scratchings-out, no hesitations that she could see. The handwriting looked even and firm. *Myth, fear, myth, fear. I see that I have signed up for 19 washing days so far this year. My dream – to the extent I even have one right now – would be to work as an anonymous suburban general practitioner. Do northern towns have suburbs?*

That was where the text ended. Not a word about the man she saw through the hotel window. Not a word or a hint. Nothing. But isn't that exactly the kind of thing diaries are for?

She looked further back in the book. From time to time her own name appeared. *Linda is a true friend*, she had written on July 20, in the middle of an entry about her mother. She and her mother had *argued over nothing* and later that evening she was planning *to go to Malmö to see a Russian movie.*

Linda sat with the journal for almost an hour, struggling with her conscience. She looked for entries about herself. *Linda can be so demanding*, she found on August 4. What did we do that day? she wondered. She couldn't remember. It was a day like any other. Linda didn't even have an organiser right now. She scribbled appointments on scraps of paper, and wrote phone numbers on the back of her hand.

Finally, she closed the journal. There was nothing there, just the strange nonsense words at the end. It's not like her, Linda thought. The rest of the entries are the work of a balanced mind. She has no more problems than most people. But the last day, the day she thinks she just saw her father in Malmö for the first time in 24 years, she three times writes the phrase, *myth, fear*. Why doesn't she write something about seeing her father? Why write something that doesn't make any sense?

Linda felt her anxiety revive. Had there perhaps been something serious behind Anna's talk about losing her mind? Linda walked to the window where Anna often stood during their conversations. The sun was reflected from a window in the building opposite and she had to squint to see anything. Could Anna have been overcome by a temporary derangement? She *thought* she had seen her long-lost father – could this have disconcerted her so she lost her bearings and began behaving in an erratic manner?

Linda gave a start. There, in the car park behind the building, was Anna's car, the red VW Golf. If she had left for a few days the car would be gone too. Linda hurried down to the car park and checked the car doors: locked. The car was clean and shiny, and this surprised her. Anna's car is generally dirty, she thought. Every time we go out her car is covered in dust. Now it's squeaky clean – even the wheel trims have been polished.

She went back up to the flat, sat in the kitchen and tried to come up with a plausible explanation. The only thing she knew for certain was that Anna had not been at home to meet Linda as they had arranged. It wasn't a misunderstanding; there was nothing wrong with her memory. She had chosen not to stay at home that day. Something else had come up that was more important, something for which she didn't need her car. Linda turned on the answering machine and listened to the messages but heard only her own bellowing of Anna's name. She let her gaze wander to the front door. Someone rang the bell, she thought. Not me, not Zeba, not Henrietta. Who else then? Anna had broken up with her boyfriend in April, someone Linda had never met called Måns Persson. He too was a student in Lund, studying electromagnetics, and he had turned out to be less faithful than Anna would have liked. She had been deeply hurt by the breach in their relationship and she had told Linda on several occasions that she was going to take her time before letting herself get that close to a man again.

Linda had also recently had a Måns Persson experience, a man she had kicked out of her life in March. His name had been Ludwig and he had seemed uniquely suited to that name. His personality was part emperor and part impresario. He and Linda had met at a pub when Linda had been out drinking with some student colleagues. Ludwig had been with another group and they had simply ended up squeezed next to each other in a crowded booth. Ludwig was in the sanitation business; he operated a rubbish truck and made his pride in his work seem like the most natural thing in the world. Linda had been attracted by his huge laugh, his happy eyes and the fact that he never interrupted her when she talked, actually straining to catch every word although the noise around them had been deafening.

They had started seeing each other and for a while Linda dared to think that at last she had found a real man. But then, purely by accident, she had heard from a friend of a friend that when Ludwig wasn't working or spending time with Linda, he was spending time with a young woman who ran a catering business in Vallentuna. They had had a heated confrontation. Ludwig pleaded with her, but Linda sent him packing and wept for a whole week. She hadn't thought about it in those terms, but perhaps she too was waiting to recover from the pain of this break-up before she let herself look for someone new. She knew that the rapid succession of boyfriends in her life worried her father, though he never asked her about it.

Before leaving Anna's flat, Linda went to the kitchen, where she had spotted the spare keys to the car in a drawer. She had borrowed the car on a few occasions. It won't matter if I do it again, she thought, I'm just going to take the car to visit her mother. She left a note saying what she had done and that she would be back in a couple of hours. She didn't write anything about being worried.

First Linda stopped on Mariagatan to change into cooler clothing, since it was getting very warm. Then she drove out of town, took the turn to Kåseberga and parked at the harbour. The surface of the water was like a mirror, the only disturbance a dog swimming around next to the boats. An old man sitting on a bench outside the smoked-fish shop nodded kindly to her. Linda smiled in return, but she had no idea who he was. A retired colleague of her father, perhaps?

She got back in the car and continued on her way. Henrietta Westin lived in a house that seemed to crouch among the tall stands of trees posted on all sides like sentries. Linda had to turn back several times before she found the right driveway. Finally she drew up next to a rusty harvester. The heat outside reminded her of the holiday she and Ludwig had taken in Greece before they broke up. She shook off those thoughts and made her way through the massive trees. She stopped to listen to an unusual noise, a furious hammering. Then she saw a woodpecker high on her right. Maybe he has a part in her music, she thought. Anna has said her mother doesn't shy away from using any kind of noise. His input might very well be crucial to the percussion section.

She left the woodpecker and walked past a run-down old vegetable garden. It had clearly not been tended for many years. What do I know about her? Linda thought. And what am I doing here? She stopped again and listened. At that particular moment, in the shade of the high trees, she was no longer worried about Anna. There was surely a sensible explanation for her staying away. Linda turned and started walking back to the car.

The woodpecker was gone, or had finished his work. Everything changes, she thought. People and woodpeckers, my dreams and all the time I thought I had, that keeps slipping away through my fingers, despite my best efforts to keep it dammed up. She pulled on her invisible reins and came to a halt. Why was she walking away? Now that she had come this far in Anna's car, the least she could do was say hello to Henrietta. Without betraying her own anxiety, without making pressing enquiries about Anna's whereabouts. She might just be in Lund and I don't have her number there. I'll ask Henrietta for it.

She followed the path through the trees again and came at last to a half-timbered whitewashed house covered in wild roses. A cat lay on the stone steps and studied her movements warily as Linda approached. A window was open, and just as she bent down to stroke the cat she heard noises from inside. Henrietta's music, she thought.

Then she stood up and caught her breath. What she had heard wasn't music. It was the sound of a woman sobbing.

CHAPTER 9

Somewhere inside the house a dog started to bark. Linda felt as though she had been caught in the act and quickly rang the doorbell. It took a while for Henrietta to open the door. When she did she was restraining an angry grey dog by its collar.

"She won't bite," Henrietta said. "Come in."

Linda never felt entirely at ease with strange dogs and so she hesitated before crossing the threshold. As soon as she did so the dog relaxed, as if Linda had crossed over into a no-barking territory. Henrietta let go of the dog's collar. Linda hadn't remembered Henrietta so thin and frail. What was it Anna had said about her? That she wasn't even 50. It was true that her face looked young, but her body looked much older. The dog, Pathos, sniffed Linda's legs then retreated to her basket and lay down.

Linda thought about the sobbing that she had heard through the window. There were no traces of tears on Henrietta's face. Linda looked past her into the rest of the house, but there was no sign of anyone else there. Henrietta caught her gaze.

"Are you looking for Anna?"

"No."

Henrietta burst out laughing. "Well, I'm stumped. First you call and then you drop by for a visit. What's happened? Is Anna still missing?"

Linda was taken aback by Henrietta's directness, but welcomed it.

"Yes."

Henrietta shrugged then directed Linda into the big room – the result of many walls being removed – that served as both living room and studio.

"My guess is that she must be in Lund. She holes up there from time to time. The theoretical component of her studies is apparently very demanding, and Anna is no theoretician. I don't know who she takes after. Not me, not her father. Herself."

"Do you have a number for her in Lund?"

"No, I'm not sure she even has a phone there. She rents a room in a house, but I don't know the address."

"Isn't that a bit odd?"

"Why? Anna is secretive by nature. If you don't leave her alone she can get very angry. Didn't you know that about her?"

"No. And she doesn't have a mobile phone either?"

"She's one of the few people still holding out," Henrietta said. "Even I have one. In fact, I don't see any more need for the old-fashioned kind. But that's neither here nor there. No, Anna doesn't have a mobile phone."

Henrietta stopped as if she had suddenly thought of something. Linda looked around the room. Someone had been crying. It hadn't occurred to her that it might have been Anna until Henrietta asked her if that was what she was doing, looking for her here. But it couldn't have been Anna, she thought. Why would she be crying? She's not a person who cries very much. Once, when we were girls, she fell off the jungle gym and hurt herself. She cried then, but it's the only time I remember. When we both fell in love with Tomas, I was the one who cried; she was only angry.

Linda looked at Henrietta, standing in a beam of light in the middle of the polished wooden floor. She had an angular profile, just like Anna.

"I don't very often get visitors," she said suddenly, as if that was what had been foremost in her mind. "People avoid me just as I avoid them. I know they think I'm eccentric. That's what comes of living alone, out in the country with only a greyhound for company, composing music no-one wants to listen to. It doesn't help matters that I'm still legally married to the man who left me twenty-four years ago."

Linda picked up the bitterness and loneliness in Henrietta's voice.

"What are you working on right now?"

"Please don't feel you have to make polite conversation. Why did you drop by? Was it really that you're still worried about Anna?"

"I borrowed her car. My grandfather used to live in these parts and I thought I would take a drive. I get a little bored these days."

"Until you get to put on your uniform?"

"Yes."

Henrietta set out a thermos and some cups on the table.

"I don't understand why an attractive girl like you would choose to become a police officer. Breaking up fights on the street, that's what I imagine it to be. I know there must be other aspects to the job, but that's what always comes to mind." She poured the coffee. "But perhaps you're going to sit behind a desk," she added.

"No, I've been assigned to a patrol car and will probably be doing a lot of the work you would expect. Someone has to be prepared to jump into the fray."

Henrietta leaned to the side with her hand tucked under her chin.

"And that's what you're going to dedicate your life to?"

Her comments put Linda on the defensive, as if she was in danger of being contaminated by Henrietta's bitterness.

"I don't know what looks have got to do with it. I'm almost thirty and on good days I'm generally happy with how I look, but I've never dreamed of being Miss Sweden. But, more to the point, what would happen to our society if there were no police? My father is a policeman and I've never had cause to be ashamed of him."

Henrietta shook her head. "I didn't mean to hurt your feelings."

Linda still felt angry. She felt a need to strike back, though she wasn't really sure why.

"I thought I heard the sound of someone crying in here when I walked up to your house."

Henrietta smiled. "It's a recording I have. I'm working on a requiem and I mix my music with the sound of someone crying."

"I don't even know what a requiem is."

"A funeral mass. That's almost all I write these days."

Henrietta got up and walked to the grand piano by the window, which overlooked fields and then the rolling hills leading down to the sea. Next to the piano there was a table with a tape recorder as well as a synthesiser and other electronic equipment. Henrietta turned on the tape player. A woman's voice came on, wailing and sobbing. It was the one Linda had heard through the window. Her curiosity about this strange woman increased.

"Where did you get that recording?"

"It's from an American film. I often record the sound of crying from films, or from programmes on the radio. I have a collection of forty-four crying voices so far, everything from a baby to a very old woman I recorded secretly at a rest home. Would you like to donate a sample to the archive sometime?"

"No thanks."

Henrietta sat at the piano and played a few haunting chords. Linda went and stood next to her, while Henrietta continued to play. The room was filled with a powerful surge of music which then faded into silence. Henrietta gestured for Linda to sit next to her on the piano stool.

"Tell me again why you came here. Seriously. I've never even felt you really liked me."

"When I was little, I was afraid of you."

"Of me? No-one is afraid of me!"

That's where you're wrong, Linda thought. Anna was afraid of you too, sometimes she had nightmares about you.

"It was pure impulse, nothing more. I wonder where Anna is, but I'm not as worried as I was last night. You're probably right that she's in Lund . . ."

Linda broke off.

"What is it you aren't saying? Should I be worried about her too?"

"Anna thought she saw her father on a street in Malmö a couple of days ago. I shouldn't be telling you this. You should hear it from her."

"Is that all?"

"Isn't that enough?"

Henrietta touched the keys as if sketching out a few more bars of music.

"Anna is forever catching glimpses of her father. She's told me stories like this since she was a little girl."

Linda raised her eyebrows. Anna had never mentioned one of these sightings before and Linda was certain she would have. When they were younger they told each other everything. Anna was one of the few people Linda had told about standing on the railing of the bridge. What Henrietta said didn't match this picture.

"Anna is never going to relinquish her hope," Henrietta said. "The hope that Erik will one day come back. Even that he is still alive."

"Why did he leave?"

"He left because he was disappointed. He had such marvellous ambitions when he was younger. He seduced me with those dreams, if you must know. I had never met a man who had the same wonderful visions as Erik. He was going to make a difference to the world, to our generation. He knew without a doubt that he had been put on this earth to do something on a grand scale. We met when he was sixteen and I was fifteen. Young as I was, I knew I had never met anyone like him: he radiated dreams and life force. At that time he was still looking for his niche – was it art, sports, politics or another arena in which he would leave his mark? He had decided to give himself until he was twenty to work it out. I remember no self-doubt in him until then. But when he turned twenty he started to worry. There was a restlessness in him. Until then he had had all the time in the world. When I started making demands of him, that he help to support the family after Anna was born, he would get impatient and scream at me. He had never done that before. That was when he began making his sandals, he was good with his hands. He called them 'sandals of lättja' as a kind of protest, I think, for the fact that they were taking up his valuable time. It was probably then that he started planning his disappearance, or should I say, escape. He wasn't running away from me or Anna, he was running away from himself, from his disappointment in life. I wonder if he managed it – I've never been able to ask him, of course. One day he was just gone. It took me by surprise. It was only with hindsight that I realised how carefully he must have planned it. I can forgive him the fact that he sold my car. What I'll never understand or accept is that he left Anna. They were so close. I know he loved her. I was never as important to him, or at least not after the first couple of years while I was still a part of his dreams. How could he leave her – how can a person's disappointment – stemming as it did from an unattainable dream – conceivably weigh more heavily than the most important person in his life? I think that must be a contributing factor to his death, at least to the fact that he never returned."

"I didn't think anyone knew what happened to him."

"He has to be dead. He's been missing for all of these years. Where could he otherwise be?"

"Anna's convinced she saw him."

"She sees him on every street corner. I've tried to talk her out of it and make her face the truth. No-one knows what happened, but he has to be dead by now."

Henrietta paused. The greyhound sighed.

"What do you suppose happened?" Linda said.

"I think he gave up – when he realised the dream was nothing more than that. And that the Anna he left behind was real. At that point it was too late. He would always have been plagued by his conscience." Henrietta closed the lid over the piano keys with a thud and stood up. "More coffee?"

"No thanks. I have to get going."

Henrietta seemed anxious and Linda watched her closely. She grabbed Linda's arm and started to hum a melody that Linda recognised. Her voice alternated between high, shrill tones and softer, cleaner ones.

"Do you know that song?" she asked when she had finished.

"I recognise it, but I don't know what it is."

"'Buona Sera'."

"Is it Spanish?"

"Italian. It means 'Good Night'. It was popular in the fifties. So many people today borrow or steal or vandalise old music. They make pop songs out of Bach. I do the reverse. I take songs like 'Buona Sera' and turn them into classical music."

"How do you do that?"

"I break down the structure, change the rhythm, replace the guitar sound with a flood of violins. I turn a banal song about three minutes long into a symphony. When it's ready, I'll play it for you. Then people will finally understand what I've been trying to do all these years."

Henrietta followed her out. "Come back sometime."

Linda promised she would, then drove away. She saw storm clouds heaped up in the distance out over the sea in the direction of Bornholm. She pulled over after a while and got out of the car. She had a sudden desire to smoke. She had stopped smoking three years ago, but the need still sucked at her from time to time.

There are some things mothers don't know about their daughters, she thought. Henrietta doesn't know that Anna and I used to tell each other everything. If she had known that, she would never have told me about Anna forever seeing her father on the street. There are a lot of things I'm not sure of, but I know Anna would have told me that.

There was only one possible explanation. Henrietta had not been telling her the truth about Anna and her missing father.

CHAPTER 10

She opened the curtains a little after 5 a.m. and looked at the thermometer. It was 9°C, the sky clear with little or no wind. What a wonderful day for an expedition, she thought. She had prepared everything the night before and it didn't take her long to leave her flat across from the old railway station in Skurup. Her 40-year-old Vespa was waiting in the yard under a custom-made cover. She was the first and only owner, and since she had taken such good care of it, it was in mint condition. In fact, word of it had spread to the factory in Italy and she had received several solicitations over the years asking her if she would consider letting them put it in their museum. In return, the company would supply her with a new Vespa every year.

Year after year she had declined the offer, meaning to keep this Vespa, which she had bought when she was 22, for as long as she lived. She didn't care what happened to it after that. One of her four grandchildren might want it, but she wasn't about to write a will for the sake of an ancient Vespa.

Adjusting her backpack, she strapped on the helmet and kick-started the old machine. It roared into instant life. Half an hour later she arrived at the small car park by Led Lake. She walked the Vespa in behind some bushes beside a large oak tree. A car drove past on the main road, then silence returned.

As she prepared to walk into the forest and become invisible to the outside world, she wondered if this wasn't the most satisfying way of expressing one's independence; by daring to abandon the well-trodden path. To step into the underbrush and vanish from the eyes of the world.

Her brother Håkan was the person who had taught her that there were two kinds of people in the world: the ones who always chose the shortest distance between two points, and the ones who sought out the scenic route where the curves, slopes and vistas were to be found. They had played in the forests around Älmhult when they were growing up. After her father was severely injured by a fall from a telegraph pole while repairing a phone line, they moved to Skåne. Her mother got a job at the Ystad hospital. That was where she spent her adolescence and forgot all about exploring, until the day she stood outside the gates to Lund University and realised she had no idea what to do with her life. She turned to her childhood memories for inspiration.

It was a day during that first difficult autumn term, when she had enrolled as a law student for lack of a more interesting alternative. She had cycled out on the road to Staffanstorp and found a small unpaved road quite by chance. She left her bike there and continued on foot until she came to the ruins of a mill. That was when the idea had come to her – or rather, went through her mind like a bolt of lightning. What was a path? Why does a path go this side of a tree and not the other? Who was the first person to walk here?

She straight away knew that it would become her life's mission to chart old trails. She would become the protector and historian of old Swedish paths and walkways. She ran back to her bike, and the same day resigned from the law faculty and started instead to study history and cultural geography. She had the good fortune to meet a sympathetic professor who appreciated the originality of her interests and supported her in her cause.

She started along the path which curved gracefully around Led Lake. The tall trees shaded her from the sun. She had mapped this particular path quite a while ago. It was a standard walking path that could be traced back to the 1930s, when Rannesholm Manor was owned by the Haverman family. One of the counts, Gustav Haverman, had been an enthusiastic runner and had cleared bushes and undergrowth away from the edge of the lake to establish this trail. But a little further on, she thought, up ahead in this old forest where no-one else sees anything but moss and stone, I am going to turn off from this trail and follow the path I found just a few days ago. I have no idea where it leads, but nothing is as tempting, as magical, as following an unknown path. I still hope that one day I'll find a path that is a work of art. A path that has been created without a destination in mind, just for its own sake.

She paused at the top of a hill to catch her breath, looking down between the trees at the glassy surface of the lake. She was 62 now, and according to her own calculations she needed five more years to complete her life's work: *The History of Swedish Walkways*. In this book she would reveal that paths were among the most important clues to ancient settlements and their way of life. Paths were not laid out only for the simplest way of getting from A to B. She had ample evidence for the numerous religious and cultural factors that determined where and how paths made their way across the landscape. Over the years she had published regional studies and maps, but the conclusions of her many years of research had yet to be set down in a final form.

She slowed down when she found the spot. Where the untrained eye saw nothing but grass and moss growing at the edge of the path, she spotted the clear outline of a path that had been out of use for many years. She began to climb up the side of the hill, careful where she put her feet. Last year she had broken a leg when she was exploring a trail to the south of Brösarp. The accident had forced her to take a long rest, which stood out in her mind as a particularly difficult time. Even though it had given her more opportunity to write, she had become restless and irritated, especially without her husband to care for her. He had died shortly before the accident occurred, and he had always been the one to look after things in the home. She had sold the house in Rydsgård after that and moved to the small flat in Skurup.

She pushed aside some branches and moved in under the trees. She had once read about a meadow in the forest that could only be found by someone who had lost their way. To her mind this captured some of the mystical dimension of human existence. If one only dared to get lost, one could find the unexpected. There was a whole other world beyond the highways and byways – if you dared to take the road less travelled. And I'm the caretaker of these old forgotten paths, she thought. Sleeping beauties waiting for someone to wake them from their slumber. If paths remain unused for too long, they die.

She was deep in the forest now, a long way from the main trail. She stopped and listened. A branch broke some distance away, then all was quiet. A bird flapped noisily and flew away. She walked on, hunched over, moving very slowly. The path was almost indecipherable. She had to search for its contours under the moss, the grass and the fallen branches.

Soon she started to feel disappointed: this wasn't an old path after all. When she first saw it she had been hoping it was a section of the ancient pilgrims' trail that was thought to have passed close to Led Lake. On the north side of the Rommele Hills it was still visible. It disappeared around Led Lake only to pop up again north-west of Sturup. Sometimes she was tempted to think the pilgrims had used a tunnel, but she knew that was not their custom. They used trails and one day she hoped to find this one. It would not be today, unfortunately. After a mere 100 metres she was convinced that the path was recently established, no more than 10 or 20 years old. She would expect to be able to say why it had been abandoned once she discovered where it led. By now the trees and the undergrowth stood so thick and close together that it was all but impossible to make her way through.

Suddenly she stopped and squatted. She saw something that confused her. She picked at the moss with her finger. She had seen something white lying there: a feather. A dove, she supposed. But were there white forest doves? They were usually brown or blue. She stood up and studied the feather. Finally she recognised it as a swan's feather. But how odd that it should have fallen here, so deep in the forest. Swans came ashore from time to time but never this far inland, not in thick woodland.

A few more steps and she stopped again, this time because the ground in front of her was surprisingly flattened. Someone must have walked here only days before. But where did the prints begin? She examined the area for ten minutes and decided that someone had walked through the forest and joined the path at this point. She continued slowly. She was no longer so curious about the path as when she had thought it might be the pilgrim trail. This path was no doubt simply an extension of the paths Count Haverman had put in to satisfy his outdoor tastes, but one that had fallen out of use since his time. The prints she was following probably belonged to a hunter.

After another 100 metres she arrived at a shallow ravine, a crack in the earth covered over by bushes and undergrowth. The path ran straight down into it. She removed her backpack, tucked a torch into her pocket and scooted gingerly down into the ravine. She started lifting branches to get past them and saw to her astonishment that several of them had been cut and placed here in order to conceal the entrance to the hollow. Boys, she thought. Håkan and I often made forts in the forest. She pressed on past the undergrowth and sure enough, there was a small hut. It was unusually large to be the work of children. She was reminded of a news item Håkan had showed her from a magazine, with pictures of a shack in a forest that served as the hide-out for a wanted criminal by the wonderful name of "Beautiful Bengtsson". He had lived in his hide-out for a long time and had been found out only when someone stumbled upon it by sheer chance.

She walked up for a closer look. The hut had been built of planks of wood, with a sturdy aluminium roof. To the rear it bordered a steep part of the ravine. She checked the handle of the door – it wasn't locked. She knocked and felt an idiot. If someone was there they would have heard her by now. She was feeling more and more confused. Could someone be hiding here?

Warning bells started going off inside her head. At first she dismissed these, for she was never one to get scared easily. She had run across unpleasant men in remote areas before now, and although it had sometimes frightened her she had always managed to control her fear and put up a tough front. Nothing had ever happened, and nothing was going to happen today. Even so, she could not help feeling that she was ignoring common sense by investigating this hut on her own. Only someone with a powerful motive to hide from prying eyes would have chosen a place like this. On the other hand, she did not want to turn back without finding out what was in here. Her path had indeed had a destination. No-one without her trained eyes would have spotted it. But the person who used the hut had not used the old path. That was strange. Was the old path she had found simply a back-up, the way fox holes had more than one exit? Her curiosity got the better of her.

She opened the door of the hut and looked in. There were two small windows on either side, but they let in only a little light. She turned on the torch. There was a bed on one side, and on the other, a small table with a chair, two gas lamps and a camp stove. Who lived here? And how long had it been empty? She leaned over and felt the blanket on the bed. It wasn't damp. Someone has been here recently, she thought. In the last couple of days. Again she felt she had better leave. A person who made his home here was not the kind to welcome visitors.

She was about to turn and leave when the light of her torch fell on a book lying by the bed. She bent down. It was a Bible. She opened it and saw a name had been written on the inside then scratched out. The book was well used and in places the pages were torn. Some verses had been underlined and annotated. Carefully she put the book back where she had found it. She turned off the torch and realised at once that something had changed. There was more light now. Someone must have opened the door behind her. She turned, but it was too late. The blow to her face came with the force of a charging predator. She was plunged into a deep and bottomless darkness.

CHAPTER 11

After her visit to Henrietta, Linda sat up waiting for her father to come home, but by the time he softly pushed the front door open at 2 a.m., she had fallen asleep on the sofa with a blanket pulled up to her ears. When she woke it was from a nightmare. She couldn't remember what she had dreamed just that she felt as if she were being suffocated. Low snores rolled through the flat, sounding like breakers on the shore. Her father's bedside light was still on, and he lay on his back, wrapped in a sheet, not unlike a walrus comfortably stretched out on a rock. She leaned over him and checked his breath between snores. Definitely alcohol.

She wondered who he could have been drinking with. The trousers which lay beside the bed were dirty, as if he had walked through mud. He's been in the country, she thought. That means a night of drinking with Sten Widén. They've sat out in the stables and shared a bottle of vodka.

Widén was one of her father's oldest friends and now he was seriously ill. Her dad had a habit of talking about himself in the third person when it came to expressing something emotional, and he had taken to saying: "When Sten dies Kurt Wallander will be a lonely man." Widén had lung cancer. Linda was familiar with the story of how he had raised fine racehorses at the stable by the ruins of Stjärnsund Castle. A few years ago he had sold the ranch, but just as the buyer was about to close on the deal, Widén changed his mind and backed out, using a clause in the contract that allowed this. He had bought more horses and then, shortly afterwards, received his diagnosis. It had been a year since then, a period of grace given the severity of his condition. Now he was once more selling his horses and his stable. He had arranged for himself a bed at a hospice in Malmö. This time there was no backing out of the deal.

Linda went to her room, put on her pyjamas and climbed into bed. She lay staring at the ceiling and reproached herself for being so hard on her father. Why shouldn't he be allowed to enjoy a night of drinking with his old friend, especially since he is dying? I've always thought of Dad as a good friend to the few he has. It's only right that he stay up late drinking in the stables. She felt like waking him and apologising for her disapproval. But he wouldn't appreciate being woken. It's his day off tomorrow, she remembered. Maybe we'll do something fun.

As she lay there, she was thinking about Henrietta and the fact that she hadn't been telling the truth. What was she hiding? Did she know where Anna was, or was there some other reason? Linda curled up on her side. Soon she was going to miss not having a boyfriend to cuddle up to. But where am I going to find one in this town? She pushed her thoughts aside and fell asleep.

Wallander shook her awake at 9.00. Linda jumped out of bed. Her father didn't seem hung-over. He was dressed and had even combed his hair.

"Breakfast," he said. "Time is ticking, life is fleeting."

Linda showered and dressed. Her father was playing patience at the kitchen table when she came in and sat down.

"You were with Sten last night, were you not?"

"Right."

"And you drank too much."

"Wrong. We drank far, far too much."

"How did you get home?"

"Taxi."

"How is he doing?"

"I wish I could be sure of being able to face the end with the same equanimity. He simply says, 'You only have so many races in your life. You have to try to win as many of them as you can.'"

"Do you think he's in pain?"

"I'm sure he is, but he doesn't say anything. He's like Rydberg."

"I don't really remember him."

"He was an old policeman with a mole on his cheek. He was the one who made a policeman out of me when I was young and didn't understand anything. He died much too early, but never a single word of complaint. He had also run his races and accepted the fact that his time was up."

"Who's going to be that kind of mentor to me?"

"I thought you had been assigned to Martinsson."

"Is he any good?"

"He's an excellent policeman."

"You know, I have memories of Martinsson from when I was a kid. I don't know how many times you came home angry about something he had said or done."

Wallander gathered up his cards.

"I was the one who trained Martinsson, just as Rydberg trained me. Of course, I am sure I did come home and complain about him. He can be damned thick-headed. But once he gets something into his head he never forgets it."

"So that makes you my mentor indirectly."

Wallander got up. "I don't even know what that word means. Come on, we're leaving."

She looked at him with surprise. "Did I forget something?"

"We decided we would go out – not where. It's going to be a beautiful day and before you know it the fog will be here to stay. I hate the fog in Skåne. It creeps right into one's head. I can't think straight when it's grey and misty everywhere. But you're right: we do have a goal today."

He sat down again and filled his cup with the last of the coffee before continuing. "Hansson. Do you remember him?"

Linda shook her head.

"I didn't think you would. He's one of my colleagues. Now he's about to sell his parents' house outside Tomelilla. His mother has been dead a long time, but his father turned a hundred and one before he went. According to Hansson he was clear-headed and mean-spirited to the end. But the house is up for sale and I want to take a look at it. If Hansson hasn't been exaggerating, it may just be what I'm looking for."

There was a breeze, but it was warm. When they drove past a long caravan of well-polished vintage cars Linda surprised her father by recognising most of the models.

"Since when do you know about cars?"

"Since my latest boyfriend, Magnus."

"I thought his name was Ludwig?"

"You have to keep up, Dad. Anyway, isn't Tomelilla all wrong for you? I thought you wanted to sit on a bench with your faithful dog, looking out over the sea."

"I don't have that kind of money. I'll have to settle for the next best thing."

"You could get a loan from Mum. Her golf-playing banker is pretty loaded."

"Never in a million years."

"I could borrow it for you."

"Never."

"No view of the ocean for you, then."

Linda glanced at her father. Was he angry? She couldn't decide. But she realised this was also something they had in common, flare-ups of irritation, a tendency to be hurt by almost nothing. Sometimes we are so close and other times it's like a crevasse has opened up between us. And then we have to build the rickety bridges that usually manage to connect us again.

He took a piece of folded paper from his pocket.

"Map," he said. "Give me the directions. We're going to get to the roundabout soon that is at the head of the page. I know we go in the direction of Kristianstad, but you'll have to guide me from there."

"I'm going to trick you into Småland," she said and unfolded the paper. "Tingsryd? Does that sound good? We'll never find our way back from there."

The house was attractively situated on a little hill, surrounded by a strip of forest, beyond which there were open fields and marshes. A bird – a kite, by the look of it – hung in the air currents above the house, and in the distance there was the rising and falling sound of a tractor at work. Linda sat down on an old stone bench between some red-currant bushes. Her father squinted at the roof, tugged at the drainpipes and tried to peer into the house. Then he disappeared to the other side.

The moment he was gone, Linda started thinking about Henrietta.

Now that some time had passed since the visit, her intuition had solidified into certainty: Henrietta had been lying. She was hiding something about Anna. Linda dialled Anna's number on her mobile phone and got the answering machine. She didn't leave a message. She put the phone away and walked round the house to find her father. He was pulling at an old water pump that squeaked and sprayed brown water into a bucket. He shook his head.

"If I could move this house down to the sea I'd take it in a minute," he said. "Here there's just too much forest for my taste."

"What about living in a trailer?" Linda suggested. "Then you could camp on the beach. Lots of people would be happy to let you stay on their land."

"And why is that?"

"Who wouldn't want police protection for free?"

He grimaced and walked back to the car. Linda followed. He's not going to turn around, she thought. He's already put this place behind him.

Linda watched the kite swoop over the fields and disappear on the horizon.

"What now?" he asked her.

Linda immediately thought of Anna. She realised she wanted most of all to talk to her father about it, about the worry she felt.

"I'd like to talk," she said. "But not here."

"I know just the place."

"Where?"

"You'll see."

They drove south, turning left towards Malmö and leaving the main road at a turn-off for Kade Lake. The forest around the lake was one of the most beautiful Linda had ever seen. She had had a feeling her father was going to bring her to this place. They had taken many walks here when she was younger, especially when she was about ten or eleven. She also had a vague memory of being here with her mum, but could not recall the whole family coming here together.

They left the car by a stack of timber. The huge logs gave off a fresh scent, as if they had been felled recently. They walked through the forest, on a path leading to the strange metal statue erected to the memory of a visit the warrior King Charles XII was rumoured to have made to Kade Lake. Linda was about to start talking about Anna when her father raised his hand. They had stopped in a narrow glen surrounded by tall trees.

"This is my cemetery," he said.

"Your *what*?"

"It's one of my secrets, maybe the most secret, and I'll probably regret telling you this tomorrow. I've assigned all the trees you see here to the friends I've had who've died. Your grandfather is here, my mother and my old relatives."

He pointed to a young oak.

"This tree I've given to Stefan Fredman, the desperate Indian. Even he belongs in my collection of the dead."

"What about the other one you were talking about?"

"Yvonne Ander? Yes, she's over there." He pointed to another oak, one with an extensive network of branches.

"I came here a week or so after your grandfather died. I felt as if I had completely lost my footing in life. You were much stronger than I was. I was sitting at the station, trying to sort out a case involving a brutal assault. Ironically it was a young man who had damn near killed his father with a sledgehammer. The boy lied about everything and suddenly I couldn't take it any more. I stopped the interrogation and came here and that's when I felt that these trees had become headstones for everyone I knew who had died. That I should come to visit them here, not where they are actually buried. Whenever I'm here I feel a calm I don't feel anywhere else. I can hug the dead here without them seeing me."

"I won't tell anyone," she said. "Thanks for sharing it with me."

They lingered a while longer. Linda wanted to ask about the identity of a few more of the trees, but she said nothing. The sun was shining through the leaves, but the wind picked up and it immediately became colder. Linda took a deep breath and launched into the topic of Anna's disappearance.

"It'll drive me up the wall if you shake your head and tell me I'm imagining things. But if you can explain to me exactly why I'm wrong, I promise I'll pay attention."

"There's something you'll find out when you become a police officer," he said. "The unexplainable almost never happens. Even someone disappearing into thin air turns out to have a perfectly reasonable explanation. You will learn to differentiate between the unexplained and the merely unexpected. The unexpected can look baffling until you have the necessary background information. This is generally the case with disappearances. You don't know what's happened to Anna and it's only natural that it would worry you, but my intuition tells me you should draw on the highest virtue of our profession."

"Patience?"

"Exactly."

"For how long?"

"A few more days. She'll have turned up by then, or at least been in touch."

"All the same, I'm certain her mother was lying to me."

"I am not sure your mother and I always stuck to the truth when we were asked about you."

"I'll try to be patient, but I do feel that there's more to this. It's just not right."

They returned to the car, It was past 1.00 and Linda suggested they stop for lunch. They chose a roadside restaurant with a funny name, My Father's Hat. Wallander had a fleeting recollection of having lunched there with his father and their ending up in a heated argument, but what the argument had been about he couldn't remember.

They were drinking their coffee when a phone rang. Linda fumbled for hers, but it turned out to be Wallander's. He answered, listened and made a few notes on the back of the bill.

"What was that?" Linda asked.

"Someone reported missing."

He put money on the table, tucking the bill into his pocket.

"What do you have to do now? Who's disappeared?"

"We'll go back by way of Skurup. A widow by the name of Birgitta Medberg – her daughter is worried."

"What were the circumstances of her disappearing?"

"The caller wasn't sure. Apparently the woman is a historian who maps old walkways and often does extensive fieldwork, sometimes in dense forest. An unusual occupation."

"So she may simply be lost."

"My first thought. We'll soon find out." Wallander called the daughter of Birgitta Medberg to tell her he was on his way, and then they drove to Skurup. The wind was now blustering. It was 3.08 on Wednesday, August 29.

CHAPTER 12

They stopped in front of a two-storey brick building quintessentially Swedish, Linda thought. Wherever you go in this country the houses all look the same. The central square in Västerås could be replaced with the one in Örebro, this Skurup apartment building could as easily be in Sollentuna.

"Where have you ever seen a building like this before?" she asked her father as they got out of the car and he was fumbling with the lock. He looked up at the brick façade.

"Looks like the place you had in Sollentuna, before you moved to the dorms at the police training college."

"Well remembered. So what do I do now?"

"Come with me. You can treat this as a warming-up exercise to real police work."

"Aren't you breaking some rules by doing this? No-one should be present at an interrogation without relevant cause – something like that?"

"This isn't an interrogation session, it's a conversation. Let's hope it will serve to put someone's mind at rest."

"But still."

"No buts. I've been breaking rules since I first started work. According to Martinsson's calculations I should have been locked up for a minimum of four years for all the things I've done. But who cares, if you're doing a good job? That's one of the few points Nyberg and I can agree on."

"Nyberg? The head of forensics?"

"The only Nyberg I know. He's retiring soon and in one sense no-one will be sorry to see the back of him. On the other hand, despite his terrible temper, maybe all of us will miss him."

They crossed the street. A bicycle missing its back wheel was propped up by the front door. The frame was bent as if it had been the victim of a violent assault. They walked into the entrance and read the names of the people who lived there.

"Birgitta Medberg. Her daughter's name is Vanya. From the phone call, I gather that she has a tendency to hysteria. She also has a very shrill voice."

"I am not hysterical!" a woman yelled from above. She was leaning over the railing of the staircase, watching them.

"Remind me to keep my voice down in stairwells," Wallander muttered.

They walked up to her landing.

"Just as I thought," Wallander said in a friendly voice to the hostile woman waiting for them. "The boys at the station are too young to know the difference between hysteria and a normal level of concern."

The woman, Vanya, was in her forties, heavy, with yellow stains around the neck and wrists of her blouse. Linda thought it was probably a long time since she had washed her hair. They walked into the flat and Linda immediately recognised the strong scent that hung in the air. Mum's perfume, she thought. The one she wears when she's upset or angry. She had another she preferred when she was happy.

They were shown into the living room. Vanya dropped into an armchair and pointed her finger at Linda. "Who's that?"

"An assistant," Wallander said in a firm voice. "Please tell us what happened, starting at the beginning."

Vanya told them in a nervous, jerky style. She seemed to have trouble finding the right words even though it was clear that she was not the kind of person who spoke in long sentences. Linda immediately understood her concern was genuine, and compared it to the way she felt about Anna.

Vanya told them that her mother was a cultural geographer whose principal work was tracing and mapping old roads and walkways in southern Sweden. She had been widowed for a year and had four grandchildren, two of whom were Vanya's daughters. On this particular day Vanya and her daughters were supposed to have visited her at noon. Birgitta had arranged to be out on one of her short excursions before then. But when Vanya arrived, Birgitta had not yet returned. Vanya waited for two hours, then called the police. Her mother would never have disappointed her grandchildren like this, she said. Something must have happened.

When Vanya had finished her story, Linda tried to guess what question her father would ask first. Perhaps something along the lines of: Where was she going?

"Do you know where she was going this morning?" he said.

"No."

"She has a car, I take it."

"Actually she has a red Vespa. Forty years old."

"Really?"

"All Vespas used to be red, my mother tells me. She belongs to an association for owners of vintage mopeds and Vespas. The office is in Staffanstorp, I think. I don't know why – why she wants to be with those people, I mean. But she seems to like them."

"You said she became a widow about a year ago. Did she show any signs of depression?"

"No. And if you think she's committed suicide you're wrong."

"I'm not saying she did. But sometimes even the people closest to us can be very good at hiding their feelings."

Linda stared at her father. He glanced briefly in her direction. We have to talk, she thought. It was wrong of me not to tell him about the time I stood on the bridge and was going to jump. He thinks the only time was when I slashed my wrists.

"She would never hurt herself. She would never do that to us."

"Is there anyone she may have gone to visit?"

Vanya had lit a cigarette. She had managed already to spill ash on her blouse and the floor.

"My mother is the old-fashioned type. She never drops in on someone without calling."

"My colleagues have confirmed that she hasn't been admitted to a hospital in the area, and there are no reports of an accident. Does she have a medical condition we should know about? Does she have a mobile phone?"

"My mother is a very healthy woman. She takes care of herself –

not like me, though it's hard to get enough exercise when you're in the grocery business." Vanya made a gesture of disgust at her body.

"A mobile phone?"

"She has one, but she keeps it turned off. My sister and I are always on at her about it."

There was a lull in the conversation. They heard the low sound of a radio or television coming from the flat next door.

"So, let's get this straight. You have no idea where she may have gone. Is there anyone who could have more specific information regarding her research? Is there a journal or working papers of some kind we could look at?"

"Not that I know of, and my mother works alone."

"Has this happened before?"

"That she's disappeared? Never."

Wallander took a notebook and a pen out of his jacket pocket and asked Vanya for her full name, home address and telephone number. Linda noticed that he reacted to her last name, Jorner. He stopped writing and looked up.

"Your mother's surname is Medberg. Is Jorner your husband's name?"

"Yes, Hans Jorner. My mother's maiden name was Lundgren. Is this important?"

"Hans Jorner – any connection to the gravel company in Limhamn?"

"Yes, he's the youngest son of the company director. Why do you ask?"

"I'm curious, that's all."

Wallander stood up and Linda followed suit.

"Would you mind showing us the flat? Does she have a study?"

Vanya pointed to a door and then put her hand to her mouth to smother an attack of smoker's cough. They walked into a study in which the walls were covered with maps. Stacks of papers and folders were neatly arranged on the desk.

"What was all that about Jorner?" Linda asked in a low voice.

"I'll tell you later. It's an unpleasant story."

"And what was it she said? She's a grocer?"

"Yes."

Linda leafed through a few papers. He stopped her immediately.

"You can come along, listen and look to your heart's content. But don't touch anything."

Linda left the room in a huff. He was right, of course, but his tone was objectionable. She nodded politely to Vanya, who was still coughing, and left the flat. As soon as she was down on the street, she regretted her childish reaction.

Her father emerged ten minutes later.

"What did I do? Is there something wrong?"

Linda made an apologetic gesture. "It's nothing. I've forgotten already."

He unlocked the car while the wind pulled and tugged at their clothing. They got into the car, but he didn't start the engine right away.

"You noticed my reaction when she said her name was Jorner," Wallander said and squeezed the steering wheel angrily. "When Kristina and I were little there were times when no art buyers had come to my father's studio in their fancy cars for a while. So we had no money. At those times Mother had to go to work. She had no education, so the only available occupations were on the assembly line or housekeeper. She chose the latter and landed a position with the Jorners, though she came home each night. Old man Jorner – Hugo was his name – and his wife Tyra were terrible people. As far as they were concerned, there had been no social change over the past fifty years. In their eyes the world was upper class and lower class and nothing in between. He was the worst.

"One time Mother came home completely devastated. Even your grandfather, who never talked to her much, wondered what had happened. I hid behind the sofa and will never forget what I overheard. There had been a dinner party at the Jorners', perhaps eight people. My mother served the food, and when the guests were ready for coffee, Hugo asked her to bring in a stool from the kitchen. They were all a bit tipsy by this point, and when she came in with it he asked her to climb up on it. She did as he asked and then he said that from her present vantage point she should be able to see that she had forgotten to lay a coffee spoon for one of the guests. Then he dismissed her, and she heard how everyone laughed as she left the room.

"I still remember it word for word. When she had finished she started crying, and said she would never go back. My dad was so upset he was ready to grab the axe from the woodshed and smash Jorner's head. But she managed to calm him. I'll never forget it. I was ten, maybe twelve, at the time. And now I meet one of his daughters-in-law."

He started the car, still angry.

"I often wonder about my grandmother," Linda said. "I think what I wonder about most was how anyone could stand to be married to my grandfather."

Wallander laughed. "She always used to say that if she just rubbed him with a little salt he did what she told him. I never really understood that – I remember thinking: how do you rub a person with salt? The secret was her patience. She had an infinite fund of patience."

Wallander stepped on the brake and swerved as a dashing convertible overtook them on a blind bend. He swore.

"I should pull them over."

"Why don't you?"

"My mind's on other things."

Linda looked over at her father, who did appear tense.

"There's something about this missing woman that bothers me," he said. "I think Vanya Jorner was telling us the truth, and I think her anxiety is genuine. My feeling is that Birgitta Medberg became sick or temporarily confused, either that or something has happened to her."

"Something criminal?"

"I don't know. But I think my day off is over. I'll take you home."

"I'll come with you to the station. I'll walk from there."

Wallander parked in the police-station garage and Linda started walking home in the wind that had become surprisingly cold. It was 4.30. She started in the direction of Mariagatan, but she changed her mind and went to Anna's building instead. She waited a minute after ringing the doorbell, then let herself in.

* * *

It took her only a few seconds to recognise that something was different. Certainly someone had been in the flat since she was here last. But by what evidence did she know this? Was something missing? She scrutinised the living-room walls and the bookcase. Nothing apparently changed there.

She sat in the chair that Anna preferred. Something *had* changed. But what? She got up and stood by the window to see the room from a different angle. That was when she saw it. There used to be a large blue butterfly in a frame hanging on the wall between a Berlin exhibition poster and an old barometer. The butterfly was gone. Linda shook her head. Was she imagining things? No, she was sure she remembered it being here last time. Could Henrietta have come and picked it up? On the face of it, it didn't make sense. She took off her coat and methodically went through the flat.

When she opened Anna's wardrobe she knew immediately that someone had been there too. Several items of clothing were gone, as well as a carrier bag. Linda could tell because Anna often left the wardrobe doors open. She sat on the bed and tried to think. Her gaze fell on Anna's journal lying on the desk. She must have left it behind, she thought and then corrected herself: Anna would never have left it behind. She might have taken the clothes, perhaps even the butterfly. But she would never have left her journal, not in a million years. Whoever it was who had been here, it wasn't Anna.

CHAPTER 13

Walking into an empty room was like dipping below the mirror-like surface of a still lake and sinking into the silent and alien underwater landscape. She tried to remember everything she had been taught. Rooms always bore the traces of what had happened in them. But had anything of note happened here? There were no blood stains, indeed, no sign of a struggle, nothing. A framed butterfly was missing, as well as a bag and some clothes. That was all. But even if it had been Anna who had stopped by to pick them up, she would have left the same number of traces as an intruder. All Linda had to do was find them. She walked through the flat again, but didn't see anything else.

Finally she stopped at the answering machine and played the messages. Anna's dentist had called to ask her to reschedule her annual check-up, Mirre had called from Lund to ask if Anna was going to go to Båstad or not, and finally there was Linda's own booming voice, urging Anna to pick up.

Linda took the address book from the telephone table and looked up the number of the dentist, Sivertsson.

"Dr Sivertsson's office."

"My name is Linda Wallander. I'm returning Anna Westin's calls as she's out of town for the next few days. Would you mind telling me the exact day and time of her appointment?"

The receptionist put her on hold for a few moments, then came back on the line.

"September the tenth at nine a.m."

"Thank you. She probably has it written down somewhere."

"I don't remember Anna ever missing an appointment."

Linda hung up and tried to find a phone number for Mirre. She thought about her own over-stuffed address book which she was forever patching with tape. Somehow she could never bring herself to buy a new one, it stored too many memories. All the crossed-out numbers were for her markers in a private graveyard. That led her thoughts away from Anna and to the moment in the forest with her father. A tenderness for him welled up inside her. She sensed what he had been like as a boy. A small kid with big ideas, maybe too big for his own good. There's so much I don't know about him, she thought. What I think I know often turns out to be wrong. I used to look on him only as a big friendly man who was not too sharp, but stubborn and with a pretty good intuition about the world. I've always thought he was a good policeman. But now I suspect he's much more sentimental than he appears, that he takes pleasure in the little romantic coincidences of everyday life and hates the incomprehensible and brutal reality he confronts through his work.

Linda pulled up a chair and turned the pages of a book about Alexander Fleming and the discovery of penicillin. Anna had obviously been reading it. It was in English and it surprised her that Anna was up to the challenge. They had talked about doing a language programme in England when they were younger. Had Anna gone and done it on her own? She put the book back and took up Anna's address book again. Every page was covered in numbers, like a blackboard during a lecture on advanced mathematics. There were numbers scratched out and changes on every page. Linda smiled nostalgically at a couple of her own old numbers, as well as the names and numbers of two of her ex-boyfriends. What am I looking for? I guess I'm trying to find traces of Anna that would explain what's happened. But why would they be here?

She kept going through the address book, still feeling that she was trespassing on Anna's most private self. I've climbed her fence, she thought. I'm doing it for her sake, but it still feels wrong.

Then the word "Dad" caught her eye, written boldly in red. The phone number was 19 digits long, all ones and threes. A number that doesn't exist, Linda thought. A secret number to the unknown city where all missing persons go.

She wanted to put the book away, but forced herself to look all the way to the end. The only other entry of any interest was a number to

"My Room in Lund". Linda hesitated, then dialled the number. Almost at once, a man answered.

"Peter here."

"I'd like to speak to Anna. Is she in?"

"I'll check."

Linda waited. She heard music in the background, but she couldn't put a name to the singer.

"No, she's not in. Can I take a message?"

"Do you know when she'll be back?"

"I don't even know if she's around. I haven't seen her for a while. But I can ask the others."

She waited again.

"No-one's seen her here for a couple of days."

Before Linda had a chance to get the address he cut the connection. She was left holding the receiver to her ear. No Anna, she thought. The man called Peter had clearly not been worried and Linda was starting to feel foolish. She thought about herself and of her own readiness to take off without leaving a note of where she was going. Her father had many times been on the verge of reporting her missing when she was younger. But I always sense when I'm gone too long and I always call in, she thought. Why wouldn't Anna do the same?

Linda rang Zeba and asked if she had heard from Anna. Zeba said no, there had been no news, no call. They agreed to meet for coffee the following day.

As she put the phone down, Linda thought: I'm staging this as a disappearance so as to have something to do. As soon as I can put my uniform on and actually start working, she'll turn up. It's like a game.

She went into the kitchen and made herself a cup of tea, then took it with her to Anna's bedroom. She sat on the side of the bed away from where Anna normally slept. Putting her tea down on the bedside table, she stretched out – just for a moment – but in the twinkling of an eye she had fallen asleep.

She didn't know where she was at first, when she awoke. She looked at the time – she had been asleep for an hour. The tea was cold. She drank it anyway because her mouth was dry, then stood up and straightened the bedspread. That was when she saw it.

On Anna's side. There was an indentation still visible: someone had lain there and not smoothed the bed afterwards. That couldn't have been Anna. She was the kind of person who never left crumbs on the table.

Linda lifted the bedspread on impulse and found a T-shirt, size XXL, dark blue, with the Virgin Airlines logo. She sniffed carefully and confirmed that it didn't smell like Anna – it had the masculine scent of aftershave or perhaps very strong deodorant. She laid the T-shirt on the bed. Anna preferred nightgowns, and classy ones at that. Linda was willing to bet that she would not have used a Virgin Airlines T-shirt even for one night.

The phone in the living room rang. Linda flinched, but walked into the living room and looked at the answering machine. Should she answer it? She stretched out her hand, then pulled it back. The machine answered after the fifth ring. *Hi, it's Mum. Your friend Linda – the one who wants to become a policewoman for some strange reason – came out here today looking for you. Just wanted to let you know. Call me when you get back. Bye.*

Linda ran the message again. Henrietta's voice sounded calm. There seemed to be no unvoiced anxiety, nothing hidden between the words. She picked up the implicit criticism of her choice of career. That bothered her. Did Anna share her mother's dismissive attitude? To hell with them, Linda thought. Anna can carry on her disappearance act without me. She walked around one last time, watering the plants, then left the flat.

By the time Wallander came home, around 7 p.m., she had cooked and eaten dinner. She heated up the food she had kept for him while he changed. She sat in the kitchen while he ate.

"What happened with the missing woman?"

"Svartman and Grönkvist are in charge of it. Nyberg is examining her flat. We decided to take her disappearance seriously. Now we can only wait."

"And what do you think?"

Wallander pushed his plate away.

"Something about it still worries me, but I could be wrong."

"What worries you?"

"Certain people ought not to go missing, that's all. It's not something they do – if it happens, it means something is wrong. I guess that's been my experience."

He got up and put on a pot of coffee.

"We had an estate agent once who went missing, about ten years ago. Maybe you remember it? She was religious, something evangelical. They had small children. The moment the husband came in to notify us she was missing I knew something bad had happened. And I was right. She had been murdered."

"But Birgitta Medberg is a widow and she doesn't have small children. She's probably not even religious – I certainly can't imagine that stout daughter of hers being religious, can you?"

"Her waistline's got nothing to do with it. You can't tell just by looking at someone. But I'm talking about something else, something unexpected."

Linda told him about her latest discovery in Anna's flat. She watched her father's face take on a strong look of disapproval.

"You shouldn't be getting yourself mixed up in this," he said. "If anything's happened to Anna it's a case for the police."

"I'm almost police."

"You're still a rookie, and the appropriate line of work for you is breaking up drunken brawls in town."

"I just think it's strange she's gone, that's all."

Wallander carried his plate and his coffee cup to the sink.

"If you're genuinely concerned you should go to the police."

He left the kitchen. Linda stayed behind. His ironic tone irritated her, not least because he was right.

She sulked in the kitchen until she felt ready to see her father again. He was in the living room, fast asleep in a chair. Linda shook his arm when he started to snore. He jerked awake and raised his arms as if to ward off an attack. Just like me, she thought. That's another thing we have in common. He went to the bathroom, then got ready for bed. Linda watched a film on television without really concentrating. By midnight she too was in bed. She dreamed about her former boyfriend, Herman Mboya, who was back in Kenya.

The buzzing of her mobile phone woke her. It vibrated next to the bedside lamp. She answered it at the same time as she checked her watch. Three fifteen. There was no voice on the other end, only breathing. Then the line went dead. Linda knew it had something to do with Anna. It was some sort of message, even if it only consisted of a few breaths. It had to mean something.

Linda never managed to fall back to sleep. Her father got up at 6.15. She let him shower and change in peace, but when he started making noise in the kitchen she joined him. He was surprised to see her up and dressed at that hour.

"I'm coming with you."

"Why?"

"I thought about what you said, that if I was worried about Anna I should raise the matter with the police. Well, I *am* worried, and I'm going to report her disappearance. I do think something is badly wrong."

CHAPTER 14

Linda had never learned to predict when her father would fall into one of his rages. She remembered with painful clarity how she and her mother would cringe when he got like this. Her grandfather was the only one who simply shrugged it off or gave as good as he got.

By now she had learned to look for certain signs; the tell-tale red patch on the forehead, the nervous pacing. But this morning she was once more taken by surprise at the vehemence of his reaction to her decision to report Anna's disappearance. He started by throwing a pack of paper napkins to the floor. There was a comical element to this gesture since the anticipated violence of the crash never came, and the white papers fluttered softly across the kitchen. Yet it was enough to reawaken Linda's childhood fear. She recalled what Mona had said after the divorce: "He can't see it himself. He doesn't know how intimidating it is to be met with a raging temper when you least expect it. Others probably think of him as a friendly, slightly eccentric perhaps, but capable policeman, which is probably a fair assessment of him in the workplace. But at home he let his temper run loose like a wild animal. He became a terrorist in my eyes. I feared him, and I also grew to hate him."

Linda thought of her mother's words as she sat across from her giant of a father, still furious, now kicking at the napkins.

"Why don't you listen to me?" he was saying. "How are you ever going to be a respected police officer if you think a crime has been committed every time one of your friends doesn't pick up the phone?"

"Dad, it's not like that."

He swept the rest of the napkins off the table. A child, Linda thought. A big child chucking all his toys to the ground.

"Don't interrupt me! Didn't they teach you anything at the training college?"

"I learned to take things seriously."

"You're going to be laughed out of the force."

"So be it, but Anna has disappeared."

His rage subsided as suddenly as it had started. There were still a few drops of sweat on his cheek. That was short, Linda thought. And not as volcanic as I remember. Maybe he's more afraid of me now, or else he's getting old. I bet he even apologises this time.

"I'm sorry," he said.

Linda didn't answer. She picked up a few napkins from the floor and put them in the bin. Her heart was still pounding with fear. I'll always feel this way when he's angry, she thought.

"I don't know what gets into me."

Linda stared at him, waiting to speak until he actually looked at her.

"You just need to do a bit more exercise."

He flinched as if she had struck him, then he blushed.

"You know I'm right," she said. "Anyway, you should get going. I'll walk so you don't have to be embarrassed."

"I was planning to walk myself, as it happens."

"Do it tomorrow. I need some space. I don't like it when you scream like that."

Wallander set off, meekly, as instructed. Linda was drenched in sweat so she changed her top. By the time she left the flat she had still not made up her mind whether she was going to report Anna's disappearance.

The sun was shining and there was a brisk breeze. Linda paused out on the street, unsure of what to do next. She prided herself on being as a rule a decisive person, but being around her father sometimes sapped her will power. She could hardly wait until she was able to move into her flat behind Mariakyrkan. She couldn't stand living with him much longer.

Finally she did steer her course to the police station. If something really had happened to Anna, she would never forgive herself for not reporting it. Her career as a police officer would be over before it had begun.

She walked past the People's Park and thought about a magician she had seen there as a child when she had been out with her father. The magician had taken gold coins out of children's ears. This memory gave rise to another, one that had to do with a fight between her parents. She had woken up in her room to the sound of their angry voices. They had been arguing about money, some money that should have been in the account, that was gone, that had been frittered away. When Linda had carefully tiptoed to her door and peered into the living room she had seen her mother with blood running from her nose. Her father had been looking out of the window, his face sweaty and flushed. She immediately realised that he had hit her mother, on account of the money that wasn't there.

Linda stopped walking and squinted up at the sun. The back of her throat was starting to constrict. She remembered looking at her parents, thinking that she was the only one who could solve their problems. She didn't want Mona to have a bloody nose. She had gone back into her bedroom and taken out her piggy bank. She walked into the living room and put it on the table. The room fell silent.

She kept squinting up at the sun, but the tears came anyway. She rubbed her eyes and changed direction, as if this would force her mind to change track. She turned on to Industrigatan and decided to postpone reporting Anna's disappearance. Instead she would drop into the flat one last time. If anyone's been there since last night, I'll know, she thought. She rang the doorbell – no answer. When she opened the front door her whole body was tense, every antenna alert. But there was nothing.

She walked around in the flat, looking at the bed where she had lain the night before. She sat down in the living room and went through what had happened. Anna had now been gone for three days.

Linda shook her head angrily and went back into the bedroom. She apologised to the air and started looking through the journal again. She flipped back about 30 days. Nothing. The most notable entry was an aching tooth on August 7 and 8 and a resulting appointment with Dr Sivertsson. Linda remembered those days and furrowed her brow.

On August 8 she, Zeba and Anna had taken a long walk out at Kåseberga. They had taken Anna's car. Zeba's boy was cooperative for once and they had taken turns carrying him when he was too tired to walk.

But a toothache?

Linda had the feeling again that there was some kind of double language in Anna's journal, perhaps a code. But why? And what could an entry about toothache possibly signify?

She kept reading and looked closely at the handwriting itself. Anna frequently changed her pen, even in the middle of a sentence. Perhaps she was interrupted by the phone and couldn't find the same pen when she came back. Linda put the journal down and went to get a glass of water from the kitchen.

When she turned the next page she drew a breath. At first she was sure she was getting it confused. But no, there it was: on August 13 Anna had written *Letter from Birgitta Medberg.*

Linda read it again, this time by the window with the sun on the page. Birgitta Medberg was not a common name. She put the book down on the windowsill and picked up the directory. It took her just a few minutes to confirm that there was only one Birgitta Medberg in this whole area of southern Sweden. She called enquiries and asked about Birgitta Medbergs in the rest of the country; there were only four others with that name. Of the five, one was listed as a cultural geographer in Skåne.

Linda returned to the journal and read impatiently through the text until she reached the odd message at the end: *myth, fear, myth, fear*. But there was no other reference to Birgitta Medberg.

Anna disappears, she thought. A few weeks earlier she received a letter from Birgitta Medberg who has now also gone missing. In the middle of all this is Anna's father whom she thinks has just reappeared on a street in Malmö after a 24-year absence.

Linda began to search the flat for Birgitta Medberg's letter. She no longer felt guilty for violating Anna's privacy. She found a number of letters over the next three hours. Unfortunately, the letter from Medberg was not one of them.

Linda took Anna's car keys and left the flat. She drove down to the Harbour Café and had a sandwich and a cup of tea. A man her own age in oil-spattered overalls smiled at her as she was getting ready to leave. It took her a while to recognise him as a contemporary from school. She stopped and they said hello. Linda struggled in vain to remember his name. He stretched out his hand after first wiping it clean.

"I'm sailing," he said. "I have an old boat with a dud motor. That's why I'm covered in grease."

"I've only just moved back to town," Linda said.

"What do you do?"

Linda hesitated. "I've just graduated from the police training college."

His name suddenly came back to her: Torbjörn. He smiled at her again.

"I thought you were into old furniture."

"I was, but I changed my mind."

He stretched out his hand again. "Ystad is pretty small. I'm sure I'll see you around."

Linda hurried to the car, which was parked behind the old theatre. I wonder what they'll think, she thought. I wonder if they'll be surprised that Linda became a policewoman.

She drove to Skurup, parked on the main square and then walked to the house where Birgitta Medberg lived. There was a strong smell of cooking in the stairwell. She rang the doorbell, there was no answer. She listened, then called through the letter box. When she was sure no-one was there, she took out her pass keys and opened the door. I'm starting my police career by breaking and entering, she thought. She was sweating and her heart was thumping. Alert for any noise, she searched through the flat. She was all the time afraid someone was going to come in. She didn't know exactly what it was she was hoping to discover here, just that it would be something to confirm the connection between Anna and Birgitta Medberg.

She was about to give up when she found a paper under the green writing pad on the desk. It was a photocopy of an old surveyor's map on which the lines and words were hard to make out.

Linda turned on the desk lamp and was able to make out the writing on the bottom of the page: Rannesholm Estate. She recognised the name, but where was it exactly? She had seen a map of southern

Sweden in the bookcase. She took it out and found Rannesholm. It was a few miles north of Skurup. Linda looked at the older map again. Even though it was a poor copy she thought she could see the outlines of some notes and arrows. She tucked both maps into her coat, turned off the light and checked for noise through the letter box before letting herself out of the flat.

It was 4.00 by the time she reached a public car park in the nature reserve at Rannesholm. What am I doing here? she asked herself. Am I just playing a game to pass the time? She locked the car and walked down to the lake. A pair of swans was out in the middle where the wind sent ripples across the surface of the water. Rain clouds were moving in from the west. She zipped up her jacket. It was still summer, but there was an unmistakable feeling of autumn in the air. She looked back at the car park. It was empty except for Anna's car. She tossed a few pebbles into the lake. There is a connection between Medberg and Anna, she thought. But what could they have in common? She threw another stone into the water. The only thing I can think of that links them is the fact that both of them have disappeared. The police are investigating one case, but not the other.

The rain came sooner than she had expected. Linda ducked under a tall oak tree next to the car park. Raindrops fell all around and the whole situation seemed completely idiotic. She was about to brace herself for a run through the rain to the car when she saw something glittering between the wet branches of a bush nearby. At first she thought it was a discarded beer can. She pulled one of the branches aside and saw a black tyre. She started pulling the branches away with both hands and her heart beat faster. Then she ran back to the car to get her phone. For once her father had his mobile phone with him and turned on.

"Where are you?" he said.

His voice was unusually gentle. She could tell that he was still trying to make up for the morning.

"I'm at Rannesholm Manor," she said. "In the car park."

"What are you doing there?"

"Dad, there's something you need to see."

"I can't. We're about to have a meeting about crazy new directives from Stockholm."

"Skip it. Just get over here – I've found Birgitta Medberg's Vespa."

She heard her father's sharp intake of breath on the other end.

"Are you sure?"

"Yes."

"And how did this happen?"

"I'll tell you when you get here."

There was a noise on the line and the connection was broken, but Linda didn't bother calling him again. She knew he was on his way.

CHAPTER 15

It was raining even harder now. Linda saw something flashing through the windscreen and turned on the wipers. It was her father's car. He parked, ducked out and jumped into the passenger seat. He was impatient, clearly in a hurry.

"Tell me."

Linda told him about the journal. His impatience made her nervous.

"Do you have it with you?" he interrupted.

"No. Why? I've given you what it says word for word."

He said no more and she continued her account. When she had finished, he sat and stared out at the rain.

"A strange story," he said.

"You always say to watch for the unexpected."

He nodded, then looked her over.

"Did you bring a waterproof?"

"No."

"I have one you can borrow."

He pushed the door open and ran back to his car. Linda was amazed to see her large, heavy-set father move so quickly, with such agility. She got out into the rain.

He was at the back of his car, putting on his gear. When he saw her, he handed her a waterproof coat that came all the way down to her ankles. Then he fished out a baseball cap with the logo of a local car-repair shop and planted it on her head. He looked up at the sky. The rain poured down his face.

"It's Noah and the flood all over again," he said. "I don't remember rain like this since I was a child."

"It rained a lot when I was young," Linda said.

He nudged her on and she led the way over to the oak tree and pushed the bushes away so he could see. Wallander took out his mobile phone and she heard him call the police station. He grumbled when they didn't pick up right away. Wallander read out the number plate and waited for confirmation. It was her Vespa. Wallander put the phone back in his pocket.

The rain stopped at that exact moment. It happened so fast it took them a while to register what had happened. It was like rain on a film set being turned off after the take.

"God has decided to take pity on us," Wallander said. "You've found Birgitta Medberg's Vespa."

He looked around.

"But no Birgitta Medberg."

Linda hesitated, then pulled out the Xerox copy of the old map that she had found in Birgitta Medberg's flat. She regretted it as soon as she had taken it out, but it was too late.

"What's that?"

"A map of the area."

"Where did you find it?"

"Here on the ground."

He took the dry piece of paper from her and gave her a searching look. Here comes the question I won't be able to answer, she thought.

But he didn't ask her. Instead he studied the map, looked down to the lake and the road, at the car park and the various paths that branched out from it.

"She was here," he said. "But this is a big park."

He studied the area right around the Vespa. Linda watched him, trying to read his mind.

Suddenly he looked at her. "What's the first question we should be asking?"

"Whether she hid the Vespa for the long term, or was she only trying to protect it from being vandalised while she was doing her work?"

He nodded. "There's a third alternative, of course."

Linda understood what he was getting at. She should have thought of it right away.

"That someone else hid it."

"Exactly."

* * *

A dog came running out from one of the paths. It was white with little black spots. Linda couldn't remember what that breed was called. Then another and finally a third appeared out of the forest, followed by a woman in rain gear from head to toe. She was walking briskly and put all three dogs on leads when she caught sight of Linda and Wallander. She was in her forties, tall, blonde and attractive. Linda saw her father react instinctively to the presence of a good-looking woman: he stood up straight, pulled up his head to make his throat appear less wrinkled and held in his stomach.

"Excuse me," he said. "My name is Wallander and I'm with the Ystad police."

The woman looked at him sceptically. "May I see your identification?"

Wallander dug out his wallet and presented his ID card, which she studied closely.

"Has anything happened?"

"No. Do you often walk your dogs in this area?"

"Twice a day, actually."

"You must know these paths very well."

"Yes, I would say I do. Why?"

He ignored her last question. "Do you meet many people in the forest?"

"Not during the autumn. Spring and summer there are a fair number of people in the park, but soon it will only be dog owners who make the effort. That's always a relief. Then I can let the dogs off the lead."

"But aren't they supposed to stay on the lead all year round? That's what the sign says."

He pointed at a sign a few feet away. She raised her eyebrows.

"Is that why you're here? To catch people who let their dogs run loose?"

"No. There's something I'd like to show you."

The dogs strained on their leads while Wallander drew away some branches to reveal the Vespa.

"Have you ever seen this scooter before? It belongs to a woman in her mid-sixties by the name of Birgitta Medberg."

The dogs wanted to pull forward and sniff it, but were firmly restrained by their owner. Her voice was steady and without hesitation.

"Yes," she said. "I've seen both the Vespa and the woman. Quite a few times."

"When did you see her last?"

She thought about it.

"Yesterday."

Wallander threw a quick glance at Linda, who was standing to one side, listening intently.

"Are you sure?"

"No, not completely. But I think it was yesterday."

"Why can't you be sure?"

"I've seen her so often over the last few weeks."

"The last few weeks – can you be more precise?"

She thought again before answering.

"I suppose all through July, perhaps the last week of June. That was when I first saw her. She was walking on a path on the other side of the lake and we stopped and chatted for a bit. She told me she was mapping old walking trails around Rannesholm. I saw her again from time to time after that. She had many interesting things to tell. Neither I nor my husband had any idea that there were pilgrim trails on our property. We live in the manor," she said. "My husband manages an investment fund. My name is Anita Tademan."

She looked at the Vespa again and her expression became anxious.

"Is something wrong?"

"We don't know. I have one more question for you. When last you saw her, which path was she on?"

Anita Tademan pointed over her shoulder.

"That one I was just on. It's a good one when it rains because the canopy is so thick. She found a completely overgrown path in there which starts about five hundred metres into the forest next to a fallen beech. That was where I last saw her. Can't you tell me what this is all about?"

"She may have disappeared. We're not sure."

"How awful . . . That nice woman."

"Was she always on her own?" Linda said.

Wallander looked at her with surprise, but did not look angry.

"I never saw her with anyone," Anita Tademan said. "And if that's all your questions, I must be on my way."

She let the dogs off their leads and started up the road to the castle. Linda and her father stood watching her for a while.

"A beauty."

"Snobby and rich," Linda said. "Hardly your type."

"Never say never," he said. "I know how to behave in polite society. Both your mother and your aunt have taught me well."

He looked down at his watch and then up at the sky.

"We'll go the five hundred metres and see if we find anything."

He started down the path at a rapid pace. She followed him and had to half run to keep up. A strong scent of wet earth rose from the forest floor. The path wound around boulders and the exposed roots of old trees. They heard a pigeon fly off from a branch, and then another.

Linda was the one who spotted it. Wallander was walking so fast he hadn't seen where a thin path forked off. She called out to him and he backtracked.

"I was counting," she said. "This is about four hundred and fifty metres in."

"The woman said five hundred."

"If you don't count every step, five hundred can feel like four or six hundred metres, depending."

"I know how to judge distances," he said, irritated.

They started following the new path, that was only barely visible. But both of them noted soft imprints. One pair of boots, Linda thought. One person.

The path led them deep into a part of the forest that looked untouched. They stopped at the edge of a shallow ravine that cut through the forest. Wallander crouched down and picked at the moss with his finger. They made their way carefully into the ravine. At one point Linda's foot caught in some roots and she fell. A branch broke and sounded like a gunshot. They heard birds fly up all around them, although they couldn't see them.

"Are you all right?"

Linda brushed the damp dirt from her clothes. "I'm fine."

Wallander made his way through the brush and Linda followed closely. He parted a few of the branches in front of them and she saw a small hut. It was like something out of a fairy tale, the house of a witch. The shack leaned up against the rock face. A pail lay half buried in the earth outside the door. Both of them listened attentively for sounds, but there were none. Only raindrops tapping leaves.

"You wait here," Wallander said, and walked up to the hut.

Wallander opened the door and looked in and flinched, at the same time stepping back. Linda caught up with him and pushed past him, peering into the interior. At first she didn't know what she was looking at. Then she realised that they had found Birgitta Medberg. Or, more precisely, what remained of her.

PART 2

The Void

CHAPTER 16

What Linda saw through the open door, what had caused her father to flinch and stumble backwards, resembled something she had once seen as a child. The image flickered to life in her mind. She had seen it in a book Mona had inherited from her mother, the other grandmother Linda had never met. It was a large book with old-fashioned type, a book of Bible stories. She remembered the full-page illustrations, protected by a translucent sheet of tissue paper. One of the pictures depicted the scene she was now witnessing at first hand, with only one difference: in the book the picture had shown a man's head with eyes closed, placed on a gleaming tray, a woman dancing in the background. Salome with her veils. That picture had made an almost unbearably strong impression on her.

Perhaps it was only now, when the picture had escaped from the page, her memory resurrected in the guise of a woman, that the moment of childhood horror was fully replaced. Linda stared at Birgitta Medberg's severed head on the earth floor. Her clasped hands lay close by, but that was all. The rest of her body was missing. Linda heard her father groan in the background, then she felt his hands on her back as he dragged her away.

"Don't look!" he shouted. "You shouldn't see this. Turn back."

He slammed the door shut. Linda was so scared she was shaking. She scuttled back up the side of the ravine, ripping her trousers in the process. Her father was at her heels. They ran until they reached the main path.

"What is going on?" she heard him mutter under his breath. "What's happened?"

He called the station and raised the alarm, using code words that she knew were designed to slip under the noses of journalists and inquisitive amateurs listening in on police radio frequencies. Then they returned to the car park and waited. Fourteen minutes went by until they heard the first siren in the distance. They had said nothing to each other during their wait. Linda was shaken and wanted to be with her father, but he turned his back and walked a few steps away. Linda had trouble making sense of what she had seen. At the same time another fear was mounting, a fear that this was somehow connected with Anna. What if there is a connection? she thought despairingly. And now one of them is dead, butchered. She interrupted her train of thought and crouched down, suddenly faint. Her father looked over at her and started towards her. She forced herself to stand and shook her head at him as if to say it was nothing, a momentary weakness.

Now she was the one who turned her back. She tried to think clearly – slowly, deliberately, but above all clearly. An officer who can't think clearly cannot do her job. She had written this statement out and pinned it to the wall next to her bed. She knew she always had to keep her cool, but how was she supposed to do that when right now she felt like bursting into tears? There was no trace of calm in her mind, only terrible flashes of the severed head and the clasped hands. And even worse, the question of what had happened to Anna. She couldn't keep new images from forming in her mind: Anna's head, Anna's hands. John the Baptist's head on a platter and Anna's hands, her head and Birgitta Medberg's hands.

The rain had started again. Linda ran over to her father and showered his chest with blows.

"Now do you believe me? Don't you realise something must have happened to Anna?"

Wallander grabbed her shoulders, trying to keep her at arm's length.

"Calm down. That was the Medberg woman back there, not Anna."

"But Anna wrote in her journal that she knew her. And now Anna is gone. Don't you get it?"

"You have to calm down. That's all."

Linda slowly regained control of herself, or rather, felt a paralysis settle over her. Three, then four police cars came slipping and sliding into the muddy car park. The police officers got out and gathered around Wallander after quickly having thrown on the rain gear that they all seemed to keep stashed in the back of their cars. Linda stood outside the circle, but no-one tried to stop her when she eventually joined them. Martinsson was the only one who acknowledged her, with a nod, but even he never asked her what she was doing there. At that moment, in the rainy car park by Rannesholm Manor, Linda cut the cord to her life at the police training college. She fell in line behind the others and followed them in their long train into the forest. When a crime technician dropped a lamp stand she picked it up and carried it for him.

She stayed there while day turned into dusk and finally evening. Rain clouds came and went, the ground was saturated with moisture, the lights erected around the site casting strong shadows. The crime technicians painstakingly marked out a working path to the hut. Linda took care not to get in their way and she never put her foot down without placing it in someone else's footprint. Sometimes her father met her gaze, but it was as if he could not really see her. Ann-Britt Höglund was always at his side. Linda had bumped into her from time to time since coming back to Ystad, but she had never liked her. In fact she felt her father would do as well to stay away from her. Höglund had barely greeted her today, and Linda sensed she would not be an easy person to work with, if that ever became the case. Höglund was a detective inspector, and Linda was a rookie who hadn't even started working as yet and who would be busy breaking up street fights until she might one day earn the opportunity to apply for a more specialised line of work.

She watched her future colleagues go about their business, noting the order and discipline that seemed always to be on the verge of giving way to sheer chaos. From time to time someone raised their voice, especially the irritable Nyberg, who often swore at his team for not watching where they put their feet. Three hours after they had arrived, the human remains were removed from the scene, enclosed in thick plastic. Everyone stopped working as they were carried away. Linda could see the contours of Birgitta Medberg's head and hands through the plastic casing.

Then everyone resumed their work. Nyberg and his men crawled around on hands and knees; someone was sawing off branches and clearing away the underbrush, others were setting up lamps or repairing generators. People came and went, phones rang, and in the middle of all this her father stood rooted to one spot as if restrained by invisible cords. Linda felt sorry for him, he looked so lonely standing there, always available to answer a steady stream of questions, making snap decisions so that the investigation could proceed. He's walking a tightrope, Linda thought. That's how I see him. A nervous tightrope walker who should go on a diet and address the issue of his loneliness once and for all.

He only realised much later that she was still there. He finished talking to someone on his phone, then turned to Nyberg, who was holding out an object for him to look at. He held it in the beam of one of the strong lights that attracted insects and burned them to death. Linda took a step closer to see what it was. Nyberg handed Wallander a pair of rubber gloves that he pulled on to his big hands with some difficulty.

"What's this?"

"If you weren't completely blind you would see it was a Bible."

Wallander didn't seem to take any notice of Nyberg's tone.

"A Bible," Nyberg repeated. "It was on the ground next to the hands. There are bloody fingerprints. But they could belong to someone else, of course."

"The murderer?"

"Possibly. The whole hut is spattered with blood. It was quite a gory scene. Whoever did this must have been drenched in it."

"No weapons?"

"Nothing at this point. But this Bible is worth a closer look, even apart from the fingerprints."

Linda drew nearer as her father put on his glasses.

"Open it to the Book of Revelations," Nyberg said.

"I don't know my way around this thing. Just tell me what's in there."

But Nyberg would not let himself be hurried.

"Who knows their Bible any more? But the Book of Revelations is an important chapter, or whatever the parts are called."

He turned to Linda. "Do you know? In the Bible, is it called a chapter?"

Linda gave a start. "No idea."

"You see, the young are no better than we. Whatever. The thing is that someone has written comments between the lines. See?"

Nyberg pointed to a page. Wallander held it closer to his eyes.

"I see some grey smudges. Is that what you mean?"

Nyberg called out to someone named Rosén. A man with mud up to his chest came clomping over with a magnifying glass. Wallander tried again.

"Yes, someone has been writing between the lines. What does it say?"

"I've made out two of the lines," Nyberg said. "It seems as if whoever wrote in here wasn't happy with the original. Someone has taken it upon themselves to improve on the word of God."

Wallander removed his glasses. "What does that mean, 'the word of God'? Can you try to be more specific?"

"I thought the Bible *was* the word of God. How much more specific do I have to get? I just think it's interesting that someone should rewrite passages in the Bible. Is that something a normal person does? A person in basic possession of his or her senses?"

"A lunatic, then. But what is this hut? A place someone is living in or just hiding in?"

Nyberg shook his head. "Too early to say. But can't they be the same thing for someone who wants to stay out of the public eye?" Nyberg gestured out to the forest, impenetrably dark beyond the spotlights. "We've had dogs searching the area and I think they're still out. The units claim the terrain is all but impassable. If you needed a hide-out you couldn't pick a better place."

"Any clues as to who might have been using it?"

Nyberg shook his head. "There are no personal effects, no clothes. We can't even be sure if it was a man or a woman."

A dog started barking somewhere in the darkness as a light rain began to fall. Höglund, Martinsson and Svartman emerged from various directions and gathered around Wallander. Linda hovered in the background, part participant, part spectator.

"Give me a scenario," Wallander said. "What happened here? We know a repulsive murder took place – but why? Who did it? Why did Medberg come here? Did she plan to meet someone? Was she even killed here? Where is the rest of the body? Tell me."

The rain continued to fall. Nyberg sneezed. One of the spotlights went out. Nyberg kicked the light over then helped set it up again.

"A picture of what happened," Wallander said.

"I have seen a lot of things that qualify as repulsive," Martinsson said. "But nothing like this. Whoever did this was truly fucked up. But where the rest of the body is, or who used this hut, we don't know yet."

"Nyberg found a Bible," Wallander said. "We'll run prints on everything, of course, but it turns out someone's written new text between the lines in the book. What does that tell us? We have to see if the Tademans ever visited this place. We may have to go door-to-door for answers. We'll maintain an investigation with a broad front, working round the clock."

No-one spoke.

"We have to get this lunatic," Wallander said. "The sooner the better. I don't know what this is, but I'm scared."

Linda stepped into the light. It was like stepping on to the stage without having learned your lines.

"I'm scared too."

Wet, tired faces turned to her. Only her father looked tense. He's going to explode, she thought. But she had to do this.

"I'm scared too," she repeated. And then she told them about Anna. She made a point of not looking at her father as she spoke. She tried to remember all the details – omitting the parts about her intuitive fears – and to present all the facts.

"We'll look into it," her father said when she had finished. His voice was ice cold.

Linda instantly regretted what she had done. I didn't want to, she thought. I did it for Anna's sake, not to get back at you.

"I know," she said. "I'm going home now. There's no reason for me to be here."

"You found the Vespa, didn't you?" Martinsson asked. "Isn't that right?"

Wallander nodded and turned to Nyberg.

"Can you spare someone who can escort Linda to her car?"

"I'll do it," Nyberg said. "I have to use the toilet up there anyway. Can't do my business in the forest – the dogs' noses are far too sensitive."

Linda clambered up out of the ravine, only now realising how tired and hungry she was. Nyberg's strong torch lit up the path for her. They ran into a dog unit on the way, the dog's tail drooping behind him. Other lights glimmered among the trees. Night orienteering, Linda thought. Police officers hunting for clues in the dark. Nyberg muttered something unintelligible when they reached the car park, then he was gone. Linda got into her car, someone lifted the yellow tape to let her past, and she was out on the road. There were onlookers all along the road to the main road, people in parked cars waiting for something to happen, to see something. She felt as if her invisible uniform was back on. Go home! she thought. There's a brutal murder to be solved and you're getting in the way of our work. But then she shook the thoughts away. She wasn't a policewoman, not quite yet.

After a while she noticed she was driving too fast and slowed down. A hare sprang out on to the road. For a brief moment his eye was frozen in her headlights. She slammed on the brakes. Her heart was beating hard. She took a few deep breaths. Lights from other cars came at her and she decided to turn into a car park. She turned off the lights first, then the engine. Darkness settled in all around her. She got out her mobile phone, but it rang before she had a chance to dial the number. It was her father. He was furious.

"Do you know what you did back there? You were telling me I didn't know how to do my job."

"I didn't say anything about you," she said. "I'm just afraid that something's happened to Anna."

"Don't ever do anything like that again. Ever. If you do, I'll make sure your stint in Ystad is over before you know it."

She didn't have a chance to answer. He hung up. He's right, she thought. I didn't think, I just started to talk. She was about to dial his number to apologise or at least explain herself, but she realised there was no point. He was still angry and it would take a couple of hours before he'd be ready to hear her out.

Linda needed to talk to someone, and she dialled Zeba's number. The line was busy. She counted slowly to 50 and dialled again. Still busy. Then without knowing why, she dialled Anna's number. Busy. Linda was startled and tried again. Still busy. She breathed a sigh of relief. Anna's back, she thought. She started the engine, turned on the headlights and swung out on to the road. Good God, she thought. I'll have to tell her everything that happened just because she didn't turn up that night.

CHAPTER 17

Linda got out of the car and stared up at the windows of Anna's flat. They were dark. Her fear returned; the phone had been busy. Linda called Zeba again. She answered right away, as if she had been waiting by the phone. Linda talked in a hurry, stumbling over her words.

"It's me. Were you talking to Anna just now?"

"No."

"Are you sure?"

"Of course I'm sure! Have you been trying to call? I was explaining to my brother why I'm not going to lend him any money. He's a spendthrift. I have four thousand kronor in the bank and that's the extent of my fortune. He wants to borrow the whole lot to buy a share in a trucking venture involving cargo transports to Bulgaria . . ."

"To hell with him," Linda interrupted. "Anna's vanished. She's never stood me up before."

"Well, there has to be a first time."

"That's what my father says, but Anna's been gone for three days."

"Maybe she's in Lund."

"No. It doesn't even matter where she is. It's just not like her to be gone like this. Has she ever done this to you – not been on time for a date or not been at home when she had invited you over?"

Zeba thought it over. "Actually, no."

"See."

"Why are you so worked up over this?"

Linda almost told her about the severed head and hands. But that would mean breaking her professional code.

"I don't know. You're right – I'm worked up over nothing."

"Come over."

"I don't have time."

"I think you're going crazy with all this free time on your hands. But I have something for you, a mystery that needs solving."

"What is it?"

"A door I can't get to open."

"Can't do it, sorry. Call the property manager."

"You need to slow down."

"I will. See you."

Linda rang the doorbell in the hope that the windows were dark because Anna was asleep. But the flat was empty still and the bed untouched. Linda looked at the phone. The receiver was in place and the message light wasn't blinking. She sat down and turned over in her mind everything that had happened in the last three days. Every time an image of the severed head flashed through her mind she felt ill. Or were the hands worse? What kind of a maniac cut off a person's hand? Cutting off a person's head was a way to kill them, but their hands . . . She wondered if the medical people would be able to determine if Birgitta Medberg's hands had been cut off before or after she died. And where was the rest of the body? Suddenly her nausea got the better of her. She only just made it to the lavatory before she has sick. Afterwards she lay on the bathroom floor. A little yellow rubber duck was stuck under the bathtub. Linda stretched out her hand to touch it, remembering when Anna had got the duck.

It had been a long time ago. They had been maybe twelve. She couldn't remember whose idea it was, but between them they had decided to go to Copenhagen together. It was spring and they were bored and restless at school. They covered for each other when one of them cut class, which happened more and more frequently. Mona had given her permission, but her father wouldn't hear of it. She heard him describe Copenhagen as a den of sin and iniquity, a beast waiting to consume two very young girls who knew nothing about life. In the end Anna and Linda had gone anyway. Linda knew there would be trouble waiting for her when she returned, so as a kind of advance revenge she lifted

100 kronor from her father's wallet before she left. They took the train to Malmö and the ferry to Copenhagen. To Linda it seemed like their first serious excursion into the adult world.

It had been breezy but sunny, a happy, giggly day. Anna won the rubber duck at an amusement stand at the Tivoli and at first all of their experiences were transparent, joyous ones. They had their freedom and their adventure. Invisible walls fell down around them wherever they went. Then the image darkened. Something happened that day that was the first real blow to their friendship. We were sitting on a green bench, Linda remembered. Anna had been borrowing money from me all day because she was broke. She had to go to the toilet and asked me to hold her handbag. Somewhere in the background a Tivoli orchestra was playing. The trumpet was out of tune.

Linda was thinking of all this lying on the bathroom floor. The warmth from the heating system under the tiled floor felt good against her back.

It was a green bench and a black bag. After all these years she couldn't say what had made her open the handbag. There had been two crisp 100-kronor notes inside, not even crumpled or hidden within a secret compartment. She had stared at the money and felt a stab of betrayal. She closed the bag and decided she wouldn't say anything, but when Anna came back and asked if Linda would buy her a Coke, something exploded inside her. They stood there shouting at each other. Linda had forgotten what Anna had said in her defence, but they had gone their separate ways and had sat apart on the return ferry to Malmö. It took them a long time to start speaking to each other again. They never talked about what had happened in Copenhagen, but they did eventually manage to resume their friendship.

Linda sat up. There are lies at the heart of this, she thought. I'm sure Henrietta concealed something from me when I was there, and I know Anna is capable of lying. I discovered that in Copenhagen and I've found her out on later occasions as well. But with her at least I know her so well that I can tell when she is telling the truth. The story she told me about seeing her father – or a doppelgänger – in Malmö is true. But what's behind all this? What didn't she tell me? Sometimes the part that's left out is the biggest part of the lie.

Her mobile phone rang. She knew it was her father. She got to her feet to be well balanced in case he was still angry, but his voice only told her that he was tense and tired. Her father had more voices than other people had, it seemed to her.

"Where are you?" he said.

"In Anna's flat."

She could hear that he was still in the forest. There were voices of people walking past, the scrape of walkie-talkies and a dog barking sharply.

"What are you doing there?" he said after a while.

"I'm more afraid now than I was before."

To her surprise he said: "I know. That's why I'm calling. I'm on my way over. I need to hear about this in more detail. There's no reason for you to worry, of course, but I'm taking this matter seriously now."

"How could I not worry? It's not natural for her to be gone like this, that's what I've been trying to tell you all along. If you don't understand that, then you can't possibly know why I'm afraid. Also, her phone line was busy, but then when I got here she wasn't in. Someone was here, I'm sure of it."

"I'll get the whole story when I get there. What's the address?"

Linda gave it to him.

"How's it going there?" she said.

"I've never seen anything like this."

"Have you found the body?"

"Not yet. We haven't found anything at all, least of all a clue to what happened here. I'll sound the horn when I arrive."

Linda stood over the sink and rinsed her mouth. To get rid of the taste, she brushed her teeth with one of Anna's toothbrushes. She was about to leave the bathroom when, on a whim, she opened the cabinet above the sink. What she saw surprised her. This is just like leaving the journal behind, she thought.

Anna from time to time suffered from eczema on her throat. She had spoken of it only a few weeks earlier when they were at Zeba's place, talking about their dream holidays. Anna had said that the first thing she would pack was the prescription-strength cream that kept her eczema under control. Linda remembered her saying that she bought this cream only one tube at a time so as to have it as fresh as possible. And yet here it was, among the bottles and toothbrushes on the shelf. Anna had a thing about toothbrushes. Linda counted 19 in the cabinet, eleven of them unused. She looked at the cream again. Anna would not have left this behind, Linda thought. Not willingly. Neither this cream nor her journal. She closed the cabinet and went into the living room. So what could have happened? There was no indication that Anna had been removed by force, at least not in the flat itself. Perhaps something had happened on the street. She could have been knocked over or forced into a car.

Linda stood at the window and waited for her father. She was tired and she felt cheated. She had had no preparation for this day's discoveries during her time at the college. She had not dreamed that she would ever find herself confronting a severed grey-haired, female head, or a pair of clasped hands cut off at the wrists.

Not simply clasped, she thought. Hands knitted together in prayer before they were cut. She shook her head. What was going on in those last moments, in the dramatic pause while the axe was raised above those hands? What was Birgitta Medberg seeing? Had she looked into another's eyes and understood what would happen? Or had she been spared that hideous knowledge? Linda stared at a street lamp swaying in the wind. She sensed what must have happened: the hands were clasped together pleading for mercy. The executioner denies the plea. She must have known, Linda thought. She knew what was coming and she pleaded for her life.

Headlights lit up the façade of the building. Her father parked the car, then got out and looked around for the right entrance until he saw Linda in the window gesturing to him. She threw the keys down to the street and heard him come up the stairs. He's going to wake up every last neighbour, she thought. I have a father who thunders through life like an infantry platoon. He was sweat-stained and tired, his clothes wet through.

"Is there anything to eat?"

"I think so."

"And a towel?"

"The bathroom is over there. There are towels on the bottom shelf."

When he came back to the kitchen he had taken off all his clothes except his vest and pants. The wet clothes were hanging on the hot pipes in the bathroom. Linda had set the table with all the food she could find in the refrigerator. She knew he wanted to eat in peace. When she had been growing up, it had been forbidden to talk or make noise around the table at breakfast. His silence had driven Mona up the wall – she always waited until after he had left for work before she had her breakfast. But Linda had often sat there sharing the silence with him. Sometimes he lowered the paper, usually the *Ystad Allehanda*, and winked at her. Silence at breakfast was sacred.

"I should never have brought you along," he said suddenly, a sandwich halfway to his mouth. "There's no excuse for it. You should never have had to see what was in that hut."

"How is it going?"

"It isn't going anywhere. We have no clues, no explanation for what happened."

"But what about the rest of the body?"

"There's no sign of it. The dogs can't pick up a scent. We know that Birgitta Medberg was mapping trails in that part of the forest, so it seems reasonable to assume she came up on the hut by accident. But who was hiding there? Why this murder, why mutilate the body and dispose of it in this way?"

Wallander finished the sandwich, made a new one and left it half eaten.

"So tell me. Anna Westin, this friend of yours, what does she do? She's a student – but of what exactly?"

"Medicine. You know that."

"I never rely on my memory. You had arranged to meet her, you said. Was that here?"

"Yes."

"And she wasn't here when you arrived?"

"Correct."

"Is there any possibility of a misunderstanding?"

"No."

"Tell me the part about her father again. He's been gone for twenty-four years and has never once in any way communicated with her. And then she's in a hotel in Malmö and she sees him through a window, right?"

Linda told him everything again in as much detail as she could muster. He was quiet when she finished.

"We have one person who turns up after having been missing," he said finally. "And the following day the person who saw that person goes missing herself. One appears, the other one disappears."

He shook his head. Linda told him about the journal and the eczema cream. And about her visit to Henrietta. He listened attentively to everything.

"What makes you think she was lying?"

"If Anna thought she regularly saw her father, she would have told me long before now."

"How can you be so sure?"

"I know her."

"People change. You can never know everything about people, even friends."

"Is that true of me too?"

"Of me, of you, your mother, Anna, everyone. Then, of course, there are people who are totally incomprehensible. My father was an outstanding example of the latter."

"I knew him."

"You think you did."

"Just because the two of you didn't get along doesn't mean I felt the same way. And we were talking about Anna."

"I gather you didn't report her missing."

"I followed your advice."

"For once."

"Oh shit, give me a break."

"Show me the journal."

Linda went to get it, and opened it to the page where Anna wrote about the letter from Birgitta Medberg.

"Did she ever mention her name to you?" Wallander said.

"Not that I can remember."

"Did you ask her mother if she had any connection to Birgitta Medberg?"

"I saw Henrietta before I knew about this."

Wallander went to the bathroom to get his notepad from his jacket.

"I'll have someone talk to her again tomorrow."

"I can do it."

Wallander sat down. "No," he said sternly. "You can't do it. You're not a police officer yet. I'll get Svartman or someone else to do it. You aren't going to be doing any more investigating on your own."

"Do you always have to sound so pissed off?"

"I'm not pissed off, I'm tired. And worried. I don't know why what happened in that hut happened, only that it was horrifying. And I don't know if it marks an end or a beginning." He looked at his watch and got up again.

"I have to go back there," he said.

Then he stopped in the middle of the kitchen, undecided.

"I find it hard to believe it was a coincidence," he said. "That the Medberg woman by pure misfortune ran into the wicked witch who lives in the gingerbread house. I can't see that you'd get murdered for knocking on the wrong door. There are no monsters in Swedish forests. No trolls even. She should have stuck to butterflies."

Wallander walked back to the bathroom and put on his clothes. Linda tagged along. What was it he had said? The door to the bathroom was slightly ajar.

"What did you just say?"

"That no monsters live in Swedish forests?"

"After that."

"I didn't."

"You did. After the monsters and trolls, the last thing?"

"She should have stuck to butterflies and not started in on mapping ancient trails."

"Why butterflies?"

"Höglund talked to the daughter – someone had to inform the relatives. The daughter said Medberg had had a large butterfly collection. She sold it a few years ago to help Vanya and her children buy a flat. Vanya always felt guilty about it because she thought her mother missed the butterflies. People often have these kinds of reactions when someone dies. I was the same way when Dad died. I could start crying at the thought of how he wore unmatching socks."

Linda held her breath. He noticed something was up.

"What is it?"

"Come with me."

They went into the living room. Linda turned on a lamp and pointed to the wall.

"I've tried to keep an eye out for things that are different, I've already told you that. But I forgot to say that something was taken."

"What?"

"A butterfly case. You know, a butterfly in a frame. It disappeared the day after Anna went missing."

Wallander frowned. "Can you be sure?"

"Yes," she said. "And that the butterfly was blue."

CHAPTER 18

It seemed to Linda that it took a blue butterfly to convince her father to take her seriously. She wasn't just a kid any more, not just an officer-in-training with potential, but a fully-grown adult with judgment and keen powers of observation.

She was sure of herself. The butterfly in a frame had been removed at the same time as or shortly after Anna's disappearance. That settled it. Wallander called his team in the field and asked Höglund to come to the flat. He asked how things were going at the crime scene. Linda heard Nyberg's irritated voice in the background, then Martinsson, who was sneezing violently, and finally Lisa Holgersson, the Chief of Police. Wallander put the phone down.

"I want Ann-Britt to be here," he said. "I'm so tired I'm not sure I can trust my own judgment any more. Are you positive you have told me all the relevant facts?"

"I think so."

Wallander shook his head.

"It seems too much of a coincidence."

"A day or so ago you said one always has to be prepared for the unexpected."

"I talk a lot of crap," he said. "Is there any coffee in the house?"

The water had just boiled when Höglund sounded her horn on the street below.

"She drives much too fast," Wallander said. "She has two young children – what is she thinking of? Throw her the keys, will you?"

Höglund caught the keys in one hand and walked briskly up the stairs. Linda noted that she had a hole in her sock but that her face was made up – heavily made up. When did she have time to do that? Did she sleep with her make-up on?

"Would you like some coffee?"

"Please."

Linda thought her father would be the one to talk, but when she came in with the coffee cup and put it on the table in front of Höglund, Wallander nodded at her to begin.

"It's better for her to hear it from the horse's mouth," he said. "Don't leave out any details, you can count on Höglund to be a good listener."

Linda picked up her story with both hands and unfolded it as carefully as she was able; all in the right order. Then she showed Höglund the journal with the page that mentioned Medberg. Wallander broke in only when she started talking about the butterfly. Then he took over, changing her story to something that would perhaps form the basis of an investigative narrative. He got up from the sofa and tapped the wall where the butterfly had been.

"This is where the lines meet," he said. "Two points, perhaps three. Medberg's name is in Anna's journal, and she wrote Anna at least one letter, although we haven't found it. Butterflies figure in both of their lives, although we don't yet know what the significance of this is. And then there is the most important point in common: they have both been missing."

Someone cried out down on the street, a drunk man who was speaking Polish or Russian.

"It's certainly a strange coincidence," Höglund said. "Who knows Anna best?"

"Difficult to say."

"Does she have a boyfriend?"

"Not right now."

"But she's had one?"

"Doesn't everyone? I think probably her mother knows her best."

Höglund yawned and ruffled her hair.

"What about all this business with her father? Why did he leave home? Had he done something?"

"Anna's mother seems to think he was running away."

"From what?"

"Responsibility."

"And now he's back and Anna disappears. And Medberg is murdered."

"No," Wallander broke in, "not murdered. That isn't an adequate description of it. She was slaughtered, butchered. Hands clasped as in prayer, head severed, torso and limbs missing. Martinsson has found his way to the Tademans, by the way. Herr Tademan was very intoxicated, according to Martinsson, which is interesting. Fru Tademan – whom Linda and I met – seems to have been much easier to talk to. They haven't seen any unusual persons in the area, apparently no-one knew about the hide-out in the forest. She called someone she knows who rough-shoots around there, but he hadn't seen any hut or even the ravine, strangely enough. Whoever used the place knew how to keep a low profile while still relatively close to people."

"Close to what or whom?"

"We don't know."

"We'll have to start with the mother," Höglund said. "Should we call her right now or wait until morning?"

"Wait until morning," Wallander said after hesitating. "We have our hands full right now as it is."

Linda felt her face flush. "What if anything happens to Anna in the meantime?"

"What if her mother forgets to tell us something important because we get her out of bed in the middle of the night? We'll scare her half to death."

He walked to the door.

"That's how it's going to be. Go home and get some sleep. But you'll be coming with us to see Anna's mother tomorrow morning."

Höglund and Wallander put on their boots and rain gear and left. Linda watched them from the window. The wind was blowing harder, coming in strong gusts from the east and south. She washed the cups and thought about the fact that she needed to sleep. But how was she supposed to do that? Anna was gone, Henrietta had lied, Birgitta Medberg's name was inscribed in the journal. Linda started to look through the flat again. Why couldn't she find Medberg's letter?

She searched more energetically this time, pulling bookshelves away from the wall and backings from paintings to make sure nothing was hidden inside them. She continued with this until the doorbell rang. Linda stopped. It was after 1.00. Who rang a doorbell in the middle of the night? She opened the door and found a man outside, in thick glasses, brown dressing gown and pink, worn slippers on his feet. He said his name was August Brogren.

"There's a great deal of noise coming from this flat," he said angrily. "Be so kind, Miss Westin, as to keep it down."

"I'm sorry," Linda said. "I'll be quiet from now on."

August Brogren took a step closer.

"You don't sound like Miss Westin," he said. "In fact you aren't Miss Westin at all. Who are you?"

"A friend."

"When one has bad eyesight one learns to differentiate people's voices," Brogren explained sternly. "Miss Westin has a gentle voice, but yours is hard and rasping. It is like the difference between soft white bread and hard tack."

Brogren fumbled his way back to the handrail of the staircase and started back down. Linda thought about Anna's voice and understood Brogren's description. She closed the door and got ready to leave. Suddenly she was close to tears. Anna is dead, she thought. But then she shook her head. She didn't want to believe that, didn't want to imagine a world without Anna. She laid the car keys on the kitchen table, locked the door and walked home through the deserted streets. When she got home she wrapped herself in a blanket and curled up on her bed.

Linda woke with a start. The hands on the alarm clock glowed in the dark and showed 2.45 a.m. She had hardly been asleep an hour. What had caused her to wake up? She had dreamed something, sensed a danger approaching from afar like an invisible bird diving soundlessly towards her head. A bird with a beak as sharp as a razor. The bird had woken her.

Even though she had slept so briefly she felt clear-headed. She thought about the investigators still out at the crime scene, people moving back and forth in the strong spotlights, insects swarming in the light beams and burning themselves on the bulbs. It seemed to her that she had woken up because she didn't have time to sleep. Was Anna calling out to her? She listened, but the voice was gone. Had it been there in her dreams? She looked at the time. It was now 2.57. Anna called out to me, she thought again. And she knew what she was going to do. She put on her shoes, took her coat and ran down the stairs.

The car keys were still lying on Anna's kitchen table. In order to avoid having to pick the lock every time, she had pocketed a spare set of keys to the flat usually stored in a box in the hallway. As she drove out of town it was 3.20. She swung north and ended up parking on a small overgrown track that lay out of sight of Henrietta's house. Stepping out of the car she listened for any noise, then gently closed the door. It was chilly. She pulled her coat tightly around her body and chastised herself for not having brought a torch.

She started along the track, taking care not to trip. She didn't know exactly what she was planning to do, but Anna had called out to her and she felt compelled to respond. She followed the dirt track until she came to the path leading to the back of Henrietta's house. Three windows were lit. The living room, she thought. Henrietta is still up – although she could have gone to bed and left the lights on.

Linda walked towards the light, cutting a wide swath around a rusty harrow, and getting closer to the garden. She stopped and listened. Was Henrietta in the middle of composing? She made it to the fence and climbed over it. The dog, she thought. Henrietta's dog. What am I going to do if it starts barking? And what am I doing out here in the first place? Dad, Höglund and I are coming back here in a few hours. What do I think I can find out on my own now? But it wasn't really about that. It was about waking up from a nightmare that seemed like a cry for help from a friend.

She approached the lighted windows pace by watchful pace, then stopped short. Voices. At first she could not determine where they were coming from, then she saw that one of the windows was pushed open. Anna's neighbour had said that her voice was gentle. But this wasn't Anna's voice, it was Henrietta's. Henrietta and a man. Linda listened, trying to will her ears to send out invisible antennae. She walked even closer and was now able to see in through the glass. Henrietta sat in profile, the man was on the sofa with his back to the window. Linda couldn't hear what the man was saying. Henrietta was talking about a composition, something about twelve violins and a lone cello, something about a last communion and apostolic music. Linda didn't understand what she meant. She tried to be absolutely quiet. The dog was in there somewhere. She tried to work out who Henrietta was talking to, and why they were talking in the middle of the night.

Suddenly, very slowly, Henrietta turned her head and looked straight at the window. Linda jumped. It seemed like Henrietta was looking directly into her eyes. She can't have seen me, Linda thought. It's impossible. But there was something about her gaze that frightened her. She turned and ran, accidentally stepping on the edge of the water pump, causing a clang from within the pump structure. The dog started to bark.

Linda ran back the way she had come. She tripped and fell, got up and stumbled on. She heard a door open somewhere behind her as she threw herself over the fence and ran down the path, aiming to make her way back to the car. At some point she made a wrong turn. She was lost. She stopped, gasping for air, and listened. Henrietta had not set the dog loose. It would have found her by now. She listened. There was no sound of pursuit, but she was still so frightened she was shaking. After a while she began making her way cautiously back to the path, but she couldn't see where she was going because it was so dark. The darkness alarmed her, making shadows into trees and trees into shadows. She stumbled and fell.

When she stood up, she felt a searing pain in her left leg. She felt she had been stabbed with a knife. She screamed and tried to get away from the pain, but she couldn't move. It was as if an animal had sunk its teeth into her, except that this animal didn't breathe or make any noise. Linda groped down her leg until her hand hit something cold and metallic connected to a chain. Then she understood. She was caught in a hunter's trap.

Her hand was wet with blood. She continued to cry out, but no-one heard her, no-one came.

CHAPTER 19

Linda tried to free herself from the trap. She didn't remotely like the idea of calling her father, but the trap was impossible to budge. She took out her mobile and dialled his number. She explained that she needed help and where she was.

"What's happened?"

"I'm caught in a trap."

"What on earth are you talking about?"

"I have a steel trap cutting into my leg."

"I'm on my way."

Linda waited, shivering. It seemed an eternity before she saw the headlights from a car in the distance. She heard the car stop. Linda called out. And then the sounds of doors opening and the dog barking. She called and called. They walked over in the first light of morning, a torch lighting their way. Her father, Henrietta and the dog. There was a third person with them, but he hung back.

"You're caught in an old fox trap. Who is responsible for putting this here?"

"Not me," Henrietta said. "It must be the man who owns the land around here."

"We'll have a word with him." Wallander forced the trap open. "We'd better get you to the hospital," he said.

Linda tried to put some weight on the foot. It hurt, but she was able to steady herself with it. The man in the shadows now came closer.

"This is a colleague you haven't met yet," Wallander said. "Stefan Lindman. He started with us a couple of weeks ago."

Linda looked at him. His face was partly lit by the torch and she liked what she saw.

"What are you doing here?" Henrietta asked.

"I can explain," Lindman said.

He spoke with some kind of dialect. But which was it? She asked her father later when they were driving back to Ystad.

"He's from western Götaland," Wallander said. "A strange language. They have trouble commanding respect, as do people from eastern Götaland and the island of Gotland. The ones who command the most respect are northerners, apparently. I don't know why."

"How is he going to account for me being out there tonight?"

"He'll think of something. But maybe you can tell me yourself what you thought you were doing."

"I had a dream about Anna."

"What sort of dream?"

"She called out to me. I woke up and came here. I didn't know what I was planning to do. I saw Henrietta inside talking to a man. Then she looked over in my direction and I ran and then I got caught in that trap."

"At least I know you're not sneaking out for private assignations."

"This is serious, don't you understand?" she screamed. "Anna is missing!"

"Of course I take it seriously. I take her disappearance seriously. I take my whole life and yours seriously. The butterfly was the clincher."

"What are you doing about it?"

"Everything that can and should be done. We're turning every stone, chasing every lead. And now we're not going to discuss this further until we've had your leg checked out at the hospital."

It was an hour before anyone could deal with her. Wallander dozed in an uncomfortable chair. And the process of cleaning the wound and bandaging it seemed to take for ever. As they were leaving at last, Lindman walked in. Linda saw now that he had closely cropped hair and blue eyes.

"I said that you had terrible night vision," he said cheerfully. "It doesn't make a lot of sense, but it will have to do as an explanation for what you were doing out there."

"I saw a man in the house with her," Linda said.

"Fru Westin told me she had a visit from a man who wants her to set music to his dramatic verse. It was late, to be sure, but it didn't sound a suspicious or unsatisfactory explanation."

Linda put her jacket on against the morning cold. She regretted having yelled at her father in the car. It was a sign of weakness. Never scream, always keep your cool. But she had done something stupid and had needed to turn the attention to someone else's shortcomings. She also felt a huge wave of relief. Anna's being missing was no longer a figment of her imagination. A blue butterfly made all the difference. The price was a painful ache in her leg.

"Stefan will take you home. I have to get back to the station."

Linda went into the ladies' room and combed her hair. Lindman was waiting for her in the corridor. He was wearing a black leather jacket and was sloppily shaven on one side of his face. She chose to walk on his good side.

"How does it feel?"

"What do you think?"

"It must hurt. I know something about that."

"What do you mean?"

"Pain."

"You have had your leg caught in a bear trap?"

"A fox trap. No, I haven't."

"Then you don't know how it feels."

He held the door open for her. She was still put off by his unshaven cheek and didn't say anything else. They came out into a car park at the back of the hospital. It was broad daylight. He pointed to a rusty Ford. As he was unlocking the door an ambulance driver came over and demanded to know what he meant by blocking the emergency entrance.

"I came to pick up a wounded police officer," Lindman said, nodding in Linda's direction.

The ambulance driver accepted this and left. Linda manoeuvred herself into the passenger seat.

"Your dad said you live on Mariagatan. Where is that?"

Linda explained, and wondered about the strong smell in the car.

"It's paint," Lindman said. "I'm renovating a house in Knickarp."

They turned on to Mariagatan and Linda pointed out her doorway. He opened the passanger door for her.

"It was nice to meet you," he said. "By the way, how I know what it's like to be in pain is that I've had cancer. Steel trap or a tumour – it's much the same thing."

Linda watched his car drive off. She had forgotten his last name.

She let herself into the flat and felt fatigue set in. She was about to collapse on to the sofa when the phone rang.

"I heard you made it home." It was her father.

"Who was the man who drove me home?"

"Stefan."

"No – his last name."

"Lindman. He's from Borås, I think. Or else it was Skövde. It's time you were resting."

"I want to know what Henrietta said to you."

"I don't have time to go into that right now."

"You have to. Just tell me the important bits."

"Wait a minute."

He broke off. Linda guessed he was on his way out of the station. She heard doors closing, phones ringing, and then the sound of an engine starting up.

He came back on the line, his voice tense. "Are you there?"

"I'm here."

"OK, I'll make this quick. She doesn't know where Anna is. She hasn't heard from her recently. Knows nothing to suggest that Anna is depressed. Anna has apparently not said anything about seeing her father. On the other hand, Henrietta claims that this happened all the time during Anna's childhood. So it's the mother's word against yours. She can't give us any leads, nor does she know anything of Medberg. So, as you see, not very productive."

"Did you notice that she was lying?"

"How would I have noticed that?"

"You always say that all you have to do is breathe on someone to know if they're telling the truth or not."

"I didn't have the impression that she was lying."

"She *was* lying."

"I have to go now. But Lindman – the one who gave you the lift – is working on the connection between Medberg and Anna. We've sent out missing-persons reports on her. That's all we can do for the moment."

He hung up. Linda didn't feel like being alone so she called Zeba. She was in luck: Zeba's son was at her cousin Titchka's house and she was free to come over.

"Buy some breakfast on the way," Linda said. "I'm starving. The Chinese restaurant by Runnerströmstorg, for example. I know it's out of your way, but I'll make it up to you the next time you find yourself caught in a fox trap."

Linda told Zeba what had happened. Zeba had heard the news on the radio about the head which had been found, but she did not believe that anything bad had happened to Anna.

"Even if I were a professional villain I'd think twice before picking on Anna. Don't you know she did martial-arts courses? I don't remember which kind, but I think it's the one where everything is allowed – short of killing someone, of course. No-one messes with Anna and gets away with it."

Linda regretted having said anything to Zeba, but she was glad that she had come. Zeba stayed another hour before it was time to pick up her son.

Linda woke up when the doorbell rang. At first she was going to ignore it, but then she changed her mind and limped out into the hallway. Stefan Lindman was standing outside the door.

"I'm sorry if I woke you."

"I wasn't sleeping."

Then she looked at herself in the hall mirror. Her hair was standing on end.

"Actually I *was* sleeping," she said. "I don't know why I said that. My leg hurts."

"I need the keys to Anna Westin's flat," he said. "I heard you tell your father you had a spare set."

"I'll come along."

He seemed surprised by this. "I thought you were in pain?"

"I thought that too. What are you going to do over there?"

"Try to create a picture for myself."

"If it's a picture of Anna, then I'm the person you should be talking to."

"I'd like to have a look by myself first. Then we can talk."

Linda pointed to a set of keys on a table. The key ring had a profile of an Egyptian pharaoh.

"What was that you said about your having cancer?" she asked.

"I had cancer of the tongue, if you can believe it. Things looked bad for a while, but I survived and there's been no recurrence."

He looked her in the eye for the first time.

"I still have my tongue. I wouldn't be able to speak without it! But my hair has never recovered. Soon it'll all be gone."

He walked away down the stairs and Linda returned to bed. Cancer of the tongue. She shuddered at the thought. Her fear of death came and went, though right now her life force was strong. But she had never forgotten what went through her mind when she balanced on the railing of the bridge. Life wasn't just something that took care of itself. There were big black holes you could fall into with long, sharp spikes at the bottom, monstrous traps.

She turned over on her side and tried to sleep. Right now she didn't have the energy to think about black holes. Then she was startled out of her half-awake state. It was something to do with Lindman. She sat up. She had finally caught hold of the thought that had been bugging her. She dialled a number on her mobile phone. Busy. On the third try her father answered.

"It's me."

"How do you feel?"

"Better. There was something I wanted to ask about the man who was at Henrietta's house tonight. The one who was said to be commissioning a composition. Did she say what he looked like?"

"Why would I have asked her that? She only gave me his name. I made a note of the address. Why?"

"Do me a favour. Call her and ask about his hair."

"Why would I do that?"

"Because that's what I saw."

"I will, but I really don't have the time for this. We're up to our knees in rain over here."

"Will you call back?"

"If I reach her."

Twenty minutes later her phone rang.

"Peter Stigström – the man who wants Henrietta to set his verse to music – has shoulder-length dark hair with a few grey streaks. Will that do?"

"That will do just fine."

"Are you going to explain yourself now or when I get home?"

"That depends on when you were planning to come home."

"Pretty soon. I have to get out of these clothes."

"Do you want something to eat?"

"No, we've been catered for out here in fact. There are some enterprising types from Kosovo who make a living putting up food stands around crime scenes and fires. I have no idea how they hear about our work, but there's probably a mole at the station who gets a commission. I'll be home in an hour."

When the conversation was over, Linda sat looking down at the phone for a few minutes. The man she had seen through the window had not had shoulder-length dark hair with a few grey streaks. His hair had been short, neatly trimmed.

CHAPTER 20

Wallander came bounding in, his clothes again soaked, his boots filthy with mud, but with the cheerful news that the weather would very soon clear up. Nyberg had called air-traffic-control at Sturup, he said, and received the report that the next 48 hours would be free of rain. Wallander changed, declined Linda's offers of food and made himself an omelette.

Linda waited for the right moment to tell him about the conflicting descriptions of Henrietta's visitor. She didn't know exactly why she was waiting. Was it a lingering childhood fear of his temper? She didn't know, she just waited. And then, when he pushed away his plate and she plopped into the chair across from him and was about to launch into her story, he started talking.

"I've been thinking about your grandfather," he said.

"What about him?"

"What he was like, what he wasn't like. I think you and I knew him in different ways. That's as it should be. I was always looking for bits of myself in him, worried about what I would find. I've grown more like him the older I get. If I live as long as he did, maybe I'll find myself a ramshackle, leaky house and start painting pictures of grouse and sunsets."

"It'll never happen."

"Don't be too sure."

Linda broke in and told him about the man she had seen whose close-cropped head didn't match Henrietta's description. He didn't ask her if she was sure of what she had seen. He had known from the start that she was sure. He reached for the phone and dialled a number from memory – first incorrectly, then getting it right. Lindman answered. Wallander told him succinctly that in the light of what Linda had observed, they had to make another visit to Henrietta Westin.

"We have no time for lies," he said. "No lies, no half-truths, no incomplete answers."

Then he put the phone down and said to Linda: "This is unorthodox at best. Not even necessary, strictly speaking, but I'm still going to ask you to come along. If you feel up to it, that is."

Linda felt a surge of pleasure.

"I'll do it."

"How's the leg?"

"It's fine."

She saw that he didn't believe her.

"Does Henrietta know why I was there last night?" she asked. "She can hardly have believed what Stefan told her."

"All we want to know is who was there with her. We have a witness, we don't have to tell her it's you."

They walked down to the street and waited for Lindman. The air-traffic controllers had been right, the weather was changing. Drier winds were blowing in from the south.

"When will it snow?" Linda asked.

He looked at her in amusement.

"Not for a while, I hope. Why do you ask?"

"Though I was born and raised here, I can't remember when it comes."

Stefan Lindman pulled up in his car. Linda climbed into the back seat, her father sat in the front. His seat belt was snagged on something and he had trouble getting it on.

They drove towards Malmö and Linda saw the sea shimmering on her left. I don't want to die here, she thought. The thought came out of nowhere. I don't only want to exist here. Or be like Zeba. Or be a single mother like her, or thousands of others whose lives become one long damned struggle to pay the rent and the babysitters and to get there on time. I don't want to be like Dad, who can never find the right house and the right dog and the wife he needs.

"What was that?" Wallander said.

"Nothing."

"That's funny. I could have sworn you were swearing."

"I didn't say anything."

"I have a strange daughter," Wallander said to Lindman. "She curses without even knowing it."

They turned on to the road to Henrietta's house. The memory of being caught in the trap made Linda's leg throb. She asked what would happen to the man who had set the trap.

"He went a little pale when I told him he had snared a police cadet. I assume he'll have a hefty fine to pay."

"I have a good friend in Östersund," Lindman said. "A policeman. Giuseppe Larsson is his name . . ."

"He sounds Italian."

"He's from Östersund, but he has a connection of sorts with an Italian lounge singer."

"What's that supposed to mean?" Linda leaned forward between the seats. She had a sudden urge to touch Lindman's face.

"His mother had a dream that his father was not her husband but an Italian singer she had heard perform at an outdoor concert. It's not just us men who have these fantasies."

"I wonder if Mona has ever had the same thoughts," Wallander said. "In your case it would be a black dream-father, Linda, since she worshipped Hosh White."

"Josh," Lindman said. "Not Hosh."

Linda wondered vaguely what it might have been like to have a black father.

"Anyway," Lindman said. "This Larsson has an old bear trap on the wall at his place. It looks like an instrument of torture from the Middle Ages. He says that if a human got caught in one the steel teeth would cut all the way through the bone. Animals that get trapped in them have been known to gnaw their own legs off in desperation."

Lindman stopped the car and they climbed out. The wind was gusty. They walked up to the house, in which several of the windows were lit up. When they entered the front garden all three of them wondered why the dog hadn't started to bark. Lindman knocked on the door, but no-one answered. Wallander peeked in through a window. Lindman felt the door. It was unlocked.

"We can say we thought we heard someone call 'Come in!'," he said tentatively.

They walked in. Linda's view was blocked by the broad backs of the two men. She tried standing on tiptoe to see past them but winced with the pain.

"Anybody home?" Wallander called out.

"Doesn't look like it," Lindman said.

They proceeded through the house. It looked much as it had when Linda was there last: papers, sheet music, newspapers and coffee cups scattered all about. But she recognised that this superficial impression of disarray only disguised a home comfortably arranged to meet Henrietta's every need.

"The door was unlocked," Lindman said, "and her dog is gone. She must be out on an evening walk. Let's give her a quarter of an hour. If we leave the door open she'll know someone's inside."

"She may call the police if she thinks the house is being burgled," Linda said.

"Burglars don't leave the front door wide open for everyone to see," her father said.

He sat in the most comfortable armchair, folded his hands over his chest and closed his eyes. Lindman put his boot in the front door to keep it open. Linda picked up a photo album that Henrietta had left lying on the piano. The first pictures were from the early '70s. The colours were starting to fade. Anna sat on the ground surrounded by chickens and a yawning cat. Anna had told her about the commune near Markaryd where she had spent the first years of her life. In another picture Henrietta was holding her, in baggy clothes, clogs and a Palestinian shawl around her neck. Who is behind the camera? Linda wondered. Probably Erik Westin, the man who was about to vanish.

Lindman walked over to her and she pointed to the pictures, explaining what she knew about them: the commune, the green wave, the sandal maker who vanished into thin air.

"It sounds like something out of a story," he said. "Like *A Thousand and One Nights.* I mean the part about 'The Sandal Maker Who Vanished into Thin Air'."

They kept turning the pages.

"Is there a picture of the husband?"

"I've seen a few at Anna's place, but that was a long time ago. I've no idea where they'd be."

Pictures of life in the commune gave way to images of an Ystad flat. Grey concrete, a wintry playground.

"By this time he had been gone for some years," Linda said. "The person taking the pictures is closer to Anna now. The photographs in the commune are always taken from a greater distance."

"Her father took the earlier pictures and now Henrietta is the one taking them. Is that what you mean?"

"Yes."

They flipped through to the end of the album, but there was no picture of Westin. One of the last pictures was of Anna's final school speech day. Zeba was at the edge of the group. Anna had been there too, but wasn't in the photograph.

Linda was about to turn the page when the lights flickered and went out. The house was plunged into darkness and Wallander woke with a start. They heard a dog barking. Linda sensed the presence of people out there in the night, people who did not intend to show their faces, but rather shied away from the light and were retreating even further into the world of shadows.

CHAPTER 21

He felt secure only in total darkness. He had never understood why there was always this talk of light in connection with mercy, eternity, images of God. Why couldn't a miracle take place in total darkness? Wasn't it harder for the Devil and his demons to find you there, in the shadows, than on a bright plain where white figures moved as slowly as froth on the crest of a wave? For him, God had always manifested Himself as an enveloping, deeply comforting darkness.

He felt the same way now as he stood outside the house with the shining bright windows. He saw people moving around inside. When all the lights suddenly went out and the last door of darkness was sealed he took it as a sign from God. I am his servant in the darkness, he thought. No light escapes from here, but I shall send out holy shadows to fill the void in the souls of the lost. I shall open their eyes and teach them the truth of the images that reside in the shadow world. He thought about the lines in John's second epistle: "For many deceivers are entered into the world, who confess not that Jesus Christ is come in the flesh. This is a deceiver and an antichrist." It was the holiest key to his understanding of God's word.

After the terrible events in the jungles of Guyana, he could recognise a false prophet: a man with raven-black hair, even white teeth, who surrounded himself with light. Jim Jones had feared the dark.

He had cursed himself countless times for not seeing through the guise of this false prophet who would lead them so astray – even to their deaths. All of them except himself. This had been the first task God had assigned him: to survive in order to tell the world about the false prophet. He was to preach about the kingdom of darkness, which would become the fifth gospel, which he would write to complete the holy writings of the Bible. This too was foretold at the end of John's letter: "Having many things to write unto you, I would not write with paper and ink: but I trust to come unto you, and speak face to face, that our joy may be full."

This evening he had been thinking about all the years that had gone by since he was here. Twenty-four years, a large part of his life. He was a young man when he left. Now age had started to claim his body. He took care of himself, chose his food and drink sensibly, kept himself constantly in motion, but the process of growing old had begun. No-one could escape it. God lets us age the better for us to understand that we are completely in His hands. He gives us this remarkable life, but as a tragedy so that we will understand that only He has the power to grant us mercy.

Everything had been what he had dreamed of until he followed Jim Jones to Guyana. Though he missed those he had left behind, Jim had managed to convince him that the loss of them was necessary to prepare him for the higher purpose God held in store for him. He had listened to Jim and sometimes he had not thought about his wife and his child for weeks at a time. It was only after the massacre, when the whole community lay rotting on the fields, that they returned to his consciousness. By then it was too late. The void created by the God whom Jim had killed in him was so huge that he could not think of anyone but himself.

He had retrieved the money and papers he had stored in Caracas, then took the bus to Colombia, to the city of Barranquilla. He remembered the long night he spent in the border station between Venezuela and Colombia, the city of Puerto Paez where armed guards watched over the travellers like hawks. Somehow he had managed to convince these guards that he was John Clifton – as his documents stated – and he even managed to convince them that he didn't have any money left. He had slept with his head on the shoulder of an old Indian woman who had a small cage with two chickens on her lap. They had exchanged no words, only a look. She had seen his suffering and his exhaustion and offered him her shoulder and wrinkled neck to rest his head. He dreamed that night of those he had left behind. He woke up, the sense of loss almost physical. The old woman was awake. She looked at him and he lay back against her shoulder. When he woke in the morning she was gone. He felt inside his shirt to touch the wad of dollar bills.

It was still there. He wanted her back again, the old woman who had let him sleep. He wanted to lean his head against her shoulder and neck and stop there for the rest of his life.

From Barranquilla he took a flight to Mexico City. He washed off the worst of his filth in a public toilet. He bought a new shirt and a small Bible. It had been confusing to see so many people milling about in rushing crowds again, this life that he had left behind when he followed Jim. He walked past the news stands and saw that what had happened had made the front-page headlines. Everyone had died, he read. No-one was thought to have survived. That meant they must think he was dead too. He existed, but he had stopped living, since he was presumed to be one of the bloated bodies found in the jungle.

He still didn't have a clear plan. He had $3000 after paying for the fare to Mexico City, and if he was frugal he could get by on that for quite a while. But where should he go? Where could he find the first step back to God, out of this unbearable emptiness? He didn't know. He stayed in Mexico City, in a pension, and spent his days attending various churches. He deliberately avoided the large cathedrals as well as the neon-lit tabernacles run by greedy and power-hungry clergy. Instead he sought out the small congregations where the love and the passion were palpable, where the ministers were hard to tell apart from those who came to listen to their sermons. That was the way he had to find for himself. Jim hid himself in the light, he thought. Now I want to find the God who can lead me to the holy darkness.

One day he woke up with the overwhelming presentiment that it was time to leave. He took a bus going north that same day, and to make the journey as cheap as possible he took local buses. Sometimes he hitched a lift with a truck driver. He crossed the border into Texas at Laredo, where he checked into the cheapest motel he could find. He spent a week in the public library, reading every article there was on the catastrophe. To his alarm, he learned that former members of the People's Temple were accusing the FBI and CIA or the American government of fostering hostility towards Jim Jones and his movement, thereby inciting the mass suicide. How could they defend the false prophet?

During long sleepless nights it occurred to him that he should write about what had happened. He was the only living witness. He bought a notebook and started to write, but he was overcome with doubt. If he was going to tell the real story, he would have to reveal his true identity: not John Clifton, as his documents claimed, but another man with another name and nationality. Did he want that? He hesitated.

Then he read an interview with a woman named Mary-Sue Legrande in the *Houston Chronicle*. There was a photo of her: a woman in her forties with dark hair and a thin, pointed face. She talked about Jim Jones and claimed to know his secrets. She was a distant spiritual relative of Jim. She had known him at the time he had the series of visions that would later lead him to found his church, the People's Temple.

I know Jim's secrets, said Mary-Sue Legrande. But what were they? She didn't say. He stared at the photograph. Mary-Sue was looking right at him. She was divorced with a grown son and she owned a small mail-order company in Cleveland. Her company sold something they called "manuals for self-actualisation".

He put the newspaper back on the shelf, nodding to the friendly librarian, and walked out on to the street. It was a mild December day. He stopped in the shade of a tree. If Mary-Sue Legrande can tell me Jim Jones's secrets, I will understand why I was taken in by him. Then I will never suffer this same weakness again.

He got off the train in Cleveland on Christmas Eve after a journey of more than 30 hours. He found a cheap hotel close to the railway station and ate his dinner at a Chinese grocery. There was a green plastic Christmas tree with flashing lights in the lobby of the hotel. He lay on the bed in the dark hotel room. Right now, I'm nothing more than the person who is registered in this hotel room, he thought. If I were to die now, no-one would miss me. They would find enough money in my sock to cover the costs of the room and a funeral – that is, if no-one stole the money and they had to dump me in a pauper's grave. Perhaps someone would discover that I was not John Clifton. But the case would probably be put on the back burner. That would be the extent of it. Right now, I'm nothing more than a traveller in a hotel room, and I can't remember the name of the hotel.

Snow fell over the city on Christmas Day. He ate warm noodles, fried vegetables and rice at the Chinese grocery and then returned to lie motionless on his bed. The following day, December 26, the snowy weather had passed. A thin white powder had dusted the streets and pavements and it was –3°C. There was no wind and the water on Lake Erie was icy calm. He had located Mary-Sue Legrande with the help of a telephone directory and a map. She lived in a neighbourhood in the south-western part of the city. He thought it was God's will that he meet her this day. He washed, shaved and put on the clothes he had bought in a second-hand shop in Laredo. What will she see when she opens her door? he wondered. A man who hasn't given up, a man who has suffered greatly. He shook his head at his reflection in the mirror. I don't inspire fear, he thought. Perhaps pity.

He left his hotel and took a bus along the lake shore. Mary-Sue Legrande lived on 1024 Madison, in a stone house partially hidden behind tall trees. He hesitated before he walked up to the door and rang the bell. Mary-Sue Legrande looked exactly like her photograph in the *Houston Chronicle*, save that she was even thinner. She regarded him with suspicion, ready to slam the door in his face.

"I survived," he said. "Not every one of us died in Guyana. I survived. I've come because I want to know Jim Jones's secrets. I want to know why he betrayed us."

She looked at him for a long time before answering. When she did, she didn't show any sign of surprise, or of any emotion whatsoever.

"I knew it," she said at last. "I knew someone would come."

She opened the door wide. He followed her in and stayed in her house for almost 20 years. With her help he came to know the real Jim Jones, the man he had not been able to see through. Mary-Sue told him in her mild voice about Jim Jones's dark secret. He was not the messenger of God he had given himself out to be; he had taken God's place. Mary-Sue claimed that Jim Jones knew deep down that his vanity would one day be the destruction of everything he had built. But he had never been able to overcome this flaw and change course.

"Was he deranged?" he had asked.

No, Mary-Sue insisted, Jim Jones had been far from deranged. He had meant well; he had wanted to start a Christian awakening around the world. It was his vanity and pride that had prevented him from succeeding, that had turned his love into hate. But someone needed to take up where he left off, she told him. Someone who was strong enough to resist the pitfall of pride yet at the same time be merciless when the need arose. The Christian awakening would only come to pass through the shedding of blood.

He stayed and helped her run the mail-order business she called God's Keys. She had written all the self-help manuals herself, with their blend of vague suggestions and inaccurate Bible quotations, and he soon came to understand that she knew Jim Jones well because she was a kind of charlatan herself. But he stayed with her because she let him. He needed time to plan what would become his life's mission. He was going to be the one to take over where Jim Jones had gone astray. He would evade the pitfalls of pride and vanity, and he would never forget that the Christian rebirth would demand sacrifice and blood.

The mail-order company did well, especially with a product she called "The Aching Heart Package", priced at $49, not including tax and shipping. They started to get rich, leaving the house on Madison for a larger one in Middleburg Heights. Mary-Sue's son Richard came home after completing his studies in Minneapolis and settled in a house nearby. He was a loner but always friendly. He seemed grateful not to have to take on his mother's loneliness.

The end arrived quickly and unexpectedly. One day Mary-Sue came back from a trip into Cleveland and sat across from him at his desk. He thought she had been running errands.

"I have cancer and I'm going to die," she announced. She said the words with a strange air of relief, as if telling the truth lifted a great burden from her shoulders.

She died on the 87th day after she came back from the doctor with the news. It was in the spring of 1999. Richard inherited all her assets, since she had never married. They drove out to Lake Erie for a walk the evening after the funeral. Richard wanted him to stay; he suggested that they continue running the mail-order business and share the profits. But he had already made up his mind. The void in him had been assuaged by living with Mary-Sue, but he had a mission to accomplish. His thinking and his plans had matured over the years. He didn't say any of this to Richard. He simply asked him for some money – only as much as Richard could comfortably part with. Then he would make his preparations. Richard asked no questions.

He left Cleveland on May 19, 2001, flying to Copenhagen via New York. He arrived in Helsingborg on the south coast of Sweden on the evening of May 21. He paused after stepping on to Swedish soil. He had left all his memories of Jim Jones behind him at last.

CHAPTER 22

Wallander was looking for the number of the electricity company when the power came back on. Seconds later they all gave a start as Henrietta and the dog walked in. The dog jumped up on Wallander with its muddy paws. Henrietta ordered it to its basket and it obeyed. Then she threw its lead furiously aside and turned to Linda.

"I don't know what gives you the right to walk into my house when I'm not here. I don't like people sneaking around."

"If the power hadn't gone out we would have walked right out again," Wallander said. Linda could tell he was losing his temper.

"That's not an answer to my question," Henrietta said. "Why did you come in the first place?"

"We just want to know where Anna is," Linda said.

Henrietta didn't seem to listen to her. She walked around the room, looking carefully at her things.

"I hope you didn't touch anything,"

"We haven't touched anything," Wallander said. "We simply have a few questions to ask and then we'll be on our way."

Henrietta stopped and stared at him. "What is it you need to know? Kindly tell me."

"Should we sit down?"

"No."

This is when he explodes, Linda thought and closed her eyes. But her father managed to control himself, perhaps because she was there.

"We need to be in touch with Anna. She's not in her flat. Can you tell us where she is?"

"No, I can't."

"Who would know?"

"Linda is one of her friends, have you asked her? Or maybe she doesn't have time to talk to you since she spends all her time spying on me."

This sent Wallander over the edge. He yelled so loudly even the dog sat up. I know all about that voice, Linda thought, the yelling. God knows, it's one of the earliest memories I have.

"You will answer my questions clearly and truthfully. If you won't cooperate, we will bring you down to the station. We need to find your daughter because we believe she may have some information regarding Birgitta Medberg." Wallander made a short pause before continuing. "We also need to assure ourselves that nothing has happened to her."

"And what could possibly have happened to her? Anna studies in Lund. Linda knows that. Why don't you talk to her housemates?"

"We will. Is there anywhere else she could be, in your opinion?"

"No."

"Then we'll move on to the question of the man who was in your house last night."

"Peter Stigström?"

"Could you describe him for us, beginning with his hair?"

"I already have."

"We can call on Mr Stigström in person, of course, but perhaps you could humour us."

"He has long hair, about shoulder length. It's dark brown, with grey streaks. Will that do?"

"Can you describe his neck?"

"Good grief – if you have shoulder-length hair it covers your neck. How would I know what it looks like?"

"Are you sure of this?"

"Of course I'm sure."

"Then I'll thank you for your time."

He got up and left, slamming the front door behind him. Lindman hurried out after him. Linda was confused. Why hadn't he confronted Henrietta with the fact that she had seen a man with short hair? As she made ready to leave, Henrietta blocked her way.

"I don't want anyone coming in here when I'm gone. Is that clear? I don't want to feel I have to lock the door every time I take the dog out."

"Yes."

Henrietta turned her back to her.

"How is your leg?"

"It's better, thanks."

"Sometime maybe you'll tell me what you were doing out there."

Linda left the house. Now she understood why Henrietta wasn't worried about Anna, even though an appalling murder had been committed which had some connection with her daughter's life. She wasn't worried because she knew very well where Anna was.

Lindman and Wallander were waiting in the car.

"What is it she does?" Lindman asked. "All that sheet music. Does she write popular stuff?"

"She composes the kind of music no-one wants to hear," Wallander said. He turned to Linda. "Isn't that right?"

"Something like that."

A mobile phone rang. All three clutched at their pockets. It was Wallander's. He listened to the caller and checked his watch.

"I'll be right there."

"We're heading out to Rannesholm," he said. "Apparently there's some information about people who have been seen in the area over the past few days. We'll take you home first."

Linda asked him why he hadn't confronted Henrietta about the conflicting descriptions of the Stigström man's hair.

"I decided to sit on it," he said. "Sometimes these things need time to ripen."

Then they talked about Henrietta's apparent lack of concern for her daughter's safety.

"She knows where Anna is," Wallander said. "There's no other way to account for it. Why she's lying is a mystery, though I expect we'll find out sooner or later if we keep pushing. It's just not top priority for us at this point."

They drove on in silence. Linda wanted to ask more about the investigation at Rannesholm, but she felt it would be wiser to wait. They stopped outside the flat on Mariagatan.

"Could you switch off the engine for a minute?" Wallander turned round so he could see Linda. "Let me repeat what I just said: I'm satisfied that no harm has come to Anna. Her mother knows where she is and why she's staying away. We don't have the manpower to investigate this any further. But there is nothing to stop you from going to Lund and talking to her friends there. Just do me a favour, don't pretend to be a police officer."

Linda got out and waved them off. As she was opening the front door, she had one of those flashes of memory of something that Anna had said. Was it the last time before she disappeared? It came and at once was gone. Linda scoured her memory, but she couldn't recover it.

The next day Linda got up early. The flat was empty and her father had clearly not been home at all since the day before. She left shortly after 8 a.m. The sun was shining and it was warm. Because she had plenty of time she decided to take the coast road to Trelleborg and then turn north to Lund when she got to Anderslöv. She listened to the news on the radio, but there was nothing about Birgitta Medberg.

Then her mobile phone rang. It was her father.

"Where are you?"

"On my way to Lund. Why are you calling?"

"I just wanted to see if I needed to wake you up."

"You didn't have to do that. By the way, I saw you never made it home last night."

"I slept awhile in a room up at the manor. We've staked out a few rooms for the time being."

"How is it going?"

"I'll tell you later. Bye."

She put the phone back in her pocket. She found the address in Lund's inner city, looked for a place to park and bought an ice cream. Why had her father called? He's trying to control me, she thought.

The house which Anna shared was a two-storey wooden building with a small garden in front. The gate was rusty and about to fall off its hinges. Linda rang the doorbell but no-one answered. She rang again and strained her ears. She didn't hear a ringing inside, so she started to knock loudly. Eventually a shadow appeared on the other side of the glass panel. The man who opened the door was in his twenties, his face covered in spots. He was wearing jeans, an undershirt and a large brown dressing gown with big holes. He reeked of sweat.

"I'm looking for Anna Westin," Linda said.

"She's not here."

"But she lives here?"

The man stepped aside so that Linda could come in. She felt his eyes on the back of her head when she walked past him.

"She has the room behind the kitchen," he said.

They walked into the kitchen, which was a mess of dirty dishes and leftover food. How can she live in this pigsty? Linda thought.

She reluctantly stretched out her hand to shake his, shuddering at his limp and clammy handshake.

"Zacharias," he said. "I don't think her door's locked, but she doesn't like anyone to go in there."

"I'm one of her close friends. If she hadn't wanted me to go in she would have locked the door."

"How am I supposed to know that you are her friend?"

Linda felt like shoving him out of the kitchen, but pulled herself together.

"When did you last see her?"

He stepped back. "What is this – a cross-examination?"

"Not at all. I've been trying to get in touch with her and she hasn't got back to me."

Zacharias kept staring at her. "Let's go into the living room," he said.

She followed him into a room full of shabby, ill-matched furniture. A torn Che Guevara poster hung on one wall, a tapestry embroidered with some words about the joys of home on another. Zacharias sat at a table with a chess set. Linda sat as far away from him as she could get.

"What do you study?" she said.

"I don't. I play chess."

"And you make a living from that?"

"I don't know. I just know I can't live any other way."

"I don't even know how the pieces move."

"I can show you, if you like."

Not a chance, Linda thought. I'm getting out of here as soon as I can.

"How many of you live here?"

"It depends. Right now there's four of us: Margareta Olsson, who studies economics; me; Peter Engbom, who is supposed to be studying physics, but is currently mired in the history of religion; and then Anna."

"Who is studying medicine," Linda said.

The gesture was almost imperceptible, but she had seen it. He had registered surprise. At that moment, she caught hold of the thought she had lost last night.

"When did you see her last?"

"I don't have a good memory for these things. It may have been yesterday or a week ago. I'm in the middle of a study of Capablanca's most accomplished endgames. Sometimes I think it should be possible to transcribe chess moves like music. In which case Capablanca's games would be fugues or enormous masses."

Another unplayable music nut, she thought.

"That sounds interesting," she said and got up. "Is anyone else home now?"

"No, just me."

Linda walked back to the kitchen, with Zacharias at her heels.

"I'm going in now, whatever you say."

"Anna won't like it."

"You can always try to stop me."

He watched her as she opened the door and walked in. Anna's room had been a maid's room at one time. It was small and narrow. Linda sat on the bed and looked around. Zacharias appeared in the doorway. Linda had the feeling he was going to throw himself on top of her. She got up and he took a step back, but he went on watching her. It was no use. She wanted to look in all the desk drawers, but as long as he was standing there she couldn't bring herself to do that.

"When do the others get home?"

"I don't know."

Linda walked into the kitchen again. He smiled at her, revealing a row of yellow teeth. She was starting to feel sick and decided to leave.

"I can show you all the chess moves," he said.

She opened the front door and paused on the steps.

"If I were you, I'd spend some much-needed time in the shower," she said, and turned on her heel.

She heard the door slam shut behind her. What a waste of time, she thought angrily. The only thing she had managed to do was to demonstrate her weaknesses. She kicked open the gate. It hit the letter box attached to the fence. She stopped and turned. The front door was closed, and she couldn't see anyone looking out of a window. She opened the letter box. There were two letters. She picked them out. One was addressed to Margareta Olsson from a travel agency in Göteborg. The other was addressed, by hand, to Anna. Linda hesitated for a moment, then took it with her to the car. First I read her journal, then I open her post, she thought. But I'm doing it because I'm worried about her. Inside the envelope was a folded piece of paper. She flinched when she opened it; a dried, pressed spider fell on to her lap.

The message was short, apparently incomplete and with no signature: *We're in the new house, in Lestarp, behind the church, first road on the left, a red mark on an old oak tree, back there. Let us never underestimate the power of Satan. And yet we await a mighty angel descending from the heavens in a cloud of glory . . .*

Linda put the letter on the passenger seat. She thought back to the insight she had had in the house. It was the one thing she could thank that noxious chess player for. He had listed what everyone who lived in the house studied, as well as their names. But Anna was just Anna. She was studying medicine, ostensibly to become a physician. But what had she said when she told Linda about the day she saw her father in Malmö? She had seen a woman who had collapsed in the street, someone who needed help. And she had said that she couldn't stand the sight of blood. Linda had been struck by the incongruity of this coming from someone who professed to want to be a doctor.

She looked at the letter beside her. What did it mean? . . . *we await a mighty angel descending from the heavens in a cloud of glory.*

The sun was strong. It was the beginning of September, but it was one of the warmest days of the summer. She took a map of Skåne out of the glove compartment. Lestarp was between Lund and Sjöbo. Linda pushed down the sun visor. It's so childish, she thought. The business with the dried spider, the kind that falls out of lamp shades. But Anna is missing. This childishness exists alongside the reality, the reality of a little gingerbread house in the forest. Hands at prayer and a severed head.

It was as if it was only now that she fully understood what she had seen in the hut that day. And Anna was no longer the person she thought she knew. Maybe she isn't even studying medicine. Perhaps this is the day on which I realise that I know nothing about Anna Westin. She's dissolving in an unfathomable fog.

Linda was not aware of formulating a plan as such, she just started driving towards Lestarp. It was 29°C in the shade.

CHAPTER 23

She parked beside the church in Lestarp. It had been recently renovated. Newly painted doors gleamed. A small black and gold plaque above them was inscribed with the year 1851. Linda remembered her grandfather saying something about his own grandfather drowning in a storm at sea that very year. She thought about him as she looked for a toilet in the porch. It was located in the crypt. The cool air felt good to her after the heat outside. I only remember important years, her grandfather had said. A year when someone drowns in a terrible accident, or when someone, like you, is born.

When she had finished, she washed her hands thoroughly as if she were washing off the remains of the chess-playing Zacharias's handshake. She looked at her face in the mirror. It passed muster, she decided. Her mouth was stern as always, her nose a little big, but her eyes were arresting and her teeth were good. She shuddered at the thought of the chess player trying to kiss her, and hurried back up the stairs. An old man was carrying in a box of candles. She held open the doors for him. He put the box down and then placed his hands on his back.

"You would think God could spare his devoted servant from the trials of back pain," he said in a low voice. Linda realised he was keeping his voice down because someone was sitting in the pews. She thought at first it was a man, then saw she was mistaken.

"Gudrun lost two children," the old man whispered. "She comes here every single day."

"What happened?"

"They were run over by a train, a terrible tragedy. One of the ambulance drivers who took care of their remains lost his mind."

He picked up the box again and continued along the aisle. Linda walked out into the sun. Death is all around me, she thought, calling out to me and trying to deceive me. I don't like churches, or the sight of women crying. How does that square with my wanting to become a police officer? Does it make any more sense than Anna not being able to stand the sight of blood? Maybe you can want to become a doctor or a police officer for the same reason: to see if you have what it takes.

Linda wandered into the little cemetery attached to the church. Walking along the row of headstones was like perusing the shelves of a library. Every headstone like the cover of a book. Here lay householder Johan Ludde and his wife Linnea. They had been buried for 96 years, but he was 76 when he died and she was only 41. There was a story here, in this poorly tended grave. She wondered what her own headstone would look like. One which was overgrown caught her eye. She crouched down and cleared moss and soil from its face. Sofia, 1854–1869. Fifteen years old. Had she too teetered on the rail of a bridge, but with no-one to save her?

She left the car where it was and followed the narrow road to the back of the church. She came upon the tree with the red mark almost immediately and turned on to a road down a small hill. The house was old and worn, the main part whitewashed stone with a slate roof, an addition built of rustic, red-painted wood. Linda stopped and looked around. It was absolutely quiet. A rusty, overgrown tractor stood on one side, by some apple trees. Then the front door opened and a woman in white clothes started walking out to greet Linda, who didn't understand how she had been spotted. She hadn't seen anyone and she was still partly hidden by the trees. But the woman was making her way briskly straight for her, smiling. She was about Linda's age.

"I saw that you needed help," she said when she was close enough. She spoke a mixture of Danish and English.

"I'm looking for a friend of mine," Linda said. "Anna Westin."

The woman smiled. "We have no use for names here. Come with me. You may find the friend you are looking for."

The mildness of her voice made Linda suspicious. Was she walking into a trap? She followed her into the cool interior of the main house. It took some time for Linda to see clearly. The slowness of her eyes to adjust from bright outdoor light to dim interiors was one of her few physical weaknesses, one she had discovered during her time at the training college.

All of the walls on the inside were whitewashed and there were no rugs on the bare, broad planks of the floor. There was no furniture, but a large black wooden cross hung between two arched windows. People sat on the floor along the walls. Many with their arms wrapped around their knees, all silent. They were of all ages and in different styles of dress. One man with short hair was wearing a dark suit and tie; by his side was an older woman in very simple clothes. Linda looked around but could not see Anna among them. The woman who had come out to greet her looked enquiringly at her, but Linda shook her head.

"There's one more room," the woman said.

Linda followed. The wooden walls of the next room were also painted white. The windows there were much less elaborate. Here, too, people were sitting along the walls, but Linda did not see anyone who looked like Anna. What was going on in this house? What had the letter said? *. . . a mighty angel in a cloud of glory*?

"Let us go out again," the woman said.

She led the way across the lawn, around the side of the house to a collection of stone furniture in the shadow of a beech. Linda's curiosity was now fully engaged. Somehow these people had something to do with Anna. She decided to come clean.

"The friend I'm looking for is missing. I found a letter in her letter box which described this place."

"Can you tell me what she looks like?"

I don't like this, Linda thought. Her smile, her calm. It's completely disingenuous and makes my skin crawl. Like when I shook that boy's hand.

Linda gave her Anna's description. The woman's smile never wavered.

"I don't think I've seen her," she said. "Do you have the letter with you?"

"I left it in the car."

"And where is the car?"

"I parked it by the church. It's a red Golf. The letter is on the front seat. The car is unlocked, actually, which I know is careless of me."

The woman was silent. Linda felt uncomfortable.

"What do you do here?"

"Your friend must have told you. Everyone who is here has the mission of bringing others to our temple."

"This is a temple?"

"What else would it be?"

Of course, Linda thought sarcastically. What was I thinking? This is clearly a temple and not simply the somewhat dilapidated remains of a humble Swedish farmstead where the owners once struggled to put food on the table.

"What is the name of your organisation?"

"We don't use names. Our community comes from within, through the air we share and breathe."

"That sounds very deep."

"The self-evident is always the most mysterious. The smallest crack in a musical instrument alters its timbre completely. If a whole panel falls out, the music ceases. So it is with human beings. We cannot fully live without a higher purpose."

Linda did not understand the answers she was getting, and didn't like this feeling. She stopped asking questions.

"I think I'll leave now."

She walked away quickly without turning around, nor did she stop until she reached the car. Instead of leaving right away, she sat and looked out at the trees. The sun was shining through the leaves and into her eyes. Just as she was about to start the engine, she saw a man cross the gravel yard in front of the church.

At first she saw only his outline, but when he crossed into the shade of the tall trees she felt as if she had just taken a gulp of frozen air. She recognised his neck, and not just that. During the seconds before he walked out into the blinding sun, she heard Anna's voice reverberate inside her head. The voice was very clear, telling her about the man she had seen from the hotel window. I am also sitting by a window, Linda thought. A car window. And I'm convinced that I've just seen Anna's father. It is utterly unreasonable, but that's what I think.

CHAPTER 24

Is it absurd to think that you can identify a person by their neck? Linda wondered. What had convinced her about something she had no grounds for knowing? You can't recognise someone you've never met, let alone someone you've only ever seen in snapshots and only heard about from a person who anyway hasn't herself seen him in more than 20 years.

She shook off the thought and drove back to Lund. It was early afternoon and the sun was still strong. The great heat hung oppressively over the day. She parked outside the house she had visited just a few hours earlier and prepared herself for another meeting with Zacharias the chess player. But the door was opened by a girl younger than Linda with blue streaks in her hair and a chain suspended from her nostril to her cheek. She was wearing black clothes in a combination of leather and vinyl. One of her shoes was black, the other white.

"There are no rooms available," she said brusquely. "If there's still a notice up at the Student Union, it's a mistake."

"I don't need a room. I'm looking for Anna Westin. I'm a friend of hers – my name's Linda."

"I don't think she's here, but you can take a look."

She let Linda pass her. Linda cast a quick glance into the living room. The chess set was still there, but not the player.

"I was here a few hours ago," Linda said. "I talked to the boy who plays chess."

"You can talk to whomever you like."

"Are you Margareta Olsson?"

"That's my assumed name."

Linda was taken aback. Margareta looked amused.

"My real name is Johanna von Lööf, but I prefer simple names. That's why I call myself Margareta Olsson. There's only one Johanna von Lööf in this country, but a couple of thousand Margareta Olssons. Who wants to be unique?"

"Beats me. You study law, right?"

"No. Economics."

Margareta pointed to the kitchen. "Are you going to see if she's in or not?"

"You know she isn't here, don't you?"

"Of course I know. But there's nothing stopping you from checking it out for yourself."

"Do you have some time to chat?"

"I have all the time in the world, don't you?"

They sat in the kitchen. Margareta was drinking tea, but she didn't offer Linda any.

"Economics. That sounds hard work."

Margareta tossed her head with irritation.

"It is hard. Life should be hard. What did you want to know?"

"I'm looking for Anna. She's my friend, and I want to make sure nothing has happened to her. I haven't heard from her for a while and that's not like her."

"And what can I do for you?"

"You can tell me when you last saw her."

Margareta's answer was caustic.

"I don't like her. I try to have as little to do with her as possible."

Linda had never heard that before – someone not liking Anna. She thought back to their school days. Linda had often fought with other students, but she couldn't remember Anna doing so.

"Why?"

"I think she's stuck up. I can generally tolerate this in others since I'm as bad myself. But not in her case. There's something about her that drives me up the wall."

She got up and rinsed her cup.

"It probably bothers you to hear me say this about your friend."

"Everyone has a right to their opinion."

Margareta sat down again. "Then there's another thing. Or two, more precisely. She's stingy and she doesn't tell the truth. You can't trust her. Either what she says or that she won't use all your milk."

"That doesn't sound like Anna."

"Maybe the Anna who lives here is a different person. All I'm saying is, I don't like her, she doesn't like me. We cope. I don't eat when she's eating and there are two bathrooms. We rarely bump into each other."

Margareta's mobile phone rang. She answered and then left the kitchen. Linda thought about what she had just been told. More and more she was starting to realise that the Anna she had become re-acquainted with was not the same as the Anna she had grown up with. Even though Margareta – or Johanna – didn't make the best impression, Linda instinctively felt that she had been telling the truth.

I have nothing more to do here, she thought. Anna has chosen to stay away. She has some reason for it, just as there will turn out to be a reason why she and Birgitta Medberg were in contact.

Linda got up to leave as Margareta came back into the kitchen.

"Are you angry?"

"Why would I be angry?"

"Because I've told you unflattering things about your friend."

"I'm not angry."

"Then maybe you'd like to hear more?"

They sat down at the table again. Linda noticed that she was tense.

"Do you know what she studies?"

"Medicine."

"That's what I thought, too, we all did. But then someone told me she had been expelled from the medical school. There were rumours about plagiarism – I don't know if that was true or not. Maybe she simply gave up. But she never said anything to us about it. She pretends that she's still studying medicine, but she's not."

"What *does* she do?"

"She prays."

"Prays?"

"You heard me," Margareta said. "Prays. What you do when you go to church."

Linda lost her temper. "I know what it is. Anna prays, you say. But where? When? How? Why?"

Margareta did not react to her outburst. Linda was grudgingly impressed by this display of self-control that she herself lacked.

"I think it's genuine. She's searching for something. I can understand her in a way. Personally, I'm on a quest for material wealth; other people are looking for the spiritual equivalent."

"How do you know all of this if you don't even talk to her?"

Margareta leaned over the table. "I snoop, I eavesdrop. I'm the person who hides behind curtains and hears and sees everything that goes on. I'm not kidding."

"So she has a confidante?"

"That's a strange word, isn't it? 'Confidante' – what does it really mean? I don't have one, I doubt if Anna Westin does either. To be completely honest, I think she's unusually dim-witted. God forbid I would ever be diagnosed and treated by a physician like her. Anna Westin talks to anyone who will listen. I think all of us here find her conversation a series of naïve and worthless sermons. She's always lecturing us on moral topics. It's enough to drive anyone off their rocker, except perhaps our dear chess player. He cherishes vain hopes about getting her into bed."

"Any chance?"

"Nil."

"What do her lectures consist of?"

"She talks about the poverty of our daily existence. That we don't nurture our inner selves. I don't know exactly what she believes in other than that she's Christian. I tried to discuss Islam with her one time and she went ballistic. She's a conservative Christian. More than that I don't know. But there's something genuine about her when she talks about her religious views. And sometimes I hear her when she's in her room. It sounds real. That's when she isn't lying or stealing. She's being herself. Beyond that, I can't say."

Margareta looked at her. "Has something happened?"

Linda shook her head. "I don't know. Maybe."

"But you're worried?"

"Yes."

Margareta got up. "Anna Westin's God will protect her. At least that's what she always brags about. Her God and some earthly angel named Gabriel. I think it was an angel. I can't remember exactly. But with that kind of protection she should be fine."

She stretched out her hand. "I have to go now. Are you also a student?"

"I'm a police officer. Or will be soon, that is."

Margareta took a closer look at her.

"I'm sure you will, the number of questions as you've been asking."

Linda realised she had one more.

"Do you know anyone called Mirre? She left a message on Anna's answering machine."

"No. But I can ask the others."

Linda gave her her phone number and left the house. She was still vaguely envious of Margareta Olsson's poise, her self-confidence. What did she have that Linda didn't?

The following morning, Monday, Linda was awakened by the sound of the front door slamming shut. She sat up in bed. It was 6 a.m. She lay down and tried to fall back to sleep. Raindrops were splattering against the windowsill. It was a sound she remembered from childhood. Raindrops, Mona's shuffling, slippered gait, and her father's firm footsteps. Once upon a time these sounds had been her greatest source of security. She shook off her thoughts and got up. Her father had forgotten to turn off the stove and he hadn't finished his coffee. He's nervous and he left in a hurry, she thought.

She pulled the paper towards her and leafed through it until she saw an article about the developments in the Rannesholm case. There was a short interview with her father. It was early, he had said, and although there were almost no clues, they had some leads. But no, he was not able to comment further for the time being. She put the paper away and thought about Anna. If Margareta Olsson was right – and she had no reason to doubt she was telling the truth – Anna had become a very different person. But why had she left the flat in Ystad? And why did she claim to have seen her father? Why wasn't Henrietta telling the truth? And that man Linda had seen walk past the church – why was she convinced it was Anna's father?

And the other crucial question: what was the connection between Anna and Birgitta Medberg?

Linda had trouble separating all these thoughts. She heated the coffee and wrote everything down on a piece of paper. Then she crumpled it up and threw it away. I have to talk to Zeba, she thought. I'll tell her everything. She's smart. She never loses touch with reality. She'll give me some good ideas. Linda showered, put her clothes on and then called Zeba. Her answering machine cut in. Linda tried her mobile, but it was out of range. Since it was raining, she could hardly have taken her boy out for a walk. Maybe she was with her cousin.

Linda was impatient and irritated. She thought of calling her father, possibly even her mother, just to have someone to talk to. She decided she didn't want to interrupt her father. And a conversation with Mona could drag on for ever. She didn't need that. She pulled on her boots and an anorak and walked down to the car. She was getting used to having a car. That was dangerous. When Anna came back she would have to start walking again. When she couldn't borrow her father's car. She drove out of the city and stopped at a petrol station. A man at the next pump nodded to her. She recognised his face without being able to place him until she was standing in the queue at the cashier's window. It was Sten Widén, her father's friend.

"It's Linda, isn't it?" His voice was hoarse and weak.

"Yes. Sten, right?"

He laughed, something that seemed to cost him an effort.

"I remember you as a little girl. And suddenly you're all grown up. A police officer no less."

"How are the horses?"

He didn't answer until she had finished paying and they were walking back to their cars.

"Your dad has probably told you what's going on," Widén said. "I have cancer and I'm going to die soon. I'm selling the last of the horses next week. That's how it is. Good luck with your life."

He didn't wait for an answer, just got into his muddy Volvo and drove away. Linda watched him leave and could think only one thing – how grateful she was that she wasn't the one selling her last horses.

She drove to Lestarp and parked by the church. Someone has to know, she thought. But if Anna isn't here, where is she? Linda pulled up the hood of her yellow anorak and hurried down the road at the back of the church. The garden was deserted. The old tractor was wet and shiny from the rain. She banged on the front door and it swung open. But no-one had opened it, it hadn't been properly closed. She called out, but no-one answered. The house was abandoned. Nothing was left. She saw that they had taken the black cross on the wall. It felt as if the house had been empty for a long time.

Linda stood in the middle of the room. The man in the sun, she thought. The one I saw yesterday and thought was Anna's father. He came here, and today everyone is gone. She left the house and drove to Rannesholm. There she was told that Inspector Wallander was up at the manor, holding a meeting with his associates. She walked over in the rain and settled down in the big hall, to wait for him. She thought about the last thing Margareta Olsson had said, something about Anna Westin not having to worry about her safety because she had God and an earthly guardian angel named Gabriel for protection. It seemed important, but she just couldn't think how.

CHAPTER 25

Linda never ceased to be surprised by her father, by his rapid mood changes, that is. When she saw him come through a door in the large hall at Rannesholm Manor she expected him to seem tired, anxious and downcast. But he was in good spirits. He sat down next to her and launched into a long-winded story about a time he had left a pair of gloves at a restaurant and been offered a broken umbrella in their stead. Is he getting soft in the head? she wondered. Then her father left to go to the toilet and Martinsson stopped on his way out. He told her Wallander had been in better humour ever since she had moved back to town. Martinsson hurried on when Wallander returned. Linda remembered that there had been some trouble between them not so long ago.

Wallander sat down so heavily on the old sofa that the springs groaned. She told him about running into Sten Widén at the petrol station.

"He's remarkably stoical," Wallander said. "He's very like Rydberg – the same calm attitude. I hope that will turn out to be true for me one day, that I'll be stronger than I think."

Some officers walked past carrying cases of equipment. Then the room was silent.

"Are you making progress?" Linda said.

"Not much, or slowly, I should say. The worse the crime, the more impatient one becomes about solving it, even though in these cases patience is critical. I once knew an officer in Malmö – Birch – who used to compare our investigative work to that of a surgeon facing a complicated operation. The calm, time and patience needed for such procedures are key ingredients even for us. Birch is dead now, as it happens. He drowned in a tiny lake. He was swimming, must have got cramp, no-one heard him. He should have known better, of course, but now he's dead. I feel as if people are dying all around me, although I know it's irrational. People are being born and dying off all the time. But the dying seems more pronounced when you reach the front of the queue. Now that my father is dead there's no-one ahead of me any more."

Wallander looked down at his hands. Then he turned to her and smiled. "What was it you asked me?"

"How is the investigation going?"

"We haven't found a single trace of the murderer. We have no idea who was living in that hut."

"What do you think?"

"You know you should never ask me that. Never what I think, only what I know or what I suspect."

"I'm curious."

He sighed. "I'll make an exception. I think Birgitta Medberg came upon the hut accidentally in her search for the pilgrim trail. The person who was there panicked or became enraged and killed her. But the fact that he dismembered the body complicates the picture."

"Have you found the rest of it?"

"We have divers in the lake and a dog unit combing the forest. They haven't come up with anything yet."

He got ready to launch himself out of the sofa.

"I take it there's something you want to tell me."

Linda told him in great detail about her visits to Anna's house in Lund as well as to the house in Lestarp.

"Too many words," he said when she had finished.

"I'm working on it. But you got the gist?"

"Yes."

"Then it couldn't have been too bad."

"I'd give it a beta query," Wallander said.

"What's a beta query?"

"When I was at school, anything less than a beta query was considered a failing grade."

"So what do you think I should do?"

"Stop worrying. You haven't been listening to me. What happened to Medberg was a mishap, one of almost biblical proportions. She went down the wrong path. Unless I'm mistaken, Medberg had excruciatingly bad luck. Therefore there's no longer any reason to think Anna is in any danger. The journal shows there is a connection between the two of them, but it's no longer of concern to us."

Ann-Britt Höglund and Lisa Holgersson came walking briskly past. Holgersson nodded kindly to Linda. Höglund didn't seem to notice her. Wallander got up.

"Go home now," he said.

"We could have used an extra set of hands," Holgersson said. "I wish the money was there. When is it you start?"

"Next Monday."

"Oh, good."

Linda watched them go out together, then she too left the manor. It was raining now and much colder, as if the weather couldn't make up its mind. She walked back to the car. The house behind the church had sparked her curiosity. Why were they all gone? I can at least find out who the owner of the house is, she thought. I don't need a permit or a police uniform for that. She drove back to Lestarp and parked in her usual spot. The doors to the church were half open. After hesitating for a moment, she walked in. The old man she had met before was in the porch. He recognised her.

"Can't stay away from our beautiful church?"

"I wanted to ask you something."

"Isn't that why we all come here? To find answers to our questions?"

"That wasn't quite what I meant. I was thinking about the house down the hill behind the church. Do you know who owns it?"

"It's been in many different hands. When I was young, a man with one leg shorter than the other lived there. His name was Johannes Pålsson. He worked as a day labourer at Stigby farmstead and was good at mending china. The last few years he lived alone. He moved the pigs into the main room and the chickens into the kitchen. That kind of thing went on in those days. When he was gone, someone else used the place as storage for grain. Then there was a horse breeder and after that, sometime in the 1960s, the house was sold to someone whose name I've forgotten."

"You don't know who owns it now?"

"Oh, I've seen people come and go lately. They're peaceful and discreet. Some say they use the house for meditation. They've never bothered us. But I don't know who the owner is. You should be able to find out through the property tax records."

Linda thought for a moment. What would her father have done?

"Who knows all the gossip in this village?"

He looked at her with a smile. "That would be me, wouldn't it?"

"But apart from you. If there's anyone who might know who owns the house, who would it be?"

"Maybe Sara Edén. She was the school teacher; she lives in the little house next to the car-repair place. She devotes her time to talking on the phone. She knows everything that's going on, and fills in the rest as needed. She's a good sort, just insatiably curious."

"What happens if I ring her doorbell?"

"You'll make a lonely old woman's day."

The front door opened wider and the woman Linda had seen the previous day walked in. She met Linda's gaze before walking to her pew.

"Every day," the old man said. "The same time, the same face, the same grief."

Linda left the church and walked down to the house. It was still empty. She returned to the church, decided to let the car stay where it was and walked down the hill to 'Rune's Auto and Tractor'. On one side of the shop there was a ramshackle pile of spare parts, on the other there was a high fence. Linda supposed that the retired teacher didn't care for a view of a car-repair shop. She opened the gate and stepped into a well-tended garden. An elderly woman was kneeling over a flower bed. She stood up when she heard Linda.

"Who are you?" she asked sternly.

"My name is Linda. Do you mind if I ask you some questions?"

Sara Edén came over to where Linda was standing, holding a garden shovel aggressively outstretched. It occurred to Linda that there were people who were the human equivalent of ill-tempered dogs.

"Why would you want to ask me questions?"

"I'm looking for a friend who's disappeared."

Sara Edén seemed sceptical.

"Isn't that something for the police? Looking for missing persons?"

"I am from the police."

"Then perhaps you'll show me your ID. That's my right, my older brother once informed me. He was the headmaster at a school in Stockholm. He lived to be one hundred and one years old despite his bothersome colleagues and even more bothersome students."

"I don't have an ID card yet. I'm still in training."

"I'll have to take your word for it, then. Are you strong?"

"Fairly."

Sara Edén pointed to a wheelbarrow filled to the brim with plants and weeds.

"There's a compost heap on the other side of the house, but my back has been giving me a bit of trouble. I must have slept in a strange position."

Linda took hold of the wheelbarrow. It was very heavy, but she managed to coax it around to the compost heap. When she had emptied it, Sara Edén showed a kindlier side. There were some chairs and a table tucked into a little arbour.

"Do you want a cup of coffee?" she asked.

"Yes, please."

"Then you'll have to fetch one yourself from the vending machine by the furniture warehouse on the road to Ystad. I don't drink coffee – or tea for that matter. But I can offer you a glass of mineral water."

"No, thank you."

They sat down. Linda had no trouble imagining fru Edén as a school teacher. She probably saw Linda as an unruly schoolgirl.

"Well? Tell me what happened."

Linda gave her an outline of the events and said she had traced Anna to the house behind the church.

"We were supposed to meet," she said. "But something happened."

The old lady looked doubtful. "And how do you think I can be of assistance?"

"I'm trying to find out who owns the house."

"In the olden days one always knew who was who and who owned what. But in this day and age there's no way of telling. One day I'll find out I've been living next to an escaped criminal."

"I thought perhaps in such a small place people still knew these things."

"There have been a great many comings and goings in that house during the past while, but nothing that caused any disturbance. If I have understood it correctly, the people there now are involved in some sort of health organisation. Since I take good care of myself and am not planning to give my departed brother the satisfaction of dying at a younger age, I watch what I eat and drink, and am curious about this new so-called alternative medicine. I went up to the house one time and spoke to a very friendly English-speaking lady. She gave me a pamphlet. I don't remember what the organisation was called, but they espoused meditation and certain natural juices for promoting health."

"Did you ever go back?"

"The thing was far too vague for my taste."

"Do you still have the pamphlet?"

Edén nodded towards the compost.

"I doubt there's anything left of it by now."

Linda tried to think of something else to ask, but didn't see the point of pursuing it further. She got up.

"No more questions?"

"No."

They walked back to the front of the house.

"I dread the autumn," fru Edén said. "I'm afraid of the creeping fog and the rain and the noisy crows in their treetops. The only thing that keeps my spirits up is the prospect of the spring flowers I'm planting now. Ah, yes, there may be something else."

They were standing now on either side of the gate.

"There was a Norwegian," the old lady said. "I sometimes go into Rune's shop and complain if they're making too much of a racket on a Sunday. I think Rune is a little afraid of me. He's the kind of person who never grows out of the respect he had for his teachers. The noise usually stops. One day he told me about a Norwegian who had just filled up his car and who paid with a thousand-kronor note. Rune isn't used to notes that big. He said something about the Norwegian owning the house there."

"So I should ask Rune?"

"Only if you have time on your hands. He's on holiday in Thailand. I don't even want to think about what he might be getting up to."

Linda thought for a moment. "A Norwegian. Did he say at all what he looked like?"

"No. If I were in your place I would ask the people who most likely handled the sale of the house. That would be the Sparbanken property sales division. They have an office in town. They may know."

Linda left. She thought that Sara Edén was a person she would have liked to know more about. She crossed the street, passed a hair salon and stepped into the tiny Sparbanken office. There was only one person inside; he looked up when she came in. Linda asked him her question and the answer came without his having to consult any binders or notes.

"That's right," he said. "We handled the sale of that house. The seller was a dentist from Malmö by the name of Sved. He had used the house as a summer retreat for a while but grown tired of it. We advertised the property online and in the *Ystad Allehanda.* A Norwegian came in and demanded to see the place. I asked one of the Skurup estate agents to take care of him. That's fairly normal, since I run the branch by myself and can't always take on the extra responsibility for property sales. Two days later, the sale was finalised. As far as I recall, the Norwegian paid cash. They've got money coming out of their ears these days."

The last comment revealed his grumbling displeasure at the vibrant Norwegian economy. But Linda was more interested in the Norwegian's name.

"I don't have the papers here, but I can call the Skurup office."

A client entered the branch, an old man who walked with the help of two canes.

"Please excuse me while I attend to herr Alfredsson," the man behind the counter said.

Linda waited impatiently. It took what seemed like an age before the old man was finished. Linda held the door open for him. The man behind the counter placed his call, and after about a minute he received an answer that he wrote on a piece of paper. He finished the conversation and pushed the note over to Linda. She read: Torgeir Langas.

"It's possible he spells the last name with a double 'a'; that would be Langaas."

"What's his address?"

"You only asked me for his name."

Linda nodded.

"If you need more information, you can ask the Skurup office directly. Do you mind my asking why you so urgently need to contact the owner of the house?"

"I may want to buy it," Linda said, and left.

She hurried to the car. Now she had a name. As soon as she opened the car door she noticed that something was amiss. A receipt that had been on the dashboard was on the floor, a matchbox had been moved. She had left the car unlocked and someone had been in it while she was gone.

Hardly a thief, she thought. The car radio is still here. But who's been in the car? Why?

CHAPTER 26

The first thought that ran through Linda's head was purely irrational: Mum did this. She's been rifling through my stuff again like she used to. Another thought ran through her like electric current: a bomb. Something was going to explode and tear her to pieces. She got hesitantly into the car. And of course there was no bomb. A bird had left a big dropping on the windscreen, that was all. Now she noticed that the seat had been pushed back. The person who had got into the car was taller than she was. So tall that he or she had had to adjust the seat to sit behind the wheel. She sniffed for new scents, but she couldn't pick anything out, no aftershave or perfume. She looked everywhere. Something was different about the black plastic cup of loose change that Anna had taped behind the gear stick, but it was not clear what.

Linda's thoughts returned to her mother. The game of cat-and-mouse had gone on for most of her childhood. She couldn't remember exactly when she realised that her mother was forever rifling through her things in search of who knew what secrets. Maybe it had started when Linda was eight or nine and she could tell that something had changed about her room when she came home from school. At first she had assumed she must be wrong. The red cardigan had been lying over the green jumper like that, not the other way around. She had even asked Mona, who had snapped at her. That was when the suspicion had been born and the game started in earnest. She left traps in her clothes, among her toys and books. But it seemed as if Mona sensed what was going on. Linda laid increasingly elaborate traps and even recorded in a notebook the precise arrangement of her things so that she could catch her mother.

Linda went on studying the interior of the car. Some kind of mother has been here, she thought, one who may be a man or a woman.

Snooping in kids' stuff is more common than you'd think. Most of my friends had at least one parent who did it. She thought about her father. He had never been through her things. Sometimes she had seen him peer through her half-open door to make sure she was there, but he never made surreptitious expeditions into her life. That had always been her mother.

She concentrated on what sort of person it had been in the car. Taking the radio would have been a simple way to cover his or her tracks. That way Linda would have assumed there had been a straightforward break-in. So this is not a particularly cunning mum, she thought.

She got no further. There was no conclusion to be drawn, no answer to who or why. She re-adjusted her seat, got out and looked around. A man had walked past in the blinding sun. She had seen his back and thought it was Anna's father. Linda scratched her head in irritation. Anna had just been imagining things when she said she had seen her father in the street. Maybe her acute disappointment was what had made her take off. She had done that before – taken a trip without warning – but Zeba said she had always let at least one person know where she was going.

Who did she tell this time? Linda wondered.

She walked across the gravel yard in front of the church, glancing up at some pigeons circling the bell tower, and then carried on down to the empty house. A man named Torgeir Langaas bought this house, she thought. He paid cash.

She walked round to the back of the house, looking thoughtfully but absent-mindedly at the stone furniture. There were several blackcurrant and red-currant bushes. She picked a few strands of berries and ate them. Her memories of Mona returned. Linda did not think she had snooped out of curiosity, rather it seemed that she was prompted by fear. Of what had she been so afraid? Was she afraid I wasn't who she thought I was? A nine-year-old can play roles and have her secrets, but hardly of a magnitude that requires continuous snooping in order to truly understand her, especially if she is your own child.

Open warfare had broken out only when Mona started reading Linda's journal. Linda had been 13 by then and had been keeping her journal hidden behind a loose panel at the back of her wardrobe. One day she discovered that it had been pushed back a few centimetres too far. She could still remember her rage. That time she really hated her mother.

There was an epilogue to that memory. Linda had decided to set another trap for her mother. She wrote a message on the first blank page in the journal stating that she knew her mother was reading it, that she was snooping in all her things. She put the journal back in its hiding place and set off for school. About halfway there, she decided to cut class. She knew she would not be able to concentrate anyway. She spent the day wandering around the shops in town. When she came home she broke out in a cold sweat, but her mother greeted her as if nothing had happened. After they had all gone to bed Linda took out the journal and saw that her mother had written, without apology or explanation: *I won't read it again, I promise.*

Linda picked a few more berries. We never talked about it, she thought. I think she stopped snooping altogether after that, but I could never be sure. Maybe she got better at covering her tracks, maybe I stopped caring as much. But we never talked about it.

The estate agent's name was Ture Magnusson and he was in the middle of the sale of a house in Trunnerup to a retired German couple. Linda skimmed through a folder of houses for sale while she waited. Ture Magnusson spoke very bad German. Finally he got up and walked over to her. He smiled.

"They want a moment alone," he said, sitting down as he introduced himself. "These things take time. What can I do for you?"

Linda gave him her story without playing the policewoman this time. Magnusson nodded even before she was finished. He remembered the sale without having to look it up.

"That house was indeed bought by a Norwegian," he said. "A pleasant sort, quick to make up his mind. He was what you would call an ideal client, paid in cash, no hesitation, no second thoughts."

"How can I get in touch with him? I'm interested in the house."

Magnusson leaned back and seemed to take stock of her. His chair creaked as he pushed it on to its back legs and balanced it up against the wall.

"To be perfectly honest, he paid a very high price for the house. I shouldn't tell you that, but it's true. I can show you three other places in better condition, in more beautiful surroundings, all going for less."

"This is the house I want. I'd like at least to ask the owner if he would consider selling."

"Of course. I understand. Torgeir Langaas was his name," Magnusson said, singing the last sentence. He had a good voice. He went into the next room, and soon reappeared with an opened folder.

"Langaas," he read. "He spells the name with a double 'a'. He was born in somewhere called Baerum, forty-three years old."

"Where in Norway does he live?"

"He lives in Copenhagen."

Magnusson put the document down in front of her so that she could see. *Nedergade 12.*

"What sort of man would you say he was?"

"Why do you ask?"

"I want to know if there's any point my looking him up – in your opinion."

Magnusson leaned against the wall.

"It was clear from the moment I laid eyes on him that Langaas meant business. He was courteous and he had picked out the house he wanted so we drove out there together and inspected the property. He asked no questions, I remember. When we returned, he pulled the cash out of his shoulder bag. I don't think that's ever happened to me except on one other occasion when one of our young tennis stars came with a suitcase full of bank notes and bought a large estate in West Vemmenhög. He's never once been there since, so far as I know."

Linda wrote down Langaas's address in Copenhagen and prepared to leave.

"Come to think of it," Magnusson said, "there was something else about him. It was nothing extraordinary, just that he turned around a lot, as if he was afraid of running into someone he knew. He also excused himself a number of times and when he returned from the toilet the last time his eyes were glazed over."

"Had he been crying?"

"No, I would say rather that he seemed high."

"Alcohol?"

"He might have been drinking vodka."

Linda tried to think of something else to ask.

"But respectful, pleasant," Magnusson said again. "Perhaps he will sell you the house. Who knows?"

"What does he look like?"

"He has a pretty normal-looking face. What I remember best is his eyes, not simply because they were glazed over, but because there was something disturbing about them. Some people would perhaps have found his look menacing."

"And yet his manner struck you as pleasant?"

"Oh very. The ideal client, as I said. In fact, I bought myself a rather nice bottle of wine that evening. Just to celebrate such an easy day's work."

Linda left the estate agent's office. This is another step on the way, she thought. I can go to Copenhagen and find this Torgeir Langaas. I don't know exactly why – perhaps it helps to diminish my anxiety – I'm treating this as if Anna decided on the spur of the moment to go away and forgot to mention it to me.

Linda drove towards Malmö. Just before the turn-off to Jägersro and the Öresund Bridge she decided to make an unexpected visit. She pulled up outside the house in Limhamn, parked and walked in through the gate. A car was parked in the driveway. She stopped herself as she was about to ring the doorbell; why, she couldn't say. Instead, she walked to the back of the house, to the glassed-in porch. The garden was neatly tended. The gravel path was raked. The door to the porch was ajar. She pushed it open and listened. All was quiet, but she was sure that someone was there. These people spent far too much time locking doors and checking their alarm systems. She walked into the living room, looking at the painting over the sofa. It was a picture she had often looked at as a child, fascinated and disturbed by the brown bear shot through with flames and seeming about to explode. It still disturbed her. Her father had won it in a lottery and given it to her mother as a birthday present.

Linda heard a noise in the kitchen and walked to the door. Her "hello" stuck in her throat. Mona was standing at the kitchen counter. She was naked and drinking vodka straight from the bottle.

CHAPTER 27

Afterwards, Linda would think that it had been like staring at an image from her past. An image that reached beyond the reality of her mother standing there naked to something else, an impression, a memory she only managed to grasp when she drew a deep breath. She had experienced the same thing herself once.

She had been only 14 at the time, in the midst of those fearful teenage years when nothing seems possible or comprehensible, but when everything is at the same time straightforward, easy to see through. All parts of the body vibrate with a new hunger. It had happened during a brief period in her life when not only her father but even her mother disappeared off to work all day, pulling herself out of an unfulfilling stay-at-home existence to work for a shipping company. This finally allowed Linda to be alone for a few hours after school, or to bring friends home with her. She was happy.

That was the time Torbjörn came into her life. He was her first real boyfriend, one whom Linda imagined looked much like Clint Eastwood would have looked at 15. Torbjörn Rackestad was half Danish, a quarter Swedish and a quarter American Indian, which gave him not only a beautiful face, but also a dash of exoticism.

It was with him that Linda started a serious investigation into all that went under the rubric of love. They were slowly approaching the moment of truth, although Linda prevaricated. One day, when they were sprawled half-naked on her bed, her door opened. It was Mona. She had had a fight with her boss and left work early. Linda still broke into a sweat when she remembered the shock she had felt. At the time, she had started laughing hysterically. She had buried her face in her hands so she didn't know exactly how Torbjörn had reacted, but he must have pulled his clothes on and left very soon afterwards.

Mona hadn't lingered in the doorway. She had simply given her a look that Linda could never quite describe. There had been everything in that look from despair to a kind of smug triumph in finally having her worst fears about her daughter's nature confirmed. When Linda eventually went out into the living room they had had a shouting match. Linda could still recall Mona's repeated war cry: "I don't give a shit what you do as long as you don't get pregnant." Linda could also hear echoes of her own shouts, their sound, not the words. She remembered the embarrassment, the fury, the humiliation.

All these thoughts ran through her head as she stared at the nude older woman by the sink. It occurred to her, too, that she hadn't seen her mother naked since she was a little girl. Mona had put on a lot of weight, the flesh spilled out in unappealing bulges. Linda's face registered disgust, a brief unconscious expression but distinct enough for Mona to see it and snap out of her initial shock. She slammed the bottle down on the counter and pulled the refrigerator door open as a kind of shield for her body. Linda couldn't help giggling at the sight of her mother's head sticking up over the door.

"What do you mean by sneaking in like this? Why can't you ring the doorbell?"

"I wanted to surprise you."

"But you can't just barge into a person's house!"

"How else would I find out that my mother spends her days getting pissed?"

Mona slammed the refrigerator door shut.

"I am not a drunk!" she screamed.

"You were swigging vodka from the bottle, Mum."

"It's water. I chill it before I drink it."

They lunged for the bottle at the same time, Mona to hide the truth, Linda to uncover it. Linda got there first and sniffed it.

"This is pure, undiluted vodka. Go and put something on. Have you taken a look at yourself recently? Soon you'll be as big as Dad. You're all blubber, he's just heavy."

Mona grabbed the bottle out of her hands. Linda didn't fight her. She turned her back to Mona.

"Mum, put your clothes on."

"I can be naked in my own house if I want to be."

"It's not your house, it's the banker's."

"His name is Olof and he happens to be my husband. We own this house together."

"You do not. You have a pre-nuptial agreement. If you get divorced he keeps the house."

"Who told you that?"

"Grandpa."

"That old bastard. What did he know?"

Linda turned and slapped her face. "Don't say that about him."

Mona took a step back, unbalanced more by the alcohol than the blow.

"You're just like your father. He hit me too."

"Put some clothes on, for God's sake."

Linda watched as her mother took one more long swallow from the bottle. This isn't happening, she thought. Why did I come here? Why didn't I go straight to Copenhagen?

Mona tripped and fell. Linda wanted to help her up, but her attempts were pushed aside. Mona finally pulled herself into a chair.

Linda went into the bathroom and brought a dressing gown, but Mona refused to put it on. Linda began to feel sick to her stomach.

"Can't you cover yourself up?"

"All my clothes feel too tight."

"Then I'm leaving."

"Can't you at least stay for a cup of coffee?"

"Only if you put something on."

"Olof likes to see me naked. We always walk around the house naked."

Now I'm becoming a mother to my mother, Linda thought, firmly guiding her into the dressing gown. Mona put up no resistance. When she reached for the bottle, Linda moved it away. Then she set about making coffee. Mona followed her movements with dull eyes.

"How is Kurt?"

"He's fine."

"That man has never been fine in his entire life."

"Right now he is. He's never been better."

"Then it must be because he's rid of his old man – who loathed him."

Linda held her hand up as if to strike and Mona shut up. She lifted her palms in apology.

"You have no idea how much he misses him. No idea."

Mona got up from the chair, swaying but staying on her feet. She disappeared into the bathroom. Linda pressed her ear against the door. She heard a tap running, no bottles being taken from a secret stash.

When Mona reappeared she had combed her hair and washed her face. She looked around for the vodka that Linda had poured down the sink, then served the coffee. Linda suddenly felt a wave of pity for her. I never want to be like her, she thought. Not this snooping, nervous, clinging woman who never really wanted to leave Dad, but who was so insecure that she ended up doing the very things she didn't want.

"I'm not usually like this," Mona said.

"Just now I thought you said you and Olof always walk around naked."

"I don't drink as much as you think."

"Mum, you used to drink next to nothing. Now I catch you stark naked in your kitchen, tossing back vodka in the middle of the day."

"I'm not well."

"You mean you're sick?"

Mona started to cry, to Linda's dismay. When had she last seen her mother cry? She would sometimes fall into a nervous, almost restless sobbing if a meal didn't turn out well or if she had forgotten something, and she had cried when she had fought with Linda's father. But these tears were different. Linda decided to wait them out. The sobbing stopped as suddenly as it had started. Mona blew her nose and drank her coffee.

"I'm sorry."

"I'd rather you told me what was bothering you."

"What would that be?"

"Only you know that, not me. But obviously there is something on your mind."

"I think Olof has met another woman. He denies it, but if there's one thing life has taught me it's to tell when a man is lying. I learned that from your father."

Linda immediately felt the need to jump to his defence.

"I don't think he tells lies more than anyone else. No more than I do."

"Oh, the things I could tell you."

"And you can't know how little I care about that."

"Why do you have to be so mean?"

"I'm telling you the truth."

"Right now I could actually do with some plain old-fashioned kindness."

Linda's feelings had always oscillated between pity and anger, but now they seemed to have reached an unprecedented intensity. I don't like her, she thought. My mother asks for a love I'm incapable of giving her. I need to get out of here. She put down her cup.

"Are you leaving already?"

"I'm on my way to Copenhagen."

"What for?"

"I don't have time to go into it."

"I hate Olof for what he's doing."

"I'll come back another time when you're sober."

"I can't live like this any more."

"Then leave him. You've done it before."

"You don't need to tell me what I've done." Her voice was full of aggression again.

Linda turned and walked out. She heard Mona's voice behind her: "Stay a little longer." And then, just as she was about to close the door: "All right, then – go. But don't you dare show your face here again!"

Linda reached the car sweating and furious. Bitch, she thought, but she knew that before she was halfway across the Öresund Bridge her anger would switch to guilt: a good daughter would have stayed with her mother, listening to her troubles.

The guilt had already started to take over as she paid her toll for the bridge and she wished she were not an only child. I'm the one who will have to take care of them one day. She shivered in dismay and made up her mind to tell her father what had happened. He would know if Mona had ever had problems with alcohol in the past, if there was something Linda didn't know about.

She reached Denmark and started to feel better. The decision to talk to her father made her feel less guilty. Leaving Mona was the only thing she could have done. If she had stayed they would have been yelling at each other until Mona sobered up.

Linda drove to a car park and got out. She sat on a bench facing the sound, and stared over the water at the misty outline of Sweden. Somewhere there were her parents. They had enveloped her whole childhood in a strange mist. My dad was worse, she thought. The talented but gloomy policeman – who had a sense of humour but never let himself laugh. My father, who never found a new woman to share his life since he still loves Mona. Baiba tried to explain it to him, but he wouldn't listen. According to Baiba, he said, "Mona belongs to the past." But he hasn't got over her and he never will. She is his one great love. Now I've seen her wandering around naked, knocking back the hard stuff in the middle of the day. She too is lost in the foggy gloom. I'm almost 30 and I haven't managed to free myself from it.

Linda kicked angrily at the gravel, picked up a pebble and threw it at a seagull. The eleventh commandment is the most important, she thought, the one that reads: "Thou shalt never become like thy parents." She got up and returned to the car. She stopped in Nyhavn and bought a city map.

Darkness was falling by the time she reached Nedergade. It was a street in a shabby neighbourhood, of tall, identical apartment buildings. Linda felt unsafe and would have preferred to come back in broad daylight, but the bridge toll was too expensive to waste the journey. She locked the car and stamped her foot on the pavement as a way of rousing her courage.

She tried to make out the names of the people who lived in the building, although it was difficult to see in the dim light. The front door opened and a man with a scar across his brow walked out. He was startled when he saw her. She caught the door and walked back in before it closed behind him. Inside there was another noticeboard with names, but no-one by the name of Langaas or Torgeir. A woman walked by carrying a bag of rubbish. She was about Linda's age, and smiled at her.

"Excuse me," Linda said. "I'm looking for a man by the name of Torgeir Langaas."

The woman stopped and put the bag down. "Does he live here?"

"He gave this as his address."

"What was his name? Torgeir Langaas? Is he Danish?"

"Norwegian."

She shook her head. It seemed to Linda she genuinely wanted to help.

"I don't know of any Norwegians around here. We have a couple of Swedes and some people from other countries, but that's all."

The front door opened and a man walked in, dressed in a hooded sweatshirt. The woman with the rubbish bag asked him if he knew of a Torgeir Langaas. He shook his head. The hood was pulled up and Linda couldn't see his face.

"Try fru Andersen on the first floor. She knows everything about everyone in this building. I'm sorry I can't help you myself."

Linda thanked them and started up the stairs. Somewhere above her a door was pulled open, and loud Latin American music reverberated in the stairwell. Outside fru Andersen's door there was a small stool with an orchid. Linda rang the bell. Immediately a dog started to bark in the flat. Fru Andersen, shrunken and hunched over, was one of the smallest women Linda had ever seen. The dog, still barking beside her slippered feet, was also one of the smallest Linda had seen. She asked fru Andersen her question. The old lady pointed to her left ear.

Linda shouted out her question again.

"I may hear badly, but there's nothing wrong with my memory," frun Andersen said. "There's no-one by that name living here."

"Could he be staying with someone?"

"I know everyone who lives here, whether they be on the contract or not. It's been forty years, believe it or not, since they built this block. Now there are all kinds of people here, of course." She leaned closer to Linda and lowered her voice. "They sell drugs here. And no-one does anything about it."

Fru Andersen insisted on inviting her in and serving her coffee that she poured from a pot in the narrow kitchen. Linda managed to leave after half an hour. By then she knew all about what a wonderful husband herr Andersen had been, a man who had died far too young.

The Latin American music had stopped. Instead there was the sound of a child wailing. Linda walked out of the front door and looked each way before crossing the street. She sensed someone's presence in the shadows and turned her head. It was the man with the hooded sweat-shirt. He grabbed her by her hair. She tried to get away, but the pain was too great.

"There is no Torgeir," he said through clenched teeth. "No Torgeir Langaas. Drop it."

"Let me go!" she screamed.

He let go of her hair, and punched her hard in the temple. Everything went black.

CHAPTER 28

She was swimming as fast as she could, but the great waves had almost caught up with her. Suddenly she saw rocks in front of her, big black prongs sticking out of the water ready to spear her. Her strength ebbed away and she screamed. Then she opened her eyes.

Linda felt a sharp pain in her head and wondered what was wrong with the bedroom light. Then she saw her father's face looming over her and wondered if she had slept in. What was she supposed to do today? She had forgotten.

Then she remembered. What caught up with her was not the great waves but the memory of what had happened right before she plunged into darkness. The stairwell, the street, the man who stepped out of the shadows, delivered his threat and hit her. She winced. Her father laid a hand on her arm.

"It's OK. Everything's going to be OK."

She looked around the hospital room, the dim lighting, screens and the rhythmic hissing of medical equipment.

"I remember now," she said. "But how did I get here? Am I hurt?"

She tried to sit up while at the same time testing all her limbs to make sure nothing was broken. Wallander tried to restrain her.

"They want you to stay lying down. You were knocked unconscious, though there doesn't appear to have been any internal damage, not even a concussion."

"How did you get here?" she asked and closed her eyes. "Tell me."

"If what I've heard so far from my Danish colleagues and one of the emergency-room physicians here at the Rikshospital is correct, you were extremely lucky. A patrol car was driving past and actually saw a man knock you down. It only took minutes for the ambulance to arrive. The officers found your driver's licence as well as your ID card from the training college. They had contacted me within half an hour. I drove over as soon as I heard about it. Lindman is here too."

Linda opened her eyes and looked at her father. She thought in a fuzzy way that she might be a little in love with Lindman, even though she had hardly had anything to do with him. Am I delirious? I return to consciousness after some lunatic has knocked me out and the first thing I think is that I've fallen in love, and much too quickly at that.

"What are you thinking about?"

"Where's Lindman now?"

"He went to get a bite to eat. I told him to go home, but he wanted to come with me."

"I'm thirsty."

Wallander gave her some water. Linda's head was clearer now; images from the moments before the assault were coming back.

"What happened to the creep who assaulted me?"

"They arrested him."

Linda sat up so quickly her father couldn't stop her.

"Lie down!"

"He knows where Anna is. Or perhaps he doesn't know that – but he does know something."

"Stay calm."

Reluctantly she stretched out on the bed again.

"I don't know his name, he could be Torgeir Langaas. But he knows something about Anna."

Her father sat down on a chair beside the bed. She looked at his watch. It was 3.15.

"Where are we? In the night or the day?"

"It's night. You've been sleeping like a baby."

"He grabbed my hair and he threatened me."

"I don't understand what you were doing here in the first place. Why Copenhagen?"

"The bastard who attacked me may know where Anna is. Maybe he assaulted her too. Or he may have something to do with Birgitta Medberg."

Wallander shook his head. "You're tired. The doctor said your memory would come back in bits and pieces, and things may be jumbled for a while."

"Don't you understand what I'm saying?"

"I do. As soon as the doctor checks you again we can go home. Lindman can drive your car."

The truth was starting to dawn on her.

"You don't believe a word I've told you, do you? That he threatened me?"

"No, I know he threatened you. He's admitted that."

"He admits what exactly?"

"That he threatened you because he wanted the drugs that he assumed you bought while you were in the apartment building."

Linda stared at her father while her mind was trying to absorb this information.

"He threatened me and told me to stop looking for Torgeir Langaas. He never said a word about drugs."

"We should be grateful the matter has been cleared up, and that the police were nearby at the time. He's going to be charged with assault and attempted robbery."

"There was no robbery. It's all about the man who owns the house behind the church in Lestarp."

Wallander frowned. "What house is that?"

"I haven't had time to tell you about this before. I went to Anna's house in Lund and found a lead that pointed to a house behind the church in Lestarp. After I was there – asking about Anna – everyone in the house left, vanished. The only thing I managed to find out was that the house is owned by a Norwegian called Torgeir Langaas, and his address is here in Copenhagen."

Her father looked at her for a long time, then took out his notebook and started reading from one of the pages.

"The man they arrested is Ulrik Larsen. If my Danish colleague is to be believed, Larsen is not the kind of man who owns a country house in Sweden."

"Dad, you're not listening to me!"

"I am listening, but this is a man who has confessed to trying to steal drugs from you."

Linda shook her head desperately. Her left temple throbbed. Why didn't he understand what she was trying to tell him?

"My mind is completely clear. I know I was knocked out, but I'm telling you what actually happened."

"You *think* you are. But I don't see the connection between what you were doing in Copenhagen and your barging in on Mona and upsetting her as you did."

Linda went cold. "How do you know that?"

"She called me. She was in a terrible state, crying so hard she couldn't speak clearly. At first I thought she was drunk."

"She *was* drunk, dammit. What did she say?"

"That you had accused her of all manner of things and complained about both her and me. She's crushed. And her banker husband was apparently not there to comfort her."

"I found Mum naked in her kitchen, with not much left of a bottle of vodka."

"She said you sneaked into the house."

"I walked in through the veranda doors, which hardly qualifies as sneaking in. She was as high as a kite, whatever she may have told you."

"We'll talk about this later."

"Thanks."

"So what were you doing in Copenhagen?"

"I've told you."

Wallander shook his head. "Explain to me why a man has been arrested for trying to rob you."

"You explain to me why you refuse to believe me."

He leaned over. "Do you understand what I went through when they called me? When they told me you had been admitted to a hospital in Copenhagen after an assault – do you know what that felt like?"

"I'm sorry that you had to worry about me."

"Worry? I was scared out of my mind – more frightened than I've been in years."

Maybe you haven't been so scared since I tried to kill myself, she thought. She knew his greatest fear was that something would happen to her.

"I'm sorry, Dad."

"I wonder what it's going to be like when you start working," he said. "If that will turn me into a sleepless old man when you're working the night shift."

She tried to tell her story again, painstakingly slowly, but still he seemed not to believe her. She had just finished when Lindman walked into the room. He nodded happily at her when he saw she was awake. He had brought a bag of sandwiches.

"How are you doing?"

"I'm fine."

Lindman handed the bag to Wallander, who immediately started to eat.

"What kind of car do you have? I'm going to get it for you," Lindman said.

"A red Golf. It's parked across the street from the apartment block, Nedergade twelve. I think it's in front of a tobacconist's."

He held up the key.

"I took this from your coat pocket. You were lucky, you know. Desperate drug addicts are about the worst thing you can run into."

"He wasn't a drug addict."

"Tell Lindman what you told me," Wallander said, between bites.

She proceeded through her account again, calmly and methodically, just as she had been taught.

"This doesn't sit very neatly with what our Danish colleagues reported," Lindman said when she was done. "Nor with what the mugger admitted to."

"I'm only telling you what really happened."

Wallander wiped his hands with a paper napkin.

"Let me put it this way," he said. "It's unusual for people to confess to crimes they haven't committed. It happens, admittedly, but not very often, and least of all with addicted drug users, since what they fear most is incarceration and the possibility that they will be cut off from their drug supply. Do you see what I'm saying?"

Linda didn't answer. A doctor walked into the room and asked her how she felt.

"You can go home," he said. "But take it easy for a few days, and call your doctor if the headache doesn't subside."

Linda sat up. Something had just occurred to her.

"What does Ulrik Larsen look like?"

Neither Lindman nor her father had seen him.

"I'm not leaving until I know what he looks like."

Her father was instantly livid. "Haven't you caused enough trouble? We are going home – *now*."

"Surely it can't be hard to get a description of him. Can't you telephone one of these Danish colleagues you keep referring to?"

Linda realised she was almost shouting herself. A nurse popped her head round the door and gave them a stern look.

"We need this room for another patient," she said.

There was a woman lying on a stretcher in the corridor, a head bandage soaked in blood, banging her fist against the wall. They found an empty waiting room.

"The man who hit me was about one hundred and eighty centimetres tall. I couldn't see his face because he was wearing a sweatshirt with the hood pulled up. The sweatshirt was either black or dark blue. He had dark trousers and brown shoes. He was thin. He spoke Danish and had a high-pitched voice. He also smelled of cinnamon."

"Cinnamon?" Lindman said.

"Maybe he had been eating a cinnamon bun, how should I know? Anyway, call your colleagues and find out if the man they have in custody matches this description. If I can just find that out I'll keep my mouth shut, for the time being anyway."

"I said no," Wallander said. "We're leaving now."

Linda looked at Lindman. He nodded carefully after Wallander had turned his back.

The doorbell rang. Linda sat up in a daze and looked at the clock. It was 11.15. She climbed out of bed and put on her dressing gown. Her head was sore but the throbbing was gone. She opened the front door. It was Lindman.

"I'm sorry if I woke you."

She let him in. "Wait in the living room. I'll be right there."

She ran into the bathroom, splashed water on her face, brushed her teeth and combed her hair. When she came back he was standing in front of the balcony door, which was open.

"How are you feeling today?"

"I feel OK. Would you like some coffee?"

"I don't have time. I just wanted to tell you about a phone call I made an hour ago."

Linda waited. He must have believed what she told him back at the hospital.

"What did they say?"

"It took a while to get to the right officer. I had to wake somebody called Ole Hedtoft, who had been on the night shift. He was one of the patrol officers who found you, and who arrested the mugger.

Lindman took a piece of paper out of the pocket of his leather jacket and looked at her.

"Give me Ulrik Larsen's description again."

"I don't know if his name is Ulrik Larsen, but the man who attacked me was one hundred and eight centimetres, thin, with a black or dark blue hooded sweatshirt, dark trousers and brown shoes."

Lindman nodded and then rubbed the bridge of his nose with his thumb and index finger.

"Hedtoft described the same individual. But perhaps you misunderstood the threat he made."

Linda shook her head. "That's not possible. He actually used the name of the man I was looking for, Torgeir Langaas."

"Well, somebody must have misunderstood something."

"Why do you keep going on about a misunderstanding? I know what happened, and I'm more than ever afraid that Anna is in danger."

"Report it, then, and tell her mother. Why doesn't she report her missing herself?"

"I don't know."

"Shouldn't she be worried?"

"I can't explain why she doesn't seem to be worried, but I do know that Anna may be in danger."

Lindman started for the front door. "Report it to the police and let us take care of it."

"You haven't done much up to this point."

Lindman stopped dead. "We are working around the clock," he said angrily. "We're investigating something that has actually taken place: a murder, and an exceptionally repulsive and baffling one at that."

"Then we're in the same boat," she said calmly. "My friend Anna isn't there when I call or knock on her door. And I am baffled by that too."

She opened the door for him. "Thanks for at least believing part of what I told you."

"This is between us. There's no need to mention it to your father."

Lindman ran down the stairs. Linda ate a rapid breakfast, got dressed then called Zeba. She didn't answer. Linda drove to Anna's flat. This time there were no signs that the place had been disturbed in any way. Where are you? Linda called out silently. You'll have a lot to explain when you get back.

She opened a window, pulled up a chair and opened Anna's journal. There has to be a clue somewhere, she thought. Something that can explain what's happened.

Linda started reading from early August 2001. Suddenly she stopped. There was a name scribbled in the margin, as if a reminder by Anna to herself. Linda frowned. She had seen or heard the name recently, but where? She put down the journal. The heat was oppressive. There was a distant rumble of thunder. A name that she had seen or heard, the question was where or from whom. She put on some coffee and tried to distract herself enough to make her brain relax and shake out the source of the name. Nothing happened.

It was only later, when she was about to give up and go out, that she remembered. It was the name of one of the people living in the apartment block on Nedergade.

CHAPTER 29

Vigsten. She was sure she was right. She couldn't be sure if it was someone living on the street side or in the inner building over the courtyard, nor if it had been a D or an O as a first initial, but she knew she was right about the surname. What do I do now? she thought. I'm on to something that is actually starting to hang together. But I'm the only one taking it seriously. I haven't managed to convince anyone else. Of course, I have no idea what is actually going on. Anna thought she saw her father and then she disappeared. Two disappearances that camouflage each other, cancel each other out or complete each other?

Linda felt a sudden need to talk to someone, and there was no-one to turn to except Zeba. She ran down the stairs of Anna's building and drove to Zeba's place. She was just on her way out with her son. Linda tagged along. They went to a playground nearby, where the boy ran off to the sand pit. There was a bench too, but it was littered with dirt and chewing gum.

They sat on the edge of the sand pit while the boy threw sand around and let out whoops of joy. Linda looked at Zeba and felt the usual sting of envy: Zeba was extravagantly beautiful. There was something arrogant and inviting about her at the same time, the kind of woman Linda had once dreamed of becoming. But I became a policewoman, she thought. A policewoman who hopes she won't turn out to be a scaredy-cat.

"I've been trying to reach Anna," Zeba said. "She hasn't been home. Have you seen her?"

This infuriated Linda. "Hello? Hello? Have you been listening to a single word I've told you? About her being gone, that I'm worried sick about her, that I think something's happened to her?"

"But you know what she's like, don't you?"

"Do I? Apparently I don't. What *is* she like?"

Zeba frowned. "Why are you so worked up?"

"I'm frightened for her."

"What do you think has happened? And what is that bruise on your head, for heaven's sake?"

Linda decided to tell her all she knew. Zeba paid attention in silence. The boy played.

"I could have told you", Zeba said when Linda finished, "the part about Anna being religious."

Linda looked at her. "Religious?"

"Yes."

"She never said anything to me."

"You only just met up again, after a long gap. And Anna is the kind of person who says different things to different people. She tells a lot of lies."

"Really?"

"I had been meaning to warn you, but I thought it would be better if you found out for yourself. She is a compulsive liar."

"She didn't used to be like that."

"People change, don't they?" Zeba said in a mocking tone. "I'm friends with Anna because of her good qualities. She's cheerful, nice to my son, helpful. But when she starts telling one of her stories, I don't bother to listen. Do you know that she spent last Christmas with you?"

"I was in Stockholm last winter."

"She said she had been up to see you. Among the many things you did together, apparently, was take a trip over to Helsinki on the ferry."

"That's absurd."

"Of course. But that's what Anna told me. I don't know why she lies, perhaps it's an illness, or else she's just bored."

"Do you think she was lying when she said she had seen her father in Malmö?"

"Of course. It's typical of her to invent an amazing reappearance like that. Her father has probably been dead for a long time."

"So you don't think anything bad has happened to her?"

Zeba looked at her with an amused expression. "What could possibly have happened to her? She's gone off like this many times before. She comes back when she's had enough, and she has a fantastic and completely made-up story about where she's been."

"And nothing of what she says is true?"

"Compulsive liars are only successful if they weave in enough threads that are true. Then we believe it, then the lie sails by, until eventually we find that their whole world is built out of lies."

Linda shook her head in bewilderment. "And the medical studies?"

"I never believed a word of it."

"But where does she get her money? What does she do?"

"I've wondered about that. Sometimes I think she might be a professional con artist, but really I don't have a clue."

Zeba's son called out to her and she joined him in the sand pit. Linda watched her as she walked over. A man on the street turned to look at her. Linda thought about what Zeba had said. It explains a part of it, she thought. And it reduces my anxiety and above all infuriates me, since I now know that Anna has been feeding me lies. But it doesn't explain everything.

When they had gone their separate ways, Linda walked into the centre of town and took some money from an ATM. She was careful with money since she worried about finding herself without. I'm like my father, she thought. We're both thrifty to the point of being miserly.

She walked home, cleaned the flat and then called the housing agency. After several tries she managed to reach the man in charge of her case. She asked if she could move into the flat earlier than planned. Apparently that was not possible. She lay on the bed in her room and thought about the conversation with Zeba. Her concern for Anna's well-being had been replaced by a feeling of unease over the fact that she hadn't seen through her lies. But how was one to see through someone who did not concoct remarkable, fantastical stories but lied about everyday events?

Linda got up and called Zeba.

"We didn't finish talking about Anna's religious beliefs."

"Why don't you ask her about it when she comes back? She does believe in God."

"Which one?"

"The Christian one. She goes to church occasionally, or she says she does. But I know she does pray because I've caught her at it a couple of times. She gets down on her knees."

"Do you know if she belongs to a particular congregation or sect?"

"No. Do you?"

"I don't know. Have the two of you ever talked about it?"

"She's tried to a couple of times, but I've always put a stop to it. Me and God never got along."

Linda heard a howl in the background.

"Whoops . . . he just hurt himself. Bye."

Linda went back to her bed and continued staring up at the ceiling. What do we really know about people? An image of Anna floated through her mind, but it was like looking at a stranger. Mona was there too, naked, with her bottle. Linda sat up. I'm surrounded by a bunch of crazy people, she thought. The only normal one is Dad.

She walked on to the balcony. It was still warm. I'm going to drop this thing here and now, she said to herself. I should concentrate on something important, like enjoying this weather for instance.

Linda read the newspaper story about the Medberg investigation. Her father was interviewed again. She had read the same thing many times. No crucial leads . . . the investigation goes on on many fronts . . . results may take time. She put the paper down and thought about the name in Anna's journal. Vigsten. The second person there after Birgitta Medberg to have crossed Linda's path.

One more time, she thought. One more trip across the bridge even though it's expensive. I should present Anna with a bill for all the worry she's caused.

This time I'm not going to walk about on Nedergade in the dark, she thought as she drove across the bridge to Denmark. I'm going to look up the man – I'm assuming it's a man – whose name is Vigsten and ask him if he knows where Anna is. That's all. Then I'm going to go home and cook dinner for my father.

Linda parked in the same place as on the previous day and was struck by a sense of dread as she got out of the car. It was as if she hadn't fully grasped it before now: she had been attacked on this street only last night.

She got back into her car and locked the doors. Take it easy, she thought. I'm going to get out of the car; no-one's going to knock me down. I'll just walk into that building and find this Vigsten person. And that will be that.

Linda kept telling herself to stay calm, but she ran across the street. A cyclist veered sharply to avoid her and nearly fell, yelling an obscenity after her. The front door opened when she pushed it. She saw the name almost immediately. On the fourth floor, F. Vigsten. She hadn't remembered the initial correctly. She started up the stairs. Fredrik Vigsten, she thought. It'll be Fredrik if it's a man, that's typically Danish. Or Frederike for a woman. She stopped and caught her breath on the third floor. Then she rang the bell, a short melody. She waited and counted slowly to ten. Then she rang again, and the door opened at about the same time. An old man with ruffled hair and glasses hanging from a cord round his neck gave her a stern look.

"I can't walk any faster," he said. "Why don't you young people have some patience?"

He stepped back without asking for her name or why she was there, ushering her into the hall.

"I sometimes forget when I have a new student," he said. "I don't make a note of everything as conscientiously as I should. Please feel free to hang up your things. I'll wait in there."

He shuffled down a long corridor with short, almost springy steps. A student of what? Linda wondered. She took off her jacket and followed him. It was a large flat, perhaps the result of knocking down a wall between two smaller ones. In the room furthest from the front door there was a baby grand piano. The white-haired man was by the window, leafing through a monthly planner.

"I don't see an appointment for today," he complained. "What did you say your name was?"

"I'm not a student," Linda said. "I just want to ask you some questions."

"I've been answering questions my whole life," he said. "I've told them why it's so important to sit correctly when playing the piano. I've tried to explain to countless young pianists why not everyone can learn to play Chopin with the required combination of caution and power. Above all, I try to get my impatient opera singers to stand properly, and not to attempt the hardest pieces without wearing good shoes. Have you got that? Opera singers need good shoes, and pianists need to take care not to develop haemorrhoids. What did you say your name was?"

"It's Linda, and I'm neither a pianist nor an opera singer. I've come to ask you about something that has nothing to do with music."

"Well, you must have the wrong man, because I can only answer questions about music. The world beyond that is incomprehensible to me."

Linda was momentarily confused.

"Your name is Fredrik Vigsten, isn't it?"

"Not Fredrik: Frans. But the last name is right."

He sat at the piano and began turning over the paper of some music. Linda had the feeling that he had forgotten she was there.

"Your name appears in the journal of my friend Anna Westin," she said.

Vigsten tapped rhythmically on the paper and did not seem to hear her.

"Anna Westin," she repeated, louder.

He looked up abruptly. "Who?"

"Anna Westin. A Swedish girl."

"I've had many Swedish pupils," he said. "Of course, now it is as if everyone has forgotten about me, and . . ."

He interrupted himself and looked at Linda. "Did you tell me your name?"

"I'm happy to tell you again. It's Linda."

"And you are not a pupil? Not a pianist? Not an opera singer?"

"No."

"You're asking about someone called Anna?"

"Anna Westin."

"I don't know an Anna Westin. Vest-in. My wife was a ves*tal*, but she died thirty-nine years ago. Have you any idea what it is like to be a widower for almost four decades?"

He stretched out a thin hand with finely etched blue veins and touched her wrist.

"Alone," he said. "It was one thing when I had my day job doing rehearsals at Det Kongelige. But one day they told me I was too old. Maybe it was that I insisted on doing things the old-fashioned way. I didn't tolerate sloppiness."

"I found your name in my friend's journal," Linda interrupted him.

She took his hand. The fingers that held hers so hungrily were surprisingly strong.

"Anna Westin, is that right?"

"Yes."

"I've never had a pupil of that name. My memory is not what it once was, but I remember all their names. They are the only ones who have given my life meaning since Mariana was taken by the gods."

Linda didn't think there was any point in pursuing the conversation. There was really only one thing left to ask.

"Do you know someone called Torgeir Langaas?"

But Vigsten was lost in dreamland again. He picked out some notes on the piano with his free hand.

"Torgeir Langaas," she repeated. "A Norwegian."

"I have had many Norwegian pupils. The one I remember best was Trond Ørje. He was from Rauland and a wonderful baritone, but he was so shy that he could only pull it off in the recording studio. The most remarkable baritone and the most remarkable person I ever met in my life. He cried with alarm when I told him he had talent. A remarkable man. There are others . . ."

Linda got up. She was never going to get a sensible answer out of him. Nor did it seem likely that Anna had had any contact with him.

She left without saying goodbye. As she walked to the front door she heard him start to play the piano. She glanced into the other rooms on her way out. The flat was a mess and the air was stale. A lonely man who has only his music, she thought. Like my grandfather and his painting. What am I going to do when I get old? What about Dad? And what about Mum? Will it be the bottle?

She lifted her jacket from the hook in the hall. Music filled the flat. She stood motionless and studied the clothes hanging by the door.

Vigsten might be a lonely old man, but he had a coat and a pair of shoes that did not belong to an old man. She looked back into the flat. There was no-one there, but she knew now that Frans Vigsten did not live alone. Fear came over her so swiftly that she jumped. The music stopped and she listened for any other sound. Then she fled, back to the street, running across to the car, and driving away as fast as she could. She only started to calm down when she was on the Öresund Bridge heading home.

At the same time as Linda was crossing the bridge, a man broke into the pet shop in Ystad. He doused the shelves of caged birds, hamsters and mice with petrol, threw a match on the floor and left just as the animals caught fire.

PART 3

The Noose

CHAPTER 30

He chose the locations for his ceremonies with great care. This was something he had learned as early as during his flight from Jonestown: where could he rest, where would he feel safe? There were no ceremonies as such in his world in those days. That came later, once he had reestablished a connection to God which was finally going to help to fill the emptiness that threatened to consume him from inside.

It was more important than ever now not to make any mistakes in choosing the places where his assistants prepared their assignments. Everything had gone well until the unfortunate incident where a woman accidentally lit upon one of their hide-outs and was killed by his disciple, Torgeir Langaas.

I never saw Langaas's weakness in its totality, he thought. The spoiled brat that I plucked from a sewer in Cleveland had a temper I never succeeded in taming. I treated him with infinite patience and listened to him talk. But he carried a powerful rage inside him, beyond mending.

Why? he had asked in vain. Why this senseless fury at a woman stumbling down the wrong path? They had once even discussed what they should do in the event that someone one day chose to follow the abandoned path. They had to remain alert to the possibility of the unexpected. The agreed response had been to meet all visitors in a cordial fashion, then remove themselves from the area as soon as practical. Langaas had adopted the polar-opposite response. A fuse had blown in his brain. Instead of giving the woman a friendly welcome, he had reached for an axe. Why he had cut off her head he couldn't say, nor why he saved the head and arranged the hands as if in prayer. The remaining body parts he had put in a sack with a very big stone, then he had removed his clothes and swum to the middle of the nearest lake, and let the sack sink to the bottom.

* * *

Langaas was strong. That had been one of his first impressions when he stumbled over the drunk crawling in the gutter in one of Cleveland's ugliest slums. He was going to walk on, but he heard the man moaning some slurred words in a language that sounded like Danish or Norwegian. He had stopped and bent down, understanding that God had put this man in his path. Langaas had been close to death. The physician who later examined him and prepared the rehabilitation programme had been very clear on this point: there was no room left for any alcohol or drugs in his body. Only his physical strength had saved him to this point, but his organs were drawing on their last reserves. His brain was damaged, would perhaps never recover the portions of memory that had been destroyed.

He would always remember the moment when a homeless Norwegian by the name of Torgeir Langaas had looked up at him with eyes so bloodshot they glowed like a rabid dog's. But it wasn't the look that had made such an impression, it was what he said, because in Langaas's confused mind the face bending towards him belonged to God. He had grabbed his redeemer's coat with his massive hands and directed his terrible breath at his face.

"Are you God?" Langaas said.

Everything that had been unresolved in his life – his failures, his hopes, his dreams – was reduced to a single point, and he answered: "Yes, I am your God."

In the following moment he had been beset with doubts, although there was no reason his first disciple couldn't be one of the lowest. But who was he? How had he ended up here?

He had walked away and left Langaas there, but his curiosity was not extinguished. He returned to the slum the very next day. It was like descending into hell, he thought. The lost souls crawled all about. In looking again for Langaas he had been close to being mugged several times, but at last an old man with a stinking pus-filled sore where his left eye should have been told him there was a Norwegian with large hands who sometimes crawled into a rusting bridge pillar when there was rain or snow. That was where he found him. Langaas was sleeping, snoring. His clothes reeked of sweat and urine and his face was badly cut.

It didn't take more than a couple of sessions under the crumbling bridge to get Langaas's life story. He was born in Baerum in 1948, heir to the Langaas Shipping Company, which specialised in oil and cars. His father, Captain Anton Helge Langaas, had studied the industry during his own years at sea. Langaas Shipping was an offshoot of the established Refsvold Shipping Company. The parting was not amicable. Nor was it known where Captain Langaas had made the fortune which obliged the unwilling board members of the Refsvold Shipping Company to admit him into their midst. Rumours abounded.

Captain Langaas waited until his company was in the black and he himself was financially secure before marrying. In a gesture of contempt for the shipping aristocracy, he chose his wife as far from the sea as it was possible in Norway, from a village in the forests east of Roros. There he found a woman called Maigrim who delivered post to the isolated farmhouses of the region. They built a large house in Baerum outside Oslo and had three children one after the other. Torgeir and then two girls, Anniken and Hege.

Torgeir Langaas sensed what his parents wanted of him from an early age, but at an equally early age realised that he would never be able to live up to their expectations. He started rebelling in early adolescence. Captain Langaas fought a battle that was doomed from the outset. Finally he capitulated and accepted that Torgeir would never take his place in the family business. He turned instead to his daughters. Hege resembled her father, showing a focused determination even as a child, and earned an executive position within the company at the age of 22. Torgeir had already started his long slide into oblivion with a focused determination of his own. He had developed several addictions, and none of the expensive clinics nor Maigrim's best efforts did any good.

The final breakdown came one Christmas. Torgeir gave his family presents of rotting meat, old car tyres and dirty cobblestones. Then he tried to set fire to himself, his siblings and his parents. He ran away, having no intention of ever returning, a fat bank balance to his name.

When his passport expired and was not renewed he was wanted by Interpol, but no-one looked for him in the streets of Cleveland. He kept his assets hidden from those he lived among, changing his bank, changing everything except his name. He still had five million Norwegian crowns to his name when the man he came to see as his saviour turned up in his life.

I did not take adequate stock of his weakness, he thought again. The rage that grew into uncontrollable violence. Langaas was so blinded by his fury that he hacked the woman to pieces. But there was something of value even in this unexpected reaction and capacity for brutality. To set light to animals was one thing, killing a person quite another. Apparently Langaas would not hesitate to commit such an act. Now that all the animals had been sacrificed, he would lift him to the next level of human sacrifice.

They met at the railway station in Ystad. Langaas had come by train from Copenhagen since occasionally he lost his concentration when driving. Langaas had bathed – that was part of the purification process that preceded the ritual sacrifice. It was important to be clean. Jesus even washed his disciples' feet. He had explained to Langaas that everything was there in the Bible. It was their map, their guide.

Langaas carried a small black bag. He did not have to ask what was in it. The Norwegian had long since proved himself to be reliable – save in the case of the woman in the forest. That had caused a most unnecessary amount of publicity and activity. Newspapers and television stations were still broadcasting the news. The act they were preparing now had had to be postponed for two days, and he had felt it best that Langaas use his Copenhagen safe house while they waited it out.

They walked towards the centre of town, turned when they reached the post office and continued on to the pet shop. There were no customers inside. The woman behind the counter was young. She was busy putting out cat food on a shelf when they arrived. There were hamsters, kittens and birds in the cages. Langaas smiled but said nothing; there was no point in letting her hear his Norwegian accent.

While Langaas walked around the shop and made a mental note of how he would carry out his task, his saviour selected and bought a packet of bird seed. Then they left the shop, walked past the theatre and towards the harbour. It was a warm day still with a number of yachts coming and going.

That was the second element of their preparations, to be close to the water. Once they had met by the shores of Lake Erie, and from then on they always sought out a body of water when they had important work to do.

"The cages are close together," Langaas said. "I'll spray with both hands in either direction, throw in the match and run. Everything will be on fire within a few seconds."

"And then?"

"Then I say: 'The Lord's will be done.'"

"And then?"

"I go left, then right. Not too fast, not too slow. I stop on Stortorget and make sure no-one's on my tail. Then I walk to the newsstand by the hospital, where you'll be waiting."

They broke off to watch a small boat on its way into the harbour. The engine sound was loud and hacking.

"These are the last animals. We have reached our first goal."

Langaas was about to kneel right then and there on the pier. He dragged him up by the arm.

"*Never* in public."

"I forgot."

"Are you calm?"

"Yes."

"Who am I?"

"My father, my shepherd, my saviour, my God."

"Who are you?"

"The first disciple. Found on a street in Cleveland, saved and helped back into life. I am the first apostle."

"What else?"

"The first priest."

Once I made sandals for a living, he thought. I dreamed about bigger things and had to run away to escape my shame, my sense of failure, my sense of having destroyed the dreams by my inability to live up to them. Now I fashion people in the same way that I once cut out soles, insteps and straps.

It was 4 p.m. They walked around the city and passed the time on various park benches. They were beyond words now. From time to time he looked at Langaas. He seemed calm and focused on the task at hand.

I've made him happy, he thought. A man who grew up spoiled but also stifled and desperately unhappy. Now I bring joy to his life by taking him seriously and giving him a purpose.

They progressed from bench to bench until it was 7.00. The pet shop closed at 6.00. Many people were out in the streets in the warm evening. That was to their advantage.

They went their separate ways. He walked to Stortorget and turned around. Their plans ticked like a timer in his head. Langaas was breaking down the front door with the crowbar. Now he was on the inside, closing the shattered door behind him, listening for signs of anyone in the shop. Now he was dropping the bag, taking out the bottles of petrol and the lighter.

He heard the boom and thought he saw a flash of light reflected in some windows a long way down the street. A narrow plume of smoke rose up to the sky. He turned and walked away. He heard the first sirens before he had made it to the appointed meeting place.

It's over, he thought. We are reviving the Christian faith, the Christian dictates of a righteous life. The long years in the desert have come to an end. No longer are we concerned with the simple beast who feels pain but lacks comprehension. Now we turn to the human being.

CHAPTER 31

When Linda got out of the car at Mariagatan she smelled something that reminded her of a week's holiday she had spent with Herman Mboya in Morocco. They had chosen the cheapest package and stayed in a hotel infested with cockroaches. During that week she had begun to think they didn't have a future together. The following year they had parted; Herman had returned to Africa and she had started along the way that finally led her to the police training college.

The smell was what had triggered the memory. Mounds of rubbish were burned every night in Morocco. But no-one burns their rubbish in Ystad, she thought. Then she heard the fire engine and police sirens. There must be a fire somewhere in the centre of town. She started to run.

It was still burning when she arrived, panting like a house-bound old woman. When had she become so out of shape? She saw tall flames leaping up through the roof. The people who lived in the upper storeys had been evacuated. A fire-damaged pushchair had been abandoned in the street. Firemen were securing the surrounding buildings. Linda made her way up to the police tape.

Her father was quarrelling with Svartman about a witness who had not been thoroughly interviewed and to make it worse had been allowed to leave.

"We'll never get this madman if we can't even follow the simplest of routines."

"Martinsson was in charge."

"He's told me twice that he handed it over to you. Now you'll have to track the witness down."

Svartman left, clearly upset. They're like angry bulls, Linda thought. All this time and energy spent marking their territory.

* * *

A fire engine which was reversing in the narrow street knocked a hose loose. It started whipping about, spraying water. Wallander jumped to the shelter of a doorway and saw Linda at the same time.

"What happened here?" she asked.

"One or more fire bombs in the shop. Petrol, same as the calf."

"Any clues?"

"One witness, but no-one seems to know where the person went."

Wallander was so angry he was shaking. This is how he'll die, Linda thought. Exhausted, outraged by a careless error in an urgent investigation.

"We absolutely have to get these bastards," he said, interrupting her chain of thought.

"I think this is different."

"In what way different?" He looked at her as if she knew the answer.

"I don't know. It's as if it were really about something else."

Höglund called out to Wallander.

Linda watched him walk away, a large man with his head pulled down into his shoulders, stepping carefully over the hoses and past the smoking remains of what had once been a pet shop. Linda's gaze fell on a tearful young woman who was watching the blaze. The owner, she supposed. Or simply someone who loves animals. There were a number of spectators, all silent. Burning buildings always inspire dread, she thought. A house fire is a frightening reminder that our own homes could one day be razed to the ground.

"Why aren't they asking me questions? I don't get it."

Linda turned and saw a young woman, pressed up against the shop front. She was talking to a friend. A swirl of thick smoke made them both pull back further.

"Why don't you just go over and tell them what you saw?" her friend said.

"I'm not going out of my way for the police."

The witness, Linda thought, and approached the woman.

"What did you see?" she said.

The woman peered at her and Linda saw that she was slightly wall-eyed.

"Who are you?"

"My name is Linda Wallander. I'm a police officer." It's almost true, she thought.

"How can someone kill all those animals? Is it true there was a horse in there too?"

"No," Linda said. "What did you see?"

"A man."

"What was he doing?"

"He started the fire that burned all the animals. I was coming from the direction of the theatre. To post some letters, which I do several times a week. I was about halfway to the post office, about a block from the pet shop, when I noticed that someone was walking behind me. I jumped since he had been walking almost soundlessly. I let him pass. I followed him, trying to walk as quietly as he did, I don't know why. But then I realised I had left a letter in the car so I turned around and went back for it."

Linda raised her hand. "How long did it take you to go back and get the letter?"

"Three or four minutes. The car is parked by the delivery entrance at the theatre."

"What happened when you came back this way? Did you see the man again?"

"No."

"And when you walked past the pet shop, what did you do?"

"I looked in the window – but I'm not so interested in hamsters and turtles."

"Did you see anything inside?"

"A blue light. It's always on. It's some kind of heat lamp, I think."

"And then?"

"I posted my letters – that took another three minutes or so – and I was walking back to the car, and the shop exploded, or it felt that way. I had just walked past it. There was a sharp light all around me. I threw myself down on to the pavement. The shop was all in flames. An animal must have escaped, it ran past me with its fur on fire. It was horrible."

"What did you do then?"

"It all happened so fast. But I saw a man standing on the other side of the street. The light was so strong that I was positive it was the man who had overtaken me on the street. He was carrying a bag."

"Had he been carrying it before?"

"Yes. I didn't mention that. A black bag, an old-fashioned doctor's bag."

"What did you do?"

"I called to him to help me."

"Were you hurt?"

"I thought so. It was such a loud bang and such a terrible light."

"Did he help?"

"No, he looked at me and walked away."

"In which direction?"

"Towards Stortorget."

"Had you seen him before this evening?"

"Never."

"How would you describe him?"

"He was tall and strong-looking. Maybe bald, or with very short hair. He had a dark blue coat, dark trousers. His shoes I had looked at when he walked past me and I wondered how he could walk so quietly. They were brown and had a thick rubber sole, but they weren't sneakers."

"Can you remember anything else?"

"He shouted something."

"Who was he shouting to?"

"I don't know."

"Did you see anyone else there?"

"No."

"What did he say?"

"It sounded like, 'The Lord's will be done'."

"'The Lord's will be done'? Is that it?"

"I'm sure of 'Lord', but the word 'will' sounded like it was pronounced in a foreign language. Danish, maybe. Or Norwegian, more like. Yes, that's it. He sounded like he was speaking Norwegian."

Linda's heart beat a little faster. It has to be the same man, she thought. Unless there's a conspiracy involving a whole pack of Norwegians. But that's preposterous.

"Did he say anything else?"

"No."

"Can I have your name, please?"

"Amy Lindberg."

Linda fished a pen out of her pocket and wrote the phone number on her wrist. They shook hands.

"Thanks for listening to me," the woman said, and she turned to rejoin her friend.

Torgeir Langaas. He keeps cropping up like some kind of shadow.

Linda could tell that the fire-fighting operation had reached a new phase. The firemen were moving more slowly, a sure sign that the blaze would soon be contained. She saw her father talking to the fire chief. When his head moved in her direction she pulled herself back even further, although it was impossible that he could see her in among the shadows. Lindman walked past with the young woman she had seen earlier, the one who had watched the fire and cried. It suits him to comfort crying women, she thought. I, on the other hand, almost never cry. I stopped all that while I was still little. She watched Lindman escort the woman to a patrol car. They said a few words to each other, then he opened the door for her and she climbed in.

The conversation with Amy Lindberg kept coming back to her. The Lord's will be done. What did the Lord want, exactly? That a pet shop burn to the ground, that some helpless animals die in terror and pain? First it was swans, she thought. Then the calf: singled out, charred, dead. And now a whole shop full of pets. It was obviously the same man, one who had calmly observed his work and pronounced: "As God decreed."

Linda walked over to Lindman, who looked at her with surprise.

"What are you doing here?"

"I'm just a curious onlooker, but I need to talk to you."

"What about?"

"The fire."

He thought for a moment. "I have to go home and eat something anyway," he said. "I am finished here. You can come along."

Lindman's flat was in one of three high-rises planted arbitrarily, it seemed, across an area with a few other houses and a paper-recycling centre.

His was the middle building. The glass in the front door had been broken and replaced by a piece of cardboard, in which someone had also kicked a hole. A message was scribbled on the wall: *Life is for sale. Alert the news media.*

"I read that every day," Lindman said. "Makes you think, doesn't it?"

He unlocked the flat and handed her a hanger for her coat. They walked into the living room, which was furnished with a few simple pieces, randomly located around the room.

"I don't have anything to offer you except water or beer," he said. "This is just a place for me to camp out."

"Where are you moving? You said something about Knickarp."

"To a house there with a large garden. I'm looking forward to it."

"I'm still at home," Linda said. "I'm counting the days until I get out."

"You have a good father."

She was taken aback. "What do you mean?"

"Just that. You have a good father. I didn't."

There were some newspapers on a side table. She picked up a copy of the *Borås Daily*.

"It's not that I'm nostalgic," he said. "I enjoy reading about everything I've managed to escape."

"Was it that awful?"

"I had to leave when I knew I was going to survive the cancer."

He fell silent. Linda wasn't sure how she was going to change the subject.

"I'll get the beer and some sandwiches," he said. Linda declined the sandwich.

He came back with two glasses.

She told him of the conversation she had overheard, and what Amy Lindberg had told her. Lindman began to take notes. Linda reverted to the incident when Anna thought she saw her father in Malmö, and she told him all she knew of this shadowy figure of a Norwegian, who was perhaps named Torgeir Langaas, and who kept figuring in her enquiries. She also told him of her second visit to Nedergade in Copenhagen.

She said in conclusion, "Someone has killed a woman; someone is killing animals; and Anna has disappeared."

"I fully appreciate your concern," Lindman said. "Not only because of the vaguely disturbing possibility of a return by Anna's father. We also have the menacing presence of a person unknown, someone who says: 'As God decreed'. You have also learned that your friend Anna is religious. These random facts are starting to look like pieces of a grotesque puzzle, not least in the horrific evil of allowing the severed hands to go on begging for mercy. From everything you have told me, and which I see now in what I know, it's clear that there's a religious dimension to all of this that we haven't taken as seriously as we perhaps should have."

He drank the last of his beer. There was a rumble of thunder in the distance.

"That's out over Bornholm," Linda said. "There are often thunderstorms out there."

"And an easterly wind, which means it's on its way here."

"What do you think about what I just told you?"

"That it's true. And that what you've told me will have a real impact on our investigation."

"Which investigation?"

"Medberg. Anna's disappearance is not yet a priority, but I think that will change now."

"Am I right to be scared?"

"I don't know. I'm going to write up everything you've told me, and it would be a good idea for you to do the same. I'll let my colleagues know about this tomorrow."

Linda shivered. "Dad will be furious that I went to you first and not to him."

"Why don't you blame it on the fact that he was so busy with the fire?"

"He keeps saying that he's never too busy when it comes to me."

Lindman helped her on with her coat. She thought that she was genuinely attracted to him. His hands on her shoulders were gentle.

She returned to Mariagatan. Her father was waiting for her at the kitchen table and she could tell from his face that he was angry. Stefan, you bastard, she thought. Couldn't you at least have waited until I got home?

She sat down across from him and braced herself.

"If you're going to rant and rave I'm going to bed. No, I'll leave. I'll sleep in the car."

"You could at least have talked to me. This amounts to a breach of trust, Linda. A huge breach of trust."

"For Christ's sake – you were up to your neck in a pet massacre. A street was going up in flames."

"You shouldn't have taken it upon yourself to talk to that girl. What gave you the right to do that? How many times do I have to tell you this is not your business? You haven't started working yet."

Linda pulled up her sleeve and showed him Amy Lindberg's phone number.

"Will that do? I'm going to bed."

"I find it deeply disturbing that you don't even have enough respect for me not to go behind my back."

Linda's eyes widened. "Go behind your back? Who said anything about going behind your back?"

"You know what I'm saying."

Linda swept a salt cellar and a vase of withered roses to the floor. He had gone too far. She ran into the hallway, grabbed her coat and slammed the front door behind her. I hate him, she thought, fumbling in her pocket for the car keys. I hate his endless nagging. I'm not spending another night there.

She tried to calm herself in the car. He thinks I'm going to feel guilty, she thought. He's waiting for me to go back inside and tell him that Linda Caroline rebelled a little but takes it all back now.

"Well, I'm not going back," she said aloud. "I'll stay with Zeba." She was about to start the car when she changed her mind. Zeba would talk, ask questions, discuss. She didn't have the energy for that. She drove to Anna's flat instead. Her father could sit at the kitchen table until the end of the world as far as she was concerned.

She put the key in the lock and pushed the door open. Anna was standing in the hall with a smile on her face.

CHAPTER 32

"I knew it had to be you. No-one else would drop by like this, like a thief in the night. You probably intuited that I had come back and woke up. Isn't that it?" Anna said brightly.

Linda dropped her keys.

"I don't understand. Is it really you?"

"Of course it's me."

"I can't tell you how relieved I am."

Anna frowned. "Why are you relieved?"

"I've been worried sick about you."

Anna raised her hands in apology. "I'm guilty, I know. Do you want me to apologise or to tell you what happened?"

"You don't have to do either right now. It's enough that you're here."

They went into the living room. Even though Linda was struggling to come to terms with the fact that Anna was back and sitting there in her usual chair, she noticed that the framed blue butterfly was still missing.

"I came over because I had a fight with my father," Linda said. "I thought I would sleep on your sofa since you were away."

"You can still sleep here even though I'm back."

"He made me so mad. My father and I are like two fighting cocks. We were arguing about you, as it happens."

"About me?"

Linda stretched out her hand and brushed Anna's arm with her fingertips. Anna was wearing a dressing gown from which the sleeves had been cut off for some reason. Her skin was cold. There was no doubt that it really was Anna who had come back and not an impostor. Anna's skin was always cold. Linda could remember that from their childhood when they – with the tingling feeling of exploring forbidden territory – had played dead. The game had made Linda warm and sweaty, but Anna had been cold, so cold, in fact, that they had stopped playing. Her cold skin scared them both.

"I was so worried about you," Linda said. "It's not like you to disappear and not be home when we had agreed to meet."

"You have to remember my world was turned upside down. I thought I had seen my father. I was convinced he had come back."

She paused and looked down at her hands.

"What happened?" Linda said.

"I went to look for him," she said. "I didn't forget about our plans, but I thought you would understand. I had seen my father and I had to find him. I was so worked up I was shaking and didn't dare drive. I took the train to Malmö and set about looking for him. It was an absolutely indescribable experience. I walked up and down the streets using all my senses, thinking there had to be a trace of him somewhere, a scent, a sound.

"It took me several hours to get from the station to the hotel where I had seen him. When I walked into the hotel lobby a fat lady was half sleeping in the chair I had been sitting in. I became furious; she had taken my place! No-one had the right to sit in the holy chair where I had seen my father and he had seen me. I walked up to her and shook her arm. I told her she had to move because the furniture was going to be replaced. She did as I asked, although I can't imagine how she could think I was one of the hotel staff in my raincoat and with wet hair stuck to my cheeks. I sat in the chair when she had left. I thought that if I just stayed there long enough he would return."

Anna stopped talking and left to go to the toilet. Thunder rumbled in the distance. She came back and continued:

"I sat in the chair until the receptionists started looking at me suspiciously. I booked a room, but tried to spend as little time there as possible. On the second day the fat lady came back. She said, 'You thief – you stole my seat!' She was so worked up I thought she was going to hyperventilate. I thought that no-one would make up a story about sitting in a particular chair in the hope of catching a glimpse of a father they hadn't seen in over twenty years. So I told her the truth and she believed me. She sat down in the chair next to me and said she'd be happy to keep me company while I waited. It was crazy. She talked non-stop, mainly about her husband, who was at a men's headwear conference. You can laugh – I didn't, of course. She told me about it in excruciating detail, about the rows of sombre men in airless conference rooms deciding which hats to order for the new season. She talked until I was ready to strangle her. But then her husband appeared. He was as fat as she was, and was wearing a broad-brimmed and probably very expensive hat. She and I had never actually introduced ourselves. As she was about to leave with her husband, she said to him: 'This young lady is waiting for her father. She's been waiting for him a long time.' And the man asked, 'How long?' 'Almost twenty-five years,' she told him. He looked at me thoughtfully but also with great respect. And the hotel lobby with its polished, sterile surfaces and strong smell of commercial-grade cleaning agents was transformed into a church. He said, 'One can never wait too long'. Then he put on his hat and I watched them leave the hotel. The whole situation was absurd, almost unbelievably so, but that's what made it so real.

"I stayed in that chair for close to two days before I realised that my father was not going to reappear. I decided to go out and look for him, though I kept the room. There was no master plan to my search. I walked through the parks, along the canals and the various harbours. My father had left me and Henrietta because he sought a freedom he couldn't have while he was with us. Therefore I looked for him in the open spaces. There were times when I thought I had found him. I would get so dizzy I'd have to lean against a wall or a tree, but it was never him. All the longing I had been bottling up for so long finally turned to rage. There I was, still looking for him, still wanting so badly to find him, and he had simply chosen to humiliate me by showing himself to me once and then disappearing again. Naturally I started to doubt myself. How could I have been so sure that it was him? Everything spoke against it. The last night I was there I ended up in Pildamm Park. I called out into the darkness: 'Daddy, where are you?' But no-one answered. I stayed in the park until dawn and then I suddenly felt as though I had been through the final trial in my relationship with him, as though I had been wandering in a fog of delusion, thinking that he was going to show himself to me, and when at last I emerged into the light I accepted that he didn't exist. Well, maybe he does exist, maybe he's not actually dead. But for me, from that point on, he was going to be a mirage, a dream that I could evoke from time to time at will, nothing more. For all of these years I had believed, deep down, that he was out there somewhere. Now, at the moment I thought he had finally returned, I knew he was never coming back at all. Now that I could no longer hold on to the idea of him as a living, breathing person, someone to be mad at, to keep waiting for, he was finally gone for real."

The storm clouds had moved on to the west. Anna stopped talking and looked down at her hands. To Linda it seemed that she was making sure none of her fingers were missing. She tried to imagine what it would be like if her own father had disappeared when she was a child. It was an impossible thought. He had always been there, a huge enveloping shadow, sometimes warm, sometimes cold, encircling her and keeping his eye on her. Linda wondered if following in his footsteps and becoming a police officer was going to be the greatest mistake of her life. Why did I do it? she asked herself. He's going to crush me with all the kindness, understanding and love – even jealousy – he should really be giving to another woman and not to his own daughter.

Anna looked up. "It's over," she said. "It was no more than a reflection in the glass. I can return to my studies. Let's not talk about it any more. I'm sorry I worried you so much."

Linda wondered if she had heard about the death of Birgitta Medberg. That was an unanswered question – what connection was there between her and Anna? And what about Vigsten in Copenhagen? Was the name Torgeir Langaas in any of her diaries? I should have ploughed through them while I had the chance, Linda thought callously. Reading one page or a thousand makes no difference, once you've crossed the line.

Somewhere inside her the sliver of anxiety was still there, gnawing away at her. But she decided that these questions would have to wait until later.

"I went to see your mother," Linda said. "She didn't seem particularly worried. I took that as a sign that she knew where you were. But she didn't seem to want to tell me anything."

"I didn't tell her I thought I saw my father."

Linda thought about what Henrietta had claimed, that Anna was regularly reporting sightings of her father. Who is lying, or not telling the whole truth? Linda decided that it wasn't important for the moment.

"I went to see Mona yesterday," she said. "I was going to surprise her, and in fact I did."

"Was she happy?"

"Not particularly. I found her in the kitchen, stark naked, drinking vodka."

"Is she an alcoholic?"

"That remains to be seen. I suppose anyone can have a bad day."

"You're right," Anna said. "Well, I need to get some sleep. Do you want me to make up the sofa for you?"

"No, I'm going home," Linda said. "Now that I know you're back, I can sleep in my own bed, even though I'll probably have another fight with my father first thing tomorrow morning."

Linda got up and walked into the hall. Anna stood in the doorway to the living room. The storm had passed.

"I didn't tell you what happened at the end of my trip," she said. "I saw someone I wasn't expecting. This morning I was having a cup of coffee at the railway station while I waited. Someone came over to my table. You'll never guess who it was."

"Since I'll never guess, it must have been the fat lady."

"Right. Her husband was standing guard over one of those huge old-fashioned trunks. It was probably full of hats all set to become the latest fashion. The fat lady was sweating and her cheeks were flushed. She leaned over to me and asked me if I had seen him. I didn't want to disappoint her, so I said yes, I had seen him. Everything had gone well. Her eyes filled with tears, then she said, 'May I tell my husband? We are returning to Halmstad now, and meeting a young woman who has been reunited with her father is a memory to cherish for life.' "

"I'll come round tomorrow," Linda said. "Let's go out as we were planning a week ago."

They agreed to meet around noon. Linda gave Anna the car keys.

"I borrowed your car when I was looking for you. I'll fill it up for you tomorrow."

"There's no need to do that. You shouldn't have to pay for being worried about me."

* * *

Linda walked home. The storm clouds were gone, but there was a light rain. She drew in the smell of asphalt and damp earth. Everything is all right, she thought. I was wrong. Nothing has happened. But the niggling splinter of anxiety survived. Anna had said, "I saw someone I wasn't expecting."

CHAPTER 33

Linda came awake with a start. Her curtain was askew, letting in a ray of sunlight reflected off the roof of a building across the street. She stretched her arm out into the light. When does the day start? she wondered. Every morning she had the feeling that she had had a dream just before waking which told her that the day was about to begin.

She sat up. Anna was back. Linda held her breath for a moment to allay the fleeting suspicion that she had dreamed it. But Anna had really been there, in that funny dressing gown with the sleeves cut off. Linda lay down again and put her hand back in the ray of sun. Summer will be over soon, she thought. I start work in five days. Then I get a new flat and my father and I won't rub each other raw any more. Soon it will be autumn and one morning there will be frost on the ground. She looked at her arm bathed in sunlight. We're still in the time before the frost.

She got up when she heard her father rattling around in the bathroom. She couldn't help laughing – no-one else could cause such a racket in a bathroom. It was as if he were doing fierce battle with soaps, taps and towels. She put on her dressing gown and went to the kitchen. It was 7.00. Her father appeared, still drying his hair.

"I'm sorry about last night," he said.

Without waiting for a response, he walked over to her and bent his head.

"Can you tell if I'm losing my hair?"

She flicked through his wet hair.

"There's a tiny bald spot right here."

"Damn it. I don't want to go bald."

"Grandpa didn't have much hair. It must run in the family. You'd look like an American army officer if you cut it all off."

"I don't want to look like an American officer."

"Anna's back."

Wallander stopped in the middle of filling a kettle. "Anna Westin?"

"She's the only Anna I know who's been missing. Yesterday when I left I went over to her place to sleep. And there she was in her hall."

"Where had she been?"

"She had gone to Malmö and stayed in a hotel. She was looking for her father."

"Did she find him?"

"No. She finally realised that she had only imagined seeing him. So she came back. That was yesterday."

Wallander sat down. "She spends a few days in Malmö looking for her father. She stays in a hotel and tells no-one – neither a friend nor her mother – where she is. Is that right?"

"Yes."

"Do you have any reason to doubt her word?"

"Not really."

"What does that mean? Yes or no?"

"No."

Wallander filled the rest of the kettle.

"So I was right. Nothing had happened."

"Birgitta Medberg's name is in her journal. As is that man named Vigsten. I don't know how much Lindman told you during your gossip yesterday."

"It was no gossip. He was very thorough – I shouldn't wonder if he's going to be the new Martinsson when it comes to making clear, concise reports. I'm going to have Anna come down to the station so she can answer a few questions for us. You can tell her that, but don't mention Medberg, and no more independent investigating from your side, understood?"

"Now you're starting to sound like a patronising chief inspector," Linda said.

He looked surprised. "I am an inspector, in case you didn't know," he said. "But I don't think I've ever before been accused of being patronising."

They ate their breakfast in silence, each with a section of the *Ystad Allehanda*. At 7.30 Wallander got up to leave, but changed his mind and sat down again.

"You said something the other day," he said tentatively.

Linda immediately knew what he was thinking of. It amused her to see him so embarrassed. "You mean what I said about you needing a little exercise?"

"What did you really mean by that?"

"What do you think? Isn't it self-explanatory?"

"My sex life is my business."

"You don't have a sex life."

"It's still my business."

"Even if it's non-existent? Well, I don't think it's good for you to be alone. Every week that passes you put on more weight. All those extra pounds scream out how lonely you are – you might as well hang a sign around your neck saying, 'I need to get laid'."

"You don't have to raise your voice."

"Who could possibly hear us?"

Wallander got up, quickly. "Forget it," he said. "I'm going in."

She watched him as he rinsed out his coffee cup. Am I too hard on him? she thought. But if I don't tell him, who will?

Linda called Anna at around 10.00.

"I just want to make sure I didn't dream the whole thing."

"And I realise now how much worry I caused. I called Zeba, so she knows I'm back."

"And Henrietta?"

"I'll talk to her later. Are you still coming over at noon?"

"I'll be there."

Linda didn't put the receiver down right away after they ended the conversation. That little splinter was still inside her somewhere. It's a message, she thought. My body is trying to tell me something, like dreams where everything leads back to you even though it may seem that you're dreaming about someone else. Anna has come back. She's unhurt and nothing seems out of the ordinary, but I can't forget the two names in her journal: Medberg and Vigsten. And then there's a third person, the Norwegian, Langaas. I won't be able to batten down the hatches on these worries until I get some answers.

She sat on the balcony. The day was cool and fresh after the night's thunderstorm. The paper said the rain had caused sewers to overflow in Rydsgård. A butterfly lay dead on the balcony floor. That's another question I need answered, Linda thought.

She put her legs up on the balcony railing. Only five more days, she thought. Then I'll no longer be in limbo.

Linda didn't know where the thought had come from, but she went inside and called enquiries. The hotel was one of the Scandic group. She was put through and a cheerful man's voice answered. She sensed the trace of a Danish accent.

"I'd like to speak to one of your guests, Anna Westin."

"One moment."

The first lie is easy, she thought. Then it gets harder.

The cheerful voice returned. "I have no-one registered under that name."

"Perhaps she's checked out already. I know she was there yesterday."

"Westin, Anna, you said?"

"Yes."

"One moment, please." This time he returned almost immediately. "There's been no-one of that name registered in the last two weeks. Can I check the spelling of the name?"

"She spells it with a W."

"We've had a Wagner, Werner, Wiktor with a W, Williamsson, Wallander . . ."

Linda squeezed the receiver. "Excuse me. What was the last name?"

"Williamsson?"

"No, the next one."

"Wallander." The cheerful voice took on a steely edge. "I thought you said you were looking for a fru Westin?"

"Her husband's name is Wallander. Perhaps she had booked them under his name?"

"Please hold the line, madam."

It can't be, she thought. This isn't happening.

"I'm afraid that isn't correct either. The only Wallander we've had was a woman who was staying alone here in a single room."

Linda couldn't speak.

"Hello? Are you still there?"

"Was her first name Linda, by any chance?"

"Yes, as a matter of fact, it was. I'm sorry I can't do anything more for you. Perhaps your friend was staying elsewhere. We also have a wonderful establishment outside Lund."

"Thank you."

Linda almost slammed the phone down. At first she had felt surprise; now it was anger. She ought to speak to her father and not go on with this on her own. Right now this is the only question that matters to me, she thought: Why would Anna go to Malmö to look for her father and book a room under my name?

She tore a piece of paper from a pad on the kitchen table and crossed out the word "asparagus" which was written on it. He doesn't even eat asparagus, she thought irritably. By the time she was ready to start jotting down all of the names and events associated with Anna's disappearance, she no longer knew where she should begin. So she drew the outline of a butterfly and started filling it in with blue. Then the pen ran out and she got another. One of the wings was blue, the other black. This is a butterfly that doesn't exist anywhere save in the realm of imagination, she thought. Like Anna's father. Reality is full of other things, such as burning swans, a butchered body in the forest, a mugger in Copenhagen.

At 11.00 she walked to the harbour, strolling out on to the pier and sitting down on a bollard. She tried to think of a reasonable explanation for Anna using her name. A dead wild duck floated in the oil-slicked water. When Linda finally got up, she had still not thought of a reasonable explanation. It must exist, she thought. I just can't think of it.

She rang Anna's bell at exactly noon. The anxiety she had felt earlier was gone. Now she was simply on her guard.

CHAPTER 34

Langaas opened his eyes, surprised as always that he was still alive. His life should have ended in that Cleveland gutter, his body disposed of by the state of Ohio.

He lay still in what had once been the maid's room off the kitchen – a room Vigsten had forgotten all about – and listened to noises issuing from the flat. A piano tuner was working on the baby grand. He came every Wednesday. Langaas had enough of an ear to know that the tuner needed to make only very minor adjustments to the pitch. He imagined old Vigsten sitting on a chair by the window, his eyes following the tuner in his work. Langaas stretched out. Everything had gone according to plan yesterday evening. The pet-shop premises had burned to the ground; not a mouse nor a hamster had escaped. Erik had stressed how important this last animal sacrifice was, how crucial that nothing go wrong. Erik came back to this point over and over, that God allowed no mistakes.

Every morning Langaas recited the oath that Erik had taught him, the first and foremost disciple: "It is my duty to God and my Earthly Master to follow the orders I receive without hesitation and undertake all actions necessary to teach the people what will happen to those that turn away from Him. Only by accepting the Lord through the words of His one true prophet will redemption be possible and the mercy of being counted among those who will return after the great transformation."

He folded his hands in prayer and mumbled the verses from Jude that Erik had taught him: "And the Lord, having saved the people from the land of Egypt, afterward destroyed them that believed not." You can turn every room into a cathedral, Erik had told him. The church you seek is here and everywhere.

Langaas whispered his oath, closed his eyes and pulled the blankets up to his chin. The piano tuner hit the same high note again and again.

Erik's words were what had sparked the memory of his grandfather. Despite his diminishing comprehension, he had spent his last years alone in his house by Femunden. One of Langaas's sisters had spent a whole week with him without his registering the fact. Langaas had told Erik about his idea and received his cautious blessing. Old Frans Vigsten had popped up as if from nowhere. Langaas sometimes wondered if Erik had steered him in his direction. Langaas had been at a café in Nyhavn, testing himself to see if he could resist the multitude of temptations that came his way. The old man had been there drinking wine. Out of the blue he had come over to Langaas and asked: "Could you tell me where I am?"

Langaas had realised that he was senile rather than drunk.

"At a café in Nyhavn."

The old man had lowered himself on to a chair across from him, and after a long silence asked: "Where is that exactly?"

"Nyhavn? In Copenhagen."

"I can't seem to remember where I live."

They found the address on a piece of paper in the man's wallet. Nedergade.

"My memory comes and goes," he said. "But this may be where I live, where my piano is, and where I receive my students."

Langaas had helped him into a cab and then accompanied him to Nedergade. The name Vigsten appeared on the list of residents in the entrance. Langaas followed him up to the flat. Vigsten recognised the smell of stale air.

"This is where I live," he said. "This is what it smells like."

Then he had wandered off into the recesses of the large flat and appeared completely to forget about the man who had helped him home. Langaas found and pocketed a spare set of keys before he left. A few days later he returned and made the maid's room into another of his temporary residences. Vigsten had still not grasped that he was host to a man who was waiting to be transported to a higher state. He had long since forgotten about their meeting in Nyhavn; he assumed Langaas was a pupil. When Langaas said he was there to service the radiators, Vigsten had turned his back and blanked him out in the same instant.

Langaas looked at his hands. They were large and strong, and they shook no longer. It had been many years since he had been lifted from the gutter, and he had not had a drop of alcohol or any drugs since then. Erik had always been there, supporting him. He could never have done it without him. It was through Erik that he had his faith, the strength he needed to continue living.

I am strong, he thought. I wait in my hiding places for my instructions. I follow them to the letter and return into hiding. Erik never knows exactly where I am, but I can always sense when he needs me.

I have received this strength from Erik, he thought. And I have only one small weakness left that I have not been able to shake off. The fact that he kept a secret from Erik was a source of great shame. The prophet had always spoken openly to him, the man from the gutter. He had not concealed any part of himself, and he had demanded the same from the man who would be his disciple. When Erik had asked him if he was free of all secrets and weaknesses he had answered yes. But it had been a lie. There was one link to his old existence. For the longest time he had resisted the task that awaited him. But when he woke this morning he knew he could no longer put it off. Setting fire to the pet shop last night had been the final step before he was lifted to the next level. He could wait no longer. If Erik did not discover his weakness, then surely God would turn His anger upon him. This fury would also strike Erik, and that was an unbearable thought.

He got up and dressed. He saw that it was overcast and windy. He hesitated between the leather jacket and his long coat, but decided on the jacket. He fingered the feathers from pigeons and swans that he picked up from the streets when he walked. Perhaps this collecting of feathers is also a form of weakness, he thought. But it is a weakness for which God forgives me.

He got off his bus at the town hall, walked to the railway station and bought the morning paper. The pet-shop arson in Ystad was front-page news. A police officer was reported as saying, "Only a sick person could do something like this, a sick person with sadistic tendencies."

Erik had taught him to keep his cool, whatever happened. But reading that what he had done was regarded as a twisted kind of sadism outraged him. He crumpled the newspaper and threw it into a rubbish bin. As penance for this weakness he gave 50 kronor to a drunk who was asking for any spare change. The man stared after him, slack-jawed. I'll come back one day and beat you to a pulp, Langaas thought savagely. I'll crush your face with a single blow, in the name of the Lord, in the name of the Christian uprising. Your blood will be spilled and join the river that will one day lead us to the promised land.

It was 10.00. He went to a café and ate breakfast. Erik had ordered him to lie low this day. His instructions were simply to seek out one of his hiding places and wait. Perhaps Erik knows I still have a weakness, he thought. Maybe he's known all along, but wants to see if I have the strength to deliver myself of it on my own.

God makes His plans well, he thought. God and Erik, His servant, are no dreamers. Erik has explained how God organises everything down to the very last detail of a person's life. This is why this day has been granted to me, in order that I should rid myself of my one remaining weakness, and stand prepared at last.

Sylvi Rasmussen had come to Denmark in the early 1990s, along with a boat-load of other illegal immigrants in a ship that had made a landfall on the west coast of Jutland. At that point she had already undertaken a long and at times terrifying journey from her home in Bulgaria. She had travelled in trucks, in trailers hitched to tractors, and had even spent two terrible days sealed in an increasingly airless container. Her name wasn't Sylvi Rasmussen then, it was Nina Barovska. She had borrowed the money for her journey, and when she came ashore on that deserted beach in Jutland two men had been waiting for her. They had taken her to a flat in Århus where they had raped her and beaten her again and again for a week, and then – when they had crushed her will – they had taken her to a flat in Copenhagen where they forced her to work as a prostitute. She had tried to escape after a month, but the two men cut off her little finger from each hand and threatened to do something worse if she ever tried to escape again. She didn't. To make her existence more bearable she started using drugs and hoped she would not have to live too long.

One day a client whose name was Torgeir Langaas had come to her.

He became a regular and she would try to talk to him, desperately wanting to make the time they spent together more human, less cold. But he always shook his head and mumbled unintelligible responses. Although he was gentle, she would sometimes break into uncontrollable shivering after one of his visits. There was something threatening about him, something uncanny, though he was among her most loyal and generous clients. His large hands touched her carefully, but still he frightened her.

He rang her doorbell at 11.00. He invariably came to her in the mornings. Since he wanted to spare her the moment of realisation that she was to die this day at the beginning of September, he took hold of her from behind as they were on their way into the bedroom. His large hands reached for her forehead and her neck and he snapped her spine. He put her body on the bed, roughly pulled her clothes off and did what he could to make it look like a sex crime. When he had finished, he looked around and thought that Sylvi had deserved a better fate. If circumstances had been different he would have wanted to bring her with him to the promised land. But Erik set the rules, and he demanded that his disciples be free of all worldly weakness. Now he had achieved this state. Women, desire, were gone from his life.

He left the flat. He was ready. Erik was waiting for him. God was waiting.

CHAPTER 35

Her grandfather had often complained about difficult people, a category which included almost everyone. Accordingly, he did his best to minimise contact with other people. However, as he said, one could never avoid them altogether. Linda had been particularly struck by the image he had used.

"They're like eels," he said. "You try to keep a hold of them, but they wriggle free of your grasp. The thing about eels is also that they swim at night. By that I don't mean you only meet difficult people at night – if anything they seem more likely to come up with their idiotic suggestions in the morning. Their darkness is of a different order; it's something they carry inside. It's their total obliviousness to the difficulty they cause others by their constant meddling. I have never meddled in other people's lives."

That was the biggest lie of his life. He had died without recognising the extent to which he himself had meddled in the lives of those around him, especially in trying to bend his two children to his will.

These musings about difficult people came unbidden to Linda as she was about to ring Anna's bell. She paused, her finger hovering a few centimetres in front of the buzzer. Anna is a difficult person, she thought. She doesn't seem to understand the worry she caused, or how her actions affect me.

When she rang the bell, Anna opened, smiling, dressed in a white blouse and dark trousers. She was barefoot and had pulled her hair back into a loose knot.

Linda had decided to bring it up straightaway to clear the air. She threw her jacket over a chair and said: "I have to tell you that I read the last few pages of your journal. I only did it to see if there was any explanation to your disappearance."

Anna flinched. "Then that was what I sensed," she said. "It

was almost as if there was a different smell when I opened the pages."

"I'm sorry, but I was so worried. I read the last couple of pages, nothing more."

We lie to make our half-truths seem more plausible, she thought. Anna may see through me. The journal will always be between us now. She'll be asking herself what I did and didn't read.

Anna stood by the window in the living room, her back to Linda. Linda looked at her friend, thinking she might as well be looking at an enemy. She realised she no longer knew Anna.

"There's one question you still need to answer." She waited for Anna to turn around, but she didn't. "I hate talking to people's backs."

Still no reaction. You may be a difficult person, Linda thought. But sometimes difficult people go too far. Grandpa would have thrown an eel like this into the fire and let it writhe in the flames.

"Why did you check into that hotel under my name?"

Linda tried to read Anna's back while she wiped the sweat from her neck. This will be my curse, she had thought in the first month of her police training. There are laughing policemen and crying policemen, but I'm going to be known as the perspiring policewoman.

Anna burst into laughter and turned around. Linda tried to judge if her laughter was genuine.

"How did you find out?"

"I called the hotel."

"May I ask why?"

"I don't know."

"What did you ask them exactly?"

"It's not so hard to work out."

"Tell me."

"I asked if an Anna Westin was still there or if she had checked out. They didn't have an Anna Westin, but they did have a Wallander, they said. It was that simple. But why did you do it?"

"What would you say if I told you I don't know why I used your name? Maybe I was afraid my father would run away again if he found out that I had checked into the hotel where we saw each other. If you want the truth, it's that I don't know."

The phone rang, but Anna made no move to pick it up. The answering machine switched on and Zeba's chirpy voice filled the room. She was calling for no reason, she informed them happily.

"I love people who call for no reason with so much positive energy," Anna said.

Linda didn't answer. She had no room to think about Zeba.

"I read a name in your journal: Birgitta Medberg. Do you know what's happened to her?"

"No."

"Don't you read the papers?"

"I was looking for my father."

"She's been found murdered."

Anna looked closely at her. "Why?"

"I don't know why."

"What are you saying?"

"I'm saying she was murdered. The police don't know who did it, but they're going to want to talk to you about her."

Anna shook her head. "What happened? Who would want to kill her?"

Linda decided not to reveal any details about the murder. She simply sketched out the news in broad brushstrokes. Anna's dismay looked completely genuine.

"When did this happen?"

"A few days ago."

Anna shook her head again, left the window and sat down in a chair.

"How did you know her?" Linda said.

Anna looked narrowly at her. "Is this an interrogation?"

"I'm curious."

"We rode horses together. I don't remember the first time we met, but there was someone who had two Norwegian Fjord horses that needed exercising. Birgitta and I volunteered to ride them. I didn't know her at all. She never said very much. I know she studied pilgrimage trails. We also shared an interest in butterflies, but I don't know anything more about her. She wrote to me fairly recently and suggested we buy a horse together. I never replied."

Linda tried to remain alert for an indication that Anna was lying.

I'm not the person who should be doing this, she thought. I should be driving a patrol car and picking up drunks. My father should be talking to Anna, not me. It's just that damn butterfly. It should be hanging on the wall.

Anna had already followed her gaze and read her mind.

"I took the butterfly with me when I went to look for my father. I was going to give it to him, but then when I realised it was all my imagination I threw it in the canal."

It could be true, Linda thought. Or she lies so convincingly that I can't tell.

The phone rang again. Ann-Britt Höglund's voice came on the answering machine. Anna looked at Linda, who nodded. Anna picked up the receiver. The conversation was brief and Anna didn't say much. She hung up.

"They want me to come to the police station now," she said.

Linda got up. "Then you'd better go."

"I want you to come with me."

"Why?"

"I'd feel more secure."

Linda hesitated. "I'm not sure it's appropriate."

"But I'm not accused of anything. They just want to have a conversation with me, at least that's what the woman said. And you're both a police officer and my friend."

"I'm happy to go down there with you, but I'm not sure they'll let me stay in the room when they talk to you."

Höglund came into the reception area at the police station to meet Anna. She looked disapprovingly at Linda. She doesn't like me, Linda thought. She's the kind of woman who prefers young men with piercings and an attitude. Höglund had put on weight. Soon you'll be dumpy, Linda thought with satisfaction. I still wonder what my father saw in you when he courted you a few years ago.

"I want Linda to be there," Anna said.

"I don't know if that will be possible," Höglund said. "Why do you want her to be there?"

"I have a tendency to make things more complicated than they are," Anna said. "I just want her there for support, that's all."

Höglund shrugged and looked at Linda.

"You'll have to ask your father if it's OK," she said. "You know where his office is. He's waiting in the small conference room two doors down from there."

Höglund left them and marched off.

"Is this where you'll be working?" Anna said.

"Hardly. I'll be spending time in the garage and in the front seat of patrol cars."

The door to the small conference room was half open. Wallander was leaning back in his chair, a cup of coffee in his hand. He's going to break that chair, Linda thought. Do police officers have to get so fat? I'll have to take early retirement. She pushed the door open. Wallander didn't seem particularly surprised to see her with Anna. He shook Anna's hand.

"I would like Linda to stay," she said.

"Of course."

Wallander threw a glance behind them into the corridor.

"Where's Höglund?"

"I don't think she wanted to come along," Linda said, seating herself as far away from her father as possible.

That day Linda learned something important about police work both from Anna and from her father. Her father impressed her by steering the conversation with imperceptible yet total control. He never confronted Anna directly, he approached her from the side, listening to her answers, encouraging her even when she contradicted herself. He gave the impression of having all the time in the world, but never let her off the hook.

What Anna taught her was through her lies. She appeared to be trying to keep them to a minimum, but without success. Once, when Anna bent down to pick up a pencil that had rolled off the table, Linda and her father exchanged a look.

When it was over and Anna had gone home, Linda sat at the kitchen table at home and tried to write down the conversation exactly as she remembered it, like a screenplay. How was it that Anna had begun? Linda started to write and the exchange slowly reproduced itself on paper.

KW: Thanks for coming. I'm glad that nothing serious happened to you. Linda was very worried, and I was too.
AW: I suppose I don't need to tell you about the person I thought I saw in Malmö.
KW: No, you don't. Would you like something to drink?
AW: Juice, please.
KW: I'm afraid we don't have any. There's coffee, tea or plain water.
AW: I'll pass.

Slowly but surely, Linda thought. He has all the time in the world.

KW: How much do you know about what happened to Birgitta Medberg?
AW: Linda told me she was killed. It's horrible. Incomprehensible. I know you saw her name in my journal.
KW: Not us. Linda was the one who saw it when she was trying to work out what had happened to you.
AW: I don't like people reading my journal.
KW: Of course not. But Birgitta's name was there, wasn't it?
AW: Yes.
KW: We're trying to build a picture from all the people she came in contact with. The conversation we're having is identical to those my colleagues are having with others, all round us.
AW: We rode a pair of Norwegian Fjord horses together. They're owned by a man called Jörlander. He lives on a small farm near Charlottenlund. He was a juggler in an earlier life. He has something wrong with his leg and can't ride now. We exercised the horses for him.
KW: When did you first meet Birgitta?
AW: Seven years and three months ago.

KW: How come you remember it so precisely?
AW: Because I've thought about it. I knew you would ask me that.
KW: Where did you meet?
AW: In the stables. She had also heard that Jörlander needed volunteers. We rode two or three times a week. We always talked about the horses, that was all.
KW: You never met each other apart from the riding?
AW: I thought she was boring, to be perfectly honest. Except for the butterflies.
KW: Which butterflies are they?
AW: One day when we were riding we realised we both had a passion for butterflies. Then we had a new topic of conversation.
KW: Did you ever hear her express any fears?
AW: She seemed nervous whenever we had to take the horses across a busy road, I remember that.
KW: And other than that?
AW: No.
KW: Did she ever have anyone with her?
AW: No, she would always be by herself on her little Vespa.
KW: So you had no other contact with each other?
AW: No. Just a letter she wrote to me once. Nothing else.

A slight hesitation, Linda thought as she wrote. An imperceptible tremor at times, but here she actually stumbled. What was she hiding? Linda thought about what she had seen in the hut and broke out into a sweat.

KW: When did you last see Birgitta?
AW: Two weeks ago.
KW: In what context was that?
AW: For heaven's sake, how many times do I have to repeat myself? Riding.
KW: This is the last time, I assure you. I just want to make sure I have all the facts straight. What happened in Malmö, by the way? When you were looking for your father.

AW: How do you mean?

KW: I mean, who rode the horses for you? Who filled in for Birgitta and for you?

AW: Jörlander has some reserves, young girls mostly. He doesn't like to use them because of their age, but he must have had to. You can ask him.

KW: We will. Do you remember if there was anything different about the last time you met?

AW: Who? The young girls?

KW: No, I was thinking of Birgitta.

AW: She was her normal self.

KW: Do you remember what you talked about?

AW: I've told you several times now that we didn't talk very much. A little about the horses, the weather, butterflies. That was about it.

And right here he had suddenly sat up in his chair, Linda thought, a tactical manoeuvre telling Anna to be on her guard.

KW: We have another name from your journal: Vigsten. Of Nedergade in Copenhagen.

Anna had looked over at Linda in surprise, then narrowed her eyes. There goes that friendship, Linda had thought at the time. If it wasn't gone already, that is.

AW: Obviously someone has read more of my journal than I realised.

KW: That's as may be. Vigsten. What can you tell me about this person?

AW: Why is this important?

KW: I don't know if it's important.

AW: Does he or she have anything to do with Birgitta?

KW: Perhaps.

AW: He's a piano teacher. He was my teacher for a while, and we've kept in touch since then.

KW: Is that it?
AW: Yes.
KW: When was he your teacher?
AW: It was during the autumn of 1997.
KW: And only then?
AW: Yes.
KW: Can I ask why you stopped going to him?
AW: I wasn't good enough.
KW: Did he tell you that?
AW: I did. Not to him, to myself.
KW: It must have cost a great deal of money to have a piano teacher in Copenhagen, with all that travel.
AW: You have to set your priorities.
KW: You're going to be a doctor, I understand.
AW: Yes.
KW: How is that going?
AW: What do you mean?
KW: Your studies.
AW: Fine.

At this point Wallander's manner changed. He leaned towards Anna, still friendly, but now he clearly meant business.

KW: Birgitta Medberg was murdered in Rannesholm in an unusually brutal way. Someone cut off her head and hands. Can you imagine anyone who could do such a thing?
AW: No.

Anna was very calm, Linda thought. Too calm. Calm in the way that only someone who knows what's coming can be. But then she retracted her conclusion. It was possible, but she shouldn't make the leap prematurely.

KW: Can you understand how anyone could do this to her?
AW: No.

Then came the abrupt finish. After her last answer his hands came down on the table.

KW: Thank you for your time. You've been very helpful.
AW: But I haven't actually been able to help you with anything.
KW: Oh, I wouldn't say that, Anna. Thank you again. You may hear more from us at some point.

He had escorted them both back to reception. Linda could tell that Anna was tense. She must be wondering what she said without knowing. My father is still questioning her, but he's doing it inside her head, waiting to see what she's going to say.

Linda pushed the paper away and stretched her back. Then she called her father on his mobile.

"I don't have time to talk, but I hope you found it instructive."

"Absolutely. But I don't think she was telling the truth."

"I think we can safely assume she wasn't telling us the whole truth. But the question is why. Do you know what I think?"

"No, tell me."

"I think her father has actually returned. But we can talk more about that tonight."

Wallander came back to the flat just after seven o'clock. Linda had cooked dinner. They sat at the kitchen table and he had just begun to set out the grounds he had for thinking that Anna's father had returned when the phone rang.

CHAPTER 36

They had arranged to meet in a parking place between Malmö and Ystad. Even a car park could become a cathedral if you chose to see it that way. The balmy September air rose from the ground like pillars for this towering, yet invisible, church.

He had told them to be there at 3 p.m., instructing them to wear normal clothes as they would be impersonating tourists from Poland on a shopping trip to Sweden. Alone or in small groups. They would be arriving from different directions and would receive their final instructions from Erik, who would have Torgeir Langaas at his side.

Erik had spent the last weeks in a caravan in a campsite in Höör. He had given up the flat in Helsingborg and bought an inexpensive used caravan in Svedala. His elderly Volvo had towed it to the campsite. Apart from his meetings with Langaas and the carrying out of their plans, he had spent all his time in the caravan, praying and preparing for the task ahead. Every morning he looked into the little mirror on the wall and asked himself if he was staring into the eyes of a madman. No-one could become a prophet without a great deal of natural humility, he would think to himself. To be strong was to be able to ask oneself the hardest questions. Even if his commitment to the task God had assigned him never wavered, he still needed to be sure that he was not carried away with pride. But the eyes that gazed steadily at him from the mirror confirmed what he already knew: that he was the anointed leader of the new age. There was nothing ill-conceived about the great task that lay before them. Everything was spelled out in the Holy Book. The Christian world had become mired in a bog of misconceptions and had tried God's patience to the point where He had simply given up, waiting for the one who was prepared to act as His true servant to step in and set things right.

"There is only one God," Erik Westin said at the beginning of all his prayers. "One God and His only son, whom we crucified. This cross is the symbol of our only hope. The cross is plainly made – of wood, not gold or precious marble. The truth lies in poverty and simplicity. The emptiness we carry inside can be filled only by the Holy Ghost, not by material goods or riches, however tempting they may appear to us."

He had carried on long conversations with God. He had also thought a great deal about Jim Jones, the false prophet, the fallen angel. He thought about the exodus from the United States to Guyana, the initial period of joy and then the terrible betrayal that had led to murder. In his thoughts and prayers there was always a place for those who had died in the jungle. One day they would be set free from the evil that Jim Jones had committed and uplifted to the highest realms, where God and the angels awaited them.

During this last little while he had also felt affirmed and accepted by those he had once left behind. They had not forgotten him. They understood why he had left and why he had now returned. One day when everything was over he would withdraw from the world and take up the life he had left so long ago: making sandals. He would have his daughter by his side and all would be fulfilled.

The time had come at last. God had appeared to him in a vision. All sacrifice is made for the creation of life, he thought. No-one knows if they have been chosen to live or to die. He had re-instituted the ritual sacrifices with their origins in the earliest days of Christianity. Life and death went hand in hand: God was both logical and wise. Killing in order to sustain life was an important practice in combating the emptiness that existed inside man. Now the moment was here.

On the morning of the day that they were to meet in the parking place, Erik Westin went down to the dark lake, which still retained some of the summer's heat. He washed himself thoroughly, clipped his nails and shaved. He was alone in the remote camping area. After Langaas called, Erik threw his mobile phone into the lake. Then he put on his clothes, took his Bible and money with him to the car and drove a short distance up the road. There was one thing left to take care of. He set fire to the caravan and drove away.

* * *

Altogether there were 26 of them, 17 men and 9 women, and all had a cross tattooed on their chest above the heart. The men were from Uganda, France, England, Spain, Hungary, Greece, Italy and the United States. The women were American and Canadian, with the exception of one British woman who had lived in Denmark for a long time. It had taken Erik four years to build the core group of the Christian army he planned to lead into battle.

They were meeting each other for the first time. A light rain fell as they assembled. Erik had parked his car on a hill overlooking the parking place. He kept an eye on the proceedings with the help of a telescope. Langaas was there to receive them. He had been instructed to say that he didn't know where Erik was. Erik had often explained to him that secret agreements of this nature could strengthen people's belief in the holy task that awaited them. Erik looked into the telescope. There they were, some in cars, some on foot, two on bikes, one on a motorcycle, and a few more walked out from a small forested area next to the parking place as if they had been camping there. Each one carried only a small rucksack. Erik had been very strict on this point: no-one was to have a large amount of luggage or wear unusual attire. Nothing that would attract attention to God's undercover army.

He trained the lens on Langaas's face. He was leaning against the glass-fronted information board. It would not have been possible without him, Erik thought. If I hadn't stumbled across him in that dirty Cleveland street and managed to transform him into an absolutely ruthlessly devoted disciple, I would not yet be ready to give my army their marching orders.

Langaas turned his head in the direction they had agreed. Then he stroked his nose twice with his left index finger. All was ready. Erik put away his telescope and started walking down to the parking place. There was a dip beside the road which meant he could walk right up to them without being seen. That way he would seem to appear from nowhere. When he walked among them everyone stopped what they were doing, but no-one talked, as he had instructed.

Langaas had come in a truck, into which they now loaded the bicycles and motorcycle, letting the people climb in after them. The cars would have to be left behind. Erik drove and Langaas sat in the front with him. They found their way to Mossbystrand, where they stopped and everyone got out. They walked down to the beach. Langaas carried two large baskets of food. They sat closely pressed together among the sand dunes, like tourists who found the weather a little too cold.

Before they started to eat, Erik said the necessary words: "The Lord commands our presence. He decrees the battle."

They unpacked the baskets and ate. When the food was finished, Erik ordered them to rest. Then he walked with Langaas to the water's edge. They went through the plan one last time. A large cloudbank moved in from over the sea, darkening the sky.

"We're getting just what we wanted," Langaas said. "It would be a grand night for catching eels."

"We are getting what we need, for we are the righteous and the just," Erik said.

They waited until it was evening, then climbed back into the truck. It was 7.30 when Erik swung back on to the road and drove east. He turned north just after Svarte, passing the main road from Malmö to Ystad, and then continued on a road that went west, past Rannesholm Manor. Two kilometres past Harup he turned on to a small dirt road, switching off the headlights and the engine. Langaas climbed out of the cab. In the wing mirror Erik watched two of the American men climbing out of the truck: Peter Buchanan, a former hairdresser from New Jersey, and Edison Lambert, a jack-of-all-trades from Des Moines.

Erik felt his pulse quicken. Was there anything that could go wrong? He regretted even thinking the question. I'm not crazy, he thought. I place my trust in God and His plan. He started the engine and pulled back out on to the road. One motorcycle overtook him, then another. He continued driving north, throwing a glance at Hurup Church where Langaas and the two Americans were headed. Half a mile north of Hurup he turned left towards Staffanstorp, then left again after ten minutes, drawing up in front of an abandoned farmhouse. He climbed out of the truck and motioned for those still in the back to follow him.

He checked the time: exactly on schedule. They walked slowly in order to accommodate the few who were older, or less fit, like the

British woman, who had been operated on for cancer six months before. Erik had debated whether to include her, but after consulting with God he received the answer that she had survived her illness precisely so that she could complete her mission. They followed a road that led to the back of Frennestad Church. Erik felt in his pocket for the key that Langaas had made him. Two weeks ago he had tried it, and it turned without a squeak. He stopped them when they reached the churchyard. No-one said anything and all he could hear was breathing. Only calm breaths, he noted. No-one is panting, no-one seems anxious, not even she who is going to die.

Erik looked again at his watch. In 43 minutes Langaas, Buchanan and Lambert would set fire to the church in Hurup. They started walking again. The gate opened without a sound. Langaas had oiled it. They walked in single file up to the church. Erik unlocked the doors. It was cool inside; one person shivered. He turned on the torch and looked around. Everyone seated themselves in the front pews, as they had been instructed. The last missive Erik had distributed included 123 instructions which were to be memorised to the letter. He knew already that they had done so.

Erik lit the candles that Langaas had placed near the altar. In the dim light he could see Harriet Bolson, the woman from Tulsa, seated on the far right. She was completely calm. The ways of God are inscrutable, he thought. But only to those who do not need to understand them. He looked at his watch. It was important that the two actions, the burning of Hurup Church and that which was to take place in Frennestad Church, be synchronised. He looked at Harriet Bolson again. She had a thin, worn face even though she was only 30 years old. Perhaps her face shows the traces of her sin, he thought. She could only be cleansed through fire. He turned off the torch and walked into the shadows by the pulpit. He reached into his rucksack and took out one of the ropes that Langaas had bought in a chandler's in Copenhagen. He placed it in front of the altar, then checked his watch again. It was time. He turned and motioned for everyone to stand. He called them up one by one. He handed one end of the rope to the first person.

"We are irrevocably bound together," he said. "From now on, from this day forward, we will never need a rope again. We are bound by our loyalty to God and our task. We cannot tolerate that the Christian world sink any deeper into degradation. The world will be cleansed through fire, and we must start with ourselves."

While he was uttering the last words he had slowly moved so that he stood in front of Harriet Bolson. At the same moment that he tied the rope around her neck, she understood what was about to happen. It was as if her mind went blank from the sudden terror. She didn't scream or struggle. Her eyes closed. *All my years of waiting are finally over.*

The church in Hurup began to burn at 9.15. As the fire engines were on their way they received reports that Frennestad Church was also on fire.

Langaas and the Americans had already been picked up. Langaas took Erik's place and drove the truck to the new hide-out.

Erik remained behind in the darkness. He sat up on a hill close to Frennestad Church. He could see the firemen trying in vain to staunch the blaze. He wondered if they or the police would make it inside before the roof caved in.

He sat there in the darkness and watched the flames. He thought about how he would one day watch the fires burning with his daughter by his side.

CHAPTER 37

That night two churches in roughly the same area, a triangle bounded by Staffanstorp, Anderstorp and Ystad, burned to the ground. The heat was so intense that at dawn only the bare, smoking skeleton of the buildings remained. The bell tower of Hurup Church collapsed, and those who heard it said it sounded like a howl of bottomless despair.

The churchwarden of Frennestad Church was the first to make it into the burning building, in hopes of saving its unique mass staves dating from the Middle Ages. Instead he made a gruesome discovery that would haunt him for the rest of his life. A woman in her thirties lay in front of the altar. She had been strangled by a thick rope pulled so tightly it had almost removed her head from her body. He rushed screaming from the scene, and fainted on the front steps.

The first fire engine arrived a few minutes later. It had been on its way to Hurup when it had received fresh instructions. None of the firefighters understood fully what had happened, whether the first alarm had been a mistake or if both churches were indeed on fire.

There was a similar state of confusion at the police station in Ystad during the first few minutes when the two calls came in. When Wallander got up from the dinner table he was under the impression that he was going to Hurup where a woman had been reported dead. Since he had drunk some wine with dinner he asked for a patrol car to collect him.

It was only as they were leaving Ystad that he learned of the misunderstanding: the church in Hurup was on fire, but the dead woman had been found in Frennestad Church. Martinsson, who was driving, started shouting at the switchboard operator to try to determine once and for all how many damned churches were on fire.

Wallander sat quietly for the duration of the journey, not only because Martinsson was driving with his usual recklessness, but because he sensed that his worst fears were being confirmed. The animals that had been killed were only the beginning. Lunatics, he thought, satanists, fanatics. As they drove through the darkness he thought he was beginning to discern a logic to these events, if only dimly.

By the time they pulled up outside the burning church in Frennestad they at least had a clearer idea of the current situation. The two churches had caught fire at almost exactly the same time. In addition there was a woman dead in Frennestad. They sought out the fire chief, Mats Olsson, to whom, as it turned out, Martinsson was distantly related. In the midst of the intense heat and chaos, Wallander heard them issue greetings to their respective wives. Then they went into the inferno that was the church. Martinsson let Wallander take the lead, as he usually did at a crime scene, and as he was particularly willing to do in the devastating heat. But the aisle provided them a route and a fireman preceded them with a hose. The dead woman lay in front of the altar with a rope around her neck. Wallander tried to imprint the scene on his memory. It was staged, it had to be. He turned to Olsson.

"How long can we stay?"

"The roof is going to cave in. We're not going to be able to put the fire out in time to save it."

"When?"

"Soon."

"How long?"

"Ten minutes. I can't let you stay any longer."

No technicians would be able to make it to the scene in time. Wallander put on a helmet that someone handed him.

"Go out and see if anyone in the crowd has a camera, or better yet a video camera," he said. "Confiscate it. We're going to need to document this."

Martinsson left. Wallander started to examine the dead woman. The rope was thick, like a ship's hawser. It lay around her neck with the ends outstretched. Two people pulling in different directions, he thought. Like the olden days when criminals were ripped apart by tying them to two horses that were sent off in different directions.

He glanced at the ceiling. Flames were starting to come through.

There were people running all around him, carrying objects from the church. An elderly man in wellington boots and pyjamas was straining to rescue a beautiful old altar cabinet. There was something touching about their struggle. These people have realised they're losing something irreplaceably precious, he thought.

Martinsson returned with a video camera.

"Do you know how to use it?"

"I think so," Martinsson said.

"Then you be our photographer. Take full shots, details, from all angles."

"Five minutes," Olsson said. "That's all you have."

Wallander crouched down beside the dead woman's body. She was blonde and bore an uncanny resemblance to his sister Kristina. An execution, he thought. First animals, now people. What was it the Lindberg woman had claimed she heard? "The Lord's will be done"?

He rapidly searched the woman's pockets. Nothing. He looked around. There was no handbag. He was about to give up when he saw a breast pocket on her blouse. Inside was a piece of paper with a name and address: Harriet Bolson, 5th Avenue, Tulsa.

"Time's up," Mats Olsson said. "Let's go."

He rounded up the people still in the church and hurried them out. The body was carried away and Wallander took the rope.

Martinsson called in to the station.

"We need information on a woman from Tulsa, that's Oklahoma, United States," he said. "All registers, local, European, international. Highest priority."

Linda turned off the television impatiently. She knew that the spare keys to her father's car were on the bookshelf in the living room. She picked them up, then headed out the door and jogged to the police station.

Wallander's car was parked in the corner. Linda recognised the car next to it as Höglund's. Linda fingered the Swiss army knife in her pocket, but this was not a night for slashing tyres. She had heard him mention Hurup and Frennestad. She unlocked the car door and drove as far as the water tower. There she pulled over and got out a map. She knew where Frennestad was, but not Hurup. She found it, turned off the light and drove out of Ystad. Halfway to Hörby she turned left and after a few kilometres she could see the smoke from Hurup Church. She drove as close as she could, then parked and walked up to the church. Her father wasn't there. The only police officers were young cadets and it struck her that if the fire had started a few days later she could have been one of them. She told them who she was and asked where her father was.

"There's another church on fire," she was told. "Frennestad Church. They have a casualty."

"What's going on?"

"It looks like arson – two churches don't just catch fire at the same time. But we don't know what happened in Frennestad Church, only that there's a body."

Linda thanked them and walked away. A sudden noise made her turn. Parts of the church roof collapsed and a shower of sparks shot up into the sky. Who would burn a church? she wondered. But she couldn't find an answer to that question any more than she could imagine what kind of person would set fire to swans or cattle or a pet shop.

She got back in the car and drove to Frennestad. She saw the burning church from a distance. Burning churches I associate with war, she thought. But here there are churches burning in peacetime. Can a country be engaged in an invisible war against an unseen enemy? She was unable to pursue this thought any further. The road to the church was blocked by cars. When she caught sight of her father in the light of the fire, she stopped. He was talking with a firefighter. She tried to see what he was holding. A hose? She walked closer, pushing past people who were crowded together outside the restricted area. He was holding a rope, she realised. A hawser.

Nyberg walked up to Wallander and Martinsson, who were standing outside the church. He looked irritable as usual.

"I thought you should take a look at this," he said, holding out his hand.

It was a small necklace. Wallander took out his glasses. One side of the frame broke when he was putting them on. He swore and had to hold the glasses in place.

"It looks like a shoe," he said.

"She was wearing it," Nyberg said. "Or had been. The chain broke when the rope was pulled tight. The necklace fell inside her blouse. The doctor found it."

Martinsson took it and turned towards the fire to get more light.

"An unusual motif for a pendant," he said. "Is it really a shoe?"

"It could be a footprint," Nyberg said. "Or the sole of a foot. Once I saw a pendant in the shape of a carrot. A diamond was placed where the greens would have been. That carrot cost four hundred thousand kronor."

"It may help us identify her," Wallander said. "That's what counts right now."

Nyberg walked to the low wall backing on to the graveyard and started yelling at a photographer who was taking pictures of the burning church. Wallander and Martinsson walked down to the barricades.

They saw Linda and waved her over.

"Just couldn't stay away?" her father said. "You can come with us."

"How is it going?"

"We don't know what we're looking for," Wallander said slowly. "But these churches didn't set fire to themselves, that much is sure."

"They're working on tracing Harriet Bolson," Martinsson said. "They'll let me know the minute they find something."

"I'm trying to understand the significance of the rope," Wallander said. "Why a church and why an American woman? What does it mean?"

"A few people, at least three but maybe more, come to a church in the middle of the night," Martinsson said.

Wallander stopped him. "Why more than three? Two who commit the murder and one victim. Isn't that enough?"

"Theoretically, yes. But something tells me there were more, maybe many more. They unlocked the door. There are only two existing keys. The minister has one and the churchwarden who fainted has the other. They have both confirmed possession of their keys. Therefore we have to assume they used a sophisticated pass key or a copy," Martinsson said. "A group, a society, a band of people who chose this church to execute Harriet Bolson. Is she guilty of something? Is she a victim of religious convictions? Are we dealing with satanists or some other kind of lunatic fringe? We don't have the answers."

"Another thing," Wallander said. "What about the note I found on her body? Why was it left behind?"

"So that we would be able to identify her. Perhaps it was a message to us."

"We have to confirm her identity," Wallander said. "If she so much as visited a dentist in this country we'll know."

"They're working on it." Martinsson sounded affronted.

"I don't mean to get on your case. What's the word?"

"Nothing, as of yet," Martinsson said. "Then there's another thing. Whoever saw a pendant necklace shaped like a shoe? Or a sandal?"

He shook his head and walked away.

Linda held her breath. Had she heard him correctly?

"What was it he said? What have you found?"

"A note with a name and address."

"Apart from that. Something else."

"A pendant necklace."

"That looked like something?"

"A footprint, a shoe. Why do you ask?"

She ignored the question.

"What kind of shoe?"

"Maybe a sandal."

The light from the fire grew brighter in spurts as gusts of wind caught the flames.

"May I remind you that Anna's father was a sandal maker before he disappeared. That's all."

It took a moment for it to sink in.

"Good," he said. "Very good. That may be the opening we need. The question is just where it leads us."

CHAPTER 38

Wallander had tried to send Linda home to get some sleep, but she had insisted on staying. She had curled up in the back seat of a patrol car and woke up only when he rapped sharply on the window pane. He's never learned the art of waking a person gently, she thought. My father doesn't simply wake people up, he tears them from their dreams.

She got out of the car and shivered. Shreds of fog drifted over the fields. The church had burned to the ground and only the gaping, sooty walls remained. Thick smoke still rose through the caved-in roof. Most of the fire engines were gone; only two crews were needed for the mop-up. Martinsson had left, but she could see Lindman in the distance. He came over to her and handed her a cup of coffee. Her father was speaking to a journalist on the other side of the police line.

"I've never seen anything like this landscape before," Lindman said. "Not in the west, not up in Härjedalen. Here, Sweden simply slopes down into the sea and ends. All this mud and fog. It's very strange. I'm trying to find my feet in a landscape that's completely alien to me."

Linda mumbled that fog was fog, mud was mud. What could possibly be strange about something so ordinary?

"Anything new on the woman?" she said.

"Not yet. But she's definitely not a Swedish citizen."

"Any reason to think she's not the person whose name is on the note?"

"No. It's far-fetched to think the murderer would leave a false name."

Wallander came walking over. The journalist disappeared down the hill.

"I've talked to Chief Holgersson," he said. "You are already involved in the fringes of this investigation, so we may as well let you in on the whole thing. I'd better get used to having you around. It'll be a little like having a ball constantly bouncing up and down by my side."

Linda thought he was making fun of her.

"At least I can still bounce. More than some people I know."

Lindman laughed. Wallander looked cross, but kept himself in check.

"Don't ever have children, Lindman," he said. "You see what I have to deal with."

A car swung on to the road leading up to the church. It was Nyberg.

"He's freshly showered," Wallander said. "Ready for another day of unpleasantness, no doubt. He'll keel over and die the day he retires and no longer has to be digging in the mud with rainwater up to his knees."

"He's like a dog," Lindman said in a low voice. "Have you noticed? It's almost as if he's sniffing around, and wishes he could just get down on all fours."

Linda had to agree: Nyberg really did look like an animal intent on picking up a scent.

Nyberg joined their group, seeming not to notice Linda. He smelled strongly of aftershave.

"Do we have any idea how the fires started?" Wallander said. "I talked to Olsson and he said the churches were both set alight in several places. The churchwarden who came on the scene early said that it looked as if the fire was burning in a circle, which would imply that it caught hold in more than one spot."

"We haven't found any evidence yet," Nyberg said. "But it's clearly arson."

"There's a difference between the two cases," Wallander said. "The fire in Hurup seems to have started more in the manner of an explosion. Someone in one of the nearby houses said it sounded as if a bomb had gone off. The blazes were started in different ways, but synchronised."

"It's a definite pattern," Lindman said. "Starting a fire to detract attention from the murder."

"But why a church?" Wallander asked. "And why would you strangle a person with a hawser?"

He looked at Linda. "What do you see in all this?"

She felt herself blushing. The question had come so suddenly she was unprepared for it.

"The site has been chosen deliberately," she started hesitantly. "Strangling someone with a rope seems akin to torture. But this has also to do with religion, like an eye for an eye, death by stoning or living burial. Why not strangle someone with a hawser?"

Before anyone had a chance to respond, Lindman's mobile rang. He listened, then held it out to Wallander.

"We're starting to get information from the States," he said. "Let's go back to Ystad."

"Do you need me?" Nyberg asked.

"I'll call if we do," Wallander said. Then he turned to Linda. "But you should be there," he said. "Unless you want to go home and sleep first."

"You know there's no need to even ask."

He threw a glance at her. "I'm trying to be considerate."

"Think of me as a police officer and not your daughter."

They were silent in the car, from both lack of sleep and a fear of saying something that would irritate the other.

Once they had parked in front of the station, Wallander walked off to the town prosecutor's office. Lindman caught up with Linda just outside the front door.

"I remember my first day as a police officer," he said. "I was in Borås and had been to a party with friends the night before. The first thing I did when I walked through the front doors of the station was rush into the toilet and throw up. What do you plan to do?"

"Not that, at any rate," Linda said.

Höglund was standing by the reception desk. She still only barely registered Linda's presence and Linda decided to treat her the same way from now on.

There was a message for Linda: Chief Holgersson wanted to speak to her.

"Have I done anything wrong?" Linda said.

"I wouldn't think so," Lindman said, then left.

I like him, in spite of his betraying my confidence, Linda thought. More and more, actually.

Holgersson was on her way out when Linda walked down the corridor to her office.

"Kurt has explained the situation to me," Holgersson said. "We're going to let you sit in on this one. It's a strange coincidence that one of your friends is involved."

"We don't know that for sure," Linda said. "She might be."

The door to the conference room was closed at 9.00. Linda sat in the seat her father had pointed out to her. Lindman sat next to her. She looked at her father at the head of the table, drinking mineral water. He looked the way she had always imagined him in these situations: thirsty, his hair standing on end, prepared to jump into yet another day of a complicated criminal investigation. But it was an overly romanticised image and therefore a false one, she knew. She shook it off with a grimace.

She had always been under the impression that he was good at his job, a skilful investigator, but today she realised he had talents she hadn't even imagined. Among other things, she was impressed by his ability to keep so many facts in his head, scrupulously arranged according to time and place. While she listened to him, something stirred in her at a much deeper level. It was as if she only now understood why he had had so little time for her or Mona. There had simply been no room for them. I have to talk to him about this, she thought. When all of the events have been explained and everything is over we have to talk about the fact that he prioritised work over us.

Linda stayed behind in the room when the meeting was over. She opened a window and thought about everything that had been said. Her father had set his bottle of mineral water down and summarised the very unclear situation they were in: "Two women have been murdered. Everything starts with these two. Maybe I'm being too presumptuous in assuming the same killer is responsible for both deaths since there is no obvious connection, no motive, not even any similarities. Medberg was killed in a hut hidden deep inside Rannesholm Forest, and now we find another woman, most probably a foreigner, strangled with a thick rope in a burning church. The only connections we have found between these events are tenuous, accidental – not really connections at all. On the outskirts of this is another murky series of events. That is why Linda is here."

Wallander slowly picked his way across the terrain, which involved everything from swans set on fire to severed hands. It was as if he proceeded with antennae stretched out in every direction at once. It took him one hour and twelve minutes without a break or retakes to reach his conclusion: "We don't know yet what has happened. Behind the two dead women, the burning animals and the torched churches lies something else that we can't quite put our finger on. And we don't know if what we have here marks the culmination of something, or simply the beginning."

At the words "simply the beginning", Wallander sat down, but continued to speak.

"We're still waiting for information regarding the person we believe to be called Harriet Bolson. While we wait I'm going to open this up for general discussion, but before I do I'd like to make a final comment. I have a feeling that the animals were not burned to satisfy the perverted instincts of a sadist. These atrocities may each have been a form of sacrifice, or acts with their own twisted logic. We have Medberg's praying hands and a Bible that someone sat and wrote commentary in. And now something that looks like a ritual killing in a church. We have an eyewitness who claims she heard the man who set fire to the pet shop shouting the words 'As decreed by God' or something along that line. All of these things may point to a religious message, perhaps the work of a sect or a handful of crazed individuals. I doubt the latter. There is an organised quality to this cruelty that speaks against it being the work of a single individual. But are we talking about two or a thousand? We don't know. That's why I want us to take the time to discuss the matter without prejudice before we go on with our investigation. I think we'll be more effective if we allow ourselves to push everything else aside and concentrate on this point for a moment."

But this discussion was averted by a door opening and a woman announcing that American faxes about Harriet Bolson had started to come in. Martinsson left and returned with a few papers, among them a blurred photograph of a woman. Wallander held his broken glasses in front of his face and nodded. The dead woman was Harriet Bolson.

"My English is not quite what it should be," Martinsson said and passed the papers over to Höglund, who started to read aloud.

Linda had picked up a notebook as she walked into the room. Now she started making notes, without being clear about why she was doing so. She was involved in something without being fully involved, but she sensed her father had an assignment for her, one that he would present her with when the time was ripe.

Höglund said the American police seemed to have covered the case thoroughly, but perhaps that hadn't been so hard since Harriet Jane Bolson had been registered as a missing person since January 12, 1997. That was when her sister, Mary Jane Bolson, had gone to the Tulsa police and filed the report. She had initially tried to reach her sister on the phone for a week without success. Then she had got in her car and driven the 300 kilometres to Tulsa where her sister lived and worked as archivist and secretary to a private art collector. Mary Jane had found her sister's flat empty. She was not at her place of work. She seemed in fact to have disappeared without trace. Mary Jane and all of Harriet's friends had described her as a reserved but conscientious and friendly woman who had had neither a drug addiction nor any other vice which might help explain her disappearance. The police in Tulsa had completed a preliminary investigation and maintained a current file on her case, but during the last four years no addition to it had been made. No clue, no sign of life, nothing.

"A police officer by the name of Clark Richardson is eagerly awaiting our reply and confirmation of the fact that the woman we've found really is Bolson. He would like the information as soon as possible."

"We can supply him with that information immediately," Wallander said. "It's her, there's no doubt about it. Is there really no theory about her disappearance?"

Höglund scoured the documents.

"Harriet was unmarried," she said. "She was twenty-six when she disappeared. She and her sister were daughters to a Methodist pastor in Cleveland, Ohio. Prominent, it says. They had a happy childhood, no evidence of trouble, studies at various universities. She had a position in Tulsa with a good salary. She lived simply with regular habits. She worked hard all week and went to church on Sundays."

"Is that it?" Wallander asked when Höglund had finished reading.

"That's it."

He shook his head.

"There has to be something more to her story," he said. "We need to know everything about her. That will be your job. Pour on the charm. Give Officer Richardson the idea that this is the most important murder investigation in Sweden right now. Which it probably is, by the by."

This was followed by a short period of open discussion. Linda listened attentively. After half an hour her father tapped the table with his pencil and ended the meeting. Everyone except Linda and her father left the room.

"I want you to do me a favour," he said. "Talk to Anna, hang around, but don't ask any questions. Try to work out why Medberg's name was really in her journal. And Vigsten. I've asked my Danish colleagues to look a little closer at him."

"Not the old man," Linda said. "He's senile. But there was someone else there, someone who kept himself hidden."

"We don't know that for sure," he said impatiently. "Have you understood what I've asked you?"

"Act normal," Linda said, "but try to get answers to these questions."

He nodded and stood up. "I'm worried," he said. "I don't know what's happening, and I'm fearful of what may be next."

Then he looked at her, stroked her swiftly, almost shyly, on the cheek, and left the room.

Linda invited Zeba and Anna to join her for coffee down at the harbour the same day. They had just sat down when it started to rain.

CHAPTER 39

Zeba's boy played happily with a toy car that squeaked because it was missing two of its wheels. Linda looked at him. Sometimes he could be almost unbearably needy and attention-seeking. At other times, like now, he was peaceful, lost in thought about the invisible roads his little yellow roadster was travelling.

The café was almost empty at this time of day. Three Danish sailors in one corner were hunched over a chart. The young woman behind the counter yawned.

"Girl talk," Zeba said suddenly. "Why don't we have more time for that?"

"Talk away," Linda said. "I'm listening."

"What about you?" Zeba asked, turning to Anna. "Are you listening?"

"Of course."

They were quiet. Anna pushed a teaspoon around in her cup, Zeba folded a pinch of snuff into her upper lip. Linda sipped her coffee.

"Is this all there is?" Zeba wondered aloud. "In life, I mean."

"What are you thinking of?" Linda said.

"All our dreams. What became of them?"

"You dreamed of having children," Anna said. "At least, that seemed to be your main goal."

"You're right. But all the other stuff. I was such a dreamer! Especially when I was drunk out of my mind, you know the way you drink when you're a teenager, when you end up on your hands and knees, throwing up in a bush, having to fight off some swain who's looking to take advantage of the situation. But I never even realised any of my dreams. I drank them away, you could say. When I think of all the things I was going to do: be a fashion designer, rock star – fly a jumbo jet, for God's sake."

"It's not too late," Linda said.

Zeba put her chin on her hands and looked at her.

"Of course it is. Did you really dream about becoming a policewoman?"

"Never. In my dreams, if you can call them that, I was always going to devote my life to the theatre or to old furniture. Neither very exciting."

Zeba turned to Anna. "What about you?"

"I wanted to find a meaning in my life."

"Did you find it?"

"Yes."

"And?"

Anna shook her head. "It's not the kind of thing you can talk about. Either you find it or you don't."

Linda thought Anna seemed to be on her guard. From time to time she looked at Linda as if she was thinking: "I know you're trying to see through me." But I can't be sure, Linda thought.

Two of the sailors got up to leave. One of them patted Zeba's boy on the head.

"His existence hung by a hair for a while," Zeba said.

Linda raised her eyebrows. "How do you mean?"

"I was close to having an abortion. Sometimes I wake up in the middle of the night in a cold sweat and think I really did it, that he doesn't exist."

"I thought you wanted a baby."

"I did, but I was scared. I didn't think I'd be up to it."

"Thank God you didn't," Anna said.

Zeba and Linda were both taken aback by her emphatic declaration. She sounded stern, almost angry. Zeba was immediately on the defensive.

"God makes no sense – not to me – in that context. Maybe you'll understand when you get pregnant one day."

"I'm against abortion," Anna said. "That's just the way it is."

"Having an abortion doesn't mean you're 'for' abortion," Zeba said calmly. "There can be other reasons for it."

"Like what?"

"Like being too young. Or too sick."

"I'm against abortion full stop," Anna said.

"I'm happy I had my boy," Zeba said. "But I don't regret the abortion I had when I was fifteen."

Linda was astonished, and so was Anna. She seemed to stiffen and stared at Zeba.

"For God's sake, why are you staring at me like that?" Zeba said. "I was fifteen years old – what would you have done?"

"Probably the same thing," Linda said.

"Not me," Anna said. "It's a sin."

"Now you're sounding like a priest."

"I'm just telling you what I think."

Zeba shrugged. "I thought this was girl talk. If I can't talk about my abortion with my friends, who am I supposed to talk to?"

Anna stood up. "I have to go now," she said. "I forgot about something I need to do."

She disappeared out of the door. Linda thought it was odd that she would leave without saying goodbye to Zeba's son.

"What got into *her*?" Zeba said. "It's enough to make you think she had an abortion herself and can't stand to talk about it."

"Maybe she did," Linda said. "You think you know everything about a person, but the truth often comes as a surprise."

Zeba and Linda ended up staying longer than they had planned. With Anna gone, the atmosphere became more light-hearted. They giggled like teenagers. Linda followed Zeba home, saying goodbye outside her building.

"What do you think Anna will do?" Zeba said. "Say that we can't be friends any more?"

"I think she'll realise she overreacted."

"I'm not so sure about that," Zeba said. "But I hope you're right."

Linda went home. She lay on her bed, closed her eyes and drifted off. She was walking to the lake again, where someone had seen burning swans and called the police. Suddenly she opened her eyes. Martinsson had said they would check the phone log of calls to the station that night. That meant the conversation was preserved on a cassette tape.

Linda couldn't recall anyone commenting on what the man had sounded like. Suppose it was a Norwegian by the name of Torgeir Langaas? Amy Lindberg had heard someone who spoke either Norwegian or Danish. She got up. If the man who called in had an accent we may be able to determine a link between the burning animals and the man who bought the house behind the church in Lestarp.

She walked on to the balcony. It was 10 p.m. and the air was chilly. It will be autumn soon, she thought, the frost is on its way. It will crunch under my feet by the time I become a police officer.

The phone rang. It was her father. "I just wanted to let you know I won't be home for dinner."

"It's ten o'clock, Dad. I ate dinner hours ago."

"Well, I'll be here for another couple of hours."

"Do you have time to talk?"

"What's up?"

"I was thinking of taking a walk down to the station."

"Is it important?"

"Maybe."

"I can't give you more than five minutes."

"I only need two. Correct me if I'm wrong, but don't all emergency calls to the police get recorded and stored?"

"Yes. Why?"

"How long are they kept?"

"For a year. Why are you asking?"

"I'll tell you when I get there."

It was 10.40 p.m. when Linda walked into the station. Her father came out into the deserted reception to greet her. His office was full of cigarette smoke.

"Who's been here?"

"Boman."

"Who's that?"

"He's our prosecutor."

Linda was suddenly reminded of another prosecutor.

"Where did she go?"

"Who?"

"The one you were in love with? She was a prosecutor."

"That was a long time ago. I made a mess of my chances."

"How?"

"One's worst embarrassments should be kept to oneself. Anyway, there are other prosecutors here now, and Boman is one of them. I'm the only one who lets him smoke."

"You can't breathe in here now!"

She opened the window.

"What was it you wanted?"

Linda explained.

"You're right," he said, when she had finished.

Wallander stood up and motioned for her to follow. They bumped into Lindman in the corridor. He was carrying a file of folders.

"Put those down and come with us," Wallander said.

They went to the archive where the tapes were stored. Wallander gestured for one of the officers on duty to come over and talk to him.

"The evening of August the twenty-first," he said. "A man called to report seeing swans burning at Marebo Lake."

"I wasn't on that night," the officer said after studying a log book. "It was Undersköld and Sundin."

"Call them."

The officer shook his head. "Undersköld is in Thailand and Sundin is at a satellite intelligence conference in Germany. It won't be easy to get hold of either of them."

"What about the tape?"

"I'll find it for you."

They gathered around a cassette player. Between a call about a car theft and a drunk calling for help in "looking for Mum" was the call about the swans. Linda flinched when she heard the voice. It sounded as if he was trying to speak Swedish without an accent, but he couldn't disguise his origins. They played the tape several times.

Police: Ystad Police Station.

Man: I would like to report that burning swans are flying over Marebo Lake.
Police: Burning swans?
Man: Yes.
Police: Can you repeat that? What is burning?
Man: Burning swans are flying over Marebo Lake.

That was the end of the call. Wallander was listening through headphones that he then passed to Lindman.

"He has an accent, no doubt about it. I think he sounds Danish."

Or Norwegian, Linda thought. What's the difference?

"I'm not sure it's Danish," Lindman said and passed the headphones to Linda.

"The word he uses for 'burning,'" she said. "Is it the same in Norwegian and Danish?"

"We'll find out," Wallander said. "But it's embarrassing that a police cadet has to be the one to bring this up."

They left the room, after Wallander had given instructions about keeping the tape readily available. He led the others to the canteen. A group of patrol officers sat around one table, Nyberg and some technicians around another. Wallander poured himself a cup of coffee, then sat down by a phone.

"For some reason I still remember this number," he said.

He put the receiver close to his ear and asked the person he was speaking to to come down to the station as soon as possible. It was clear that whoever it was was reluctant to do this.

"Perhaps you would rather I send a patrol car with sirens blaring," Wallander said. "And have them handcuff you so your neighbours will wonder what you've been up to."

He hung up.

"That was Christian Thomassen," he said. "He's first mate on one of the Poland ferries. He's also an alcoholic, though currently on the wagon. He's Norwegian and should be able to give us a positive identification, as it were."

* * *

Seventeen minutes later one of the largest men Linda had ever seen was escorted into the canteen. He had huge feet stuffed into enormous rubber boots, was close to two metres tall, had a beard down to his chest and a tattoo on his bald pate. When he sat down, Linda discreetly stood up to see the tattoo more clearly. It depicted a compass card. Thomassen smiled at her.

"It's pointing south-south-west," he said. "Straight into the sunset. That way the Grim Reaper will know which way to take me when the time comes."

"This is my daughter," Wallander said. "Do you remember her?"

"Maybe. I don't remember too many people, to be honest. I've survived my drinking, but most of my memories haven't."

He stretched out his hand so she could shake it. Linda was afraid he would squeeze too hard. His accent reminded her of the man on the tape. She offered to fetch him a cup of coffee, but he said no.

"Let's go in," Wallander said. "I want you to listen to a recording for us."

Thomassen listened carefully. He asked to hear the conversation four times, but stopped Lindman when he was about to play it for a fifth time.

"He's Norwegian," Thomassen said. "Not Danish. I was trying to hear where in Norway he's from, but I can't pinpoint it. He's probably been away for a long time."

"Do you think he's been here a long time?"

"Not necessarily."

"But no question that he's Norwegian?"

"No. Even if I've lived here for nineteen years and drunk myself silly for eight of those, I haven't forgotten where I came from."

"That's all we wanted to know," Wallander said. "Do you need a lift back?"

"I came down on the bike," Thomassen said, smiling. "I can't when I've been drinking. I just fall over and hurt myself."

"A remarkable man," Wallander said to Linda after he left. "He has a beautiful bass voice. If he hadn't been so lazy and drunk so much, he could have been an opera singer. I suspect he would have become world famous, for his bulk if nothing else."

They went back to Wallander's office.

"So he's Norwegian," Wallander said. "Now we can be sure that the man who set fire to the swans was the same as the one who set fire to the pet shop, as we suspected. It will probably turn out to be the same man who set fire to the calf. The question is whether he was the one who was hiding in the hut in the forest."

"The Bible," Lindman said.

Wallander shook his head. "Swedish. They've managed to decipher a lot of what's been written in the margins and it's all in Swedish."

Linda waited.

Lindman shook his head. "I have to sleep," he said. "I can't think clearly any more."

"Eight o'clock tomorrow," Wallander said.

Lindman's steps died away in the corridor. Wallander yawned.

"You should get some sleep too," Linda said.

He nodded, then stood up. "You're right. We need to sleep. I need to sleep. It's already midnight."

There was a knock on the door. One of the officers on phone duty looked in.

"This just came," he said, handing a fax to Wallander. "It's from Copenhagen. Someone called Knud Pedersen."

"I know him," Wallander said.

The officer left. Wallander skimmed the fax, but then sat down again at his desk and read it more carefully.

"Strange," he said. "I know from way back that Knud Pedersen is a policeman who keeps his eyes open. They've had a murder there recently, a prostitute by the name of Sylvi Rasmussen. She was found with her neck broken. The unusual thing is that her hands were clasped in prayer – not severed this time, but Pedersen has read about our case and thought we should know about this one."

Wallander let the fax fall to the desk. "Copenhagen again," he said.

Linda was about to ask a question, but he raised his hand.

"We should get some sleep," he said. "Tired policemen always end up giving the criminal a chance to slip away."

* * *

Wallander suggested they go home on foot.

"Let's talk about something totally different," he said. "Something to clear our thoughts."

They walked back to Mariagatan without saying a single word.

CHAPTER 40

Each time he saw his daughter it was as if the ground disappeared beneath his feet. It might take several minutes before he regained his equilibrium. Images from his younger life flickered through his mind. Normally he bore his memories with calm; he checked his pulse and it was always steady no matter how upset he felt. "Like the feathered animal, you should shake hate, lies and anger from your body," God had said to him in a dream. It was only when he met his daughter that he was overcome with weakness. When he saw her face, he also saw the others: Maria and the baby left behind to rot in the jungle marshland that madman Jim Jones had chosen for his paradise. Sometimes he longed passionately for those who had died, and also felt guilty that he hadn't been able to save them. God demanded this sacrifice of me in order to test me, he thought.

He always varied the times and places he met his daughter. Now that he had stepped out of his former state of invisibility and shown himself to her, he made sure in turn that she did not disappear from him. He often tried to surprise her. Once, just after they had been reunited, he washed her car. He sent a letter to her address in Lund when he had wanted her to come to their hide-out behind the church in Lestarp. He had several times visited her flat without her knowledge, using her phone line to make important calls and once spending the night there.

I left her behind once, he thought. Now I have to be the stronger so that she doesn't do the same to me. He had prepared himself for the possibility that she wouldn't want to follow him. Then he would have disappeared again. But already after the first three days he decided he would be able to make her one of the chosen. The fact that convinced him was the unexpected coincidence that she knew the woman whom

Langaas had happened upon and killed in the forest. He had understood then that she had been waiting for his return all these years.

This time he was going to see her in the flat. She had put a flowerpot in the window as a sign that the coast was clear. A few times he had got in with the set of keys she had given him without waiting for the flowerpot because God told him when the coast was clear. He had explained to her that it was important that she act natural in front of her friends. Nothing has happened on the surface, he told her. Your faith grows deep inside you for now, until the day I call it forth from your body.

Each time they met he did something that Jim Jones had taught him – the only lesson that was not spoiled by betrayal and hatred. Jones had taught him how to listen to a person's breath, especially of those who were new and who perhaps had not yet found the proper humility to lay their lives in their leader's hands.

He walked into the flat. She kneeled on the floor of the hall and he laid his hand on her forehead and whispered the words that God demanded he say to her. He reached for a vein in her throat where he could feel her pulse. She trembled, but she was less afraid now. It was starting to become more familiar to her, all these elements of her new life. He kneeled in front of her.

"I am here," he whispered.

"I am here," she replied.

"What does the Lord say?"

"He demands my presence."

He stroked her cheek, then they stood up and walked into the kitchen. She had put out the food he requested: salad, crispbread, two slices of meat. He ate slowly, in silence. When he had finished, she came over with a bowl of water, washed his hands and gave him a cup of tea. He looked at her and asked her if anything had happened since they had last met. He was interested in hearing about her friends, especially the one who had been looking for her.

He sipped the tea and listened to her first words, noticing that she was nervous. He looked at her and smiled.

"What is troubling you?"

"Nothing."

He grabbed her hand and forced two of her fingers into the hot tea. She flinched, but he held her hand there until he was sure she had scalded herself. She started to cry. He let go.

"God demands the truth," he said. "You know I am right when I say that something is troubling you. You have to tell me what it is."

Then she told him what Zeba had said when they were at the café and her little boy played under the table. He noticed that she wasn't sure she was doing the right thing. Her friends were still important to her. That wasn't unusual, in fact he had been surprised at the speed with which he had been able to convert her.

"Telling me about this was the right thing to do," he said when she was finished. "It is also only appropriate that you hesitated in this. Hesitation is a way to prepare to fight for the truth and not take it for granted. Do you understand what I am saying?"

"Yes."

He looked at her for a long time, scrutinising her. She is my daughter, he thought. She gets her seriousness from me.

He stayed a while and told her about his life, wanting to bridge the long years of his absence. He would never be able to convince her to follow him if she did not fully understand that his absence had been decreed by God. It was my time in the desert, he had said repeatedly. I was sent out not for thirty days, but twenty-four years.

When he left her flat he was sure she was going to follow him. And even more significantly, she had given him yet another possibility to punish a sinner.

Langaas was waiting for him at the post office, since they always tried to meet in public. They had a brief conversation, then Langaas leaned forward so that his pulse could be checked. It was normal.

Later that same day they met at the parking place. It was a mild, cloudy evening with rain likely at night. Langaas had replaced the truck with a bus that he had stolen from a company in Malmö, being careful to put on new plates. They drove east, passing Ystad and continuing on minor roads towards Klavestrand, where they stopped at the church. It stood on a hill, some 400 metres from the nearest house. No-one would notice the bus where it was parked. Langaas unlocked the church door with the key that he had copied. They used shielded torches as they erected the ladders and covered with large black plastic rubbish bags the windows facing towards the road. Afterwards they lit the candles on the altar. Their footsteps made no sound; all was silent.

Langaas came to him in the vestry where he was making his preparations.

"Everything is ready."

"Tonight I will let them wait," Erik said.

He gave the remaining hawser to Langaas.

"Put this on the altar. The hawser inspires fear, fear inspires faith."

Langaas left him alone. Erik sat down at the pastor's table with a candle in front of him. When he closed his eyes he was back in the jungle. Jim Jones came walking out of his hut, the only one that was supplied with electricity from a small generator. Jim was always so well groomed. His teeth were white, his smile carved into his face. Jim was beautiful, he thought, even if he was a fallen angel. I cannot deny that there were moments with him when I was completely happy. I also cannot deny that what Jim gave me, or what I believed he gave me, is what I am trying to give the people who now follow me. I have seen the fallen angel; I know what to do.

He folded his arms and let his head come to rest on them. He was going to let them wait for him. The hawser on the altar would be a stimulus of the fear they should feel for him. If the ways of God were inscrutable, so too would be the ways of His servant. He knew Langaas would not disturb him again. He started to dream. It was like stepping down into the underworld, a world where the heat of the jungle penetrated the cold stone walls of the church. He thought about Maria and the child. He slept.

He woke with a start at 4 a.m. At first he wasn't sure where he was. He stood up and shook life back into his stiff body. After a few minutes he walked out into the church. They were sitting, all of them, in the first few pews, frozen, fearful, waiting. He stopped and looked at them before letting them see him. I could kill them all, he thought. I could get them to cut off their hands and eat themselves. Because I too have a weakness. I do not completely trust my followers. I am afraid of the thoughts they think, thoughts I cannot control.

He walked out and stood in front of the altar. This night he was going to tell them about the great task that awaited them, the reason they had made the long journey to Sweden. Tonight he would pronounce the first words of the text that would become the fifth gospel.

He nodded to Langaas, who opened the old-fashioned brown trunk on the floor next to the altar. Langaas walked down the row of people, handing out the death masks. They were white, like masks in a pantomime, devoid of expression, of joy or sorrow.

God has made man in His image, Erik thought. But no-one knows the face of God. Our lives are His breath, but no-one knows His face. We have to wear the white mask in order to obliterate the ego and become one with our Creator.

He watched while they put on the death masks. It filled him with a sense of power and strength to see them cover their faces.

Finally Langaas put on a mask. The only one not wearing one was Erik.

This too he had learned from Jim. The disciples always have to know where to find their Master. He is the only one who should not be masked.

He pressed his right thumb against his left wrist. His pulse was normal. Everything was under control. In the future this church may become a shrine, he thought. The first Christians who died in the catacombs of Rome have returned. The time of the fallen angels is finally over.

The day he had chosen was September 8. This had come to him in a dream. He had found himself in a deserted factory with puddles of rainwater and dead leaves on the floor. There had been a calendar on the wall. When he woke up he remembered that the date in the dream was September 8. That is the day everything ends and everything begins again.

He took a step towards them and started to speak.

"The time has come. I had not intended that we should meet before the day that you undertake your great task, but God has spoken to me tonight and told me that yet another sacrifice is necessary. When we meet again another sinner will die."

He picked up the hawser and held it above his head.

"We know what God demands of us," he intoned. "The old scriptures teach us the law of an eye for an eye and a tooth for a tooth. He who kills must himself be killed. We must remove all doubt from our minds. God's breath is steel and he demands hardness from us in return. We are like the snake who wakes from his winter sleep, we are the lizards who live in the crevices of the rock and change colour when threatened. Only through complete devotion and ruthlessness will we conquer the emptiness that exists inside men. The great darkness, the long days of degeneration and impotence are over."

He paused and saw that they understood. He walked along the row and stroked their foreheads, then gave the sign that they were to stand. Together they said the holy words that he told them had come to him in a vision. They did not have to know the truth, that it was something he had read when he was a young man. Or had the words in fact come to him in a dream? He could not be sure, but it was of no importance.

And in our redemption we are lifted high on wings of might
To join Him in His power and shine with His holy light.

* * *

Later they left the church. Langaas locked up and they drove away in the bus. A woman who came in to clean in the afternoon did not notice that anyone had been there at all.

PART 4

The Thirteenth Tower

CHAPTER 41

The telephone woke her. She checked the time: 5.45 a.m.

There were noises coming from the bathroom. Her father was already up, but hadn't heard the phone. Linda ran out into the kitchen and answered.

"May I please speak to Inspector Wallander?" a woman's voice said.

"Who is speaking?"

"May I please speak to him?"

The woman spoke in a cultured way. Not a cleaning lady at the station, Linda thought.

"He's busy right now. Who may I say is calling?"

"Anita Tademan from Rannesholm Manor."

"We've met, actually. I'm his daughter."

Anita Tademan ignored her. "When will I be able to speak to him?"

"As soon as he gets out of the bathroom."

"It's very important."

Linda wrote down the number and put some water on for coffee. The kettle had just started to boil when Wallander appeared in the kitchen. He was so wrapped up in his own thoughts that it did not even strike him as strange to see her up so early.

"Anita Tademan just called," Linda said. "She said it was important."

Wallander looked at his watch. "It must be, at this hour."

She dialled the number for him and held out the phone.

While he was speaking with fru Tademan, Linda looked through the cupboards and discovered that there were no more coffee beans.

Wallander hung up. Linda had heard him agree to a time.

"What did she want?"

"For me to come and talk to her."

"What about?"

"To tell me something she heard from a distant relative who lives in a house on the Rannesholm grounds. She didn't want to elaborate on the phone, and insisted I come up to the manor. I'm sure she thinks she's too important to come down to the station like a regular person. But that's when I put my foot down. Maybe you heard that part?"

"No – why?"

Wallander muttered something unintelligible and started to rifle through the cupboard.

"It's all gone," Linda said.

"Do I have to be the only one around here who takes responsibility for keeping coffee in the house?"

That immediately infuriated her.

"You can't know how incredibly relieved I will be to move out. I should never have come back here."

He threw out his arms in apology. "That might have been best, but we don't have time to argue about it now. Parents and children shouldn't live on top of each other."

They drank tea and leafed through their respective parts of the previous day's paper. Neither one could concentrate on what they were reading.

"I want you to come along," he said. "Get dressed."

Linda showered and dressed as quickly as she could. But when she was ready he had already left. He had jotted something down in the margin of the newspaper. She took that as a sign that he was in a hurry. He's as impatient as I am, she thought.

She looked out of the window. The thermometer said it was 22°C – still summer. It was raining. She half ran, half walked to the station. It was as if she were hurrying to school, with the same anxiety about being on time.

Wallander was talking on the phone when she came in. She sat in the chair across from his desk until he put the phone down and stood up.

"Come with me."

They walked into Lindman's office. Höglund was leaning against the wall, a mug of coffee in her hand. For once she acknowledged Linda's presence. Someone's mentioned it to her, Linda thought. Hardly my father. Maybe Lindman.

"Where is Martinsson?" Höglund asked.

"He just called," Wallander said. "He has a sick child on his hands, so he'll be in a little later. But he was going to make some calls from home and find out more about this Sylvi Rasmussen."

"Who?" Höglund asked.

"Why are we all crowding around in here anyway?" Wallander said. "Let's go to the conference room. Does anybody know where Nyberg is?"

"He's still working on the two fires."

"What does he think he's going to find there?"

The last comment came from Höglund. Linda sensed that she was one of those who looked forward to his retirement.

They discussed the case for three hours and ten minutes, until someone knocked on the door and said that an Anita Tademan had arrived to speak to Inspector Wallander. Linda wondered if the discussion had really come to its natural conclusion, but no-one made any objection when her father stood up. He stopped beside her chair on his way out.

"Anna," he said. "Keep talking to her, keep listening to her."

"I don't know what we should talk about. She's going to see through me, that I'm keeping an eye on her."

"Just be natural."

"Shouldn't you talk to her again yourself?"

"Yes, but not just yet."

Linda left the station. The rain had turned into a thin drizzle. A car honked its horn, so close to her that she jumped. It was Lindman. He pulled over and opened the door.

"Jump in. I'll take you home."

"Thanks."

There was music on. Jazz.

"Do you like this stuff?" she said.

"Yes. A lot, actually."

"Jazz?"

"Lars Gullin. A sax player, one of Sweden's best jazz musicians ever. He died much too young."

"I've never heard of him, but I don't like this kind of music."

"In my car I play what I like."

He seemed stung, and Linda instantly regretted what she'd said. Unfortunately, she thought, one of the many things I've inherited from my father is this ability to make thoughtless, hurtful comments.

"Where are you going?" she said.

His answer was curt. "Sjöbo. To see a locksmith."

"Is it going to take long?"

"I don't think so. Why?"

"Maybe I could come along. If you'll have me."

"Only if you can put up with the music."

"From now on I love jazz."

The tension was broken. Lindman laughed and drove north. He drove fast. Linda had the urge to touch him, to stroke her fingers over his shoulder or his cheek. She felt more desire than she could remember feeling in a long time. She had a silly thought, that they should check into a hotel in Sjöbo. Not that there was a hotel in Sjöbo. She tried to shake off the thought, but it stayed with her. Rain splattered the windscreen. The saxophone poured out some high, insistent, quick notes. Linda tried to pick out the melody line, without success.

"If you're talking to a locksmith in Sjöbo it must have something to do with the investigation. One of them. How many are there exactly?"

"Medberg is one. Bolson is another, the burned animals, and the two church fires. Your father wants them all treated under the rubric of one investigation, and the prosecutor has agreed. At least for now."

"And the locksmith?"

"His name is Håkan Holmberg. He's not your run-of-the-mill locksmith; he makes copies of very old keys. When he heard that the police were wondering how the arsonists broke into the churches he remembered that he made two keys a few months ago that might very well have been old church keys. I'm on my way to see if he remembers anything else. His workshop is in the centre of Sjöbo. Martinsson had heard of him before. He's won prizes for his craftmanship. He's also studied philosophy and teaches in the summer."

"In his workshop?"

"In another part of the farmstead. Martinsson has thought about signing on sometime. The students work in the smithy half the time and explore philosophical issues the rest of the time."

"Not something for me," Linda said.

"What about your father?"

"Even less so."

They arrived in Sjöbo and stopped outside a red-brick house with a giant iron key hanging outside the door.

"Maybe I shouldn't go in with you."

"If I understood matters correctly, you've started work now."

They walked in. It was very hot. A man working at the forge nodded at them, then took out a piece of glowing iron and started hammering it.

"I need to finish this key," he said. "You can't interrupt this kind of work once you've started. It lets a kind of hesitation into the iron. That happens and the key will never sit well in its lock."

They watched him with fascination. At last the key lay finished on the anvil. Holmberg wiped the sweat from his brow and washed his hands. They followed him into a courtyard with tables and chairs. A coffee pot and some cups had been put out. They shook hands. Linda felt foolishly flattered by Lindman's introduction of her as "a colleague". Holmberg put on an old straw hat and served the coffee. He noticed Linda looking at it.

"One of the few crimes I've ever committed," he said. "I take a trip overseas every year. A few years ago I was in Lombardy. One afternoon I was somewhere close to Mantua, where I had spent a few days in honour of the great Virgil, who was born there. I caught sight of a scarecrow out in a field. I don't know what crop he was supposed to be protecting. I stopped and thought that for the first time in my life I wanted to commit a crime, become a dishonest blacksmith, in a word. So I snuck out on to the field and stole his hat. Sometimes in my dreams it isn't a scarecrow at all but a living person. He must have realised I was a harmless coward who would never steal from anyone, that's why he let me take the hat. Perhaps he was the remains of a Franciscan monk hoping to do one last good deed on this earth. In any case, it was an overwhelming, tumultuous experience for me, to commit this crime."

Linda glanced at Lindman and wondered if he knew who Virgil was. And Mantua? Where was that? It had to be Italy, but she had no idea if it was a region or a city. Zeba would have known; she could sit for hours over her maps and books.

"Tell me about the keys," Lindman said.

Holmberg rocked back in his chair and fished a pipe out of the breast pocket of his overalls.

"It happened by accident, in a way," he said, after lighting the pipe. "I don't watch or read any kind of news in the summer – it's a way to rest my mind. But one of my customers came to pick up a key. It was the key of an old seaman's chest that had once belonged to a British admiral's ship in the eighteenth century. He told me about the fires and the police's suspicion about copied keys. I recalled that I had made two keys a few months ago that looked like church keys. I'm not saying it was definitely the case, but I suspected it was."

"Why?"

"Experience. Church keys often look a certain way. And there aren't very many other doors these days that still use the locks and keys of the old masters. I decided to call the police."

"Who ordered these keys?"

"He said his name was Lukas."

"Lukas . . ."

"Herr Lukas. An uncommonly well-mannered sort. He was in a hurry and made a generous deposit."

Lindman took a packet out of his pocket, which he unwrapped. Holmberg immediately recognised the contents.

"Those are the keys I made copies of."

He stood up and walked into the smithy.

"This could be something," Lindman said. "A strange old man. But his memory seems good."

Holmberg returned with an old-fashioned ledger in his hand, turning the pages until he found the right one.

"It was June the twelfth. Herr Lukas left two keys. He wanted the copies made by the twenty-fifth at the latest. That didn't leave me very long since I had a lot to do, but he paid well and even I need money, to keep up the forge and take my holiday."

"What address did he give you?"

"No address."

"Telephone number?"

Holmberg turned the ledger around so that Lindman could see. He dialled the number on his mobile phone, listened, then turned it off again.

"That was a florist in Bjärred," he said. "I think we can safely assume that herr Lukas doesn't have anything to do with them. What happened after that?"

Holmberg flipped forward a few pages.

"He fetched the keys on June the twenty-fifth. That was all."

"How did he pay?"

"Cash."

"Did you write a receipt for him?"

"No. I rely on my bookkeeping. I take great pains to pay my share of taxes, though this kind of case is ideal for the less scrupulous."

"How would you describe Mr Lukas?"

"Tall, light hair, maybe losing a little of it in front. Courteous, polite. When he first came in, he was wearing a suit. Same when he picked up the keys, though not the same one that time."

"How did he get here?"

"I can't see the road from the workshop, but I assume he drove a car."

Linda saw Lindman gather himself for the next question, intuitively sensing what it must be.

"Can you describe the way he spoke?"

"He had an accent."

"What kind of accent?"

"Something Scandinavian. Not Finnish, nor Icelandic. That would leave Danish or Norwegian."

"Do you have anything else to say about him?"

"Not that I can think of."

"Did he say that these keys were for church doors?"

"He said they were for some kind of storage facility, in an old manor house, come to think of it."

"Which manor?"

Holmberg knocked some ash out of his pipe and wrinkled his forehead.

"He may have told me the name, but I've forgotten."

They waited. Holmberg shook his head.

"Could it have been Rannesholm?" Linda asked.

The question simply sprang out of her, like the last time.

"Right," Holmberg said. "That was it. Rannesholm. An old brewery at Rannesholm."

Lindman got up, as if he was suddenly in a hurry. He finished the rest of his coffee.

"Thank you," he said. "This has been valuable."

"Working with keys is always meaningful," Holmberg said, and smiled. "Locking and opening is, in a sense, man's very purpose on this earth. Key rings rattle throughout history. Each key, each lock has its tale. And now I have yet another to tell."

He followed them out.

"Who was Virgil?" Linda said.

"Dante's guide," he answered. "And a great poet."

He raised the old straw hat, which was starting to come apart, and went back inside. They got into the car.

"So often you meet fearful, angry, shaken people," Lindman said. "But sometimes there are moments of light. Like this man. I'm filing him away in my archive of interesting people I'll remember when I'm old."

They left Sjöbo. Linda saw a sign for a hotel and giggled. He looked at her, but didn't ask anything. His mobile rang. He answered, listened, hung up and increased his speed.

"Your father has finished talking to fru Tademan," he said. "Apparently something important has come to light."

"Better not tell him that I was with you today," she said. "He had something other in mind for me today."

"What was that?"

"Talking to Anna," she said.

"Maybe you'll have time for both."

Lindman let her off in the centre of town. When she made it to Anna's flat and was greeted by her at the door, she immediately realised something was wrong. Anna had tears in her eyes.

"Zeba is gone," she said. "Her boy was screaming so loudly that the neighbours were worried. He was home alone. And Zeba was gone."

Linda held her breath. Fear overwhelmed her like a violent pain. Now she knew she was close to a terrible truth she should already have grasped.

She looked into Anna's eyes and there she saw only her own fear.

CHAPTER 42

The situation was at once both crystal clear and confusing. Linda knew Zeba would never have abandoned her son of her own free will, or forgotten about him. Clearly something had happened. But what? It was something she felt she should know, something that was almost within her grasp and yet eluded her. The big picture. Her father always talked about looking for the way events came together. But she saw nothing.

Since Anna seemed even more confused than she did, Linda forced her to sit down in the kitchen and talk. Anna spoke in unconnected fragments, but it did not take Linda more than a few minutes to piece together what had happened.

Zeba's neighbour, a woman who often watched the boy for her, had heard him crying through the thin walls. Since he cried for an unusually long time without Zeba seeming to intervene, she went round and rang the bell. When there was no answer she let herself in with the key Zeba had given her and found the boy alone. He stopped crying when he saw her.

This neighbour, whose name was Aina Rosberg, had noticed nothing unusual about the flat. It was messy as usual, but there were no signs of commotion. That was the phrase she had used, "no signs of commotion". Fru Rosberg had called one of Zeba's cousins, Titchka, who wasn't home, and then Anna. That was what Zeba had instructed her to do if anything ever happened: first call Titchka, then Anna.

"How long ago did this happen?" Linda asked.

"Two hours ago."

"Has fru Rosberg called again?"

"I called her back. But Zeba still hadn't returned."

Linda thought for a moment. Most of all she wanted to talk to her father, but she also knew what he would say. Two hours was not a long time, there was no doubt a natural explanation for Zeba's absence.

But what could it possibly be?

"Let's go over to her flat," Linda said. "I want to take a look at it."

Anna made no objections. Ten minutes later fru Rosberg let them in.

"Where can she be?" the neighbour said. "This isn't like her. Nobody would leave such a young child alone, least of all her. What would have happened if I hadn't heard him crying?"

"I'm sure she'll be back soon," Linda said. "But it would be best if the boy could stay with you until then."

"Of course he can," fru Rosberg said, and left to go back to her flat.

When Linda walked into Zeba's flat she picked up a strange smell. Her heart grew cold with fear; she knew something serious had happened. Zeba had not left of her own free will.

"Can you smell that?" she asked.

Anna shook her head.

"That sharp smell. Like vinegar."

"I don't smell anything."

Linda sat in the kitchen, Anna in the living room. Linda could see her through the open door. Anna was nervously pinching herself in the arm. Linda tried to think clearly. She walked over to the window and looked out. She tried to imagine Zeba walking on to the street. Which way had she gone? To the left or to the right? Had she been alone? Linda looked at the little tobacconist's shop across the street. A tall, heavily built man was standing in the doorway, smoking. When a customer came by he walked back in, then resumed his station at the doorway. Linda thought he was worth a try.

Anna still sat on the sofa, lost in thought. Linda patted her on the arm.

"I'm sure she'll turn up," she said. "Probably nothing has happened. I'm going down to the tobacconist's. I'll be back soon."

* * *

There was a sign welcoming customers to "Yassar's shop". Linda bought some chewing gum.

"Do you know Zeba?" she asked. "She lives on the other side of the street."

"Zeba? Sure. I give her little one sweets when they come in."

"Have you seen her today?"

His answer came without hesitation.

"A few hours ago. I was putting up one of the flags that had come down outside. I don't understand how a flag can fall down when there is no wind . . ."

"Was anyone with her?" Linda interrupted.

"She was with a man."

Linda's heart beat faster. "Have you seen him before?"

Yassar looked anxious. Instead of answering her question, he started asking his own.

"Why do you want to know? Who are you?"

"You must have seen me before. I'm a friend of Zeba's."

"Why are you asking all these questions?"

"I need to know."

"Has anything happened?"

"No. Have you ever seen the man before?"

"No. He had a small grey car, he was tall, and later I thought about how strange it was that Zeba was leaning on him."

"How do you mean, 'leaning on him'?"

"Just that. She was leaning, clinging. As if she needed support."

"Can you describe the man?"

"He was tall. That's about it. He had a hat on, a long coat."

"A hat?"

"A grey hat. Or blue. A long grey coat. Or blue. Everything about him was either blue or grey."

"Did you see the number plate?"

"No."

"What about the make of the car?"

"I don't know. Why are you asking all these questions? You come into my shop and make me as worried as if you were a police officer."

"I am a police officer," Linda said, and left.

When she came back to the flat, Anna was sitting where she had left her on the sofa. Linda had the same feeling that there was something she should be seeing, realising, seeing through, although she didn't know what it was. She sat down next to Anna.

"You have to go back to your place, in case Zeba calls. I'm going down to the police station to talk to my father. You can drop me off there."

Anna grabbed Linda's arm so roughly that Linda jumped. Then, just as abruptly, she let go. It was a strange reaction. Perhaps not the action itself, but the intensity of it.

When Linda walked into the reception, someone called out to her that her father was at the prosecutor's office, on the other side. She went over. The outer door was locked, but an assistant, recognising her, let her in.

"Are you looking for your father? He's in the small conference room."

She pointed down a corridor. A red light was on outside one of the rooms. Linda sat down and waited.

After ten minutes Ann-Britt Höglund came out, saw her and looked surprised. Then she turned back to the room.

"You have an important visitor," she said, and went her way.

Wallander came out with the very young-looking prosecutor. He introduced Linda and the young lawyer left. Linda pulled him down into a chair and told him what had happened, not even trying to be systematic about the order of her telling. Wallander was quiet for a long time after she finished. Then he asked a few questions, primarily about Yassar's observations. He returned several times to the issue of Zeba "leaning" on the man.

"Is Zeba the touchy-feely kind?"

"No, I'd say the opposite, actually. It's normally the man who is all over her. She's tough and avoids showing any weakness, although she has several."

"If she was being taken away against her will, why didn't she cry out?"

Linda shook her head. Wallander answered his own question as he stood up.

"Maybe she wasn't able to."

"And she wasn't leaning on the man? She was drugged and would have fallen down if he hadn't held her up?"

"That's exactly what I'm thinking."

He walked quickly to his office. Linda had difficulty keeping up. On the way Wallander knocked on Lindman's door and pushed it open. The office was empty. Martinsson walked past, carrying a large teddy bear.

"What the hell is that?" Wallander asked irritably.

"It was made in Taiwan. There's a large package of amphetamines inside."

"Get someone else to take care of it."

"I was about to hand it over to Svartman," Martinsson said, not hiding his own irritation.

"Try to round everyone up. I want a meeting in half an hour."

Martinsson left.

Wallander sat down behind his desk, then leaned towards Linda.

"You didn't ask Yassar if he heard the man say anything."

"I forgot."

Wallander handed her the phone.

"Call him."

"I don't know what his number is."

Wallander dialled enquiries for her. Linda asked to be put through. Yassar answered. He hadn't heard the man say anything.

"I'm starting to worry," Yassar said. "What has happened?"

"Nothing that we know of," Linda said. "Thanks for your help."

She put the phone down.

"He didn't hear anything."

Her father rocked back and forth on his chair and looked at his hands. She heard voices come and go in the corridor.

"I don't like it," he said finally. "Her neighbour is right. No-one leaves such a young child alone."

"I keep having the feeling that I'm overlooking something," Linda said. "Something I should see, something that's staring me in the face. There's a connection – the kind you're always talking about – but I can't think what it is."

He looked at her keenly.

"As if part of you already knows what's happened? And why?"

She shook her head.

"It's more as if I've kind of been waiting for this to happen. And as if Zeba isn't the one who's disappeared, but Anna. A second time."

He looked at her for a long time.

"Can you explain what you mean?"

"No."

"We'll give Zeba a few more hours," he said. "If she's not back by three o'clock we'll have to do something. I want you to stay here."

Linda followed him to the conference room. When everyone was gathered and the door closed, Wallander began by telling them about Zeba's disappearance. The tension in the room mounted.

"Too many people are disappearing," Wallander said. "Disappearing, reappearing, disappearing again. By coincidence or because of factors as yet unknown, all this seems to involve my daughter, a fact that makes me like this even less."

He tapped a pencil on the tabletop and continued.

"I talked to fru Tademan. She is not what you would call a particularly pleasant woman. In fact she's about as good an example of an arrogant, conceited Scanian aristocrat as I've ever had the misfortune to meet. But she did the right thing in getting in touch with us. A distant cousin who lives on the Rannesholm Estate saw a band of people near the edge of the forest. There were at least twenty of them and they vanished as soon as they were seen. They could have been a group of tourists, but the fact that they were apparently so anxious not to attract attention means they could also have been something else."

"Such as?" Höglund said.

"We don't know. But we found a hut in the forest there, you will bear in mind, where a woman was murdered."

"That hut could hardly house twenty people or more."

"I know. Nonetheless, this is important information. We have suspected that there were several people involved in the Frennestad

Church fire and murder. Perhaps what we have here is an indication that there are even more."

"Are we dealing with a kind of gang?" Martinsson said.

"Possibly a sect," Lindman said.

"Or both," Wallander said. "That's something we don't yet know. This piece of information may turn out to lead us in the wrong direction, but we're not drawing any conclusions. Not yet, not even provisional ones. Let us leave fru Tademan's information to one side for the moment."

Lindman reported on his meeting with Håkan Holmberg and his keys. He didn't mention the fact that Linda had been with him.

"The man with an accent," Wallander mused. "Our Norwegian or Norwegian-Danish link. He turns up again. I believe we can safely accept Mr Holmberg's assurance that the keys he copied were those to both Hurup and Frennestad Churches. A Norweigian orders copies of some church keys. An American woman is later strangled in the church. By whom and why? That's what we need to find out." He turned to Höglund. "What do our Danish colleagues say about Frans Vigsten?"

"He's a piano teacher. He was a rehearsal pianist at Det Kongelige Theatre and apparently very much admired as such. Now he's increasingly senile and has trouble taking care of himself. But no-one has any information indicating that anyone else lives in the flat, least of all Vigsten himself."

Wallander threw a hasty glance at Linda before continuing.

"Let's stay in Denmark for a moment. What about this woman Sylvi Rasmussen? What do we have on her?"

Martinsson rifled through his papers.

"Her original name was something else. She came to Denmark as a refugee after the collapse of Eastern Europe. Drug addict, homeless, the same old story leading to prostitution. She was well liked by clients and friends. No-one has anything bad to say about her. There was nothing else unusual about her life, not even the sheer predictable tragedy of it." Martinsson looked through the papers again before putting them down. "No-one knows who her final client was, but he must be the murderer."

"Any indication of who it was?"

"There are the prints of twelve different people in her flat. They're being examined and the Danes will let us know what they find."

Linda noticed that Wallander was picking up the pace of the meeting. He tried to interpret the information that was brought in, never receiving it passively, always looking for the underlying message.

Finally he opened the meeting for general discussion. Linda was the only one who didn't say anything. After half an hour they took a break. Everyone left to stretch their legs or to get some coffee, except Linda, who was assigned to guard the windows.

A gust of wind blew some of Martinsson's papers onto the floor. Linda gathered them up and saw a picture of Sylvi Rasmussen. She studied her face, seeing fear in her eyes. She shivered when she thought of her life and fate.

She was about to put the papers back when a detail caught her eye. The pathologist's report stated that Sylvi Rasmussen had had two or three abortions. Linda stared at the paper. She thought of the Danish sailors who had been sitting in the corner, the boy playing on the floor and Zeba telling them about her abortion. She also thought about Anna's unexpected reaction. Linda froze, holding her breath and Sylvi Rasmussen's photograph.

Wallander came back into the room.

"I think I get it," she said.

"Get what?"

"I have one question. That woman from Tulsa."

"What about her?"

Linda shook her head and pointed to the door.

"Close it."

"We're in the middle of a meeting."

"I can't concentrate if everyone comes back in. But I think I'm on to something important."

He saw that she was in deadly earnest, and went to close the door.

CHAPTER 43

Wallander put his head out of the door and told someone that the end of the meeting would be postponed a little while. Someone started to protest, but he shut the door.

They sat across from each other.

"What did you want to ask?"

"Did the Bolson woman ever have an abortion? Did Medberg? If I'm correct, the answer will certainly be yes for Bolson, but for Medberg, most probably no."

Wallander frowned, at first perplexed, then simply uncomprehending. He pulled his stack of papers over and started looking through them with growing impatience. He tossed the file to the side.

"Nothing about an abortion."

"Are all the facts there?"

"Of course not. A full description of a person's life, however uneventful or uninteresting, fills a much larger folder than this. Harriet Bolson does not seem to have had an exciting life, and certainly there's nothing as dramatic as an abortion in the material we received from the force in Tulsa."

"And Medberg?"

"I don't know, but that information should be easier to come by. All we have to do is talk to her unpleasant daughter – although perhaps it's not the kind of thing mothers tell their children. I don't think Mona ever had an abortion. Do you know?"

"No."

"Does that mean that you don't know if she did or that she never had one?"

"Mum never had an abortion. I would know."

"I don't understand what you're getting at. Why is this important?"

Linda tried to clear her head. She could be wrong, but every instinct told her she was right.

"Can you find out about the abortions?"

"I'll do it when you've told me why it's important."

Something inside her burst. Tears started to run down her face and she banged her fists into the table. She hated crying in front of her father. Not just in front of him, in front of anybody. The only person she had ever been able to cry in front of was her grandfather.

"I'll ask them to do it," Wallander said, and stood up. "But I expect you to tell me what this is all about when I get back. People have been murdered, Linda. This isn't an exercise at the training college."

Linda grabbed an ashtray from the table and threw it at him, hitting him right above the eyebrow. Blood ran down his face and dripped on Harriet Bolson's file.

"I didn't mean to do that."

Wallander pressed a fistful of napkins against the gash.

"I just can't stand it when you needle me," she said.

He left the room. Linda picked up the ashtray from the floor, still trembling with agitation. She knew he was furious with her. Neither of them could stand to be humiliated. But she didn't feel any regret.

He came back after 15 minutes with a makeshift bandage over his wound and dried blood still smeared across his cheek. Linda expected him to yell at her, but he simply sat down in his chair.

"Does it hurt a lot?" she asked.

He ignored her question.

"Höglund called fru Jorner, Medberg's daughter. She found the question deeply insulting and threatened to call the evening papers and complain, but Höglund did establish that she has no knowledge of any abortion."

"That's what I thought," Linda said. "And what about the other one? The one from Tulsa?"

"Höglund is contacting the US," he said. "We're not entirely sure about the time difference, but she's sending a fax, which will no doubt be waiting for the day shift whenever it starts."

Wallander felt the bandage with his fingertips.

"Your turn," he said.

Linda started speaking slowly to keep her voice from wobbling, but also so she wouldn't skip anything.

"There are five women," she said. "Three of them are dead, one of them has disappeared, and the last one disappeared and then returned. I'm starting to see a connection between them, apart from Medberg, who we're assuming was killed because she found herself in the wrong place at the wrong time. But what about the rest? Sylvi Rasmussen was murdered; she had also had two or three abortions. Let's assume that information from Tulsa will confirm that Bolson had had an abortion. It's true, too, for the person who's just gone missing: Zeba. She told me only a few days ago that she had one. I think this may be the connection between these women."

Linda paused and drank some water. Wallander tapped his fingers and stared at the wall.

"I still don't get it," he said.

"I'm not finished yet. Zeba didn't just tell me about her abortion, she told Anna as well. And Anna had the strangest reaction. She was upset by it in a way I couldn't relate to, nor could Zeba. To say that Anna strongly disapproves of women who have had abortions would be an understatement. She walked out on us. And when Anna later discovered that Zeba was missing she clung to my arm and cried. But it was as if she wasn't so much afraid for Zeba as for herself."

Linda stopped. Her father was still fingering his bandage.

"What do you mean, she was afraid for herself?"

"I'm not sure I know."

"Try."

"I'm telling you all I can."

Wallander gazed absently at the wall. Linda knew that staring at a blank surface was a sign of intense concentration on his part.

"I want you to tell the others," he said.

"I can't."

"Why not?"

"I'll get nervous. I might be wrong. Maybe that woman from Tulsa never had an abortion."

"You have an hour to prepare," Wallander said, and stood up. "I'll tell them."

He walked out and closed the door. Linda had the feeling that she was imprisoned, not with a lock and key, but by the imposed time limit. She decided to write down what she was going to say in a notebook, and pulled one across the table towards her. When she flipped it open she was confronted with a bad sketch of a seductively posed naked woman. To her surprise she saw it was Martinsson's notepad. But why should that surprise me? she thought. All the men I know spend an enormous amount of mental energy undressing women in their minds.

She reached for an unused notepad beside the overhead projector and jotted down the five women's names.

After 45 minutes the door opened and everyone marched in like a delegation, led by her father. He waved a piece of paper in front of her.

"Harriet Bolson had two abortions."

He sat down, as did everyone else.

"The question, of course, is why this matters to our investigation. That's what we're here to discuss. Linda is going to present us with her ideas. Over to you, Linda."

Linda drew a deep breath and managed to present her theory without faltering. Wallander took over when she finished.

"I think that Linda is on to something that may be very important. The terrain is still far from mapped, but there is enough substance here to merit our attention, more substance, in fact, than we have managed to uncover thus far in other facets of the investigation."

The door opened and Lisa Holgersson slipped in. Wallander put his papers down and lifted his hands as if he were about to conduct an orchestra.

"I think we can glimpse the outline of something that we do not yet understand but is there nonetheless."

He stood up and pulled over a flip chart set on an easel, with the legend HIGHER WAGES DAMMIT scrawled across it. Chuckles broke out across the room. Even Holgersson laughed. Wallander turned to a clean sheet.

"As usual I ask that you hold your thoughts until I'm done," he said. "Save the rotten tomatoes and catcalls."

"Looks like your daughter's already been taking potshots," Martinsson said. "Blood is seeping through the bandage. You look like the old Döbeln at Jutas, to use a literary analogy."

"Who's that?" Lindman said.

"A man who stood guard over a bridge in Finland," Martinsson said. "Didn't they teach you anything when you were in school?"

"We had to read that when I was a girl, but you're getting them confused," Högland said. "The man standing guard had a different name. It's a book by some Russian author."

"No, Finnish," Linda heard herself say. "Sibelius, isn't it?"

"For the love of God," Wallander said.

"I'll call my brother Albin," Martinsson said, standing up. "We have to get to the bottom of this."

He left the room.

"I don't think it was Sibelius," Holgersson said after a moment. "He was a composer. But something similar."

Martinsson returned after a few minutes of silence.

"Topelius," he said. "Or possibly Runeberg. And Döbeln did have a large bandage, I was right about that."

"He didn't guard the bridge, though, did he?" Höglund muttered.

"I'm trying to create an overview here," Wallander interrupted them, and proceeded one by one to touch on all the known facts of the case.

After the rather lengthy overview he sat down.

"There's one thing we've neglected: why haven't we brought in the estate agent in Skurup, the one who sold the house in Lestarp, to listen to the burning-swans tape? We need to take care of that as soon as possible."

Martinsson got up again and left the room. Lindman opened a window.

"Have we talked to Norway about Torgeir Langaas?" Holgersson asked.

Wallander looked at Höglund.

"No word yet," she said.

Wallander looked down at his watch in a way that indicated the meeting was drawing to a close.

"It's too early to arrive at any definitive conclusions," he said. "It's too early and yet we have to work with two assumptions. Either that all this hangs together. Or that it doesn't. And yet the first alternative is compelling. What do we have? Sacrifices, fires and ritual murder, a Bible in which someone has changed the text. It's easy for us to see this as the work of a madman, but maybe that isn't the case. Maybe we're dealing with a group of very deliberate, methodical people, with a twisted and ruthless agenda. We need to work quickly. There's a gradual increase in tempo in these events, an acceleration. We have to find Zeba, and to talk to Anna Westin again."

He turned to Linda.

"I thought you could bring her in. We're going to have a friendly but necessary conversation. We're simply worried about Zeba, that's all you have to say."

"Who's taking care of her son?" Höglund asked Linda directly, without the superior air she normally adopted.

"Zeba's neighbour."

Wallander hit the table with the flat of his hand, marking the end of the meeting.

"Torgeir Langaas," he said as everyone stood. "Lean on our Norwegian colleagues. The rest of us will look for Zeba."

Linda and her father went to get a cup of coffee, without exchanging a word. Then they went to his office. Martinsson knocked on the door half an hour later, coming in before Wallander answered. He stopped when he saw Linda.

"Sorry," he said.

"What is it?"

"Ture Magnusson is here to listen to the tape."

Wallander jumped out of his chair, grabbing Linda by the arm and pulling her along. Ture Magnusson seemed nervous. Martinsson went to fetch the tape. Wallander received a call from Nyberg and immediately launched into an argument with him, so Linda was left to take care of Magnusson.

"Have you found the Norwegian?"

"Not yet."

"I'm not sure I will be able to recognise his voice."

"We'll just hope for the best."

Wallander hung up. At the same time Martinsson came back with a worried look on his face.

"The tape must still be here," he said. "It's not in the archive."

"Didn't anyone put it back?" Wallander asked crossly.

"Not me," Martinsson said.

He looked through the shelf behind the tape recorder. Wallander stuck his head into the call centre.

"Can we get a little help here?" he shouted. "We're missing a tape!"

Höglund joined them, but no-one could find the tape. Linda watched her father get increasingly red in the face. But in the end it was Martinsson who exploded.

"How the hell are we supposed to do our work when archived tape can go missing like this?"

He picked up a booklet of instructions for the tape recorder and threw it against the wall. They kept looking for the tape. Linda finally had the feeling that the whole police district was looking for it, but it didn't turn up. She looked at her father. He seemed tired and despondent, but she knew it would pass.

"We owe you an apology," Wallander said to Magnusson, "for bringing you down here. The tape appears to be misplaced. There's nothing for you to do."

"I have a suggestion," Linda said.

She had been debating with herself whether to suggest this.

"I believe I can imitate the voice," she said. "He's a man, I know, but I'd like a shot at it."

Höglund gave her a disapproving look. "What makes you think you could possibly imitate his voice?"

Linda could have given her a long answer, about how she had discovered a talent for imitation at parties. How her friends had been impressed, and she had assumed it was a one-off success, but how she soon realised she simply had a knack for it. There were voices she couldn't imitate at all, but most of the time she was spot on.

"It's not as if we have anything to lose," she said.

Lindman had come back into the room. He nodded encouragingly.

"I guess since we're all here anyway . . ." her father said hesitantly.

He waved to Ture Magnusson.

"Turn around. Don't look, just listen. If you have even the slightest doubt, tell us."

Linda quickly decided on a plan. She was not going to do the voice right away, but work up to it. It would be a test for everyone in the room, not just Magnusson.

"Who remembers what his exact words were?" Lindman said.

Martinsson had the best memory. He repeated the text. Linda made her voice as deep as possible, and found the right accent.

Magnusson shook his head.

"I'm not sure. I almost think I recognise it, but it's not quite right."

"I'd like to do it again," Linda said. "It didn't come out the way I wanted it to."

No-one objected. Again Linda only approximated the right intonation and phrasing. Again, Magnusson shook his head.

"I don't know," he said. "I really couldn't say for sure."

"One last time," Linda said.

This was the time that counted. She took a deep breath and repeated the text, this time getting as close as possible to the original.

"Yes," Magnusson said. "That's what he sounded like. That's his voice."

"But that was on the third try," Höglund said. "What's that worth?"

Linda couldn't quite hide her satisfaction. Her father saw it at once.

"Why did he only recognise it on the third attempt?" he said.

"Because the first two times I didn't sound like him," she said. "It was only the third time that I did the voice exactly."

"I didn't hear a difference," Höglund said suspiciously.

"When you imitate someone's voice all the ingredients have to be right," Linda said.

"That's quite something," Wallander said. "Are you serious about this?"

"Yes."

Wallander looked straight at Ture Magnusson.

"Are you sure?"

"Absolutely."

"Then we thank you for taking the trouble of coming in."

Linda was the only one who shook Magnusson's hand. She followed him out.

"Great job," she said. "Thank you for coming in."

"How could you do that so well?" he said. "It was almost as if I could see him in front of me."

"Anna," Wallander said. "We need to talk to her now."

Linda rang the bell to Anna's flat, but no-one opened. Anna wasn't home. Linda shivered as she stood in the stairwell. She was afraid that Anna had decided to disappear again.

CHAPTER 44

It had been Langaas's task to collect Anna from by the boarded-up pizzeria in Sandskogen. Erik had been planning to get her himself to make sure she was completely willing, but he decided she was so dependent on him she wasn't likely to put up any resistance. Since she had no idea what had happened to Harriet Bolson – Langaas had strict instructions to say nothing – she had no reason to try to get away. The only thing he feared was her intuition. He had tried to gauge it and had concluded it was almost as strong as his own. Anna is my daughter, he thought. She is careful, attentive, constantly receptive to the messages of her subconscious.

Langaas had been briefed on how to handle the situation, though it was unlikely that Anna had been frightened by Zeba's disappearance. There was a chance that she would talk to Linda, the girl Erik judged to be her closest confidante, though he had warned her and thereafter forbidden her to have intimate conversations with anyone but himself. She could be led astray, he had told her, now that she had found the right path. He was the one who had been gone for so long, but she was the prodigal child, she was the one who was finally coming home. What was happening now was necessary. Her father was the one who was going to hold people responsible for turning their backs on the Lord and for building cathedrals where they worshipped at the altar of their own egos rather than humbling themselves before their true maker. He had seen the bewitched look in her eye and known that with time he would be able to erase all doubt from her mind. The problem was that he didn't have this time. It was a mistake, he acknowledged. He should have contacted her long before he showed himself to her in Malmö. But he had had all the others to work on, the members of his army who were one day to open the gates and take their place in his plan. Harriet Bolson's death had been their biggest challenge to date. He had told Langaas to observe their reactions over the next few days, in case anyone seemed liable to break down, or even so much as waver in their conviction. No-one had shown any such signs. On the contrary, Langaas reported a growing impatience among them to undergo the ultimate sacrifice that lay ahead.

Before Langaas went to pick up Anna, Erik made sure that he understood he was to use force if she did not come willingly. That was why he had chosen such an out-of-the-way place for the rendezvous. He had watched Langaas's reaction when he spoke of the possible use of force. A momentary hesitation, a glimpse of anxiety had flickered in his eyes. Erik had made his voice as mild as possible while he leaned forward and put a hand on his shoulder. What was it that worried him? Had Erik ever played favourites among his disciples? Had he not plucked Langaas from the gutters of Cleveland? Why shouldn't his daughter be treated like everyone else? God created a world where everyone was equal, a world people had turned their backs on and destroyed. Was it not that world they were trying to recover?

If all went well and Anna showed herself worthy, she would one day be his successor. God's New Kingdom on Earth could not be left without a ruler, as in the past. There had to be a leader, and God Himself had told him it was to be a position that would go from father to child.

There were times when he thought that Anna was not the one. If that proved to be the case, he would have to have more children and select his successor from among them.

Erik wasn't sure how Langaas had found these houses that stood empty and unattended, but it was a matter of trust between them. He had also found a villa in Sandhammaren which was conveniently isolated from its neighbours and belonged to a retired sea captain who was in hospital with a broken leg. This house had the advantage of a basement room. The house had thick concrete walls and the room in the basement was well constructed and had a small window in the sturdy door. When Langaas first showed it to him they agreed that it seemed as if the sea captain had his very own prison cell. Langaas had suggested it was originally a bomb shelter during the war. But why the thick glass window in the door?

He stopped and listened. In the beginning, when the drugs had worn off, Zeba had screamed, punched the door and attacked the walls with the bucket they had put in for her use as a toilet. When she was quiet, he had looked in through the window. She had been curled up on the bed. They had put a sandwich and a plastic cup of water on a table, but she had touched neither. He hadn't expected her to.

When he looked at her through the window a second time, she was on the bed with her back to him, sleeping. He watched her for a long time until he was sure that she was breathing. Then he went back upstairs and sat down on the veranda, waiting for Langaas to arrive with Anna.

There was still one problem, the question of what was to be done about Henrietta. So far, it seemed that Anna and Langaas had been able to convince her that all was well, but Henrietta was moody and unreliable. Once upon a time he had loved her – although that time lay wrapped in a haze of unreality – and if it were possible, he would spare her life.

He looked out to sea. People were walking on the beach. One of them had a dog, one was carrying a small child on his shoulders. I am doing this for your sakes, he thought. It is for you that I have gathered the martyrs, for your freedom, to fill the emptiness you may not even realise you carry within yourselves.

The walkers on the beach vanished beyond his sight. He looked at the water. The waves were almost imperceptible. A faint wind blew from the south-east. He went into the kitchen and poured himself a glass of water. Langaas and Anna would arrive in half an hour. He returned to the veranda and watched a ship making its slow way west on the horizon.

True Christian martyrs were so rare now that people hardly thought they existed. Some priests had died for the sake of their fellow men in concentration camps during the Second World War, and there had been many other holy men and women since then. But in general the act of martyrdom had slipped from Christian culture. It was the Muslims now who called on the faithful to make the ultimate sacrifice. He had studied their preparations on video, how they documented their intention to die the death of a martyr. In short, he had learned his craft at the hands of those he hated most, his chief enemy, the people he had no intention of making room for in the New Kingdom. Ironically, the dramatic events that were about to take place would in all probability be attributed to Muslims. A welcome benefit of this would be to provoke greater hatred of their faith, but it was regrettable that it would take the world a while fully to understand that the Christian martyrs had returned. This would be no mere isolated phenomenon, no Maranatha, but a wave of true evangelical power that would continue until the New Kingdom of the Lord was fully realised on Earth.

He studied his hands. Sometimes when he contemplated what lay before him they would shake. Now they were steady.

For a short while they will see me as a madman, he thought. But when the martyrs march forth in row upon row, people will understand that I am the apostle they have been waiting for. I could not have managed this without the help of Jim Jones. He taught me how to overcome my fear of death, of urging others to die for the greater good. He taught me that freedom and redemption come only through bloodshed, through death, that there is no other alternative and that someone must lead the way.

Someone must lead the way. Jesus had done so, but God had forsaken him because he had not gone far enough. Jesus had a weakness, he thought. He did not have the strength I possess. We will complete what he lacked the strength to do.

Erik scanned the horizon. The ship he had been watching was gone, and the soft breeze had died down. Soon they would be here. The rest of the day and night he would concentrate on her. It had been a big step for her to lie about her relationship to the man Vigsten in Copenhagen who was Langaas's unwitting host. Anna had never taken a piano lesson in her life, but she had managed to convince the policeman she had talked to. Erik again felt irritation at the fact that he had underestimated the time needed to work on her. But it was too late. Not everything could go according to his plan, and the important thing was that the larger events not be affected.

* * *

The front door opened. He strained to hear them. During the past long and difficult years he had trained all of his senses. It was as if he had sharpened the blades of his hearing, sight and smell. Sometimes he thought of them like finely crafted knives hanging from his belt. He listened to the footsteps. Langaas's were heavy, Anna's lighter. She was moving at her own speed, which indicated that she had come gladly. They walked on to the veranda.

Erik stood and embraced Anna. She was anxious, but not so much that he would be unable to comfort her. He asked her to sit while he followed Langaas to the door. They spoke in low tones. The report Langaas gave him was reassuring. The equipment was stored safely, the others were waiting in two separate houses. No-one showed any signs of anything except impatience.

"They are hungry now," Langaas said.

"We are approaching the fiftieth hour. Two days and two hours until we come out of hiding and make the first strike."

"She was completely calm when I collected her. I felt her pulse and it was normal."

Erik's rage boiled up as if from nowhere.

"Only I have the right to feel a person's pulse! Not you, never you."

Langaas turned pale. "I should not have done it."

"No. But there is something you can do for me to make up for it."

"What is it?"

"Anna's friend. The one who has been too curious, too interested. I am going to talk to Anna now. If it turns out that this friend suspects anything, she should disappear."

Langaas nodded.

Erik signalled for him to leave, and quietly returned to the veranda. Anna was sitting in a chair against the wall. She always keeps her back to the wall, he thought. He kept watching her. She appeared relaxed, but somewhere he had doubts. There is one sacrifice I do not want to make, he thought. A sacrifice I fear. But I must be prepared even for this. Not even my daughter can expect to go free. No-one can expect to do that, except me.

When he sat down, the unexpected suddenly happened. A scream came up through the floor. The walls were simply not thick enough.

He cursed the sea captain silently. Anna froze. The scream modulated into something like the roar of a desperate animal.

Zeba's voice, Zeba's scream. Anna stared at him, the man who was her father and so much more. She bit her lower lip so hard it started to bleed.

He wasn't sure if Anna had abandoned him or if Zeba's scream would only have thrown her off track for a moment. It would be a long and difficult night.

CHAPTER 45

Linda stared at Anna's door, thinking she should kick it open. But why – what was it she thought she would find there? Not Zeba, who was the only one she cared about right now. Standing outside the door she broke into a cold sweat as she felt she understood the gist of what was happening, without being able to translate her insight into words. She pushed her hands into her pockets. She had returned all of Anna's keys, except the ones to the car. But what good will they do me? she thought. Where would I go? Is her car even there? She walked down to the car park and saw that it was. Linda tried to think clearly, but fear blocked her thoughts. To start with she had been worried about Anna. Now it was Zeba who had disappeared. Then she grasped something which had been confusing her. It was about Anna. At first she had been afraid that something had happened to her, but now she was afraid of what she could do.

I'm imagining things, she thought. What is it I think Anna could do? She started walking in the direction of Zeba's house, then turned around and hurried back to Anna's car. Normally she would at least write a note, but there was no time for that. She drove to Zeba's house at high speed. The neighbour who had Zeba's son was out, but her daughter was at home and she gave her the key to Zeba's flat. Linda let herself in and picked up the strange smell again. Why is no-one testing this? she thought.

She walked into the middle of the living room, breathing quietly as if hoping to trick the walls into thinking no-one was there. Someone comes here. Her boy is here, but he can't talk. Zeba is drugged and carried away. Her boy starts to cry and eventually the neighbour comes to check on him.

Linda looked around her, but could see no trace of what had happened. All I see is an empty flat, and I can't interpret emptiness.

She stubbed her toe on the way out. As she was walking to her car, Yassar came out of his shop.

"Did you find her?"

"No. Have you thought of anything else?"

Yassar sighed. "Nothing. My memory is not so good, but I'm sure she was clinging to his arm."

Linda felt a need to defend Zeba. "She wasn't clinging to him, she was drugged."

Yassar looked worried. "You may be right," he said. "But do things like that really happen in a town like Ystad?"

Linda heard only a part of what Yassar had to say. She was already on her way to see Henrietta. She had just started the engine when her mobile rang. It was the police station, but not her father's office number. She hesitated, then answered. It was Lindman. She was happy to hear his voice.

"Where are you?"

"In a car."

"Your father asked me to call. He wants to know where you are. And where is Anna Westin?"

"I haven't found her."

"What do you mean?"

"What do you mean, what do I mean? I went to her flat and she wasn't there. Now I'm trying to work out where she could be. When I've found her, I'll bring her back to the station."

Why don't I tell him the truth? she wondered. Is it something I learned because I had two parents who never told me what was going on, who always chose to skirt their way around the truth of every issue?

It was as if he saw through her.

"Is everything all right with you?"

"Apart from the fact that I haven't found Anna – yes."

"Do you need any help?"

"No."

"That didn't sound very convincing. Do remember you aren't a police officer yet."

"How can I forget it when you're all always bringing it up?"

She switched the phone off and threw it on to the passenger seat. She had only turned one corner before she stopped and switched it on again. Then she drove straight to Henrietta's house. The wind had picked up and the air was brisk when she got out of the car and walked to the house. She looked towards the place where she had been caught in the animal trap. In the distance, on one of the narrow dirt roads between the fields, a man was burning rubbish next to his car. The thin spiral of smoke was torn apart by the gusts of wind.

Autumn was just around the corner, the first frost not far off. She walked into the garden and rang the doorbell. The dog started to bark. She drew a deep breath and shook out her body as if she were about to crouch down into the starting blocks. Henrietta opened the door. She smiled. Linda was immediately suspicious; it seemed as if Henrietta had been expecting her. Linda also noted that she had put on make-up, as if she wanted to make a good impression on someone, or to conceal the fact that she was pale.

"This is unexpected," Henrietta said, and stepped aside.

Not true, Linda thought.

"You're always welcome. Come in."

The dog sniffed her, then returned to his basket. Linda heard a sigh. She looked around, but no-one was there. Sighs seemed to emanate from the thick stone walls themselves. Henrietta put out a coffee pot and two mugs.

"What's that sound?" Linda asked.

"I'm playing one of my oldest compositions," Henrietta said. "It's from 1987, a concert for four sighing voices and percussion. Listen!"

Linda heard a single voice sigh, a woman.

"That's Anna. I managed to persuade her to participate. She has a melodious sigh, full of sadness and vulnerability. There is always a somewhat hesitant quality to her speaking voice, but never to her sigh."

Henrietta walked over to the tape recorder and turned it off. They sat down. The dog had started snoring and it was as if this sound drew Linda back to reality.

"Do you know where Anna is now?"

Henrietta looked down at her nails, then at Linda, who sensed a moment of doubt in her eyes. She knows, and she's prepared to deny it, Linda thought.

"My mistake, then. Each time I think you're here to see me and what you're really after is to find out where my daughter is."

"*Do* you know where she is?"

"No."

"When did you talk to her last?"

"She called yesterday."

"From where?"

"From her flat."

"She doesn't have a mobile phone?"

"No, she doesn't, as you must know. She resists joining the ranks of those who are always available."

"So she was home last night?"

"Are you interrogating me, Linda?"

"I need to know where Anna is, and what she's up to."

"I don't know where she is – presumably in Lund. She's in medical school, you know."

No she isn't, Linda thought. Maybe Henrietta didn't know that Anna had taken a break from her studies. That will be my trump card. But not now – later.

She chose another route.

"Do you know Zeba?"

"Little Zeba? Of course."

"She's disappeared, just like Anna."

Not a twitch or a quiver betrayed that Henrietta knew anything. Linda felt as if she had been floored by a punch she never saw coming. That had happened during her time at the training college. She had been in a ring, and suddenly found herself face-down on the floor without knowing how she got there.

"And maybe she'll reappear, just as Anna did."

Linda sensed more than saw her opportunity and she went in with her fists held high.

"Why didn't you tell me the truth, Henrietta? Why didn't you say you knew where she was?"

It hit the mark. Beads of sweat broke out on Henrietta's forehead.

"Are you saying that I lied to you? If that is the case, I want you to leave right now. I will not be called a liar in my own home. You are poisoning me. I cannot work, the music is dying."

"Yes, I *am* saying that you lied, and I won't leave until you answer my questions. I have to know where Zeba is because I think she's in danger. Anna is mixed up in this somehow, maybe you are too. And for sure, you know a lot more than you're telling me."

"Go away! I don't know anything!" Henrietta yelled. The dog got up and started to bark.

Henrietta walked to a window, absently opening it, then closing it, then leaving it slightly ajar. Linda didn't know how to continue, but knew she couldn't let go. Henrietta seemed to have calmed down. She turned around.

"I'm sorry I lost my temper, but I don't like being accused of lying. I don't know where Zeba is, and I have no idea why you seem to think Anna is involved."

Her indignation seemed genuine, or else she was a better actress than Linda imagined. She was still speaking with a raised voice, and she was still by the window.

"The night I got caught in the animal trap," Linda said, "who were you talking to?"

"Were you spying on me?"

"Call it what you like. Why else would I have been here? I needed to know why you didn't tell me the truth when I came to ask you about Anna."

"The man who was here had come to talk to me about a composition we are planning together."

"No," Linda said, forcing her voice to remain steady. "It was someone else."

"You are accusing me of lying again."

"I know you are."

"I always tell the truth," Henrietta said. "But I prefer never to reveal any part of my private life."

"You lied, Henrietta. I know who was here."

"You know who was here?" Henrietta's voice was high and shrill again.

"It was either a man called Torgeir Langaas, or it was Anna's father."

Henrietta flinched.

"Torgeir Langaas," she almost screamed. "I don't know anyone called Torgeir Langaas. And Anna's father has been gone for twenty-four years. He's dead. Anna is in Lund and I have no idea where Zeba might be."

She went out into the kitchen and returned with a glass of water. She moved some cassette tapes out of the way and sat down on a chair next to Linda, who had to turn her body to look at her. Henrietta smiled. When she spoke again her voice was soft, almost careful.

"I didn't mean to get so carried away."

Linda looked at her and somewhere in her head a warning light came on. There was something she should be seeing. The conversation had been a failure. The only thing she had achieved was to put Henrietta even more on her guard. An experienced officer should have been in charge of this questioning, she thought. Now it would be even harder for her father, or whomever it would fall to, to get Henrietta to reveal whatever it was she was hiding.

"Is there anything else you think I've been lying about?"

"I believe hardly anything you say, but I can't force you to stop telling lies. I just want you to know that I'm asking these questions because I'm worried about Zeba."

"What could possibly have happened to her?"

Linda drew a deep breath. "I think someone, perhaps more than one person, is killing women who have had abortions. Zeba has had an abortion. The woman who was found dead in that church had had one – two, actually. You've heard about the dead woman and the churches set on fire?"

Henrietta sat absolutely still, which Linda took as a yes.

"What has Anna to do with all this?"

"I don't know, but I'm frightened."

"What frightens you?"

"The thought that someone might try to kill Zeba. And that Anna is somehow involved."

Something in Henrietta's face changed. Linda couldn't say what it was, but it flickered there for a moment. She decided she wasn't going to get any further and bent to pick her jacket off the floor. There was a mirror next to the table. She glanced in it as she reached down and saw Henrietta's face. She was looking past Linda.

Linda grabbed her jacket and sat up. Henrietta had been looking at the open window.

She started putting on her jacket and stood up, turning around. There was no-one outside, but Linda knew that someone had been there. She froze. Henrietta's loud voice, the window that was opened for no reason, her repetitions of the names Linda had given her, and her vehement rejections of the accusations. Linda finished putting on her jacket. She didn't dare turn around and look Henrietta in the eye since she was afraid her realisation would be spelled out in her face.

Linda quickly made her way to the front door and bent down to pat the dog. Henrietta followed her into the hallway.

"I'm sorry I couldn't help you."

"You could have helped," Linda said. "But you chose not to."

She opened the door and walked out. When she reached the end of the path she turned and looked around. I don't see anyone, she thought, but someone can see me. Someone watched me in the house and – more to the point – heard what we said. Henrietta repeated my questions and the person outside now knows what I know and what I believe and fear.

She walked swiftly to the car. She was scared, but she also berated herself for making a mistake. The point at which she was patting the dog and leaving was the point at which she should have started her questions in earnest. But she had chosen to leave.

She kept checking the rear-view mirror as she drove away.

CHAPTER 46

As Linda was walking into the police station she tripped and split her lip on the hard floor. For a moment she was dizzy, then she managed to get up and wave away the receptionist who was on her way over to help her. When she saw blood on her hand she walked to the ladies', washed her face with cold water and waited for the bleeding to stop. Lindman was on his way in through the front doors as she came back out into reception. He looked at her with an amused expression.

"You make quite a pair," he said. "Your father claims he walked into a door. What about you? That same door been making trouble for you too? Maybe we should call you Black-Eye and Fat-Lip, to avoid any confusion with the two of you having the same name."

Linda laughed, which caused the wound to reopen and bleed. She went back into the ladies' and got more tissue paper. They walked down the corridor together:

"It wasn't a door. I threw an ashtray at him."

They stopped outside Wallander's office.

"Did you find Anna?"

"No."

Lindman knocked on the door.

"You'd better go in and tell him."

Wallander had his feet on the desk and was chewing on a pencil. He raised his eyebrows at her.

"Did you bring Anna?"

"I can't find her."

"What do you mean?"

"What do you think I mean? She's not at home."

Wallander didn't conceal his impatience. Linda prepared for the onslaught, but then he noticed her swollen lip.

"What happened to you?"

"I tripped in reception."

He shook his head, then started to laugh. Linda was glad of this shift in his mood, but she found his laugh hard to take. It sounded like a horse neighing and was far too high-pitched. If they were ever out together and he started to laugh, people would turn round to see who could possibly be responsible for such sounds.

Wallander threw his pencil down and took his feet off the desk.

"Have you called her place in Lund? Her friends? She has to be somewhere."

"Nowhere that we can reach her, I think."

"You've called her mobile number, at least?"

"She doesn't have one."

He was immediately interested in this piece of information.

"Why not?"

"Apparently she doesn't want one."

"Is there any other reason?"

Linda knew that there was more than idle curiosity behind these questions.

"Everyone has a mobile phone these days, especially you young people. But somehow not Anna Westin. How do you explain that?"

"I can't. According to Henrietta, she doesn't want to be 'always available'."

Wallander thought about this.

"Are you sure you know everything? Maybe she does have a phone, but she hasn't told you about it?"

"How could I know that?"

"Exactly."

Wallander pulled his phone over and dialled a number. Höglund came into the office, looking tired and scruffy. Her hair was messy and her shirt was soiled. Linda was reminded of fru Jorner, Medberg's daughter. The only difference that she could see between them was that Höglund was not so fat.

Linda heard her father ask Höglund to see if a mobile phone was registered under Anna's name. She was irritated that she hadn't thought of that herself. Before leaving the room, Höglund gave Linda a smile that was more like a grimace.

"She doesn't like me."

"If memory serves, you don't care much for her either. It all evens itself out in the end. Even in a small police station like this people don't always get along."

He stood up.

"Coffee?"

They went to the canteen and Wallander was immediately drawn into an evidently exasperating exchange with Nyberg. Martinsson came in waving a piece of paper.

"Ulrik Larsen," he said. "The one who tried to mug you in Copenhagen."

"Not mug me," Linda said sharply. "The one who threatened me and told me to stop asking questions about a man called Torgeir Langaas."

"That's exactly what I was going to tell you about," Martinsson said. "Larsen has withdrawn his story. The problem is that he won't let them have a new version. He continues to deny that he threatened you, and he maintains he doesn't know anyone by the name of Langaas. Our Danish colleagues are convinced he's lying, but they can't get him to tell the truth."

"Is that it?"

"Not completely. But I want Kurre to hear the rest."

"Don't call him that," Linda warned. "He hates it."

"Tell me about it," Martinsson said. "He likes it about as much as I like being called 'Marta'."

"Who calls you that?"

"My wife. When she's in a bad mood."

Wallander and Nyberg finished discussing whatever it was they disagreed about, and Martinsson relayed the news about Larsen.

"There's one more thing," he said, "which is really the most significant. Our Danish colleagues have run a background check on Larsen and he has no previous criminal record. In fact, it turns out in all other respects that he's a model citizen: thirty-seven years old, married, three children, and with an occupation where practitioners rarely turn to criminal activity."

"What is it?" Wallander said.

"He's a minister."

They stared at Martinsson.

"What do you mean, a minister?" Lindman said. "I thought he was a drug addict."

Martinsson looked through his papers.

"Apparently he played the role of a drug addict, but he's a minister in the Danish state church, with a parish in Gentofte. There have been all kinds of headlines over there about the fact that a minister of the church has been accused of assault and robbery."

"It turns up again, then," Wallander said softly. "Religion, the church. This Larsen is important. Someone has to go over and assist our colleagues in their investigation. I want to know how he fits in."

"*If* he fits in," Lindman said.

"He does," Wallander said. "We just need to know how. Ask Höglund to do it."

Martinsson's telephone rang. He listened and then finished his cup of coffee.

"The Norwegians are stirring," he said. "We've received some information about Langaas."

"Let's have a look at it."

Martinsson went to get the faxes. There was a fuzzy version of a photograph.

"This was taken more than twenty years ago," Martinsson said. "He's tall. More than one hundred and ninety centimetres."

They studied the snapshot. Have I seen this man before? Linda wondered. But she couldn't be sure.

"What do they say?" Wallander asked.

Linda noticed he was getting more and more impatient. Just like me, she thought. The anxiety and impatience go hand in hand.

"They found our man Langaas as soon as they started to look. It would have come through sooner if the officer in charge had not misdirected our urgent query. In other words, the Oslo office is plagued by the same problems as we are. Here, tapes from the archives go missing, there, requests from other stations. But it all got sorted out in the end, and Langaas is involved in an old missing-persons case, as it turns out."

"In what way involved?" Wallander said.

"You won't believe me when I tell you."

"Try me."

"Torgeir Langaas disappeared from his native Norway nineteen years ago."

They looked at each other. Linda felt as if the room itself was holding its breath. She saw her father sit up in his chair as if readying himself to charge.

"Another disappearance," he said. "Somehow all of this is about disappearances."

"And reappearances," Lindman said.

"Or a resurrection," Wallander said.

Martinsson kept reading, picking his way through the text as if there were landmines hidden between the words: "Torgeir Langaas was the heir of a shipping magnate. His disappearance was unexpected and sudden. No crime was suspected since he left a letter to his mother, Maigrim Langaas, in which he assured her that he was not depressed and had no intention of committing suicide. He left because he – and I quote – 'couldn't stand it any longer'."

"What couldn't he stand?"

It was Wallander who interrupted him again. To Linda it seemed as if his impatience and worry came out of his nostrils like invisible smoke.

"It's not clear from this report, but he left with quite a bank balance. Several bank accounts, in fact. His parents thought he would tire of his rebellion after a while. They didn't go to the police until two years had passed. The reason they gave, it says here in the report from January the twelfth, 1984, was that he had stopped writing letters, that they had had no sign of life from him for four months, and that he had emptied each of his accounts. Since which time, no-one has heard from him."

Martinsson let the page fall to the table.

"There's more, but those are the main points."

Wallander raised his hand. "Does it say where the last letter was posted? And when were the bank accounts emptied?"

Martinsson looked through the papers for these answers, but without success. Wallander reached for his mobile.

"What's the number?"

He dialled the number that Martinsson read out. The Norwegian officer's name was Hovard Midstuen. Once he was on the line, Wallander asked his two questions, gave his phone number and hung up.

"He said it would only take a few minutes," Wallander said. "We'll wait."

Midstuen called back after 12 minutes. Wallander pounced on the telephone and scrawled a few notes. He thanked his Norwegian colleague and triumphantly shut off the phone.

"This might be starting to hang together."

He read from his notes: Langaas's last letter was posted in Cleveland, Ohio. It was also from there that the accounts were emptied and closed.

Not everyone made the connection, but Linda saw what he was getting at.

"The woman who was found dead in Frennestad Church came from Tulsa," he said. "But she was born in Cleveland, Ohio. I still don't understand what's happening, but there's one thing I know, and that's that Linda's friend Zeba is in mortal danger. It may be that her other friend, Anna Westin, is in danger too." He paused. "It might also be that Anna Westin is part of this. That's why we need to concentrate on these two and nothing else for the moment."

It was 3 p.m. and Linda was scared. She could think only of Zeba and Anna. A thought flew through her mind: she would start her real work as a police officer in three days. How would she feel about that if something happened to either of her friends in the meantime?

CHAPTER 47

When Anna recognised the scream as Zeba's, Erik knew that God was testing him in the same way He had tested Abraham. He perceived her every reaction. She had merely flinched and then she had carefully adopted an expression which lacked all emotion. A moment of doubt, a series of questions – was that a kind of animal or, in fact, a human scream? Could it be Zeba? She was searching for an answer that would satisfy her, and at the same time she was waiting to hear the scream again. What Erik didn't understand was why she didn't ask him about it. In a way, it was as well that Zeba had made her presence known. Now there was no turning back.

He would see soon if Anna was worthy of being called his daughter. And what would he do if it turned out she did not possess the strength he expected of her? It had taken him many years to travel the road his inner voices had told him to follow. He had to be prepared to sacrifice even that which was most precious to him, and it would be up to God whether he, Erik, would be granted a stay at the last minute.

I won't talk to her, he thought. I must preach to her, as I preach to my disciples. She broke in during a pause. He let her speak, because he knew he could best interpret a person's state of mind at such a moment of vulnerability.

"Once upon a time you were my father. You lived a simple life."

"I had to follow my calling."

"You abandoned me, your daughter."

"I had to. But I never left you in my heart. And I came back to you."

She was tense, he could see that, but still her sudden loss of control surprised him. Her voice rose to a shriek.

"That screaming I heard was Zeba! She's here somewhere below where we are sitting. What is she doing here? She hasn't done anything."

"You know what she has done. It was you who told me."

"I should never have told you!"

"She who commits a sin and takes the life of another must bear the wrath of God. This is justice, and the word of the Lord."

"Zeba didn't kill anyone. She was only fifteen at the time. How could she have cared for a child at that age?"

"She should not have allowed a child to be conceived."

Erik could not calm her, and felt a wave of impatience. This is Henrietta, he thought. She's too much like her.

"Nothing is going to happen to Zeba," he said.

"Then what is she doing here, in a basement?"

"She is waiting for you to make up your mind. To decide."

This confused her, and Erik smiled inwardly. He had spent many years in Cleveland poring over books about the arts of warfare. That work was paying off now. It was she who was on the defensive now.

"I don't understand what you mean. I'm frightened."

Anna started to sob, her body shook. He felt a lump in his throat, remembering how he had comforted her as a child when she cried. But he forced the feeling away, and asked her to stop.

"Of what are you scared?"

"Of you."

"You know I love you. I love Zeba. I have come to join the earthly and the divine in transcendent love."

"I don't understand you when you talk like that!"

Before he had a chance to say anything else there was a new cry for help from the basement and Anna flew from her chair.

"I'm coming!" she cried, but he grabbed her before she could leave the veranda. She struggled, but he was too strong for her. When she continued to struggle, he hit her with an open hand. Once, then again, and then a third time. She fell to the floor after the third blow, her nose bleeding. Langaas appeared at the French windows and Erik motioned for him to go down into the basement. Langaas understood and left. Erik pulled Anna up on to a chair and felt her forehead with his fingertips. Her pulse was racing. His own was only somewhat accelerated. He sat down and waited. Soon he would have broken her will.

This was the last set of defences. He had surrounded her and was attacking from all sides. He waited.

"I didn't want to do that," he said after a while. "I only do what is necessary. We are about to embark on a war against emptiness, soullessness. It is a war in which it is not always possible to be gentle, or merciful. I am joined by people who are prepared to give their lives for this cause. I myself may have to give my life."

She didn't say anything.

"Nothing will happen to Zeba," he repeated. "But nothing in this life comes to us without a price."

Now she looked at him with a mixture of fear and anger. She held a handkerchief to her nose. He explained what it was he wanted her to do. She stared at him with wide eyes. He moved his chair closer to her and placed his hand over hers. She stared back, but she did not pull it away.

"I will give you one hour," he said. "No door will be locked, no guard will watch over you. Think about what I have said, and come to your decision. I know that if you let God into your heart and mind, you will do what is right. Do not forget that I love you very much."

He stood up, traced a cross on her brow with his finger, and left.

Langaas was waiting in the hallway.

"She settled down when she saw me. I don't think she'll do it again."

They walked through the garden to an outhouse that had been used for storing fishing equipment. They stopped outside the door.

"Has everything been prepared?"

"Everything has been prepared," Langaas said.

He pointed to four tents that had been erected beyond the outhouse, lifting the flap to one of them. Erik looked in. There were the boxes, piled one on top of the other. He nodded. Langaas tied the tent flap shut.

"The cars?"

"The ones that will drive the greatest distances are waiting on the road. The others have been parked in the positions we agreed."

Erik Westin looked down at his watch. The many, often difficult, years he had spent laying the groundwork had seemed without end. Now time was passing too fast. From this point on everything had to conform exactly to the plan.

"It's time to start the countdown," he said.

He threw a glance at the sky. Whenever he had thought forward to this moment he had always imagined that the heavens would mirror its dramatic import, but in Sandhammaren on this day, September 7, 2001, there were no clouds and almost no breeze.

"What is the temperature?" he said.

Langaas looked at his watch, which had a built-in thermometer, as well as a pedometer and compass.

"Eight degrees," he said.

They walked into the outhouse, which still smelled strongly of tar. Those who were waiting for him sat in a semicircle on low wooden benches. Erik had planned to perform the ceremony with the white masks, but he decided to wait. He did not yet know if the next sacrifice would be Zeba or the policeman's daughter. They would perform the ceremony then. Now they only had time for a shorter ritual; God would not accept anyone who arrived late for their appointed task. Not to be mindful of one's time was like denying that time was the gift of the Lord. Those who needed to travel to their destinations would shortly have to leave. They had calculated how much time was needed for each leg of their journey, and had followed the checklists in the carefully prepared manuals. In short, they had done everything in their power, but there was always the possibility that the dark forces would prevent them from achieving their goals.

When the cars with the three groups who had to travel had left, and the others had returned to their hide-outs, Erik remained in the outhouse. He sat motionless in the dark with the necklace in his hand – the golden sandal that was now as important to him as the cross. Did he have any regrets? That would be blasphemy. He was no more than an instrument, but one equipped with a free will to understand and then to dedicate himself to the path of the chosen. He closed his eyes and breathed in the tar. He had spent a summer on the island of

Öland as a child, visiting a relative who was a fisherman. The memories of that summer, one of the happiest of his childhood, were nestled in the scent of tar. He remembered how he had crept out in the light summer night and run down to the boat shed in order to draw the smell more deeply into his lungs.

Erik opened his eyes. He was past the point of no return. The time had come. He left the outhouse and took a circuitous route to the front of the house. He looked at the veranda from the cover of a large tree. Anna was in the same chair. He tried to interpret her decision from the way she was sitting, but he was too far away.

Suddenly there was a rustling behind him. He took fright. It was Langaas. Erik was furious.

"Why are you sneaking around?"

"I didn't mean to."

Erik struck him hard in the face, below the right eye. Langaas accepted the blow and lowered his head. Then Erik stroked his hair and they walked together to the house. He made his way soundlessly to the veranda until he was standing behind her. She noticed his presence only when he bent over and she felt his breath on the nape of her neck. He sat down beside her, pulling his chair closer until their knees touched.

"Have you made your decision?"

"I will do as you ask."

He had expected that she would say this, but still it came as a relief.

He collected a shoulder bag lying by the wall, and pulled out a small, thin and extremely sharp knife. He lowered it gently into her hands, as if it were a kitten.

"The moment she reveals that she knows things she shouldn't, I want you to stab her – not once, but three or four times. Strike her in the chest and force the blade up before you pull it out. Then call Torgeir and stay out of sight until we get you. You have six hours to do this, no more. You know I trust you, and love you. Who could love you more than I do?"

She was about to say something, but stopped herself. He knew she had been thinking of Henrietta.

"God," she said.

"I trust you, Anna," he said. "God's love and my love are one and the same. We are living in a time of rebirth. A new kingdom. Do you understand this?"

"Yes."

He looked deep into her eyes. He was still not utterly certain about this, but he had to believe that he was doing the right thing.

He followed her out.

"Anna is going home now, Torgeir."

They got into a car that was parked by the front door. Erik tied the kerchief over her eyes himself to make sure that she saw nothing.

"Take a detour," he said in a low voice to Langaas. "Make her think it is further than it really is."

The car came to a stop at 5.30. Langaas took out Anna's earplugs, then told her to keep her eyes closed and count to 50 after he had taken off her blindfold.

"The Lord watches you," he said.

He helped her to step out on to the pavement. Anna counted to 50, then opened her eyes. She didn't know where she was at first. Then she recognised it. She was on Mariagatan, outside Linda's flat.

CHAPTER 48

During the afternoon and evening of September 7, Linda once again watched her father try to bring all the threads into one whole and come up with a plan for how they should proceed. During those hours she became aware that the praise he often received from his colleagues and at times in the media – when they were not chastising him for his dismissive attitude towards them at press conferences – was justified. Her father was not only knowledgeable and experienced, but he possessed a remarkable ability to focus and to inspire his colleagues.

During her time at the training college, the father of one of her friends was an ice-hockey coach for a top team in the second league. She and her friend had once been allowed into the changing room before a game, during the breaks and after the game was over. This coach had the ability she had just witnessed in her father, an ability to motivate people. After two periods the team was losing by four goals, but the coach didn't let up. He urged them on, cajoled them not to let themselves be beaten, and in the last period the players had stormed back on to the ice and very nearly managed to turn the game around.

Will my father manage to turn this game around? she wondered. Will he find Zeba before it is too late? Over the course of the day, during a meeting or briefing when she stood at the back of the room, she several times had to rush out to go to the toilet. Her stomach had always been her weakest point; fear gave her diarrhoea. Her father, on the other hand, sometimes bragged about having the stomach lining of a hyena – apparently their stomach acid was the strongest in all the animal kingdom. His weakness was his head, and sometimes when he was under a great deal of stress he would get tension headaches that might last days and could be cured only by prescription-strength pills.

Linda was afraid, and she knew she wasn't the only one. There was an unreal quality to the calm and concentration at the police station. She understood something which no-one had mentioned at the training college: that sometimes the most important task of a police officer was keeping their own fear in check. If it got out of control, all this concentration and focus would crumble.

Shortly after 4.00 Linda saw her father pacing up and down the corridors like a caged animal. The press conference was about to take place. He kept sending in Martinsson to see how many journalists were assembled, and how many television cameras. From time to time he asked Martinsson about individuals by name, and from the tone of his voice it was clear he was hoping they would not be there. She watched him walking anxiously to and fro. He was the animal waiting to be sent into the arena. When Holgersson came to announce that it was time, he lunged into the conference room. The only thing missing was a roar.

During the 30-minute press conference, Wallander concentrated on Zeba. Photographs were passed out, a slide was projected on to the screen. Where was she? Had anyone seen her? Skilfully he side-stepped being pulled into detailed explanations, keeping his remarks concise and ignoring questions he did not want to answer.

"There is still a dimension here that we do not understand," he said in closing. "The church fires, the two dead women and the burned animals. We cannot be entirely sure that there is a connection, but what we do know is that this young woman may be in danger."

What danger? Who posed this danger? Could he add anything? The room buzzed with dissatisfaction. Linda imagined him lifting an invisible shield and simply letting the questions bounce back unanswered. Chief Holgersson said nothing during the proceedings, except to chair the question-and-answer session. Svartman mouthed answers to Wallander when there were data that escaped him.

Suddenly Wallander stood up as if he couldn't take it any longer, nodded and left the room. He shook off the reporters who threw themselves after him. Afterwards he left the station without saying another word.

"That's what he always does," Martinsson said. "He takes himself out for some air, as if he were his own dog. Walks around the water tower. Then he comes back."

Twenty minutes later he came storming down the corridor. Pizzas were delivered to the conference room. Wallander told everyone to hurry up, shouting at a young woman from the office who had not provided them with the paper he had asked for, and then he slammed the door. Lindman, who was sitting beside her, whispered:

"One day I think he's going to lock the door and throw away the key. We'll turn into pillars of salt. If we're lucky, we'll be excavated in a thousand years."

Ann-Britt Höglund had just returned from a quick investigation in Copenhagen.

"I met this Ulrik Larsen," she said and pushed a photograph over to Linda. She recognised him immediately. Yes, he was the one who had warned her not to look for Torgeir Langaas, and who had then knocked her down.

"He's changed his mind," Höglund continued. "There's no more talk of drugs. He still denies having threatened Linda, but gives no alternative explanation. He is evidently a controversial minister. His sermons have become increasingly fire-and-brimstone of late."

Linda saw her father's arm shoot out and interrupt.

"This is important. How do you mean, 'fire-and-brimstone', and be specific about 'of late'."

Höglund flipped through her notebook.

"I was led to believe that 'of late' meant this last year. The fire-and-brimstone is shorthand for the fact that he has started preaching about the Day of Judgment, the crisis of Christianity, ungodliness and the punishment that will be meted out to all sinners. He has been admonished both by his own congregation and by the bishop, but refuses to change the tenor of his sermons."

"I take it you asked the most important question?"

Linda wasn't sure what that was, and when Höglund gave her answer, she felt stupid.

"His views on abortion? I was actually able to ask him myself."

"The answer?"

"There was none. He refused to speak to me. But from the pulpit he has stated that abortion is a despicable crime that deserves the severest punishment."

At that moment Nyberg opened the door. "The theologian is among us."

Linda looked around the room and saw that only her father knew what Nyberg was talking about.

"Show him in," Wallander said.

Nyberg went out, and Wallander explained for whom they were waiting.

"Nyberg and I have been trying to make sense of that Bible that was left, or maybe deliberately left in the hut where Medberg was murdered. Someone has written their own version of the text, notably in the Book of Revelation, in Romans and parts of the Old Testament. But what kind of changes? Is there a logic there? We talked to the state crime people, but they had no experts to send. That's why we contacted the Department of Theology at Lund University and established contact with Dr Hanke, who has come here today."

Dr Hanke, to everyone's surprise, turned out to be a pretty young woman with long blonde hair, dressed in black leather trousers and a low-cut top. Linda saw that it discomfited her father. Hanke walked around the room shaking hands and then sat down on a chair that was pulled up next to Lisa Holgersson.

"My name is Sofia Hanke," she said. "I'm a lecturer and have written a dissertation on the Christian paradigm shift in Sweden after the Second World War."

She opened her portfolio and took out the Bible that had been found in the hut.

"This has been fascinating," she said. "But I know you don't have a lot of time, so I'll try to make it brief. The first thing I want to say is that I believe this is the work of one person, not because of the handwriting, but because there is a kind of logic to what is written here."

She looked in a notebook and continued, "I've chosen an example to illustrate what I mean, from Romans Chapter Seven. By the way, how many of you know the Bible? Perhaps it's not part of the current curriculum at the training college . . ."

Everyone who met her gaze shook their head, except Nyberg, who surprised them by saying, "I read from the Bible every night. Foolproof way to induce sleep."

Everyone laughed, including Sofia Hanke.

"I can relate to that experience," she said. "I ask mainly because I'm curious. In any case, Romans Chapter Seven discusses the human tendency to sin. It says, among other things: 'Yes, the good that I wish to do, I do not; but the evil that I do not wish to do, I do.' Between these lines our writer has rearranged good and evil. The new version reads: 'Yes, the evil that I wish to do, I do; but the good that I do not wish to do, I do not do.' St Paul's message is turned upside down. One of the grounding assumptions of Christianity is the idea that humans want to do what is right, but always find reasons to do evil instead. But the altered version says that humans do not even want to do what is right. This happens again and again in the changes. The writer turns texts upside down, seemingly to find new meanings. It would be easy to assume this is the work of a deranged soul, but I don't think that's what we're dealing with. There is a strained logic to these emendations. I think the writer is hunting for a significance he or she believes is concealed in the Bible, something that is not immediately apparent in the words themselves. He or she is looking between the words."

"Logic," Wallander said. "What kind of logic is there in something this absurd?"

"Not everything is absurd; some of it is straightforward, even simple. There are also other texts in the margin. For example: 'All the wisdom life has taught me can be summed up in the words "he who loves God, is blessed".'"

Linda saw that her father's patience was beginning to be tried.

"Why would someone do this? Why do you think we find this Bible in a miles-from-anywhere hut in which a woman has been the victim of an abominable murder?"

"It could be a case of religious fanaticism," Hanke said.

Wallander leaned forward. "Tell us more."

"I normally refer to something I call Preacher Lena's tradition. A long time ago a milkmaid in Östergötland had mystical visions and started preaching. After a while she was taken to a lunatic asylum, but such people have always been around. Religious fanatics who either choose to live as lone preachers or who try to assemble a flock of devotees. Most of these people are honest to the extent that they act out of a genuine belief in their divine inspiration. Of course, there have always been con artists, but these are in the minority. Most of them preach their beliefs and start their sects from a genuine desire to do good. If they commit crimes or evil deeds they often try to legitimise these acts in the eyes of their God, for example by the interpretation of Bible verses."

The discussion with Dr Hanke went on, but Linda could tell that her father was thinking about other things. These scribbles between the lines of the Bible found in the Rannesholm hut had yielded no clues. Or had they? She tried to read his thoughts – she had been practising since early childhood – but there was a big difference between being alone with him at home and being in a conference room full of people at the police station, as now.

Nyberg escorted Dr Hanke out and Holgersson opened a window. The pizza cartons were starting to empty. Nyberg returned. People walked out and in, talked on the phone, went to get cups of coffee. Only Linda and her father stayed at the table. He looked at her vaguely and then retreated into his own thoughts.

When they started their long meeting Linda was quiet and no-one asked her any questions. She sat there like an invited guest. Her father looked at her a few times. If Birgitta Medberg had been a person who mapped old, overgrown paths, then her father was a person who was looking for passable roads to travel. He seemed to have an endless patience even though he had a clock inside him ticking crossly and loudly. That was what he had told her once in Stockholm when he met Linda and a few of her student friends and told them about his work. During times of enormous pressure, as when he knew that a person's life was in danger, he had a feeling that there was a clock ticking away on the right side of his chest, parallel to his heart. Outwardly, however, he was patient, and displayed signs of irritation only if anyone started to veer away from the subject: where was Zeba? The meeting went on, but from time to time someone made or received a call or else someone left and returned with some documentation or a photograph that was immediately made into a thread of the investigation.

Chief Holgersson closed the door at 8.15 p.m. after a short break. Now no-one was allowed to disturb them. Wallander took off his coat, rolled up the sleeves of his dark blue shirt and walked to the flip chart. On a blank sheet he wrote Zeba's name and drew a circle around it.

"Let's forget about Medberg for the moment," he said. "I know it may be a fatal mistake, but right now there is no logical connection between her and Harriet Bolson. It may be the same killer or killers, we don't know. But my point is that the motive seems different. If we leave Medberg out, we see that it is easier to find a connection between Harriet Bolson and Zeba. Abortion. Let us assume that we are dealing with a number of people – we don't know how many – who with some religious motivation judge and punish women who have had abortions. I use the word 'assume' here since we don't know. We know that people have been murdered, animals killed and churches burned to the ground. Everything that has happened gives us the inescapable impression of thorough planning."

Wallander looked at the others, then went back to his place at the table.

"Let us assume everything is part of a ceremony," he said. "Fire is an important symbol in similar cases. The burning of the animals may have been a sacrifice of some kind. Harriet Bolson was executed in front of the altar in a way that could be interpreted as ritual sacrifice. We found a necklace with a sandal pendant around her neck."

Lindman raised his hand and interrupted him.

"I've been wondering about that note with her name on it. If it was left there for us, then why?"

"I can't tell you. What do *you* think?"

"Doesn't it suggest that we are dealing with a lunatic who is challenging us, who wants us to try to catch him?"

"It could be. But that's not the important thing right now. I think these people are planning to do to Zeba what they did to Harriet Bolson."

The room grew quiet.

"This is where we are," he said. "We have no suspect, no sure-fire motive, no definite direction. In my opinion, we're at a stalemate."

No-one disagreed.

"We keep working," he said. "Sooner or later we'll find our direction."

The meeting was over. People left in different directions. Linda felt in the way, but she had no thoughts of leaving the station. In three days, on Monday, she would at last be able to pick up her uniform and start working in earnest. But the only thing that meant anything right now was Zeba. She went to the toilet. On her way back her mobile rang. It was Anna.

"Where are you?"

"At the station."

"Is Zeba back yet? I called her flat, but there's no answer."

Linda was immediately on her guard. "She's still missing."

"I'm so worried about her."

"Me too."

She must really be worried, Linda thought. She couldn't lie that well.

"I need to talk," Anna said.

"Not now," Linda said. "I can't get away right now."

"Not even for a few minutes? If I come up to the station?"

"You aren't allowed in."

"But can't you come out? For a few minutes?"

"Are you sure this can't wait?"

"Of course it can."

Linda heard that Anna was disappointed. She changed her mind.

"A few minutes, then."

"Thanks. I'll be there in ten."

Linda walked down the corridor to her father's office. Everyone seemed to have vanished. She wrote on a note that she left on the desk: I've gone out for some air and to talk to Anna. Back soon. Linda.

She put on her jacket and left. The corridor was empty. The only person she passed on the way out was the cleaning lady with her trolley. The police officers manning the incoming calls were busy and did not look up. No-one noticed her walk through reception.

The cleaning woman, Lija, who was from Latvia, normally started at the far end of the corridor where the criminal police had their offices. Since several rooms there were occupied, she started with Inspector Wallander's office. There were always loose pieces of paper under his chair that he hadn't managed to throw into the waste-paper basket. She swept up everything that was under the chair, dusted here and there, and then left the room.

CHAPTER 49

Linda waited outside the station. She was cold and pulled the jacket tightly across her body. She walked down to the poorly lit car park and spotted her father's car. She pushed a hand into her pocket and confirmed that she still had the spare keys. She checked her watch. More than ten minutes had gone by. Why wasn't Anna here?

Linda waited at the front entrance of the police station. No-one was around. In other parts of the building there were shadows behind the lit windows. She walked back over to the car park. Suddenly, something made her feel ill at ease and she stopped short, looking around, listening. The wind rustled through the trees as if to catch her attention. She turned round quickly, adopting a defensive posture as she did so. It was Anna.

"Why on earth did you sneak up on me like that?"

"I didn't mean to frighten you."

"Where did you appear from?"

Anna pointed vaguely in the direction of the entrance to the car park.

"I didn't hear your car," Linda said.

"I walked."

Linda was more than ever on her guard. Anna was tense, her face troubled.

"Tell me what is so important."

"I just want to know about Zeba."

"We talked about that on the phone." Linda gestured towards the glowing windows of the station. "Do you know how many of them are working in there right now?" she said. "People with only one thing on their mind: finding Zeba. You can think what you like, but I'm part of that team and I really don't have time to stand here talking to you."

"I'm sorry, I should go."

This isn't right, Linda thought. Her whole inner alarm system was ringing. Anna was acting confused. Her sneaking manner and her unconvincing apology didn't add up.

"Don't go," Linda told her sharply. "Now that you're here, you might as well tell me why."

"I've already told you."

"If you know anything about where Zeba is, you have to tell me."

"I don't know where she is. I just came to ask you if you've found her, or at least have any clues."

"You're lying."

Anna's reaction was so surprising that Linda didn't have time to prepare for it. It was as if she underwent a sudden transformation. She shoved Linda in the chest and shouted at her, "I never lie! And you don't understand what's happening!"

Then she turned and walked away. Linda didn't say anything. She watched, speechless, as Anna went. Anna had one hand in her pocket. She has something in there, Linda thought. Something she's clinging to like a lifebelt. But why is she so upset? Linda wondered if she should run after her, but Anna was already far away.

She walked back up to the front doors of the station, but something stopped her. She tried to think fast. She shouldn't have let Anna go. If it was true, as she thought, that Anna was acting strangely, then she should have brought her into the station and asked someone else to talk to her. She had been given the task of staying close to Anna. She had made a mistake and brushed her away too soon.

Linda tried to decide. She wavered between going back in, and trying to stop Anna. She chose the latter and decided to borrow her father's car since that would be faster. She drove the way that Anna should have walked, but she didn't find her. She drove the same way again but still saw nothing. She tried another possible route – same result. She drove to Anna's apartment building and stopped. The lights were on in her flat. On her way to the front door, Linda saw a bicycle. The tyres were wet and the water-splashed frame had not yet dried. It wasn't raining, but the streets were full of puddles. Linda shook her head. Something warned her against ringing the bell.

Instead, she returned to the car and backed it up until it was in shadow.

She felt that she needed to consult with someone and so she dialled her father's mobile. No answer. He must have left it somewhere again, she thought. She dialled Lindman's number. Busy. Like Martinsson's, which she tried next. She was about to try all three again when a car turned in to the street and stopped outside Anna's door. It was dark blue or black, maybe a Saab. The light in Anna's flat went out. Linda's body was tense, her hands holding the mobile were sweating. Anna appeared and climbed into the back seat, then the car drove away. Linda followed. She tried again to call her father, but still no answer. On Österleden she was overtaken by a speeding truck. She stayed behind the truck but pulled out from time to time to make sure that the dark car was still there. It turned off to Kåseberga.

Linda kept as great a distance between herself and the other car as she dared. She tried to make another call, but she succeeded only in dropping the phone between the seats. They passed the road to Kåseberga harbour and kept on east. It was only when they reached Sandhammaren that the car in front of her made a right turn. The turn seemed to come out of nowhere, as it had not indicated. Linda continued past the turning and stopped only when she had gone over a brow and round a corner. She found a bus stop and turned round, then drove back. She did not dare to take the same turning.

Instead she chose a dirt road to the left. It came to an end by a broken-down gate and a rusting combine harvester. Linda got out of the car. There was a stronger wind here by the sea. She searched for her father's black-knit watch cap. She pulled it over her head and felt as if it made her invisible. She wondered if she should try to call again, but when she saw that her battery was running low, she put the phone in her pocket and started walking back the way she had come. It was only a few hundred metres to the other road. She walked so fast she broke into a sweat. The road was dark. She stopped and listened, but could hear only the wind and the roar of the sea.

She searched among the driveways of the houses in the area for about 45 minutes, and had almost given up when she spotted the dark blue car between some trees. There was no building nearby, not that she could see. She listened, but everything was quiet. She shielded the torch with her hand to hide the light, then shone it into the car. There were a scarf and some earplugs in the back seat where Anna had been. Then she directed the beam of light on to the ground. There were paths leading in several directions, but one had a multitude of footprints.

Linda thought again about calling her father, but she reminded herself that the battery was low. Instead she sent him a text message: WITH ANNA. WILL CALL LATER. She switched off the torch and started along the sandy path. She was surprised that she wasn't scared, even though she was breaking the golden rule so often repeated during her training. Never work alone, never go into the field alone. She stopped, hesitating. Perhaps she should turn back. I'm just like Dad, she thought, and inside she felt a gnawing suspicion that this was about showing him she was good enough.

Suddenly she caught sight of a light between the trees and the sand dunes up ahead. She listened. There was still only the sound of the wind and the sea. She took a few steps in the direction of the light. There were several lighted windows. It was a house standing alone, without neighbours. There was a fence and a gate. The garden was large, and she knew the sea must be close, although she couldn't see it. She wondered who it was who had such a large house near the shore and what Anna was doing there, if that was where she was. Then her phone rang. She was startled, but answered it quickly. It was one of her fellow students from the training college, Hans Rosquist, who now worked in Eskilstuna. They hadn't talked since the graduation ball.

"Is this a bad time?" he said.

Linda could hear music, the clinking of glasses and bottles in the background.

"Sort of," she said. "Call me tomorrow. I'm working."

"You can't talk even for a few minutes?"

"No. Let's chat tomorrow."

She kept a finger on the off button in case he called again. When she had waited for two minutes without anything happening she tucked the phone back into her pocket. Cautiously she climbed the fence. There were cars parked in front of the house and there were also tents on the lawn.

Someone opened a window close to where she was. She jumped back and dropped to her knees. There was a shadow behind a curtain and the sound of voices. She waited. Then, noiselessly, she made her way to the window. The voices had stopped. The sense that there were eyes out here in the darkness was very powerful. I should run away from this place, she thought, her heart pounding. I shouldn't be here, at least not alone. A door opened, she couldn't see exactly where, but she saw the long patch of light it cast on to the lawn. She held her breath. Now she caught the whiff of tobacco smoke on the wind. At the same time the voices through the window started up again.

The patch of light on the grass disappeared and the unseen door was closed. The voices became clearer. It took a few minutes for Linda to realise that there was only one speaker, a man. But the pitch of his voice varied so much that she had thought it was several people. He spoke in short sentences, paused and then continued. She strained to hear the language he was using. It was English.

She didn't understand what he was talking about at first. It was simply an incoherent jumble: the names of people, of cities: Luleå, Västerås, Karlstad. It was part of a briefing, she realised. Something was being arranged to happen in these places, at a time and on a date which were repeated over and over. Whatever it was, it would happen in 26 hours' time. The voice spoke methodically, slowly, and could occasionally become sharp, almost shrill, before falling again to a mild tone.

Linda tried to imagine what the man looked like. She was very tempted to stand on tiptoe and try to see into the room, but she stayed in her uncomfortable position crouched next to the wall. Suddenly the voice started to talk about God. Linda felt a contraction in her stomach.

She didn't have to think about what the alternatives were. She knew she should make her way back and contact the station. Possibly they were even wondering where she had gone. But she also felt that she couldn't leave just yet, not while the voice was talking about God and the thing that was to happen in 26 hours. What was the message between the lines of what he was saying? He talked about a special grace that awaited the martyrs. Martyrs? Who was he talking about? There were too many questions and not enough room in her head. What was going on, and why was his voice so mild?

How long did she listen until she grasped what he was saying? It might have been half an hour or just a few minutes. The terrifying truth slowly dawned on her and at that point she had already started to sweat, in spite of the cold. Here in a house in Sandhammaren a group of people were preparing a terrible attack – no, 13 attacks – and some of those who would set the catastrophe in motion had already left.

She heard a number of phrases repeated: "located by the altars and towers". Also, "the explosives", and "at the corners of the structures". Linda was suddenly reminded of her father's annoyance when someone was trying to inform him of an unusually spectacular theft of dynamite. Was there a connection with what she was hearing through the window? The man was talking about how important it was to attack the foremost symbols of the false prophets, and that this was why he had chosen the 13 cathedrals as targets.

Linda was so cold that her legs were stiff and her knees ached. She knew that she had to get away immediately. What she had heard, what she now knew to be true, was so terrifying that she could hardly keep it in her head. This isn't really happening, she thought, not in Sweden. These kinds of things happen far away.

She straightened her back. It was quiet inside now. Then, just as she was about to leave, another voice started. She stiffened. The man who was speaking now said: "All is ready", only that: "All is ready". But he wasn't speaking a true Swedish; it was as if she were hearing a voice inside herself and on the tape that had disappeared from the station's archive. She shivered and waited for Torgeir Langaas to say something else, but the room was silent.

Linda felt her way back to the fence and climbed over. She didn't dare to turn on her torch. She walked into branches and stumbled over rocks.

After a while, she realised she was lost. She couldn't find the path and she had ended up in some sand dunes. Wherever she turned, she could see no lights except from a ship far out to sea. She took off her hat and stuffed it into her pocket, as if her bare head would help her to find her way. She tried to work out where she was from her relation to the sea and the direction of the wind. Then she started walking, pulling out the hat again and putting it on.

Time was of the essence. She couldn't keep wandering aimlessly in these sand dunes. She had to make a call. But the phone wasn't in her pocket. She felt through all her pockets. The hat, she thought. It must have fallen out when I took out the hat. It fell on to sand and I didn't hear it.

She crawled back along her tracks, with the torch on, but she didn't find it. I'm so incompetent, she thought furiously. Here I am crawling around without a ghost of an idea how to get away from here. But she forced herself to regain her composure. Once more she tried to determine the right direction. From time to time she stopped and let the torch cut through the dark.

At long last she found the path she had walked in on. The house with the brightly lit windows was on her left. She veered as far away from it as she could, then broke into a run towards the dark blue car. It was a moment accompanied by a rush of relief. She looked at her watch: 11.15. The time had flown by.

The arm came out of the darkness, from behind, with no warning and gripped her tightly. She couldn't move, the force holding her was too great. She felt breath against her cheek. The arm turned her around and a torch shone into her face. Without him saying a word, she knew that the man looking at her was Torgeir Langaas.

CHAPTER 50

Dawn came as a slowly creeping shade of grey. The blindfold over Linda's eyes let in some light and she knew the night was coming to an end. But what would the day bring? It was quiet all around her. One stroke of good fortune was that her bowels had not betrayed her. It was a stupid thought, but when Langaas had grabbed her it had sped through her mind like a little sentry, screaming: Before you kill me you have to let me go to the toilet. If there isn't one around then leave me for a minute. I'll crouch in the sand, I always have toilet paper in my pocket, and then I'll kick the sand over my shit like a cat.

But of course she hadn't said anything. Langaas had breathed on her, the torch had blinded her. Then he had pushed her aside, put the blindfold over her eyes and tightened it. She had banged her head when he shoved her into the car. Her fear was so great that it could only be compared to the terror she felt when she was balancing on the railing of the bridge and arrived at the surprising insight that she didn't want to die. It had been quiet all around her, just the wind and the bellow of the sea.

Was Langaas still there by the car? She didn't know, nor did she know how much time passed before the doors were opened. But she deduced from the motion of the car that two people had climbed in, one behind the wheel and the other in the passenger seat. The car jerked into action. The person driving was careless and nervous, or simply in a hurry.

She tried to follow the route they were taking. They came out on the main road and turned to the left – that was towards Ystad. She also calculated that they drove through Ystad, but at some point on what was probably the road to Malmö she lost control of her inner map. The car changed direction several times, tarmac gave way to gravel, and that in turn gave way to tarmac. The car stopped, but no doors opened. It was still quiet. She didn't know how long she sat there, but it was towards the end of this phase of waiting that the grey light of morning started to trickle in through her blindfold.

Suddenly the peace was broken by the sound of the car doors being thrown open, and someone pulled her out. She was led along a paved road and then on a sandy path. She was ushered up four stone steps, noting that the edges were uneven. She thought the steps were old. Then she was surrounded by cool air, an echoing coolness. She was in a church. The fear that had grown numb during the night returned with full force. She saw in her mind's eye what she had only heard about: Harriet Bolson strangled at the altar.

Steps echoed on the stone floor, a door was opened and she tripped over the ledge. Her blindfold was removed. She blinked in the grey light and turned in time to see Langaas's back as he walked out and locked the door behind him. A lamp was lit in the room. It was a vestry, with oil paintings of stern ministers from yesteryear. Shutters were closed over the windows. Linda looked around for a door to a toilet, but there was none. Her bowels were still calm, but her bladder was about to burst. There were some tall goblets on a table. She hoped the Lord would forgive her and used one of them as a chamber pot. She looked at her watch: 6.45, Saturday September 8. She heard a plane passing right over the church.

Linda cursed the mobile she had managed to lose during the night. There was no phone in the vestry, of course. She searched the cupboards and drawers. Then she started to work on the windows. They opened, but the shutters were tightly sealed. She looked through the vestry a second time, but she did not find any tools.

The door opened and a man walked in. Linda recognised him at once even though he was thinner than in the pictures Anna had showed her, the pictures she had kept hidden in her desk. He was dressed in a suit with a dark blue shirt buttoned all the way up. His hair was combed back and long in the neck. His eyes were light blue, just like Anna's, and it was even more clear than from the photographs how much they resembled each other. He stopped in the shadows by the door and smiled at her.

"Don't be afraid," he said kindly, and approached her with his arms outstretched, as if he wanted to demonstrate that he was unarmed.

A thought flashed through Linda's head when she saw his open outstretched arms. Anna must have had a weapon in her coat pocket. That's why she came to the station. To kill me. But she couldn't bring herself to do it. The thought made Linda weak at the knees. She staggered to one side and Westin helped her sit down.

"Don't be afraid," he said again. "I regret that you had to wait in the car, and I am sorry that I am forced to detain you for a few hours more. Then you will be free to go."

"Where am I?"

"That I cannot tell you. The only thing that matters is that you not be afraid. I also need you to answer one question."

His tone was concerned, the smile seemed genuine. Linda was confused.

"You have to tell me what you know," Westin said.

"About what?"

He fixed her with his gaze, still smiling. "That wasn't very convincing," he said softly. "I could ask my question more directly, but that won't be necessary since you understand full well what I mean. You followed Anna last night and you found your way to a house by the sea."

Most of what I tell him has to be true, otherwise he'll see through me. There is no alternative, she thought, giving herself more time by blowing her nose.

She recognised his voice now. He was the one who had been preaching to an invisible congregation in the house by the sea. Although his voice and presence gave an impression of a gentle calm, she could not forget what he had said during the night.

"I didn't make it to a house," she said. "I found a car under the trees. But it is true that I was looking for Anna."

Westin seemed lost in thought, but Linda knew he was weighing her answer. He looked at her again.

"You did not find your way to a house?"

"No."

"Why were you looking for Anna?"

No more lies, Linda thought.

"I am worried about Zeba."

"Who is that?"

Now he was the one who was lying and she the one trying to conceal the fact that she saw through it.

"Zeba is a friend Anna and I have in common. I think she was abducted."

"Why would Anna know where she is?"

"She has seemed awfully tense lately."

He nodded. "You may be telling the truth," he said. "Time will tell."

He stood up without taking his eyes off her.

"Do you believe in God."

No, Linda thought. But I know the answer you're looking for.

"I believe in God."

"We shall soon see the measure of your faith," he said. "It is as is written in the Bible: 'Soon our enemies will be destroyed and their excesses consumed by fire'."

He walked to the door and opened it.

"You won't have to wait by yourself."

Zeba came in, followed by Anna. The door closed behind Westin and a key turned in the lock. Linda stared at Zeba, then at Anna.

"What are you *doing*?" Linda asked.

"Only what needs to be done." Anna's voice was steady, but forced and hostile.

"She's off her head," Zeba said. She collapsed on to a chair. "Completely insane."

"Anyone who kills an innocent child is insane. It is a crime that must be punished."

Zeba leaped up from her chair and grabbed Linda's arm.

"She's barking mad," she shouted. "She is saying I should be punished for the abortion."

"Let me talk to her," Linda said.

"You can't reason with insane people."

"I don't believe she's insane," Linda said as calmly as she could manage.

She walked over to Anna and looked her straight in the eye, feverishly trying to order her thoughts. Why had Westin left Anna with her and Zeba?

"Don't tell me you're part of this," Linda said.

"My father has returned. He has restored the hope I had lost."

"What kind of hope is that?"

"That there is a meaning to life, that God has a meaning for each of us."

That's not true, Linda thought. She saw in Anna's eyes what she had seen in Zeba's: fear. Anna had turned her body so that she could see the door. She's afraid it will open, Linda thought. She's terrified of her father.

"What is he threatening you with?" she said, almost in a whisper.

"He hasn't threatened me."

Anna had also lowered her voice to a whisper. It can only mean she's listening, Linda thought. That gives us a possibility.

"You have to stop telling lies, Anna. We can get out of this, if you'll just stop lying."

"I'm not lying."

Linda didn't want to argue with her. Time was too short. If she didn't want to answer a question, or answered one with a lie, Linda could only press on.

"Believe what you like," Linda said. "But you won't make me responsible for people being murdered. Don't you understand what's going on?"

"My father came back to get me. A great task awaits us."

"I know what task it is you're talking about. Is that what you want? That more people lose their lives, more churches burn?"

Linda saw that Anna was close to breaking point. She had to keep going, not relax her grip.

"And if Zeba is punished, as you call it, you will for all eternity have her son's face in front of you, an accusation you will never escape. Is that what you want?"

They heard the sound of the key in the lock. They had run out of time. But just before the door opened, Anna pulled a mobile phone out of her pocket and passed it to Linda. Westin was in the doorway.

"Have you said your goodbyes?"

"Yes," Anna said. "I've said goodbye."

Westin stroked her forehead with his fingertips. He turned to Zeba and then to Linda.

"Only a little while longer," he said. "An hour or so."

Zeba lunged at the door. Linda grabbed her and forced her into the chair. She kept her there until Zeba started to calm herself.

"I have a phone now," Linda whispered. "We'll get through this."

"They're going to kill me."

Linda pressed her hand over Zeba's mouth.

"If I'm going to get us out of this, you have to help me by being quiet."

Zeba did as she was told. Linda was shaking so hard she twice dialled the wrong number. The phone rang again and again without her father picking up. She was just going to shut it off when he answered. When he heard her voice he started to shout. Where was she? Didn't she understand how worried he was?

"We don't have time," she whispered. "Listen to me."

"Where are you?"

"Be quiet and listen."

She told him what had happened when she left the station, after leaving the note on his desk.

He interrupted. "There's no note. I stayed the whole night waiting for you to call."

She was about to cry. He didn't interrupt her again, only breathing heavily as if each breath were a difficult question he needed to find an answer to, an important decision that needed to be made.

"Is this true?" he said at last. "They're stark raving mad."

"No," Linda said. "It's something else. They believe in what they're doing."

"Whatever, we'll alert every force in the country," he said, clearly aghast. "I believe we have fifteen cathedrals."

"I only heard mention of thirteen," Linda said. "Thirteen towers. The thirteenth tower is the last one and marks the onset of the great cleansing process, whatever that means."

"And you don't know where you are?"

"No. I'm pretty sure we drove back through Ystad, the roundabouts matched up. I don't think we could have gone as far as Malmö."

"Can you think of anything else you were aware of when you were in the car?"

"Different kinds of road. Tarmac, gravel, sometimes dirt roads."

"Did you go over any bridges?"

She thought hard. "I don't think so."

"Did you hear any sounds at all?"

She thought of it immediately. The aircraft flying in. She had heard them several times.

"I heard aeroplanes. One was pretty close."

"How do you mean, 'close'?"

"It seemed to be coming in to land. Or it was taking off."

"Wait," her father said.

He called to someone behind him.

"We're getting out a map," he said when he came back on the line. "Can you hear an aeroplane now?"

"No."

"Were they big or little planes?"

"Jets, I think. Big."

"Then it has to be Sturup."

Paper rustling in the background. Linda heard her father tell someone to call air-traffic-control at Sturup, to patch the call into his line with Linda.

"We have a map here," he said. "Can you hear anything now?"

"Aircraft? No, nothing."

"Can you tell me anything about where you are in relation to the aircraft noise?"

"Are church towers towards the east or west?"

"How would I know that?"

Wallander shouted to Martinsson.

"Martinsson says the tower is always in the west, the altar in the east. It has to do with the resurrection."

"Then the planes have been coming from the south. If I'm facing east, the planes have been coming from the south towards the north. Maybe north-west. They have been passing almost directly overhead."

There was mumbling and scraping at the other end. Linda felt the sweat running on her body. Zeba sat, apathetic, cradling her head in her hands.

Her father came back on the line. "I'm going to let you talk to a flight controller called Janne Lundwall. I'll be able to hear your conversation and may jump in from time to time. Do you understand?"

"I'm not stupid, Dad. But you have to hurry."

His voice wavered when he answered. "I know. But we can't do much if we don't know where you are."

Janne Lundwall's voice came on. "Let's see if we can work out where you are," he said cheerily. "Can you hear any planes right now?"

Linda wondered what her father had told him. The flight controller's bright tone only increased her anxiety.

"I can't hear anything."

"We have a KLM flight due in five minutes. As soon as you hear it, you let me know."

The minutes passed painfully slowly. Finally she heard the faint sound of an approaching plane.

"I can hear something now."

"Are you facing east?"

"Yes. The noise is on my right."

"Good. Now, you tell me when the plane is right above you or if it's in front of you."

There was a noise at the door. Linda switched off the mobile and shoved it in her pocket. Langaas came in. He stopped, looked at both of them and then left without a word. Zeba sat curled up in her chair. Only when Langaas had closed the door behind him did Linda realise that the plane had come and gone.

She dialled her father's number again. He was upset. He's just as scared as I am, Linda thought. Just as scared, and he has as little idea of where I am as I do. We can talk to each other, but we can't find each other.

"What happened?"

"Someone came in. The one called Langaas. I had to shut down."

"Good God . . . Here's Lundwall again."

The next flight in was due in four minutes, a charter flight from Las Palmas, 14 hours behind schedule.

"A whole lot of grumpy, pissed-off passengers on their way in for landing," Lundwall said happily. "Sometimes I'm grateful I'm tucked up here in my tower. Can you hear anything?"

Linda told him when she heard the plane.

"Same as before. Tell me when it's above you or in front."

The noise grew louder. At the same time the mobile started to beep. Linda looked at the display. The battery was almost exhausted.

"The phone is dying," she wailed.

"We have to know where you are," her father shouted.

It's too late, Linda thought. She cursed the phone and pleaded with it not to die on her just yet. The plane came closer and closer, the phone still beeped. Linda called out when the whine of the engines was right between her ears.

"We have a pretty good idea where you are," Lundwall said. "Just one more question . . ."

What he wanted to know, Linda never discovered. The phone cut out. Linda switched it off and hid it in a cupboard with robes and mantles. Did they have enough information to identify the church? She could only hope so. Zeba looked at her.

"It's going to be OK," Linda said. "They know where we are."

Zeba didn't answer. She was glassy-eyed and took hold of Linda's wrist so hard that her nails dug into the skin and drew blood. We're equally frightened, Linda thought. But I'm pretending not to be. I have to keep Zeba calm. If she goes into a panic our waiting period may be cut short. What were they waiting for? She didn't know, but if the truth was that Anna had told her father about Zeba's abortion, and if an abortion was the grounds for Harriet Bolson's execution in Frennestad Church, then there was no doubt what was going to happen.

"It's going to be all right," Linda whispered. "They're on their way."

* * *

They waited. It could have been half an hour, maybe it was more. Then, as if lightning had struck, the door flew open and three men came in and grabbed Zeba. Two more followed and grabbed Linda. They were pulled out of the room. Everything went so fast that it never even occurred to Linda to resist. The arms that held her were too strong. Zeba screamed, the howl of an animal. Erik Westin was waiting with Langaas in the church. There were two women and a man in the front pew. Anna was there too, but she sat further back. Linda tried to intercept her gaze, but Anna's face was like a mask. Or was she actually wearing a mask? Linda couldn't tell. The people sitting in the front pew had what looked like white masks in their hands.

Linda was filled with a paralysing fear when she saw the hawser in Westin's hands. He's going to kill Zeba, she thought desperately. He's going to kill her and then he's going to kill me because I have seen too much. Zeba struggled to free herself.

Then it was as if the walls collapsed. The church doors burst open, and four of the stained-glass windows, two on either side of the church, were shattered. Linda heard a voice shouting through a megaphone, and it was her father. He shouted as if he didn't trust the megaphone's amplifying capacity. Everyone in the church froze.

Westin gave a start, then ran to grab hold of Anna and use her as a shield. She tried to pull away. He shouted at her to be calm, but she kept resisting. He dragged her towards the west doors of the church. Again she tried to escape his grasp. A shot rang out. Anna jerked and collapsed. Westin had a gun in his hand. He stared in disbelief at his daughter, then ran out of the church. No-one dared to stop him.

Wallander and a large number of armed officers – Linda didn't recognise most of them – stormed into the aisle through the side doors. Langaas started to shoot. Linda pulled Zeba with her into a pew and pushed her on to the floor. The exchange of fire went on. Linda didn't look to see what was happening. And then it grew quiet. She heard Martinsson's voice. He shouted that a man had gone out through the front doors. That must be Langaas, she thought.

She felt a hand on her shoulder and flinched. Perhaps she even screamed without realising. It was her father.

"You have to get out," he said.

"How is Anna?"

He didn't answer, and Linda knew she was dead. She and Zeba scurried out into the day. A long way down the road leading to the church, she saw the dark blue car hurtling away with two police cars in distant pursuit. Linda and Zeba sat on the ground on the other side of the cemetery wall.

"It's over," Linda said.

"Nothing is over," Zeba whispered. "I'm going to live with this for the rest of my life. I'm always going to feel something pressing around my throat."

Suddenly there was one more shot, then two more. Linda and Zeba crouched behind the low wall. There were voices, orders, cars that took off at high speed with their sirens going. Then silence again.

Linda told Zeba to stay put. She got to her knees and looked over the wall. There were a lot of officers surrounding the church, but everyone was still. It was like looking at a painting. She saw her father and went over to him. He was pale and grabbed her arm hard.

"Both of them got away," he said. "Westin and Langaas."

He was interrupted by someone who handed him a mobile phone. He listened, then handed it back without a word.

"A car loaded with dynamite has just driven into Lund Cathedral, clear through the poles with iron chains, and crashed into the left tower. There's chaos there. No-one knows how many are dead. We seem to have averted attacks against the other cathedrals. Twenty people have been arrested so far."

"Why did they do it?" Linda said.

Wallander thought for a long time.

"Because they believe in God and love him," he said. "I can't think their love is reciprocated."

"Was it hard to find us?" Linda said. "There are a lot of churches hereabouts."

"Not really," he said. "Lundwall was able to locate you. We had two churches to choose from. Whose was the mobile?"

"Anna's. She felt terrible about what she had done."

They walked over to Zeba. A black car had arrived and Anna's body was carried from the church.

"I don't think he meant to shoot her," Linda said. "I think the gun went off in his hand."

"We'll catch him," Wallander said. "And then we'll find out."

Zeba stood up as they approached. She was frozen and shivering.

"I'll go with her," Linda said. "I did almost everything wrong, I know, and I'm sorry."

"I'll be able to relax when I've got you wearing a uniform and know you're securely in a patrol car, circling the streets of Ystad," her father said.

"My mobile is somewhere in the dunes at Sandhammaren."

"We'll send someone out there who can call your number. Maybe the sand will answer."

Svartman was standing by his car. He opened the back door and swept a blanket around Zeba, who crawled in and made herself small in the corner.

"I'll come with her," Linda said.

"How are you doing, Linda?" Svartman asked.

"I don't know. The only thing I'm sure of is that I'm going to start work on Monday."

"Put it off for a week," her father said. "There's no hurry."

Linda sat in the car and they drove away. A plane flew low over their heads, coming in to land. Linda looked at the landscape. It was as if her gaze was being sucked into the brown-grey mud. And there was the sleep that she needed more than anything else right now. After that there was one more day of what had been a long wait before she started working. Very soon now she would throw away her invisible uniform. She thought about asking Svartman if he reckoned they would catch Westin and Langaas, but she didn't say anything. Right now, she didn't want to know.

Later, not now. Frost, autumn and winter; time enough later for thinking. She leaned her head on Zeba's shoulder and closed her eyes.

Suddenly she saw Westin's face in front of her. That last moment when Anna fell slowly towards the floor. Now she realised the despair that had been in his face, the vast loneliness. The face of a man who has lost everything.

She looked out at the landscape again. Slowly Erik Westin's face fell away into the grey clay.

Zeba was asleep by the time they drew up outside Yassar's shop. Linda gently shook her awake.

"We're here," she said. "We're here and everything is over."

CHAPTER 51

Monday September 10 was a cold and blustery day in Skåne. Linda had tossed and turned and only managed to sleep at dawn. She was woken by her father coming in and sitting down on the side of her bed. Just like when I was little, she thought. He was always the one who would sit on the side of my bed, never my mother.

He asked how she had slept and she told him the truth: not well, and she had been plagued by nightmares.

The previous evening, Lisa Holgersson had called to say that Linda could wait a week before starting work. But Linda had said no. She didn't want to put it off any longer, even after everything that had happened. They finally agreed that she would take one extra day and start work on Tuesday.

Wallander got to his feet. "I've got to go," he said. "What do you have planned?"

"I'll see Zeba. She needs someone to talk to, and so do I."

Linda spent the day with Zeba. Her son was back with fru Rosberg for the day. The phone rang and rang, mostly reporters. Finally she and Zeba escaped to Mariagatan. They went over what had happened again and again, especially the part that Anna had played. Could they understand it? Could anyone understand it?

"She missed her father all her life," Linda said. "When he finally turned up she refused to believe anything except that he was right, whatever he said and whatever he did."

Zeba often fell silent. Linda knew what she was thinking, about how close to death she had been, and that not only Anna's father but Anna herself had been to blame.

Mid-morning Wallander called and told her that Henrietta had collapsed and been taken to the hospital. Linda remembered Anna's sighs that Henrietta had incorporated in one of her compositions. It's all she has left, she thought. Her dead daughter's sighs.

"She left a letter in the house," Wallander said. "In which she tried to explain what she had done. She didn't tell us about Westin returning because she was afraid. He had threatened her and said that she and Anna would die if she said anything. There's no reason not to believe her, but she surely could have found a way to let *someone* know what was going on."

"Did she say anything about my last visit?" Linda said.

"Langaas was in the garden. She opened the window so that he would hear that she didn't give anything away."

"Westin used Langaas to frighten people."

"He knew a lot about people, we shouldn't forget that."

"Is there any trace of them?"

"We will find them, since this matter is top priority, all over the world. But maybe they'll get new hiding places, new followers. No-one knows how many places Langaas had up his sleeve, and no-one will know for sure until the two of them are found."

"Langaas is gone, Erik Westin is gone, but the most gone of all is Anna."

Linda and Zeba talked about the fact that maybe Erik Westin was already building a new sect. There were many gullible people out there who would follow him. One such was Ulrik Larsen, the minister who had threatened Linda. He was one of Westin's followers, waiting to be called to action. Linda thought about what her father had said. They couldn't be sure of anything until Westin was caught. One day maybe a new assault would be launched, like the one in Lund.

Afterwards, when she had followed Zeba home, after first making sure she was up to being on her own with her boy, Linda took a walk and sat on the pier by the Harbour Café. It was cold and windy, but she found a sheltered spot out of the wind. She didn't know if she missed Anna or if what she felt was something else. We were only true friends as children, she thought.

* * *

That evening Wallander came home and reported that Langaas had driven into a tree. Everything pointed to suicide. Westin was still at large. Linda wondered if she would ever know whether it was Westin she had seen in the sunlight outside Lestarp Church.

But there was one question she had found the answer to herself. The nonsense words in Anna's journal: *myth, fear, myth, fear*. It was so simple, Linda thought: *my father, my father*. That was all it was.

Linda and her father sat up and talked. The police in Cleveland were reconstructing Erik Westin's life and had found a connection to the minister Jim Jones and his sect who had been massacred in the jungle in Guyana. Westin might be a person whom it would never be possible fully to understand, but it was important to realise that he was by no means a madman. His self-image, not least as expressed in the holy pictures he asked his disciples to carry with them, was of a humble person carrying out God's work. He wasn't insane so much as a fanatic, prepared to do whatever it took to realise his beliefs. He was ready to sacrifice people if need be, kill anyone who stood in his way, and punish those whom he deemed to have committed mortal sins. He sought his justifications in the Bible.

Westin was also a desperate man because he saw only evil and decay around him. Not that this in any way justified his actions. The only hope of preventing something like this in the future, of identifying people prepared to blow themselves up in something they claimed was a Christian effort, was not to dismiss Westin as a madman plain and simple.

There was not much to add. Everyone who was to have carried out the well-planned bombings was awaiting trial. Police, if necessary all over the world, would be looking for Westin, and soon autumn would come with frosty nights and cold winds from the north-east.

They were about to go to bed when the phone rang. Wallander listened in silence, then asked a few short questions. When he hung up, Linda did not want to ask him what had happened. She saw the glimmer of tears in his eyes and he told her that Sten Widén had died. The woman who had called was a girlfriend, possibly the last one he had lived with. She had promised Widén that she would contact Wallander and tell him that everything was over and that it had "gone well".

"What did she mean by that?"

"We used to talk about it when we were younger, Sten and I. That death was something one could face like an opponent in a duel. Even if the outcome was a given, a skilful player could hold off and tire death out so that it only had the power to deliver a single blow. That was how we wanted our deaths to be, something we could take care of so they would 'go well'."

He was sad, she could see that.

"Do you want to talk some more?"

"No. This is something I have to work through on my own."

They were quiet for a while, then he stood up and went to bed without a word. Linda didn't manage to sleep many hours that night either. She thought about all the people out there prepared to blow themselves up, and the churches they hated. From what her father and Lindman had said, and from what she had read in the papers, these people were a far cry from monsters. They spoke of their good intentions, their hopes to pave the road for the true Kingdom of God on Earth.

One day she had been prepared to wait, but no more. Therefore she walked up to the station on the morning of September 11. It was a cold, dreary day after a night that had left traces of the first frost. Linda tried on her uniform and signed receipts for her equipment. Then she had a meeting with Martinsson which lasted an hour, and received her first shift assignment. She was free for the rest of the day, but she didn't feel like sitting alone at home at Mariagatan, so she stayed at the station.

At 3 p.m. she was drinking coffee in the canteen, talking to Nyberg, who had sat at her table of his own accord and was showing his friendly side. Martinsson came in and shortly thereafter her father. Martinsson turned on the television.

"Something's happened in the States," he said.

"What sort of thing?" Linda said.

"We'll have to wait and see."

There was an image of a clock, counting down the seconds to a special news report. More and more people filtered into the canteen. By the time the news report came on, the room was almost full.

EPILOGUE

The Girl on the Roof

The call had come into the station shortly after 7.00 on Friday night, November 23, 2001. Linda, who that evening was partnered up with an officer named Ekman, answered the police dispatch department's broadcast. They had just resolved a family conflict in Svarte and were driving back to Ystad.

A young woman had climbed on to the roof of an apartment building to the west of the city and was threatening to jump. To make matters worse, she was armed. Head of operations wanted as many patrol cars to the scene as possible. Ekman turned on the siren and put his foot down.

Curious onlookers had already gathered by the time they arrived. Spotlights illuminated the girl, sitting up on the roof with a shotgun in her arms. Ekman and Linda were briefed by Sundin, who was responsible for getting her down. A ladder truck, provided by Rescue Services, was also in place. But the girl had threatened to jump if the ladder was driven any closer.

The girl, Maria Larsson, was 16 and had been treated for several episodes of mental illness. She lived with her mother, who was a substance abuser. This particular evening something had gone wrong. Maria had rung a neighbour's doorbell and when the door opened she had rushed in and grabbed a shotgun and some cartridges that she knew were kept in the flat. The owner of the flat could count on serious trouble, since he had obviously stored both the weapon and the cartridges in an insecure manner.

But this was about Maria. She had threatened first to jump, then to shoot herself, then to shoot anyone who tried to approach her. The mother was too far gone to be of any use, and there was also the chance that she would start to shout at her daughter and incite her to carry out her threats.

Several officers had tried to speak to the girl through a trapdoor 20 metres from the drainpipe where she was sitting. Right now an old minister was trying to talk to her, but she aimed the weapon at his head and he ducked out of sight. They were working on finding a close friend of Maria's who would perhaps be able to get through to her. No-one doubted that she was desperate enough to do what she had threatened.

Linda borrowed a pair of binoculars and looked at the girl. Even when the call came through she had thought of the time she had stood on the bridge railing. When she saw Maria shaking on the roof, her cramped hold on the shotgun and the tears that had frozen on her face, it was like looking at herself. Behind her she could hear Sundin, Ekman and the minister talking. No-one knew what to do. Linda lowered the binoculars and turned to them.

"Let me talk to her," she said.

Sundin shook his head doubtfully.

"I was in the same situation once," she said. "And she might listen to me since I'm not that much older than she is."

"I can't let you take that risk. You're not experienced enough to judge what you should and shouldn't say. And her weapon is loaded. She's showing signs of an increasing desperation. Sooner or later she'll use the weapon."

"Let her try." It was the old minister. He sounded very firm.

"I agree," Ekman said.

Sundin wavered. "Shouldn't you at least call your dad first and talk to him?"

Linda almost lost her composure.

"For goodness' sake, this has nothing to do with him. This is between me and Maria Larsson. Nothing to do with him."

Sundin consented, but he made her put on a bullet-proof vest and helmet before he let her go up. She kept the vest on, but removed the helmet before sticking her head up through the trapdoor. The girl on the roof heard the screech of the metal. When Linda peeked out Maria had the gun aimed at her head. She almost ducked under again.

"Don't come near me!" the girl shrieked. "I'll shoot and then I'll jump!"

"Take it easy," Linda said. "I'm not going to move an inch. But will you let me talk to you?"

"What could you have to say to me?"

"Why are you doing this?"

"I want to die."

"I wanted to die too, once. That's what I have to say to you."

The girl didn't answer, and Linda waited. Then she started to tell her about how she had stood on the bridge railing, what had led up to it, and about the person who had finally been able to talk her down.

Maria listened, but her initial reaction was anger.

"What do you think that has got to do with me? My story is going to end down on the street. Go away! Leave me alone!"

Linda wondered what she should do next. She had hoped her story would be enough. What a naive assessment that was. I've watched Anna die, she thought. But I have also witnessed Zeba's joy at still being alive.

She decided to keep talking.

"I want to give you something to live for," she said.

"There is nothing."

"Hand me the gun and come here. For my sake."

"You don't know me."

"No, but I've teetered on a bridge railing. Sometimes I have nightmares where I throw myself off."

"When you're dead, you don't dream anything. I don't want to live."

The conversation went back and forth. After a while – how long it was Linda couldn't say, because time seemed to have been suspended when she first poked her head up out of the roof and faced the barrels of the shotgun – she could tell that the girl was fully engaged in what they were saying. Her voice was less shrill. This was the first step. Now she held an invisible lifeline of sorts around Maria's body. But nothing was resolved until the moment when Linda had used up all her words and started to cry. And that was when Maria finally gave in.

"All right," she said. "I just want them to turn off all the fucking lights. I don't want to see my mother. I only want to talk to you. And I won't come down right away."

Linda hesitated. What if it was a trap? What if she had decided to jump as soon as the lights were turned off?

"Why won't you come down with me now?"
"I want ten minutes."
"What for?"
"Ten minutes to see what it feels like to have decided to live."

Linda climbed down and all the lights were turned off. Sundin kept an eye on the time. Suddenly it was as if all of the events from the dramatic days at the beginning of September came out of the darkness at her with full force. She had been so grateful for her work and the new flat had taken so much of her attention that she had not yet had the opportunity to slow down enough to take on the impact of what she had been through. Even more important was the time she had been spending with Stefan Lindman. They had started seeing each other, and sometime in the middle of October Linda had realised she wasn't alone in having fallen in love. He had been getting over a lost love in the north. Now, as she stood here trying to pick out the outline of the girl on the roof who had decided to live, it was as if the moment had arrived for a kind of resolution to all that had happened.

Linda stamped her feet to stay warm and looked up again at the roof. Had the girl changed her mind? Sundin mumbled that there was only a minute left. Then the time ran out. The ladder truck drove up to the edge of the building. Two firemen helped the girl down, a third went up and collected the weapon. Linda had told Sundin and the others what she had promised and she insisted that her side of the bargain be upheld. Therefore she was the only one of them there when Maria reached the bottom of the ladder. Linda hugged her and suddenly both of them started to sob. Linda had the strangest feeling that she was hugging herself.

An ambulance was near by. Linda helped Maria into it and waited until it drove away, the gravel crunching under its wheels. The frost had arrived; it was already below freezing. Officers, the old minister, the firemen: everyone came up and shook her hand.

* * *

Linda and Ekman stayed until the fire engines and the patrol cars had left, the yellow tape had been taken down and the crowd had gone. Then there was a radio message about a suspected drunk driver on Österleden. They left, and Linda swore under her breath. Most of all she would have liked to go back to the station for a cup of coffee.

But that would have to wait, like so many other things. She leaned against Ekman to read the thermometer for the temperature outside.

It was –3°C. Winter had arrived in Skåne.